Foundations of **DISCRETE MATHEMATICS**

THE PRINDLE, WEBER & SCHMIDT SERIES IN MATHEMATICS

Althoen and Bumcrot, *Introduction to Discrete Mathematics*
Boye, Kavanaugh, and Williams, *Elementary Algebra*
Boye, Kavanaugh, and Williams, *Intermediate Algebra*
Buchthal and Cameron, *Modern Abstract Algebra*
Burden and Faires, *Numerical Analysis, Fourth Edition*
Cass and O'Connor, *Fundamentals with Elements of Algebra*
Cullen, *Linear Algebra and Differential Equations, Second Edition*
Cullen, *Mathematics for the Biosciences*
Dick and Patton, *The Oregon State University Calculus Curriculum Project*
Eves, *In Mathematical Circles*
Eves, *Mathematical Circles Adieu*
Eves, *Mathematical Circles Revisited*
Eves, *Mathematical Circles Squared*
Eves, *Return to Mathematical Circles*
Fletcher, Hoyle, and Patty, *Foundations of Discrete Mathematics*
Fletcher and Patty, *Foundations of Higher Mathematics*
Gantner and Gantner, *Trigonometry*
Geltner and Peterson, *Geometry for College Students, Second Edition*
Gilbert and Gilbert, *Elements of Modern Algebra, Second Edition*
Gobran, *Beginning Algebra, Fifth Edition*
Gobran, *Intermediate Algebra, Fourth Edition*
Gordon, *Calculus and the Computer*
Hall, *Algebra for College Students*
Hall and Bennett, *College Algebra with Applications, Second Edition*
Hartfiel and Hobbs, *Elementary Linear Algebra*
Kaufmann, *Algebra for College Students, Third Edition*
Kaufmann, *Algebra with Trigonometry for College Students, Second Edition*
Kaufmann, *College Algebra, Second Edition*
Kaufmann, *College Algebra and Trigonometry, Second Edition*
Kaufmann, *Elementary Algebra for College Students, Third Edition*
Kaufmann, *Intermediate Algebra for College Students, Third Edition*
Kaufmann, *Precalculus, Second Edition*
Kaufmann, *Trigonometry*
Laufer, *Discrete Mathematics and Applied Modern Algebra*
Nicholson, *Elementary Linear Algebra with Applications, Second Edition*
Pence, *Calculus Activities for Graphic Calculators*
Powers, *Elementary Differential Equations*
Powers, *Elementary Differential Equations with Boundary-Value Problems*

Powers, *Elementary Differential Equations with Linear Algebra*
Proga, *Arithmetic and Algebra, Second Edition*
Proga, *Basic Mathematics, Third Edition*
Rice and Strange, *Plane Trigonometry, Fifth Edition*
Schelin and Bange, *Mathematical Analysis for Business and Economics, Second Edition*
Strnad, *Introductory Algebra*
Swokowski, *Algebra and Trigonometry with Analytic Geometry, Seventh Edition*
Swokowski, *Calculus, Fifth Edition*
Swokowski, *Calculus of a Single Variable*
Swokowski, *Calculus with Analytic Geometry, Fourth Edition (Late Trigonometry)*
Swokowski, *Fundamentals of Algebra and Trigonometry, Seventh Edition*
Swokowski, *Fundamentals of College Algebra, Seventh Edition*
Swokowski, *Fundamentals of Trigonometry, Seventh Edition*
Swokowski, *Precalculus: Functions and Graphs, Sixth Edition*
Tan, *Applied Calculus, Second Edition*
Tan, *Applied Finite Mathematics, Third Edition*
Tan, *Calculus for the Managerial, Life, and Social Sciences, Second Edition*
Tan, *College Mathematics, Second Edition*
Trim, *Applied Partial Differential Equations*
Venit and Bishop, *Elementary Linear Algebra, Third Edition*
Venit and Bishop, *Elementary Linear Algebra, Alternate Second Edition*
Wiggins, *Problem Solver for Finite Mathematics and Applied Calculus*
Willard, *Calculus and Its Applications, Second Edition*
Wood and Capell, *Arithmetic*
Wood, Capell, and Hall, *Developmental Mathematics, Fourth Edition*
Wood and Capell, *Intermediate Algebra*
Zill, *A First Course in Differential Equations with Applications, Fourth Edition*
Zill, *Calculus with Analytic Geometry, Second Edition*
Zill, *Differential Equations with Boundary-Value Problems, Second Edition*

THE PRINDLE, WEBER & SCHMIDT SERIES IN ADVANCED MATHEMATICS

Brabenec, *Introduction to Real Analysis*
Ehrlich, *Fundamental Concepts of Abstract Algebra*
Eves, *Foundations and Fundamental Concepts of Mathematics, Third Edition*
Keisler, *Elementary Calculus: An Infinitesimal Approach, Second Edition*
Kirkwood, *An Introduction to Real Analysis*
Ruckle, *Modern Analysis: Measure Theory and Functional Analysis with Applications*

Foundations of DISCRETE MATHEMATICS

Peter Fletcher
Virginia Polytechnic Institute and State University

Hughes Hoyle
The Citadel

C. Wayne Patty
Virginia Polytechnic Institute and State University

PWS-KENT PUBLISHING COMPANY
Boston

PWS-KENT
Publishing Company

20 Park Plaza
Boston, Massachusetts 02116

PWS-KENT Publishing Company is a division of Wadsworth, Inc.

Library of Congress Cataloging-in-Publication Data

Fletcher, Peter.
Foundations of discrete mathematics / Peter Fletcher, Hughes Hoyle, C. Wayne Patty.
p. cm.
ISBN 0-534-92373-9
1. Mathematics. 2. Electronic data processing—Mathematics. I. Hoyle, Hughes. II. Patty, C. Wayne. III. Title.
QA39.2.F588 1990
510—dc20 90-30756
CIP

Printed in the United States of America.
90 91 92 93 94 — 10 9 8 7 6 5 4 3 2 1

International Student Edition
ISBN 0-534-98381-2

Sponsoring Editor: *Steve Quigley*
Production Supervisor: *Elise Kaiser*
Production: *Del Mar Associates*
Composition: *J.M. Post Graphics, Corp.*
Interior Illustration: *Kristi Mendola*
Interior Design: *John Odom*
Cover Design: *Elise Kaiser*
Manufacturing Coordinator: *Margaret Sullivan Higgins*
Cover Printer: *Henry N. Sawyer Company, Inc.*
Text Printer/Binder: *R. R. Donnelley & Sons, Inc.*
Cover Illustration: *Technical Ecstasy* by Steven Hunt; courtesy of The Image Bank.

CONTENTS

PREFACE

Foundations of Discrete Mathematics introduces students to basic ideas and techniques of discrete mathematics. The text presumes no background in calculus or computer science, and is designed for a one- or two-semester course to be taken by students with the same mathematical maturity that is expected of calculus students. Two years of high school algebra is the only prerequisite, and students whose mathematical background exceeds our minimum requirements should be able to skim, or skip entirely, much of the material of the first four chapters.

Paul Halmos once observed: "Calculus books are bad because there is no such subject as calculus; it is not a subject because it is many subjects." We believe that Halmos could have leveled this same charge at discrete mathematics texts. Certainly to most college students the distinction between discrete mathematics and other mathematics is not well understood, and so there is the danger that a course in discrete mathematics will be perceived as a hodgepodge of unrelated topics or as a mathematical muddling of computer science. For this reason, we have taken care to explain what discrete mathematics is and to highlight some common threads that unify our subject. We have in mind the ground shared by algorithms and recursion, techniques of counting inherent in combinatorics and probability, and the interest in (discrete) structures common to the study of directed graphs, graphs, trees, and Boolean algebras.

As a mathematics text, our book strives to teach mathematical reasoning and an appreciation of the need to read and write mathematics with care. Because we assume no background in calculus, some compromises are inevitable. We do not define real numbers or logarithms, and our definition of a matrix is the usual doodle-definition, a rectangular array of numbers. We hope, however, that we have treated the reader honestly. Knowing that our students may soon be considering formal languages, we have tried to give an elementary explanation of the difference between a proposition and a propositional expression. We are uncomfortable with the definition found in other discrete mathematics texts that a tautology is a compound proposition that is always true. After all, there is no true proposition, compound or otherwise, which, like Cinderella's coach-and-four, suddenly turns false at the stroke of midnight. The notion of recursive definition is fundamental to discrete mathematics, and we have no qualms in taking as an axiom that this method of definition really does define a sequence. While it is true that our axiom is a consequence of the axiom of induction, it is not fair to invite readers to prove this consequence knowing full well that they will fall into the trap of saying "It is defined for 1; if it is defined for n, it is defined for $n + 1$; so by induction it is defined." It is also possible to hoodwink readers by saying nothing, where in all fairness something needs to be said. For instance, the equivalence classes of an equivalence relation form a partition, and the natural relation formed from a partition is an equivalence relation. Enough said? No. We still need to know that, going from equivalence relation to partition to equivalence relation, we get home again, and that the trip from partition to equivalence relation to partition also brings us back where we started.

One misconception about our subject is that there is no room in mathematics for trial and error, as if somehow proof precedes conjecture. We therefore give the student a chance to experiment.

For the most part we have followed the guidelines presented by the Mathematical Association of America in the report from *Committee on Discrete Mathematics in the First Two Years* (1986). In particular, our book presents a beginning discrete mathematics course, and, as recommended by the MAA, it can be used for a one-year course, at the level of the calculus but independent of it. However, with reality in mind, we also designed the text so that it can be used readily in a one-semester course. This text is intended to be comprehensive, and students (particularly those who are taking computer science courses) may find it a handy reference at a later date.

It is a little intimidating for authors, and we suspect for publishers alike, to look about and see how many discrete mathematics texts are already on the market. As we have pointed out, we believe that discrete mathematics, like calculus, is not one subject but many. Perhaps this explains why there are many different books and different opinions. Some texts, it seems to us, are written more as discrete engineering than as discrete mathematics. These texts provide a vast array of techniques and problems to attack, but the student in going from skirmish to skirmish gains little understanding of the underlying mathematics. It is as if Donald Knuth had begun his book *Surreal Numbers* by saying: "Here is what we mean by a number and here are three examples of the use of numbers, which the reader should study carefully before attempting the exercises."

The goal of our text is to present discrete mathematics as mathematics.

In Chapter 1 we explain the logic behind direct proofs, proofs by contradiction, and proofs by contrapositive; in Chapter 4 we present induction, not as a Peano postulate, but as a powerful method of proof. The instructor has a choice of how much emphasis to place on proofs, and this text is not designed to be an introduction to proofs.

We introduce the basic concepts of sets and functions in Chapters 2 and 3. Even if the instructor chooses to skim or omit much of the material in the first four chapters, these chapters are available as a reference when needed in Chapters 5 through 15. These last eleven chapters are written with as much independence as possible. They contain references to other chapters, but in most cases these references are designed to link topics and can be omitted without loss of continuity.

The text can be used in courses with differing points of view. One approach is to design a course that begins with Chapter 7 (Boolean Algebra and Gate Networks) so that Boolean algebra can be used to unify such topics as circuit design, logic, set theory, and formal languages. On the other hand, the text can be used in a course that never mentions Boolean algebra. A course might also begin with Chapter 5 (Algorithms), Chapter 6 (Counting), or Chapter 11 (Graph Theory).

Our goal is to present a text that a beginning student will read, for the student's ultimate objective is to gain an understanding and an appreciation of a new way of thinking about mathematics. Naturally, because our text is designed for a beginning course, it is sometimes necessary to sacrifice rigor for clarity. But we warn the reader when such sacrifices are made, and so we hope to provide a course that is accessible to the large and diverse group of students who need discrete mathematics and that is of equal standing with calculus in the college curriculum.

TERMINOLOGY

In this text we use two conventions. The first is that we never rely on a plural form to establish that one object is different from another; so we may say "let x and y be apples" without implying that x is different from y. The second convention is similar. When we give a grammatical construction such as there "are two" square roots of 9, or 9 "has two" square roots, we always mean "exactly two." Otherwise we add some qualifying expression such as "at least" or "at most."

Let n be an integer. If $n > 0$, n is called a *natural number*, a *counting number*, or a *positive integer;* in this text, we use the terms *natural number* and *positive integer*. If $n < 0$, n is called a *negative integer*. Note that 0 is neither positive nor negative. A *rational number* is the quotient of an integer by a non-zero integer. A real number that is not rational is called an *irrational number*. Finally a real number x is called *positive* if $x > 0$ and *negative* if $x < 0$.

It is convenient to give a few important number systems their own names. We let $\mathbb{N}$ denote the set of all natural numbers and $\mathbb{Z}$ denote the set of all integers—positive, zero, or negative. We let $\mathbb{R}$ denote the set of all real numbers and $\mathbb{Q}$ denote the set of all rational numbers. Although our choice of notation is standard, the reader may be a bit puzzled by the choice of $\mathbb{Z}$ to denote the set of all integers. This letter is taken from the German word for number, which is *Zahl.* We indicate especially difficult exercises with an asterisk.

The *absolute value* of a real number x, denoted by $|x|$, is defined by $|x| = x$ if $x \geqslant 0$ and $|x| = -x$ if $x < 0$. It can be shown that the absolute value of x is the larger of x and $-x$, and we often use this characterization of absolute value. The absolute value of a real number x can be $-x$, but this can happen only when x is not a positive number. Thus, for each real number x, $|x| \geqslant 0$.

Let n be an integer. A non-zero integer d *divides* n provided that there is an integer m such that $dm = n$. When d divides n, we say that d is a *divisor* of n or a *factor* of n and that n is a *multiple* of d. An integer greater than 1 whose only positive divisors are itself and 1 is called a *prime number*. An integer greater than 1 that is not a prime number is called a *composite*. Note that 1, which is neither prime nor composite, divides every integer.

ACKNOWLEDGMENTS

It is a pleasure to acknowledge the advice we have received from our colleagues: George Crumley, Charles Feustel, Margaret Francel, and Spencer Hurd. We have incorporated the ideas and insights of our reviewers and thank them for sharing their experiences with us: Nancy Baxter, *Dickinson College*; Robert Beezer, *University of Puget Sound*; Carole Bernett, *Harper College, TMPS Division*; Paul M. Cook, *Furman University*; Larry Dornhoff, *University of Illinois*; Frederick Hoffman, *Florida Atlantic University*; Francis Masat, *Glassboro State College*; Dix Pettey, *University of Missouri*; George Schultz, *St. Petersburg Junior College, Clearwater Campus*; and Diane M. Spresser, *James Madison University*.

We thank Steve Quigley and Dave Geggis, our editors at PWS–KENT, for their professional support and encouragement. Thanks also to Elise Kaiser, our PWS–KENT production supervisor, and Nancy Sjoberg, our production coordinator.

Peter Fletcher
Hughes Hoyle
C. Wayne Patty

INTRODUCTION

It is easy to pick up a text with the catchwords "discrete mathematics" in its title, but often it is impossible to find a concrete definition of discrete mathematics in such texts. Those who work in this field of mathematics are equally evasive. One practitioner told us, "I feel about discrete mathematics the way Supreme Court Justice Stewart felt about pornography. I don't know what it is, but I know it when I see it." One view is that discrete mathematics is just finite mathematics, which is the study of mathematical problems concerning finite sets, with a few bits from computer science thrown in to make the subject appear up to date. It is true that discrete mathematics subsumes finite mathematics, but in our view it is something more.

The adjective *discrete* means "separate, detached, discontinuous," and these synonyms capture some of the intent of the mathematical definition, which is taken from calculus. In calculus, one defines a set D of real numbers to be discrete provided that for each x belonging to D there is a positive number ε such that no other member of D lies between $x - \varepsilon$ and $x + \varepsilon$. This definition is not so formidable as it first appears. It merely suggests, for example, that we can center a coin along the number line at each point of a discrete set in such a way that no two coins overlap. Perhaps a few of the points are so far from the other points that we can center half-dollars over them, others may be close enough to require quarters, still others may require nickels or dimes, or even some imaginary coins of smaller size; but we can always find small enough coins so that no coin covers two or more points of our discrete set. In other words, each point has its own private bit of territory into which no other member of D intrudes.

In more advanced study, the concept of a discrete set is generalized to any geometric space in which there is a sensible notion of distance between points. In this setting, a set D of points is *discrete* provided that for each point p of D there is a positive number ε such that no other point of D lies within ε of p. Once again, the idea is just that each point of D may be assigned its own private territory. For example, in the Cartesian plane the set D of all points whose coordinates are both integers is a

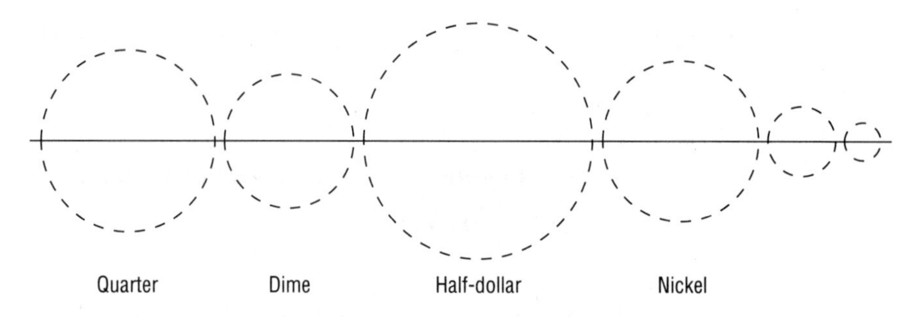

discrete set because we can find a circle with center at each such point whose interior does not contain any other point of D.

In any geometric space, each finite set of points is a discrete set. Suppose that we are given a finite set F of points. We first compute the distance between each pair of points of F. Since there are only finitely many points, there are only finitely many distances. In particular, F has a closest pair of points. We can use this closest pair of points to choose an appropriately sized coin to center over each point of F such that no two coins overlap. Thus any finite set is discrete, and so our definition includes all finite mathematics.

The ascendency of discrete mathematics in the second half of the twentieth century can be attributed in large measure to two machines, the telephone and the computer. The telephone industry is faced with problems involving switching, relay circuits, networking, and coding. Because these problems often turn out to be problems in discrete mathematics, much recent research in discrete mathematics has been motivated by problems involving the telephone and has been published by Bell Laboratories. In this text, however, we emphasize applications of discrete mathematics to computer science. In part, this emphasis reflects our interests and experience, but we also believe that many students first become interested in discrete mathematics because of a prior interest in the computer. Like the telephone, the computer has provided a rich source of problems in discrete mathematics. Its ability to perform calculations with speed has made practical some techniques of discrete mathematics which would have otherwise remained esoteric and useless.

The present-day standards of the mathematical community allow a computer program as part of a proof in discrete mathematics, and so a knowledge of computers has become just as important in studying discrete mathematics as discrete mathematics has become in studying the art of computing. Indeed, the use of computers within discrete mathematics is becoming commonplace, and computer-aided proofs include solutions to such famous and long-standing problems as the four-color conjecture, which was first posed by Auguste Möbius around 1840. Although the reader of this text is never required to write a program or to use a computer, we give a brief introduction to the computer in Appendix A.

Chapter

1

LOGIC

Logic has important practical applications. We use its rules, for example, in writing computer programs and in designing computer circuits. Logic also plays a major role in the construction of mathematical proofs. In such contexts, we deal with sets of propositions, the connectives *and* and *or*, which correspond to *intersection* and *union* in set theory, and the notion of *negation*, which corresponds to *complement*.

1.1 PROPOSITIONS AND CONNECTIVES

A *proposition* (or *statement*) is a sentence that is either true or false but not both. We illustrate this idea in Examples 1 and 2.

Example 1

The following sentences are propositions:

a. William Shakespeare was president of the United States.
b. $3 + 2 = 5$.
c. Las Vegas is the capital of Nevada.
d. If x is a real number, then $x^2 < 0$.
e. $2^{1999} - 1$ is a prime number.
f. There is intelligent life outside our solar system.

Sentence (e) is certainly true or false, although we do not happen to know which. Sentence (f) is certainly true or false, but no one inside our solar system knows which. ■

In Example 2 we list some sentences that it could be argued are neither true nor false. Hereafter, all sentences we use to study logic are presumed to be either true or false. Thus, if we later consider the sentence "If it is a cloudy day, then Sally Lou has blue eyes," we expect that you will not question what it means to say a day is cloudy or the eyes of Sally Lou are truly blue.

Example 2

a. Shall we go to your place or mine?
b. Will you marry me?
c. Sally Lou is beautiful.
d. This sentence is false. ■

We are all quite adept at building new propositions from old ones in ordinary English. There are three basic ways of building new propositions: we can connect two propositions with "and," we can connect two propositions with "or," or we can negate a proposition. In this text we use capital letters P, Q, R, and so forth to represent propositions, $P \wedge Q$ to represent P and Q, $P \vee Q$ to represent P or Q, and P' to represent the denial of proposition P.

Suppose that we are given boxes □ and ◇ into which we may place propositions. Since, by definition, any proposition placed in the box can have only two truth values, we can tabulate the behavior of $\wedge$, $\vee$, and $'$ and thereby define these signs by listing all possible inputs and the possible outputs in tabular form. The resulting tables are called *truth tables* for $\wedge$, $\vee$, and $'$. There are two commonly accepted ways to indicate the truth values in these truth tables. One is to use T or F (for true or false) and the other is to use 1 or 0 (for true or false). The definitions of $\wedge$, $\vee$, and $'$ are given by the truth tables in Figure 1.1.

Figure 1.1

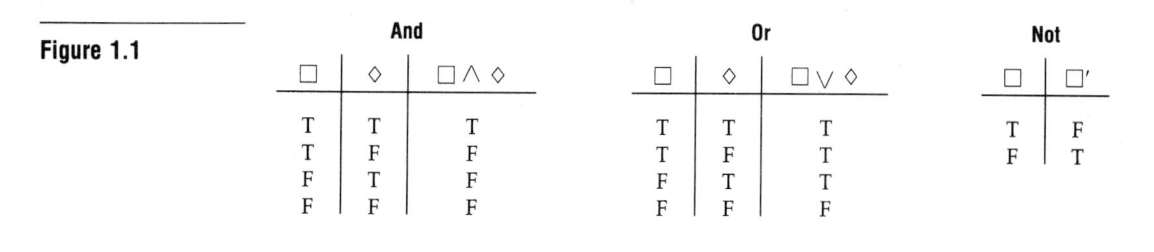

And

□	◇	□ ∧ ◇
T	T	T
T	F	F
F	T	F
F	F	F

Or

□	◇	□ ∨ ◇
T	T	T
T	F	T
F	T	T
F	F	F

Not

□	□′
T	F
F	T

Example 3

Assume that the proposition "Jo owns a car" is true and that the proposition "Jo lives in Texas" is false. Under these assumptions,

a. "Jo owns a car and Jo lives in Texas" is false.
b. "Jo owns a car and Jo does not live in Texas" is true.
c. "Jo owns a car or Jo lives in Texas" is true.
d. "Jo owns a car or Jo does not live in Texas" is true.
e. "It is not the case that Joe does not own a car and Jo lives in Texas" is ambiguous.

If we mean that it is not the case that Jo does not own a car, and furthermore that Jo does live in Texas, then sentence (e) is false. But since this sentence has another reasonable interpretation, you should see that we often need parentheses in forming compound propositions. ■

The definitions of two other connectives, $\rightarrow$ and $\leftrightarrow$, called the *conditional* and the *biconditional*, are given by the truth tables in Figure 1.2.

Figure 1.2

Conditional (implies)

□	◇	□ → ◇
T	T	T
T	F	F
F	T	T
F	F	T

Biconditional (if and only if)

□	◇	□ ↔ ◇
T	T	T
T	F	F
F	T	F
F	F	T

Given propositions P and Q, the conditional proposition $P \rightarrow Q$ crops up in our daily lives, and the English language has many ways to express this proposition. Here are some of the ways that are used to express that $P \rightarrow Q$ is true.

P implies Q.

If P, then Q.

Q if P.

If P is true, then Q is true.

Q is true whenever P is true.

P is true only if Q is true.

For P to be true it is necessary that Q be true.

For Q to be true it is sufficient that P be true.

In a conditional proposition $P \to Q$, P is called the *antecedent*, or *hypothesis*, and Q is called the *consequent*, or *conclusion*. "Antecedent" and "hypothesis" (and "consequent" and "conclusion") are often used interchangeably, but the term "hypothesis" has the popular interpretation that it is something one assumes true. As we see in the discussion of the direct method of proof in Section 1.4, this is the correct interpretation when one is proving a theorem. It is not, however, the correct interpretation in a conditional proposition. The distinction is that a conditional is true or false and a proof is valid or invalid. The truth table of the conditional is easy to remember because there is only one case in which the implication is false—namely, when the antecedent is true and the consequent is false. In law, this corresponds to a breach of contract in which one person has agreed to a consequence Q provided that a condition P is met. If P is true and yet Q is false, the first party to the contract has met the required condition P and yet the second party to the contract has failed to fulfill the promised consequence Q. Note that in the three remaining cases there has been no breach of contract. In the first line of the truth table, the first party has met the required condition and sure enough the second party has paid off by fulfilling the agreed upon consequence. In the last two lines of the table, the first party has failed to meet the conditions of the contract and so there is no breach of contract regardless of whether the second party has fulfilled the consequence.

We illustrate the truth value combinations for the conditional with the following four compound propositions:

P: If 7 is a positive integer, then Atlanta is the capital of Georgia.

Q: If Des Moines is the capital of Iowa, then 4 is a prime number.

R: If $4 > 7$, then George Washington was the first president of the United States.

S: If 5 is an even integer, then no one will pass this course.

Proposition P is true, since it is composed of two propositions each of which is true. Proposition Q is false, since it is composed of a true hypothesis and a false conclusion. Propositions R and S are true, since the hypothesis of each is false. Note that, even though S is true, S does not say that no one will pass this course.

We see from the truth table in Figure 1.2 that, for propositions P and Q, the biconditional $P \leftrightarrow Q$ is true when P and Q are both true and when P and Q are both false but is false in the remaining two cases. Thus we can think of $P \leftrightarrow Q$ as follows:

P if and only if Q.

If P then Q, and if Q then P.

P implies Q, and Q implies P.

P is a necessary condition for Q, and P is a sufficient condition for Q.

Just as with the conditional, one should be careful to distinguish a biconditional proposition $P \leftrightarrow Q$ and a proof of a theorem of the form "P if and only if Q." Again, in the direct proof of the theorem one assumes that P is true and proves that Q is true and then assumes that Q is true and proves that P is true. One does not assume that P or Q is true in a biconditional proposition.

We illustrate the truth value combinations for the biconditional with the following four compound propositions:

P: George Washington was the first president of the United States if and only if $3 < 4$.

Q: $2 < 5$ implies that $13 < 4$, and $13 < 4$ implies that $2 < 5$.

R: For $2 = 1$, it is necessary and sufficient that $3 < 4$.

S: The moon is made of cheese if and only if $4 > 7$.

Proposition P is true, since it is composed of two simple propositions each of which is true. Propositions Q and R are false, since each is composed of one proposition that is true and one that is false. Proposition S is true, since it is composed of two simple propositions each of which is false.

Exercises 1.1

1. In Example 2 we consider the sentence "This sentence is false." Discuss the fairness of the following true–false question:

   ________ This sentence is false.

2. Identify the antecedent and the consequent in each of the following propositions.

   a. If $3 > 4$, then $\sqrt{2}$ is irrational.

   b. John plays golf whenever it is not cloudy.

   c. When the Boston Celtics win, the mayor of Boston gets free tickets.

d. In order to pass this course, it is sufficient for the student to pass the final exam.

e. In order to pass this course, the student must do all the homework assignments.

3. Let P be the proposition "$2 + 3 = 5$" and let Q be the proposition "The moon is made of cheese." Find the symbolic form of the following statements.

a. $2 + 3 \neq 5$.

b. $2 + 3 \neq 5$ and the moon is not made of cheese.

c. $2 + 3 \neq 5$ or the moon is made of cheese.

d. If $2 + 3 \neq 5$, then the moon is made of cheese.

e. That the moon is made of cheese implies that $2 + 3 \neq 5$.

4. Let P be the proposition "$\sqrt{2}$ is irrational" and let Q be the proposition "The moon is made of green cheese." Translate the following symbolic statements into grammatically correct English sentences.

a. $P' \wedge Q$ **b.** $(P \wedge Q)'$ **c.** $P' \wedge (P \vee Q')$
d. $Q' \rightarrow P'$ **e.** $P' \rightarrow (P \wedge Q')$ **f.** $P \leftrightarrow (Q' \rightarrow P)$

5. Assume that "Mary is a girl" is a true statement and that "Mary is ten years old" is a true statement. Which of the following are true?

a. Mary is ten years old or Mary is a girl.

b. If Mary is ten years old, then Mary is a girl.

c. Mary is ten years old and Mary is a girl.

d. Mary is ten years old if and only if Mary is a girl.

e. Mary is not a girl.

f. Mary is not a ten-year-old girl.

6. Assume that "Joe is a girl" is a false statement and that "Mary is ten years old" is a true statement. Which of the following are true?

a. Joe is a girl or Mary is ten years old.

b. Joe is a girl and Mary is ten years old.

c. If Mary is ten years old, then Joe is a girl.

d. If Joe is a girl, then Mary is ten years old.

e. Joe is a girl if and only if Mary is ten years old.

7. Assume that "Joe is a girl" is a false statement and that "Joe is ten years old" is a false statement. Which of the following are true?

a. Joe is ten years old or Joe is a girl.

b. If Joe is ten years old, then Joe is a girl.

c. Joe is ten years old and Joe is a girl.

d. Joe is ten years old if and only if Joe is a girl.

e. Joe is not a girl.

f. Joe is not a ten-year-old girl.

1.2

EXPRESSIONS AND EQUIVALENT EXPRESSIONS

In the previous section we considered several fundamental ways to combine propositions to form compound propositions and we used squares and diamonds as placeholders in writing the associated truth tables. There is an obvious disadvantage to our placeholder notation. What other figures can we use? We could use pentagons, hexagons, and so on, but if we are considering a compound proposition made up of very many simple propositions it would be annoying counting up sides. And although the human eye can readily distinguish between a square and a pentagon, the difference between a small nonagon and a small decagon is not readily observable. We therefore introduce the idea of a *propositional expression*. We are given an alphabet of *variables*, and we have $\vee$, $\wedge$, $'$, and parentheses. We add the signs $\rightarrow$ and $\leftrightarrow$. Using variables, these signs, and parentheses, we form expressions. We rely on your intuition rather than attempt a formal definition of a propositional expression. Each variable is a propositional expression, and if the variables and symbols are combined in a way that appears meaningful, the result is an expression. For example, $(X \vee Y)' \wedge (X' \rightarrow (Y \leftrightarrow Z))$ is an expression, as is $(X \vee Y) \leftrightarrow (X \wedge Y)$; we hope you do not believe that $X\ X \rightarrow\ )) \wedge (Y \leftrightarrow$ is also an expression. Note that it makes no sense to ask if the expression $(X \vee Y) \leftrightarrow (X \wedge Y)$ is true. As yet, this expression has no meaning whatsoever, but when we replace the variables X and Y with propositions P and Q we obtain the proposition $P \vee Q \leftrightarrow P \wedge Q$, which like any other proposition is either true or false.

What does it mean to say that two propositional expressions are equivalent? We give one answer now and another answer in Chapter 7.

Definition

Two expressions are **equivalent** provided that they have the same truth values for all possible values of true or false for all variables appearing in either expression.

When two expressions X and Y are equivalent, we write $X \Leftrightarrow Y$ and say that X and Y are *logically equivalent*. The signs $\leftrightarrow$ (biconditional) and $\Leftrightarrow$ (logical equivalence) should not be confused. The first is one of the signs we may use to *form* expressions, but the second sign cannot appear *in* any propositional expression. In other words, an expression may contain the biconditional ($\leftrightarrow$), but logical equivalence ($\Leftrightarrow$) tells something about the relationship between two expressions. We consider some examples to illustrate the notion of equivalent expressions.

Example 4

Are the expressions $X' \vee Y$ and $X \rightarrow Y$ equivalent?

Analysis

We have only two variables, and so the truth tables of each expression have four rows each:

X	Y	$X \rightarrow Y$
T	T	T
T	F	F
F	T	T
F	F	T

X	Y	X'	$X' \vee Y$
T	T	F	T
T	F	F	F
F	T	T	T
F	F	T	T

In each of the four possible cases, the resulting values for $X \rightarrow Y$ and $X' \vee Y$ are the same. Therefore the expressions are equivalent. ■

Example 5

Are $X \wedge Y$ and $(X \wedge Y) \vee (W \wedge W')$ equivalent expressions?

Analysis

There are eight possible combinations of truth values for the three variables X, Y, W.

X	Y	W	$X \wedge Y$	$W \wedge W'$	$(X \wedge Y) \vee (W \wedge W')$
T	T	T	T	F	T
T	T	F	T	F	T
T	F	T	F	F	F
T	F	F	F	F	F
F	T	T	F	F	F
F	T	F	F	F	F
F	F	T	F	F	F
F	F	F	F	F	F

In each of the eight possible combinations of truth values, the resulting values for $(X \wedge Y) \vee (W \wedge W')$ and $X \wedge Y$ are the same. Therefore the expressions are equivalent. ■

Example 6

Are the expressions $W \vee X$ and $Y \vee X$ equivalent?

Analysis

Substituting a false proposition for W, a false proposition for X, and a true proposition for Y yields a false proposition for the first expression and a true proposition for the second expression. Therefore the two expressions are not equivalent. We write the truth tables of these two expressions to show that one must be careful to consider all variables when determining equivalence of expressions.

W	X	$W \vee X$
T	T	T
T	F	T
F	T	T
F	F	F

Y	X	$Y \vee X$
T	T	T
T	F	T
F	T	T
F	F	F

When we list all the variables appearing in either expression in tabular form, we see that the two expressions do not have the same truth values for all possible values of true or false for *all* variables appearing in either expression. Thus the expressions are not equivalent.

W	X	Y	$W \vee X$
T	T	T	T
T	T	F	T
T	F	T	T
T	F	F	T
F	T	T	T
F	T	F	T
F	F	T	F
F	F	F	F

W	X	Y	$Y \vee X$
T	T	T	T
T	T	F	T
T	F	T	T
T	F	F	F
F	T	T	T
F	T	F	T
F	F	T	T
F	F	F	F

■

Definition

If P and Q are propositions, the **converse** of $P \rightarrow Q$ is $Q \rightarrow P$.

Example 7

The converse of the conditional proposition "If $4 > 3$, then the moon is made of cheese" is "If the moon is made of cheese, then $4 > 3$."

Observe that the propositional expression $X \rightarrow Y$ and its converse $Y \rightarrow X$ are not logically equivalent. In particular, if X and Y have different truth values, then $X \rightarrow Y$ and $Y \rightarrow X$ have different truth values. ■

Definition

If P and Q are propositions, the **contrapositive** of $P \rightarrow Q$ is $Q' \rightarrow P'$.

The contrapositive of "If $4 > 3$, then the moon is made of cheese" is "If the moon is not made of cheese, then $4 \leqslant 3$." Observe that, in writing the contrapositive of a conditional proposition, the denial of the conclusion becomes the hypothesis, and the denial of the hypothesis becomes the conclusion. In Example 7, both the conditional proposition and the contrapositive of this conditional proposition are false. By constructing a truth table, we can see that $X \to Y$ and the contrapositive $Y' \to X'$ are logically equivalent expressions.

X	Y	$X \to Y$	Y'	X'	$Y' \to X'$
T	T	T	F	F	T
T	F	F	T	F	F
F	T	T	F	T	T
F	F	T	T	T	T

To lessen the use of parentheses, we agree that in the absence of parentheses "negation" takes precedence over "and" and "or," and "and" and "or" take precedence over the conditional and biconditional. Thus $X' \vee Y'$ means $(X') \vee (Y')$ rather than $(X' \vee Y)'$.

Exercises 1.2

8. Write the converse and contrapositive of the following propositions.
   a. If today is Friday, then I will party tonight.
   b. If $3 > 4$, then George Bush was the thirty-first president of the United States.
   c. The number x is positive whenever it is the square root of a positive number.
   d. In order for the number x to be negative, it must be less than its absolute value.
9. Construct truth tables for each of the following expressions.
   a. $(X \wedge Y)'$ b. $X' \wedge Y'$ c. $(X \to Y) \leftrightarrow (Y' \to X')$
   d. $X \to (Y \to Z)$ e. $(X \to Y) \to Z$ f. $(X \vee Y)' \to Z$
   g. $((X \vee Y) \to Z)'$
10. Let P be the proposition $2 = 5$. If possible, find a proposition Q such that $P \vee Q \leftrightarrow P \wedge Q$ is true. If possible, find a proposition Q such that $P \vee Q \leftrightarrow P \wedge Q$ is false.
11. a. Write the truth tables for the expressions $(X \leftrightarrow Y) \to Z$ and $X \leftrightarrow (Y \to Z)$.
    b. Find propositions P, Q, and R such that one of the two compound propositions $(P \leftrightarrow Q) \to R$ and $P \leftrightarrow (Q \to R)$ is true and the other is false.
12. Under the assumption that $P \to Q$ is false, give the truth values for the following expressions.
    a. $Q \to P$ b. $P \leftrightarrow Q$ c. $P \vee Q$ d. $P \wedge Q$
13. Denote "Mary is a girl" by P and "Joe is a boy" by Q.
    a. Complete the following truth table:

	P	Q	$P \to Q$
(1)	T	T	
(2)	T	F	
(3)	F	T	
(4)	F	F	

b. Which line of the truth table has P and $P \to Q$ both true?

c. Assume that "Mary is a girl" and "If Mary is a girl then Joe is a boy" are true statements. Can you conclude that Joe is a boy?

14. Denote "Mary is a girl" by P, "Joe is a boy" by Q, and "Mary is ten years old" by R.

a. Assume that P is true, that $P \to Q$ is true, and that $Q \to R$ is true. Using a method similar to that used in Exercise 13, determine if Mary is ten years old.

b. Assume that $P \vee Q$ is true and that P' is true. Determine if Joe is a boy.

c. Assume that $P \wedge Q$ is true and that P' is true. Determine if Mary is ten years old.

15. Which of the following statements are true?

a. $(P \leftrightarrow Q) \Leftrightarrow (Q \leftrightarrow P)$

b. $(P \leftrightarrow Q)' \Leftrightarrow (P' \to Q')$

c. $(P' \leftrightarrow Q) \Leftrightarrow (P \leftrightarrow Q')$

d. $(P' \to Q) \Leftrightarrow (P' \wedge Q')$

1.3

TAUTOLOGIES AND CONTRADICTIONS

In the preceding section we used an alphabet of variables consisting of W, X, Y, and Z and reserved the letters P, Q, R, and S to name propositions. The laws of logic, however, are usually stated using the letters P, Q, R, and S, and from now on we use the letters P, Q, R, and S both for the names of propositions and as variables of our expressions. The result is instantaneous ambiguity. If we write $P \to Q$, we may be referring to the expression $P \to Q$, in which case P and Q are variables, or we may have in mind two propositions such as P = "George Washington slept at the Waldorf Astoria" and Q = "The Waldorf Astoria is a historic site," in which case $P \to Q$ is a proposition.

Because we must determine from context whether we are considering a proposition or an expression, it may be worthwhile reconsidering the features that distinguish the two. A proposition is true or false. An expression has no meaning until its variables are replaced by propositions; consequently, an expression cannot be true or false. Each expression has a truth table. This table lists the truth or falsity of those propositions that can be obtained by replacing the variables in the expression by all allowable combinations of true and false. A proposition has no truth table, because by definition a proposition has only one truth value. Finally, two expressions can be logically equivalent, but it does not make sense to talk about two propositions being logically equivalent.

Some compound expressions always yield a true proposition no matter what propositions replace their variables, and some compound expressions always yield a false proposition. Consider, for example, the truth tables for the expressions $P \vee P'$ and $P \wedge P'$:

P	P'	$P \vee P'$
T	F	T
F	T	T

P	P'	$P \wedge P'$
T	F	F
F	T	F

We notice that $P \vee P'$ is always true and that $P \wedge P'$s is always false.

Definition

A propositional expression is called a **tautology** if it yields a true proposition regardless of what propositions replace its variables. A propositional expression is called a **contradiction** if it yields a false proposition regardless of what propositions replace its variables.

Recall that two expressions P and Q are logically equivalent provided that they have the same truth values for all possible assignments of truth values to their variables. Thus all tautologies are equivalent, as are all contradictions. We reemphasize the difference between $\Leftrightarrow$ and $\leftrightarrow$. $P \leftrightarrow Q$ is an expression and so has no truth value, but the proposition $P \Leftrightarrow Q$ is true if and only if the expression $P \leftrightarrow Q$ is a tautology.

We have seen that the conditional $P \rightarrow Q$ and its contrapositive $Q' \rightarrow P'$ are logically equivalent expressions, and the tautology $(P \rightarrow Q) \leftrightarrow (Q' \rightarrow P')$ is among the best-known tautologies in mathematics. In Example 4 we established that $(P \rightarrow Q) \leftrightarrow (P' \vee Q)$ is a tautology, as well.

We have given one example of a contradiction, $P \wedge P'$. The following example provides another contradiction.

Example 8

The expression $(P \rightarrow Q) \wedge (P \wedge Q')$ is a contradiction.

Analysis

The following truth table presents all possible truth values of P and Q.

P	Q	$P \to Q$	Q'	$P \wedge Q'$	$(P \to Q) \wedge (P \wedge Q')$
T	T	T	F	F	F
T	F	F	T	T	F
F	T	T	F	F	F
F	F	T	T	F	F

Since $(P \to Q) \wedge (P \wedge Q')$ is false regardless of what truth values are assigned to P and Q, it is a contradiction. ■

Let P, Q, and R be variables; let t be any tautology and c be any contradiction. The following theorem can be established by constructing appropriate truth tables (see Exercise 16c for an example).

Theorem 1.1

a. $P \vee Q \Leftrightarrow Q \vee P$ — commutative laws
$P \wedge Q \Leftrightarrow Q \wedge P$

b. $P \vee (Q \vee R) \Leftrightarrow (P \vee Q) \vee R$ — associative laws
$P \wedge (Q \wedge R) \Leftrightarrow (P \wedge Q) \wedge R$

c. $P \wedge (Q \vee R) \Leftrightarrow (P \wedge Q) \vee (P \wedge R)$ — distributive laws
$P \vee (Q \wedge R) \Leftrightarrow (P \vee Q) \wedge (P \vee R)$

d. $c \vee P \Leftrightarrow P$ — identity laws
$t \wedge P \Leftrightarrow P$

e. $P \vee P' \Leftrightarrow t$
$P \wedge P' \Leftrightarrow c$

If we agree that all logically equivalent expressions are the same, we can glean a number of important results in symbolic logic for free. Any of these results can also be established by constructing appropriate truth tables.

Theorem 1.2

Let P, Q, and R be variables; let t be any tautology and c be any contradiction. Then the following logical equivalences hold:

a. $P \vee t \Leftrightarrow t$
b. $P \wedge c \Leftrightarrow c$
c. $P \vee P \Leftrightarrow P$
d. $P \wedge P \Leftrightarrow P$
e. $t' \Leftrightarrow c$
f. $c' \Leftrightarrow t$
g. $P'' \Leftrightarrow P$
h. $(P \vee Q)' \Leftrightarrow P' \wedge Q'$ — de Morgan's laws
i. $(P \wedge Q)' \Leftrightarrow P' \vee Q'$ — de Morgan's laws

Theorem 1.3

If P and Q are variables, then $(P \to Q)' \Leftrightarrow (P \wedge Q')$.

Proof

Recall that in Example 4 we showed that $P \rightarrow Q$ and $P' \vee Q$ are equivalent expressions. Therefore $(P \rightarrow Q)' \Leftrightarrow (P' \vee Q)'$. By (g) and (h) of Theorem 1.2, $(P' \vee Q)' \Leftrightarrow (P \wedge Q')$. Hence $(P \rightarrow Q)' \Leftrightarrow (P \wedge Q')$. □

Example 9

Write the negation of "If $|2| = |-2|$, then $2 = -2$."

Analysis

By Theorem 1.3, the negation is "$|2| = |-2|$, but $2 \neq -2$." ■

Exercises 1.3

16. Let P, Q, and R be variables. Prove that the following expressions are tautologies.

a. $P \rightarrow P$
b. $(P')' \leftrightarrow P$
c. $P \wedge (Q \vee R) \leftrightarrow (P \wedge Q) \vee (P \wedge R)$
d. $(P \vee Q)' \leftrightarrow P' \wedge Q'$

17. Write a useful negation of each of the following propositions.

a. FORTRAN is a language, or SUPERCALC is a branch of mathematics.
b. The course is easy and I will pass.
c. If you can write Pascal programs, then you can easily learn FORTRAN.
d. If you know the password, then you can log onto the computer.

18. Prove that the following propositional expressions are contradictions.

a. $((P \vee Q) \wedge P') \wedge Q'$ b. $(P \wedge Q) \wedge P'$

19. Write the contrapositive and a nontrivial negation of each of the following propositions. (A negation that begins "It is not true that . . ." is trivial.)

a. If IBM is a museum or $3 < 4$, then $2^2 = 4$ and Raleigh is the capital of North Carolina.
b. If Digital Equipment manufactures brass instruments and Texas Instruments manufactures guitars, then $3 < 2$ or Mozart was a musician.
c. If $|-3| = |3|$ or the Mississippi is a river, then Buffalo is the capital of New York or Mickey Mantle was a movie star.
d. If Michael Jackson was president and Bob Hope was a pro golfer, then Elvis Presley was a rock star and Marilyn Monroe was a redhead.

20. Write, in symbols, the converse, contrapositive, and negation of each of the following propositional expressions.

a. $P \rightarrow (Q \vee R)$ b. $P \rightarrow (Q \wedge R)$ c. $(P \vee Q) \rightarrow R$
d. $(P \wedge Q) \rightarrow R$

21. Which of the following expressions are logically equivalent to $P \rightarrow Q$?

a. $P \rightarrow Q$ b. $P' \rightarrow Q$ c. $Q' \rightarrow P$ d. $P' \rightarrow Q'$
e. $Q' \rightarrow P'$ f. $P \vee Q'$ g. $P \wedge Q'$ h. $P' \vee Q$
i. $P' \wedge Q$ j. $P \vee Q$ k. $P \wedge Q$ l. $Q \rightarrow P$

22. Construct a truth table to prove or disprove each of the following statements.

 a. $[P \wedge (P \vee Q)] \Leftrightarrow P$
 b. $[(P \rightarrow Q) \rightarrow R] \Leftrightarrow [(P \vee R) \wedge (Q \rightarrow R)]$
 c. $(P \rightarrow Q) \Leftrightarrow (P \vee Q)$
 d. $[(P \rightarrow Q) \rightarrow (P \rightarrow R)] \Leftrightarrow [P \rightarrow (Q \rightarrow R)]$

23. Suppose that "John is smart," "John or Mary is ten years old," and "If Mary is ten years old, then John is not smart" are each true statements. Which of the following statements are true?

 a. John is not smart.
 b. Mary is ten years old.
 c. John is ten years old.
 d. Either John or Mary is not ten years old.

*24. On one side of a 3×5 card write the sentence "The sentence on the other side of this card is true." On the other side write the sentence "The sentence on the other side of this card is false." Argue that neither sentence can be either true or false.

*25. It is rumored that on an island in the Pacific there are two tribes of people, the Truth Tellers and the Liars. The Truth Tellers always tell the truth and the Liars always lie. A visitor hiking across the island comes to a fork in the road, and he knows that one branch leads to the Truth Tellers' village and the other to the Liars' village. Of course he wishes to go to the Truth Tellers' village. A man is standing at the fork of the road, but the visitor does not know if he is a Truth Teller or a Liar. What one question can the visitor ask the man to discover for certain the way to the Truth Tellers' village?

1.4 METHODS OF PROOF

Let us examine a proposition of the form $P \rightarrow Q$. We know that there are three conditions under which this proposition is true: (1) P and Q are true; (2) P is false and Q is true; or (3) P is false and Q is false. We observe that, if P is false, then $P \rightarrow Q$ is true. Therefore, to prove that P implies Q, it is sufficient to assume that P is true, and, under this assumption, to prove that Q is true. The *direct method* of proof is precisely this method; that is, we assume that P is true and, under this assumption, we proceed through a logical sequence of steps to arrive at the conclusion that Q is true. Before we illustrate the direct method with a simple theorem, we remind you that an integer n is *odd* provided that there is an integer k such that $n = 2k + 1$, and that n is *even* provided that there is an integer q such that $n = 2q$. So, 19 is odd since there is an integer 9 such that $19 = 2(9) + 1$, and 46 is even since there is an integer 23 such that $46 = 2(23)$.

Theorem 1.4 If n is an odd integer, then n^2 is odd.

Proof

The hypothesis is "n is an odd integer," and the conclusion is "n^2 is odd." So we begin by assuming that n is odd. Then, by definition, there is an integer k such that $n = 2k + 1$. So

$$n^2 = (2k + 1)(2k + 1) = 4k^2 + 4k + 1 = 2(2k^2 + 2k) + 1$$

Therefore, there is an integer m (namely, $m = 2k^2 + 2k$) such that $n^2 = 2m + 1$. So, by definition, n^2 is odd. We have proved that if the hypothesis is true then the conclusion is true; so the proof is complete. □

In Section 1.2, we saw that the proposition $P \to Q$ is equivalent to its contrapositive $Q' \to P'$. Thus, since an integer is either odd or even, Theorem 1.4 may be restated as follows: if n^2 is an even integer, then n is even. In general, another way of proving any proposition of the form "If P, then Q" is to assume that Q is not true, and, under this assumption, to proceed through a logical sequence of steps to arrive at the conclusion that P is not true. This is the *contrapositive method* of proof. We can illustrate the contrapositive method with another simple theorem. Recall that a nonzero integer a *divides* an integer b provided that there is an integer q such that $b = aq$. A natural number p is a *prime* if $p \geqslant 2$ and p has exactly two positive divisors, itself and 1. Our example of a contrapositive proof involves the concept of a perfect number. A natural number n is a *perfect number* provided that it is equal to the sum of its smaller positive divisors. Thus 6 is a perfect number, since $6 = 1 + 2 + 3$, and $28 = 1 + 2 + 4 + 7 + 14$ is another perfect number.

Theorem 1.5 If n is a perfect number, then n is not a prime.

Proof

The contrapositive of this theorem is "If n is a prime, then n is not a perfect number." Suppose that n is a prime number. Then $n \geqslant 2$ and n has only one positive divisor other than n, namely, 1. Therefore n is not equal to the sum of its smaller positive divisors, and hence n is not a perfect number. We have proved that if n is a prime number then n is not a perfect number. Since the contrapositive is known to be equivalent to the original implication, we have proved the original implication. □

A third method of proof is the *contradiction method*. Recall that $(X \to Y)'$ and $X \wedge Y'$ are logically equivalent expressions. In a proof by contradiction of a proposition $P \to Q$, we begin by assuming that the proposition

P is true and that the proposition Q is false. We wish to find a proposition R such that $P \wedge Q' \rightarrow R \wedge R'$. If we find such a proposition R, we say that we have reached a contradiction. Since a true statement never implies a false statement, if $(P \wedge Q') \rightarrow R \wedge R'$, then $(P \wedge Q')$ is a false proposition and $(P \wedge Q')'$ is a true proposition. Because $(X \wedge Y')'$ and $X \rightarrow Y$ are logically equivalent expressions, they have the same truth value for all possible substitutions of propositions for the variables X and Y. In particular, $(P \wedge Q')'$, which is true, has the same truth value as $P \rightarrow Q$. Thus we know that $P \rightarrow Q$ is a true proposition.

The most difficult step in the contradiction method is to decide which contradiction to look for. There are no specific guidelines, since each proof gives rise to its own contradiction, but *any* contradiction will do.

The advantage of the contradiction method over the contrapositive method is that we get two statements from which to reason rather than just one. The disadvantage is that we have no definite knowledge of where the contradiction will arise. In general, the best rule of thumb is to use contradiction when the statement "not Q" gives some useful information. We consider a simple example in Theorem 1.6. Recall that a real number r is *rational* provided that there are integers m and n, with $n \neq 0$, such that $r = m/n$, and that r is *irrational* if it is not rational.

Theorem 1.6 If r is a real number such that $r^2 = 2$, then r is irrational.

Proof

Suppose $r^2 = 2$ and r is not irrational. Then r is rational, so there are integers m and n such that $r = m/n$. We may assume that m and n do not have common divisors greater than 1, because if they did we could divide both numerator and denominator by the greatest common divisor. So $r^2 = m^2/n^2$, and hence $m^2 = r^2n^2$. Since $r^2 = 2$, we have $m^2 = 2n^2$. Hence m^2 must be even. By Theorem 1.4, m must be even. (Why?) Therefore there is an integer p such that $m = 2p$. So $2n^2 = m^2 = 4p^2$, and hence $n^2 = 2p^2$. It follows that n^2 is even, and, again by Theorem 1.4, n must be even. Since m and n are both even, they have a common divisor greater than 1. This is a contradiction to our assumption that m and n have no common divisors greater than 1, so we have established the theorem. □

The three methods of proof we have discussed are reviewed in Table 1.1. In each case, we are concerned with proving that P implies Q. The advantage and the difficulty of the contradiction method are illustrated by the table. We have more information to work with since we are assuming both P and Q'. On the other hand, R suddenly appears (that is, we do not know R before we begin the proof).

Table 1.1

Direct method	Contrapositive method	Contradiction method
Assume P	Assume Q'	Assume P and Q'
⋮	⋮	⋮
(logical sequence of steps)	(logical sequence of steps)	(logical sequence of steps)
⋮	⋮	⋮
Conclude Q	Conclude P'	Conclude R and R'

Suppose that P and Q are propositions and that we want to prove a theorem that states "P if and only if Q." Then we really have two theorems to prove: we must prove that if P is true then Q is true, and we must prove that if Q is true then P is true. Consider the following examples.

Theorem 1.7 An integer n is even if and only if n^2 is even.

Proof

The contrapositive of Theorem 1.4 tells us that if n^2 is even then n is even. So, to complete the proof of the theorem, it is sufficient to show that if n is even then n^2 is even.

Suppose n is even. Then, by definition, there is an integer k such that $n = 2k$. Therefore $n^2 = (2k)^2 = 4k^2 = 2(2k^2)$. Hence, by definition, n^2 is even. □

Theorem 1.8 Let x be a real number. Then $x = 1$ if and only if $x^3 - 3x^2 + 4x - 2 = 0$.

Proof

We know that x is a real number, and we must prove two things: (1) if $x = 1$, then $x^3 - 3x^2 + 4x - 2 = 0$, and (2) if $x^3 - 3x^2 + 4x - 2 = 0$, then $x = 1$.

First, suppose $x = 1$. Then $x^3 - 3x^2 + 4x - 2 = 1 - 3 + 4 - 2 = 0$. This proves the first part of the theorem—that, if $x = 1$, then $x^3 - 3x^2 + 4x - 2 = 0$.

Now let us prove the second part. We assume that $x^3 - 3x^2 + 4x - 2 = 0$. Since $x^3 - 3x^2 + 4x - 2 = (x - 1)(x^2 - 2x + 2)$, either $x - 1 = 0$ or $x^2 - 2x + 2 = 0$ (using the property that, if the product of two real numbers is 0, then at least one of the numbers must be 0). If $x^2 - 2x + 2 = 0$, the quadratic formula yields $x = 1 \pm i$. This contradicts the hypothesis that x is real. Therefore $x^2 - 2x + 2 \neq 0$. Hence $x - 1 = 0$, and therefore $x = 1$.

In the second half of this proof, we have two propositions $x - 1 = 0$ and $x^2 - 2x + 2 = 0$ such that $(x - 1 = 0) \vee (x^2 - 2x + 2 = 0)$ is true. We show that $x^2 - 2x + 2 = 0$ implies that x is not a real number. Therefore, by the contrapositive, since x is a real number, $x^2 - 2x + 2 \neq 0$. Therefore, for $(x - 1 = 0) \vee (x^2 - 2x + 2 = 0)$ to be true, $x - 1 = 0$ must be true. □

We give another example of a direct proof in Theorem 1.9.

Theorem 1.9 If a, b, and c are integers and a divides b, then a divides bc.

Proof

Since a divides b, there is an integer q such that $b = aq$. So $bc = (aq)c$, and hence qc is an integer such that $bc = a(qc)$. Therefore a divides bc. □

The proof of Theorem 1.10 illustrates that a theorem can be proved by considering more than one possibility.

Theorem 1.10 If $-x^2 - x + 2 > 0$, then $-2 < x < 1$.

Proof

Let x be any real number such that $-x^2 - x + 2 > 0$. Note that $-x^2 - x + 2 = -(x + 2)(x - 1)$. So $-x^2 - x + 2 > 0$ if and only if $(x + 2)(x - 1) < 0$. Now the product of two real numbers is negative if and only if one of the numbers is negative and the other one is positive, so we have two cases to consider. We consider the first case, $x + 2 > 0$ and $x - 1 < 0$. In this case, x must be greater than -2 and less than 1; that is, $-2 < x < 1$. In the second case, $x + 2 < 0$ and $x - 1 > 0$. Thus x must be less than -2 and greater than 1. No real number satisfies both conditions; that is, there is no real number x such that $x < -2$ and $x > 1$.

We have now shown that any real number x that satisfies the condition $-x^2 - x + 2 > 0$ also has the property that $-2 < x < 1$. Since the proof is independent of the real number x chosen, as long as it satisfies the condition $-x^2 - x + 2 > 0$, we have proved the theorem for all real numbers x. □

We give another example of proof by contradiction in Theorem 1.11.

Theorem 1.11 There do not exist prime numbers a, b, and c such that $a^3 + b^3 = c^3$.

Proof

The proof is by contradiction. We want to show that if a, b, and c are primes then $a^3 + b^3 \neq c^3$. Suppose that there are primes a, b, and c such that $a^3 + b^3 = c^3$. If a and b are odd, then a^3 and b^3 are odd, so c^3 is even. Therefore c is an even prime, so $c = 2$, but this is impossible since a and b are both greater than 2. So at least one of a and b is even, and hence we may assume that $b = 2$. Therefore $b^3 = 8 = c^3 - a^3 = (c - a)(c^2 + ca + a^2)$. Now a and c are primes, so each of c^2, ca, and a^2 is greater than or equal to 4 and hence the sum is greater than or equal to 12. Thus $|c^3 - a^3| \geqslant 12$. This contradicts the assumption that $c^3 - a^3 = 8$. □

We have been considering how to prove a given statement, but you may encounter a conjecture that is false. For example, suppose someone asks you to prove that all odd numbers greater than 100 are prime. Just because one or more attempted proofs breaks down, you cannot be sure that you have not just overlooked some clever argument that establishes the desired result. But if you note that 105 is an odd number greater than 100 and that 105 is not prime, then you have an example to show that the statement is false. Such an example is called a *counterexample.* (The formal definition is given in Section 1.5.) Notice that 105 is by no means the only counterexample to our rather naive conjecture. Indeed, any number greater than 100 that ends in a 5 provides a counterexample. But there is no glory in pointing out a second example. Once you give one counterexample, the issue is settled.

"Like . . . is this a special case or does it hold for every letter in the alphabet?"

Before considering the exercises that follow, try the theorem on the blackboard in the preceding cartoon. This is a good example of a theorem for which you must be careful not to assume what you are proving, but you need to assume $a \neq -b$.

Exercises 1.4

26. Let a, b, and c be integers. Prove that, if a divides b and b divides c, then a divides c.

27. Let a, b, c, m, and n be integers. Prove that, if a divides b and c, then a divides $nb + mc$.

28. Let a be an even integer and b be an odd integer. Prove that $a + b$ is odd and ab is even.

29. Let x be a real number. Prove that $x = 2$ if and only if $x^3 - 2x^2 + 4x - 8 = 0$.

30 Let n be an integer such that n^2 is even. Prove that n^2 is divisible by 4.

31. Let x and y be positive real numbers such that $x \neq y$. Show that $x + y > 4xy/(x + y)$.

32. Let a, b, and c be integers such that $a^2 + b^2 = c^2$. Prove that at least one of a and b is even.

33. Let n be an integer. Prove that $n^3 - n$ is even.

34. Give a counterexample to each of the following statements.

a. For each $n \in \mathbb{N}$, $n^2 + n + 1$ is a prime number.

b. For each $x \in \mathbb{R}$, $x^3 < 0$ or $x^3 > x^2$.

c. For each $n \in \{2,3,5,7,11,13\}$, $2^n + 1$ is a prime number.

***d.** For each $n \in \{2,3,5,7,11,13\}$, $2^n - 1$ is a prime number.

1.5

QUANTIFIERS

Although we have implied the existence of infinitely many expressions, we have as yet no way of talking about more than a finite number of them at one time. In mathematics it is frequently necessary to talk about an infinite number of propositions in one fell swoop (even though it is not permissible to form a compound propositional expression with infinitely many components). It is common to have a statement $Q(x)$ for every element x of some set and to need to say something about all these statements at once. For example, we might have a statement $Q(n)$ for every natural number n. If for each element x of some set we have a statement $Q(x)$, we think of Q as a machine that picks out a proposition $Q(x)$ for each element x of the set. We call the machine Q a *predicate*.

Here are some examples of predicates defined on the set of integers: let $E(n)$ be the statement "n is an even integer"; let $O(n)$ be the statement "n is an odd integer"; let $P(n)$ be the statement "n is a prime"; and let $S(n)$ be the statement "n is a perfect square." Then $E(2)$ and $P(2)$ are true, but $O(2)$ and $S(2)$ are false. Also $O(3)$ and $P(3)$ are true, but $E(3)$ and $S(3)$ are false. Likewise $E(4)$ and $S(4)$ are true, but $O(4)$ and $P(4)$ are false.

We can use logical connectives to combine statements involving predicates just as we did for propositions. For example, $E(n) \vee O(n)$ says that n is even or n is odd. Thus for each integer n, $E(n) \vee O(n)$ is a true statement. On the other hand, $E(n) \wedge P(n)$ says that n is even and n is prime. Therefore $E(n) \wedge P(n)$ is true whenever $n = 2$ and it is false whenever $n \neq 2$.

Let U be a set called the *universe of discourse.* Statements such as "for each x belonging to U, $Q(x)$" are common, and they have a special notation. In particular, we use $(\forall x \in U)(Q(x))$ to mean "for each x belonging to the set U, $Q(x)$." We also use $(\exists x \in U)(Q(x))$ to mean "there exists an x belonging to the set U such that $Q(x)$." Thus $(\forall n)(E(n))$ means "for each integer n, n is an even integer." In this case, it is understood that n is an integer since we said that the predicate E is defined on the set of integers. Therefore $(\forall n)(E(n))$ is false. To see that this is the case, we have only to find at least one integer n such that $E(n)$ is false, and 1 is such an n. On the other hand, $(\exists n)(P(n))$ means "there is an integer n such that n is prime." Therefore $(\exists n)(P(n))$ is true.

Phrases such as "for some x" and "there exists an element x" are called *existential quantifiers,* and as we have seen $\exists$ is used to denote an existential quantifier. We also refer to the symbol $\exists$ as an existential quantifier. The sentences "there is an integer n such that n is prime" and "n is prime for some integer n" are two different ways of saying the same thing.

Phrases such as "for each x," "for any x," and "for all x" are called *universal quantifiers,* and $\forall$ is used to denote a universal quantifier. We also refer to the symbol $\forall$ as a universal quantifier. The sentences "for each integer n, n is a perfect square," "n is a perfect square for all integers n", and "for any integer n, n is a perfect square" are three different ways of saying precisely the same thing.

The notation $(\exists n)(E(n) \wedge P(n))$ says "there is an integer n such that n is even and n is a prime." Thus $(\exists n)(E(n) \wedge P(n))$ is true. The notation $(\forall n)(O(n) \vee S(n))$ says "for each integer n, n is odd or n is a perfect square." Thus $(\forall n)(O(n) \vee S(n))$ is false.

Example 10

For each real number x, let $Q(x)$ be the statement $|x| = x$. Then $(\forall x \in \mathbb{R})(Q(x))$ is false because, if x is any negative real number, then $|x| \neq x$. However, $(\exists x \in \mathbb{R})(Q(x))$ is true because, if x is any positive real number, then $|x| = x$. ■

Example 11

For each natural number n, let $Q(n)$ be the statement $n + 2 = 5$. Then $(\exists n \in \mathbb{N})(Q(n))$ is true because $Q(3)$ is true. Observe that $Q(n)$ is false whenever $n \neq 3$. So $(\forall n \in \mathbb{N})(Q(n))$ is false. ■

Example 12

For each real number x, let $Q(x)$ be the statement $x^2 \geqslant 0$. Then $(\forall x \in \mathbb{R})(Q(x))$ is true. If $(\forall x \in \mathbb{R})(Q(x))$ is true, then $(\exists x \in \mathbb{R})(Q(x))$ is also true. ■

Often statements involve both "for each" and "there exists," and it is important to realize that $\forall$ and $\exists$ do not commute. Consider these statements:

a. For each real number x there is a real number y such that $x < y$. $(\forall x \in \mathbb{R})(\exists y \in \mathbb{R})(x < y)$.
b. There is a real number y such that, for each real number x, $x < y$. $(\exists y \in \mathbb{R})(\forall x \in \mathbb{R})(x < y)$.

The first of these statements is true but the second is not.

Besides statements involving both "there exists" and "for each," it is also common to see statements of the form "there does not exist an x belonging to a set U such that $Q(x)$." This is the *denial*, or *negation*, of "there exists an x in U such that $Q(x)$." We use $[(\exists x \in U)(Q(x))]'$ to denote "there does not exist an x belonging to the set U such that $Q(x)$." Similarly, the *denial*, or *negation*, of "for each x belonging to the set U, $Q(x)$" is denoted $[(\forall x \in U)(Q(x))]'$. Theorem 1.12 tells us how to write the denial of a statement involving quantifiers.

Theorem 1.12

Let Q be a predicate and let U be a set. Then,
a. $[(\exists x \in U)(Q(x))]'$ is a true proposition if and only if $(\forall x \in U)[Q(x)']$ is a true proposition.
b. $[(\forall x \in U)(Q(x))]'$ is a true proposition if and only if $(\exists x \in U)[Q(x)']$ is a true proposition.

Proof

a. Suppose that $[(\exists x \in U)(Q(x))]'$ is a true statement. Then there is no x belonging to U such that $Q(x)$ is true. Consequently, for each x in U, $Q(x)$ is false. Therefore $Q(x)'$ is true for every x in the set U, and $(\forall x \in U)[Q(x)']$ is a true statement.

Now suppose that $(\forall x \in U)[Q(x)']$ is a true statement. Then $Q(x)'$ is true for every x in the set U. So $Q(x)$ is false for every x in the set U; that is, there does not exist an element x of the set U such that $Q(x)$ is true. Hence $[(\exists x \in U)(Q(x))]'$ is a true statement.

b. We can use part (a) to prove part (b). But it is probably easier (even though longer) to prove part (b) directly. Suppose that $[(\forall x \in U)(Q(x))]'$ is true. Then it is not the case that, for each element x of U, $Q(x)$ is true. Therefore there is some element x of U such that $Q(x)$ is false. So there is an element x of U such that $Q(x)'$ is true. Hence $(\exists x \in U)[Q(x)']$ is true.

Now suppose that $(\exists x \in U)[Q(x)']$ is true. Then there is an element x of U such that $Q(x)'$ is true. So there is an element x of U such that $Q(x)$ is false. Hence it is not the case that, for each element x of U, $Q(x)$ is true. Therefore $[(\forall x \in U)(Q(x))]'$ is true. □

The rules given in Theorem 1.12 are called *de Morgan's laws for quantifiers.* De Morgan's laws are really quite simple. To deny "there exists x such that $Q(x)$," we replace "there exists" with "for each" and deny $Q(x)$. For example, the denial of "there exists a real number x such that $x^2 - 2x + 6 = 0$" is "for each real number x, it is not the case that $x^2 - 2x + 6 = 0$." Perhaps a more useful English sentence is "If x is any real number, then $x^2 - 2x + 6 \neq 0$."

Likewise, to deny "for each x, $Q(x)$," we replace "for each" with "there exists" and deny $Q(x)$. For example, the denial of "for each integer n, n is even" is "there exists an integer n such that n is not even."

How do we deny the proposition $(\forall x \in \mathbb{R})(\exists y \in \mathbb{R})(y > x)$? By Theorem 1.12b, the statement $[(\forall x \in \mathbb{R})(\exists y \in \mathbb{R})(y > x)]'$ is true if and only if $(\exists x \in \mathbb{R})[(\exists y \in \mathbb{R})(y > x)]'$. By Theorem 1.12a, $[(\exists y \in \mathbb{R})(y > x)]'$ is true if and only if $(\forall y \in \mathbb{R})[(y > x)']$. So $[(\forall x \in \mathbb{R})(\exists y \in \mathbb{R})(y > x)]'$ is true if and only if $(\exists x \in \mathbb{R})(\forall y \in \mathbb{R})\ [(y > x)']$. The denial of $y > x$ is $y \leq x$, so $[(\forall x \in \mathbb{R})(\exists y \in \mathbb{R})(y > x)]'$ is true if and only if $(\exists x \in \mathbb{R})(\forall y \in \mathbb{R})(y \leq x)$. Therefore, in English, the denial is "There exists a real number x such that, if y is any real number, then $y \leq x$." Note that we simply replace "for each" with "there exists" and "there exists" with "for each" and deny the statement $y > x$. Note also that $(\forall x \in \mathbb{R})(\exists y \in \mathbb{R})(y > x)$ simply says "There is no largest real number," so the denial says "There is a largest real number."

There is one more existential quantifier that appears frequently in mathematics. Consider the statement "There is a real number x such that $x^2 = 0$." This statement is true, but even more is true. In fact there is *only one* real number x such that $x^2 = 0$. We write $(\exists! x \in \mathbb{R})(x^2 = 0)$. The quantifier $\exists!$ is read "There exists uniquely. . . . " The denial of $(\exists! x \in U)(P(x))$ is somewhat more involved than the denial of $(\exists x \in U)(P(x))$ (see Exercise 40). Since $(\exists! x \in U)(P(x))$ states that there exists *one and only one* $x \in U$ such that $P(x)$, the denial of $(\exists! x \in U)(P(x))$ states either that there does not exist $x \in U$ such that $P(x)$ or that there is more than one $x \in U$ such that $P(x)$.

Example 13

Write a denial of $(\exists! n \in \mathbb{N})(n^2 - 5n + 6 = 0)$.

Analysis

We want to write the denial of "There is one *and* only one $n \in \mathbb{N}$ such that $n^2 - 5n + 6 = 0$." This denial is "There is no $n \in \mathbb{N}$ such that $n^2 - 5n + 6 = 0$ *or* there is more than one $n \in \mathbb{N}$ such that $n^2 - 5n + 6 = 0$." Another way of saying this is "For each $n \in \mathbb{N}$, $n^2 - 5n + 6 \neq 0$ or there exist at least two natural numbers n such that $n^2 - 5n + 6 = 0$." ■

Definition

If Q is a predicate and x is a member of a set U, then a **counterexample** to $(\forall x \in U)(Q(x))$ is a member t of the set U such that $Q(t)$ is false.

The natural number 3 is a counterexample to $(\forall n \in \mathbb{N})(E(n))$.

Exercises 1.5

35. Write a nontrivial negation of each of the following statements.
    - a. Every integer is a real number.
    - b. There is a real number that is not an integer.
    - c. Some integers are prime.
    - d. Not all integers are prime.
    - e. All houses are made of brick.
    - f. No integer is prime.
    - g. Every car is either silver or blue.
    - h. Some cars are silver on the outside and maroon on the inside.
    - i. There is an integer that is even or a perfect square.
    - j. Every integer is even and a perfect square.
36. Using the predicates E, O, P, and S, which are defined on the set of integers, logical connectives, and universal and existential quantifiers, express each of the following sentences in symbols.
    - a. Every perfect square is an integer.
    - b. There is an integer that is prime but not a perfect square.
    - c. If n is an integer, then n is even or a prime.
    - d. There is an integer that is either a perfect square or is both even and a prime.
    - e. Every integer is either not positive but even or is both odd and a perfect square.
37. Write a useful negation of each of the following statements.
    - a. If x is any real number, then there exists a real number y such that $2^x = y$.

b. There exists a real number x such that, if y is any real number, then $2^x = y$.

c. For each pair of rational numbers a and b with $a < b$, there is an irrational number c such that $a < c < b$.

38. Under the agreement that we are considering members of the set of all integers, determine which of the following statements are true and which are false. Explain your answers.

a. $[(\forall x)(\forall y)(x = y)]'$
b. $(\forall x)(\forall y)(xy = yx)$
c. $(\forall x)(\exists y)(xy = 1)$
d. $(\forall x)(\exists y)(xy = x)$
e. $(\exists x)(\forall y)(xy = y)$
f. $(\forall x)(\exists y)(\forall z)(xy = z)$
g. $(\forall x)(\forall y)(\exists z)(xy = z)$

39. The concepts "all" and "some" can be represented by pictures: "All *A*s are *B*s" means that the circle of *A*s is contained in the circle of *B*s, and "Some *A*s are *B*s" means that there is something (p) that is in both the circle of *A*s and the circle of *B*s.

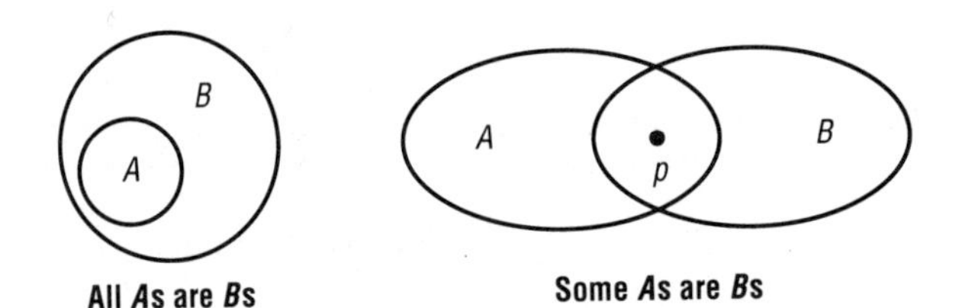

All *As* are *Bs* **Some *As* are *Bs***

a. Draw a picture of the following two statements: (1) "All students love mathematics," (2) "Harry is a student."

b. Using the picture from (a) and assuming that the statements in (a) are true, what can you say about the statement "Harry loves mathematics"?

c. Draw a picture of the following two statements: (a) "All mathematics teachers are dull," and (2) "Some teachers who hold an advanced degree are dull."

d. Using the picture from (c) and assuming that the statements in (c) are true, what can you say about the statements "Some mathematics teachers hold an advanced degree," and "Some dull people have an advanced degree"?

40. Under the agreement that we are considering members of the set of all integers, write a denial of statements (a), (b), and (c) without using any negative words.

a. $(\exists! n)(n^2 = 4)$

b. $(\exists! n)$(n has exactly two positive divisors)

c. $(\exists! n)$($n < 100$ and $n^2 > 50$)

d. Which of the denials you have written are true statements?

41. Let the universe of discourse be the set of all integers and let $P(x)$ be the predicate $x^3 = -x$. Which of the following are true?

    a. $P(0)$ b. $P(-1)$ c. $P(1)$

    d. $(\exists x)(P(x))$ e. $(\forall x)(P(x))$ f. $(\exists! x)(P(x))$

42. Let the universe of discourse consist only of 0 and -1 and let $P(x)$ be the predicate $x^3 = -x$. Which of the following are true?

    a. $(\exists x)(P(x))$ b. $(\exists! x)(P(x))$ c. $(\forall x)(P(x))$

Chapter 1 Review Exercises

43. Write a useful negation of each of the following propositions.

    a. You can draw a flowchart and you can write a COBOL program.

    b. If you document your programs, then they will be easy to debug.

    c. The machine is an IBM or I cannot operate it.

    d. You can log onto the computer if and only if you know the password.

44. Let P, Q, and R be variables. Prove that the following expressions are tautologies.

    a. $[P \wedge (P \rightarrow Q)] \rightarrow Q$ b. $[(P \vee Q) \wedge P'] \rightarrow Q$

    c. $(P \rightarrow Q) \rightarrow [(P \wedge R) \rightarrow (Q \wedge R)]$ d. $[(P \wedge Q) \rightarrow R] \leftrightarrow [P \rightarrow (Q \rightarrow R)]$

45. Write the contrapositive of each of the following propositions.

    a. If 1992 has 366 days, then a rainbow has five colors.

    b. The absolute value of x is equal to x whenever x is nonnegative.

    c. In order for $|x + y|$ to be equal to $|x| + |y|$, both x and y must be positive.

46. Assume that "Tom failed the course" is a true statement and that "Sue made an A" is a false statement. Which of the following are true?

    a. Tom failed the course and Sue made an A.

    b. If Sue made an A, then Tom failed the course.

    c. Sue made an A or Tom failed the course.

    d. If Tom passed the course, then Sue failed.

    e. If Sue did not make an A, then Tom passed the course.

47. Let n be a natural number. Prove that $n^2 + n$ is even.

48. Let x be a real number. Prove that $x = -1$ if and only if $x^3 + x^2 + x + 1 = 0$.

49. Let x be a real number. Prove that if $x^2 - x - 2 < 0$ then $-1 < x < 2$.

50. Write a useful negation of each of the following statements.

    a. All apples are red.

    b. Some silos are used to store wheat.

    c. There is a math book that is easy to read.

d. Each of the following is true: $2 = 3$, $5 < 4$, $2 = \sqrt{9}$.

e. For each positive real number x, there is a natural number n such that $1/n < x$.

f. There is a positive real number a such that if b is any positive real number then $a \leq b$.

*51. Three men attended an opera and checked their hats. When the opera was over, they discovered that their hats were gone. The only hats that remained were three black hats and two white hats, none of which belonged to the men. They decided to close their eyes and each chose a hat at random. As they walked down the street in single file, the last man said, "I can see both your hats but I can't tell what color my hat is." The second man said, "I can see the first man's hat but I can't tell what color my hat is." The first man said, "I can't see anyone's hat but I know what color my hat is." What color is the first man's hat and how does he know?

*52. Three men are arguing about who is the smartest, and a fourth man decides to give them a test. Each of the three is blindfolded and a black or white dot is painted on the forehead of each. The three men are put at the vertices of a triangle so that when the blindfolds are removed they will be able to see the color of the dots on the other two men; of course, they will not be able to see the color of the dot on their own foreheads. When the blindfolds are removed, if a man sees a black dot, he is to raise his hand. The first man to discover the color of the dot on his forehead is to be declared the smartest. The blindfolds are removed and each man raises his hand. After a few seconds' hesitation, one man declares that he knows what color dot is on his forehead. What color is his dot and how does he know?

Chapter

2

SETS

Set theory and logic are the two main underpinnings of mathematics, and, like logic, the theory of sets deserves to be studied as a subject in its own right. Such a study would reveal that set theory is itself a branch of mathematics with its own system of axioms and its own body of proofs and theorems. The insistence in higher-level courses that set theory be presented with a system of axioms is not merely an attempt to breathe life into an old subject; the naive and intuitive view of set theory that we present here will not stand up to scrutiny. If you are to avoid the built-in ambiguities and contradictions of intuitive set theory, you must eventually study axiomatic set theory. Presumably, however, no one has ever had the misfortune of studying axiomatic set theory without having first viewed set theory from an intuitive point of view, and since we are primarily interested in the use of sets in the study of discrete mathematics, it is the intuitive view of sets that we present here.

2.1

SET NOTATION

We think of a set as a collection of objects. We do not attempt to define the words *set, collection,* or *object,* but we assume that if we have a set S there is some rule that determines whether a given object x is a member of S. It may happen, however, that we do not know explicitly which objects belong to S. For example, if S is the set to which a natural number n belongs provided that $2^n - 1$ is prime, we do not actually know which natural numbers belong to S, even though we have a rule that describes these natural numbers. Indeed, it is not even known whether S is a finite set. Nonetheless, S is a perfectly good set, because we have an explicit rule whereby we can (at least in principle) decide whether a given natural number belongs to it.

Membership in a set is also an all-or-nothing situation. We cannot have a set S and an object x that belongs only partially to S. Just as you are either registered for this course or you are not, the assertion that a given object belongs to a given set S is a proposition, which like all propositions is true or false but not both.

If A is a set and x is an object that belongs to A, we say that x is an *element* of A or that x is a *member* of A or that x *belongs* to A and write $x \in A$. If x is not a member of A, then we write $x \notin A$. So, for example, if A is the set of all prime numbers, then $5 \in A$ but $6 \notin A$. We use the notation $x,y \in A$ to mean that $x \in A$ and $y \in A$.

Sets can be specified in a variety of ways. If a set A consists of only a small number of objects, we can describe it by simply listing its members between braces. For example, if A is the set of all prime numbers less than 25, then we can describe A by writing $A = \{2,3,5,7,11,13,17,19,23\}$. It is also common practice to use ellipses (. . .) to indicate some elements that are not made explicit. We might describe a set B by writing $B = \{1,2, \ldots ,10\}$; here we understand that the set B is really $\{1,2,3,4,5,6,7,8,9,10\}$. We read $C = \{2,3,5\}$ as "C is the set whose members are 2, 3, and 5." Observe that $2 \in \{2,3,5\}$ but $4 \notin \{2,3,5\}$.

Definition

Let A and B be sets. We say that A is a **subset** of B, written $A \subseteq B$, provided that every member of A is also a member of B. If A is not a subset of B, we write $A \not\subseteq B$. We say that the sets A and B are **equal,** written $A = B$, provided that $A \subseteq B$ and $B \subseteq A$. We say that A is a **proper subset** of B, written $A \subset B$, provided that $A \subseteq B$ and $A \neq B$.

Whenever we are given a set S, we may use *set-builder notation* to describe a subset of S. For example, we could write $A = \{n \in \mathbb{N}: n$ is prime and $n < 15\}$, which would be read "A equals the set of all n belonging to $\mathbb{N}$ such that n is prime and n is less than 15." Thus $\{n \in \mathbb{N}: n$ is prime and $n < 15\}$ describes the set $\{2,3,5,7,11,13\}$. Note that in set-builder notation $\{x \in S: \ldots\}$ is always read "The set of *all* x belonging to S such that. . . . " Thus $B = \{x \in \mathbb{N}: x^2 - 5x + 6 = 0\}$ is read "B equals the set of all x belonging to $\mathbb{N}$ such that $x^2 - 5x + 6 = 0$." In this case, $B = \{2,3\}$.

Before leaving set-builder notation, we offer two warnings. First, set-builder notation is intended as a simple and short way to describe subsets. If the subset you have in mind has only a few members, it is usually kinder to list these members rather than to describe your set in set-builder notation. We confess, therefore, that $\{x \in \mathbb{N}: x^2 - 5x + 6 = 0\}$ is a ridiculous way to describe the set $\{2,3\}$. Moreover, it is often the case that a subset can be described most easily in plain English. There is no obligation to use set-builder notation, and your choice between this notation and English should always be based on a desire for clarity and not a desire for prestige.

Our second warning is that in describing any infinite set you must use either English or set-builder notation, because it is impossible to list the members of an infinite set. This warning seems pretty obvious, but the temptation is to list a few members of a set and hope that everyone will guess correctly what set you have in mind. Unfortunately, there is no way to determine an infinite set by the behavior of any of its finite subsets (see Exercise 6). If you are told that a certain list of natural numbers begins 2,4 and is an arithmetic progression, you know that the nth member of the set is $2n$; if you are told that the list is a geometric progression, you know that the nth member of the list is 2^n; but if you are just told that the list begins 2,4, you have no way of determining any other number in the list.

We have notation for certain special subsets of $\mathbb{R}$. In particular, if $a,b \in \mathbb{R}$ and $a < b$, we define the following *intervals:*

$$[a,b] = \{x \in \mathbb{R}: a \leq x \leq b\}$$

$$(a,b) = \{x \in \mathbb{R}: a < x < b\}$$

$$[a,b) = \{x \in \mathbb{R}: a \leq x < b\}$$

$$(a,b] = \{x \in \mathbb{R}: a < x \leq b\}$$

The interval $[a,b]$ is said to be *closed,* and the interval (a,b) is said to be *open.* Brackets [,] indicate that the endpoint(s) are included, and parentheses (,) indicate that the endpoint(s) are excluded. For convenience, we use the symbols ∞ and $-\infty$ and define

$$[a,\infty) = \{x \in \mathbb{R}: x \geqslant a\}$$
$$(-\infty,a] = \{x \in \mathbb{R}: x \leqslant a\}$$
$$(a,\infty) = \{x \in \mathbb{R}: x > a\}$$
$$(-\infty,a) = \{x \in \mathbb{R}: x < a\}$$

A word of caution is in order. The symbols ∞ and $-\infty$ *do not represent real numbers.* They are simply part of the notation. We would never use the notation $[a,\infty]$, $(a,\infty]$, $[-\infty,a)$, or $[-\infty,a]$.

Observe the difference between $[2,4]$, $(2,4)$, and $\{2,4\}$. The intervals $[2,4]$ and $(2,4)$ contain infinitely many members of $\mathbb{R}$, whereas $\{2,4\}$ contains exactly two members of $\mathbb{R}$. Note that both $\{2,4\}$ and $(2,4)$ are subsets of $[2,4]$. On the other hand, $\{2,4\} \not\subseteq (2,4)$, since neither 2 nor 4 is a member of $(2,4)$.

Just as it is easy to confuse set braces, brackets, and parentheses, it is also easy to confuse the meanings of the symbols $\in$ and $\subseteq$. For example, $2 \in \{2,3\}$ but $2 \not\subseteq \{2,3\}$. On the other hand, $\{2\} \subseteq \{2,3\}$, but $\{2\} \notin \{2,3\}$. Let us prove that $\{2\} \subseteq \{2,3\}$. Let $x \in \{2\}$. Since $\{2\}$ is the set whose only member is 2, $x = 2$. Since $2 \in \{2,3\}$ and $x = 2$, $x \in \{2,3\}$. Therefore $\{2\} \subseteq \{2,3\}$. Observe that $\{\{2\},3\}$ is the set whose members are $\{2\}$ and 3; that is, one of the members of $\{\{2\},3\}$ is the set whose only member is $\{2\}$. In other words, $\{2\} \in \{\{2\},3\}$, but $2 \notin \{\{2\},3\}$.

Are the sets $\{2,3\}$ and $\{3,2\}$ the same? Using practically the same proof that was given previously, you should be able to determine that $\{2,3\} \subseteq \{3,2\}$ and that $\{3,2\} \subseteq \{2,3\}$. Hence $\{2,3\} = \{3,2\}$. We see from this simple example that, unlike lists, sets have no particular order. Is the set $\{2,3\}$ a proper subset of $\{3,2,3\}$? If not, are these sets equal (see Exercise 8)?

The unique set that has no members is called the *empty set,* and it is denoted by the symbol $\varnothing$, which is the last letter of the Danish–Norwegian alphabet. Recall that $\{2\}$ is the set whose only member is 2. What is $\{\varnothing\}$? It is the set whose only member is $\varnothing$, so $\{\varnothing\} \neq \varnothing$. Observe that $\varnothing \in \{\varnothing\}$ and $\varnothing \subset \{\varnothing\}$, but $\varnothing \notin \varnothing$.

If A is a set, we can form a new set whose members are the subsets of A. This set is called the *power set* of A and is denoted by $\mathcal{P}(A)$.

Example 1

a. $\mathcal{P}(\varnothing) = \{\varnothing\}$.
b. If $A = \{1\}$, then $\mathcal{P}(A) = \{\varnothing,A\}$.
c. If $A = \{1,2\}$, then $\mathcal{P}(A) = \{\varnothing,\{1\},\{2\},A\}$.
d. If $A = \{1,2,3\}$, then $\mathcal{P}(A) = \{\varnothing,\{1\},\{2\},\{3\},\{1,2\},\{1,3\},\{2,3\},A\}$. ■

Exercises 2.1

1. List the members of each of the following sets

a. $\{n \in \mathbb{Z}: -n \in \mathbb{N} \text{ and } -n < 5\}$

b. $\{n \in \mathbb{R}: n = m^2 \text{ for some prime natural number } m \text{ less than } 10\}$

c. $\{n \in \mathbb{Z}: n = m^3 - m$ for some even natural number m less than $10\}$

d. $\{n \in \mathbb{Z}: n = [1 - (-1)^m]/m$ for some integer m with $|m| < 4\}$

e. $\{n \in \mathbb{Z}: n^2 = 4\}$

f. $\{n \in \mathbb{N}: n^2 = 4\}$

g. $\{n \in \mathbb{R}: n^2 = 4\}$

2. How many members are there in each of the following sets?

a. $\{x \in \mathbb{Q}: x^2 = 2\}$

b. $\{x \in \mathbb{R}: x^2 = 2\}$

c. $\mathcal{P}(\{1,2,3,4\})$

d. $\{n \in \mathbb{Z}: n$ is even and $|n| < 50\}$

e. $\{n \in \mathbb{N}: n$ is even and n is a prime$\}$

3. For each of the following sets, either list all the members of the set or describe the set in plain English.

a. $\{x \in \mathbb{N}: x^2 = x\}$

b. $\{x \in \mathbb{R}: x^2 = x\}$

c. $\{n \in \mathbb{N}: -n > 3\}$

d. $\{x \in \mathbb{R}: x > 5$ and $x^2 < 17\}$

e. $\{x \in \mathbb{R}: x > 5$ or $x^2 < 17\}$

4. Determine which of the following statements are true.

a. $\varnothing \in \{\varnothing\}$

b. $\varnothing \subseteq \{\varnothing\}$

c. $4 \in \{4\}$

d. $\{4\} \subseteq \{\{4\}\}$

e. $\varnothing \in \{4\}$

f. $\varnothing \subseteq \{4\}$

g. $[2,4] \subseteq \mathbb{N}$

h. $\{2,3,4\} \subseteq \mathbb{N}$

i. $1/2 \in \{0,1\}$

j. $1/2 \in (0,1)$

k. $\{4\} \subseteq \{4,\{4\}\}$

l. $[0,1] \in \mathcal{P}(\mathbb{R})$

5. Determine which of the following statements are true.

a. The empty set is a subset of every set.

b. There is a set that is a member of every set.

c. Since $\varnothing$ is a member of $\{\varnothing\}$, $\varnothing = \{\varnothing\}$.

d. If A is a proper subset of $\varnothing$, then $A = \mathcal{P}(\mathbb{N})$.

e. If $A = B$, then $A \subseteq B$.

f. If $A \subset B$, then $A = B$.

g. If $A \subseteq \varnothing$ and $A \neq \varnothing$, then $A = [0,4]$.

6. a. Evaluate $2^{n-1} - (n-5)(n-4)(n-3)(n-2)(n-1)/120$, for $n = 1,2,3,4,5,6$.

b. For $n = 1,2,3,4,5,6$, place n dots randomly on a circle, draw all chords between pairs of dots, and count the number of pieces into which the chords divide the inside of the circle.

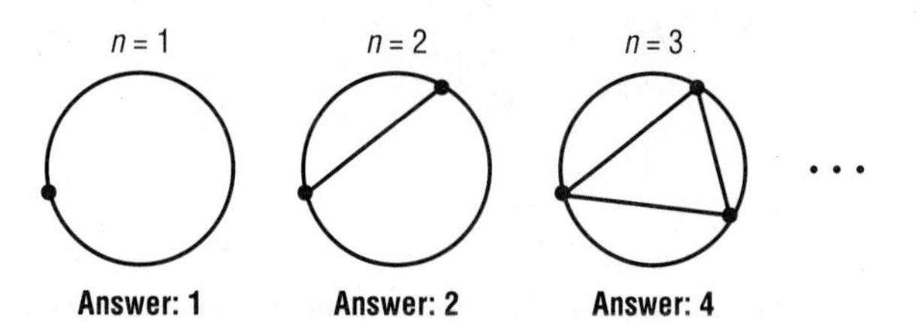

c. A certain college entrance examination asks the following question: A list of numbers begins 1,2,4,8,16, ... The next number in the list is (a) 48 (b) 32 (c) 64 (d) 31 (e) 100. Is the question fair? Explain.

7. List all members of $\mathcal{P}(\mathcal{P}(\{1\}))$.

8. a. Show that $\{1,2\} = \{2,1\}$.
   b. Show that $\{1,2,1\} = \{1,2\}$.

9. a. Find two sets A and B such that $A \not\subseteq B$ and $B \not\subseteq A$.
   b. Find two sets A and B such that $A \subseteq B$ or $B \subseteq A$.
   c. Find two sets A and B such that $A \subseteq B$ but $B \not\subseteq A$.

2.2 OPERATIONS ON SETS

In this section we discuss several ways of forming new sets from existing sets. We assume that all sets under consideration are subsets of some given set U, so that each member of any set is a member of U. The set U is called a *universal set*. Relations between sets can be illustrated by the use of *Venn diagrams*. In a Venn diagram, it is customary to represent the universal set by a rectangular region. Then an arbitrary set is represented by a figure drawn within the rectangular region. For example, the Venn diagram in Figure 2.1 represents the information that $A \subseteq B$ and $B \subseteq C$.

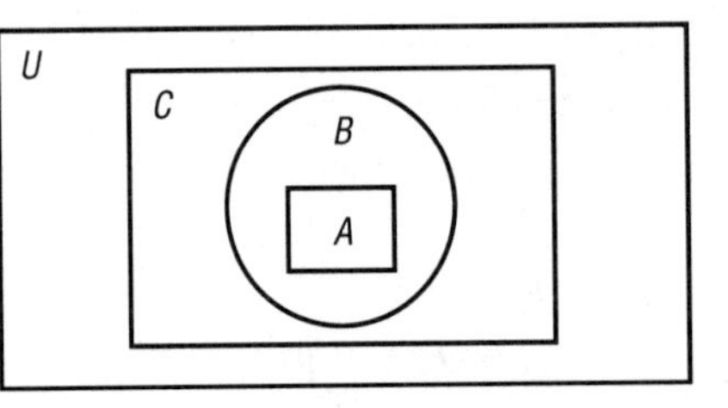

Figure 2.1

Definition

If A and B are sets, the **union** of A and B is the set of all objects that belong to A or to B. The union of A and B is denoted by $A \cup B$, which is read "A union B." In set-builder notation,

$$A \cup B = \{x \in U : x \in A \text{ or } x \in B\}$$

Recall that in the definition of the disjunction of two propositions we used the "inclusive or." In mathematics, unless explicitly specified to the contrary, we always interpret "or" as "inclusive or." Thus an element x of U belongs to the union of sets A and B in each of the following cases:

a. $x \in A$ and $x \notin B$
b. $x \notin A$ and $x \in B$
c. $x \in A$ and $x \in B$

The only case where x does *not* belong in $A \cup B$ is the case where $x \notin A$ and $x \notin B$. The shaded region in the Venn diagram in Figure 2.2 represents $A \cup B$.

Figure 2.2

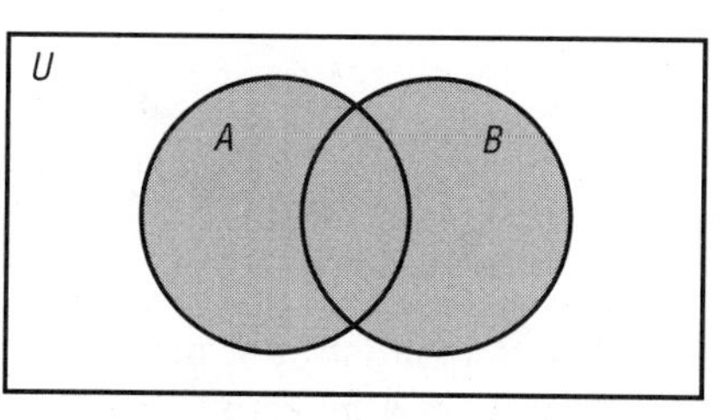

Definition

If A and B are sets, the **intersection** of A and B is the set of all objects that belong to both A and B. The intersection of A and B is denoted by $A \cap B$, which is read "A intersect B." In set-builder notation,

$$A \cap B = \{x \in U: x \in A \text{ and } x \in B\}$$

An element x of U belongs to the intersection of sets A and B only in the case where $x \in A$ and $x \in B$. The element x does *not* belong in $A \cap B$ in each of the following cases:

a. $x \in A$ and $x \notin B$
b. $x \notin A$ and $x \in B$
c. $x \notin A$ and $x \notin B$

In other words, $x \notin A \cap B$ provided that $x \notin A$ or $x \notin B$ (note the "inclusive or"). The shaded region in the Venn diagram in Figure 2.3 represents $A \cap B$.

Figure 2.3

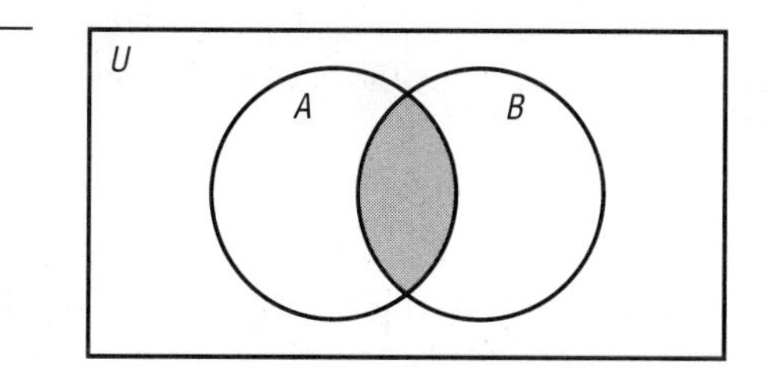

Example 2

Let $U = \{2,3,4,5,8,10,17,32,42\}$, $A = \{2,4,8,32\}$, and $B = \{5,8,10,17\}$. Then $A \cup B = \{2,4,5,8,10,17,32\}$ and $A \cap B = \{8\}$. ■

Example 3

Let $A = \{2,3,7,15,18\}$ and let $B = \{7,18,32,35\}$. Then $A \cup B = \{2,3,7,15,18,32,35\}$ and $A \cap B = \{7,18\}$. ■

Example 4

Let A be the set of all even integers and let B be the set of all odd integers. Then $A \cup B = \mathbb{Z}$ and $A \cap B = \varnothing$. ■

Definition

> If A and B are sets and $A \cap B = \varnothing$, then A and B are **disjoint sets.**

The concept of disjoint sets is illustrated in the Venn diagram in Figure 2.4. As Example 4 indicates, if A is the set of all even integers and B is the set of all odd integers, then A and B are disjoint sets. Note that if A is any set then A and $\varnothing$ are disjoint sets.

Figure 2.4

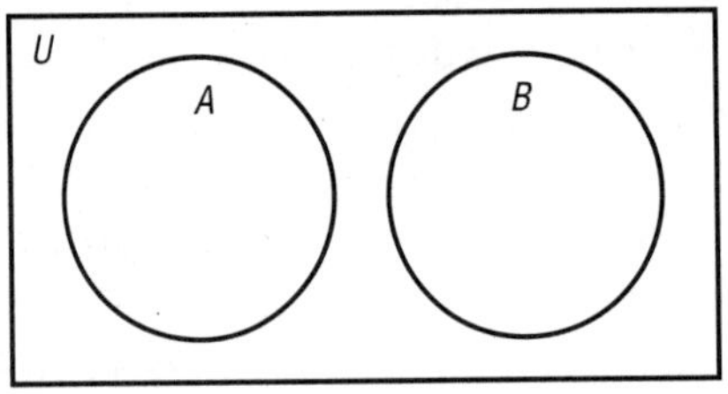

U
A
B

Example 5

Let $A = \{x \in \mathbb{R}: 2 < x \leq 4\}$ and $B = \{x \in \mathbb{R}: 3 \leq x < 7\}$. Then $A \cup B = \{x \in \mathbb{R}: 2 < x < 7\}$ and $A \cap B = \{x \in \mathbb{R}: 3 \leq x \leq 4\}$. Using the interval notation, $A = (2,4]$, $B = [3,7)$, $A \cup B = (2,7)$, and $A \cap B = [3,4]$. ■

Definition

> If A and B are sets, the **complement of B relative to A** is the set $\{x \in A: x \notin B\}$. The complement of B relative to A is denoted by $A - B$, which is read "A minus B." The complement of B relative to U is called the **complement** of B and is denoted by $B^{\sim}$.

The concept of complement is illustrated in the Venn diagrams in Figure 2.5. The shaded region in the Venn diagram on the left represents

Figure 2.5

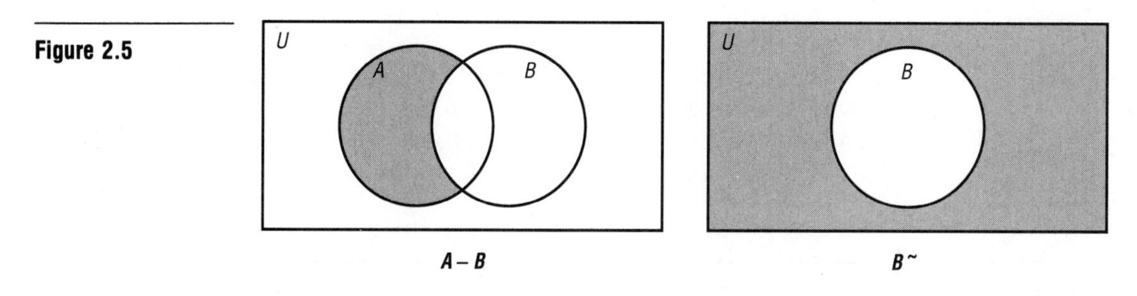

A − B **B~**

$A - B$; the shaded region in the Venn diagram on the right represents $B^{\sim}$.

Example 6

Let $U = \{1,2,3,4,5,6,7,8\}$, $A = \{3,4,5,6,7\}$, and $B = \{5,7,8\}$. Then $A - B = \{3,4,6\}$, $B - A = \{8\}$, $A^{\sim} = \{1,2,8\}$, and $B^{\sim} = \{1,2,3,4,6\}$. ■

The following observations about containment, though simple, are useful:

a. If A is a set, then $\varnothing \subseteq A$ and $A \subseteq A$.
b. Every nonempty set has at least two subsets.
c. If A, B, and C are sets, $A \subseteq B$ and $B \subseteq C$, then $A \subseteq C$.

Since $\varnothing$ is a subset of every set, whenever we are faced with the problem of showing that a set A is a subset of a set B we do not have to worry about what happens when $A = \varnothing$. Therefore we may begin such a proof by saying, "Let $x \in A$."

The following theorem is comparable to Theorem 1.1.

Theorem 2.1

Let A, B, and C be subsets of a universal set U. The following properties hold:

a. $A \cup B = B \cup A$ — commutative laws
$A \cap B = B \cap A$
b. $A \cup (B \cup C) = (A \cup B) \cup C$ — associative laws
$A \cap (B \cap C) = (A \cap B) \cap C$
c. $A \cap (B \cup C) = (A \cap B) \cup (A \cap C)$ — distributive laws
$A \cup (B \cap C) = (A \cup B) \cap (A \cup C)$
d. $\varnothing \cup A = A \cup \varnothing = A$ — identity laws
$U \cap A = A \cap U = A$
e. $A \cup A^{\sim} = U$
$A \cap A^{\sim} = \varnothing$

Proof

We prove only that $A \cap A^{\sim} = \varnothing$ and illustrate the second distributive law with Venn diagrams in Figure 2.6. Suppose $A \cap A^{\sim} \neq \varnothing$. Then there is an $x \in A \cap A^{\sim}$. Hence $x \in A$ and $x \in A^{\sim}$. Since $x \in A^{\sim}$, $x \notin A$. This is a contradiction. □

Figure 2.6

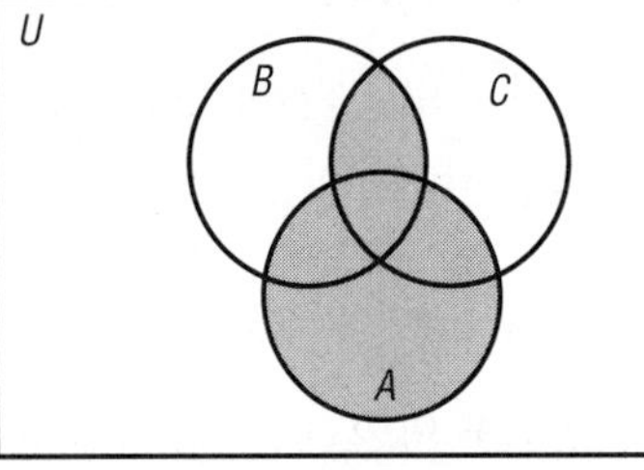

$A \cup (B \cap C)$

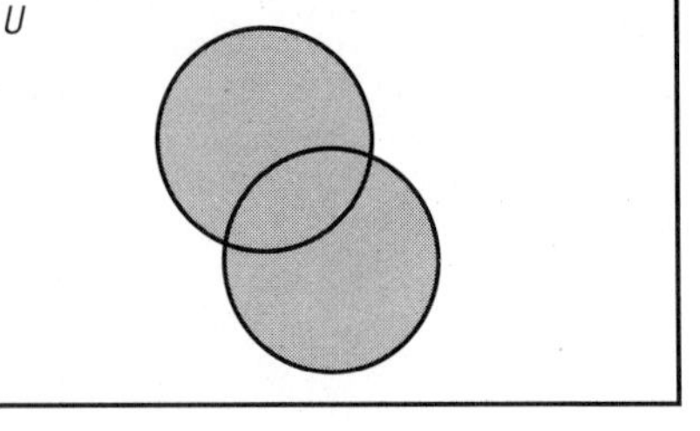

$A \cup B$

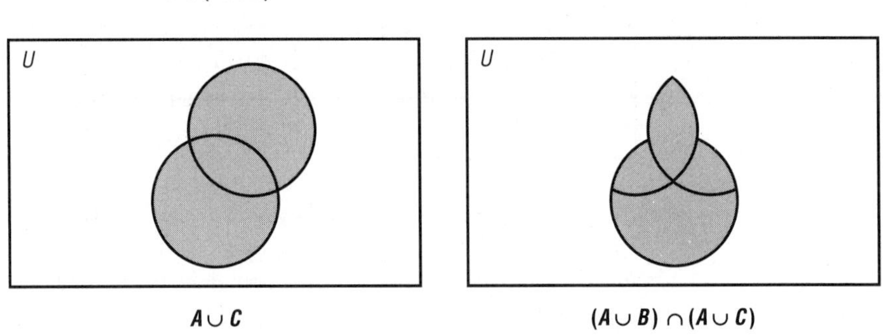

$A \cup C$ $(A \cup B) \cap (A \cup C)$

Theorem 2.2 collects some useful results concerning sets.

Theorem 2.2

Let A and B be subsets of a universal set U. The following properties hold:

a. $\varnothing^{\sim} = U$ and $U^{\sim} = \varnothing$
b. $A^{\sim\sim} = A$
c. $A \cup A = A$ and $A \cap A = A$
d. $(A \cup B)^{\sim} = A^{\sim} \cap B^{\sim}$ and $(A \cap B)^{\sim} = A^{\sim} \cup B^{\sim}$ de Morgan's laws
e. $A \cup (A \cap B) = A$ and $A \cap (A \cup B) = A$ absorption laws
f. $A - B = A \cap B^{\sim}$
g. $A \subseteq B$ if and only if $B^{\sim} \subseteq A^{\sim}$
h. $A \subseteq B$ if and only if $A \cap B = A$
i. $A \cap \varnothing = \varnothing$ and $A \cup \varnothing = A$ identity laws

Proof

We prove (f), illustrate the second part of (d) with Venn diagrams, and leave some of the remaining properties as exercises.

To prove that $A - B = A \cap B^{\sim}$, we need to prove that $A - B \subseteq A \cap B^{\sim}$ and that $A \cap B^{\sim} \subseteq A - B$. We first prove that $A - B \subseteq A \cap B^{\sim}$ by showing that each member of $A - B$ is also a member of $A \cap B^{\sim}$. Let $x \in A - B$ (that is, x is an arbitrary, but fixed, element of $A - B$). Then, by the definition of the complement of B relative to A, $x \in A$ and $x \notin B$. Therefore, by the definition of complement, $x \in A$ and $x \in B^{\sim}$. So, by the definition of intersection, $x \in A \cap B^{\sim}$. Now we show that $A \cap B^{\sim} \subseteq A - B$ by showing that each member of $A \cap B^{\sim}$ is also a member of $A - B$. Let $x \in A \cap B^{\sim}$ (once again x is an arbitrary, but fixed, element of $A \cap B^{\sim}$). Then, by the definition of intersection, $x \in A$ and $x \in B^{\sim}$. So, by the definition of complement, $x \in A$ and $x \notin B$. Therefore, by the definition of complement of B relative to A, $x \in A - B$.

We illustrate the second part of property (d) by the Venn diagrams in Figure 2.7.

Figure 2.7

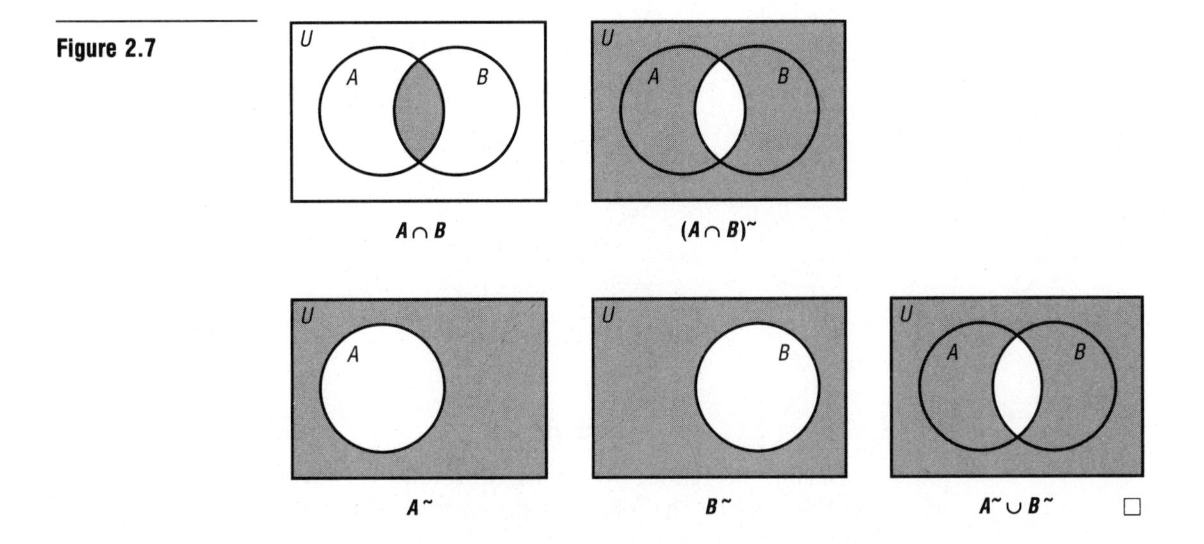

□

De Morgan's laws (Theorem 2.2d) are particularly useful and we urge the reader to remember them in plain English: the complement of the union is the intersection of the complements and the complement of the intersection is the union of the complements.

Definition

Let S be a set and let A and B be subsets of S. The **symmetric difference of A and B,** written $A \oplus B$, is defined to be $(A - B) \cup (B - A)$.

Theorem 2.3

Let S be a subset and let A and B be subsets of S. Then $A \oplus B = (A \cup B) - (A \cap B)$.

Proof

Let $U = A \cup B$ and consider A and B as subsets of the universal set U. We start with $(A \cup B) - (A \cap B)$ and use previous results to show that it is $A \oplus B$.

$$\begin{aligned}
(A \cup B) - (A \cap B) &= (A \cap B)^{\sim} \\
&= A^{\sim} \cup B^{\sim} \\
&= ((A \cup B) - A) \cup ((A \cup B) - B) \\
&= [(A \cup B) \cap A^{\sim}] \cup [(A \cup B) \cap B^{\sim}] \\
&= [A^{\sim} \cap (A \cup B)] \cup [B^{\sim} \cap (A \cup B)] \\
&= [(A^{\sim} \cap A) \cup (A^{\sim} \cap B)] \cup [(B^{\sim} \cap A) \\
&\quad \cup (B^{\sim} \cap B)] \\
&= [\varnothing \cup (A^{\sim} \cap B)] \cup [(B^{\sim} \cap A) \cup \varnothing] \\
&= (A^{\sim} \cap B) \cup (B^{\sim} \cap A) \\
&= (B^{\sim} \cap A) \cup (A^{\sim} \cap B) \\
&= (A \cap B^{\sim}) \cup (B \cap A^{\sim}) \\
&= (A - B) \cup (B - A) \\
&= A \oplus B
\end{aligned}$$
□

Despite its name, symmetric difference behaves more like addition than subtraction. The following proposition, which we state without proof, illustrates this point (see Exercise 22).

Theorem 2.4

Let S be a set and let A, B, and C be subsets of S. Then the following statements hold:

a. $A \oplus B = B \oplus A$ — commutative law
b. $A \oplus (B \oplus C) = (A \oplus B) \oplus C$ — associative law
c. $A \cap (B \oplus C) = (A \cap B) \oplus (A \cap C)$ — $\cap$ distributes over $\oplus$

Exercises 2.2

10. Let A and B be sets. Prove or find a counterexample to each of the following statements.

a. $\mathcal{P}(A \cap B) = \mathcal{P}(A) \cap \mathcal{P}(B)$ **b.** $\mathcal{P}(A \cup B) = \mathcal{P}(A) \cup \mathcal{P}(B)$

c. $\mathcal{P}(A - B) = \mathcal{P}(A) - \mathcal{P}(B)$

11. Let $U = \{1,2,3,4,5,6,7,8,9\}$, $A = \{1,4,7\}$, $B = \{2,4,5,7\}$, and $C = \{3,4,7\}$. List the members of each of the following sets.

12. **a.** $A \cup B$ **b.** $(A \cup B)^{\sim}$ **c.** $A \cap B$

d. $(A \cap B)^{\sim}$ **e.** $A^{\sim} \cap B^{\sim}$ **f.** $A^{\sim} \cup B^{\sim}$

g. $A - B$ **h.** $A - (B \cup C)$ **i.** $A - (B \cap C)$

j. $(A \cup B) \cap C$ **k.** $A \cup (B \cap C)$ **l.** $(A \cup B) \cap C^{\sim}$

12. In each of the following diagrams, shade the indicated set.

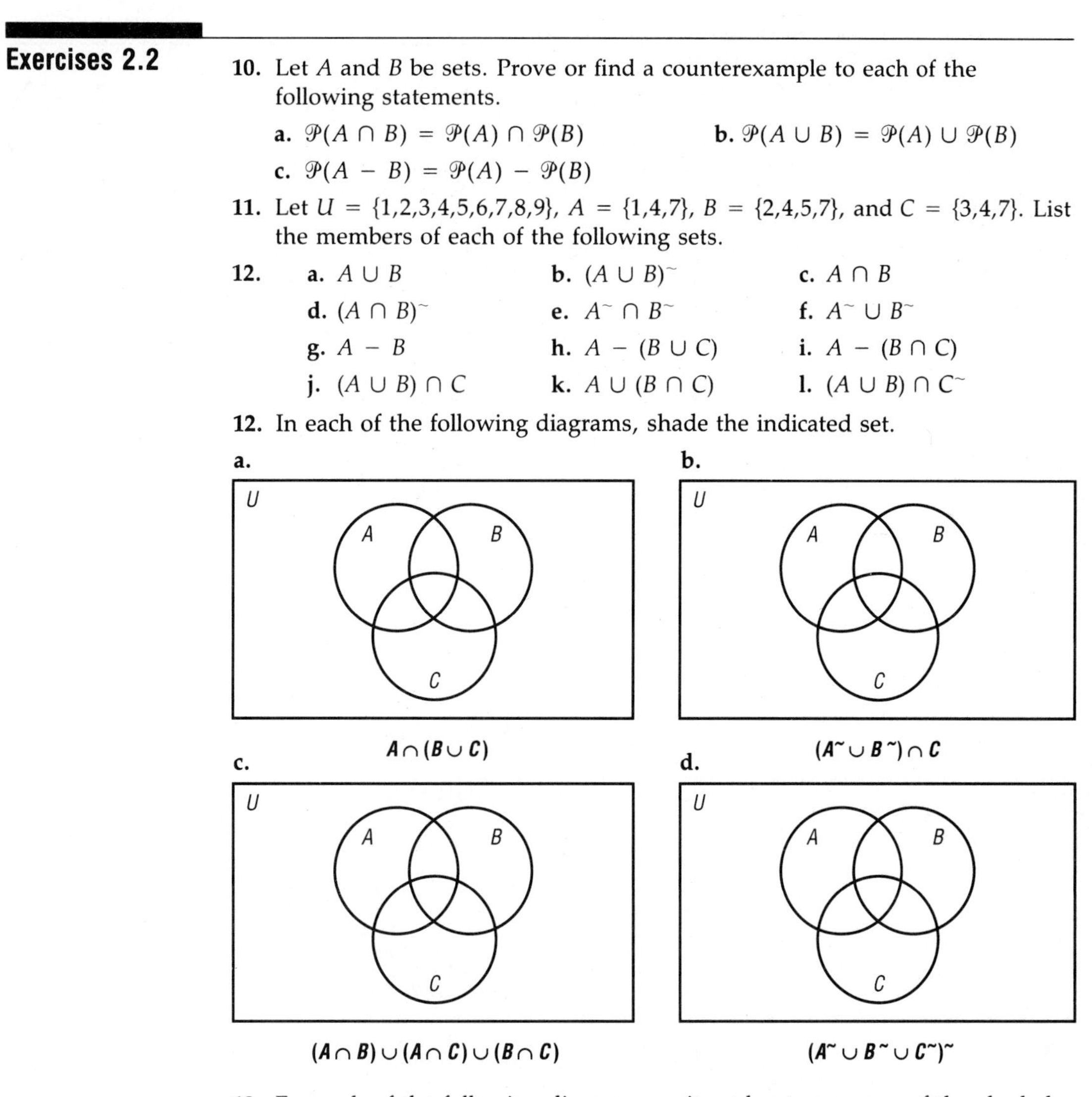

a. $A \cap (B \cup C)$

b. $(A^{\sim} \cup B^{\sim}) \cap C$

c. $(A \cap B) \cup (A \cap C) \cup (B \cap C)$

d. $(A^{\sim} \cup B^{\sim} \cup C^{\sim})^{\sim}$

13. For each of the following diagrams, write at least one name of the shaded set using only the symbols A, B, C, $\cup$, $\cap$, $^{\sim}$, (,).

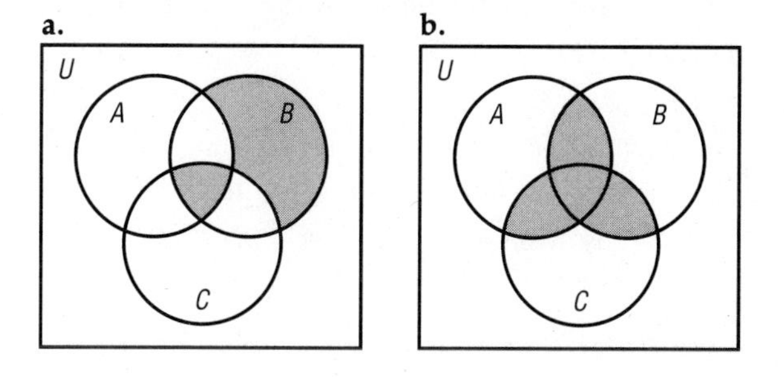

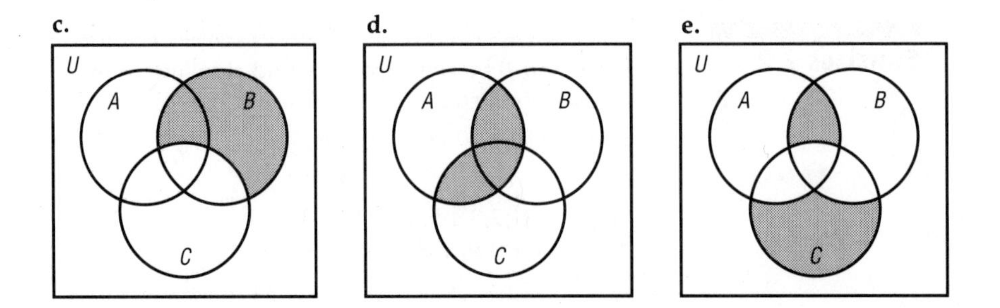

14. In the following diagram, the eight natural sections of the diagram have been numbered from 1 to 8. **a.** If any combination of these eight sections are shaded, is it always possible to find at least one name for the shaded set? **b.** Why?

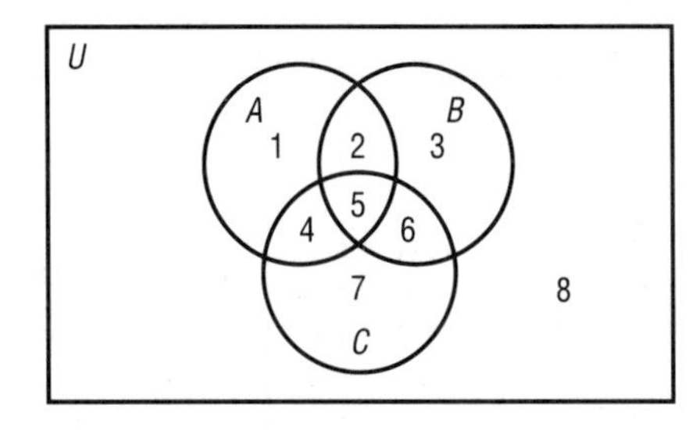

15. Let $X = \{1,2,3,4\}$. List all pairs A,B of subsets of X such that A and B are disjoint and $X = A \cup B$.

16. Let A and B be sets. Prove each of the following statements.
a. $A \subseteq B$ if and only if $B^{\sim} \subseteq A^{\sim}$ **b.** $A \cup B = A$ if and only if $B \subseteq A$
c. $A \cap B = A$ if and only if $A \subseteq B$

17. Draw Venn diagrams illustrating $A \cap [(B - C) \cup (C - B)]$ and $[(A \cap B) - (A \cap C)] \cup [(A \cap C) - (A \cap B)]$.

***18.** Prove that, if A, B, and C are sets, then $A \cup (B \cap C) = (A \cup B) \cap (A \cup C)$.

***19.** Prove that, if A and B are sets, then $A = (A - B) \cup (A \cap B)$ and $(A \cap B) \cap (A - B) = \varnothing$.

20. Let A, B, and C be sets. Find a counterexample to each of the following statements.

a. $(A \cap B) \cup C = A \cap (B \cup C)$ **b.** If $A \cap B = A \cap C$, then $B = C$
c. If $A \cup B = A \cup C$, then $B = C$ **d.** $(A - B) - C = A - (B - C)$
e. $A \cap B = A^{\sim} \cup B^{\sim}$ **f.** $(A \cup B) \cap C = A \cup (B \cap C)$
g. $(A - B) \cup C = A - (B \cup C)$ **h.** $A^{\sim} \cup C \subseteq (A^{\sim} \cup B) \cap (B^{\sim} \cup C)$

21. A student is asked to prove the following theorems:
a. For any two sets A and B, $(A \cup B) - (A \cap B) = (B - A) \cup (A - B)$.
b. For any two sets A and B, $A = (A \cap B) \cup (A - B)$.
The student's proofs are as follows:

a. Let $A \cup B$ be the universal set. Then

$$\begin{aligned}(A \cup B) - (A \cap B) &= (A \cap B)^{\sim} \\ &= A^{\sim} \cup B^{\sim} \\ &= ((A \cup B) - A) \cup ((A \cup B) - B) \\ &= ((A \cup B) \cap A^{\sim}) \cup ((A \cup B) \cap B^{\sim}) \\ &= (B \cap A^{\sim}) \cup (A \cap B^{\sim}) \\ &= (B - A) \cup (A - B)\end{aligned}$$

b. Let A be the universal set U. Then

$$\begin{aligned}(A \cap B) \cup (A - B) &= (U \cap B) \cup B^{\sim} \\ &= B \cup B^{\sim} \\ &= U = A\end{aligned}$$

What, if anything, is wrong with the student's arguments?

22. In the text we have stated that $\oplus$ is commutative and associative and that $\cap$ distributes over $\oplus$. Show by example that ordinary subtraction of integers is neither commutative nor associative. Does multiplication distribute over subtraction?

23. The terminology *symmetric difference* is in some ways unsatisfactory, and so some people call $\oplus$ "exclusive or." Briefly explain the relevance of this terminology.

24. With $U = \{1,2,3, \ldots ,9,10\}$ as a given universal set and with $A = \{1,2,3\}$, $B = \{2,4,6,8\}$, and $C = \{3,4,5\}$, list all the elements of the indicated sets.

a. $A \oplus B$ **b.** $B \oplus C$ **c.** $A \oplus (B \oplus C)$

d. $(A \oplus B) \oplus C$ **e.** $C \oplus (B \oplus A)$ **f.** $(A \cap B) \oplus (A \cap C)$

25. Prove that A and B are disjoint sets if and only if $A \cup B = A \oplus B$.

2.3

INDEXED SETS

As we discussed at the beginning of this chapter, we can describe a finite set by simply listing its members between braces. We cannot describe an infinite set in this way.

Similarly, if we have a finite number of sets and the number of sets is not too large, we can list the sets. For example, if we have five sets, we can call them A, B, C, D, and E. This method of listing sets is, at best, cumbersome. For example, if we have more than twenty-six sets we do not have enough capital letters in the English alphabet to label each set. One way to avoid this problem is to use subscripts. Often we can use members of $\mathbb{N}$ as the subscripts. So, if we have fifty sets, we can use the first fifty natural numbers as subscripts, and we can use the notation $\{A_n: n \in \mathbb{N} \text{ and } n \leq 50\}$ to indicate the set consisting of these fifty sets. We understand that for each natural number $n \leq 50$, A_n is one of our fifty sets.

Example 7

For each natural number n, let A_n be the closed interval $[0, n]$. Observe that, for each natural number k, $[0, k] \in \{A_n: n \in \mathbb{N}\}$. ■

We can use the members of any nonempty set as subscripts. Let X be any nonempty set, and for each $x \in X$ let A_x be a set. Then we have a collection of sets $\{A_x: x \in X\}$. The set X is called the *indexing set*, and the collection $\{A_x: x \in X\}$ is said to be an *indexed set*. Most of the indexing sets that we use in this book are nonempty subsets of $\mathbb{N}$, and if the members of a set $\{A_x: x \in X\}$ are themselves sets, we often call the set $\{A_x: x \in X\}$ a *family*.

Example 8

For each $n \in \mathbb{N}$, let $A_n = \{k \in \mathbb{Z}: k \text{ is divisible by } n\}$. Then $\{A_n: n \in \mathbb{N}\}$ is a family of sets, $\mathbb{N}$ is the indexing set, and for each $n \in \mathbb{N}$, A_n is a subset of $\mathbb{Z}$. In particular, $A_1 = \{k \in \mathbb{Z}: k \text{ is divisible by } 1\}$, and, since every integer is divisible by 1, $A_1 = \mathbb{Z}$. Also $A_2 = \{k \in \mathbb{Z}: k \text{ is divisible by } 2\}$, so A_2 is the set of all even integers. Observe that A_3 is the set of all multiples of 3 and that A_{10} is the set of all multiples of 10. ■

Example 9

For each natural number n, let $A_n = \{2n - 1, 2n\}$. Then $\{A_n: n \in \mathbb{N}\}$ is a family of sets, $\mathbb{N}$ is the indexing set, and, for each $n \in \mathbb{N}$, A_n is a set that consists of two members of $\mathbb{N}$. In particular, $A_1 = \{1,2\}$, $A_2 = \{3,4\}$, and $A_3 = \{5,6\}$. ■

The union and intersection of two sets, which are defined in Section 2.2, are special cases of the following definition:

Definition

Let X be a nonempty subset of $\mathbb{N}$ and, for each $n \in X$, let A_n be a set. Then $\cup\{A_n: n \in X\}$ is the set to which p belongs provided that there is an $n \in X$ such that $p \in A_n$, and $\cap\{A_n: n \in X\}$ is the set to which p belongs provided that $p \in A_n$ for each $n \in X$.

There are alternative notations for $\cap\{A_n: n \in X\}$ and $\cup\{A_n: n \in X\}$. In general, $\cap_{n\in X}A_n$ is the same as $\cap\{A_n: n \in X\}$, and $\cup_{n\in X}A_n$ is the same as $\cup\{A_n: n \in X\}$. We also use $\cup_{n=1}^{\infty}A_n$ as a substitute for $\cup\{A_n: n \in \mathbb{N}\}$ and $\cap_{n=1}^{\infty}A_n$ as a substitute for $\cap\{A_n: n \in \mathbb{N}\}$. Similarly $\cup_{n=1}^{p}A_n$ is used as a substitute for $\cup\{A_n: n \in \mathbb{N}$ and $n \leq p\}$, and $\cap_{n=1}^{p}A_n$ is used as a substitute for $\cap\{A_n: n \in \mathbb{N}$ and $n \leq p\}$.

In working with $\cup\{A_n: n \in X\}$ and $\cap\{A_n: n \in X\}$, there are two crucial facts to keep in mind:

a. $p \in \cup\{A_n: n \in X\}$ means that there is an $n \in X$ such that $p \in A_n$.
b. $p \in \cap\{A_n: n \in X\}$ means that, for each $n \in X$, $p \in A_n$.

Example 10

Let $\{A_n: n \in \mathbb{N}\}$ be the collection of sets defined in Example 7. If r is a positive real number, then there is a natural number n such that $n \geq r$. Therefore there is a natural number n such that $r \in [0, n]$. Hence $r \in \cup\{A_n: n \in \mathbb{N}\}$, so we see that $\cup\{A_n: n \in \mathbb{N}\} = \{x \in \mathbb{R}: x \geq 0\}$. If $0 \leq r \leq 1$, then $r \in A_n$ for each natural number n. However, if $r > 1$, then $r \notin A_1$. Hence $\cap\{A_n: n \in \mathbb{N}\} = \{x \in \mathbb{R}: 0 \leq x \leq 1\}$. ■

Example 11

Let $\{A_n: n \in \mathbb{N}\}$ be the collection of sets defined in Example 8. Since $A_1 = \mathbb{Z}$, $\cup\{A_n: n \in \mathbb{N}\} = \mathbb{Z}$. An integer p belongs to $\cap\{A_n: n \in \mathbb{N}\}$ if and only if p is divisible by every natural number. Since 0 is the only integer that is divisible by every natural number, $\cap\{A_n: n \in \mathbb{N}\} = \{0\}$. ■

Example 12

If $\{A_n: n \in \mathbb{N}\}$ is the collection of sets defined in Example 9, then $\cup\{A_n: n \in \mathbb{N}\} = \mathbb{N}$ and $\cap\{A_n: n \in \mathbb{N}\} = \varnothing$. ■

Definition

We say that $\{A_n: n \in X\}$ is a **pairwise disjoint family** of sets provided that, if A and B are members of $\{A_n: n \in X\}$ and $A \neq B$, then $A \cap B = \varnothing$.

The collection of sets defined in Example 9 is a pairwise disjoint family of sets. The collections defined in Examples 7 and 8 are not pairwise disjoint.

If you heeded our advice following Theorem 2.2, you have probably already memorized Theorem 2.5, which gives a generalization of de Morgan's laws.

Theorem 2.5

For each natural number n, let A_n be a set. Then $(\cup\{A_n: n \in \mathbb{N}\})^{\sim} = \cap\{A_n^{\sim}: n \in \mathbb{N}\}$ and $(\cap\{A_n: n \in \mathbb{N}\})^{\sim} = \cup\{A_n^{\sim}: n \in \mathbb{N}\}$.

Proof

We prove that $(\cup\{A_n: n \in \mathbb{N}\})^{\sim} = \cap\{A_n^{\sim}: n \in \mathbb{N}\}$ and leave the remaining equality as an exercise. Once again, to prove that these two sets are equal, we prove that each is a subset of the other. Let $x \in (\cup\{A_n: n \in \mathbb{N}\})^{\sim}$. Then, by the definition of complement, $x \notin \cup\{A_n: n \in \mathbb{N}\}$. So, by the definition of union, $x \notin A_n$ for any $n \in \mathbb{N}$. Hence, by the definition of complement, $x \in A_n^{\sim}$ for each $n \in \mathbb{N}$. Therefore, by the definition of intersection, $x \in \cap\{A_n^{\sim}: n \in \mathbb{N}\}$, and so $(\cup\{A_n: n \in \mathbb{N}\})^{\sim} \subseteq \cap\{A_n^{\sim}: n \in \mathbb{N}\}$. Now let $x \in \cap\{A_n^{\sim}: n \in \mathbb{N}\}$. Then, by the definition of intersection, $x \in A_n^{\sim}$ for each $n \in \mathbb{N}$. So, by the definition of complement, $x \notin A_n$ for any $n \in \mathbb{N}$. Hence, by the definition of union, $x \notin \cup\{A_n: n \in \mathbb{N}\}$. Therefore, by the definition of complement, $x \in (\cup\{A_n: n \in \mathbb{N}\})^{\sim}$, and so $\cap\{A_n^{\sim}: n \in \mathbb{N}\} \subseteq (\cup\{A_n: n \in \mathbb{N}\})^{\sim}$. □

Exercises 2.3

26. For each natural number n, let $A_n = \{-3, n\}$. Find $\cup\{A_n: n \in \mathbb{N}\}$ and $\cap\{A_n: n \in \mathbb{N}\}$.
27. For each natural number n, let $A_n = \{k \in \mathbb{N}: k \geq n\}$. Find $\cup_{n=1}^{5} A_n$, $\cap_{n=1}^{5} A_n$, $\cup\{A_n: n \in \mathbb{N}\}$, and $\cap\{A_n: n \in \mathbb{N}\}$.
28. For each natural number n, let $A_n = \{k \in \mathbb{N}: k \leq 2n\}$. Find $\cup_{n=1}^{4} A_n$, $\cap_{n=1}^{4} A_n$, $\cup\{A_n: n \in \mathbb{N}\}$, and $\cap\{A_n: n \in \mathbb{N}\}$.
29. For each natural number n, let $A_n = (-1/n, 2n)$. Find $\cup_{n=1}^{20} A_n$, $\cap_{n=1}^{20} A_n$, $\cup\{A_n: n \in \mathbb{N}\}$, and $\cap\{A_n: n \in \mathbb{N}\}$.
30. Let $B = \{k \in \mathbb{Z}: k$ is divisible by $3\}$. For each natural number n, let $A_n = \{n + k: k \in B\}$. Describe A_1, A_2, A_3, and A_4. If $m - n$ is divisible by 3, what is the relationship between A_m and A_n? If $m - n$ is not divisible by 3, what is the relationship between A_m and A_n?
31. For each natural number n, let A_n be a set. Prove that if $m \in \mathbb{N}$ then $A_m \subseteq \cup\{A_n: n \in \mathbb{N}\}$ and $\cap\{A_n: n \in \mathbb{N}\} \subseteq A_m$.
32. Let $p \in \mathbb{N}$ and, for each $n \in \mathbb{N}$ such that $n \leq p$, let A_n be a set. Prove that, if $A_n \subseteq A_{n+1}$ for each $n = 1, 2, \ldots, p - 1$, then $\cup_{n=1}^{p} A_n = A_p$ and $\cap_{n=1}^{p} A_n = A_1$.
33. For each natural number n, let A_n be a set. Prove that $(\cap\{A_n: n \in \mathbb{N}\})^{\sim} = \cup\{A_n^{\sim}: n \in \mathbb{N}\}$.

34. For each natural number n, let $A_n = (-1/n, 2n)$. Use Theorem 2.5 and Exercise 29 to find $\cup\{A_n^{\sim}: n \in \mathbb{N}\}$ and $\cap\{A_n^{\sim}: n \in \mathbb{N}\}$.

35. For each natural number n, let $A_n = \{x \in \mathbb{R}: -n < x \leq 1/n\}$. Find $\cup\{A_n: n \in \mathbb{N}\}$ and $\cap\{A_n: n \in \mathbb{N}\}$. Now use Theorem 2.5 to find $\cup\{A_n^{\sim}: n \in \mathbb{N}\}$ and $\cap\{A_n^{\sim}: n \in \mathbb{N}\}$.

2.4 ORDERED *n*-TUPLES AND CARTESIAN PRODUCTS

The concept of representing each point in the plane by an ordered pair of real numbers can be generalized as follows: if A and B are sets, $a \in A$, and $b \in B$, then we can form the *ordered pair* (a,b). Here a is the *first term* of the ordered pair, b is the *second term*, and the order is important. Thus $(a_1, b_1) = (a_2, b_2)$ if and only if $a_1 = a_2$ and $b_1 = b_2$. The set of all ordered pairs (a,b) such that $a \in A$ and $b \in B$ is called the *Cartesian product of A and B* (in honor of René Descartes) and is denoted by $A \times B$. Notice that the plane is the Cartesian product of $\mathbb{R}$ and $\mathbb{R}$.

Example 13

Let $A = \{1,2\}$ and $B = \{3,4,5\}$. Then

$$A \times B = \{(1,3),(1,4),(1,5),(2,3),(2,4),(2,5)\} \quad \text{and}$$
$$B \times A = \{(3,1),(3,2),(4,1),(4,2),(5,1),(5,2)\}$$

Although $A \times B$ and $B \times A$ are disjoint sets, both sets have six members. ■

Example 14

Let P denote the set of all prime numbers, and let E denote the set of all even natural numbers. Then $P \times E$ is the set of all ordered pairs (p,e), where p is a prime and e is an even natural number. Thus $(3,6) \in P \times E$. However, $E \times P$ is the set of all ordered pairs (e,p), where e is an even natural number and p is a prime. Therefore, $(3,6) \notin E \times P$. However, $(6,3) \in E \times P$ but $(6,3) \notin P \times E$. Since 2 is the only even prime, $(P \times E) \cap (E \times P) = \{(2,2)\}$. ■

Example 15

Let $A = \{1\}$, $B = \{2\}$, and $C = \{3\}$. Then $A \times (B \times C) = \{(1,(2,3))\}$, and $(A \times B) \times C = \{((1,2),3)\}$. Since ordered pairs are equal if and only if their first terms are equal and their second terms are equal, $(1,(2,3)) \neq ((1,2),3)$. Consequently, the sets $A \times (B \times C)$ and $(A \times B) \times C$ are not equal. ■

An ordered pair, as we have defined it, is intuitive, and almost everyone thinks of an ordered pair in this manner. We give a precise definition, which generalizes this concept, by considering sets that are indexed by the first n natural numbers and exploiting the ordering that such an indexed set inherits from the ordering of the natural numbers. (If $n = 2$, this definition precisely defines an ordered pair.)

Definition

A set that is indexed by the set consisting of the first n natural numbers is called an **ordered n-tuple.** It is customary to call ordered 2-tuples **ordered pairs** and ordered 3-tuples **ordered triples.** We denote the set consisting of the first n natural numbers by I_n. For each natural number $k \leq n$, the element a_k of the ordered n-tuple $\{a_i: i \in I_n\}$ is called the **kth-term** of that n-tuple.

We denote the n-tuple $\{a_i: i \in I_n\}$ by $(a_1,a_2, \ldots, a_n)$. It is even permissible to omit the parentheses, writing $a_1,a_2, \ldots, a_n$. In this case, the ordered n-tuple is often called a *list* or a *string*. Two n-tuples $(a_1,a_2, \ldots ,a_n)$ and $(b_1,b_2, \ldots, b_n)$ are equal provided that the n-tuples agree term by term. In other words, $(a_1,a_2, \ldots, a_n) = (b_1,b_2, \ldots, b_n)$ provided that the term a_k equals the term b_k for all $k \in I_n$. Again, almost everyone thinks of an ordered n-tuple in an intuitive fashion as $(a_1,a_2, \ldots, a_n)$. Each point in 3-space can be represented by an ordered triple of real numbers.

Examples 14 and 15 show that the Cartesian product is neither commutative nor associative. Because the whole idea of ordered pairs is to determine which member comes first and which comes second, it should not come as a great disappointment that the Cartesian product is not commutative. Nonassociativity is a more serious shortcoming, but the following definition provides a way of avoiding this problem altogether.

Definition

Let $A_1, A_2, \ldots, A_n$ be a list of sets. The **Cartesian product** of $A_1, A_2, \ldots, A_n$, denoted $A_1 \times A_2 \times \cdots \times A_n$, is the set of all ordered n-tuples $(a_1,a_2, \ldots, a_n)$ whose kth term a_k belongs to A_k for $k = 1,2, \ldots, n$. It is a common abuse of set-builder notation to write $A_1 \times A_2 \times \cdots \times A_n = \{(a_1,a_2, \ldots, a_n): a_k \in A_k \text{ for } k \in I_n\}$.

Example 16

Let $A = \{5,8\}$, $B = \{2,3\}$, and $C = \{1,3\}$. Then

$$A \times B \times C = \{(5,2,1),(5,2,3),(5,3,1),(5,3,3),(8,2,1),(8,2,3),(8,3,1),(8,3,3)\}$$

Notice that each member of $[A \times (B \times C)] \cup [(A \times B) \times C]$ is an ordered pair one of whose terms is an ordered pair, while each member of $A \times B \times C$ is an ordered triple. ■

Example 17

Let $A_1 = \{5,8\}$, $A_2 = \{2,3\}$, $A_3 = \{1,3\}$, and $A_4 = \{4,6\}$. Then each member of $A_1 \times A_2 \times A_3 \times A_4$ is an ordered 4-tuple. Indeed,

$$A_1 \times A_2 \times A_3 \times A_4 = \{(5,2,1,4),(5,2,1,6),(5,2,3,4),(5,2,3,6),\\ (5,3,1,4),(5,3,1,6),(5,3,3,4),(5,3,3,6),\\ (8,2,1,4),(8,2,1,6),(8,2,3,4),(8,2,3,6),\\ (8,3,1,4),(8,3,1,6),(8,3,3,4),(8,3,3,6)\}$$

■

Exercises 2.4

36. Suppose that we are given a universal set U with n elements. How many n-tuples of elements of U are there?

37. Let U be a universal set and let A be a subset of U. A *neighbor* of A is a subset B of U that can be obtained from A by adding or subtracting one element. For example, if $U = I_8$ and $A = \{2,5,7\}$, then $\{2,5\}$ and $\{2,5,7,8\}$ are neighbors of A. Let A be a five-element subset of an eight-element universal set U. How many neighbors does A have?

38. Let $A = \{1,2,3\}$, $B = \{1\}$, and $C = \{4,5\}$. Find

a. $A \times C$ **b.** $C \times A$ **c.** $A \times B$ **d.** $A \times B \times C$

39. Suppose that A has five members, B has eight members, and $A \cap B$ has two members.

a. How many members does $A \times B$ have?

b. How many members does $B \times A$ have?

c. How many members does $A \times (A \cap B)$ have?

40. Suppose that A has five members, B has ten members, and C has no members (thus $C = \varnothing$).

a. How many members does $A \times B$ have?

b. How many members does $A \times C$ have?

c. How many members does $A \times B \times C$ have?

41. Which of the following statements are true?

a. If A is a set, B is a set, and $(5,6) \notin A \times B$, then $5 \notin A$ and $6 \notin B$.

b. If A is a set, B is a set, and $5 \notin A$, then $(5,6) \notin A \times B$.

c. If A is a set, B is a set, and $(A \times B) \cap (B \times A) \neq \varnothing$, then $A \cap B \neq \varnothing$.

d. If A is a set, B is a set, and $A \cap B \neq \varnothing$, then $(A \times B) \cap (B \times A) \neq \varnothing$.

e. If $A = \varnothing$, then for each set B, $A \times B = \varnothing$.

f. There are sets A and B such that $A \times B = \{(1,1),(1,2),(2,1),(3,1),(3,2),(2,2)\}$.

g. There are sets A and B such that $A \times B = \{(5,11),(2,5),(5,2),(3,2),(3,11),(5,3),(3,5)\}$.

42. Which of the following subsets of the plane are Cartesian products of two sets A and B?

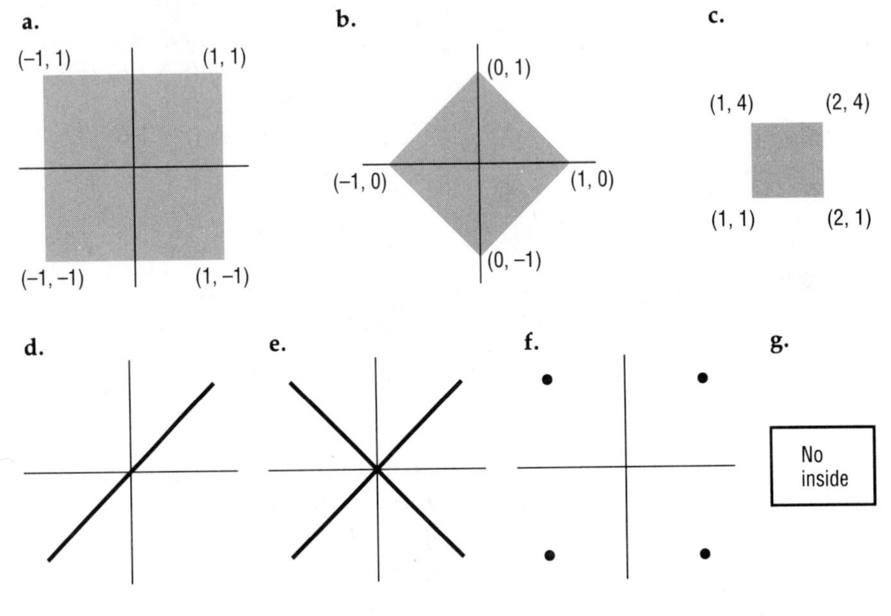

2.5 COUNTING PROBLEMS

In this section, we obtain some results about finite sets and use Venn diagrams to sort objects into pairwise disjoint collections. If A is a finite set, $n(A)$ denotes the number of elements in A. If A and B are disjoint sets, then $n(A \cup B) = n(A) + n(B)$. Theorem 2.6 provides the basis for generalizing this result.

Theorem 2.6 Let A and B be sets. Then the collection $\mathcal{A} = \{A - B, B - A, A \cap B\}$ is a pairwise disjoint family of sets and $A \cup B = (A - B) \cup (B - A) \cup (A \cap B)$.

Proof

To prove that $\mathcal{A}$ is a pairwise disjoint family of sets, we must prove that $(A - B) \cap (B - A) = \varnothing$, that $(A - B) \cap (A \cap B) = \varnothing$, and that $(B - A) \cap (A \cap B) = \varnothing$.

We first prove that $(A - B) \cap (B - A) = \varnothing$. We can use properties that we have already established in order to obtain this result.

$$\begin{aligned}
(A - B) \cap (B - A) &= (A \cap B^{\sim}) \cap (B \cap A^{\sim}) && \text{(Theorem 2.2f)}\\
&= [(A \cap B^{\sim}) \cap B] \cap A^{\sim} && \text{(Theorem 2.1b)}\\
&= [A \cap (B^{\sim} \cap B)] \cap A^{\sim} && \text{(Theorem 2.1b)}\\
&= (A \cap \varnothing) \cap A^{\sim} && \text{(Theorem 2.1e)}\\
&= \varnothing \cap A^{\sim} && \text{(Theorem 2.2i)}\\
&= A^{\sim} \cap \varnothing && \text{(Theorem 2.1a)}\\
&= \varnothing && \text{(Theorem 2.2i)}
\end{aligned}$$

By Exercise 19, $(A - B) \cap (A \cap B) = \varnothing = (B - A) \cap (A \cap B)$. We can also use previous results to complete the proof:

$$\begin{aligned}
&(A - B) \cup (B - A) \cup (A \cap B)\\
&= (A \cap B^{\sim}) \cup (B \cap A^{\sim}) \cup (A \cap B) && \text{(Theorem 2.2f)}\\
&= (A \cap B^{\sim}) \cup (B \cap A^{\sim}) \cup (B \cap A) && \text{(Theorem 2.1a)}\\
&= (A \cap B^{\sim}) \cup [B \cap (A^{\sim} \cup A)] && \text{(Theorem 2.1c)}\\
&= (A \cap B^{\sim}) \cup (B \cap U) && \text{(Theorem 2.1e)}\\
&= (A \cap B^{\sim}) \cup B && \text{(Theorem 2.1d)}\\
&= B \cup (A \cap B^{\sim}) && \text{(Theorem 2.1a)}\\
&= (B \cup A) \cap (B \cup B^{\sim}) && \text{(Theorem 2.1c)}\\
&= (B \cup A) \cap U && \text{(Theorem 2.1e)}\\
&= B \cup A && \text{(Theorem 2.1d)}\\
&= A \cup B && \text{(Theorem 2.1a)}
\end{aligned}$$

□

If $A - B$, $B - A$, and $A \cap B$ are nonempty, then $\{A - B, B - A, A \cap B\}$ is a partition of $A \cup B$. In general, a collection $\mathcal{P}$ of subsets of a set X is a *partition* of X provided that each member of $\mathcal{P}$ is nonempty and each member of X belongs to exactly one member of $\mathcal{P}$. This important concept is discussed in Chapter 10.

Theorem 2.7

If A and B are finite sets, then $n(A \cup B) = n(A) + n(B) - n(A \cap B)$.

Proof

By Exercise 19, $A = (A - B) \cup (A \cap B)$ and $(A - B) \cap (A \cap B) = \varnothing$. Therefore $n(A) = n(A - B) + n(A \cap B)$, and so $n(A - B) = n(A)$

$- n(A \cap B)$. Likewise $B = (B - A) \cup (B \cap A)$ and $(B - A) \cap (B \cap A) = \emptyset$, so $n(B) = n(B - A) + n(B \cap A)$. Therefore $n(B - A) = n(B) - n(B \cap A)$. By Theorem 2.6, $A \cup B = (A - B) \cup (B - A) \cup (A \cap B)$ and $\{A - B, B - A, A \cap B\}$ is a pairwise disjoint family of sets. Therefore $n(A \cup B) = n(A - B) + n(B - A) + n(A \cap B)$. So

$$\begin{aligned} n(A \cup B) &= n(A) - n(A \cap B) + n(B) - n(A \cap B) + n(A \cap B) \\ &= n(A) + n(B) - n(A \cap B) \end{aligned}$$ □

Theorem 2.7 is an important result, and it reappears in Chapter 6. Although Example 18 and the following exercises do not capture the importance of this theorem, they should familiarize you with the way it is applied.

Example 18

A computer distributor has 100 personal computers in stock. The owner informs the manager that 25 of the computers are ATs, 40 have 640K memory, 10 have color monitors and 640K memory but are not ATs, 55 have monochrome monitors, 10 are ATs with color monitors, 15 are ATs with 640K memory, and 5 are ATs with monochrome monitors that do not have 640K memory. To fill some orders, the manager needs to know how many personal computers have the following characteristics:

ATs, 640K memory, and color monitors

640K memory, monochrome monitors, and are not ATs

640K memory or color monitors, but are not ATs

Analysis

Observe that there are three categories of personal computers—ATs, 640K memory, and color monitors. Unfortunately, it is not clear how many computers are in each category. Five are ATs that do not have 640K memory and do not have color monitors. There are a total of 25 ATs, so 20 must be ATs with 640K memory or ATs with color monitors. If we let A denote the set of all ATs, B denote the set of all personal computers with 640K memory, and C denote the set of all personal computers with color monitors, then $n[(A \cap B) \cup (A \cap C)] = 20$. But we know that $n(A \cap C) = 10$ and that $n(A \cap B) = 15$. So by Theorem 2.7, $n(A \cap B \cap C) = 5$. This is a critical piece of information, and we can begin to complete the diagram in Figure 2.8 by entering a 5 in $A \cap B \cap C$.

Since $n(A \cap C) = 10$, we can enter a 5 in $(A \cap C) - B$, and, since $n(A \cap B) = 15$, we can enter a 10 in $(A \cap B) - C$. Since 25 are ATs, we can enter a 5 in $A - (B \cup C)$. Since 10 have color monitors and 640K memory but are not ATs, we can enter a 10 in $(B \cap C) - A$. Since 55 have monochrome monitors, 45 must have color monitors, so we can

Figure 2.8

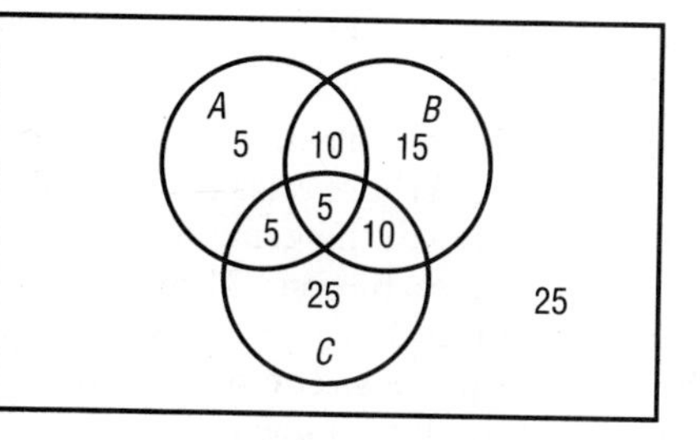

enter a 25 in $C - (A \cup B)$. Since 40 have 640K memory, we can enter a 15 in $B - (A \cup C)$. Finally, since we have 100 personal computers, we can enter a 25 in $U - (A \cup B \cup C)$. We are now prepared to supply the information the manager needs. There are 5 ATs with 640K memory and color monitors, there are 15 computers with 640K memory and monochrome monitors that are not ATs, and there are 50 computers that are not ATs but that have 640K memory or color monitors. ■

Exercises 2.5

43. **a.** Let A and B be sets such that $n(A) = 28$, $n(B) = 79$, and $n(A \cap B) = 3$. Find $n(A \cup B)$.

b. Let A and B be sets such that $n(A) = 35$, $n(B) = 96$, and $n(A \cup B) = 122$. Find $n(A \cap B)$.

44. Find sets A and B such that $A \cap B = \{1,4,7\}$, $A - B = \{3,5,6\}$, and $B - A = \{9,11\}$. Is there more than one correct answer to this exercise? Explain.

45. There are 512 computer science majors at State University, and they each own a personal computer. Of these students, 281 own a computer with a printer, 167 own a computer with a color monitor, 98 own a computer with a hard-disk drive, 75 own a computer with a printer and a color monitor, 56 own a computer with a printer and a hard-disk drive, 34 own a computer with a color monitor and a hard-disk drive, and 5 own a computer with a printer, a color monitor, and a hard-disk drive. Draw a Venn diagram like that in Example 18 and place a number in each region. How many computer science majors own a personal computer with monochrome monitor and no printer or hard-disk drive?

46. In the major leagues, there are 81 players who bat and throw right-handed and are under 30 years old, 100 who bat right-handed and are under 30, 333 who bat right-handed, 234 who bat and throw right-handed, 63 who bat and throw left-handed and are 30 or over, 162 who bat left-handed, throw right-handed, and are 30 or over, 54 who bat left-handed, throw right-handed, and are under 30, and 45 who bat and throw left-handed and are under 30. How many major league players are in each of the following categories?

a. bat right-handed

b. are under 30

c. throw right-handed

d. bat right-handed but throw left-handed

e. bat left-handed and are 30 or over

f. bat right-handed and are under 30

47. Bud, Nancy, Nero, and Jan are playing a game of Clue. There are six suspects—Mr. Green, Mrs. White, Miss Scarlet, Professor Plum, Col. Mustard, and Miss Peacock. In the game there is a playing card for each person, each weapon, and each room. Three cards, a person card, a weapon card, and a room card, are picked at random and hidden without being seen by any of the players. The rest of the cards are dealt one at a time to the players. During the course of the game the following facts are observed: (a) Bud has the gun or Mrs. White; (b) Nancy has Col. Mustard or the knife; (c) Nero has the gun or Miss Scarlet; (d) Jan has the knife and Miss Scarlet; (e) Bud has Professor Plum; (f) Nero has Mrs. Peacock or the rope; and (g) Nancy has the gun or Mr. Green. Which person is missing?

48. In the freshman class at State University, there are 810 members who play golf, 1,070 who swim, 70 who play tennis and swim but do not play golf, 975 who do not play tennis, 820 who play golf or swim but do not play tennis, 510 who play golf and tennis, 480 who play golf and swim, and 120 who play golf but neither swim nor play tennis. Draw a Venn diagram like that in Example 18 and place a number in each region that can be determined. How many freshmen play golf and tennis and swim? How many swim but play neither golf nor tennis? Which region in your Venn diagram is of indeterminable size?

49. A dean reports that there are 1,500 faculty members at State University. Of these, 1,260 are men, 1,080 earn at least $30,000 per year, 780 are over 35, 560 are males over 35, 710 are over 35 and earn at least $30,000 per year, 600 are males who earn at least $30,000 per year, and 430 are males over 35 who earn at least $30,000 per year. Draw the appropriate Venn diagram. Can these figures be correct? If not, why not? If they can be correct, how many women who make at least $30,000 per year have not celebrated their 35th birthday?

50. A dean at State University reports that, in a discrete mathematics class of 50, 8 were women who were computer science majors and who made a grade of C or better, 11 were women who were computer science majors, 17 were women who made a grade of C or better, 15 were computer science majors who made a C or better, 23 were women, 19 were computer science majors, and 34 made a grade of C or better. Draw an appropriate Venn diagram. Can these figures be correct? If not, why not? If they can be correct, how many men were in the class?

51. In a group of secretaries, it was found that 92 can climb steep mountains, 103 can leap over tall buildings, 117 can walk on water, 31 can climb steep mountains and leap over tall buildings, 21 can leap over tall buildings and walk on water, 27 can climb steep mountains and walk on water, 13 can do all three, and 11 cannot do any of the three. How many secretaries are in the group?

52. In a magazine survey of 200 people, it was found that 139 people read *Playboy*, 59 read *National Geographic*, 77 read *Golf Digest*, 19 read *Playboy* and

National Geographic, 37 read *Playboy* and *Golf Digest*, 33 read *National Geographic* and *Golf Digest*, and 9 read all three magazines.

a. How many people read at least one of the three magazines?

b. How many read none of the three magazines?

c. How many read exactly one of the three magazines?

d. How many read exactly two of the three magazines?

53. Flight 632 flies from New York to Denver with one intermediate stop in Charlotte, N.C. Is it true that the number of paying passengers on Flight 632 between New York and Charlotte plus the number of paying passengers on Flight 632 between Charlotte and Denver minus the number of passengers who bought tickets from New York to Denver equals the total number of tickets sold for Flight 632? Explain. Would it make any difference if someone bought a ticket to Charlotte, got off at Charlotte, and immediately bought a second ticket for the remainder of the flight? You may assume that everyone who bought a ticket took the flight, but do not forget that some passengers got on board at Charlotte, while others "deplaned" there.

Chapter 2 Review Exercises

54. Let A, B, and C be sets. Prove or find a counterexample to each of the following statements.

a. $(A - B) \cup (B - A) = (A \cup B) - (A \cap B)$

b. If $(A - B) \cup (B - A) = (A - C) \cup (C - A)$, then $B = C$

c. $A - B \subseteq C$ if and only if $A - C \subseteq B$

d. $(A^{\sim} \cup B) \cap (B^{\sim} \cup C) \subseteq A^{\sim} \cup C$

e. $(A^{\sim} \cup B)^{\sim} \cap A = A - B$

55. Let A, B, and C be sets. Prove that $A \cap (B \oplus C) = (A \cap B) \oplus (A \cap C)$.

56. For each $n \in \mathbb{N}$ let $A_n = [n,2n]$, and let $m \in \mathbb{N}$. Find $\bigcup_{n=1}^{10} A_n$, $\bigcup_{n=1}^{m} A_n$, and $\bigcup_{n=1}^{\infty} A_n$.

57. Let $A = \{1,2,3\}$, $B = \{4,5\}$, and $C = \{6\}$. Find $A \times (B \times C)$, $(A \times B) \times C$, and $A \times B \times C$.

58. Find the symmetric difference of $\{2,3,5\}$ and $\{3,5,7\}$.

59. Let $A = \{n \in \mathbb{N}: n \leqslant 25\}$, $B = \{n \in \mathbb{N}: 10 \leqslant n \leqslant 50\}$, and $C = \{n \in \mathbb{N}: n \leqslant 5,\ 20 \leqslant n \leqslant 30, \text{ or } 45 \leqslant n \leqslant 60\}$. Find $n(A)$, $n(B)$, $n(C)$, $n(A \cap B)$, $n(A \cap C)$, $n(B \cap C)$, $n(A \cap B \cap C)$, and $n(A \cup B \cup C)$. Calculate $n(A) + n(B) + n(C) - n(A \cap B) - n(A \cap C) - n(B \cap C) + n(A \cap B \cap C)$.

60. Let A be a set. The *successor* of A is the set $A \cup \{A\}$. Find the successor of each of the following sets:

a. $\varnothing$ **b.** $\{\varnothing\}$ **c.** $\{1,2\}$ **d.** $\{1,\{1\}\}$

61. A *multiset* is analogous to a set, but members of a multiset may occur more than once. Thus $\{a,a,a,b,b\}$ is a multiset, and we denote this set by $\{3a,2b\}$.

In general, the notation $\{n_1a_1, n_2a_2, \ldots, n_pa_p\}$ denotes the multiset in which the element a_i occurs n_i times. The number n_i is called the *multiplicity* of a_i. The *union*, $A \cup B$, of two multisets A and B is the multiset in which the multiplicity of an element is the maximum of its multiplicity in A and B. The *intersection*, $A \cap B$, of A and B is the multiset in which the multiplicity of an element is the minimum of its multiplicity in A and B. The *difference*, $A - B$, is the multiset in which the multiplicity of an element a is its multiplicity in A minus its multiplicity in B provided that this difference is positive; otherwise a does not occur in $A - B$. The *sum*, $A + B$, is the multiset in which the multiplicity of an element is the sum of its multiplicity in A and B. Let $A = \{4a,2b,5c\}$ and $B = \{5b,3c,1d\}$. Find

a. $A \cup B$ **b.** $A \cap B$ **c.** $A - B$ **d.** $B - A$ **e.** $A + B$

62. A *fuzzy set* A is analogous to a multiset. Each member of a universal set U has a *degree of membership*, d, in A, where $0 \leq d \leq 1$. Thus $A = \{.2a,.6b,.3c\}$ is a fuzzy set. The definition of union and intersection of fuzzy sets is analogous to union and intersection of multisets. The complement $A^{\sim}$ of a fuzzy set is the set in which the degree of membership of each element of U is 1 minus its degree of membership in A. Thus, if $U = \{a,b,c,d\}$, then $A^{\sim} = \{.8a,.4b,.7c,1d)$. Let $U = \{a,b,c,d\}$, $A = \{.9a,.2c\}$, and $B = \{.3b,.5c,.7d\}$. Find

a. $A \cup B$ **b.** $A \cap B$ **c.** $A^{\sim}$ **d.** $B^{\sim}$

Chapter

3

FUNCTIONS

Buying a candy bar from a vending machine is an adventure. You put your money in. Sometimes you get the candy bar of your choice, sometimes you get an entirely different candy bar, and sometimes you get no candy bar at all. In short, buying a candy bar is like playing a slot machine. This is not the way a vending machine is supposed to work. The machine has knobs 1 through 10, and it is supposed to assign to each of these ten knobs one and only one kind of candy bar.

In mathematics, especially in discrete mathematics, we are often faced with one of the tasks of a vending machine, to match up members of one set with members of another. In this chapter we study functions. Basically, a function consists of a set of inputs, like the knobs of a vending machine, a set of possible outcomes, like the set of all candy bars, and a rule that assigns to each input exactly one of the possible outcomes. Thus a function is divided into three parts: a set of inputs called the *domain,* a set of conceivable outcomes called the *codomain,* and a rule that assigns to each input one and only one of the possible outcomes. We denote the function with domain A, codomain B, and rule f by $f\colon A \to B$, which is read "f mapping A into B" or sometimes "f maps A into B."

3.1 BASIC DEFINITIONS

In this section we study the basic concepts of functions.

Definition

A function $f\colon A \to B$ consists of a set A, a set B, and a subset $G(f)$ of $A \times B$ with the property that each member of A is the first term of exactly one ordered pair belonging to $G(f)$. The set A is called the **domain** of the function (often abbreviated $\text{Dom}(f)$), the set B is called the **codomain** of the function, and the subset $G(f)$ of $A \times B$ is called the **graph** of the function. If $(a,b) \in G(f)$, we say that b is the image of a under f and write $f(a) = b$. For each subset S of A, $\{f(a)\colon a \in S\}$ is called the **image of the set S under f.** Specifically, the image of A under f is called the **range** of the function and is denoted by $\text{Rng}(f)$.

Although the range of a function is always a subset of the codomain, there is no requirement that the range equal the codomain. Moreover, although no member of the domain is the first term of two or more members of the graph, it is possible that a single member of the codomain is the image under f of two or more members of the domain. When the codomain and range of a given function $f\colon A \to B$ are equal we say that f maps A *onto* B, and when no member of the codomain is the image under f of two or more members of the domain we say that the function is *one to one.* We illustrate these ideas informally in Example 1.

Example 1

The Lone Ranger versus a gang of desperados

The cylinder of the Lone Ranger's six gun

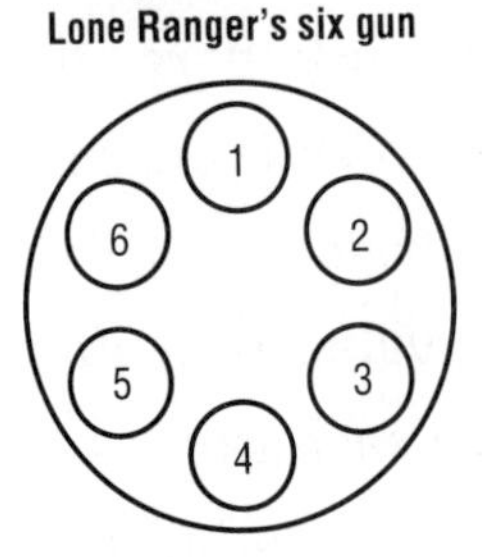

The Gang

a, Two-Gun Pete
b, Moose
c, Lefty
d, Dutch
e, Slippery Joe
f, Sloppy Joe

Let $A = \{1,2,3,4,5,6\}$ and $B = \{a,b,c,d,e,f\}$. Each part of Figure 3.1 is a candidate for a function with domain A and codomain B.

Figure 3.1

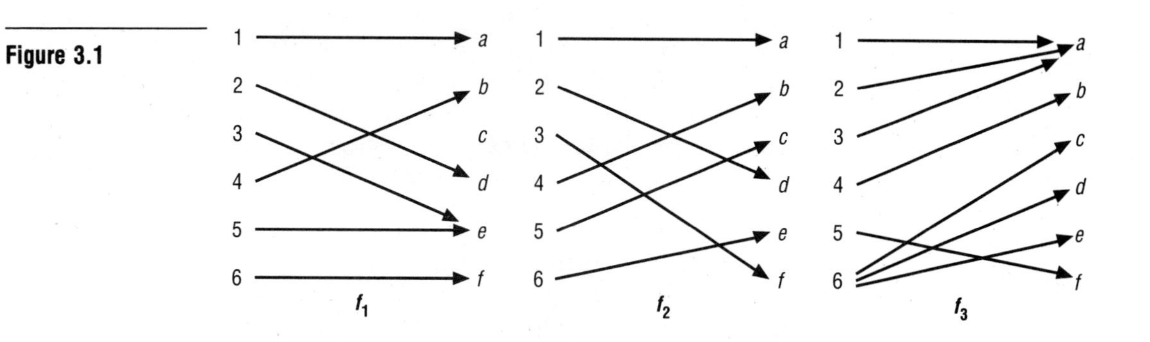

$G(f_1) = \{(1,a), (2,d), (3,e), (4,b), (5,e), (6,f)\}$. Lone Ranger's comments: "Hate to waste two silver bullets on the likes of Slippery Joe, but there was no point shooting at Lefty; he was out of range." Authors' comments: f_1: $A \to B$ is a function, but it is *not* one to one because $f(3) = f(5)$ and so Slippery was shot twice. Since Lefty does not belong to $\text{Rng}(f_1)$, the range of the function is a *proper* subset of its codomain. Hence f_1 does not map A onto B.

$G(f_2) = \{(1,a), (2,d), (3,f), (4,b), (5,c), (6,e)\}$. Lone Ranger's comments: "What a showdown. Everybody was hit once and no one was hit twice. A nice clean job!" Authors' comments: f_2: $A \to B$ is a one-to-one function mapping A onto B.

$G(f_3) = \{(1,a),(2,a),(3,a),(4,b),(5,f),(6,c),(6,d),(6,e)\}$. Movie director's comments: "Cut! Clayton, you showboat, even the kiddies in the Saturday matinee are not going to believe that you dropped Lefty, Dutch, and Slippery with one shot." Authors' comments: f_3: $A \to B$ is not a function because 6 is the first term of more than one ordered pair in $G(f_3)$. ■

Notice that in Example 1 it would not be possible to illustrate a function f: $A \to B$ that is one to one but does not map onto B without adding at least one member to the gang of desperados. Similarly, it would not be possible to illustrate a function f: $A \to B$ that maps A onto B and is not one to one without having a gang B with fewer than six members (see Exercise 5).

The definitions of a one-to-one function and of a function f mapping A onto B have been stated somewhat informally. Now we restate these definitions and give some alternative terminology. You should take the trouble to make sure that the new definitions square with the informal definitions given before Example 1.

Definition

Let $f: A \to B$ be a function. The function is called a **one-to-one function,** or **injection,** provided that, whenever a_1 and a_2 belong to A and $f(a_1) = f(a_2)$, then $a_1 = a_2$. The function is said to **map A onto B,** or to be a **surjection,** provided that each member of B is the image of at least one member of A. A function that is both an injection and a surjection is called a **bijection,** or **one-to-one correspondence.**

The prefix "bi-" of the term *bijection* is meant to suggest the possession of two properties, much as a biplane has two wings or a bicycle two wheels. The adjectives *injective, surjective,* and *bijective* have the obvious meanings, but it is not impolite to speak of a one-to-one correspondence rather than a bijection or of "one to one" rather than injective. Although we do not do so in this text, it is even commonplace to say that a function is "onto" rather than surjective.

We have already used the symbol $f(a)$ to denote the image of a under a function $f: A \to B$, and if $S \subseteq A$ we have agreed to call $\{f(a): a \in S\}$ the image of S under f. As long as no subset of A is a member of A, the notation $f(S) = \{f(a): a \in S\}$ is natural and unambiguous. In this text, we use the notation $f(S)$ to denote the image of a set S under f, and we assure you that, with the exception of Exercise 17, we never consider a function $f: A \to B$ for which $A \cap \mathcal{P}(A) \neq \varnothing$. Note that, with the convention that $f(S)$ denotes $\{f(a): a \in S\}$, the range of a function $f: A \to B$ may be denoted by $f(A)$ as well as by $\text{Rng}(f)$.

Example 2

Let $f: \mathbb{R} \to \mathbb{R}$ be defined by

$$f(x) = \begin{cases} x, & \text{if } x \leqslant 0 \\ x^2 - x, & \text{if } x > 0 \end{cases}$$

Is this function one to one? Does it map $\mathbb{R}$ onto $\mathbb{R}$? Find $f(S)$ when $S = [-1,0]$, when $S = \{-1,1,0\}$, and when $S = \mathbb{R}$.

Analysis

Since $f(0) = 0$ and $f(1) = 0$, $f: \mathbb{R} \to \mathbb{R}$ is not one to one. Let $r \in \mathbb{R}$. If $r \leqslant 0$, then $f(r) = r$. If $r > 0$, then by the quadratic formula we see that $x = (1 + \sqrt{1 + 4r})/2$ is a root of the equation $x^2 - x = r$. Since $x > 0$, $f(x) = r$. Therefore f maps $\mathbb{R}$ onto $\mathbb{R}$. Evidently $f([-1,0]) = [-1,0]$ and $f(\{-1,1,0\}) = \{-1,0\}$. Since f is a surjection, $f(\mathbb{R}) = \mathbb{R}$. ■

Example 3

Let $A = \{x \in \mathbb{R}: x \geqslant 0\}$ and let $f: A \to A$ be defined by $f(x) = (5x^2 + 3)/(3x^2 + 5)$. Is this function an injection? Is this function a bijection? Find $f(S)$ when $S = \{0,1,3,5\}$ and when $S = \{\sqrt{3}, \sqrt{5}\}$.

Analysis

To show that f is an injection, we must show that, if $f(a) = f(b)$, then $a = b$. Suppose that $f(a) = f(b)$. Then

$$\frac{5a^2 + 3}{3a^2 + 5} = \frac{5b^2 + 3}{3b^2 + 5}$$

and so

$$15a^2b^2 + 9b^2 + 25a^2 + 15 = 15a^2b^2 + 9a^2 + 25b^2 + 15$$

This equation simplifies to $16a^2 = 16b^2$, or $a^2 = b^2$. Since $a \geqslant 0$ and $b \geqslant 0$, $a = b$.

Suppose that this function is a surjection. Then for each $r \in A$ there is an $x \in A$ such that $(5x^2 + 3)/(3x^2 + 5) = r$. Thus there is an $x \in A$ such that $x^2 = (3 - 5r)/(3r - 5)$. For $r = 5/3$, no such x exists. Therefore $f\colon A \to A$ is not a surjection and hence not a bijection. It is easily verified that $f(\{0,1,3,5\}) = \{3/5,1,3/2,8/5\}$ and that $f(\{\sqrt{3},\sqrt{5}\}) = \{9/7,7/5\}$. ■

Example 4

Let X be a finite set with at least two members, and let $f\colon \mathcal{P}(X) \to \mathbb{Z}$ be defined by $f(A) = n(A)$ for each $A \in \mathcal{P}(X)$. This function is neither an injection nor a surjection. ■

Exercises 3.1

1. Let $A = \{1,2,3,4\}$ and $B = \{0,1,-1,2\}$. Which of the following sets of ordered pairs are graphs of a function $f\colon A \to B$?

   a. $\{(1,0),(2,1),(3,-1),(4,2)\}$
   b. $\{(1,0),(2,-1),(3,2)\}$
   c. $\{(1,0),(2,0),(3,1),(4,1)\}$
   d. $\{(2,0),(1,-1),(3,2),(4,1)\}$
   e. $\{(4,1),(3,2),(2,3),(1,4)\}$
   f. $\{(2,1),(3,0),(1,-1),(3,1),(4,2)\}$

2. Which of the sets of ordered pairs in Exercise 1 are graphs of one-to-one functions $f\colon A \to B$?

3. Which of the sets of ordered pairs in Exercise 1 are graphs of functions that map A onto B?

4. Each of the following rules is intended to define a function $f\colon \mathbb{R} \to \mathbb{R}$. Which of these rules does define such a function? If a given rule does not define such a function, explain why not.

   a. $f(x) = \dfrac{x}{x - 3}$
   b. $f(x) = \begin{cases} 1/x, & \text{if } x \neq 0 \\ 0, & \text{if } x = 0 \end{cases}$
   c. $f(x) = \pm\sqrt{1 + x^2}$
   d. $f(x) = \begin{cases} x, & \text{if } x \geqslant 0 \\ x^2, & \text{if } x < 0 \end{cases}$
   e. $f(x) = \begin{cases} x/2, & \text{if } x \in \mathbb{Z} \\ x, & \text{if } x \in \mathbb{R} - \mathbb{Z} \end{cases}$
   f. $f(x) = \sqrt{x + 1}$
   g. $f(x) = \begin{cases} 5, & \text{if } x^2 \geqslant x \\ -5, & \text{if } 0 \leqslant x \leqslant 1 \end{cases}$
   h. $f(x) = \begin{cases} x, & \text{if } x \geqslant 0 \\ x^2, & \text{if } x \leqslant 0 \end{cases}$

5. Let $A = \{1,2,3,4,5,6\}$ and $B = \{1,1/2,1/3,1/4,1/5,1/6,1/7\}$.
   a. Give an example of a function $f: A \to B$ that is one to one but does not map A onto B.
   b. Give an example of a function $f: B \to A$ that maps B onto A but is not one to one.
6. Let $f: \mathbb{R} \to \mathbb{R}$ be defined by $f(x) = 2x + 1$. Is f a one-to-one correspondence? Prove your answer.
7. Let $f: \mathbb{R} \to \mathbb{R}$ be defined by $f(x) = 2x + 1$. Describe the following sets explicitly: $f(\mathbb{Z})$, $f(\mathbb{N})$, $f(E)$ where E is the set of all even integers, $f(\mathbb{R})$, $\text{Rng}(f)$.
8. Each of the following functions maps $\mathbb{Z}$ into $\mathbb{Z}$. For each function determine whether it is a one-to-one function and determine whether it maps $\mathbb{Z}$ onto $\mathbb{Z}$.
   a. $f_1(n) = 2n + 3$
   b. $f_2(n) = n - 4$
   c. $f_3(n) = n^2 + 1$
   d. $f_4(n) = n^3$
9. a. Give an example of a function $f: \mathbb{N} \to \mathbb{N}$ that is not one to one and does not map $\mathbb{N}$ onto $\mathbb{N}$.
   b. Give an example of a function $g: \mathbb{N} \to \mathbb{N}$ that is one to one but does not map $\mathbb{N}$ onto $\mathbb{N}$.
   c. Give an example of a function $h: \mathbb{N} \to \mathbb{N}$ that maps $\mathbb{N}$ onto $\mathbb{N}$ but is not one to one.
10. a. Give an example of an injection $f: \mathbb{R} \to \mathbb{R}$ that is not a bijection.
    b. Give an example of a surjection $f: \mathbb{R} \to \mathbb{R}$ that is not a bijection.
    c. Give an example of a function $f: \mathbb{R} \to \mathbb{R}$ that is neither a surjection nor an injection.
11. Let $f: A \to B$ and let R and S be subsets of A.
    a. Is it necessarily true that $f(R) \cup f(S) = f(R \cup S)$?
    b. Is it necessarily true that $f(R) \cap f(S) = f(R \cap S)$?
12. Let $f: A \to B$ be a one-to-one function and let R and S be subsets of A.
    a. Is it necessarily true that $f(R) \cup f(S) = f(R \cup S)$?
    b. Is it necessarily true that $f(R) \cap f(S) = f(R \cap S)$?
13. Let $f: \mathbb{R} \to \mathbb{R}$ be defined by $f(x) = 3x^2 - 5$. Show that $f: \mathbb{R} \to \mathbb{R}$ is not a one-to-one function and that f does not map $\mathbb{R}$ onto $\mathbb{R}$.
14. Let $A = (0,\infty)$ and let $f: A \to A$ be defined by $f(x) = (7x + 8)/(8x + 7)$. Prove that $f: A \to A$ is a one-to-one function. Does f map A onto A? Explain.
15. Let $A = \{x \in \mathbb{R}: x \leqslant -1\}$ and $B = \{x \in \mathbb{R}: x \geqslant 4\}$. Define $f: A \to B$ by $f(x) = x^2 + 2x + 5$. Prove that this function is one to one.
16. Let X be a finite set such that $1 \in X$ but $2 \notin X$. Determine whether each of the following functions is an injection and whether it is a surjection.
    a. $f: \mathcal{P}(X) \to \mathcal{P}(X)$ defined by $f(A) = A - \{1\}$ for each $A \in \mathcal{P}(X)$
    b. $f: \mathcal{P}(X) \to \mathcal{P}(X \cup \{2\})$ defined by $f(A) = A \cup \{2\}$ for each $A \in \mathcal{P}(X)$

*17. Let $A = \mathbb{N} \cup \mathcal{P}(\mathbb{N})$. There are certainly functions with domain A and codomain $\mathbb{N}$. Define such a function and use it to illustrate how the notation $f(S)$ is ambiguous when $S = \{1,2,3\}$.

3.2 COMPOSITION AND INVERSES

If we are given a function $f: A \to B$, there is in general no way to run the function backward to obtain a function mapping B into A. There are two insurmountable difficulties. If f does not map onto B, there is some point b of B outside of $f(A)$ and so there is no meaningful way to take f into consideration when trying to assign a member of A to b. The other difficulty is that if $f: A \to B$ is not one to one then there is some point $b \in B$ for which there are two or more points of A that we could assign to b and no meaningful way to choose among these two or more possibilities. It is a little like the complaint of the Australian folksong: "The creeks run dry or ten feet high." If f does not map onto B, we know too little; if f is not one to one, we know too much. It is only when $f: A \to B$ is a one-to-one correspondence that we have just the right amount of information.

Definition

Let $f: A \to B$ be a one-to-one correspondence. The **inverse function** of the given function is the function $f^{-1}: B \to A$ with domain B and codomain A defined by $f^{-1}(b) = a$ if and only if $f(a) = b$.

Warning: In Section 3.4 we introduce a function $1/f$, which has nothing to do with f^{-1} (see Exercise 48).

Let $f: A \to B$ be a one-to-one correspondence. Since, for each $b \in B$, $f^{-1}(b) = a$ if and only if $f(a) = b$, $f(f^{-1}(b)) = f(a) = b$. It turns out that the equation $f(f^{-1}(b)) = b$ is useful in determining the inverse function of a given one-to-one correspondence.

Example 5

Let $f: \mathbb{R} \to \mathbb{R}$ be defined by $f(x) = -3x + 5$ for $x \in \mathbb{R}$. Find the inverse function $f^{-1}: \mathbb{R} \to \mathbb{R}$.

Analysis

The given function is a one-to-one correspondence (Exercise 22), and so it makes sense to ask for the inverse function $f^{-1}: \mathbb{R} \to \mathbb{R}$. Let $x \in \mathbb{R}$.

Then $f(f^{-1}(x)) = x$ and so $-3(f^{-1}(x)) + 5 = x$. Solving for $f^{-1}(x)$, we find that $f^{-1}(x) = (5 - x)/3$. Thus $f^{-1}\colon \mathbb{R} \to \mathbb{R}$ is determined by the assignment $f^{-1}(x) = (5 - x)/3$ for each real number x. ■

If $f\colon A \to B$ is a one-to-one function that does not map A onto B, then this function does not have an inverse function. But we may consider the one-to-one correspondence $f\colon A \to f(A)$ and determine $f^{-1}\colon f(A) \to A$.

Example 6

Let $A = (0,\infty)$ and let $f\colon A \to \mathbb{R}$ be defined by $f(x) = x/(x + 3)$ for all $x \in A$. By Exercise 23, $f\colon A \to \mathbb{R}$ is a one-to-one function but not a one-to-one correspondence, since $f(A)$ is the open interval $(0,1)$. Determine $f^{-1}\colon (0,1) \to (0,\infty)$.

Analysis

Let $x \in (0,1)$. Then $f(f^{-1}(x)) = x$ and so $f^{-1}(x)/(f^{-1}(x) + 3) = x$. Solving for $f^{-1}(x)$, we have $f^{-1}(x) = 3x/(1 - x)$. Thus $f^{-1}\colon (0,1) \to (0,\infty)$ is determined by the assignment $f^{-1}(x) = 3x/(1 - x)$ for each $x \in (0,1)$. ■

The equation $f(f^{-1}(x)) = x$, which we have been using to evaluate $f^{-1}(x)$, may be thought of as indicating that, if we first perform the function $f^{-1}\colon B \to A$ and then perform the function $f\colon A \to B$, the latter function f undoes the change inflicted by f^{-1} and so we just get back the value x with which we started. This notion of performing first one function and then another can be defined precisely, and the definition makes sense even when the functions involved are not one-to-one correspondences.

Definition

Let $g\colon A \to B$ and $f\colon B \to C$ be functions. Then $f \circ g\colon A \to C$, called the **composition of the function $g\colon A \to B$ with the function $f\colon B \to C$,** is defined by the assignment $f \circ g(a) = f(g(a))$ for each $a \in A$. The symbol $f \circ g$ is read "f composite g."

Figure 3.2

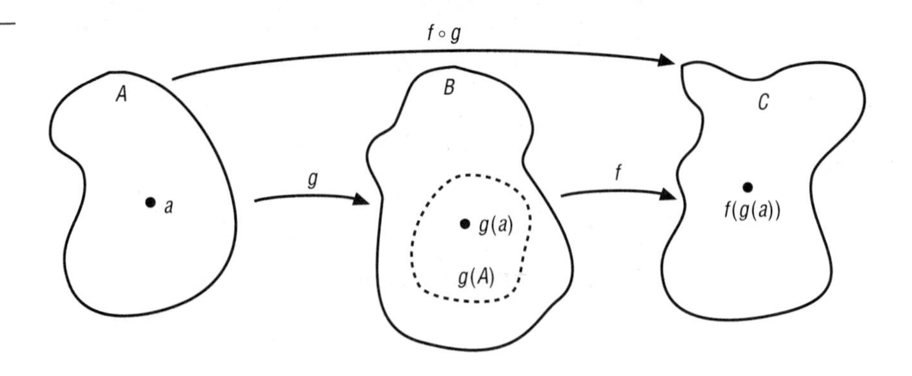

The composition of $g: A \to B$ with $f: B \to C$ requires that the range of g be a subset of the domain of $f: B \to C$. We illustrate this fact in Figure 3.2.

Example 7

Let $f: \mathbb{R} \to \mathbb{R}$ be defined by $f(x) = x^2 + 2$ and let $g: \mathbb{R} \to \mathbb{R}$ be defined by $g(x) = 3x + 4$. Then $g \circ f: \mathbb{R} \to \mathbb{R}$ is the function defined by

$$g \circ f(x) = g(f(x)) = g(x^2 + 2) = 3(x^2 + 2) + 4 = 3x^2 + 6 + 4$$
$$= 3x^2 + 10$$

and $f \circ g: \mathbb{R} \to \mathbb{R}$ is the function defined by

$$f \circ g(x) = f(g(x)) = f(3x + 4) = (3x + 4)^2 + 2$$
$$= 9x^2 + 24x + 16 + 2 = 9x^2 + 24x + 18$$

Clearly $g \circ f \neq f \circ g$, even though the domain of each is $\mathbb{R}$. ■

Example 8

Let X be a finite set such that $1 \notin X$. Let $g: \mathcal{P}(X) \to \mathcal{P}(X \cup \{1\})$ be the function defined by $g(A) = A \cup \{1\}$ for each $A \in \mathcal{P}(X)$, and let $f: \mathcal{P}(X \cup \{1\}) \to \mathbb{Z}$ be the function defined by $f(B) = n(B)$ for each $B \in \mathcal{P}(X \cup \{1\})$. Then $f \circ g: \mathcal{P}(X) \to \mathbb{Z}$ is the function defined by $f \circ g(A) = f(g(A)) = f(A \cup \{1\}) = n(A \cup \{1\}) = n(A) + 1$. ■

As we have seen, it is often important to decide whether a function is one to one or maps one set onto another. In this regard, the following definition and theorem concerning the composition of functions are useful.

Definition

Let X be a nonempty set. The bijection $i_X: X \to X$ defined by $i_X(x) = x$ for each $x \in X$ is called the **identity function on X.**

Theorem 3.1

Let A and B be sets and let $f: A \to B$ and $g: B \to A$.

a. If $f \circ g: B \to B$ is the identity function on B, then $f: A \to B$ is a surjection.
b. If $g \circ f: A \to A$ is the identity function on A, then $f: A \to B$ is an injection.
c. If $f \circ g: B \to B$ is the identity function on B and $g \circ f: A \to A$ is the identity function on A, then $f: A \to B$ and $g: B \to A$ are bijections, $f: A \to B$ is the inverse function of $g: B \to A$, and $g: B \to A$ is the inverse function of $f: A \to B$.

Proof

a. Suppose that $f \circ g: B \to B$ is the identity function on B. We must show that if $b \in B$ then there exists $a \in A$ such that $f(a) = b$. Let $b \in B$. Since $f \circ g(b) = i_B(b)$, $f \circ g(b) = b$. Let $a = g(b)$. Then $a \in A$ and $f(a) = f(g(b)) = f \circ g(b) = b$. Therefore f maps A onto B.

b. Suppose that $g \circ f: A \to A$ is the identity function on A. We must show that, if $x,y \in A$ and $f(x) = f(y)$, then $x = y$. Suppose $x,y \in A$ and $f(x) = f(y)$. Then

$$x = i_A(x) = g \circ f(x) = g(f(x)) = g(f(y)) = g \circ f(y) = i_A(y) = y$$

Therefore f is one to one.

c. Suppose that $f \circ g: B \to B$ is the identity function on B and $g \circ f: A \to A$ is the identity function on A. We first note that it follows from (a) and (b) that both $f: A \to B$ and $g: B \to A$ are bijections. We prove that $g: B \to A$ is the inverse function of $f: A \to B$, and a similar argument shows that $f: A \to B$ is the inverse function of $g: B \to A$. Let $b \in B$ and let $a = g(b)$. We must show that a is the element of A for which $f(a) = b$. But $f(a) = f(g(b)) = f \circ g(b) = i_B(b) = b$. □

Definition

A function $f: A \to B$ is **invertible** if there is a function $g: B \to A$ such that $g \circ f: A \to A$ is the identity function on A and $f \circ g: B \to B$ is the identity function on B.

Corollary 3.2 A function $f: A \to B$ is a bijection if and only if it is invertible.

Proof

Suppose that $f: A \to B$ is invertible. Let $g: B \to A$ be a function such that $f \circ g: B \to B$ is the identity function on B and $g \circ f: A \to A$ is the identity function on A. By Theorem 3.1c, $f: A \to B$ is a bijection.

Suppose that $f: A \to B$ is a bijection. To show that $f^{-1}: B \to A$ is a bijection, and to show that $f: A \to B$ is invertible, it is enough to show that $f \circ f^{-1}: B \to B$ is the identity function on B and $f^{-1} \circ f: A \to A$ is the identity function on A. We have already observed that, for each $b \in B$, $f \circ f^{-1}(b) = f(f^{-1}(b)) = b$, and so it remains to show only that, for each $a \in A$, $f^{-1} \circ f(a) = a$. Let $a \in A$ and let $f(a) = b$. Then by definition $f^{-1}(b) = a$ and so $f^{-1} \circ f(a) = f^{-1}(f(a)) = f^{-1}(b) = a$. □

As Example 9 illustrates, Theorem 3.1 and Corollary 3.2 may be used to show that a function $f: A \to B$ is a bijection.

Example 9

Suppose that $f: \mathbb{R} \to \mathbb{R}$ is the function defined by $f(x) = 3x + 4$. Define $g: \mathbb{R} \to \mathbb{R}$ by $g(x) = (x - 4)/3$. Let $x \in \mathbb{R}$. Then

$$g \circ f(x) = g(f(x)) \quad = g(3x + 4) = \frac{(3x + 4) - 4}{3} = x, \text{ and}$$

$$f \circ g(x) = f\left(\frac{x - 4}{3}\right) = 3\left(\frac{x - 4}{3}\right) + 4 = (x - 4) + 4 = x$$

Therefore $f: \mathbb{R} \to \mathbb{R}$ is invertible, and by Corollary 3.2 it is a bijection. ■

In Example 9, it may seem that $g(x) = (x - 4)/3$ appeared as if by magic. This is not the case. We found it in much the same way that we solved for $f^{-1}(x)$ in Examples 5 and 6. Since we need $f \circ g(x) = x$ for each $x \in \mathbb{R}$, the equation $f(g(x)) = x$ must hold for each $x \in X$. In other words, $3g(x) + 4 = x$, and solving for $g(x)$ we have $g(x) = (x - 4)/3$.

The following theorem, though intuitively obvious, turns out to be quite useful.

Theorem 3.3

Let A, B, and C be sets, and let $f: A \to B$ and $g: B \to C$.

a. If f maps A onto B and g maps B onto C, then $g \circ f$ maps A onto C.
b. If $f: A \to B$ and $g: B \to C$ are one to one, then $g \circ f: A \to C$ is one to one.

Proof

a. Suppose that f maps A onto B and that g maps B onto C. We need to show that if $c \in C$ then there exists $a \in A$ such that $g \circ f(a) = c$. Let $c \in C$. Since g maps B onto C, there exists $b \in B$ such that $g(b) = c$. Since f maps A onto B, there exists $a \in A$ such that $f(a) = b$. Therefore $g \circ f(a) = g(f(a)) = g(b) = c$, and hence $g \circ f$ maps A onto C.

b. Suppose that $f: A \to B$ and $g: B \to C$ are one to one. Let x and y be members of A such that $g \circ f(x) = g \circ f(y)$. Then $g(f(x)) = g(f(y))$. Since $g: A \to B$ is one to one, $f(x) = f(y)$. Since $f: A \to B$ is one to one, $x = y$. Therefore $g \circ f: A \to C$ is one to one. □

Exercises 3.2

18. Let $A = \{1,2,3,\}$ and $B = \{-1,0,1\}$. Find the graph of $f^{-1}: B \to A$ for each function $f: A \to B$ such that $G(f)$ is as follows:

a. $\{(1,-1),(2,0),(3,1)\}$ **b.** $\{(1,1),(2,-1),(3,0)\}$

c. $\{(1,1),(2,0),(3,-1)\}$

19. Let $A = \{1,2,3,\}$ and $B = \{-1,0,1\}$. What is the number of one-to-one functions that map A onto B? Write the graph of each.

20. Let $A = \{1,2,3\}$ and $B = \{-1,0,1\}$. Let $f\colon A \to B$ be the function whose graph is $\{(1,0),(2,1),(3,0)\}$ and let $g\colon B \to A$ be the function whose graph is $\{(-1,3),(1,2),(0,1)\}$.

    a. What is the graph of $f \circ g$? b. What is the graph of $g \circ f$?

21. Let $A = \{1,2,3,4\}$ and $B = \{-1,0,2,3\}$. Let $f\colon A \to B$ be the function whose graph is $\{(1,0),(2,3),(4,-1),(3,2)\}$ and let $g\colon B \to A$ be the function whose graph is $\{(0,1),(3,4),(2,1),(-1,2)\}$.

    a. What is the graph of $f \circ g$? b. What is the graph of $g \circ f$?

22. Let $f\colon \mathbb{R} \to \mathbb{R}$ be defined by $f(x) = -3x + 5$ for each $x \in \mathbb{R}$. Argue that this function is a one-to-one correspondence.

23. Let $f\colon (0,\infty) \to \mathbb{R}$ be defined by $f(x) = x/(x + 3)$ for each positive real number x.

    a. Show that f is one to one.

    b. Show that if $0 < a < 1$ then there is an $x \in (0,\infty)$ such that $f(x) = a$.

    c. Show that if $a \geq 1$ then there does not exist a positive real number x such that $f(x) = a$.

    d. What is the range of $f\colon (0,\infty) \to \mathbb{R}$?

24. Let $f\colon \mathbb{R} \to \mathbb{R}$ be defined by $f(x) = x^3 - 1$ and let $g\colon \mathbb{R} \to \mathbb{R}$ be defined by $g(x) = x^2 + 8$. Find $f \circ g\colon \mathbb{R} \to \mathbb{R}$ and $g \circ f\colon \mathbb{R} \to \mathbb{R}$.

25. Let $f\colon \mathbb{Z} \to \mathbb{Z}$ be defined by $f(x) = 2x$, let $g\colon \mathbb{Z} \to \mathbb{Z}$ be defined by $g(x) = x - 3$, and let $h\colon \mathbb{Z} \to \mathbb{Z}$ be defined by $h(x) = 0$ if x is odd and by $h(x) = 1$ if x is even. Find each of the following functions.

    a. $f \circ g\colon \mathbb{Z} \to \mathbb{Z}$ b. $g \circ f\colon \mathbb{Z} \to \mathbb{Z}$ c. $f \circ h\colon \mathbb{Z} \to \mathbb{Z}$ d. $h \circ f\colon \mathbb{Z} \to \mathbb{Z}$

    e. $g \circ h\colon \mathbb{Z} \to \mathbb{Z}$ f. $h \circ g\colon \mathbb{Z} \to \mathbb{Z}$ g. $f \circ (g \circ h)\colon \mathbb{Z} \to \mathbb{Z}$ h. $h \circ (g \circ f)\colon \mathbb{Z} \to \mathbb{Z}$

26. Show that the following functions are invertible.

    a. $f\colon \mathbb{R} \to \mathbb{R}$ defined by $f(x) = (5 - 3x)/2$

    b. $g\colon \mathbb{R} \to \mathbb{R}$ defined by $g(x) = x^3$

    c. $h\colon [0,\infty) \to [0,\infty)$ defined by $h(x) = x^2$

27. Let $f\colon (-1,1) \to \mathbb{R}$ be defined by $f(x) = x/(1 - |x|)$ for each $x \in (-1,1)$. Let $g\colon \mathbb{R} \to (-1,1)$ be defined by $g(x) = x/(1 + |x|)$ for each $x \in \mathbb{R}$.

    a. Show that $f \circ g\colon \mathbb{R} \to \mathbb{R}$ is the identity function on $\mathbb{R}$.

    b. Show that $g \circ f\colon (-1,1) \to (-1,1)$ is the identity function on $(-1,1)$.

    c. Is $f\colon (-1,1) \to \mathbb{R}$ a one-to-one function? Does f map onto $\mathbb{R}$? Justify your answers.

28. Let $X = \mathbb{R} - \{2\}$ and let $f\colon X \to X$ be defined by $f(x) = (2x + 1)/(x - 2)$ for each $x \in X$. Prove that $f\colon X \to X$ is a one-to-one correspondence.

29. Find the inverse function of the function given in Exercise 28.

3.3

GRAPHS OF FUNCTIONS

The intuitive idea of a function $f: A \to B$ is that f is a rule that assigns to each member of A one and only one member of B; in an explicit definition of a function, this intuitive notion of a rule is replaced by the set $G(f)$. There is a distinction between a rule and the graph of a function. In Example 1 we listed all members of $G(f_1)$ and $G(f_2)$, but we could give the graphs of the functions of Example 1 only because these functions were extremely simple. Most of the time we would have no way of describing a given function $f: A \to B$ if we did not have in mind some rule that determined its graph. We illustrate this point with a theorem, known as the division algorithm.

Theorem 3.4

The Division Algorithm Let n be an integer and d be a positive integer. There is one and only one pair of integers q and r such that $n = dq + r$ and $0 \leq r < d$. The number q is called the *quotient* and the number r is called the *remainder*. □

The division algorithm promises the existence of a function $g: \mathbb{Z} \times \mathbb{N} \to \mathbb{Z} \times (\mathbb{N} \cup \{0\})$ because it promises that there is, for each $(n,d) \in \mathbb{Z} \times \mathbb{N}$, one and only one pair (q,r) such that $n = dq + r$ and $0 \leq r < d$. But the division algorithm is not a recipe, and as we see in Chapter 5 it is therefore not an algorithm. Nowhere in the division algorithm are we told how to find the pair (q,r), and although the theorem promises us that a certain function exists we are never given this function's graph, nor even a rule for finding the graph. There is, however, an algorithm for finding the function promised by the division algorithm, and happily you already know it.

Example 10

Let $n = 367$ and $d = 13$. Find the numbers q and r such that $n = dq + r$ and $0 \leq r < d$.

Analysis

$$\begin{array}{r|l} & \underline{\ \ 28} \\ 13 & 367 \\ & \underline{26\ \ } \\ & 107 \\ & \underline{104} \\ & r3 \end{array}$$

Therefore $367 = 13(28) + 3$. It follows from the division algorithm that $q = 28$ and $r = 3$.

If n is a negative integer, we must make a slight modification in our procedure, as indicated in the next example.

Example 11

Let $n = -367$ and $d = 13$. Find the numbers q and r such that $n = dq + r$ and $0 \leq r < d$.

Analysis

Since $367 = 13(28) + 3$,

$$\begin{aligned} -367 &= -13(28) - 3 \qquad (\text{but } -3 < 0) \\ &= -13(28) - 13 + (13 - 3) \\ &= 13(-29) + 10 \end{aligned}$$

By the division algorithm, the quotient is -29 and the remainder is 10. ■

Although it is usually impractical, and often impossible, to list the ordered pairs that belong to the graph of a function, Pierre de Fermat and René Descartes have provided the method of graphing functions, with which you are familiar. Of course there are many functions, like the function promised by the division algorithm, that are impossible to graph. When we draw the graph of a function, much information can be gained. Although we cannot determine the codomain of a function from its graph, we can determine its domain and we can decide immediately by looking at the graph of a function if the function is one to one. If any horizontal line hits the graph twice, the function is not one to one; if each horizontal line hits the graph at most once, the function is one to one.

Example 12

Let $f\colon \mathbb{N} \to \mathbb{N}$ be defined by $f(n) = n^2$. Draw the graph of f and determine from the graph if the function is one to one.

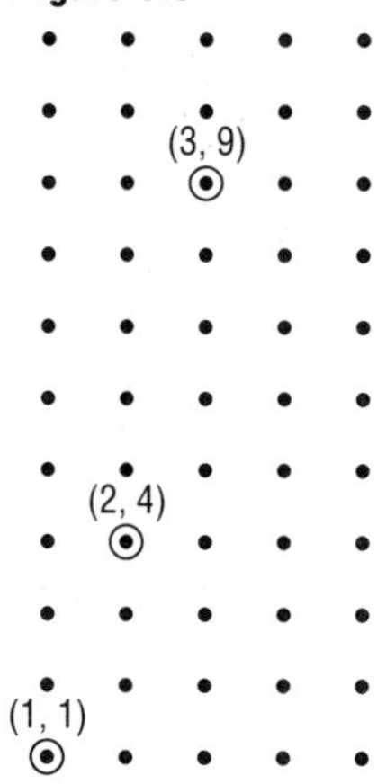

Analysis

Since $\mathbb{N}$ is an infinite set, we cannot draw the graph. As customary, we sketch part of the graph, and we use ⊙ to indicate those points of $\mathbb{N} \times \mathbb{N}$ that belong to the graph in Figure 3.3. Since 2 is a natural number and the horizontal line with equation $y = 2$ does not hit any point of the graph, the function does not map onto $\mathbb{N}$. Since each horizontal line hits the graph at most once, the function is one to one. ■

Figure 3.3

(3, 9)

(2, 4)

(1, 1)

If $f: A \to B$ is a one-to-one correspondence, we may use its graph to determine the inverse function. The reason the graph is useful is that $f(a) = b$ if and only if $f^{-1}(b) = a$; that is, $(b,a) \in G(f^{-1})$ if and only if $(a,b) \in G(f)$.

Example 13

Let $f: I_4 \to I_4$ be the function whose graph is $\{(1,2),(2,4),(3,3),(4,1)\}$. Draw the graph of f and determine that $f: I_4 \to I_4$ is a one-to-one correspondence. Also draw the graph of f^{-1}.

Analysis

Since each horizontal line hits the graph (Figure 3.4) once and only once, $f: I_4 \to I_4$ is a one-to-one correspondence. We obtain f^{-1} by flipping the graph about the indicated diagonal.

Figure 3.4

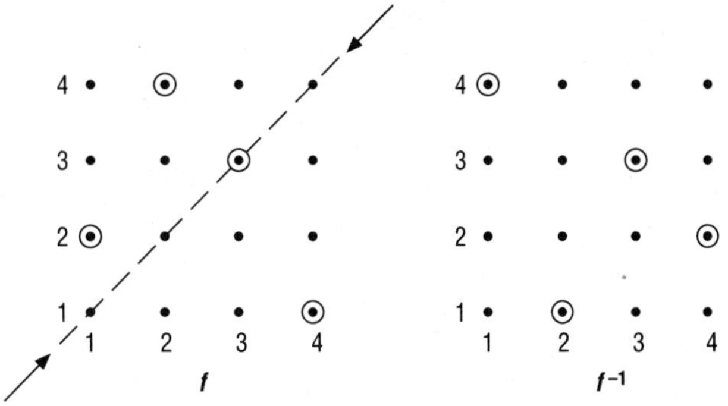

■

Definition

Let X and Y be sets, let A be a subset of X, and let $f: X \to Y$. The set $\{(a,f(a)) \in A \times Y: a \in A\}$ is denoted by $f|A$ and is called the **restriction of the graph f to A.** The function $f|A: A \to Y$ is called the **restriction to A of $f: X \to Y$.**

Example 14

Let $f: \mathbb{R} \to \mathbb{R}$ be defined by $f(x) = 3x^2 + x - 2$. The restrictions $f|[-1,2/3]$, $f|[0,2]$, and $f|\{-1,0,2/3\}$ are sketched in Figure 3.5. ■

As you might guess, and as Theorem 3.5 demonstrates, the restriction of a one-to-one function is also a one-to-one function.

Theorem 3.5

If X and Y are sets, $A \subseteq X$, and $f: X \to Y$ is one to one, then $f|A: A \to Y$ is one to one.

Figure 3.5

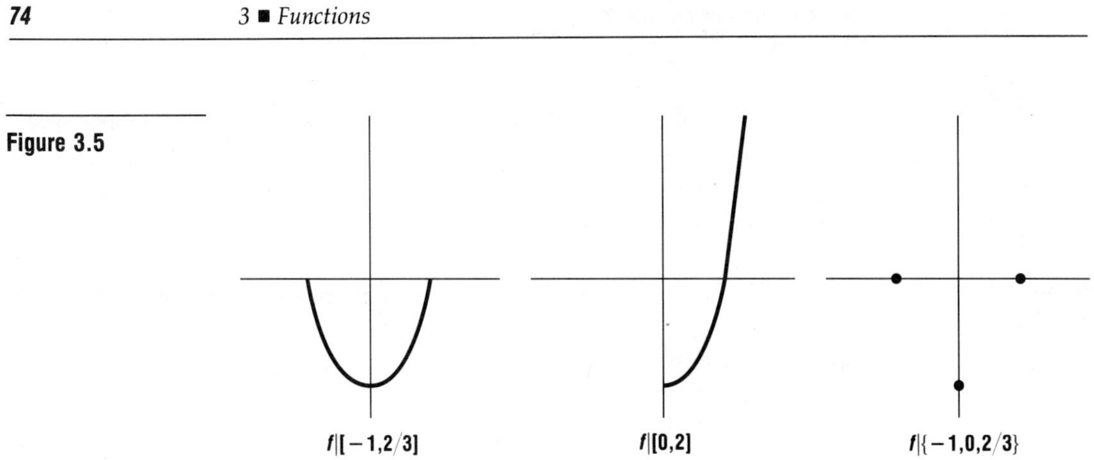

Proof

It is obvious that $f|A: A \to Y$ is a function, so it remains to show only that the function is one to one. Suppose $x, y \in A$ and $(f|A)(x) = (f|A)(y)$. Then $x,y \in X$ and $f(x) = f(y)$. Since f is one to one, $x = y$. □

Exercises 3.3

30. Each of the following graphs is the graph of a function $f: \mathbb{R} \to \mathbb{R}$. Which graphs represent injections, which represent surjections, and which represent bijections?

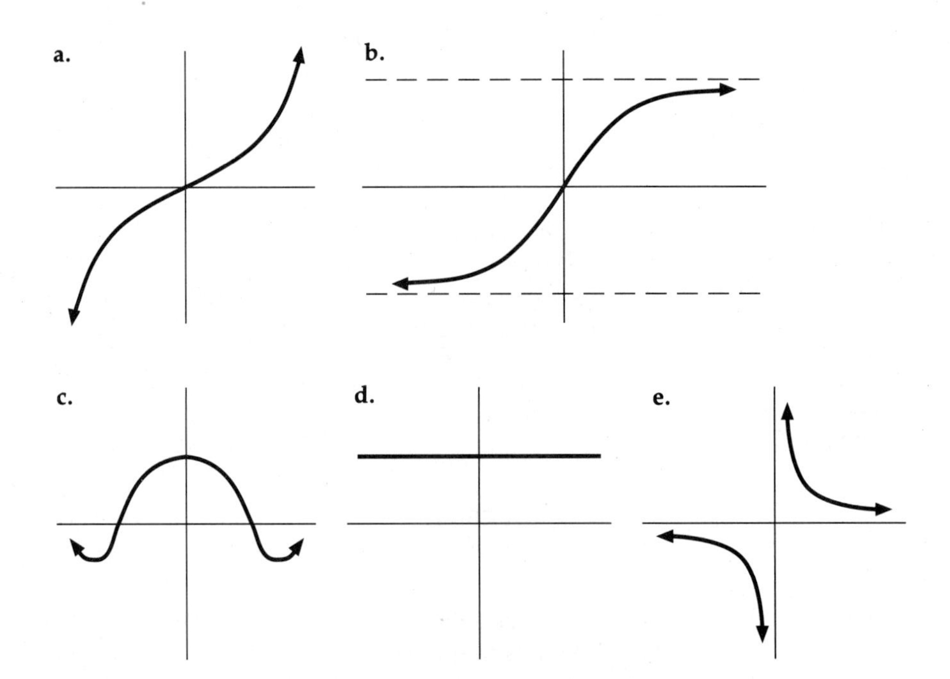

31. Sketch the graph of $f^{-1}: \mathbb{R} \to \mathbb{R}$ for each function $f: \mathbb{R} \to \mathbb{R}$ in Exercise 30 that is a bijection.

32. Give examples of two one-to-one correspondences $f: I_4 \to I_4$ and $g: I_4 \to I_4$ other than the identity function on I_4 which have the property that they are their own inverse functions.

33. Let $f: [0,1] \to [0,1]$ and $g: [0,1] \to [0,1]$ be the functions with the indicated graphs. Indicate the approximate "height" of $f \circ g(1/4)$ and $g \circ f(1/4)$. We have already indicated the approximate height of $f(1/4)$.

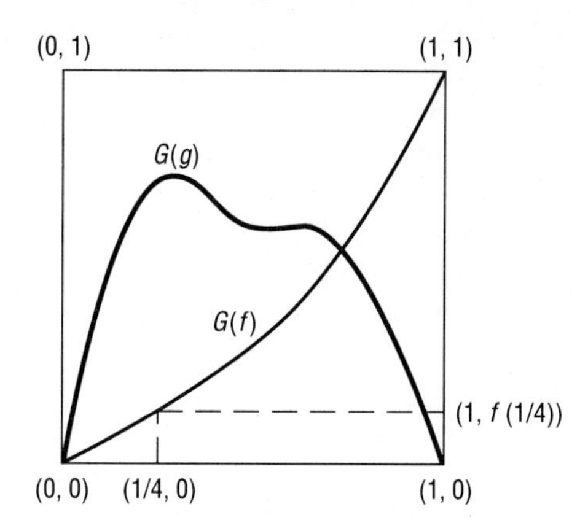

34. Let $f: \mathbb{R} \to \mathbb{R}$ be defined by $f(x) = x^2 - 3x + 4$. Sketch the graph of each of the following restrictions.

a. $f|\{0,1,2,3\}: \{0,1,2,3\} \to \mathbb{R}$ **b.** $f|[0,3]: [0,3] \to \mathbb{R}$

c. $f|\{3/2\}: \{3/2\} \to \mathbb{R}$

35. Find the integers q and r such that $n = dq + r$ and $0 \leqslant r < d$ when

a. $d = 13$ and $n = 427$ **b.** $d = 13$ and $n = -427$

c. $d = 24$ and $n = -244$

36. Let f and g be the functions mapping I_5 onto I_5 with the indicated graphs.

(5, 5) (5, 5)

f g

Draw the graphs of

a. $f \circ g$ **b.** $g \circ f$ **c.** $(g \circ f)^{-1}$ **d.** f^{-1} **e.** g^{-1} **f.** $f^{-1} \circ g^{-1}$

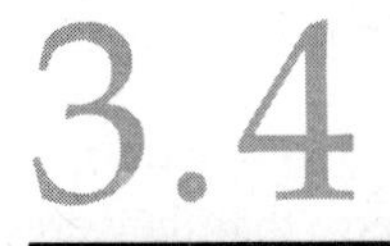

3.4 SPECIAL FUNCTIONS

As the title of this section indicates, our purpose here is to study some functions important enough to have names. Indeed, the function we consider first has two well-known names. It was originally called the *greatest-integer function* and, for each real number x, the image of x under this function was denoted by $[x]$. Although the name and the notation $[x]$ are still in use, in discrete mathematics the function is commonly called *floor* and the image of a real number x under floor is denoted by $\lfloor x \rfloor$ and called "floor x." The function floor is defined by assigning to each real number x the greatest integer that is less than or equal to x, so it seems reasonable to assume that the codomain of floor is $\mathbb{Z}$. Then $\lfloor\ \rfloor: \mathbb{R} \to \mathbb{Z}$ is a surjection, but it is certainly not an injection, since each integer is the image of infinitely many real numbers. For example, $\lfloor 5 \rfloor = 5$, $\lfloor 5.6 \rfloor = 5$, and, if $5 \leq x < 6$, $\lfloor x \rfloor = 5$. Likewise $\lfloor -4 \rfloor = -4$, $\lfloor -3.2 \rfloor = -4$, and, if $-4 \leq x < -3$, $\lfloor x \rfloor = -4$. The graph of the floor function is given in Figure 3.6. (In Figures 3.6 and 3.7, the open parenthesis at the right end of each line segment indicates that the right endpoint of the line segment is not a member of the graph.)

Figure 3.6

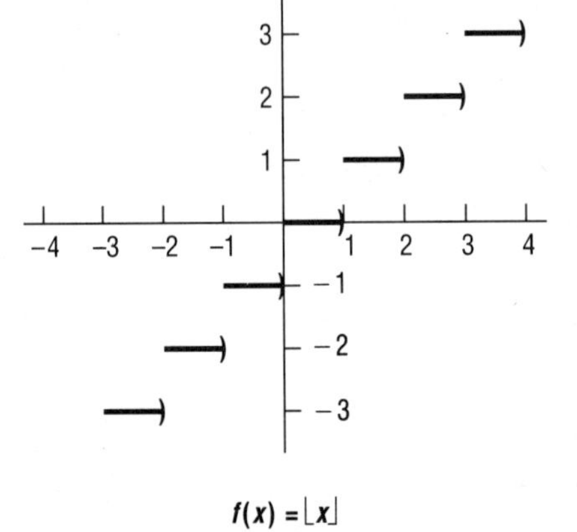

$f(x) = \lfloor x \rfloor$

Example 15

Let $f: \mathbb{R} \to \mathbb{R}$ be defined by $f(x) = \lfloor x^2 \rfloor$. Then, for example,

$$f(2.5) = \lfloor (2.5)^2 \rfloor = \lfloor 6.25 \rfloor = 6$$

$$f(1.1) = \lfloor (1.1)^2 \rfloor = \lfloor 1.21 \rfloor = 1 \quad \text{and}$$

$$f(-3.4) = \lfloor (-3.4)^2 \rfloor = \lfloor 11.56 \rfloor = 11$$

The graph of f is sketched in Figure 3.7. Notice that f is the composition $g \circ h$ where $h(x) = x^2$ and $g(x) = \lfloor x \rfloor$. ■

Figure 3.7

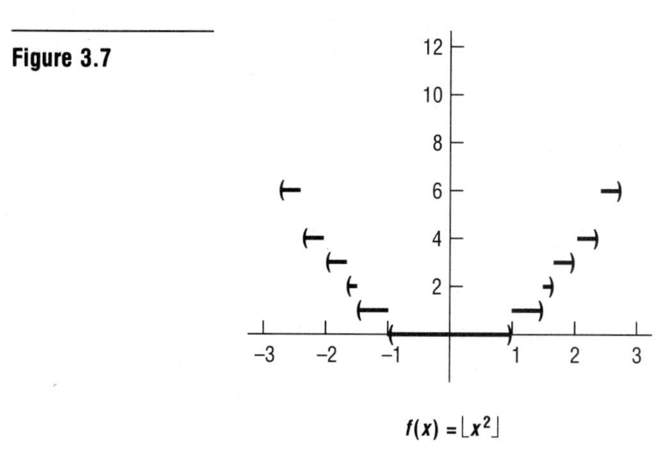

$f(x) = \lfloor x^2 \rfloor$ ■

The function floor has a companion function, called *ceiling*. As you may well guess, for each real number x, "ceiling x" is defined to be the smallest integer that is greater than or equal to x and is denoted by $\lceil x \rceil$. The two functions floor and ceiling are related by Theorem 3.6.

Theorem 3.6 For each real number x, $\lfloor x \rfloor = -\lceil -x \rceil$ and $\lceil x \rceil = -\lfloor -x \rfloor$.

Proof

Let $x \in \mathbb{R}$ and let $n \in \mathbb{Z}$ such that $n \leqslant x < n + 1$. Then $-n - 1 < -x \leqslant -n$, $\lfloor x \rfloor = n$, and $\lceil -x \rceil = -n$. Hence $\lfloor x \rfloor = -(-n) = -\lceil -x \rceil$. The proof that $\lceil x \rceil = -\lfloor -x \rfloor$ is left as Exercise 42. □

The function floor can provide functions that are difficult to graph (see Exercise 40). But beyond this, floor and ceiling crop up in discrete mathematics in surprising and important ways, as Theorem 3.7 illustrates.

Example 16

A function commonly available in high-level programming languages is the RANDOM function, usually referred to as RAN or RAND. This function is usually provided a seed at the beginning of a program, and RAN(1) produces a random number greater than or equal to 0 and less than 1. For instance, let x = RAN(1) puts some number between 0 and 1 such as 0.53678921 in the variable x. This function in conjunction with the function FLOOR can be used to produce an integer belonging to a random collection of integers. For example, if we want to simulate rolling a die in a program, then the command let x = FLOOR(6 · RAN(1)) + 1 puts one of the integers 1, 2, 3, 4, 5, or 6 in the variable x. This is true since $0 \leqslant$ RAN(1) < 1 implies $0 \leq 6 \cdot$ RAN(1) < 6. Thus FLOOR(6 · RAN(1)) is one of the integers 0, 1, 2, 3, 4, or 5. ■

Theorem 3.7

Let n and d be positive integers. The number of positive integers less than or equal to n that are divisible by d is $\lfloor n/d \rfloor$.

Proof

Suppose that $d > n$. Then there are no positive integers less than or equal to n that are divisible by d, and $\lfloor n/d \rfloor = 0$.

Now suppose that $d \leqslant n$. The list of positive integers divisible by d is d, $2d$, $3d$, $4d$, and so on, and the list of positive integers divisible by d that are less than or equal to n is $d, 2d, 3d, 4d, \ldots, kd$, where $kd \leqslant n$ but $(k + 1)d > n$. Since k is the number of positive integers less than or equal to n that are divisible by d, we need to show that $\lfloor n/d \rfloor = k$. By the division algorithm, $n = kd + r$, where $0 \leqslant r < d$. Thus

$$\lfloor n/d \rfloor = \lfloor kd/d + r/d \rfloor = \lfloor k + r/d \rfloor = k \qquad \square$$

If the range of a function is a subset of $\mathbb{Z}$, we say that the function is *integer valued*. Thus floor and ceiling are integer-valued functions. Similarly, if the range of a function is a subset of $\mathbb{R}$, we say that the function is *real valued*. Two real-valued functions may be added, subtracted, multiplied, or divided, even if their domains are sets that have no algebraic structure.

Definition

Let X and Y be sets such that $X \cap Y \neq \varnothing$, and let $f: X \to \mathbb{R}$ and $g: Y \to \mathbb{R}$. Then

a. $f + g: X \cap Y \to \mathbb{R}$ is defined by $(f + g)(x) = f(x) + g(x)$
b. $f - g: X \cap Y \to \mathbb{R}$ is defined by $(f - g)(x) = f(x) - g(x)$
c. $fg: X \cap Y \to \mathbb{R}$ is defined by $(fg)(x) = f(x)g(x)$
d. $f/g: (X \cap Y - \{x \in X \cap Y: g(x) = 0\}) \to \mathbb{R}$ is defined by $(f/g)(x) = f(x)/g(x)$

We consider two examples, one in which the domains of the functions are subsets of $\mathbb{R}$ and one in which the domains are subsets of an arbitrary set.

Example 17

Let $f\colon [0,\infty) \to \mathbb{R}$ be defined by $f(x) = \sqrt{x} + 3$, and let $g\colon (-\infty,4] \to \mathbb{R}$ be defined by $g(x) = \sqrt{4-x} - 1$. Then $(-\infty, 4] \cap [0,\infty) = [0,4]$, and $f + g\colon [0,4] \to \mathbb{R}$ is the function defined by $(f + g)(x) = \sqrt{x} + \sqrt{4-x} + 2$, $f - g\colon [0,4] \to \mathbb{R}$ is the function defined by $(f - g)(x) = \sqrt{x} - \sqrt{4-x} + 4$, $fg\colon [0,4] \to \mathbb{R}$ is the function defined by $(fg)(x) = \sqrt{x}\sqrt{4-x} + 3\sqrt{4-x} - \sqrt{x} - 3$, and $f/g\colon [0,3) \cup (3,4] \to \mathbb{R}$ is the function defined by $(f/g)(x) = (\sqrt{x} + 3)/(\sqrt{4-x} - 1)$. So, for example,

$$(f + g)(4) = \sqrt{4} + \sqrt{4-4} + 2 = 2 + 0 + 2 = 4$$

$$(f - g)(4) = \sqrt{4} - \sqrt{4-4} + 4 = 2 + 0 + 4 = 6$$

$$\begin{aligned}(fg)(4) &= \sqrt{4}\sqrt{4-4} + 3\sqrt{4-4} - \sqrt{4} - 3\\ &= 0 + 0 - 2 - 3 = -5 \quad \text{and}\end{aligned}$$

$$(f/g)(4) = \frac{\sqrt{4} + 3}{\sqrt{4-4} - 1} = \frac{2+3}{0-1} = -5$$

■

Example 18

Let X be a set, and let A, B, and C be nonempty pairwise disjoint subsets of X such that $X - (A \cup B \cup C)$ is nonempty. Define $f\colon X \to \mathbb{R}$ and $g\colon X \to \mathbb{R}$ by

$$f(x) = \begin{cases} 1, & \text{if } x \in A \\ 2, & \text{if } x \in B \\ 3, & \text{if } x \in C \\ 4, & \text{if } x \in X - (A \cup B \cup C) \end{cases}$$

and $$g(x) = \begin{cases} 0, & \text{if } x \in A \\ 5, & \text{if } x \in B \\ 6, & \text{if } x \in C \\ 7, & \text{if } x \in X - (A \cup B \cup C) \end{cases}$$

Then $f + g\colon X \to \mathbb{R}$ and $f - g\colon X \to \mathbb{R}$ are the functions defined by

$$(f + g)(x) = \begin{cases} 1, & \text{if } x \in A \\ 7, & \text{if } x \in B \\ 9, & \text{if } x \in C \\ 11, & \text{if } x \in X - (A \cup B \cup C) \end{cases}$$

$$(f - g)(x) = \begin{cases} 1, & \text{if } x \in A \\ -3, & \text{if } x \in X - A \end{cases}$$

Also, $fg\colon X \to \mathbb{R}$ and $f/g\colon (X - A) \to \mathbb{R}$ are the functions defined by

$$(fg)(x) = \begin{cases} 0, & \text{if } x \in A \\ 10, & \text{if } x \in B \\ 18, & \text{if } x \in C \\ 28, & \text{if } x \in X - (A \cup B \cup C) \end{cases}$$

$$(f/g)(x) = \begin{cases} 2/5, & \text{if } x \in B \\ 1/2, & \text{if } x \in C \\ 4/7, & \text{if } x \in X - (A \cup B \cup C) \end{cases}$$

■

Two classes of real-valued functions, the exponential functions and the logarithmic functions, play important roles in discrete mathematics. Calculus is needed to justify the definitions and prove the first principles. In the discussion that follows, therefore, we rely on your intuition and good will.

Definition

Let b be a positive real number, different from 1, and let A be a nonempty subset of $\mathbb{R}$. Then the function $f\colon A \to \mathbb{R}$ defined by $f(x) = b^x$ is called an **exponential function.**

Note that an exponential function is determined by a positive real number, *different from* 1, and a nonempty subset A of $\mathbb{R}$.

A function whose range has only one member is called a *constant function*. Although constant functions provide dramatic examples of functions that are neither injections nor surjections (provided that their domain and codomain each has more than one member), they would not be of any use to us as exponential functions. For each real number x, $1^x = 1$, and so while it is true that we can define $f\colon A \to \mathbb{R}$ by the assignment $f(x) = 1^x$, we have agreed in the previous definition that such a function is not to be considered an exponential function.

Our first example of an exponential function is the function $f\colon \mathbb{R} \to \mathbb{R}$ by $f(x) = 2^x$. It is simple enough to determine $f(3)$, $f(5)$, and $f(-5/2)$. Indeed $f(3) = 8$, $f(5) = 32$, and $f(-5/2) = 1/f(5/2) = 1/\sqrt{32}$. When x is an irrational number, however, $f(x)$ is defined using calculus. We give the graph of $f(x) = 2^x$ in Figure 3.8.

Figure 3.8

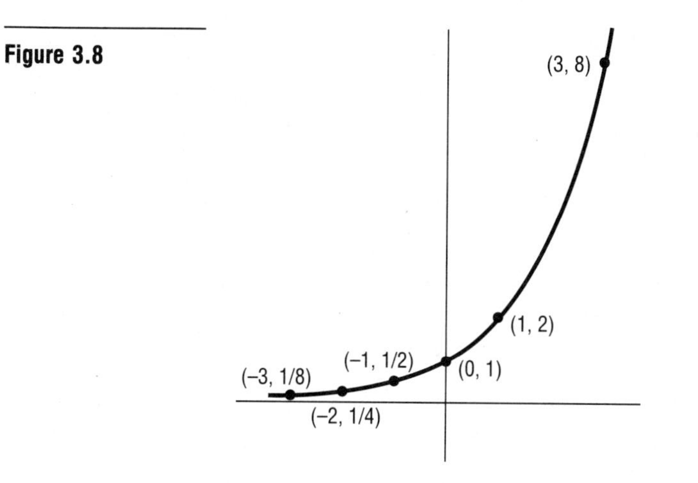

The function $f|\mathbb{Z}: \mathbb{Z} \to \mathbb{R}$, whose graph is given in Figure 3.9, is another example of an exponential function.

Figure 3.9

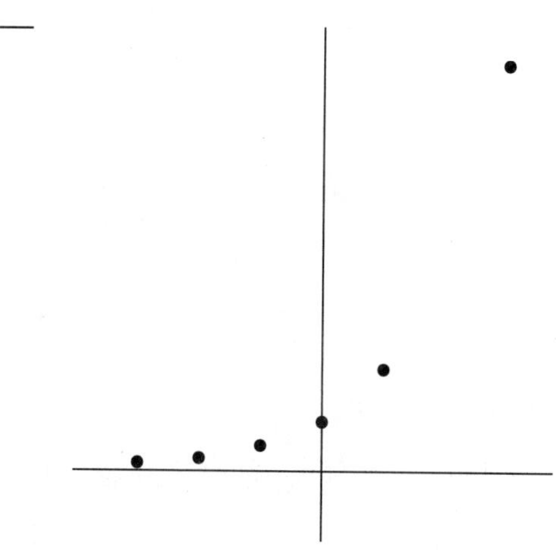

Of course the function $g: \mathbb{R} \to \mathbb{R}$ defined by $g(x) = 10^x$ is another exponential function, and because of our use of the decimal system this is the exponential function with which we are most familiar. Nonetheless, the number 2 is especially useful in discrete mathematics, and so we concentrate here on the exponential functions of Figures 3.8 and 3.9.

Let $\mathbb{R}^+$ denote the set $(0, \infty)$ of all positive real numbers. The exponential function defined by $f(x) = 2^x$ for each $x \in \mathbb{R}$ is a one-to-one function mapping $\mathbb{R}$ onto $\mathbb{R}^+$. It follows from this intuitively obvious fact, which we cannot prove without calculus, that f is invertible. Its inverse function f^{-1} is denoted by $\log_2$ and is called the *logarithm to the base 2 function*. The graph of $\log_2$ is given in Figure 3.10.

Figure 3.10

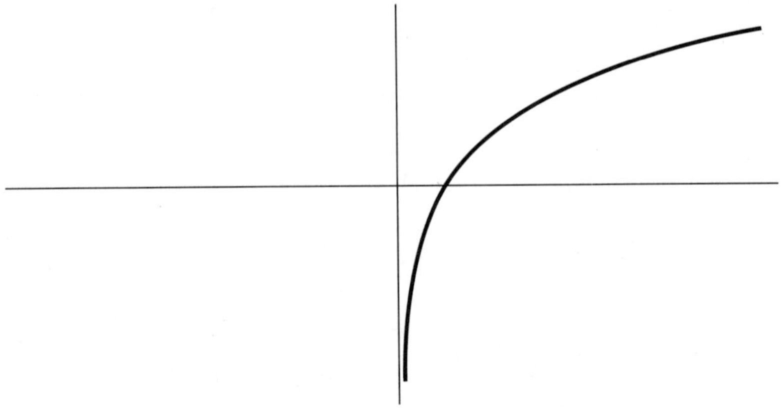

Compare the graphs in Figures 3.8 and 3.9. Note that $\log_2$ is a one-to-one function mapping $\mathbb{R}^+$ onto $\mathbb{R}$ and that $\log_2(x) = y$ if and only if $2^y = x$. Since $f \circ f^{-1}$ is the identity on $\mathbb{R}^+$ and $f^{-1} \circ f$ is the identity on $\mathbb{R}$, if $x \in \mathbb{R}^+$,

$$2^{\log_2 x} = f(f^{-1}(x)) = x$$

and, if $x \in \mathbb{R}$,

$$\log_2(2^x) = f^{-1}(f(x)) = x$$

We now give the general definition of the logarithm functions. In this definition, we assume that the exponential functions are one-to-one correspondences from $\mathbb{R}$ onto $\mathbb{R}^+$.

Definition

If b is a positive real number different from 1, and f is the function from $\mathbb{R}$ onto $\mathbb{R}^+$ defined by $f(x) = b^x$, then the **logarithm to the base *b* function** is the inverse of f.

Example 19

Let A = {poor, wealthy, millionaire}, let B = {yes, no}, and define $f: A \times B \to A$ by f((poor, yes)) = poor, f((poor, no)) = wealthy, f((wealthy, yes)) = wealthy, f((wealthy, no)) = millionaire, f((millionaire, yes)) = millionaire, and f((millionaire, no)) = poor. Then f maps $A \times B$ onto A, but f is not an injection. If A and B are nonempty sets, a function f from $A \times B$ into A is sometimes called a *transition function*. ■

Definition

A **binary operation on a set X** is a function $f: X \times X \to X$.

According to this definition, $S: \mathbb{N} \times \mathbb{N} \to \mathbb{N}$ defined by $S((m,n)) = m + n$ is a binary operation on $\mathbb{N}$ (S is addition of natural numbers). Note that subtraction of natural numbers is not a binary operation on $\mathbb{N}$ because the difference of two natural numbers is not necessarily a natural number. The function $T: \mathbb{Z} \times \mathbb{Z} \to \mathbb{Z}$ defined by $T((m,n)) = m - n$ is a binary operation on $\mathbb{Z}$ since the difference of two integers is an integer.

Definition

Let m and n be natural numbers. Then $\gcd(m,n)$ denotes the largest natural number that divides both m and n, and $\text{lcm}(m,n)$ denotes the smallest natural number divisible by both m and n. For any natural numbers m and n, $\gcd(m,n)$ is called the **greatest common divisor** of m and n, and $\text{lcm}(m,n)$ is called the **least common multiple** of m and n. If $\gcd(m,n) = 1$, then m and n are said to be **relatively prime.**

If S is the set of all natural numbers or the set of all divisors of a given natural number k, then we can define binary operations $\vee$ and $\wedge$ on S by $\vee((m,n)) = \text{lcm}(m,n)$ and $\wedge((m,n)) = \gcd(m,n)$.

The logarithmic and exponential functions have the property that they interchange the binary operations of addition and multiplication of real numbers, and this accounts for much of their importance. In the olden days of yore (before the invention of the pocket calculator), people calculated products using logarithmic tables or a slide rule to change the task of multiplying to the task of adding.

We have already observed that for appropriately behaved functions f and g we may define the composition $f \circ g$, and so it is natural to look for a set of functions X for which $\circ: X \times X \to X$ is a binary operation.

Definition

Let S be a nonempty set. A **permutation on** S is a one-to-one function that maps S onto S. The collection of all permutations on S is denoted by $\text{Sym}(S)$.

The function $f: \mathbb{Z} \to \mathbb{Z}$ defined by $f(n) = n + 3$ is a permutation on $\mathbb{Z}$, and the function f whose graph is $\{(1,2),(2,4),(3,3),(4,1)\}$ is a permutation on $\{1,2,3,4\}$.

Theorem 3.8

Let S be a nonempty set. Then α: Sym(S) $\times$ Sym(S) $\rightarrow$ Sym(S) defined by $\alpha(f,g) = f \circ g$ is a binary operation.

Proof

Let $f,g \in$ Sym(S). We must show that $f \circ g \in$ Sym(S). By Theorem 3.3, $f \circ g$ is a one-to-one function mapping S onto S. Thus by definition $f \circ g$ is a permutation on S and so belongs to Sym(S). □

Let S be a nonempty set, and let $*$: $S \times S \rightarrow S$ be a binary operation on S. If $a,b \in S$, it is customary to denote $*(a,b)$ by $a * b$.

Definition

A binary operation $*$ on a set S is **commutative** provided that $a * b = b * a$ for all $a, b \in S$. The binary operation $*$ is **associative** provided that $a * (b * c) = (a * b) * c$ for all $a, b, c \in S$.

The binary operations of addition and multiplication on $\mathbb{Z}$ are both commutative and associative, whereas the operation of subtraction on $\mathbb{Z}$ is neither commutative nor associative (see Exercise 56). The operation $*$: $\mathbb{N} \times \mathbb{N} \rightarrow \mathbb{N}$ defined by $*(a,b) = (a + b)^{a+b}$ is commutative but not associative (see Exercise 57). As we now show, if S is a set with at least three members, then composition is an associative operation on Sym(S) that is not commutative.

Theorem 3.9

Let h: $A \rightarrow B$, g: $B \rightarrow C$, and f: $C \rightarrow D$. Then $(f \circ g) \circ h$: $A \rightarrow C$ is the function $f \circ (g \circ h)$: $A \rightarrow C$.

Proof

Since the functions we are considering have the same domain and codomain, it suffices to show that, for each $a \in A$, $(f \circ g) \circ h(a) = f \circ (g \circ h)(a)$. Let $a \in A$. Then

$$(f \circ g) \circ h(a) = f \circ g(h(a)) = f(g(h(a))) \quad \text{and}$$

$$f \circ (g \circ h)(a) = f(g \circ h(a)) = f(g(h(a)))$$

□

Theorem 3.10

> Let S be a set with at least three members. Then composition is not a commutative operation on Sym(S).

Proof

Let x, y, and z be three members of S. Define f and g in Sym(S) by the following rules:

$$f(x) = y,\ f(y) = z,\ f(z) = x, \text{ and } f(s) = s \text{ if } s \notin \{x, y, z\}$$

$$g(x) = x,\ g(y) = z,\ g(z) = y, \text{ and } g(s) = s \text{ if } s \notin \{x, y, z\}$$

Then $f \circ g(x) = y$, whereas $g \circ f(x) = z$. Thus $f \circ g \neq g \circ f$. □

The final theorem of this chapter, Theorem 3.11, provides a useful connection between the inverse of a composition $(f \circ g)^{-1}$ and the composition of the corresponding inverses *in the reverse order* $g^{-1} \circ f^{-1}$. As we have just seen, if a set S has more than two members, composition is not a commutative operation on Sym(S). Thus the distinction between $f^{-1} \circ g^{-1}$ and $g^{-1} \circ f^{-1}$ made in Theorem 3.11 is a real distinction and must be remembered. Notice also that the proof of Theorem 3.11 makes repeated use of Theorem 3.9.

Theorem 3.11

> Let S be a nonempty set and let f and g be permutations on S. Then the permutation $(f \circ g)^{-1}$ is the permutation $g^{-1} \circ f^{-1}$.

Proof

Both functions have domain and codomain S, so by Theorem 3.1c it suffices to show that $(f \circ g) \circ (g^{-1} \circ f^{-1}) = i_S$ and $(g^{-1} \circ f^{-1}) \circ (f \circ g) = i_S$. But

$$\begin{aligned}
(f \circ g) \circ (g^{-1} \circ f^{-1}) &= f \circ (g \circ (g^{-1} \circ f^{-1})) \\
&= f \circ ((g \circ g^{-1}) \circ f^{-1}) \\
&= f \circ (i_S \circ f^{-1}) \\
&= f \circ f^{-1} = i_S \quad \text{and} \\
(g^{-1} \circ f^{-1}) \circ (f \circ g) &= g^{-1} \circ (f^{-1} \circ (f \circ g)) \\
&= g^{-1} \circ ((f^{-1} \circ f) \circ g) \\
&= g^{-1} \circ (i_S \circ g) \\
&= g^{-1} \circ g = i_S
\end{aligned}$$

□

The behavior of permutations with respect to composition provides a model for the behavior of matrix multiplication in Chapter 8 and a motivation for the study of directed graphs in Chapter 9. We end this section on special functions by listing those facts about the composition of permutations that we have established.

Let S be a nonempty set.

a. The identity function i_S is a permutation on S and so the set $\mathrm{Sym}(S)$ of all permutations of S is nonempty.
b. If f is a permutation on S, so is f^{-1}.
c. Composition is an associative operation on the set of all permutations on S.
d. If S has more than two members, composition is not a commutative operation on $\mathrm{Sym}(S)$.
e. If $f, g \in \mathrm{Sym}(S)$, $(f \circ g)^{-1} = g^{-1} \circ f^{-1}$.
f. If $f \in \mathrm{Sym}(S)$, then $f \circ f^{-1} = i_S$ and $f^{-1} \circ f = i_S$.
g. If $f \in \mathrm{Sym}(S)$, then $i_S \circ f = f \circ i_S = f$.

Exercises 3.4

37. Let $A = \{1,2,3,4\}$, let $f\colon A \to \mathbb{R}$ be the function whose graph is $\{(1,1),(2,3),(3,1),(4,-2)\}$, and let $g\colon A \to \mathbb{R}$ be the function whose graph is $\{(2,1),(3,-1),(1,2),(4,1)\}$. Find the graph of each of the following functions

a. $f + g\colon A \to \mathbb{R}$ **b.** $f - g\colon A \to \mathbb{R}$ **c.** $fg\colon A \to \mathbb{R}$ **d.** $f/g\colon A \to \mathbb{R}$

38. Evaluate the following integers

a. $\lfloor -6 \rfloor$ **b.** $\lfloor 6.4 \rfloor$ **c.** $\lfloor -6.4 \rfloor$ **d.** $\lceil 6.4 \rceil$ **e.** $\lceil -6.4 \rceil$

39. Find a real number x that is not an integer such that

a. $\lfloor x \rfloor = \lceil 7.2 \rceil$ **b.** $\lceil x \rceil = \lfloor -7.2 \rfloor$
c. $\lceil x \rceil = \lfloor 7.2 \rfloor$ **d.** $\lfloor x \rfloor = \lceil -7.2 \rceil$

40. Sketch the graph of each of the following functions

a. $f\colon \mathbb{R} \to \mathbb{Z}$ defined by $f(x) = \lceil -x \rceil$
b. $g\colon \mathbb{R} \to \mathbb{Z}$ defined by $g(x) = \lfloor -x \rfloor$
c. $h\colon \mathbb{R} \to \mathbb{Z}$ defined by $h(x) = \lfloor x + 1 \rfloor$

41. Sketch the graph of each of the following functions.

a. $f\colon \mathbb{R} \to \mathbb{Z}$ defined by $f(x) = \lceil 3x \rceil$
b. $g\colon \mathbb{R} \to \mathbb{Z}$ defined by $g(x) = \lfloor x/4 \rfloor$

42. Prove that, for each real number x, $\lceil x \rceil = -\lfloor -x \rfloor$.

43. Let $f\colon \mathbb{R} \to \mathbb{Z}$ be defined by $f(x) = \lfloor x/2 \rfloor$. Is the restriction of f to $\mathbb{Z}$ a one-to-one function? Does it map onto $\mathbb{Z}$? Explain.

44. **a.** How many integers less than 10,000 are divisible by 17?

b. How many integers greater than 147 and less than 8,195 are divisible by 17?

45. Define $f: \mathbb{R} \to \mathbb{R}$ and $g: \mathbb{R} \to \mathbb{R}$ by

$$f(x) = \begin{cases} -1, & \text{if } x < 0 \\ 1, & \text{if } x \geq 0 \end{cases} \quad \text{and} \quad g(x) = \begin{cases} 2x, & \text{if } x < 0 \\ 3x, & \text{if } x \geq 0 \end{cases}$$

Find $f + g$, $f - g$, fg, and f/g.

46. Define $f: \mathbb{R} \to \mathbb{R}$ and $g: \mathbb{R} \to \mathbb{R}$ by

$$f(x) = \begin{cases} 2, & \text{if } x < -1 \\ 3x, & \text{if } x \geq -1 \end{cases} \quad \text{and} \quad g(x) = \begin{cases} x^2, & \text{if } x < 0 \\ 2x, & \text{if } x \geq 0 \end{cases}$$

Find $f + g$, $f - g$, fg, and f/g.

47. Define $f: \mathbb{R} \to \mathbb{R}$ and $g: \mathbb{R} \to \mathbb{R}$ by

$$f(x) = \begin{cases} 4, & \text{if } x < -5 \\ x^2, & \text{if } -5 \leq x \leq 5 \\ -3, & \text{if } x > 5 \end{cases} \quad \text{and} \quad g(x) = \begin{cases} 2x, & \text{if } x < -2 \\ x^3, & \text{if } -2 \leq x \leq 2 \\ 3x, & \text{if } x > 2 \end{cases}$$

Find $f + g$, $f - g$, fg, and f/g.

48. Let $f: (0,\infty) \to (0,\infty)$ be defined by $f(x) = 2x + 1$ for each $x \in (0,\infty)$.

a. Find $\dfrac{1}{f(3)}$. **b.** Find $f^{-1}(3)$. **c.** Is $\dfrac{1}{f(3)} = f^{-1}(3)$?

d. Given a particular function f and an element a which is in both the domain of f and the domain of f^{-1}, is it likely that $f^{-1}(a) = 1/f(a)$?

49. Let X be an nonempty set. For each subset A of X, let $\chi_A: X \to \{0,1\}$ be the function defined by $\chi_A(x) = 1$ for each $x \in A$, and $\chi_A(x) = 0$ for each $x \in X - A$.

a. For which subsets A of X is $\chi_A: X \to \{0,1\}$ a constant function?

b. For which subsets A of X is $\chi_A: X \to \{0,1\}$ a surjection?

The function $\chi_A: X \to \{0,1\}$ is called the *characteristic function of A*, and χ_A is pronounced "chi, sub A."

50. Let A and B be nonempty subsets of a set X, and let $\chi_A, \chi_B: X \to \{0,1\}$ be the characteristic functions of A and B. Prove that if $\chi_A = \chi_B$ then $A = B$.

51. Let A be a nonempty proper subset of a set X, and let $\chi_A, \chi_{A^\sim}, \chi_X: X \to \{0,1\}$ be the characteristic function of A, $A^\sim$, and X. Prove that $\chi_A + \chi_{A^\sim} = \chi_X$.

52. Let A and B be nonempty subsets of a set X, and let $\chi_A, \chi_B: X \to \{0,1\}$ be the characteristic functions of A and B. Prove the following:

a. $\chi_{A-B} = \chi_A - \chi_A\chi_B$ **b.** $\chi_{A\cup B} = \chi_A + \chi_B - \chi_A\chi_B$ **c.** $\chi_{A\cap B} = \chi_A\chi_B$

53. a. Graph the exponential function $k: \mathbb{R} \to \mathbb{R}^+$ defined by $k(x) = 10^x$ for each $x \in \mathbb{R}$.

    b. Graph the logarithm to the base 10 function.

54. a. Graph the exponential function $r: \mathbb{R} \to \mathbb{R}^+$ defined by $r(x) = 3^x$ for each $x \in \mathbb{R}$.

    b. Graph the logarithm to the base 3 function.

55. Give an example to show that, if $*$ is a binary operation on a set S and $T \subseteq S$, then $*|(T \times T)$ is not necessarily a binary operation on T.

56. Give examples to show that the binary operation of subtraction on $\mathbb{Z}$ is neither commutative nor associative.

57. Let $*: \mathbb{N} \times \mathbb{N} \to \mathbb{N}$ be defined by $*(a,b) = (a + b)^{a+b}$ and denote $*(a,b)$ by $a * b$. Show that $(3 * 2) * 1 \neq 3 * (2 * 1)$. Is $*$ commutative? Is $*$ associative?

58. Is division a binary operation on the set of nonzero integers? on the set of nonzero real numbers?

59. Let $S = \{1,2\}$. List all the members of Sym(S). Show that composition is a commutative binary operation on Sym(S).

60. Let $S = \{1,2,3\}$. List all the members of Sym(S). Give an example of two permutations f and g on S and an $x \in S$ such that $f \circ g(x) \neq g \circ f(x)$.

61. a. How many members does Sym(I_4) have?

    b. How many members does Sym(I_5) have?

    c. Suppose that S is a nonempty set and $\mathcal{P}(S)$ has more members than Sym(S). How many members can S have?

62. For each of the following collections of integers, use the FLOOR and RAN functions (see Example 16) to write an expression that produces the collection.

    a. $\{1,2\}$

    b. $\{n \in \mathbb{N}: n \leq 36\}$

    c. $\{1,2,3,4,5,6,7,8,9,10\}$

    d. $\{2,4,6,8\}$

63. Let $A = \{\text{bat, glove, shoe}\}$ and $B = \{\text{up, down, same}\}$. Define a transition function from $A \times B$ into A.

64. Let S be the set of all positive divisors of 30. Let $\wedge: S \times S \to S$ be defined by $\wedge((a,b)) = \gcd(a,b)$, and denote $\wedge((a,b))$ by $a \wedge b$ for all $a, b \in S$. Prove that $\wedge$ is an associative operation. (*Hint:* Suppose that d divides $a \wedge b$ and that d divides c; argue that d divides a and $b \wedge c$. Suppose that d divides a and $b \wedge c$; argue that d divides $a \wedge b$ and c.)

Chapter 3 Review Exercises

65. Let $A = \{1,2,3,4\}$ and $B = \{5,6,7,8\}$. Which of the following sets of ordered pairs are graphs of a function $f: A \to B$?

    a. $\{(1,5),(2,8),(3,7),(4,6)\}$

    b. $\{(1,5),(2,7),(3,8)\}$

    c. $\{(2,5),(1,4),(3,7),(4,8)\}$

    d. $\{(2,6),(3,5),(1,6),(4,5)\}$

    e. $\{(1,5),(2,7),(1,6),(3,8),(4,7)\}$

66. Which of the sets of ordered pairs in Exercise 65 are graphs of one-to-one functions?

67. Which of the sets of ordered pairs in Exercise 65 are graphs of functions that map A onto B?

68. Let $f: \mathbb{Z} \to \mathbb{Z}$ be a function such that $f(\mathbb{N}) = f(\mathbb{Z})$. Is f one-to-one? Is f a surjection?

69. Let $A = \{1,2,3,4,5\}$ and $B = \{1,1/2,1/3,1/4,1/5,1/6\}$. Does there exist a one-to-one function $f: B \to A$? Explain your answer.

70. Let $f: \mathbb{Z} \to \mathbb{Z}$ be defined by $f(n) = 2n + 1$. Is f a one-to-one correspondence? Prove your answer.

71. Let $f: \mathbb{R} \to \mathbb{R}$ be defined by $f(x) = 2x + 7$. Find the inverse function $f^{-1}: \mathbb{R} \to \mathbb{R}$.

72. Let $A = \{1,2,3,4\}$ and $B = \{5,6,7\}$. Let $f: A \to B$ be the function whose graph is $\{(1,6),(2,6),(3,5),(4,7)\}$ and let $g: B \to A$ be the function whose graph is $\{(5,2),(6,4),(7,2)\}$. Find the graphs of $f \circ g: B \to A$ and $g \circ f: A \to B$.

73. Let $f: \mathbb{R} \to \mathbb{R}$ be defined by $f(x) = 2x + 5$ and let $g: \mathbb{R} \to \mathbb{R}$ be defined by $g(x) = x^2 + 7$. Find $f \circ g: \mathbb{R} \to \mathbb{R}$ and $g \circ f: \mathbb{R} \to \mathbb{R}$.

74. Find the integers q and r such that $n = dq + r$ and $0 \leq r < d$ when

a. $d = 17$ and $n = 639$. **b.** $d = 17$ and $n = 657$

75. Let $f: \mathbb{Z} \to \mathbb{Z}$ be defined by $f(n) = n^3 - 4n$.

a. Sketch $G(f)$ and determine if the function is one to one.

b. Do you think f maps $\mathbb{Z}$ onto $\mathbb{Z}$? Explain your answer. (*Hint:* Use the values $n = -3,-2,-1,0,1,2,3$ in order to sketch $G(f)$.)

76. Let $f: \mathbb{R} \to \mathbb{R}$ be defined by $f(x) = x^3 - 4x$. Sketch the restrictions $f|[0,2]$, $f|[-3,0]$, and $f|\{-1,0,1/2\}$.

77. Sketch the graph of the function $f: \mathbb{Z} \to \mathbb{Z}$ defined by $f(n) = \lfloor -2n + 3 \rfloor$.

78. Sketch the graph of the function $f: \mathbb{Z} \to \mathbb{Z}$ defined by $f(n) = \lceil -2n + 3 \rceil$.

79. Let $A = \{1,2,3,4,5,6,7\}$, let $f: A \to \mathbb{R}$ be the function whose graph is $\{(1,-2),(2,0),(3,1),(4,5),(5,-1),(6,0),(7,1/2)\}$, and let $g: A \to \mathbb{R}$ be the function whose graph is $\{(1,0),(2,1),(3,-1),(4,2),(5,-2),(6,3),(7,-3)\}$. Sketch the graphs of each of the following functions.

a. $f + g: A \to \mathbb{R}$ **b.** $f - g: A \to \mathbb{R}$ **c.** $fg: A \to \mathbb{R}$ **d.** $f/g: A \to \mathbb{R}$

***80.** Suppose that α is a permutation of $(0,\infty)$ such that $\alpha^{-1} = 1/\alpha$.

a. Show that $\alpha(1) = 1$.

b. Show that if $x \in (0,\infty)$ and $x \neq 1$ then $\alpha(x) \neq x$.

c. Show that if $x \in (0,\infty)$ and $x \neq 1$ then $\alpha(\alpha(x)) \neq x$.

d. Show that if $x \in (0,\infty)$ and $x \neq 1$ then $\alpha(\alpha(\alpha(x))) \neq x$.

e. Show that if $x \in (0,\infty)$ then $\alpha(\alpha(\alpha(\alpha(x)))) = x$.

Chapter

4

INDUCTION AND RECURSION

The principle of mathematical induction and the principle of recursive definition are at the heart of the proofs and definitions of much of discrete mathematics. Mathematical induction is a method of proof that is used to establish that something is true about every integer larger than some specified integer. In particular, induction can be used to prove that a formula holds for every natural number. One way it is used in computer science is to show that a program with loops performs as expected.

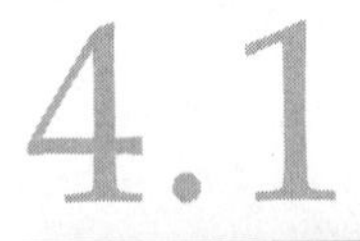

THE PRINCIPLE OF RECURSIVE DEFINITION

In order to consider the principle of recursive definition, we first introduce some terminology. A *sequence* is a function whose domain is the set of all natural numbers. For each natural number n, the set $\{1,2,3, \ldots ,n\}$ is called an *initial segment* of $\mathbb{N}$ and is denoted by I_n (see Section 2.4). A *finite sequence* is a function whose domain is an initial segment of $\mathbb{N}$. If we are given a set A and a sequence $f\colon \mathbb{N} \to A$, then for each natural number n, $f|I_n\colon I_n \to A$ is a finite sequence. For example, if we consider the sequence $f\colon \mathbb{N} \to \mathbb{N}$ defined by $f(n) = 2n$, then $f|I_3\colon \{1,2,3\} \to \mathbb{N}$ is a finite sequence. In fact, we can define (for each $n \in \mathbb{N}$) f_n to be $f|I_n$. When can we go in the opposite direction; that is, when can we build a sequence from an infinite collection of finite sequences? The principle of recursive definition provides the answer to this question.

> **Principle of Recursive Definition** Let X be a nonempty set. To define a sequence $f\colon \mathbb{N} \to X$, it is enough to define f explicitly on some initial segment I_m of $\mathbb{N}$ and to have a rule that defines, for each $n \in \mathbb{N} - I_m$, $f(n)$ in terms of $f(1), f(2), \ldots, f(n-1)$. The values of f that are given explicitly are called **initial values** and the rule that defines $f(n)$ in terms of $f(1), f(2), \ldots, f(n-1)$ is called a **recursion relation.** The sequence f is said to be defined **recursively** (or **inductively**).

Example 1

The sequence $f(n) = 2^n$ is defined recursively. There is only one initial value, $f(1) = 2$. The recursion relation is given by $f(n+1) = 2f(n)$ for all $n \in \mathbb{N}$. Notice that in our recursive definition of f we are never multiplying more than two numbers together at one time. ■

Example 2

Let f be a real-valued sequence. We define a new sequence $\Sigma_{i=1}^{n} f(i)$ recursively as follows: There is only one initial value, $\Sigma_{i=1}^{1} f(i) = f(1)$. The recursion relation is given by $\Sigma_{i=1}^{n+1} f(i) = \Sigma_{i=1}^{n} f(i) + f(n+1)$. Intuitively $\Sigma_{i=1}^{n} f(i)$ means $f(1) + f(2) + f(3) + \cdots + f(n)$, and so $\Sigma_{i=1}^{n} f(i)$ is called the *nth partial sum of f*. But addition is a binary operation and so $\Sigma_{i=1}^{n} f(i)$ defines what we mean by $f(1) + f(2) + f(3) + \cdots + f(n)$, rather than the other way around. ■

Example 3

The *Fibonacci sequence*, a famous sequence used extensively in computer science, is defined recursively. The initial values are $f(1) = 1$ and $f(2) = 1$, and the recursion relation is given by $f(n + 2) =$

$f(n+1) + f(n)$ for all natural numbers n. The members of the range of the Fibonacci sequence f are called *Fibonacci numbers*, and the number $f(n)$ is called the *nth Fibonacci number* and is denoted by f_n. We consider the Fibonacci sequence in considerable detail later in this chapter. Here we are content to evaluate the first few Fibonacci numbers:

$$f_1 = f_2 = 1$$
$$f_3 = f_2 + f_1 = 2$$
$$f_4 = f_3 + f_2 = 3$$
$$f_5 = f_4 + f_3 = 5$$
$$f_6 = f_5 + f_4 = 8$$
$$f_7 = f_6 + f_5 = 13$$

■

Example 4

Let f be a real-valued sequence. We define $\Pi_{i=1}^{n} f(i)\colon \mathbb{N} \to \mathbb{R}$ inductively. The initial condition is given by $\Pi_{i=1}^{1} f(i) = f(1)$ and the recursion relation is $\Pi_{i=1}^{n+1} f(\mathrm{i}) = (\Pi_{i=1}^{n} f(i))\, f(n+1)$.

Suppose that the sequence f we have been given is the constant sequence $f(n) = 2$ for all $n \in \mathbb{N}$. Then

$$\Pi_{i=1}^{1} f(i) = 2$$
$$\Pi_{i=1}^{2} f(i) = (\Pi_{i=1}^{1} f(i))f(2) = 2 \cdot 2 = 4$$
$$\Pi_{i=1}^{3} f(i) = (\Pi_{i=1}^{2} f(i))f(3) = 4 \cdot 2 = 8$$

and it seems that we ought to be able to prove that $\Pi_{i=1}^{n} f(i) = 2^n$ for all $n \in \mathbb{N}$. We use mathematical induction, in the next section, to prove this fact. ■

We conclude this section by stating a theorem about finite sequences.

Theorem 4.1 If $g\colon I_n \to I_n$ is an injection, then g is a surjection.

Exercises 4.1

1. Let $f\colon \mathbb{N} \to \mathbb{R}$ be defined recursively by $f(1) = 1$ and the recursion relation $f(n+1) = -f(n)$. Evaluate $f(2)$, $f(3)$, $f(4)$, $f(5)$, $f(1604)$, and $f(15{,}593)$.
2. Let $f\colon \mathbb{N} \to \mathbb{R}$ be defined recursively by the recursion relation $f(n+1) = f(n) + f(n-1) + f(n)f(n-1)$ for $n > 2$. Evaluate $f(3)$, $f(4)$, and $f(6)$ if the initial values of f are as follows.
   - **a.** $f(1) = 0$ and $f(2) = 0$
   - **b.** $f(1) = 0$ and $f(2) = 1$
   - **c.** $f(1) = 1$ and $f(2) = 0$
   - **d.** $f(1) = 1$ and $f(2) = 1$

3. The number $n!$ is thought of intuitively as the product of the first n natural numbers.
   a. Give a recursive definition of the sequence $f(n) = n!$.
   b. Find the sequence $g: \mathbb{N} \to \mathbb{N}$ such that $\Pi_{i=1}^{n} g(i) = n!$.
   c. Evaluate $f(3)$, $f(4)$, and $f(6)$.
4. Find the 12th and 14th Fibonacci numbers.
5. Find $f(4)$ and $f(6)$ if $f(1) = 1$, $f(2) = 2$, and f is defined for $n \geqslant 2$ recursively by the following.
   a. $f(n + 1) = f(n) - f(n - 1)$
   b. $f(n + 1) = f(n)f(n - 1)$
   c. $f(n + 1) = f(n)^{f(n-1)}$
   d. $f(n + 1) = 5f(n)$
   e. $f(n + 1) = f(n)^2 - f(n - 1)^3$
   f. $f(n + 1) = f(n - 1)/f(n)$
6. *Farey addition* of fractions with positive numerators and denominators is defined by $a/b * c/d = (a + c)/(b + d)$. Let $\mathbb{F}$ be the collection of all fractions. Define $f: \mathbb{N} \to \mathbb{F}$ recursively by $f(1) = 2/3$, $f(2) = 3/4$, and $f(n + 1) = f(n) * f(n - 1)$ for $n \geq 2$. Find $f(3)$, $f(4)$, and $f(6)$.

4.2

PROOF BY INDUCTION

In this section we consider the principle of mathematical induction. This principle is useful in every branch of mathematics. Moreover, its use sometimes comes as a pleasant surprise.

> **Principle of Mathematical Induction, or Finite Induction Principle**
>
> Let S be a subset of $\mathbb{N}$ such that
>
> $1 \in S$ and
> if $n \in S$, then $n + 1 \in S$
>
> Then $S = \mathbb{N}$.

The principle of mathematical induction gives conditions under which a subset S of $\mathbb{N}$ must be $\mathbb{N}$ itself, and so the principle is a set-theoretic statement. In reality, however, this principle is a powerful tool that allows us to prove infinitely many propositions in one fell swoop. A proof by induction consists of three steps:

1. Define a set S of natural numbers with the property that, if S turns out to be $\mathbb{N}$, then the desired result is known to be true.
2. Show that $1 \in S$.
3. Show that, if $n \in S$, then $n + 1 \in S$.

We illustrate the three steps of a proof by induction with a simple example. Recall that we have agreed (in Example 2) that "$1 + 2 + 3 + \cdots + n$" means $\Sigma_{i=1}^{n} i$, which has been defined recursively.

Example 5

Prove by induction that, for each natural number n, $1 + 2 + 3 + \cdots + n = n(n + 1)/2$.

Analysis

In this simple example, step 1 is easy. Let $S = \{n \in \mathbb{N}: 1 + 2 + 3 + \cdots + n = n(n + 1)/2\}$. The next step is to show that $1 \in S$. This is also easy. We simply observe that $1(1 + 1)/2 = 1$.

The third step is to show that if $n \in S$ then $n + 1 \in S$. We choose a natural number n, but instead of showing that $n \in S$, we *assume* that $n \in S$ and *show* that $n + 1 \in S$. In this example, we are assuming that $1 + 2 + 3 + \cdots + n = n(n + 1)/2$, and we want to prove that $(1 + 2 + 3 + \cdots + n) + (n + 1) = (n + 1)[(n + 1)+1]/2$. So let us proceed. Suppose $1 + 2 + 3 + \cdots + n = n(n + 1)/2$. Then

$$
\begin{aligned}
(1 + 2 + 3 + \cdots + n) + (n + 1) &= \frac{n(n + 1)}{2} + (n + 1) \\
&= (n + 1)\left(\frac{n}{2} + 1\right) \\
&= (n + 1)\frac{(n + 2)}{2} \\
&= (n + 1)\frac{(n + 1) + 1}{2}
\end{aligned}
$$

Therefore $(1 + 2 + 3 + \cdots + n) + (n + 1) = (n + 1)[(n + 1) + 1]/2$, and we have proved that if $n \in S$ then $n + 1 \in S$. Hence, by the principle of mathematical induction, $S = \mathbb{N}$. Thus, if $n \in \mathbb{N}$, then $1 + 2 + 3 + \cdots + n = n(n + 1)/2$. ■

Let us consider another example of the use of the principle of mathematical induction. The formula given in this example is the sum of the terms of a *geometric progression*, which is a sequence of the form $a, ar, ar^2, \ldots, ar^n, \ldots$, where $a, r \in \mathbb{R}$ and $r \neq 1$.

Example 6

Let $a, r \in \mathbb{R}$ and suppose $r \neq 1$. Prove that, for each $n \in \mathbb{N}$, $a + ar + ar^2 + \cdots + ar^n = a(r^{n+1} - 1)/(r - 1)$.

Proof

Let

$$S = \left\{n \in \mathbb{N}: a + ar + ar^2 + \cdots + ar^n = a\left(\frac{r^{n+1} - 1}{r - 1}\right)\right\}$$

Since $a(r^2 - 1)/(r - 1) = a(r + 1) = a + ar$, $1 \in S$. Suppose $n \in S$. Then

$$a + ar + ar^2 + \cdots + ar^n = a\left(\frac{r^{n+1} - 1}{r - 1}\right)$$

Therefore

$$\begin{aligned} a + ar + ar^2 + \cdots + ar^n + ar^{n+1} &= a\left(\frac{r^{n+1} - 1}{r - 1}\right) + ar^{n+1} \\ &= \frac{ar^{n+1} - a + (r - 1)ar^{n+1}}{r - 1} \\ &= a\left(\frac{r^{n+2} - 1}{r - 1}\right) \end{aligned}$$

Therefore $n + 1 \in S$. By the principle of mathematical induction, $S = \mathbb{N}$. ■

Our next example is the sum of the terms of an *arithmetic progression*, which is a sequence of the form $a, a + d, a + 2d, \ldots, a + nd, \ldots$.

Example 7

Let $a, d \in \mathbb{R}$. Prove that for each $n \in \mathbb{N}$,

$$a + (a + d) + (a + 2d) + \cdots + (a + nd) = \frac{(n + 1)(2a + nd)}{2}$$

Proof

Let

$$S = \left\{n \in \mathbb{N}: a + (a + d) + (a + 2d) + \cdots + (a + nd) = \frac{(n + 1)(2a + nd)}{2}\right\}$$

Since $(1 + 1)(2a + 1d)/2 = 2a + d = a + (a + d)$, $1 \in S$.

Suppose $n \in S$. Then

$$a + (a + d) + (a + 2d) + \cdots + (a + nd) = \frac{(n + 1)(2a + nd)}{2}$$

so

$$a + (a + d) + (a + 2d) + \cdots + (a + nd) + [a + (n + 1)d]$$
$$= \frac{(n + 1)(2a + nd)}{2} + [a + (n + 1)d]$$
$$= \frac{(n + 1)(2a) + (n + 1)nd + 2a + 2(n + 1)d}{2}$$
$$= \frac{2a[(n + 1) + 1] + (n + 1)d(n + 2)}{2}$$
$$= \frac{(n + 2)(2a + (n + 1)d)}{2}$$

Therefore $n + 1 \in S$. By the principle of mathematical induction, $S = \mathbb{N}$. ■

In the previous section, we promised to prove the following theorem.

Theorem 4.2

> Let $f: \mathbb{N} \to \mathbb{N}$ be defined by $f(n) = 2$. Then, for each $n \in \mathbb{N}$, $\Pi_{i=1}^{n} f(i) = 2^n$.

Proof

Let

$$S = \{n \in \mathbb{N}: \Pi_{i=1}^{n} f(i) = 2^n\}$$

Since $\Pi_{i=1}^{1} f(i) = f(1) = 2 = 2^1$, $1 \in S$.

Suppose $n \in S$. Then $\Pi_{i=1}^{n} f(i) = 2^n$. Now

$$\Pi_{i=1}^{n+1} f(i) = \Pi_{i=1}^{n} f(i) f(n + 1) = 2^n \cdot 2 = 2^{n+1}$$

Therefore $n + 1 \in S$. By the principle of mathematical induction, $S = \mathbb{N}$. □

Example 8

What is wrong with the following proof that if $n \in \mathbb{N}$ then $n > 100$?

Proof

Let $S = \{n \in \mathbb{N}: n > 100\}$. Suppose $n \in S$. Then $n > 100$. Therefore $n + 1 > 100 + 1 > 100$, so $n + 1 \in S$. By the principle of mathematical induction, $S = \mathbb{N}$. ■

Analysis

When using the principle of mathematical induction, one must always remember to prove both parts of the induction proof. Since the hard part is usually step 3, it is easy to overlook step 2, but as you can see, this sometimes leads to ludicrous results. The problem in this proof is of course that $1 \notin S$, since 1 is less than 100. ■

Let us consider another example of the use of the principle of mathematical induction. In Example 9 we make use of the numbers 7^n and 2^n, both of which are defined recursively.

Example 9

Prove that, for each natural number n, $7^n - 2^n$ is divisible by 5.

Proof

Let $S = \{n \in \mathbb{N}: 7^n - 2^n \text{ is divisible by } 5\}$. To use the principle of mathematical induction to prove that $S = \mathbb{N}$, we need to prove two things: (1) that $1 \in S$, and (2) that if $n \in S$ then $n + 1 \in S$. Since $7^1 - 2^1 = 7 - 2 = 5$, and 5 is divisible by 5, $1 \in S$.

Suppose $n \in S$. Then $7^n - 2^n$ is divisible by 5. We want to use this fact to prove that $n + 1 \in S$. Note that $n + 1 \in S$ if and only if $7^{n+1} - 2^{n+1}$ is divisible by 5. So we need to prove that, if $7^n - 2^n$ is divisible by 5, then $7^{n+1} - 2^{n+1}$ is divisible by 5. To do this, we need to write $7^{n+1} - 2^{n+1}$ in terms of $7^n - 2^n$. But how do we do this? By definition, $7^{n+1} = 7 \cdot 7^n$ and $2^{n+1} = 2 \cdot 2^n$. So we can write $7^{n+1} - 2^{n+1} = 7 \cdot 7^n - 2 \cdot 2^n$. To get the $7^n - 2^n$ term, we subtract $7 \cdot 2^n$ from $7 \cdot 7^n - 2 \cdot 2^n$ and obtain $7 \cdot 7^n - 7 \cdot 2^n - 2 \cdot 2^n = 7(7^n - 2^n) - 2 \cdot 2^n$.

Alas, we no longer have $7^{n+1} - 2^{n+1}$. To maintain equality, we add $7 \cdot 2^n$ and obtain

$$\begin{aligned} 7^{n+1} - 2^{n+1} &= 7 \cdot 7^n - 2 \cdot 2^n \\ &= 7 \cdot 7^n - 7 \cdot 2^n + 7 \cdot 2^n - 2 \cdot 2^n \\ &= 7(7^n - 2^n) + 2^n(7 - 2) \end{aligned}$$

Since $7^n - 2^n$ and $7 - 2$ are divisible by 5, (by Exercise 27 of Chapter 1) $7^{n+1} - 2^{n+1}$ is divisible by 5. Therefore $n + 1 \in S$, and we have completed step 3. ■

In Section 2.1, we defined the power set $\mathcal{P}(X)$ of a set X to be the set whose members are the subsets of X, and in Example 1 of Chapter 2 we saw that the following was true:

Number of members of X	1	2	3
Number of members of $\mathcal{P}(X)$	2	4	8

It is also true that, if X is a set with four members, then the number of members of $\mathcal{P}(X)$ is 16. Furthermore, we can use the principle of mathematical induction to prove the following theorem.

Theorem 4.3

For each natural number n, if X is a set consisting of n elements, then $\mathcal{P}(X)$ is a set consisting of 2^n elements.

Proof

Let $S = \{n \in \mathbb{N}$: if X is a set consisting of n elements, then $\mathcal{P}(X)$ is a set consisting of 2^n elements$\}$.

If X consists of one element, then, as previously noted, $\mathcal{P}(X)$ has two elements, so $1 \in S$.

Suppose $n \in S$. Let X be a set consisting of $n + 1$ elements, and choose an element x of X. Then $X - \{x\}$ consists of n elements, so $\mathcal{P}(X - \{x\})$ consists of 2^n elements. Now A is a subset of X if and only if (1) $A \subseteq X - \{x\}$, or (2) $A = B \cup \{x\}$, where $B \subseteq X - \{x\}$. We know that there are 2^n subsets of X that are also subsets of $X - \{x\}$. How many subsets of X are there of the form $B \cup \{x\}$, where $B \subseteq X - \{x\}$? There is one such subset for each subset of $X - \{x\}$, so there are 2^n subsets of X of this form. Therefore the number of subsets of X is $2^n + 2^n = 2^n(1 + 1) = 2^{n+1}$. Hence $n + 1 \in S$. By the principle of mathematical induction $S = \mathbb{N}$. □

Now that we have given several examples, which we hope have convinced you that the principle of mathematical induction is useful, we turn to the obvious question: Why should you believe that the principle of mathematical induction is true? There are several convincing arguments in support of this principle, the most famous being the domino analogy. Imagine an infinite row of dominoes labeled $1,2,3, \ldots ,n, \ldots$, where the dominoes are standing so close together that each domino that falls will knock down the domino immediately behind it. Now suppose that someone knocks over the front domino. Is it true that all the dominoes will fall? If not, consider the first domino that does not ever fall. It is not the front domino, so it makes sense to talk about the domino that immediately precedes our steadfast domino. This immediately preceding domino we know will fall, and when it does it will knock down the domino that we thought would stand forever. Thus all the dominoes must eventually fall.

At the moment, this analogy depends on vivid imaginations, but we can make another plausible argument without having to imagine infinitely many dominoes. Suppose that we have a subset S of $\mathbb{N}$ such that $1 \in S$ and if $n \in S$ then $n + 1 \in S$. Since $1 \in S$, $2 \in S$. Since $2 \in S$, $3 \in S$. Since $3 \in S$, $4 \in S$. We can obviously continue as long as we wish. If $\mathbb{N} = S$, then there is a natural number p such that $p \notin S$. Since $1 \in S$, $p \neq 1$. By repeating the argument that $n \in S$ implies that $n + 1 \in S$ $(p - 1)$ times, we arrive at the conclusion that $p \in S$. The key words to this argument are "by repeating the argument . . . $(p - 1)$ times." In mathematical writing, phrases such as "by repeating the argument" mean one of two things: (1) the author has tried many cases and is now willing to make a guess, or (2) the author sees how to construct a proof using mathematical induction but does not want to take the time to write out the details of the proof.

We give an example to illustrate the need for a proof by induction even though something may be true for as many cases as we care to check.

Example 10

Is it true that $n^2 + n + 41$ is a prime for each natural number n?

Analysis

In the following table, we give the value of $n^2 + n + 41$ for the first ten values of n:

n	1	2	3	4	5	6	7	8	9	10
$n^2 + n + 41$	43	47	53	61	71	83	97	113	131	151

Each listed value of $n^2 + n + 41$ is a prime. In fact, $n^2 + n + 41$ is a prime for $n = 11, 12, \ldots, 39$. However, if $n = 40$, then

$$n^2 + n + 41 = 40^2 + 40 + 41 = 40(40 + 1) + 41 = 41(40 + 1) = 41^2$$

■

The principle of mathematical induction is equally valid if, instead of starting with 1, we (1) start with a given integer k, (2) show that $k \in S$, and (3) show that, if $n \in S$ and $n \geqslant k$, then $n + 1 \in S$. When we do this we will know that every integer greater than or equal to k belongs to the set S. We illustrate this with Example 11.

Example 11

Prove that, for each natural number $n \geqslant 5$, $n^2 < 2^n$.

Proof

Let $S = \{n \in \mathbb{N}: n^2 < 2^n\}$. We want to show that every natural number greater than or equal to 5 belongs to S. By the principle of mathematical induction, it is sufficient to show that (1) $5 \in S$, and (2) if $n \in S$ and $n \geqslant 5$, then $n + 1 \in S$. Since $5^2 = 25$ and $2^5 = 32$, $5 \in S$.

Suppose $n \geqslant 5$ and $n \in S$. Then $n^2 < 2^n$. We want to show that $(n + 1)^2 < 2^{n+1}$. Now $(n + 1)^2 = n^2 + 2n + 1$, and $2^{n+1} = 2 \cdot 2^n$, so we want to show that $n^2 + 2n + 1 < 2 \cdot 2^n$. Since $n^2 < 2^n$, $2n^2 < 2 \cdot 2^n$. Hence it is sufficient to show that $n^2 + 2n + 1 < 2n^2$. Since $n \geqslant 5$, $n(n - 2) > 1$ (see Exercise 81). So $n^2 - 2n > 1$, or $n^2 > 2n + 1$. By adding n^2 to both sides of this inequality, we obtain $2n^2 > n^2 + 2n + 1$. This is the desired result, so $n + 1 \in S$. ■

Exercises 4.2

7. Use the formula for the sum of the terms of a geometric progression to evaluate the following sums.

   a. $2 + 2 \cdot 3 + 2 \cdot 3^2 + \cdots + 2 \cdot 3^{10}$ b. $3 + 3 \cdot 5^2 + 3 \cdot 5^4 + \cdots + 3 \cdot 5^{10}$

8. Use the formula for the sum of the terms of an arithmetic progression to evaluate the following sums.

   a. $2 + 5 + 8 + \cdots + 29$ b. $3 + 8 + 13 + \cdots + 38$

9. Write each of the following sums without using Σ notation.

   a. $\sum_{i=1}^{5} i^2$ b. $\sum_{i=1}^{6} (-1)^i$ c. $\sum_{i=1}^{5} i^i$ d. $\sum_{i=1}^{6} (-1)^i 2i$

10. Evaluate each of the following sums.

    a. $(\Sigma_{i=1}^{3} i)(\Sigma_{j=1}^{4} j)$ b. $\Sigma_{i=1}^{3} (\Sigma_{j=1}^{4} i^j)$

11. Write each of the following sums using Σ notation.

    a. $2 + 4 + 6 + 8 + 10 + 12$ b. $1 + 3 + 5 + 7 + 9 + 11$

    c. $1^2 + 2^2 + 3^2 + 4^2 + 5^2 + 6^2$

12. If $A = \{1,3,5,6,7,8,9\}$, how many members does $\mathcal{P}(A)$ have?

13. Prove by induction that, for each natural number n,

$$3 + 6 + 9 + \cdots + 3n = \frac{3n(n + 1)}{2}$$

14. Prove by induction that, for each natural number n,

$$5 + 7 + 9 + \cdots + (2n + 3) = n(n + 4)$$

15. Prove by induction that, for each natural number n,

$$1^2 + 2^2 + 3^2 + \cdots + n^2 = \frac{n(n + 1)(2n + 1)}{6}$$

16. Prove by induction that, for each natural number n,

$$1(1 + 1) + 2(2 + 1) + 3(3 + 1) + \cdots + n(n + 1) = \frac{n(n + 1)(n + 2)}{3}$$

17. Prove by induction that, for each natural number n,

$$4 + 10 + 16 + \cdots + (6n - 2) = n(3n + 1)$$

18. Prove by induction that, for each natural number n,

$$\frac{1}{1 \cdot 2} + \frac{1}{2 \cdot 3} + \frac{1}{3 \cdot 4} + \cdots + \frac{1}{n(n+1)} = \frac{n}{n+1}$$

19. Prove by induction that, for each natural number n,

$$1 + 1/2 + 1/4 + \cdots + 1/2^{n-1} = 2 - 1/2^{n-1}$$

20. Prove by induction that, for each natural number n,

$$1^3 + 2^3 + 3^3 + \cdots + n^3 = \frac{n^2(n+1)^2}{4}$$

21. The left-hand sides of the equations of Exercises 14, 15, and 18 are defined inductively as $\Sigma_{i=1}^{n} f(i)$ for some appropriate sequence f. Explain how to recognize the sequence f, and rework these exercises using Σ notation.

22. Prove by induction that, for each natural number n, $n^3 - n$ is divisible by 3.

23. Prove by induction that, for each natural number n, $3^n - 1$ is divisible by 2.

24. Prove by induction that, for each natural number n, $2^{2n-1} + 3^{2n-1}$ is divisible by 5.

25. **a.** Calculate $1 + 3 + 5 + \cdots + (2n - 1)$ for $n = 1,2,3,4,$ and 5.

b. Guess a general formula for this sum.

c. Prove, by induction, the formula in (**b**).

26. Let $S = \{n \in \mathbb{N}: n^2 + 5n + 1 \text{ is even}\}$.

a. Prove that S is an inductive set; that is, prove that if $n \in S$ then $n + 1 \in S$.

b. Which natural numbers belong to S?

c. Prove the answer given in (**b**). (*Hint:* Do not use induction. Consider two cases, one where n is even and one where n is odd.)

27. Prove by induction that, for each natural number $n \geqslant 4$, $n! > 2^n$.

28. Prove by induction that, for each natural number $n \geqslant 7$, $n! > 3^n$.

29. Prove by induction that, for each natural number $n \geqslant 2$,

$$\left(\frac{2^2}{1 \cdot 3}\right)\left(\frac{3^2}{2 \cdot 4}\right)\left(\frac{4^2}{3 \cdot 5}\right) \cdots \frac{n^2}{(n-1)(n+1)} = \frac{2n}{n+1}$$

30. Prove by induction that, for each natural number $n \geqslant 2$,

$$\left(1 - \frac{1}{4}\right)\left(1 - \frac{1}{9}\right)\left(1 - \frac{1}{16}\right) \cdots \left(1 - \frac{1}{n^2}\right) = \frac{n+1}{2n}$$

31. Let A be a set and for each natural number i let B_i be a set. Prove by induction that, for each natural number $n \geqslant 2$,

a. $A \cap (\cup_{i=1}^{n} B_i) = \cup_{i=1}^{n} (A \cap B_i)$ **b.** $\cup_{i=1}^{n} B_i^{\sim} = (\cap_{i=1}^{n} B_i)^{\sim}$

32. What is wrong with the following proof that all members of the human race are of the same sex?

Proof: Let $S = \{n \in \mathbb{N}$: in any set of n people, all the members of the set have the same sex$\}$. If we have a set consisting of one person, then clearly all the members of the set are of the same sex, so $1 \in S$. Suppose $n \in S$. Then in any set of n people, all the members of the set are of the same sex. Let A be a set consisting of $n + 1$ people. Let us call these people $P_1, P_2, P_3, \ldots, P_n, P_{n+1}$. Now remove one person, say P_{n+1}, from this set. We now have a set consisting of n people, so, since $n \in S$, all the members of this set are of the same sex. Now put P_{n+1} back in the set and remove P_n from the set. Once again, we have a set consisting of n people, so all the members of this set are of the same sex. Now observe that everyone in the original set of $n + 1$ people is of the same sex as P_1, so they all have the same sex. Therefore $n + 1 \in S$. By the principle of mathematical induction, $S = \mathbb{N}$.

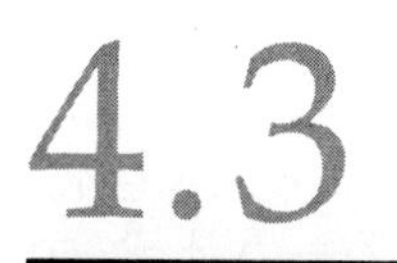

THE SECOND PRINCIPLE OF MATHEMATICAL INDUCTION

Let us continue our study of induction by considering an example that does not readily yield to any of the previous forms of induction. If $W = \{n \in \mathbb{N}: n > 5\}$, can every member of W be written as a sum of numbers each of which is a 2 or a 5?

$$6 = 2 + 2 + 2 \qquad 7 = 5 + 2$$
$$8 = 2 + 2 + 2 + 2 \qquad 9 = 5 + 2 + 2$$
$$10 = 5 + 5 \qquad 11 = 5 + 2 + 2 + 2$$
$$13 = 5 + 2 + 2 + 2 + 2 \qquad 14 = 5 + 5 + 2 + 2$$

Based on these few examples, it appears reasonable to guess that our question has an affirmative answer, but can we prove this guess by induction? Let $S = \{n \in \mathbb{N}: n > 5$ and n can be written as a sum of numbers each of which is a 2 or a 5$\}$. Since $6 = 2 + 2 + 2$, $6 \in S$. Suppose $n \in S$. We need to show that $n + 1 \in S$. It does not help to know that n can be written as a sum of 2s or 5s, since $n + 1$ involves this sum plus 1. However, if we know that $n - 1$ can be written as the sum of 2s or 5s, then we can add 2 to this sum to get $n + 1$. So instead of knowing something about n, we need to know something about $n - 1$. Fortunately, there is a form of induction, the second principle of mathematical induction, that permits us to handle such situations.

Second Principle of Mathematical Induction Let S be a subset of $\mathbb{N}$ such that

$1 \in S$ and

for each $n \in \mathbb{N}$, if $\{1,2,\ldots,n\} \subseteq S$, then $n + 1 \in S$

Then $S = \mathbb{N}$.

Just as with the principle of mathematical induction, the second principle of mathematical induction is equally valid if, instead of starting with 1, we (1) start with a given integer k, (2) show that $k \in S$, and (3) show that, if $n \geqslant k$ and $\{k, k+1, \ldots, n\} \subseteq S$, then $n+1 \in S$. Again, when we do this, we will know that every integer greater than or equal to k belongs to the set S.

Example 12

Prove that, if $n \in \mathbb{N}$ and $n > 5$, then n can be written as a sum of numbers each of which is a 2 or a 5.

Proof

Let S be defined as before and recall that $6 \in S$. Suppose $n \in \mathbb{N}$, $n > 5$, and $\{6, 7, \ldots, n\} \subseteq S$. If $n = 6$, then $n + 1 = 7 = 5 + 2$, so $n + 1 \in S$. If $n > 6$, then $n - 1 \in S$ and hence $n - 1$ can be written as a sum of 2s or 5s; that is, $n - 1 = a_1 + a_2 + \cdots + a_p$, where, for each $i = 1, 2, \ldots, p$, a_i is either 2 or 5. So $n + 1 = a_1 + a_2 + \cdots + a_p + 2$, and therefore $n + 1 \in S$. By the second principle of mathematical induction, $\{n \in \mathbb{N}: n > 5\} \subseteq S$. ■

Example 13 is another proof that uses the second principle of mathematical induction.

Example 13

Prove that the sum of the odd natural numbers up to and including the odd natural number n is $(n + 1)^2/4$.

Analysis

This example not only illustrates the use of the second principle of mathematical induction, it also illustrates a clever choice of the set S. The choice of S is the first and often the most difficult step in an induction argument. Let $S = \{n \in \mathbb{N}: n$ is even, or n is odd and the sum of the odd natural numbers up to and including the odd natural number n is $(n + 1)^2/4\}$. Note that we are interested in proving that something is true when n is odd, so we simply include all the even natural numbers in the set S. If $n + 1$ is even, there is nothing to prove, for in this case $n + 1$ is known to be a member of S. We now proceed with the proof by the second principle of induction.

Since $(1 + 1)^2/4 = 1$, $1 \in S$. Suppose $n \in \mathbb{N}$ and $\{1, 2, \ldots, n\} \subseteq S$. As we have just observed, if $n + 1$ is even, it is a member of S. Suppose $n + 1$ is odd. Then $n - 1$ is odd and $n - 1 \in S$. Therefore

$$1 + 3 + 5 + \cdots + (n - 1) = \frac{[(n - 1) + 1]^2}{4} = \frac{n^2}{4}$$

so

$$\begin{aligned}(1 + 3 + 5 + \cdots + (n - 1)) + (n + 1) &= \frac{n^2}{4} + (n + 1) \\ &= \frac{n^2 + 4(n + 1)}{4} \\ &= \frac{n^2 + 4n + 4}{4} \\ &= \frac{[(n + 1) + 1]^2}{4}\end{aligned}$$

Therefore $n + 1 \in S$, and hence by the second principle of mathematical induction $S = \mathbb{N}$.

Observe that here we need to know that $n - 1$ is a member of S in order to show that if $n + 1$ is odd then $n + 1 \in S$. It does not help to know that n is a member of S. Therefore, we use the second principle of mathematical induction rather than the principle of mathematical induction. ■

We give another example to illustrate the use of the second principle of mathematical induction:

Example 14

Every natural number other than 1 is either a prime number or the product of prime numbers.

Proof

Let $S = \{n \in \mathbb{N}: n \neq 1$, and n is either a prime number or the product of primes$\}$. Since 2 and 3 are primes, $2,3 \in S$. Suppose $n \geq 3$ and $\{2,3, \ldots ,n\} \subseteq S$. If $n + 1$ is prime, then $n + 1 \in S$. If $n + 1$ is not prime, then it can be written as the product of natural numbers p and q, where $2 \leq p \leq n$ and $2 \leq q \leq n$. Since $p,q \in S$, they are either prime or the product of primes. In either case $n + 1$ is the product of primes. Therefore $n + 1 \in S$. By the second principle of mathematical induction, $S = \mathbb{N} - \{1\}$. ■

The result established in Example 14 is part of the *fundamental theorem of arithmetic*. The remaining part, which is more difficult to prove, says that, except for the order in which the factors are written, each integer greater than 1 can be written as a prime or a product of primes in only one way.

In Section 4.2, we discussed why you should believe the principle of mathematical induction. We now ask the obvious question: What is the relationship between the second principle of mathematical induction and the principle of mathematical induction? It is clear that the second principle of mathematical induction implies the principle of mathematical induction because, if we are allowed to assume that $\{1,2, \ldots ,n\} \subseteq S$, then we are surely allowed to assume that $n \in S$. It is also true that the principle of mathematical induction implies the second principle, but the equivalence of the two principles is relatively unimportant to us. What we really need to know is that they are valid axioms. There is a third axiom, called the least-natural-number principle, that is a fundamental property of the natural numbers.

> **Least-Natural-Number Principle** Every nonempty set of natural numbers has a least member.

We have already indicated why the second principle of mathematical induction implies the principle of mathematical induction. We now prove that the least-natural-number principle implies the second principle of mathematical induction. Let S be a subset of $\mathbb{N}$ such that $1 \in S$ and, if $\{1,2, \ldots ,n\} \subseteq S$, then $n + 1 \in S$. The proof that $S = \mathbb{N}$ is by contradiction. Suppose $S \neq \mathbb{N}$. Then, by the least-natural-number principle, $\mathbb{N} - S$ has a least member p. So $\{1,2, \ldots ,p - 1\} \subseteq S$ and hence $p \in S$. This is a contradiction.

We end this section with one more warning. There is nothing wrong with guessing in mathematics, and presumably most mathematical results are first arrived at intuitively and only later established by proof as theorems. As Exercise 25 illustrates, mathematical induction can often be used to substantiate an intelligent guess. There is a natural temptation, however, to make a plausible guess and then allow that guess to stand unproved. If you were presented with a sequence beginning 1,2,4,8,16, . . . and asked to guess what number comes next, you would probably guess 32. But really you have no way of knowing the next number, and in Exercise 6 of Chapter 2 we encountered a perfectly natural sequence that began 1,2,4,8,16,31. In discrete mathematics, guesses are made with small numbers, because small numbers are easier to work with. The warning we wish to give is contained in Richard Guy's law of small numbers: There are too few small numbers to meet the many demands made of them. In other words, because we are only able to check small numbers, there are always going to be false conjectures that seem true in the beginning. If you intend to let S be the set of all even natural numbers, you cannot just write "Let $S = \{2,4,6,8, \ldots \}$." Who knows what numbers come next? No one. You may as well say to your reader, "Guess correctly the set S I am thinking of."

Exercises 4.3

These exercises are designed to illustrate the use of the second principle of mathematical induction. Although some of them yield to the principle of mathematical induction when the proper choice of S is made, their purpose is to give you some experience with the second principle. In Section 4.4 we discuss some recursion relations, and these relations provide additional examples of the use of the second principle. Thus additional exercises in the use of the second principle of mathematical induction are given at the end of the chapter.

33. Prove that every natural number $n \geqslant 14$ can be written as a sum of numbers each of which is a 3 or an 8.

34. Prove that every natural number $n \geqslant 8$ can be written as a sum of numbers each of which is a 3 or a 5.

35. Prove that, for each odd natural number $n \geqslant 3$,
$(1 + 1/2)(1 - 1/3)(1 + 1/4) \ldots (1 + (-1)^n/n) = 1$

36. Prove that, for each even natural number n,
$(1 - 1/2)(1 + 1/3)(1 - 1/4) \ldots (1 - (-1)^n/n) = 1/2$

37. Show that the sum of the first n odd natural numbers is n^2.

38. Show that the sum of the first n even natural numbers is $n^2 + n$.

4.4 RECURSIVELY DEFINED SEQUENCES AND SETS

Throughout this chapter we have stressed that mathematical induction, in all its guises, and recursive definition are allied topics. In this final section we return to the study of recursion relations, and this time we have at hand all the techniques of mathematical induction. Once we have two sequences defined recursively, we can use mathematical induction to determine readily whether the two sequences are the same. Moreover, the study of a recursively defined sequence that has more than one initial value often requires the use of the second principle of mathematical induction. We now consider some examples of recursively defined sequences and make use of induction in the study of their properties.

Example 15

A video game begins with three aliens on the screen. Each alien that is not shot down within six seconds reaches the side of the screen and splits into two aliens, which appear at different places on the screen. So, if none of the aliens is destroyed within six seconds, six aliens appear on the screen. Again, each alien not destroyed within six seconds reaches the side of the screen and splits into two aliens. In other words, every six seconds, if no aliens are destroyed, the number of aliens doubles. If a player of this video game is a novice (like one of the authors) and all the aliens survive, how many aliens are on the screen thirty seconds after the game begins?

By calculating the number of aliens at six-second intervals, we obtain the following:

Time	0	6	12	18	24	30
Aliens	3	6	12	24	48	96

Thirty seconds after the game has begun there are 96 aliens on the screen.

Suppose we want to know the number of aliens on the screen at the end of three minutes. We could extend the preceding table, but there is an easier way. We have a recursion relation. If $N(0)$ denotes the number of aliens on the screen at the beginning of the game and $N(n)$ denotes the number of aliens on the screen after n six-second intervals, then $N(0) = 3$ and $N(n) = 2 \cdot N(n - 1)$ for each $n > 0$. We prove by induction that the sequence N that we have defined recursively is the sequence $f\colon \mathbb{N} \to \mathbb{N}$ defined by $f(n) = 3 \cdot 2^n$. Let $S = \{n \in \mathbb{N}\colon \mathrm{N}(n) = 3 \cdot 2^n\}$. Since $N(1) = 3 \cdot 2 = 3 \cdot 2^1$, $1 \in S$. Suppose $n \in S$. Then

$$\begin{aligned} N(n + 1) &= 2 \cdot N(n) = 2(3 \cdot 2^n) = (2 \cdot 3)(2^n) \\ &= (3 \cdot 2)(2^n) = 3(2 \cdot 2^n) = 3(2^{n+1}) \end{aligned}$$

Therefore $n + 1 \in S$, and so $S = \mathbb{N}$. Since there are 10 six-second intervals in each minute and we are concerned with the number of aliens at the end of three minutes, $n = 30$. Therefore the number of aliens on the screen at the end of three minutes is $2^{30} \cdot 3$, or 3,221,225,472. Because this number of aliens would more than fill the video monitor, no player is allowed to play for three minutes without destroying any aliens. ■

Example 16

A game called the Tower of Hanoi, invented by the French mathematician Edouard Lucas (1842–1891), consists of a board with three upright pegs and seven disks with different outside diameters. The seven disks begin all on one peg, with the largest disk on the bottom and the smallest disk at the top (Figure 4.1). The object of the game is to transfer all the disks, one at a time, in the smallest possible number of moves, to the third peg to form an identical pyramid. During the transfers, the player is not permitted to place a larger disk on top of a smaller disk, and this is why the second peg is needed. Can you do this problem? If so, how many moves does it take to complete the transfer?

Figure 4.1

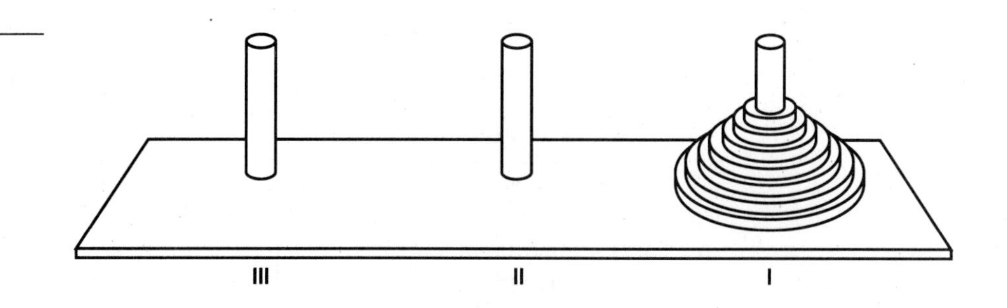

Let us start with the simplest problem, the problem in which there is only one disk on the first peg. This problem is, perhaps, too trivial. We simply move the disk from the first peg to the third peg—one move.

Suppose that we have two disks. Then we move the smaller disk from the first peg to the second peg, the larger disk from the first peg to the third peg, and finally the smaller disk from the second peg to the third peg—a total of three moves.

Suppose that we have three disks. Then we begin to see the induction, or recursion, process. We know from the previous paragraph that we can move the two smaller disks from the first peg to the second peg in three moves. Then we move the largest disk from the first peg to the third peg. Finally, we again know, from the previous paragraph, that we can move the two smaller disks from the second peg to the third peg in three moves. So we can move the three disks from the first peg to the third peg in $3 + 1 + 3 = 7$ moves. Can you now guess how many moves seven disks require?

As Lucas told his story, his game with seven disks was just a model to illustrate the legendary Tower of Hanoi, which had sixty-four disks. This tower was attended by a mystic order of monks who moved one disk a minute according to the long-established ritual that no larger disk could be placed on a smaller disk. The monks believed, so Lucas's story went, that as soon as all sixty-four disks were transferred, the earth would collapse in a cloud of dust. This legend seems somewhat alarming, so let us use induction to find a formula for the number of moves required to transfer n disks from the first peg to the third peg. For each natural number n, let s_n denote the number of moves required to move the n disks from the first peg to the third peg. Suppose that we have $n + 1$ disks. Then we can move the n smaller disks from the first peg to the second peg in s_n moves. Then we move the largest disk from the first peg to the third peg. Finally we move the n smaller disks from the second peg to the third peg in s_n moves. Therefore $s_{n+1} = s_n + 1 + s_n = 2s_n + 1$. So the recursion relation that gives the number of moves required is $s_{n+1} = 2s_n + 1$, and we have the initial value $s_1 = 1$.

It turns out that, in this particular case, there is a formula for the number of moves required to move n disks from the first peg to the third peg. Let us construct a table to see if we can guess this formula.

n = number of disks	1	2	3	4	5	6	7
s_n = number of moves	1	3	7	15	31	63	127

It appears reasonable to guess that, for each $n \in \mathbb{N}$, $s_n = 2^n - 1$, and, as we have indicated, we prove by induction that this guess is correct. Let $S = \{n \in \mathbb{N}: s_n = 2^n - 1\}$. Clearly $1 \in S$. Suppose that $n \in S$. Then $s_{n+1} = 2s_n + 1 = 2(2^n - 1) + 1 = (2^{n+1} - 2) + 1 = 2^{n+1} - 1$, and so $n + 1 \in S$. By the principle of mathematical induction $S = \mathbb{N}$. It is comforting to realize that $2^{64} - 1$ minutes is 18,446,744,073,709,551,615 minutes, or somewhat over 30 trillion years. ■

One of the applications of mathematical induction in computer science is in proving that programs do what they are supposed to do. Let us consider an example of a program that purports to calculate 2^n. The flowchart for this program is given in Figure 4.2.

Figure 4.2

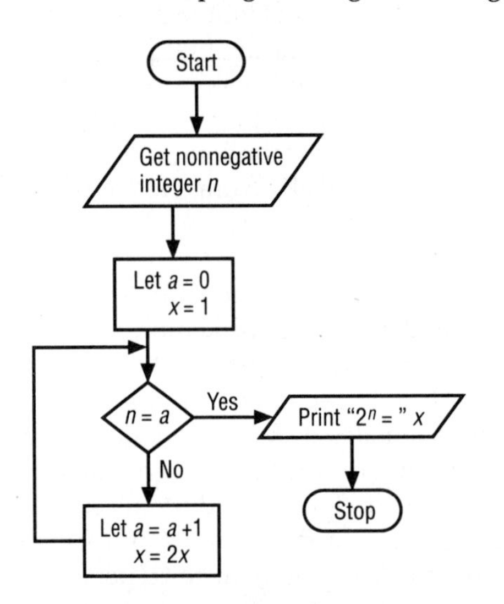

Example 17

For each nonnegative integer n, let $f(n)$ denote the output x in Figure 4.2 corresponding to the input n. Then for each nonnegative integer n, $f(n) = 2^n$.

Analysis

The analysis is by induction. Let $S = \{n \in \mathbb{N}: n$ is nonnegative and $f(n) = 2^n\}$. If $n = 0$, then $f(n) = 1$ and $2^0 = 1$, so $0 \in S$. Suppose $n \in S$. Then $f(n) = 2^n$. Now $f(n + 1) = 2f(n)$ and $2^{n+1} = 2 \cdot 2^n$, so $f(n + 1) = 2^{n+1}$.

Therefore $n + 1 \in S$. By the principle of mathematical induction, S is the set of all nonnegative integers. ■

Another example of a sequence that is defined recursively is the Fibonacci sequence, which we considered briefly in Example 3. In 1202, Leonardo de Pisa, better known as Fibonacci, published *Liber Abaci*. This book, whose title means "book of the abacus," consists for the most part of problems in business mathematics, but there is one problem that has become so famous that a mathematical journal, *Fibonacci Quarterly*, is devoted entirely to it. Suppose that newborn rabbits become adults in two months and that a pair of adult rabbits produces a pair of young rabbits each month. Starting with an adult pair of rabbits and assuming that no rabbits die, how many pairs of rabbits are there after $n (n \in \mathbb{N})$ months? Fibonacci asked for the solution for $n = 12$. Table 4.1 shows the first twelve Fibonacci numbers (second row) and the solution for $n = 12$ at the bottom right.

Table 4.1

Months	1	2	3	4	5	6	7	8	9	10	11	12
Adult pairs	1	1	2	3	5	8	13	21	34	55	89	144
Young pairs	1	2	3	5	8	13	21	34	55	89	144	233
Total pairs	2	3	5	8	13	21	34	55	89	144	233	377

It is amusing to know what it means mathematically to multiply like rabbits, but the Fibonacci sequence is far more useful than its origin might suggest. Indeed, as we indicated earlier, this sequence has important applications in computer science. Next we consider some of the properties of the Fibonacci sequence; other properties are discussed in the exercises.

Theorem 4.4

For each natural number n, let f_n denote the nth Fibonacci number. Then for each $n \in \mathbb{N}$,

a. f_n is even if and only if n is a multiple of 3.
b. $\gcd(f_n, f_{n+1}) = 1$.
c. $|f_{n+1}^2 - f_n f_{n+2}| = 1$.

Proof

a. Let $S = \{n \in \mathbb{N}: f_n$ is even if n is a multiple of 3 and f_n is odd if n is not a multiple of 3$\}$. Clearly 1, 2, and 3 belong to S. Let n be a natural number greater than 2 and suppose that $I_n \subseteq S$. Then

$$f_{n+1} = f_{n-1} + f_n = f_{n-1} + f_{n-1} + f_{n-2} = 2f_{n-1} + f_{n-2}$$

Thus f_{n+1} is even if and only if f_{n-2} is even.

Case 1: f_{n-2} is even. Since $n - 2 \in S$, there is a natural number k such that $n - 2 = 3k$. It follows that

$$n + 1 = (n - 2) + 3 = 3(k + 1)$$

In this case, $n + 1$ is a multiple of 3 and f_{n+1} is even.

Case 2: f_{n-2} is odd. Since $n - 2 \in S$, $n - 2$ is not a multiple of 3. It follows that $n + 1$ is not a multiple of 3. (Why?) In this case, $n + 1$ is not a multiple of 3 and f_{n+1} is odd.

In either case, $n + 1 \in S$. Therefore, by the second principle of mathematical induction, $S = \mathbb{N}$.

b. Let $S = \{ n \in \mathbb{N}: \gcd(f_n, f_{n+1}) = 1\}$. Clearly $1 \in S$. Let $n \in S$ and suppose that $n + 1 \notin S$. Then there is a natural number $k \neq 1$ such that k divides both f_{n+1} and f_{n+2}. Since $f_{n+2} = f_{n+1} + f_n$, it follows (from Exercise 27 of Chapter 1) that k divides f_n. Since $n \in S$ and k divides both f_n and f_{n+1}, we have reached a contradiction. Therefore $n + 1 \in S$ and, by the principle of mathematical induction, $S = \mathbb{N}$.

c. Let $S = \{n \in \mathbb{N}: |f_{n+1}{}^2 - f_n f_{n+2}| = 1\}$. Since $|f_2^2 - f_1 f_3| = |1 - (1)(2)| = 1$, $1 \in S$.

Suppose $n \in S$. Then $|f_{n+1}{}^2 - f_n f_{n+2}| = 1$, so

$$\begin{aligned} |f_{n+2}{}^2 - f_{n+1}f_{n+3}| &= |f_{n+2}{}^2 - f_{n+1}f_{n+2} - f_{n+1}{}^2| \\ &= |(f_{n+1} + f_n)f_{n+2} - f_{n+1}{}^2 - f_{n+1}f_{n+2}| \\ &= |f_{n+1}f_{n+2} + f_n f_{n+2} - f_{n+1}{}^2 - f_{n+1}f_{n+2}| \\ &= |f_n f_{n+2} - f_{n+1}{}^2| = 1 \end{aligned}$$

Therefore $n + 1 \in S$ and, by the principle of mathematical induction, $S = \mathbb{N}$. □

We have used the principle of recursive definition to define sequences, but it is possible to use this same principle to define sets. Suppose that we are given a set T and we wish to describe a subset S of T recursively.

Definition

Let T be a set and let S be a subset of T. We say that S is **defined recursively** provided that three conditions are met:

a. One or more members of S are given. (**Basis**)

b. Rules are given that allow us to determine another member of S from those members of S that have already been given or determined. (**Inductive clause**)

c. No member of T belongs to S unless it can be obtained from a member of S given by condition (a) or obtained in condition (b) in a finite number of steps. (**Extremal condition**)

Before verifying that defining sets recursively makes sense, we offer a simple and admittedly frivolous example of a set that is defined recursively:

Example 18

McDonalds sells Chicken McNuggets in boxes of 6, 9, and 20 McNuggets. What is the largest number n of Chicken McNuggets that *cannot* be bought without buying $n + 1$ McNuggets?

Analysis

The subset S of $\mathbb{N}$ to which n belongs provided that it is possible to buy exactly n McNuggets is defined recursively. The basis is the set $\{6,9,20\}$ and the rule of the inductive clause is that, whenever a and b are known to belong to S, then $a + b$ also belongs to S. This observation and simple arithmetic are all that is needed to answer the question (see Exercise 65). ■

Theorem 4.5

> Let T be a set. Suppose that we are given a recursive definition with basis, recursive clause, and extremal condition. Then there is one and only one subset S of T that satisfies the recursive definition.

Proof

Let $S(0)$ be the set of members of T given by the basis. Let $S: \mathbb{N} \to \mathbb{N}$ be the *sequence* defined recursively by $S(1) = \{x \in T: x$ can be obtained from $S(0)$ using the rules given by the recursive clause$\}$, and, for each $n \in \mathbb{N}$, let $S(n + 1) = \{x \in T: x$ can be obtained from $S(n)$ using the rules given by the recursive clause$\}$.

Let $S = \cup\{S(n): n \in \mathbb{N} \cup \{0\}\}$ and let $x \in S$. Then there is an $n \in \mathbb{N} \cup \{0\}$ such that $x \in S(n)$. If $n = 0$, x belongs to the set of members of T given in the basis. If $n = 1$, x is determined by the recursive clause from members that have already been given in the basis. If $n > 1$, then $n - 1 \in \mathbb{N}$ and so x is determined by the recursive clause from members of $S(n - 1)$, which have already been determined. Thus S is defined recursively by the given recursive definition.

Suppose that S$'$ is a subset of T that is defined by the given recursive definition. The preceding argument shows that $S \subseteq S'$, and since S' must satisfy the extremal condition, $S' \subseteq S$. □

Now we give a more serious example of a recursively defined set.

Example 19

The set P of all polynomial functions mapping $\mathbb{R}$ into $\mathbb{R}$ is a recursively defined set. As a basis we take the collection of all constant functions together with the identity function $x: \mathbb{R} \to \mathbb{R}$ defined by $x(r) = r$ for each $r \in \mathbb{R}$. The rules of the recursive clause are that if f and g belong to P then so do $f + g: \mathbb{R} \to \mathbb{R}$ and $fg: \mathbb{R} \to \mathbb{R}$. Recall that $f + g$ is defined by $(f + g)(r) = f(r) + g(r)$ for each $r \in \mathbb{R}$ and fg is defined by $(fg)(r) = f(r)g(r)$ for each $r \in \mathbb{R}$.

We illustrate this recursive definition by establishing that $(x^2 + \pi x) + 5$ belongs to P. Since π and x are basis elements, by the recursive clause both x^2 and πx are members of P. Since the recursive clause also guarantees that the sum of two members of P belongs to P, then $(x^2 + \pi x) \in P$. Finally, the basis member 5 belongs to P and so $(x^2 + \pi x) + 5$ belongs to P. Note that in this illustration both π and 5 are (constant) functions rather than numbers. We are abusing notation somewhat by using the same symbol to denote both a function and a number, but this abuse is common. And while we are at it, let us agree to write $x^2 + \pi x + 5$ even though technically we can add only two functions at a time. ■

We now introduce a concept that is designed to allow a general, but precise, mathematical treatment of languages. Of course, computer languages are of particular interest.

Definition

An **alphabet** is a nonempty finite set the members of which are called **symbols** or **characters**. We use the sign Σ to denote an alphabet. (The sign Σ should not be confused with Σ notation).

The twenty-six lowercase letters in the English alphabet constitute an alphabet according to the preceding definition. The sets {#,@,$,%} and {1,3,6} also constitute alphabets. There is a restriction on the members of Σ, which we discuss shortly.

Definition

Let Σ be an alphabet. A **nonempty word over Σ** is a finite sequence with domain I_n (for some $n \in \mathbb{N}$) and codomain Σ. A nonempty word with domain I_n is said to **have length n**. We also allow the function with domain the empty set and codomain Σ as a word. This word is called the **empty word** and is denoted by ε. We define the length of ε to be 0. Although it is true that ε is the function $\varnothing: \varnothing \to \Sigma$, ε is *not* the empty set. We denote the set of all words over Σ by Σ^*. Any nonempty subset of Σ^* is a **language over Σ**.

Now that we have introduced languages, we can give additional examples of functions. For instance, if Σ is an alphabet and L is a language over Σ, then for each $w \in L$ we can define $f(w)$ to be the length of w. Thus we obtain a function $f: L \to \mathbb{Z}$.

Example 20

Let $\Sigma = \{0,1,2,3,4,5,6,7,8,9\}$. The language consisting of all nonempty words $f: I_n \to \Sigma$ $(n \in \mathbb{N})$ for which $f(1) \neq 0$ may be interpreted as the set of all natural numbers. The word $w: I_6 \to \Sigma$ defined by $w(1) = 1$, $w(2) = 4$, $w(3) = 2$, $w(4) = 8$, $w(5) = 5$, and $w(6) = 7$ may be thought of as 142857. ■

Even the word $w: I_6 \to \Sigma$ (from Example 20) is rather tedious to write formally, and this word after all is only of length 6. For this reason, we agree to write nonempty words as we normally do. For example, we write 142857 for the word defined formally in Example 20. There is a small cost for this privilege, a price well worth paying. Suppose that the alphabet Σ of Example 20 contained 17 as well as 0,1,2,3,4,5,6,7,8, and 9. Then when we saw the word 517 we would not know if it was the word $x: I_3 \to \Sigma$ defined by $x(1) = 5$, $x(2) = 1$, and $x(3) = 7$ or the word $y: I_2 \to \Sigma$ defined by $y(1) = 5$ and $y(2) = 17$. To avoid this ambiguity, we do not allow an alphabet Σ to contain symbols that are themselves formed from symbols in Σ and that *begin* with another member of Σ. Thus, for example, we do not permit an alphabet to be the set $\{a,b,bc,c\}$, since bc begins with b, but the alphabet $\Sigma = \{a,b,cad,d,r\}$ is permitted because each a preceded by c must be the member cad of Σ and each a not immediately preceded by c must be the member a of Σ. Thus *abracadabra* is a word of length 9 over Σ, even though it is a word of length 11 over the usual English alphabet. It is this restriction that we mentioned after our definition of an alphabet.

Example 21

Let $\Sigma = \{a,b,c\}$, where we assume that a, b, and c are distinct, and let $W = \{w \in \Sigma^*\colon \text{length}(w) = 2\}$. Then the language W is $\{aa,ab,ac,ba,bb,bc,ca,cb,cc\}$. ■

Let Σ be an alphabet. It is possible (and sometimes useful) to define the set of all nonempty words over Σ recursively. We first agree that, if $a \in \Sigma$, then a also denotes the obvious one-letter word with domain I_1 and codomain Σ.

Definition

Let Σ be an alphabet. The set Σ^+ **of all nonempty words over** Σ is defined as follows:

a. If $a \in \Sigma$, then $a \in \Sigma^+$. (**Basis**)

b. Let w be a word of length n belonging to Σ^+ and let $a \in \Sigma$. Then aw: $I_{n+1} \to \Sigma$ defined by $aw(1) = a$ and, for $2 \leqslant k \leqslant n + 1$, $aw(k) = w(k - 1)$ is also a word in Σ^+. (**Inductive clause**)

c. No object belongs to Σ^+ unless it can be obtained from conditions (a) or (b) in a finite number of steps. (**Extremal condition**)

Exercises 4.4

39. Prove by induction that $(1 + 2 + \cdots + n)^2 = 1^3 + 2^3 + \cdots + n^3$ for each natural number n.

40. Define a sequence recursively by $a_1 = 3$ and $a_{n+1} = a_n + 2$ for each natural number n. Find a_2, a_3, a_4, and a_5.

41. Define a sequence recursively by $a_1 = 2$ and $a_{n+1} = (2)^{a_n}$ for each natural number n. Find a_2, a_3, a_4, and a_5.

42. Define a sequence recursively by $a_1 = a_2 = 1$ and $a_n = (a_{n-1})^2 + (a_{n-2})^3$ for each natural number $n \geqslant 3$. Find a_3, a_4, and a_5.

43. Let $b_1 = 1$ and, for each natural number $n > 1$, let $b_n = 2/b_{n-1}$. List the first eight terms of the resulting sequence.

44. A sequence f is defined recursively by the initial condition $f(1) = 1$ and the recursion relation

$$f(n + 1) = \frac{[f(n) - 8][f(n) - 4][f(n) - 2][f(n) - 1][f(n)][\pi - 32]}{96(15^3 - 15)} + 2^{n-1}$$

Find the first six values of this sequence.

45. Let f be a function such that $f(n)$ is the sum of the first n natural numbers. Give a recursive definition of $f(n)$.

46. a. Give a formula for the nth term of a sequence whose first five terms are 1, 3, 9, 27, and 81.

    b. Give a recursive definition of the sequence in (a).

47. Show that the set S defined by $2 \in S$ and $x + y \in S$ whenever $x \in S$ and $y \in S$ is the set of even natural numbers.

48. Give a recursive definition of the set of natural numbers that are multiples of 7.

49. Let $f_1, f_2, f_3, \ldots$ be the Fibonacci sequence. Prove by induction that, for each natural number $n > 1$, $f_1 + f_2 + \cdots + f_{n-1} = f_{n+1} - 1$.

50. Let $f_1, f_2, f_3, \ldots$ be the Fibonacci sequence. Prove by induction that, for each natural number n,

a. $f_1 + f_3 + f_5 + \cdots + f_{2n-1} = f_{2n}$

b. $f_2 + f_4 + f_6 + \cdots + f_{2n} = f_{2n+1} - 1$

51. Prove that every natural number can be written as a sum of distinct Fibonacci numbers.

52. Prove that, for each $n \in \mathbb{N}$, the nth Fibonacci number $f_n \leqslant 2^n$.

53. Define a sequence recursively by $F_1 = 1$, $F_2 = 1$, and $F_{n+2} = F_{n+1} + F_n + F_{n+1}F_n$ for each $n \in \mathbb{N}$.

a. Find the first five values of F.

b. Guess how to define F_n in terms of the nth Fibonacci number, and prove by the second principle of mathematical induction that your guess is correct.

54. Define a sequence recursively by $a_1 = a_2 = 1$ and $a_n = 2a_{n-1} + a_{n-2}$ for each natural number $n > 2$.

a. Prove that, for each natural number n, a_n is odd.

b. Prove that, for each natural number $n > 4$, $a_n < 6a_{n-2}$.

55. Define a sequence recursively by $a_1 = 1$, $a_2 = 2$, and $a_n = 2a_{n-1} - a_{n-2}$ for each natural number $n > 2$.

a. Calculate a_n for $n = 3,4,5$.

b. Guess the general formula for a_n.

c. Prove the guess in (**b**).

56. Define a sequence recursively by $a_1 = 1$, $a_2 = 2$, and $a_n = a_{n-1} + 2a_{n-2}$ for each $n > 2$.

a. Calculate a_n for $n = 3,4,5,6$.

b. Guess the general formula for a_n.

c. Prove the guess in (**b**).

57. Define a sequence recursively by $a_1 = a_2 = 1$ and $a_n = 3a_{n-1} - 2a_{n-2}$ for $n > 2$.

a. Calculate a_n for $n = 3,4,5$.

b. Guess the general formula for a_n.

c. Prove the guess in (**b**).

58. Define a sequence recursively by $a_1 = a_2 = a_3 = 1$ and $a_n = a_{n-1} + a_{n-2} + a_{n-3}$ for each $n > 3$. Prove that $a_n \leqslant 2^{n-2}$ for each natural number $n > 1$.

59. Define a sequence recursively by $a_1 = a_2 = 1$ and $a_n = 2a_{n-1} + 3a_{n-2}$ for each $n > 2$.

a. Prove that $a_n > 3^{n-2}$ for $n > 2$.

b. Prove that $a_n < 2 \cdot 3^{n-2}$ for $n > 2$.

60. Define a sequence recursively by $a_1 = 1$, $a_2 = 2$, and $a_n = 2a_{n-1} + a_{n-2}$ for each $n > 2$. Prove that $a_n \leqslant (5/2)^{n-1}$ for each natural number n.

61. Let $\Sigma = \{a, ba, cba, dcba, edcba\}$.

a. Give an example of a word of length 3.

b. Give an example of a word of length 5.

c. What is the length of *edcbabacbaaba*?

62. Let Σ be an alphabet, let $a \in \Sigma$, and let $w\colon I_n \to \Sigma$ be a word over Σ. Define the word $wa\colon I_{n+1} \to \Sigma$.

63. Let $\Sigma = \{0,1,2,3,4,5,6,7,8,9\}$. Define recursively the language over Σ that would be interpreted by any reasonable person as the set of all even natural numbers.

64. Let $\Sigma = \{0,1,2,3,4,5,6,7,8,9\}$. Define recursively the language over Σ that would be interpreted by any reasonable person as the set of all natural numbers that are divisible by 5.

65. What is the largest number n of Chicken McNuggets that *cannot* be bought without buying $n + 1$ McNuggets (See Example 18)?

Chapter 4 Review Exercises

66. Find $f(5)$ and $f(7)$ if $f(1) = 1$, $f(2) = 3$, and f is defined recursively by
    a. $f(n + 2) = 2f(n + 1) - 3f(n)$ **b.** $f(n + 2) = f(n + 1)/f(n)$

67. Let r be a rate of interest. (For example, if the interest rate is 8%, then $r = .08$.) Define a sequence f by the initial condition $f(1) = 1$ and the recursion relation $f(n + 1) = f(n)(1 + r)$. Find an explicit formula for f and prove that your formula is correct.

68. Prove that, for each natural number n,
    $$2 + 5 + 8 + \cdots + (3n - 1) = n(3n + 1)/2.$$

69. Prove that, for each natural number n,
    $$\frac{1}{1 \cdot 5} + \frac{1}{5 \cdot 9} + \frac{1}{9 \cdot 13} + \cdots + \frac{1}{(4n - 3)(4n + 1)} = \frac{n}{4n + 1}$$

70. Prove that, for each natural number n,
    $$3 + 3^2 + 3^3 + \cdots + 3^n = \frac{3^{n+1} - 3}{2}$$

71. Prove that, for each natural number n, $5^n - 2^n$ is divisible by 3.

72. Prove that, for each natural number $n \geqslant 5$, $(n + 1)! > 2^{n+3}$.

73. Each area of a dart board has the number 4 or the number 5. What is the largest score you cannot get using as many darts as you want?

*74. Each area of a dart board has the natural number m or the natural number n, where m and n have no common divisors greater than 1. In terms of m and n, determine the largest score you cannot get using as many darts as you want.

75. A frog falls into a 100-foot-deep well. Although the frog can jump 3 feet, the sides of the well are slippery and so the frog slides back 2 feet with each jump. How many jumps does it take the frog to escape the well?

76. Let f be the increasing sequence of natural numbers that have no other digits than 0 and 1 and do not have two consecutive 0's or two consecutive 1's. Thus $f(1) = 1$, $f(2) = 10$, $f(3) = 101$, $f(4) = 1010$, and $f(5) = 10101$. Give a recursive definition of this sequence. (*Hint:* It may be convenient to state the recursion relation by considering the case that n is odd separately from the case that n is even.)

77. Prove by induction that your answer to Exercise 76 is correct. Did you use the second principle of mathematical induction in your proof?

78. For each natural number n, let $f(n)$ denote the number of subsets of $\{1,2,3, \ldots ,n\}$ that do not contain two consecutive numbers.

a. Find a pattern for $f(n)$. (Do not forget to count $\varnothing$)

b. Use the second principle of mathematical induction to prove that your pattern is correct.

79. The offense of a professional football team can score a field goal (3 points), a touchdown (6 points), or a touchdown and point-after (7 points). (We assume that all safeties are scored by the defense.) Define recursively the set of possible points that the offense can score. Give the basis, recursive clause, and extremal condition. Is there a largest score that an offense cannot score? Explain.

80. Let a and b be real numbers such that $a \neq b$. Prove that, for each natural number n, $a^n - b^n$ is divisible by $a - b$. (*Hint:* See Example 9.)

81. Prove by induction that, for each natural number $n \geqslant 3$, $n^2 > 2n + 1$.

Chapter

5

ALGORITHMS

We begin our study of algorithms with a trip to the dictionary. *Algorithm* is a form of the Medieval Latin word "algorismus," which originated in the thirteenth century in honor of the ninth-century Persian mathematician Muhammad ibn-Mūsa al-Khwārizmi. Traditionally, algorithms were defined as procedures for calculating by means of nine figures and zero as opposed to calculating with the abacus. More recently, computer science has been defined as the study of algorithms.

5.1

THE BASIC CONCEPT

Two types of information are needed to perform our usual arithmetic calculations with whole numbers: addition and multiplication tables for one-digit numbers, and a collection of rules that describe how to add, subtract, multiply, and divide nonnegative integers with any number of digits, as in Figure 5.1.

Figure 5.1
Rules for adding 37 and 48

$$\begin{array}{r} 37 \\ +48 \\ \hline \end{array} \qquad \begin{array}{r} 1 \\ 37 \\ +48 \\ \hline 85 \end{array}$$

1. Add 7 and 8 to get 15.
2. Write the 5 and carry the 1.
3. Add 3 and 4 and the 1 you carried to get 8 and write it down.

One common way of looking at a function is to think of it as a black box with an input slot and an output slot. You are allowed to select various objects from a set, called the domain of the function, to put into the black box, and various objects, from the codomain of the function, pop out of the black box (see Figure 5.2). The one requirement of the black box is that it not be allowed to produce two different outputs from the same input.

Figure 5.2
Black box representation of a function

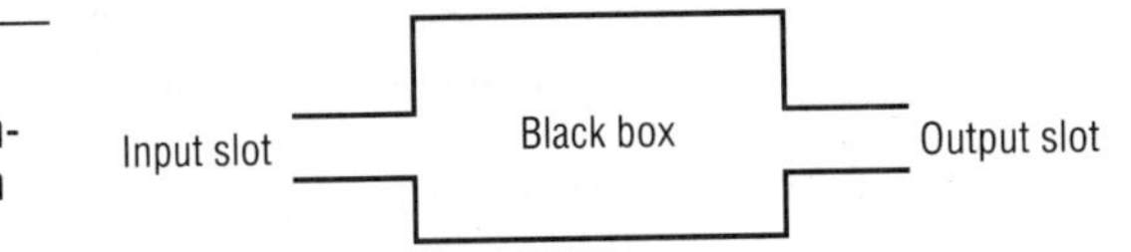

Now suppose that we put a 2 in the input slot of the black box of Figure 5.2 and observe the number 7 pop out of the output slot. Since we know that this particular black box represents a function, we are guaranteed that each time we put 2 in the input slot, 7 emerges. We have no idea what output the number 4 will produce, but we can find out—we just enter 4 and look at the number that comes out.

Although an algorithm always generates a function, it is more than just a function. An algorithm is really the plans or description of the interior of a black box that represents a function. Look at the interiors of the two black boxes A and B in Figure 5.3.

It is easy to verify that these black boxes produce identical outputs from the same input. If 2 is put in box A, then 2 times 2 is 4, and when 1 is subtracted the result is 3, which becomes the output. If 2 is put in box B, then 2 squared is 4, times 2 is 8, minus 2 is 6, divided by 2 is 3,

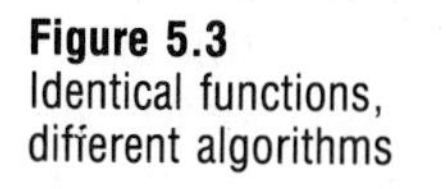

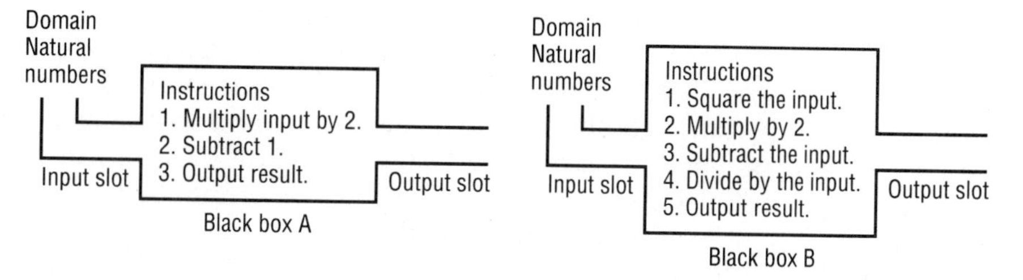

Figure 5.3 Identical functions, different algorithms

and once again the output is 3. To prove that boxes A and B produce identical outputs from identical inputs, we need only repeat the preceding argument using an arbitrary natural number x instead of the specified number 2. We leave this simple task as an exercise (see Exercise 4).

Although we choose to remain informal in this book, it is possible to give a mathematically sound definition of an algorithm (see Donald Knuth, 1973, *The Art of Computer Programming*, Vol. 1, pp. 7–8).

Definition

An **algorithm** is a linear list of instructions, together with a set of allowable inputs having the following two properties: (a) any input produces an output in a finite amount of time; and (b) the algorithm behaves like a function, in the sense that from a single input the algorithm cannot produce two different outputs.

Knuth insists that an algorithm has five important features, which he labels *finiteness, definiteness, input, output,* and *effectiveness.* Here we consider only the first four of these properties, because effectiveness involves concepts beyond our scope.

1. Finiteness Each input must result in an output in a finite number of steps. For example, if we chose $x_1 = 1$ and $x_n = x_{n-1}/2$ for $n > 1$, we would have an input set $\{1\}$ and an ordered list of instructions. But final output would not be obtained in a finite number of steps. So even if some computers and pocket calculators might eventually give up and start taking $x_n = 0$ for all $n \geqslant N$, for some large number N, these instructions would not constitute an algorithm.

It is possible to write algorithms that are completely useless even though technically they meet the requirement of finiteness. For example, the instruction "add 1 if there is a winning strategy for the white pieces in the game of chess and add 2 if there is no such strategy" meets the requirement of finiteness, because there are a finite number of different possible games of chess. In practice, however, no one knows how to follow this instruction, and any algorithm containing it is worthless.

2. Definiteness This property is guaranteed by the insistence that an algorithm behaves like a function. Instructions found in cookbooks such as "season to taste" or "add a dash of salt" have no place in an algorithm. Likewise, the outcome cannot be left to chance. For example, if we input a number, then the output cannot be determined by the flip of a coin.

3. Input Each algorithm includes a set of allowable inputs. It is not required that every allowable input be used, but we must agree at the outset what sort of objects are allowed. Imagine that you are asked to find the roots of the equation $x^4 = 16$. If you are told at the outset that the only allowable x's are natural numbers, you will find only the number 2. If x's are allowed to be integers, you will find -2 and 2. If they are allowed to be complex numbers, you will find -2, 2, $2i$, and $-2i$ (where $i^2 = -1$). Just as we cannot solve the equation $x^4 = 16$ without knowing what sort of x's are allowed, we cannot have an algorithm unless we are told in the algorithm what sort of inputs are allowed.

4. Output Algorithms must do something. It has been said that if all economists were laid end to end they would not reach a conclusion. But an algorithm always reaches a conclusion, no matter what allowable input is given, and any such conclusion is said to be an output.

An algorithm can be described in many ways. One way is to use an expression to replace the set of instructions. The algorithm for black box A in Figure 5.3 could be written simply as $2x - 1$, and the algorithm for black box B could be written as $(2x^2 - x)/x$. Other ways of describing algorithms include look-up tables, computer programs, pseudocode descriptions, and flowchart descriptions. We review these in the next section. Another way to describe an algorithm, one often overlooked, is to use plain English. As long as the description is well defined, can be executed in a finite amount of time, and represents a function, we can assume that we have an algorithm.

Exercises 5.1

1. Write an addition algorithm for two two-digit numbers in our decimal number system.
2. Write a description of a black box with domain the natural numbers that has the indicated outputs for the given inputs. (It is certainly acceptable to find an expression that solves the problem, but there are other ways as well.)

a.

Input	1	2	3
Output	2	4	6

b.

Input	2	4	6	9
Output	4	10	16	25

3. Suppose that we have a black box with domain the natural numbers that has

the indicated outputs for the given inputs. Explain why you cannot be sure that the output is 9 when the input is 9.

Input	1	2	3	4	5	6	7	8
Output	1	2	3	4	5	6	7	8

4. Prove that the algorithms in Figure 5.3 represent the same function.

5. Explain why the following list of instructions is not an algorithm.

1. Get as input a natural number x.
2. If it is raining, multiply x by 2.
3. If it is not raining, multiply x by 3.
4. Subtract 5.
5. Output x.

6. Decide whether each of the following lists of instructions is an algorithm. Justify your decision.

a. 1. Get as input a natural number x.
2. If today is Monday, Wednesday, or Friday, then add 2 to x.
3. If today is Sunday, Tuesday, Thursday, or Saturday, then add 3 to x.
4. Output x.

b. 1. Get as input a day of the week.
2. If today is Monday, Wednesday, or Friday, then let $x = 1$.
3. If today is Sunday, Tuesday, Thursday, or Saturday, then let $x = 2$.
4. Output x.

7. Assume that a certain discrete mathematics class meets on Mondays, Wednesdays, and Fridays.

a. Write an algorithm whose input set is the days of the week and whose output is the next day of the week (after the input) on which the class meets.

b. Write an algorithm whose input set is the days of the week and whose output is the number of days (after the input) before the class meets again.

5.2 DESCRIBING ALGORITHMS

The various ways of describing algorithms can be summarized as follows:

1. a list of instructions
2. an expression
3. a look-up table with instructions
4. a computer program
5. a pseudocode
6. a flowchart

We reviewed lists of instructions and expressions for algorithms in Section 5.1, so we continue our discussion here with look-up tables.

The look-up table is a seemingly simple way of describing an algorithm, but although it represents a function it is not considered an algorithm. An algorithm must not only give the table, it must also include a set of instructions for using the table. An example of a look-up table algorithm with instructions is given in Figure 5.4.

Figure 5.4 Look-up table algorithm

Find $\log_2 x$ when x is a positive integer from 1 to 7

Input	Output
1	0
2	1
3	1.585
4	2
5	2.322
6	2.585
7	2.807

Instructions

1. Get the input x.
2. If $x = 4$, then print output 2 and stop.
3. If $x > 4$, then go to step 7.
4. If $x = 2$, then print output 1 and stop.
5. If $x > 2$, then print output 1.585 and stop.
6. Print output 0 and stop.
7. If $x = 6$, then print output 2.585 and stop.
8. If $x > 6$, then print output 2.807 and stop.
9. Print output 2.322 and stop.

The instructions in Figure 5.4 constitute the binary search. With so few numbers, even a linear search would be adequate. However, as look-up tables grow larger, the look-up procedure becomes more complicated and cumbersome. For instance, the telephone directory for a city is a common look-up table where a typical input is the name of a person in that city and the corresponding output is that person's telephone number. In this example, there are a multitude of choices for instructions. Do you start at the beginning of the telephone book and look at the names in order until you get to the correct name (linear search)? Do you start at the middle of the book and discard half of the book each time (binary search)? Do you open the book randomly hoping to get lucky (chaotic search)? An algorithm for getting a person's telephone number out of a telephone book must include a systematic way of using the look-up table.

Another way of describing an algorithm is a computer program. For example, let us look at the program in Figure 5.5. The first four lines of the program get the two fractions that we wish to add. Next we cross multiply and add to get the numerator of the result and multiply to get the denominator. Lines 70–90 print the result and line 100 tells the program to end.

A fifth way of describing an algorithm is to use pseudocode. An algorithm written in pseudocode has several advantages: it is in English and thus can be understood by anyone, it can be easily translated to any programming language, and it is in a particularly good form for proving that a given algorithm works. An example of the use of pseudocode is illustrated in Figure 5.6.

Figure 5.5
Computer program algorithm

```
 10  Input "numerator one" ;A
 20  Input "denominator one" ;B
 30  Input "numerator two" ;C
 40  Input "denominator two" ;D
 50  Let X = A * D + B * C
 60  Let Y = B * D
 70  Print A;" ";C;" ";X
 80  Print "– + – = –"
 90  Print B;" ";D;" ";Y
100  End
```

Figure 5.6
Computer program algorithm

Find the larger of two numbers

1. Input two numbers x and y.
2. If $x = y$, then print x and y are the same size and stop.
3. If $x > y$, then print x is the larger of x and y and stop.
4. Print y is the larger of x and y and stop.

Figure 5.7
Find the square root of a nonnegative number

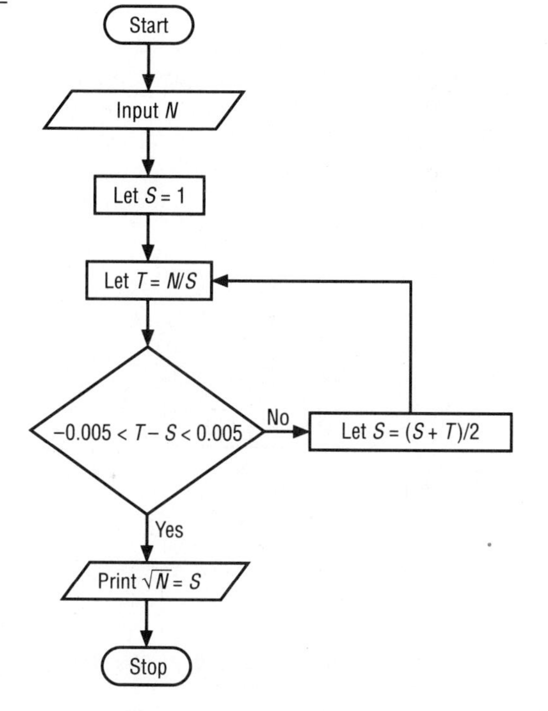

The proof that the algorithm in Figure 5.6 produces the desired result is easy to see. Input two numbers x and y. Either $x = y$, $x > y$, or $x < y$. If $x = y$, the "if" part of step 2 is true and the algorithm prints that x and y are the same size and stops. If $x > y$, the "if" part of step 2 is false and the algorithm goes to step 3. Since the "if" part of step 3 is true, the algorithm prints that x is the larger of x and y and stops. If $x < y$, the "if" part of steps 2 and 3 are false and the algorithm goes to step 4 and prints that y is the larger of x and y and stops. Since we have exhausted all the possibilities, we have shown that the algorithm produces the desired result.

A flowchart is the last method of describing an algorithm that we consider. Such a description lends itself naturally to counting the number of operations needed to complete a task by counting the number of times the algorithm visits each box and counting the number of operations performed in each box. The algorithm in Figure 5.7 on page 127 can be used to find the square root of a nonnegative number to two decimal places.

Example 1

Using the algorithm shown in Figure 5.7, find $\sqrt{3}$ to two decimal places.

Analysis

We get the value of N that is 3 and initialize S to 1, which brings us to the loop of statements shown in Figure 5.8. We proceed to traverse this loop until $-0.005 < T - S < 0.005$. The results are summarized in Table 5.1, where "number of operations" shows the number of comparisons and the number of arithmetic operations.

Figure 5.8

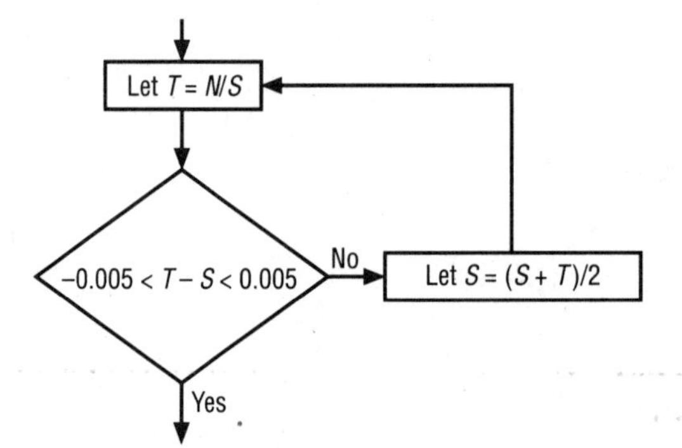

Table 5.1

Pass	S	N	$T = N/S$	$T - S$	Yes/No	$(S + T)/2$	Number of Operations
1	1	3	3	2	No	2	6
2	2	3	1.5	−0.5	No	1.75	6
3	1.75	3	1.7143	−0.0357	No	1.73215	6
4	1.73215	3	1.73200	−0.0001	Yes		4

There is one arithmetic operation needed to find T, one arithmetic operation needed to find $T - S$, two comparisons needed to determine if $T - S$ is between -0.005 and 0.005, and two arithmetic operations needed to find $(S + T)/2$. Thus this algorithm requires a total of twenty-two operations to find $\sqrt{3}$ to two decimal places. Notice that this algorithm is fast; it takes only three repetitions for completion. You do, however, have to start the fourth pass and perform the comparisons before you realize that three passes suffice. To use the algorithm in Figure 5.7 to find square roots more precisely, we have only to change the comparison condition, which bounds $T - S$. Such an algorithm, but with a much more restrictive comparison condition, is frequently used to design calculators to find the square roots of numbers. ■

Exercises 5.2

8. Write a nontechnical description that explains how to use a telephone directory.
9. Write a pseudocode description of an algorithm that subtracts two fractions.
10. Write a pseudocode description of an algorithm that finds the largest of three numbers.
11. Using the algorithm of Figure 5.7, find $\sqrt{5}$ to the nearest two decimal places.
12. Using the algorithm of Figure 5.7, find $\sqrt{10}$ to the nearest four decimal places.

5.3 BIG-O NOTATION

It is not enough to know that a computer program runs; we must also know that it runs efficiently. Imagine, for example, that we have four programs that sort a list of n numbers and that we have somehow determined the functions that give the number of steps required by each program to sort n numbers:

$$f_1(n) = n, \quad f_2(n) = n^2, \quad f_3(n) = 2^n \quad f_4(n) = n!$$

As long as we are sorting a short list, it does not matter much which program we use. For $n = 4$, $f_1(n) = 4$, $f_2(n) = f_3(n) = 16$, and $f_4(n) = 24$. But the number 4 does not provide a realistic case. Most people would sort such a short list by hand. For $n = 1000$, the third program takes 2^{1000} steps, a number that exceeds the number of atoms in our galaxy. The fourth program requires even more steps, but there is not much point

in brooding about the difference. In this section, we introduce a way of measuring how fast a function grows as the number n grows, and concurrently we introduce a way of comparing functions with respect to their growth.

The phrase "y varies directly as x" means that there is a real number k such that $y = kx$. The functions $f,g: \mathbb{R} \to \mathbb{R}$ defined by $f(x) = 2x$ and $g(x) = 3x$ are different ($f(2) = 4$ and $g(2) = 6$), and yet f and g have an important property in common. As the element in the domain of f and g varies, f and g vary in the same manner: if x doubles, $f(x)$ and $g(x)$ double; if x triples, $f(x)$ and $g(x)$ triple, and so on.

Imagine that we have three programs that sort a list of n numbers and that we have determined functions that give the number of operations required by each program to sort n numbers:

$$f(n) = 2n, \qquad g(n) = 3n, \qquad h(n) = n^2$$

when $n = 2$, $f(n) = 4$, $g(n) = 6$, and $h(n) = 4$. It would appear that the programs represented by f and h do a better job than g does. But in reality f and g vary directly as n, while h varies directly as n^2. For example, when $n = 4$, $f(n) = 8$, $g(n) = 12$, and $h(n) = 16$. When $n = 6$, $f(n) = 12$, $g(n) = 18$, and $h(n) = 36$. Somehow we are much happier with the behavior of f and g than we are with the behavior of h. From an operation-counting point of view, f and g are essentially the same. The following definition of the big-O notation is a generalization of the concept of variation.

Definition

Let S be a set of real-valued functions all of which have the same domain D, where $D = \mathbb{N}$, $\mathbb{Z}$, or $\mathbb{R}$. Let $f,g \in S$. We say that $f = O(g)$ (read "f is big oh of g") if there are positive numbers c and k such that $|f(x)| \leq c|g(x)|$ for all $x \in D$ for which $x > k$.

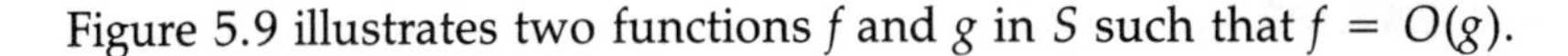
Figure 5.9 illustrates two functions f and g in S such that $f = O(g)$.

Figure 5.9

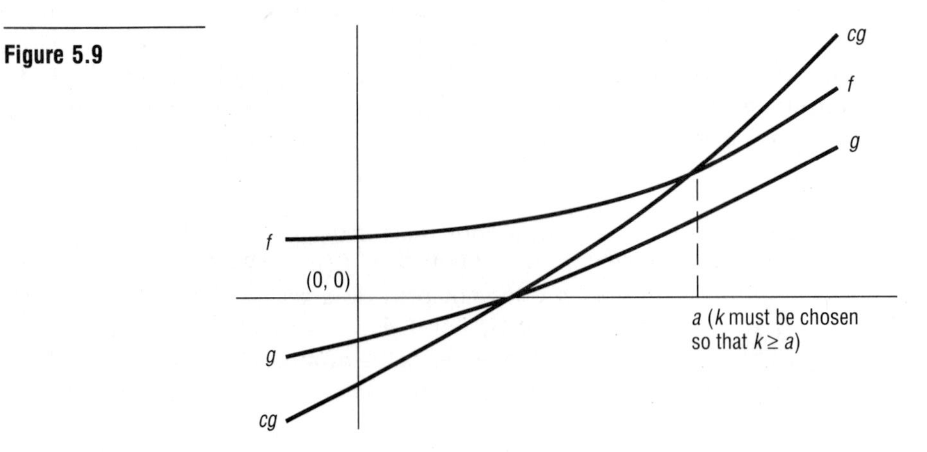

Although in this text we consider big-O relationships only between functions whose ranges contain no negative numbers, we have stated the definition of big-O so that it will apply to other real-valued functions, and we establish the basic properties of big-O using our general definition.

Example 2

Let S be the set of all real-valued functions with domain $\mathbb{R}$ and let f and g be the members of S defined by $f(x) = x^3 + 7x^2 + 11x$ and $g(x) = x^3$. Then $f = O(g)$.

Analysis

Let $c = 19$. Then for $x > 1$,

$$|f(x)| = x^3 + 7x^2 + 11x < x^3 + 7x^3 + 11x^3 = 19x^3 = c|g(x)|$$

Thus $f = O(g)$. ■

Example 3

Let $f,g\colon \mathbb{N} \to \mathbb{N}$ be defined by $f(x) = x^3$ and $g(x) = x!$. Then $f = O(g)$ but $g \neq O(f)$.

Analysis

A quick look at small values of x might lead one to think that $f(x) > g(x)$ for all $x \in \mathbb{N}$. For example,

$$f(1) = 1, \quad f(2) = 8, \quad f(3) = 27, \quad f(4) = 64, \quad f(5) = 125$$
$$g(1) = 1, \quad g(2) = 2, \quad g(3) = 6, \quad g(4) = 24, \quad g(5) = 120$$

But $f(6) = 216$ and $g(6) = 720$. Suppose $c = 1$ and $x > 6$. Then

$$2x - 2 = x + x - 2 \geqslant x + 6 - 2 \geqslant x + 4 > x \quad \text{and}$$
$$3x - 6 = x + 2x - 6 \geqslant x + 2(6) - 6 = x + 6 > x$$

Now $g(x) = x! = x(x - 1)(x - 2) \cdots 3 \cdot 2 \cdot 1$ is the product of at least six numbers. Thus

$$g(x) \geqslant x(2x - 2)(3x - 6) \geqslant x \cdot x \cdot x = x^3 = f(x)$$

Hence $|f(x)| \leqslant c|g(x)|$ and $f \in O(g)$.

To show that $g \neq O(f)$, we need a theorem about factorials—namely, that if $n \in \mathbb{N}$ and $n \geqslant 11$ then $n! > n^4$. The proof of this theorem uses mathematical induction and is left as Exercise 13. The proof that $g \neq O(f)$ is by contradiction. Suppose that $g = O(f)$. Then there are positive numbers c and k such that, if $n \in \mathbb{N}$ and $n \geqslant k$, then $|g(n)| \leqslant c|f(n)|$. Let w be a natural number such that $w \geqslant c$, $w \geqslant k$, and $w \geqslant 11$. Then $w! = |g(w)| \leqslant c|f(w)| = cw^3 \leqslant w^4$. But since $w \geqslant 11$, $w^4 < w!$. This is a contradiction, and hence $g \neq O(f)$. ■

We conclude this section with two theorems that we use throughout the remainder of this chapter.

Theorem 5.1

Let $f(x) = a_nx^n + a_{n-1}x^{n-1} + \cdots + a_1x + a_0$ be a polynomial function and let $g(x) = x^n$. Then $f = O(g)$.

Proof

Let $c = |a_n| + |a_{n-1}| + \cdots + |a_1| + |a_0|$. Let x be a real number that is greater than 1. Then,

$$\begin{aligned}|f(x)| &= |a_nx^n + \cdots + a_1x + a_0| \\ &\leqslant |a_n|x^n + |a_{n-1}|x^{n-1} + \cdots + |a_1|x + |a_0| \\ &\leqslant |a_n|x^n + |a_{n-1}|x^n + \cdots + |a_1|x^n + |a_0|x^n \\ &= (|a_n| + |a_{n-1}| + \cdots + |a_1| + |a_0|)x^n \\ &= cx^n = c|g(x)|\end{aligned}$$

Therefore $f = O(g)$. □

Theorem 5.2

Let S be a set of functions all of which have the same domain D, where $D = \mathbb{N}$, $\mathbb{Z}$, or $\mathbb{R}$, and codomain $(0,\infty)$. Let $f_1,f_2,g_1,g_2 \in S$ such that $f_1 = O(g_1)$ and $f_2 = O(g_2)$. Then the following big-O relations hold:

a. $f_1 + f_2 = O(\max\{g_1,g_2\})$ and

b. $f_1f_2 = O(g_1g_2)$

Proof

a. Since $f_1 = O(g_1)$ and $f_2 = O(g_2)$, there are positive numbers c_1, k_1, c_2, k_2 such that, if $x \in D$ and $x > k_1$, then $|f_1(x)| < c_1|g_1(x)|$ and, if $x > k_2$, then $|f_2(x)| < c_2|g_2(x)|$. Let $k = \max\{k_1,k_2\}$ and let $c = c_1 + c_2$. Let $x \in D$ such that $x > k$. Then $x > k_1$ and $x > k_2$, and hence

$$\begin{aligned}|(f_1 + f_2)(x)| &= (f_1 + f_2)(x) \\ &= f_1(x) + f_2(x) \\ &< c_1|g_1(x)| + c_2|g_2(x)| \\ &= c_1g_1(x) + c_2g_2(x) \\ &\leqslant c_1 \max\{g_1(x),g_2(x)\} + c_2\max\{g_1(x),g_2(x)\}\end{aligned}$$

$$= (c_1 + c_2)\max\{g_1(x), g_2(x)\}$$
$$= c \cdot \max\{g_1(x), g_2(x)\}$$
$$= c|(\max\{g_1, g_2\})(x)|$$

Thus $f_1 + f_2 = O(\max\{g_1, g_2\})$.

b. Since $f_1 = O(g_1)$ and $f_2 = O(g_2)$, there are positive numbers c_1, k_1, c_2, and k_2 such that, if $x \in D$ and $x > k_1$, then $|f_1(x)| < c_1|g_1(x)|$ and, if $x > k_2$, then $|f_2(x)| < c_2|g_2(x)|$. Let $k = \max\{k_1, k_2\}$ and let $c = c_1c_2$. Let $x \in D$ such that $x > k$. Then $x > k_1$ and $x > k_2$. Thus

$$|(f_1f_2)(x)| = |f_1(x)||f_2(x)|$$
$$\leqslant c_1|g_1(x)| \cdot c_2|g_2(x)|$$
$$= c_1c_2|g_1(x)g_2(x)|$$
$$= c_1c_2|(g_1g_2)(x)|$$
$$= c|(g_1g_2)(x)|$$

Hence $f_1f_2 = O(g_1g_2)$. □

Exercises 5.3

13. Show that if $n \in \mathbb{N}$ and $n \geqslant 11$ then $n^4 < n!$.
14. Let f,g: $\mathbb{R} \to \mathbb{R}$ be functions defined by $f(x) = 3x + 2$ and $g(x) = x$. Show that $f = O(g)$.
15. Let f,g: $\mathbb{R} \to \mathbb{R}$ be functions defined by $f(x) = x^2$ and $g(x) = x^2 + x$. Show that $f = O(g)$ and $g = O(f)$.
16. Let f,g: $\mathbb{R} \to \mathbb{R}$ be functions defined by $f(x) = 2x$ and $g(x) = x^2$. Show that $f = O(g)$ and $g \neq O(f)$.
17. Let f,g,h: $\mathbb{R} \to \mathbb{R}$ be defined by $f(x) = 2x$, $g(x) = 0$, and $h(x) = 17$.
    a. Show that $f \neq O(g)$.
    b. Is it true that $f = O(h)$? Explain your answer.
18. Prove that, if f,g: $\mathbb{N} \to \mathbb{N}$ are defined by $f(n) = n^2$ and $g(n) = n^2/2 - n/2$, then $f = O(g)$.

5.4 THE ANALYSIS OF ALGORITHMS

As we have seen, there can be many algorithms associated with a single function. A first course in computer programming always drives this point home. After the instructor assigns the first problem requiring the students to create a program, students compare solutions and discover that no two programs are exactly alike. An inevitable question arises: Which solution is best? A common response to this question is that any solution that works is great. But the students, without realizing it, are asking one of the hardest questions of computer science: What constitutes a good algorithm?

A similar issue arises in mathematics if we have different proofs of the same theorem. The question "Which proof is better?" has the same answer as "Which algorithm is better?" If pressed, the professor might admit that proofs are valued for their brevity, clarity, elegance, and so forth; but here the question of value is not nearly so critical. Once a theorem is proved, it never needs to be proved again. The proof is lost as the course continues, and only the theorem itself is remembered and used.

An algorithm, however, is used again and again, and so it is reasonable to ask if one algorithm is preferable to another. In a computer program setting, the two most important considerations, aside from correctness, are probably execution time and the amount of memory space needed in the computer. Other considerations include readability, provability, and elegance. In the past, the memory space provided in a computer was severely limited, and so programmers paid particular attention to the space required to execute an algorithm. With inexpensive memory now measured in megabytes, the major consideration when comparing algorithms is execution time. In an actual computer setting, each operation in a program (addition, multiplication, comparison) takes a specific number of clock cycles, and thus each operation can be timed exactly. But different computers perform operations in different ways, and it may not be the case that a particular operation always takes the same time to run on each computer. It may be that two algorithms, fairly close in execution time on two computers, have the property that the first works better on one computer while the second works better on the other.

Recall that an algorithm consists, not only of a set of instructions, but also of a set of allowable inputs. Normally, two different inputs to a particular algorithm have different execution times. We would all agree that it takes us more time to calculate 76549×47981 than to calculate 23×10. The computer, too, in much the same way, may take more time to deal with one input to an algorithm than with another. Let A be an algorithm with allowable input set D. Although the allowable input set of an algorithm may be infinite, the allowable input set of a computer algorithm is always finite. Consequently, we now assume that D is a finite set of size $n(D)$. For each $d \in D$, let $t(d)$ be the execution time of A for the input d. The *best-case time* of A is the minimum of $\{t(d)\colon d \in D\}$, the *worst-case time* of A is the maximum of $\{t(d)\colon d \in D\}$, and the *average-case time* is the average of all the numbers in the set $\{t(d)\colon d \in D\}$, that is,

$$\frac{\sum_{d \in D} t(d)}{n(D)}$$

The study of the efficiency of algorithms is called *complexity theory*. Given a particular computer and two computer algorithms, we could run all allowable inputs through both algorithms and actually calculate the best-, worst-, and average-case times for each algorithm. We could even write a program to do the testing for us. Such programs are called *bench-mark programs* and the test results are called the *bench marks* of the algorithm. Alas, this type of analysis requires the programmer to write the program before testing begins. We would like a way to estimate the execution time of an algorithm before expending the effort to write the complete program.

It is reasonable to expect the running time of an algorithm to be comparable to the number of operations performed during its execution. For the remainder of this chapter, we consider as roughly equal the running time of the operations addition, subtraction, multiplication, division, and comparison; we ignore all other operations. We also assume that the number of operations that are performed by an algorithm during its execution is a good measure of its running time.

In Figure 5.10 we look at a simple algorithm that finds the largest element of a linear list. If we look closely, we see that there are three lines of the algorithm involving manipulations that count as operations. Lines 4 and 6 involve a comparison, and line 5 involves an addition. These lines are repeated once for each natural number between 2 and $N + 1$. Since there are N natural numbers between 2 and $N + 1$, the number of operations involved when the algorithm is executed is $3N$. Here the best-, worst-, and average-case times are all the same, since the algorithm does not depend on the values in the linear list. As the number N grows larger, the number of operations required by the algorithm in Figure 5.10 grows larger. In fact, the number of operations grows in a very nice manner. If you double the number N to get $2N$ elements in the linear list,

Figure 5.10

Find the largest element of a linear list

1. Let L be a linear list of N real numbers.
2. Let $M = L[1]$.
3. Let $I = 2$.
4. If $L[I] > M$, then let $M = L[I]$
5. Let $I = I + 1$.
6. If $I > N$ then go to step 8.
7. Go to step 4.
8. Stop.

then the number of operations grows from $3N$ to $6N$. In other words, when the size of the list doubles, the number of required operations doubles. Notice that, if the original number of operations had been $5N$, doubling the size of the list would still only double the number of necessary operations. It would seem that the 3 in our answer $3N$ is not the significant factor. What really matters is N. We say that the number of operations involved is of the order N and, using the big-O notation, we say that the complexity of the algorithm in Figure 5.10 is $O(N)$.

Example 4

Let n be an arbitrary but fixed natural number and let A be the algorithm represented by the expression $2x^n + 5x^{n-1} - 3x$ with allowable input a set of real numbers appropriate to the computer on which the algorithm is to be executed. Find the number of operations performed when A is executed and find the complexity of A.

Analysis

Since x^n is x times itself $n - 1$ times, finding x^n requires $n - 1$ operations. Finding $2x^n$ requires n operations, finding $5x^{n-1}$ requires $n - 1$ operations, and finding $3x$ requires 1 operation. There is also one addition and one subtraction. So the number of operations needed for A is $n + n - 1 + 1 + 1 + 1 = 2n + 2$. It is tempting to say that the complexity of A is $O(n)$, but you must remember that the input set refers to the variable x and not to the variable n. So the complexity of A is really $O(1)$; that is, as x increases, the execution time of the algorithm remains constant. Of course, in reality one number may be more difficult for a computer to raise to the nth power than another, even though each input x requires the same number of arithmetic operations and comparisons. Our simplistic assumption that execution time depends solely on the number of comparisons and arithmetic operations sometimes leads to doubtful conclusions. ■

Example 5

Let A be an algorithm defined by the expression 2^x. Find the number of operations performed when A is executed and find the complexity of A.

Analysis

Since 2^x requires $x - 1$ multiplications, the number of operations performed when A is executed is $x - 1$. Thus the complexity of A is $O(x)$. ■

An algorithm with complexity $O(1)$ is faster than an algorithm with complexity $O(x)$. If the complexity of an algorithm is $O(x^2)$, the situation is even worse. For this case, the execution time increases as the square of the variable x, so if such an algorithm takes 3 seconds when $x = 100$, it takes approximately 9 seconds when $x = 200$.

Exercises 5.4

19. Let A be the algorithm represented by each of the following expressions, with allowable input a finite set of real numbers. How many operations does each of the following algorithms take and what is the complexity of each algorithm?

a. $2x^5 - 3x^4$

b. $x^{11} + x^{10} - x^5$

c. $x^{3n} + x^{2n}$, where n is a natural number

d. $(((x + 1)x + 2)x + 3)x + 4$

e. 4^{2x+5}

f. $2^{2x} + 5$

20. Determine the new execution time in each of the following.

a. The complexity of A is $O(1)$. The execution time is 4 seconds when $x = 100$. What is the execution time when $x = 400$?

b. The complexity of A is $O(x)$. The execution time is 2 seconds when $x = 200$. What is the execution time when $x = 500$?

c. The complexity of A is $O(x^2)$. The execution time is 25 seconds when $x = 600$. What is the execution time when $x = 300$?

5.5 SEARCHING ALGORITHMS

One important use of the computer is to store information. Telephone companies store telephone locations, interconnecting wire paths, and telephone numbers. The credit card bureau stores credit card numbers, current balances, and maximum balances on all currently active credit cards. Such information would be useless without a good retrieval system. When you use the telephone you expect instant service, and if you buy an item by credit card you will not wait for hours while the store checks on your ability to pay. Yet both these systems have millions of pieces of information stored in a computer. The retrieval of information from these systems works almost like magic. Seconds after the information is requested, you are talking to Aunt Lucy or carrying home a new television set. The process of retrieval is normally referred to as *searching*.

Two linear search algorithms A_1 and A_2 are pictured in Figure 5.11, where n is a natural number, L is a linear list of real numbers with domain I_n, and x is a real number in the codomain of L.

The best-case scenario for algorithms A_1 and A_2 is one operation when $x = L[1]$. If x is in the list, the worst-case scenario for algorithms A_1 and

Figure 5.11 Linear search algorithm

Search $L[1], L[2], \ldots, L[n]$ for x

Algorithm A_1
(It is given that x is in the range of L.)

1. Let $I = 1$.
2. If $L[I] = x$, then print x is the Ith term and stop.
3. Let $I = I + 1$.
4. Go to step 2.

Algorithm A_2
(It is known only that x is in the codomain of L.)

1. Let $I = 1$.
2. If $L[I] = x$, then print x is the Ith term and stop.
3. Let $I = I + 1$.
4. If $I > n$, then print x is not in the list and stop.
5. Go to step 2.

A_2 occurs when $x = L[n]$. In algorithm A_1, when $x = L[m]$ there are m comparisons in step 2 and $m - 1$ additions in step 3, so A_1 requires $m + m - 1 = 2m - 1$ operations to find $L[m]$. In algorithm A_2, there are m comparisons in step 2, $m - 1$ additions in step 3, and $m - 1$ comparisons in step 4, so A_2 requires $m + m - 1 + m - 1 = 3m - 2$ operations to find $L[m]$. Thus both algorithms are of complexity $O(n)$. Algorithm A_1 requires that x actually be in the list, while algorithm A_2 does not. Algorithm A_1 appears to run in two thirds the time of algorithm A_2, so it seems superior. Nevertheless, most programmers prefer to use algorithm A_2. They know from experience that someone will eventually search for a number that does not appear in the list. When this happens, algorithm A_1 never stops. The entry continues to grow, and after $L[I]$ ceases to make sense it cannot possibly equal x. So the "if" part of step 2 is never true and the algorithm goes on forever. Really, we should say that the intended algorithm A_1, which does not have x as an allowable input, is with respect to the input x no longer an algorithm; or we might say, somewhat inaccurately, that the worst case of algorithm A_1 is very bad indeed. As we have commented before in this chapter, when using the linear search we should really use the average-case execution time, since we expect to find the sought-after object after looking through roughly half the objects in the linear list. The number of operations for the average-case scenario for algorithm A_1 is $[\Sigma_{m=1}^{n}(2m - 1)]/n = n$; for algorithm A_2 it is $3n/2 - 1/2$ (see Exercises 22 and 23). The complexity for both average cases is still $O(n)$.

To get some feeling for the efficiency of the linear search, let us take n to be 100 million, which is a number of the size that the credit card bureau or a large telephone company may need to handle. Let us further suppose that our computer can perform 10,000 operations per second, that we use the average-case scenario, and that to be safe we use algorithm A_2. A little arithmetic shows that the search will take approximately 4 hours and 10 minutes, which would try the patience of most telephone callers.

The discussion of the linear search leads us, as it must have led the computer scientists before us, to wonder if there is a better way to search a linear list for a particular number. There is a better way—the binary search. Because we assume that the reader is convinced that we must be sure our search eventually ends, even if the object of our search is not present, we concern ourselves only with a so-called safe binary search—a search that ends even if the user forgets and searches for something that is not listed. Once again, we let n be a natural number and let L be a linear list with domain I_n; but this time we also require that, if I is a natural number less than n, then $L[I] < L[I + 1]$; that is, we require that the list be ordered. The binary search algorithm is given in Figure 5.12.

Figure 5.12
Binary search algorithm

Search L[1], L[2], . . . , L[n] for x
(It is known only that x is in the codomain of L.)

1. Let $A = 1$.
2. Let $B = n$.
3. Let $I = \text{FLOOR}((A + B)/2)$.
4. If $L[I] = x$, then print x is the Ith term and stop.
5. If $I = A$, then go to step 8.
6. If $L[I] < x$, then let $A = I$ and go to step 3.
7. Let $B = I$ and go to step 3.
8. Let $I = B$.
9. If $L[I] = x$, then print x is the Ith term and stop.
10. Print x is not in the list and stop.

Recall that the function FLOOR picks the largest integer less than or equal to the number input. So FLOOR(4.5) = 4. Thus the binary search algorithm of Figure 5.12 picks the number I in step 3 to be the integer in the middle of the interval $[A,B]$, and if the middle of $[A,B]$ is not an integer, the algorithm picks the largest integer less than the middle. Steps 5, 8, and 9 ensure that the algorithm ends, even if the number x does not appear in the list. Without these steps, in certain lists we might never get to check the right-most integer of an interval.

When searching for an object using the linear search, we always expect to look through roughly half the objects to find a particular object. Thus the linear search deserves to have the average-case time used when considering its running time. The probability that the binary search stumbles on the object of the search before the last possible trial is less than 1/2 and before the next-to-last trial is less than 1/4. Thus, for the binary search, it is more appropriate to consider the worst-case time than the average-case time. The binary and linear searches illustrate that it is not always fair to use the same type of comparison time when comparing algorithms.

Returning to the algorithm in Figure 5.12, we count the number of operations in a worst-case scenario. Our program ends when $I = A$ in step 5. In each loop starting at step 3, we have two operations (addition and division) that occur in step 3; one operation (comparison) in step 4; one operation (comparison) in step 5; and one operation (comparison) in step 6. The first interval we search is the interval from 1 to n, and by looking in the middle we eliminate roughly half the numbers. If we discover that our number is to the right of the middle, we put $A = I$ and leave B as it is. If we discover that our number is to the left of the middle, we put $B = I$ and leave A alone. Thus, at each stage the number x remains in the interval $[A, B]$. When A and B are one apart, the number I remains A each time step 3 is executed. Then, if our number is in the list, it must either be A or B. The last time through the loop we do not execute step 6, but we do execute step 9, which also has one operation. Thus, at each stage through the loop (even the last one) there are five operations. It remains to count the number of times the algorithm must traverse the loop. Since at each stage through the loop we cut the length of the interval in half, at the end of the first stage the length of the intervals is no more than $n/2$. At the end of the second stage the length of the interval is no more than $n/2^2$. At the end of the third stage the length of the interval is no more than $n/2^3$. At the end of the kth stage the length of the interval is no more than $n/2^k$. When the length of the interval is no more than 1, we start the final stage. So we are really asking when $n/2^k$ becomes less than or equal to 1, or what is the smallest integer k such that $n \leqslant 2^k$. For the purpose of calculating the complexity of the algorithm, we may assume that $n = 2^k$. Taking the $\log_2$ of both sides gives $k = \log_2 n$. Hence, the number of operations required is $5(\log_2 n + 1)$. Thus, for $n \geq 2$, $5(\log_2 n + 1) \leq 5(\log_2 n + \log_2 n) = 5(2 \cdot \log_2 n) = 10 \cdot \log_2 n$, and $5(\log_2 n + 1) = O(\log_2 n)$. Hence, the complexity of the algorithm in Figure 5.12 is $O(\log_2 n)$.

Once again, to get some feeling for the efficiency of the binary search, let us take the same conditions that we took for the linear search, namely that n is 100 million and that our computer can perform 10,000 operations per second. Having the term $\log_2 n$ instead of the term n makes an amazing difference. The binary search takes only 0.015 seconds. This, you will grant, is much faster than 4 hours and 10 minutes. You would not even notice 0.015 seconds when trying to make a telephone call.

There is, of course, a catch. When we use the binary search, the numbers have to be ordered. The process of putting a list of numbers in order is called *sorting*. So now the problem shifts to finding efficient ways of sorting information into some increasing order.

There is a television game show called "The Price Is Right" in which the contestants routinely search for information using a linear search. During one part of the show, the contestant is asked to guess the price of a mystery item whose dollar value is between \$0 and \$1000. After each guess, the contestant is told whether the guess is too high or too low. Let us consider two basic searches, a linear search and a binary search, and since no one is offering a brand new Corvette or a trip for two to Tahiti, let us agree in advance that the mystery item is priced at $653. As the name suggests, the linear search simply starts at the beginning and goes up a fixed amount each time. Once a certain range is established, the search is repeated starting at the beginning of the new range and the fixed amount of increase is changed accordingly. Let us imagine that a contestant uses a linear search. The conversation between the contestant and the game host goes as follows:

(1)	"$100"	"higher"	(7)	"$700"	"lower"
(2)	"$200"	"higher"	(8)	"$610"	"higher"
(3)	"$300"	"higher"	(9)	"$620"	"higher"
(4)	"$400"	"higher"	(10)	"$630"	"higher"
(5)	"$500"	"higher"	(11)	"$640"	"higher"
(6)	"$600"	"higher"	(12)	"$650"	

"Oh, I'm sorry. Time is up, but you have been a wonderfully clever contestant, and we have some parting gifts for you (a box of soap and three dozen packs of hot dogs). Now, Vanna, show our audience what Sally Lou would have won, if . . . "

Although few contestants are foolish enough to attempt a complete linear search by starting at $1 and increasing their guess $1 at a time, few contestants are clever enough to use the most efficient method of search, the binary search. The conversation between the contestant and host would have proceeded as follows had Sally Lou used a binary search:

(1)	"$500"	"higher"	(6)	"$640"	"higher"
(2)	"$750"	"lower"	(7)	"$648"	"higher"
(3)	"$625"	"higher"	(8)	"$652"	"higher"
(4)	"$688"	"lower"	(9)	"$654"	"lower"
(5)	"$656"	"lower"	(10)	"$653"	

Notice that the first response is half way between 0 and 1,000, the second is half way between 500 and 1,000, the third is half way between 500 and 750, and the fourth is a closest natural number to half way between 625 and 750. The ninth response is half way between 652 and 656, and the last response is not a guess—it is a certainty.

Exercises 5.5

21. Let L be the sorted list of numbers 1,2,3, . . . ,13,14,15.

a. Which number is found in the first trial of a binary search?

b. Which numbers are found in the second trial?

c. Which numbers are found in the third trial?

d. Which numbers are found in the last trial?

e. Show that there are more numbers found in the worst-case scenario than in all other cases combined.

22. Argue that $[\Sigma_{m=1}^{n} (2m - 1)]/n$ is the number of operations of the average-case scenario of the linear-search algorithm A_1 (see Figure 5.11). Also show that $[\Sigma_{m=1}^{n} (2m - 1)]/n = n$. (*Hint:* Recall that $2m - 1$ operations are needed to find the item $L[m]$ from our list $\{L[m]: m \in I_n\}$.)

23. Find the number of operations of the average-case scenario for the linear search algorithm A_2 (see Figure 5.11). (*Hint:* Recall that $3m - 2$ operations are needed to find the item $L[m]$ from our list $\{L[m]: m \in I_n\}$.)

24. For the linear-search algorithm L_1, does the average of the number of operations corresponding to best case and worst case equal the number of operations corresponding to the average case?

25. Suppose that a certain search algorithm searched a sorted list $L[1]$, $L[2], \ldots, L[n]$ and required, for each $i \in I_n$, i^3 operations to find the object $L[i]$.

a. Find (possibly in terms of n) the number of operations required in the best- and worst-case scenarios. Average these two numbers.

b. Find, in terms of n, the number of operations required in the average-case scenario. (*Hint:* $(\Sigma_{i=1}^{n} i)^2 = \Sigma_{i=1}^{n} i^3$.)

26. Which is faster, the average-case time of a linear search of a sorted list of 100 numbers or the worst-case time of a binary search of a sorted list of 100,000 numbers? Explain.

27. Perform the arithmetic for the linear search to support our assertion that the search of a linear list of 100 million numbers using a computer that performs 10,000 operations per second takes about 4 hours and 10 minutes.

28. Write a binary search algorithm to search an ordered list $L[1], L[2], \ldots, L[1023]$ which is simpler than the algorithm of Figure 5.12 in part because it searches only for objects that belong to the list.

5.6 SORTING ALGORITHMS

The two concepts *sorting* and *searching* provide the key to using large amounts of information entered into a computer. They are so important that book upon book has been written on different ways of sorting and searching. In the previous section we saw that the binary search, though far more efficient than a linear list, requires a sorted list. We therefore turn our attention to sorting algorithms. We suggest that the easiest way for you to understand each of our sorting algorithms is to work with a pack of 3″ × 5″ index cards. Each time we give an example of a particular sort, write one number on each card and arrange the cards following the instructions of the sort. This saves writing the numbers over and over and allows you to rerun the sort until you understand its operation.

Let n be a natural number and let L be a linear list of real numbers with domain I_n. Our first sort, a *bubble sort*, is illustrated in Figure 5.13. The steps in the bubble sort algorithm involving operations are steps 3, 7, 8, 9, and 10, and each of these steps requires one operation. When I is 1, J becomes all the natural numbers between 1 and $n - 1$ and the algorithm goes through steps 3–8, requiring three operations per loop for each J. Then steps 9 and 10 are executed, giving two more operations for a total of $3(n - 1) + 2 = 3n - 1$ operations when $I = 1$. Then I becomes 2, and J becomes all the natural numbers between 1 and $n - 2$, giving a total of $3(n - 2) + 2 = 3n - 4$ operations. When $I = 3$, $3n - 7$ operations are required. The algorithm continues through $I = n - 1$, when $3(n - (n - 1)) + 2 = 5$ operations are required, and then the algorithm stops. Thus the total number of operations needed for the bubble sort algorithm (see Exercise 30) is

$$5 + 8 + 11 + \cdots + (3n - 7) + (3n - 4) + (3n - 1) = \frac{(3n + 4)(n - 1)}{2}$$

Since $(3n + 4)(n - 1)/2 = 3n^2/2 + n/2 - 2$, the complexity of the bubble sort algorithm is $O(n^2)$.

Figure 5.13
Bubble sort algorithm

Sort $L[1], L[2], \ldots, L[n]$

1. Let $I = 1$.
2. Let $J = 1$.
3. If $L[n - J] \leq L[n - J + 1]$, then go to step 7.
4. Let $X = L[n - J]$.
5. Let $L[n - J] = L[n - J + 1]$.
6. Let $L[n - J + 1] = X$.
7. Let $J = J + 1$.
8. If $J < n - I$, then go to step 3.
9. Let $I = I + 1$.
10. If $I \leq n - 1$, then go to step 2.
11. Stop.

As always, to get a feeling for the efficiency of the bubble sort, we turn to our linear list of 100 million numbers and our computer which performs 10,000 operations per second. Here the answer is rather grim. This bubble sort would take over 480 centuries to sort our list. The linear search looks better and better. We cannot use the binary search unless the numbers are ordered, and 480 centuries is a long time to wait. But it is encouraging to note that the telephone and credit card systems function efficiently. Someone must have found a faster sort.

The *insertion sort* can be seen in Figure 5.14. The principle on which the insertion sort works is simple. Suppose we know that our list $L[1]$, $L[2], \ldots, L[k]$ is sorted and now we must arrange $L[1], L[2], \ldots, L[k]$, $L[k+1]$ into its sorted order. Since $L[k+1]$ is the only element out of order, we find its correct place. We compare $L[k+1]$ with the elements of the correctly ordered list $L[1], L[2], \ldots, L[k]$ starting with the last element. When we find where $L[k+1]$ belongs, we put it there and then rename the list. Example 6 is a simple demonstration.

Figure 5.14
Insertion sort algorithm

Sort $L[1], L[2], \ldots, L[N]$

1. Let $I = 1$.
2. Let $J = 2$.
3. Let $X = L[J]$.
4. If $X < L[I]$, then go to step 10.
5. Let $L[I + 1] = X$.
6. Let $J = J + 1$.
7. If $J > N$, then go to step 16.
8. Let $I = J - 1$.
9. Go to step 3.
10. Let $L[I + 1] = L[I]$.
11. If $I = 1$, then go to step 14.
12. Let $I = I - 1$.
13. Got to step 4.
14. Let $L[I] = X$.
15. Go to step 3.
16. Stop.

Example 6

We give one insertion in the insertion sort. $L[6]$ is the only element out of order. Compare $L[6]$ to $L[5]$, $L[4]$, $L[3]$, and $L[2]$, stopping the first time $L[6]$ is the larger number. This tells us that $L[6]$ should come after $L[2]$ and before $L[3]$. Put it there and renumber the domain of L.

$L[1] = 3$		$L[1] = 3$		$L[1] = 3$
$L[2] = 7$		$L[2] = 7$		$L[2] = 7$
$L[3] = 12$	$\rightarrow$	$L[6] = 9$	$\rightarrow$	$L[3] = 9$
$L[4] = 15$		$L[3] = 12$		$L[4] = 12$
$L[5] = 23$		$L[4] = 15$		$L[5] = 15$
$L[6] = 9$		$L[5] = 23$		$L[6] = 23$

■

In Example 7, we go through all the comparisons and insertions using the insertion sort to sort five numbers.

Example 7

Sort the following list of numbers using the insertion sort: 5, 3, 7, 2, 4.

Analysis

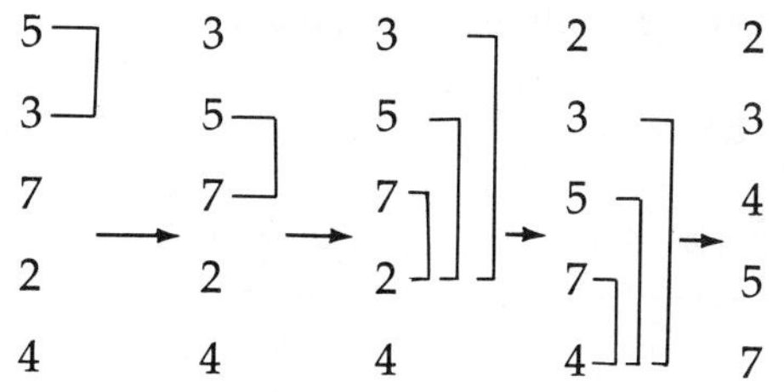

5		3		3		2		2
3		5		5		3		3
7	→	7	→	7	→	5	→	4
2		2		2		7		5
4		4		4		4		7

■

Let us count the number of operations used in the insertion sort of Figure 5.14. The best-case scenario occurs when the list is already sorted and only one comparison is needed to decide that each element is in its correct position. Steps 4, 5, 6, 7, and 8 each involve one operation, and these steps are repeated $n - 1$ times—except that on the last loop step 8 is not executed. Thus the number of operations needed in this best case is $5(n - 1) - 1 = 5n - 6$, and the complexity of the best-case scenario is $O(n)$. The worst case occurs when the list is in order from the largest element to the smallest element. Then each element in the list must be compared to all the elements before it. Suppose the list $L[1], L[2], \ldots, L[k]$ is ordered correctly and we are ready to use the algorithm to insert $L[k + 1]$ correctly into the list. Since our list started ordered backward, $L[k + 1]$ must be compared to $L[1], L[2], \ldots, L[k]$. We start through the algorithm with $J = k + 1$ and $I = k$. The algorithm circles through steps 4, 10, 11, 12, and 13, $k - 1$ times, finally ending with steps 4, 10, 11, 6, 7, and 8. Each of these steps, except step 13, involves one operation, so when $J = k + 1$ the number of operations involved is $4(k - 1) + 6$. When $k = n - 1$, the algorithm finally ends, and at this last stage steps 4, 10, 11, 6, 7, and 16 involve only five operations. Since we start at $k = 1$ and end when $k = n - 1$, the number of operations involved in the worst-case scenario of the insertion sort algorithm (see Exercise 31) is

$$(4 \cdot 0 + 6) + (4 \cdot 1 + 6) + (4 \cdot 2 + 6) + \cdots + (4(n - 2) + 6) - 1 = 2n^2 - 3$$

We note with alarm that the complexity of the insertion sort is $O(n^2)$, which is the same as the complexity of the bubble sort. In fact, with our familiar example of $n = 100$ million, the insertion sort takes over 150 centuries longer than the bubble sort. But there are times when a list is almost sorted, and then the insertion sort comes into its own. If you know that the list $L[1], L[2], \ldots, L[n-1]$ is sorted, then all you have to do is insert the number $L[n]$ into its correct place. This condition occurs frequently in real life, when you have a sorted list of numbers and want to add a new number. Just think of issuing a new credit card or installing a new telephone. This revised algorithm requires $4n - 3$ steps and has complexity $O(n)$. In this case, even for our example of 100 million numbers, the time works out to be somewhere in the neighborhood of 11 hours. If we could just get that 100 million numbers sorted once, everything would be okay. This brings us, at last, to the merge sort.

Before we try the complete merge sort, let us look at how to merge two sorted lists into one sorted list. Suppose we have two lists already sorted. We now wish to merge these two lists into one sorted list. To see how this sort works, imagine each number in two lists you are sorting to be on a separate card. The two lists are already correctly sorted, so there are two stacks of cards in front of you, face up, in the correct order, with the smaller cards on top. It is now your job to make one stack of cards face up, in the correct order, with the smaller cards on top. Look at the top two cards. Pick up the smaller of the two and place it face down, starting a third stack. Now there are two new cards up in front of you (the top card of each of the original two piles). The same rule applies. Pick up the smaller and place it face down on the third stack. Continue this until one stack is gone. Then just pick up the rest of the remaining stack and put it face down on the third stack. Turn the third stack over and you have the sorted stack. Although this process takes a while to explain, it is really like a faro shuffle of two piles of sorted cards in which the crooked dealer is always careful to let the smaller card fall first. Make yourself some 3″ × 5″ cards and try it.

Example 8

Combine the following two lists into one correctly sorted list by merging the two lists. List A: 2, 7, 13, 16; List B: 5, 8, 9, 14.

Analysis

Let C be the name of the final sorted single list.

A	B	C		A	B	C		A	B	C		A	B	C		A	B	C	
2	5			7	5	2		7	8	2		13	8	2		13	9	2	
7	8		→	13	8		→	13	9	5	→	16	9	5	→	16	14	5	→
13	9			16	9			16	14				14	7				7	
16	14				14													8	

A	B	C		A	B	C		A	B	C		A	B	C
13	14	2		16	14	2		16		2				2
16		5	→			5	→			5	→			5
		7				7				7				7
		8				8				8				8
		9				9				9				9
						13				13				13
										14				14
														16

■

The merge sort algorithm is shown in Figure 5.15. To make the bookkeeping and ending conditions easier, we assume that there is a natural number m such that $n = 2^m$. We also vary our usual way of writing algorithms and include some comments to make the merge sort easier to follow.

Figure 5.15
Merge sort algorithm

Sort $L[1], L[2], \ldots, L[n]$

1. Let Z be a list with domain $I_{n/2}$.
2. Let $I = 1$. (size of pieces to merge)
3. Let $J = 1$. (place to start moving)
4. Let $K = I + J - 1$. (place to end moving)
5. Let $W = J$. (moving variable)
6. Let $Z[W] = L[W]$.
7. Let $W = W + 1$. (moving routine)
8. If $W < K$, then go to step 6.
9. Let $P = J$. (beginning Z position)
10. Let $Q = K + 1$. (beginning L position)
11. Let $R = P + 1$. (ending Z position)
12. Let $S = Q + 1$. (ending L position)
13. Let $A = P$. (moving variable)
14. If $Z[P] < L[Q]$, then go to step 24. (moving routine)
15. Let $L[A] = L[Q]$. (L list moves)
16. Let $A = A + 1$.
17. Let $Q = Q + 1$.
18. If $Q \leq S$, then go to step 14.
19. Let $L[A] = Z[P]$. (L all moved; finish moving Z)
20. Let $A = A + 1$.
21. Let $P = P + 1$.
22. If $P \leq R$, then go to step 19.
23. Go to step 32.
24. Let $L[A] = Z[P]$. (Z list moves)
25. Let $A = A + 1$.
26. Let $P = P + 1$.

Figure 5.15 (continued)

27. If $P \leq R$,, then go to step 14.
28. Let $L[A] = L[Q]$. (*Z* all moved; finish moving *L*)
29. Let $A = A + 1$.
30. Let $Q = Q + 1$.
31. If $Q \leq S$, then go to step 28.
32. Let $J = J + 2*I$. (next chunk size *I*)
33. If $J < n$, then go to step 4. (there is another chunk)
34. Let $I = 2*I$. (on to the next size)
35. If $I < n$, then go to step 3. (not through yet)
36. Stop. (finally finished)

The merge sort algorithm pictured in Figure 5.15 appears complex, but most of the algorithm is concerned with keeping track of where we are. Something new appears in this algorithm when we introduce the linear list *Z*. We need the list *Z* because as we write the elements in the two chunks of *L* back into *L* we might write over the number we are going to use later in the first half of the list. For example, suppose we are merging *L*[1], *L*[2], . . . , *L*[8] and *L*[9], *L*[10], . . . , *L*[16]. If all the elements in the second list come before the elements in the first list, then as we write the elements of the second list into the spaces *L*[1], *L*[2], . . . , *L*[8], the original numbers in the places *L*[1], *L*[2], . . . , *L*[8] are lost.

The principle of the merge sort algorithm is easy to understand. We start with an unsorted list and merge the elements of the list two at a time to form sorted pairs. Then we merge the pairs to form sorted groups of four and merge the groups of four to form sorted groups of eight. We continue this process until we have only one large sorted group, as in Example 9:

Example 9

Sort the following list using the merge sort algorithm: 3, 12, 5, 2, 10, 8, 6, 1.

Analysis

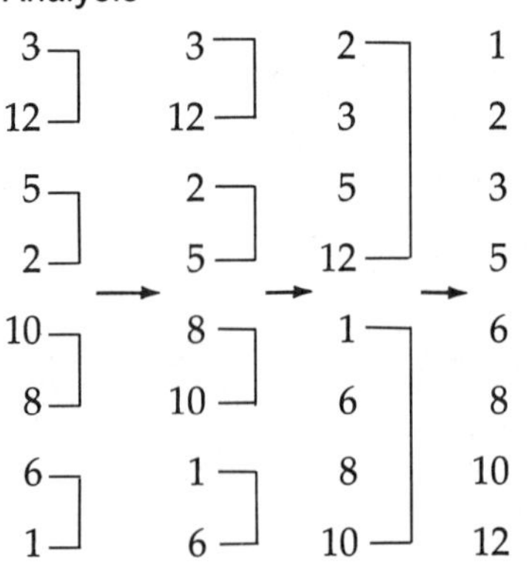

■

We now compute the complexity of the merge sort in a worst-case scenario. Such a case occurs when we have to make the maximum number of comparisons at each stage to merge two previously sorted lists. Let I be the size of each of the two groups to be merged. The number of operations occurring as the merge sort algorithm executes through one merge of two sorted lists of length I is summarized by step number in Table 5.2.

Table 5.2

Step numbers	Number of times executed	Operations per groups of steps	Total number of operations
4,5	1	2	2
6,7,8	I	2	$2 * I$
9,10,11,12,13	1	3	3
14,15,16,17,18,24, 25,26,27	$2 * I - 1$	4	$4 * (2 * I - 1)$
19,20,21,22 28,29,30,31	1	3	3
32,33	1	3	3
		Total =	$10 * I + 7$

There are n/I groups of size I and hence $(n/I)/2 \times n/I$ pairs must be merged to deal with all the groups of size I. Then there are two operations in steps 34 and 35 to get to the next size groups. The total number of operations for each I through the entire algorithm is summarized in Table 5.3.

Table 5.3

Group size	Number of operations per group	Number of groups	Number of operations to next group	Total number of operations
2^1	$10(2) + 7$	2^{m-1}	2	$10 \cdot 2^m + 7 \cdot 2^{m-1} + 2$
2^2	$10(2^2) + 7$	2^{m-2}	2	$10 \cdot 2^m + 7 \cdot 2^{m-2} + 2$
2^3	$10(2^3) + 7$	2^{m-3}	2	$10 \cdot 2^m + 7 \cdot 2^{m-3} + 2$
2^4	$10(2^4) + 7$	2^{m-4}	2	$10 \cdot 2^m + 7 \cdot 2^{m-4} + 2$
$\vdots$	$\vdots$	$\vdots$	$\vdots$	$\vdots$
2^{m-1}	$10(2^{m-1}) + 7$	2	2	$10 \cdot 2^m + 7 \cdot 2^1 + 2$

From Table 5.3 we see that the total number of operations performed in a worst-case scenario of the merge sort algorithm is

$$(10 \cdot 2^m + 7 \cdot 2^{m-1} + 2) + (10 \cdot 2^m + 7 \cdot 2^{m-2} + 2) + \cdots + (10 \cdot 2^m + 7 \cdot 2^1 + 2)$$

$$= (m - 1)10 \cdot 2^m + 7(2^{m-1} + 2^{m-2} + \cdots + 2^1) + (m - 1)2$$
$$= 10m2^m - 10 \cdot 2^m + 7(2^m - 2) + 2(m - 1)$$
$$= 10m2^m - 10 \cdot 2^m + 7 \cdot 2^m - 14 + 2m - 2$$
$$= 10m2^m - 3 \cdot 2^m + 2m - 16$$
$$= 10n\log_2 n - 3n + 2\log_2 n - 16$$
$$= (10n + 2)\log_2 n - (3n + 16)$$

Since $10n + 2 < 20n$, the complexity of a worst-case scenario of the merge-sort algorithm is $O(n\log_2 n)$.

Checking the example when n is 100 million and our computer can perform 10,000 operations per second, we find that the merge sort algorithm takes about a month. A month is a fairly long time, but it is better than 480 centuries. Once the list is sorted, searching it is fast, and additions to it, which take only about 11 hours, can be made overnight. The problem is at least solvable, and to date there is no known sort routine that has complexity less than the complexity of the merge sort.

Exercises 5.6

29. Perform the arithmetic for the bubble sort to support our assertion that the sort takes about 480 centuries to sort a linear list of 100 million numbers using a computer that performs 10,000 operations per second. Use the formula for the number of operations performed when the bubble sort is executed $3n^2/2 + n/2 - 2$. Notice, as you perform this calculation, the effect, if any, that the second and third terms of the expression have on the calculation.

30. Using a result from Chapter 4, show that

$$5 + 8 + 11 + \cdots + (3n - 7) + (3n - 4) + (3n - 1) = \frac{(3n + 4)(n - 1)}{2}$$

31. Using a result from Chapter 4, show that $\sum_{i=0}^{n-2} (4i + 6) = 2n^2 - 2$.

32. How many comparisons does it take to sort each of the following lists using the insertion sort?

a. 1, 2, 3, 4, 5 **b.** 1, 5, 2, 3, 4 **c.** 1, 3, 4, 5, 2

d. 1, 2, 4, 5, 3 **e.** 5, 4, 3, 2, 1

33. In each of the following, sort the list using the bubble sort, the insertion sort, and the merge sort. Count the number of comparisons for each.

a. 1, 2, 3, 4, 5, 6, 7, 8 **b.** 1, 3, 5, 7, 2, 4, 6, 8

c. 8, 7, 6, 5, 4, 3, 2, 1

5.7 THE EUCLIDEAN ALGORITHM

In the thirteenth century, the Middle English word *augrime* brought to mind what we would now call arithmetic algorithms. Today when the word *algorithm* is mentioned, mathematicians usually think of the Euclidean algorithm. This algorithm, though simple, gives an extremely efficient way to find the greatest common divisor of two natural numbers. The Euclidean algorithm first appeared in Euclid's *Elements* Book 7, Propositions 1 and 2, and is now a standard topic in freshman mathematics courses.

The Euclidean Algorithm

1. Input natural numbers M and N.
2. Divide M by N and let R be the remainder.
3. If $R = 0$, then print the greatest common divisor is N and stop.
4. Let $M = N$.
5. Let $N = R$ and go to step 2.

Example 10

Use the Euclidean algorithm to find the greatest common divisor of 2,574 and 1,092.

Analysis

$M = 2574 \quad N = 1092 \qquad 1092 \overline{)2574}$ quotient 2, rem = 390

$$\begin{array}{r} 2574 \\ \underline{2184} \\ 390 \end{array}$$

$M = 1092 \quad N = 390 \qquad 390 \overline{)1092}$ quotient 2, rem = 312

$$\begin{array}{r} 1092 \\ \underline{780} \\ 312 \end{array}$$

$M = 390 \quad N = 312 \qquad 312 \overline{)390}$ quotient 1, rem = 78

$$\begin{array}{r} 390 \\ \underline{312} \\ 78 \end{array}$$

$M = 312 \quad N = 78 \qquad 78 \overline{)312}$ quotient 4, rem = 0

$$\begin{array}{r} 312 \\ \underline{312} \\ 0 \end{array}$$

Thus the greatest common divisor of 2,574 and 1,092 is 78. ■

As you can see by Example 10, the Euclidean algorithm appears to be very fast. To indicate that the brevity of Example 10 is not a fluke, we next look at an example with much larger numbers. To expedite matters, we only list M, N, and R (for remainder) and do not show the actual divisions.

Example 11

Use the Euclidean algorithm to find the greatest common divisor of 167,076 and 1,928,737.

Analysis

Step			
1	$M = 167076$	$N = 1928737$	$R = 167076$
2	$M = 1928737$	$N = 167076$	$R = 90901$
3	$M = 167076$	$N = 90901$	$R = 76175$
4	$M = 90901$	$N = 76175$	$R = 14726$
5	$M = 76175$	$N = 14726$	$R = 2545$
6	$M = 14726$	$N = 2545$	$R = 2001$
7	$M = 2545$	$N = 2001$	$R = 544$
8	$M = 2001$	$N = 544$	$R = 369$
9	$M = 544$	$N = 369$	$R = 175$
10	$M = 369$	$N = 175$	$R = 19$
11	$M = 175$	$N = 19$	$R = 4$
12	$M = 19$	$N = 4$	$R = 3$
13	$M = 4$	$N = 3$	$R = 1$
14	$M = 3$	$N = 1$	$R = 0$

Thus the greatest common divisor of 167,076 and 1,928,737 is 1. ■

Example 11 shows that even with large numbers this particular algorithm works quickly. There are several observations that we can make about this algorithm.

1. If we switch the values of M and N, then the number of divisions needed changes by one (consider steps 1 and 2).
2. If M is larger than N, then after two more divisions the new M is no more than one-half the old M. The reason for this result is that the new M is the old remainder R, as is illustrated in the enclosed area in Example 11. For the old M, N, and R we have $M = NQ + R$, where $0 \leqslant R < N$. Since $R < N$, $RQ < NQ$ and so $M = NQ + R > RQ + R \geqslant 2R$. Thus $M/2 > R$ and, as we have already noted, the new M is R.
3. Thus the complexity of the Euclidean algorithm is no more than $2\log_2 M + 1$ (see Exercise 45).

A formal proof that the Euclidean algorithm produces the greatest common divisor of two natural numbers seems to require a cumbersome

use of the principle of mathematical induction, but it is somewhat easier to see why the algorithm works. The proof that follows, therefore, is intended as an informal argument.

Theorem 5.3

> Let M and N be natural numbers. Define finite sequences $M_1, M_2, \ldots,$ $N_1, N_2, \ldots, Q_1, Q_2, \ldots,$ and $R_1, R_2, \ldots$ as follows:
>
> **a.** $M_1 = M$ and $N_1 = N$
> **b.** Q_1 and R_1 are the quotient and remainder, respectively, when M_1 is divided by N_1.
> **c.** If $M_i, N_i, Q_i,$ and R_i are defined and $R_i \neq 0$, then $M_{i+1} = N_i$ and $N_{i+1} = R_i$.
> **d.** Q_{i+1} and R_{i+1} are the quotient and remainder, respectively, when M_{i+1} is divided by N_{i+1}.
>
> Then there is a smallest natural number k such that $R_k = 0$. Moreover, N_k is the greatest common divisor of M and N.

Proof

To indicate that Q_i and R_i are the quotient and remainder, respectively, when M_i is divided by N_i, we write the equation $M_i = Q_iN_i + R_i$. We are using the division algorithm, so it is understood that the nonnegative number R_i must be less than N_i (otherwise N_i would have divided into M_i at least one more time, making Q_i larger). Since $N_{i+1} = R_i$ and $R_i < N_i$, the sequence $N_1, N_2, \ldots$ is a decreasing sequence of natural numbers. Since $R_i = N_{i+1}$, the sequence $R_1, R_2, \ldots$ is a decreasing sequence of nonnegative integers. Hence there is a natural number k such that $R_k = 0$. Let d be the greatest common divisor of M and N. Now $M_1 = Q_1N_1 + R_1$ and so $R_1 = M_1 - Q_1N_1 = M - Q_1N$. Since d divides M and N, d divides R_1. But $R_2 = M_2 - Q_2N_2 = N_1 - Q_2R_1$. Since $N_1 = N$, d divides N_1 and R_1. Thus d divides R_2. Continuing in this way, we see that, if i is a natural number less than or equal to k, then d divides R_i. We have argued that d divides N_k, and so $d \leq N_k$. It remains to show that N_k divides M and N. Since $R_k = 0$, N_k divides R_k and N_k. Thus, since $M_k = Q_kN_k + R_k$, N_k divides M_k and N_k. Now $N_{k-1} = M_k$, $R_{k-1} = N_k$, and $M_{k-1} = Q_{k-1}N_{k-1} + R_{k-1}$. Therefore N_k divides M_{k-1} and N_{k-1}. Thus, working backward, we see that, if i is a natural number less than or equal to k, then N_k divides M_i and N_i. Hence N_k divides M and N and so N_k is a common divisor of M and N. The greatest common divisor of M and N is at least as large as N_k, so $N_k \leq d$. But we observed previously that $d \leq N_k$. Thus we have shown that $d = N_k$. □

The Fibonacci sequence plays an important role in the investigation of the complexity of the Euclidean algorithm. We recall the definition of the Fibonacci sequence below.

Definition

The **Fibonacci sequence** is the sequence F defined recursively by

a. $F(0) = 1$, $F(1) = 1$, and
b. If n is a nonnegative integer, then $F(n + 2) = F(n + 1) + F(n)$.

As usual, we denote $F(n)$ by f_n for each $n \in \mathbb{N}$.

Our proof of the final theorem of this chapter, Theorem 5.4, is even more informal than the proof of the Euclidean algorithm. What you should seek from this proof is an understanding of why there is a connection between the Fibonacci sequence and a 2,000-year-old algorithm for finding the greatest common divisor of two numbers.

Theorem 5.4

Let n, u, and v be natural numbers such that (a) $u > v$, (b) the Euclidean algorithm applied to u and v requires n division steps, and (c) u is the smallest natural number with properties 1 and 2.
Then $u = f_{n+2}$ and $v = f_{n+1}$.

Proof

Let us write out a table for u and v that is similar to the table of Example 11—except that whenever possible we label a number by its name as a remainder rather than by an alias such as M or N.

Step			
1	u	v	R_1
2	v	R_1	R_2
3	R_1	R_2	R_3
4	R_2	R_3	R_4
.			
.			R_{n-3}
.		R_{n-3}	R_{n-2}
$n - 1$	R_{n-3}	R_{n-2}	$R_{n-1} = \gcd(u,v)$
n	R_{n-2}	R_{n-1}	$R_n = 0$
	Begin		

For each appropriate integer k, $R_{k-2} = R_{k-1}Q_k + R_k$, and since $Q_k \geqslant 1$, $R_{k-2} \geqslant R_{k-1} + R_k$. We begin near the bottom, but we are interested only in the top two lines consisting of steps 1 and 2.

$$R_{n-3} \geqslant R_{n-2} + R_{n-1} \geqslant 1 + 1 = f_2 + f_1$$
$$R_{n-4} \geqslant R_{n-3} + R_{n-2} \geqslant R_{n-2} + R_{n-1} + R_{n-2} \geqslant 3 = f_3 + f_2$$
$$R_{n-5} \geqslant R_{n-4} + R_{n-3} \geqslant (f_3 + f_2) + (f_2 + f_1) = f_4 + f_3$$
$$\vdots$$
$$R_2 = R_{n-(n-2)} \geqslant f_{n-3} + f_{n-2} = f_{n-1}$$
$$R_1 = R_{n-(n-1)} \geqslant f_{n-2} + f_{n-1} = f_n$$

Since $v \geqslant R_1 + R_2$, $v \geqslant f_n + f_{n-1} = f_{n+1}$. □

Forget about Fibonacci for a minute. Suppose that you want to divide by a number that is at least 7, have a quotient of at least 1, and leave a remainder that is at least 4. What is the smallest number you can divide into? The answer is $4 + 7 = 11$ because 11 divided by 7 yields a quotient of 1 with a remainder of 4. Similarly, if you want to divide by a number that is at least f_{n+1}, have a quotient that is at least 1, and have a remainder that is at least f_n, the smallest number you can divide into is $f_n + f_{n+1}$. This smallest possible number is called u. Since $f_n + f_{n+1} = f_{n+2}$, $u = f_{n+2}$ and $v = f_{n+1}$.

We have not yet shown that, if you find $\gcd(f_{n+1}, f_n)$ using the Euclidean algorithm, then the algorithm will take n steps. This part of the argument is left as Exercise 44; here again, we believe that an informal argument is more enlightening than a formal proof by induction.

Theorem 5.4 has the historical distinction of being the first practical application of the Fibonacci sequence. It was first proved by G. Lame in 1845 and is called Lame's theorem. One consequence of Lame's theorem is that, when finding the greatest common divisor of two numbers using the Euclidean algorithm, the maximum number of division steps needed is no more than six times the number of digits in the larger number.

Since the Euclidean algorithm is famous and venerable, it is interesting to note an upstart algorithm discovered by J. Stein in 1961 for finding the greatest common divisor of two natural numbers. Stein's algorithm depends on binary arithmetic and usually takes more steps than the Euclidean algorithm. The advantage of Stein's algorithm is that it does not require a division step but instead halves numbers and subtracts.

Stein's Algorithm

1. Input natural numbers M and N.
2. Let $k = 0$.

3. If either M is odd or N is odd, go to step 8.
4. Let $k = k + 1$.
5. Let $M = M/2$.
6. Let $N = N/2$.
7. Go to step 3.
8. If M is odd, go to step 11.
9. Let $t = M$.
10. Go to step 13.
11. Let $t = -N$.
12. Go to step 14.
13. Let $t = t/2$.
14. If t is even, go to step 13.
15. If $t > 0$, go to step 18.
16. Let $N = -t$.
17. Go to step 19.
18. Let $M = t$.
19. Let $t = M - N$.
20. If $t \neq 0$, go to step 13.
21. Print the gcd of M and N is $M \cdot 2^k$.
22. Stop.

Although Stein's algorithm seems complex, it is simple to use. The minus sign in the algorithm keeps track of which number we are working with, M or N. If t is minus, we are working with N; otherwise we are working with M. First, all common powers of 2 are divided out of M and N, and we keep track in k of the total number of powers of 2. If either M or N is still even, then we divide out the remaining powers of 2 of that number. As soon as both numbers are odd, we let $t = M - N$. From then on, we divide t by 2 until t becomes odd, and then we replace M or N by t (the sign of t tells which). We continue letting $t = M - N$ until t is 0. The greatest common divisor is $M \cdot 2^k$.

The easiest way to see how this algorithm works is with an example. In Example 12, we use the same numbers that we used with the Euclidean algorithm in Example 10 so that you can compare the two methods.

Example 12

Using Stein's algorithm, find the greatest common divisor of 2,574 and 1,092.

Analysis

$k = 0$ $\quad$ $M = 2574$ $\quad$ $N = 1092$

$\boxed{k = 1}$ $\quad$ $M = 1287$ $\quad$ $N = 546$

M	N	$t = M - N$	$t/2$	$t/2$
1287	546	-546 $(-N)$	-273	
1287	273	1014	507	
507	273	234	117	
117	273	-156	-78	-39
117	39	78	39	
39 (boxed)	39	0		

The greatest common divisor of 2,574 and 1,092 is $39 \cdot 2^1 = 78$. ■

Exercises 5.7

34. Find the greatest common divisor of each of the following by factoring each of the two numbers and multiplying the common factors.

a. 35 and 63 **b.** 99 and 142 **c.** 312 and 396 **d.** 3,496 and 893

35. Work the problems in Exercise 34 using the Euclidean algorithm.

36. Work the problems in Exercise 34 using Stein's algorithm.

37. Find the largest natural number that divides 660, 1,386, and 14,421.

38. Find the least common multiple of each of the following numbers.

a. 6 and 9 **b.** 6, 15, and 25 **c.** 646, 11,339, and 2,091

39. Bill is going to play a game with a dart board. On the board are various sections, each having either the number 4 or the number 7. Bill can throw as many darts at the board as he wants, and there are many of both numbers on the board. Bill's score when he finishes is the sum of all the numbers that have a dart sticking in their section. It is clear that Bill's score cannot be 1, 2, or 3 but that it can be 4.

a. Can Bill's score be 5 or 6? **b.** Can Bill's score be 8?

c. Can Bill's score be 9 or 10? **d.** Can Bill's score be 11?

e. What is the largest score Bill cannot get?

40. Let us use the Euclidean algorithm to find the greatest common divisor of 35 and 20. Following the algorithm gives $35 = 1 \cdot 20 + 15$, $20 = 1 \cdot 15 + 5$, and $15 = 3 \cdot 5 + 0$. Hence the greatest common divisor of 35 and 20 is 5. It is interesting to note that performing the process of finding the greatest common divisor of 35 and 20 also gives a method of writing 5 as an integer times 35 plus an integer times 20. Starting with the equation $20 = 1 \cdot 15 + 5$, we solve for 5 in terms of 35 and 20:

$$20 = 1 \cdot 15 + 5 \qquad 5 = 20 + (-1) \cdot 15$$

$$= 20 + (-1) \cdot (35 + (-1) \cdot 20)$$

$$= 20 + (-1) \cdot 35 + 1 \cdot 20$$

$$= 2 \cdot 20 + (-1) \cdot 35$$

a. Find integers a and b such that $1 = a \cdot 73 + b \cdot 101$.

b. Find integers a and b such that $3 = a \cdot 24 + b \cdot 75$.

41. Show that, if m and n are natural numbers and d is the greatest common divisor of m and n, then there are integers a and b such that $d = am + bn$.

42. For a dart board similar to that in Exercise 39 with two numbers a and b on it, find a formula for the largest score you cannot get.

43. Find the greatest common divisor of the 10th and 11th Fibonacci numbers.

44. Show that, if n is a natural number, $u = F(n + 2)$, and $v = F(n + 1)$, then finding the greatest common divisor of u and v using the Euclidean algorithm requires n steps.

*45. Following Example 11, we listed three properties of the Euclidean algorithm. Assuming the first two of these properties, show that the complexity of the Euclidean algorithm is no more than $2\log_2 M + 1$.

46. In our informal proof of the Euclidean algorithm we omitted several arguments that would have used the principle of mathematical induction or the least-natural-number principle. Explain informally where in the proof of Theorem 5.3 these two principles are used sub rosa.

Chapter 5 Review Exercises

47. Using the usual algorithm, find $\sqrt{7}$ to the nearest three decimal places.

48. Let $f,g\colon \mathbb{R} \to \mathbb{R}$ be functions defined by $f(x) = 7x^2 - 13x + 26$ and $g(x) = x^2$. Show that $f = O(g)$.

49. Let $f,g\colon \mathbb{R} \to \mathbb{R}$ be functions defined by $f(x) = 2^x$ and $g(x) = x!$.

a. Is $f = O(g)$? b. Is $g = O(f)$?

50. Let A be the algorithm represented by each of the following expressions with allowable input a finite set of real numbers. How many operations does each take and what is the complexity of each algorithm?

a. $6x^{15} + 9x^{11} - 13x^6 + 2x^4$ b. 3^{6x} c. $x!$

51. If the complexity of an algorithm is $O(x)$ and the execution time is 3 seconds when $x = 50$, what is the execution time when $x = 250$?

52. How many comparisons does it take to sort each of the following lists using the bubble sort?

a. 3, 2, 1, 4 b. 4, 3, 2, 1, c. 1, 4, 3, 2

53. How many comparisons does it take to sort each of the lists in Exercise 52 using the insertion sort?

54. How many comparisons does it take to sort each of the lists in Exercise 52 using the merge sort?

55. Sort the list 8, 4, 6, 3, 7, 2, 1, 5,

a. using the bubble sort. b. using the insertion sort.

c. using the merge sort.

56. Find the greatest common divisor of each of the following pairs.

a. 682 and 978 b. 13,622 and 28,973 c. 6,482,905 and 12,397,406

Chapter

6

COUNTING

Learning to count is a lifelong endeavor, and those who learn to do it well see patterns and structures in mathematics that remain hidden from probes with other mathematical skills. There are four basic reasons that counting can provide mathematical insight. Counting can determine the proportion between the number of elements of a set that have a given property and the number that do not; the study of discrete probability is founded on such proportions. Counting can reveal the existence of a solution to a problem by showing that when all the nonsolutions are counted out at least one object remains. Arguments of this type, which are said to use the pigeonhole principle, are discussed in Section 6.2. Even when it is already known that solutions to a problem exist, counting solutions can sometimes show how to select a best solution. Finally, some of the most important theorems of discrete mathematics are established by counting a finite set in two different ways. Theorem 6.10 illustrates this last principle. Sometimes a theorem can be proved by mathematical induction or by counting, for both techniques are fundamental to discrete mathematics. Almost always, a counting proof is preferable, because counting often reveals why a theorem is true.

6.1 BASICS OF COUNTING

We discuss two basic counting principles in this section. The first, the sum rule, tells us the number of ways to perform either of two procedures whose outcome sets are disjoint.

> **The Sum Rule** Suppose that an operation can be performed by either of two different procedures, with m possible outcomes for the first procedure and n possible outcomes for the second. If the two sets of possible outcomes are disjoint, then the number of possible outcomes for the operation is $m + n$.

Example 1

A representative to the College Personnel Committee is to be chosen from the Mathematics Department or the Computer Science Department. How many different choices are there for this representative if there are 54 members of the Mathematics Department and 32 members of the Computer Science Department, and no faculty member belongs to both departments?

Analysis

The procedure of choosing a representative from the Mathematics Department has 54 possible outcomes, and the procedure of choosing a representative from the Computer Science Department has 32 possible outcomes. By the sum rule, there are $54 + 32 = 86$ possible choices for the representative. ■

The sum rule can be extended to more than two operations. Suppose that $n \in \mathbb{N}$ and that an operation can be performed by any of n different procedures. Suppose further that for each $i = 1,2, \ldots ,n$ there are m_i possible outcomes for the ith procedure. If the collection of n sets of possible outcomes is pairwise disjoint, then the number of possible outcomes for the operation is $\Sigma_{i=1}^{n} m_i$.

Example 2

A representative to the College Personnel Committee is to be chosen from the Biology Department, the Chemistry Department, the Geology

Department, or the Physics Department. How many different choices are there for this representative if there are 47 members of the Biology Department, 52 members of the Chemistry Department, 31 members of the Geology Department, and 38 members of the Physics Department, and no faculty member belongs to more than one of these departments.

Analysis

The procedure of choosing a representative from the Biology Department has 47 possible outcomes, the procedure of choosing a representative from the Chemistry Department has 52 possible outcomes, the procedure of choosing a representative from the Geology Department has 31 possible outcomes, and the procedure of choosing a representative from the Physics Department has 38 possible outcomes. Therefore there are $47 + 52 + 31 + 38 = 168$ possible choices for the representative. ■

Example 3

The flowchart in Figure 6.1 describes a procedure for adding the n natural numbers $m_1, m_2, \ldots, m_n$. What is the value of m when this procedure stops?

Figure 6.1

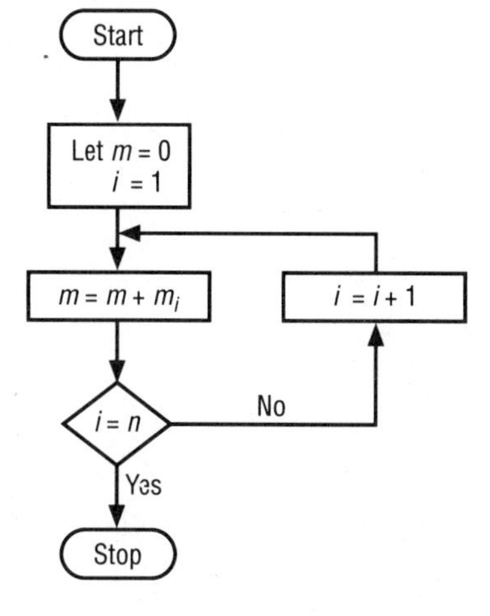

Analysis

The initial value of m is zero. When the loop is completed the first time, the value of m is m_1. When the loop is completed the second time, the value of m is $m_1 + m_2$. When the loop is completed the third time, the value of m is $m_1 + m_2 + m_3$. This process continues until the procedure stops with $m = \Sigma_{i=1}^{n} m_i$. ■

The sum rule can be stated in terms of sets. If A and B are disjoint finite sets, then the number of elements in $A \cup B$ is the sum of the number of elements in A and the number of elements in B. Notice that this result is a corollary of Theorem 2.7. According to the extended sum rule, stated in terms of sets, if $n \in \mathbb{N}$ and $\{A_i: i = 1,2, \ldots, n\}$ is a pairwise disjoint collection of sets, then the number of elements in $\cup_{i=1}^{n} A_i$ is $\Sigma_{i=1}^{n} m_i$, where, for each $i = 1,2, \ldots, n$, m_i is the number of members of A_i.

The second basic counting principle that we discuss in this section, the product rule, applies when a procedure is composed of different tasks.

> **The Product Rule** Suppose that a procedure is composed of two tasks. If m is the number of ways of performing the first task and n is the number of ways of performing the second task, then the number of ways of performing the procedure is mn.

Example 4

How many ways are there to perform an experiment that consists of rolling a die and then tossing a coin?

Analysis

A diagram can be drawn to illustrate the use of the product rule to solve this problem. The first six branches of the diagram denote the six possible outcomes of rolling a die. The two branches that originate at the end of each of these six branches denote the two possible outcomes of tossing a coin.

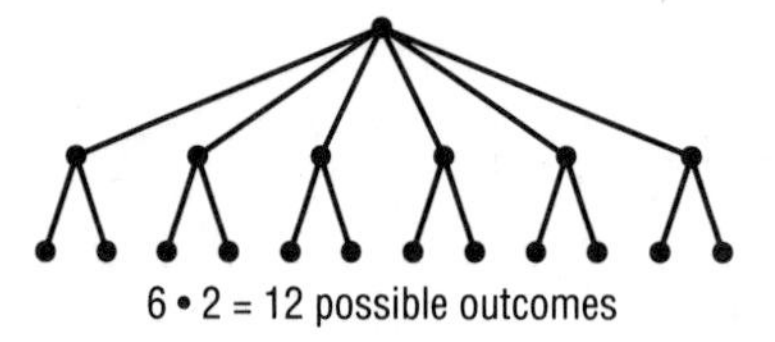

6 • 2 = 12 possible outcomes ■

Example 5

The riders in a bicycle race are identified by a single letter and a single-digit number. If all riders are to be labeled differently, how many riders are permitted in the race.

Analysis

The procedure of labeling the riders consists of two tasks—placing a single letter on each rider and then placing a single-digit number on each rider. There are 26 ways of performing the first task and 10 ways of performing the second task, so the number of riders permitted in the race is $26 \cdot 10 = 260$. ■

The product rule can be extended to more than two tasks. Suppose that $n \in \mathbb{N}$ and a procedure is composed of n tasks, $T_1, T_2, \ldots, T_n$. If, for each $i = 1,2, \ldots, n$, the number of ways of performing the ith task is m_i, then the number of ways of performing the procedure is $m_1 m_2 \cdots m_n$.

Example 6

License plates in Virginia contain a sequence of three letters followed by a sequence of three digits. How many different license plates are possible?

Analysis

There are 26 choices for each of the letters and 10 choices for each of the digits, so the number of possible license plates is $26 \cdot 26 \cdot 26 \cdot 10 \cdot 10 \cdot 10 = 17{,}576{,}000$. ■

Example 7

The flowchart in Figure 6.2 describes a procedure for multiplying the n natural numbers $m_1, m_2, \ldots, m_n$. What is the value of m when this procedure stops?

Figure 6.2

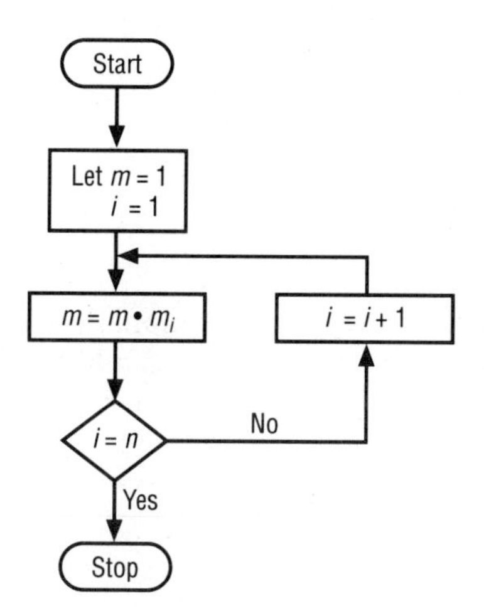

Analysis

The initial value of m is 1. When the loop is completed the first time, the value of m is m_1. When the loop is completed the second time, the value of m is $m_1 m_2$. When the loop is completed the third time, the value of m is $m_1 m_2 m_3$. The process continues until the procedure stops with $m = \prod_{i=1}^{n} m_i$. ■

Example 8

A telephone number in the United States consists of ten digits—a three-digit area code, a three-digit prefix, and a four-digit number. There are some restrictions on some of these digits. In particular, the first digit of the area code must be one of 2,3, . . . ,9, the second digit of the area code must be 0 or 1, and the first two digits of the prefix must be one of 2,3, . . . ,9. How many different telephone numbers are possible?

Analysis

The number of possible areas codes is $8 \cdot 2 \cdot 10 = 160$, the number of possible prefixes is $8 \cdot 8 \cdot 10 = 640$, and the number of four-digit numbers is $10^4 = 10{,}000$. Thus the number of different telephone numbers is $160 \cdot 640 \cdot 10{,}000 = 1{,}024{,}000{,}000$. ■

Example 9

An algorithm begins with an IF . . . THEN . . . ELSE conditional proposition. If the ELSE option occurs, the algorithm stops. If the THEN option occurs, the algorithm executes a sequence of two independent conditional propositions, each of which results in the THEN or the ELSE option. How many possible routes are there through the algorithm?

Analysis

We have two possibilities at the beginning:

THEN ELSE

If the THEN option occurs, the two conditional propositions that result are independent. Thus there are $2^2 = 4$ possibilities if the THEN option occurs. Therefore there are five possible routes:

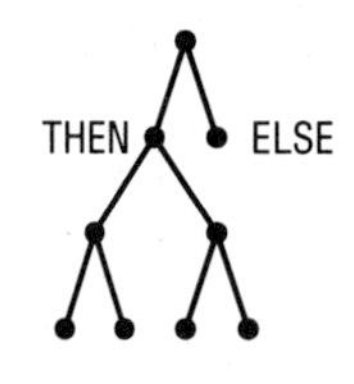

Note that five sets of data are needed to test the algorithm. ■

The product rule can also be stated in terms of sets. If, for each $i = 1,2, \ldots ,n$, A_i is a finite set, then the number of members of the Cartesian product $A_1 \times A_2 \times \cdots \times A_n$ is the product of the number of members of the sets. In other words,

$$n(A_1 \times A_2 \times \cdots \times A_n) = n(A_1) \cdot n(A_2) \cdot \quad \cdots \quad \cdot n(A_n)$$

Exercises 6.1

1. A standard deck of cards has 52 cards, four suits (spades, hearts, diamonds, clubs) each having thirteen cards (ace, king, queen, jack, 10, 9, 8, 7, 6, 5, 4, 3, 2).
   a. In how many ways can a 2 or 3 be drawn?
   b. In how many ways can a spade or heart be drawn?
   c. In how many ways can a 7, jack, or queen be drawn?
   d. In how many ways can an ace or spade be drawn?
2. A die has the numbers 1 through 6, one on each of the six sides.
   a. In how many ways can an even number be rolled?
   b. In how many ways can an even number and then an odd number be rolled?
   c. In how many ways can a number less than 6 be rolled?
   d. In how many ways can a number greater than 4 and then a number less than 4 be rolled?
   e. In how many ways can a number be rolled and then a second number rolled so that the sum of the two numbers is greater than 9?
3. There are 65 computer science majors, 48 mathematics majors, and 37 engineering majors in a large discrete mathematics class. Assume that there are no "double majors."
   a. In how many ways can a class representative who is either a computer science major or a mathematics major be chosen?
   b. In how many ways can two representatives be chosen so that one is a computer science major and one is an engineering major?
4. A man has five shirts, three pairs of pants, and two pairs of shoes. How many different outfits, consisting of one shirt, one pair of pants, and one pair of shoes, are possible?
5. A true–false test contains ten questions.
   a. In how many ways can a student answer the questions if every question is answered?
   b. In how many ways can a student answer the questions if there is a penalty for guessing (so that the student leaves some or all answers blank)?
6. Two dice, one red and one green, are rolled.
   a. How many outcomes are possible?
   b. How many outcomes have the red die showing 6?
   c. How many outcomes have exactly one die showing 6?
   d. How many outcomes have at least one die showing 6?
   e. How many outcomes have both dice showing 6?
   f. How many outcomes have neither die showing 6?
7. You roll two dice, one red and one green.
   a. In how many ways can you get two numbers whose sum is 7?

b. In how many ways can you get two numbers whose sum is 2?

c. In how many ways can you get two numbers whose sum is 12?

d. In how many ways can you get two numbers whose sum is either 2 or 12?

e. Is it more likely that the sum of the two numbers you get is 7 or that the sum is either a 2 or a 12?

8. You roll two dice, one red and one green.

a. In how many ways can you get an even sum?

b. In how many ways can you get an odd sum?

9. Seven different airlines fly from Portland to St. Louis, and twelve different airlines fly from St. Louis to Dallas. How many different possibilities are there for an airline flight from Portland to Dallas via St. Louis?

10. A menu contains four appetizers, twenty main courses, five desserts, and eight beverages.

a. How many dinners (chosen from this menu) consisting of one main course and one beverage are possible?

b. How many dinners consisting of one main course, one beverage, and one dessert are possible?

c. How many dinners consisting of one appetizer, one main course, one beverage, and one dessert are possible?

11. Mainframe IBM computers use the EBCDIC system, in which eight bits (a bit is a 0 or a 1) are used to encode a character. How many different characters can be encoded with this system?

12. In FORTRAN, a variable name must have from one to six characters, the first character must be a letter, and the remaining characters can be letters or digits. How many different variable names are possible?

13. a. When you write the numbers from 1 to 100, how many times do you write the digit 5?

b. When you write the numbers from 100 to 1,000, how many times do you write the digit 5?

14. What is the least number of area codes needed to guarantee that 40 million telephones have distinct ten-digit numbers (see Example 8)?

15. A *palindrome* is a finite sequence of letters that is the same when read forward or backward. Here are two palindromes, the first attributed to Napoleon, the second to the mathematician Peter Hilton: (1) "Able was I ere I saw Elba." (2) "Doc, note I dissent. A fast never prevents a fatness. I diet on cod." How many n-letter palindromes are there in a 2-letter alphabet? (*Hint:* Consider the 4- and 5-letter palindromes.)

6.2 THE PIGEONHOLE PRINCIPLE

According to the simplified version of Dirichlet's pigeonhole principle, named in honor of Peter Gustav Lejeune Dirichlet (1805–1859), if a flock of pigeons flies into a set of pigeonholes and there are more pigeons than pigeonholes, then at least two pigeons must fly into the same pigeonhole.

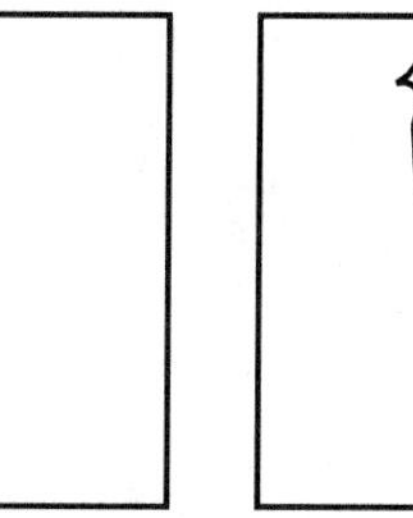

Four pigeons roosting in three pigeonholes

We state the simple version of the pigeonhole principle more precisely as follows:

Theorem 6.1

If $n \in \mathbb{N}$ and $n + 1$ or more objects are distributed among n sets, then at least one of the sets must contain at least two objects.

Proof

Suppose that none of the sets contains more than one object. Then the number of objects cannot exceed the number of sets, so the number of objects is at most n. This is a contradiction, since there are at least $n + 1$ objects. □

Example 10

If there are 367 people in a room, then at least two of them must have the same birthday. In this example, we are distributing 367 objects among 366 sets. ■

Example 11

Let m and n be natural numbers with $m > n$. If, for each $i = 1,2, \ldots ,m$, $p_i \in \mathbb{N}$ and $p_i \leq n$, then there exist i and j with $i \neq j$ such that $p_i = p_j$. In this example, we are distributing m numbers into n singleton sets with $m > n$. ■

Example 12

Show that if twenty-five PCs are interconnected then at least two PCs are directly connected to the same number of PCs.

Analysis

Label the PCs 1,2, . . . ,25, and for each $i = 1,2, \ldots ,25$ let n_i be the number of PCs that are directly connected to i. We must show that there exist i and j $(i \neq j)$ such that $n_i = n_j$. For each i, $0 \leq n_i \leq 24$. However, there does not exist i and j such that $n_i = 0$ and $n_j = 24$. Hence the set $\{0,1,2, \ldots ,23\}$ or the set $\{1,2, \ldots ,24\}$ contains all n_i's. Therefore by the pigeonhole principle there exist i and j $(i \neq j)$ such that $n_i = n_j$. ■

The pigeonhole principle states that, if there are more objects than sets, then at least one of the sets must contain at least two objects. However, even more can be said if the number of objects is more than a multiple of the number of sets. For example, an ordinary deck of cards consists of four suits—spades, hearts, diamonds, and clubs—and there are thirteen cards in each suit. A bridge hand consists of thirteen cards, so a bridge hand must contain at least four cards of at least one suit. This is true because thirteen is more than three times the number of suits.

Theorem 6.2

The Generalized Pigeonhole Principle If k and n are natural numbers and $kn + 1$ objects are distributed among n sets, then at least one of the sets must contain at least $k + 1$ objects.

Proof

Suppose that none of the sets contains at least $k + 1$ objects. Then each set contains at most k objects. Since the number of sets is n, the total number of objects is at most kn. This is a contradiction, since there are $kn + 1$ objects. □

It is easier for many people to remember the generalized pigeonhole principle when it is stated informally: If $kn + 1$ pigeons nest in n nests, then one nest has at least $k + 1$ pigeons.

Example 13

If 51 letters are distributed to 50 mailboxes, then at least one mailbox must contain at least 2 letters. If 51 letters are distributed to 10 mailboxes, then at least one mailbox must contain at least 6 letters $(k = 5, n = 10)$. ■

Example 14

If $p_1,p_2, \ldots ,p_{100}$ are natural numbers such that, for each $i = 1,2, \ldots ,100$, $1 \leq p_i \leq 11$, then at least one natural number must occur at least 10 times in the list $p_1,p_2, \ldots ,p_{100}$ $(k = 9, n = 11)$. ■

Example 15

Let $n \in \mathbb{N}$, and let $\{p_1, p_2, \ldots, p_{n+1}\}$ be a subset of $\{1, 2, \ldots, 2n\}$. Then there exist p_i and p_j $(i \neq j)$ such that one of them divides the other.

Analysis

For each $i = 1, 2, \ldots, n + 1$, there exist a nonnegative integer n_i and an odd integer q_i such that $p_i = 2^{n_i} q_i$. [If p_i is odd, then $n_i = 0$ and $q_i = p_i$. If p_i is even, divide p_i by 2 until a quotient that is odd is obtained. Then n_i is the number of divisions by 2, and q_i is the odd quotient (see Exercise 35).] Notice that, for each $i = 1, 2, \ldots, n + 1$, $q_i < 2n$. So if $A = \{q_1, q_2, \ldots, q_{n+1}\}$, then the number of members of A is less than or equal to the number of odd integers that are less than $2n$. Therefore the number of members of A is at most n. Hence there exist i and j such that $q_i = q_j$. So $p_i = 2^{n_i} q_i$ and $p_j = 2^{n_j} q_i$. If $n_i \leq n_j$, then p_i divides p_j. If $n_i > n_j$, then p_j divides p_i. ■

Example 16

If P_1, P_2, P_3, P_4, P_5 are five distinct points in the plane $(= \mathbb{R} \times \mathbb{R})$ whose coordinates are integers, then there exist P_i and P_j such that the midpoint of the line segment joining P_i and P_j has coordinates that are integers.

Analysis

There are four different parity (even–odd) patterns of the coordinates of the five points: (even, even), (even, odd), (odd, even), and (odd, odd). Therefore by Dirichlet's pigeonhole principle, at least two of the points must have the same parity pattern. Let P_i and P_j denote two points with the same parity pattern. If $P_i = (a,b)$ and $P_j = (c,d)$, the midpoint P of the line segment joining P_i and P_j is $((a + c)/2, (b + d)/2)$. Since the sum of two odd integers and the sum of two even integers are each even, the coordinates of P are integers. ■

Exercises 6.2

16. If a box contains 101 pencils of four different colors, why are there at least 26 pencils of the same color?
17. If you have black, brown, blue, white, and yellow socks in a drawer, how many socks must you choose to be certain of having two of the same color?
18. Suppose that you have eight pairs of socks in a drawer, each pair a different color. If they are unsorted, how many socks must you choose to be certain of having a matched pair?
19. How many people are in a room if at least two of them must have birthdays in the same month?
20. During the next eight days, Wayne will play five rounds of golf. But he never plays more than one round a day. Prove that he will play golf on two consecutive days.
21. Is it possible to interconnect five PCs so that exactly two PCs are directly connected to the same number of PCs? Explain your answer.

22. There are thirteen people in a room. The first name of each is Sam, Sally, or Tom, and the last name of each is Smith, Jones, Williams, or Turner. Prove that at least two people have the same first and last names.

23. How many items must be chosen from 36 ducks, 38 pitchers, 30 baskets, and 22 bears if at least 18 objects are to be of the same type? Justify your answer.

24. Let $n \in \mathbb{N}$. How many objects must be chosen from the set $\{1,2, \ldots ,2n\}$ if at least one of them is to be odd? Justify your answer.

25. Let P_1,P_2, and P_3 be integers. Must there exist P_i and P_j such that $P_i + P_j$ is even? Justify your answer.

***26.** Show that if P_1, P_2, P_3, and P_4 are integers then there exist P_i and P_j such that $P_i - P_j$ is divisible by 3.

***27.** Of six people in a room, each pair of individuals are either friends or enemies. Show that there are three mutual friends or three mutual enemies in the room. (*Hint:* There are 20 threesomes in the room. Suppose that there are neither three common friends nor three common enemies. We may assume that for any threesome there is a member, whom we call an odd man out, with the property that either (1) the two pairs containing the odd man out are friends and the pair not containing the odd man out are enemies, or (2) the two pairs containing the odd man out are enemies and the pair not containing the odd man out are friends.)

***28.** Let n be a natural number greater than 2. Show that if there are n people at a party then there are at least two people who know the same number of people at the party. Assume that if A knows B then B knows A.

29. Let $P_1,P_2, \ldots ,P_9$ be nine distinct points in Euclidean 3-space whose coordinates are integers. In other words, for each $i = 1,2, \ldots ,9$, there exist integers x,y, and z such that $P_i = (x,y,z)$. Show that there exist P_i and P_j such that the midpoint of the line segment joining P_i and P_j has coordinates that are integers.

***30.** Chuck hikes 44 miles in 10 hours. In the first hour he jogs and covers 5 miles, but in the last hour he slows to a walk and covers only 3 miles. Show that there exist two consecutive hours in which he covers at least 9 miles.

31. Let A and B be finite sets with $n(A) > n(B)$ and suppose that $f: A \to B$ is a function. Show that there are distinct members a and b of A such that $f(a) = f(b)$. Notice that this states that f cannot be one to one.

***32.** Let A and B be finite sets and suppose that $p = n(A)/n(B)$ is an integer. Show that if $f: A \to B$ is a function then there are p district members $a_1,a_2, \ldots ,a_p$ of A such that $f(a_i) = f(a_j)$ for each i and j.

***33.** Let A be a 25-member subset of $\{n \in \mathbb{N}: n \leq 100\}$. Show that there are two distinct 4-member subsets B and C of A such that the sum of the members of B is equal to the sum of the members of C.

***34.** A math professor gives seven students a pop quiz that consists of three true–false questions. Show that, given any two questions, at least two students answer these two questions in the same way.

*35. Prove by induction that if p is an even natural number then there is a natural number m such that $p/2^m$ is an odd natural number.

*36. Four people play three games of doubles tennis. Each person plays one game as a partner with each of the other three players. Show that there is a player who is on the winning team all three games or who is on the losing team all three games.

6.3 A MATCHING PROBLEM

A computer science department teaches introductory courses in seven languages: BASIC, FORTRAN, Pascal, APL, COBOL, SNOBOL, and ALGOL. Seven qualified instructors are available, and each instructor is to teach exactly one of these courses. Some of the instructors have opinions about which courses they want to teach and the department has a policy of honoring requests as far as possible. Figure 6.3 lists the courses and instructor preferences and also represents this information by a diagram in which a line appears between course and instructor if the course is on the instructor's request list. The department head would like to please as many instructors as possible. This may be thought of as an *optimization* problem.

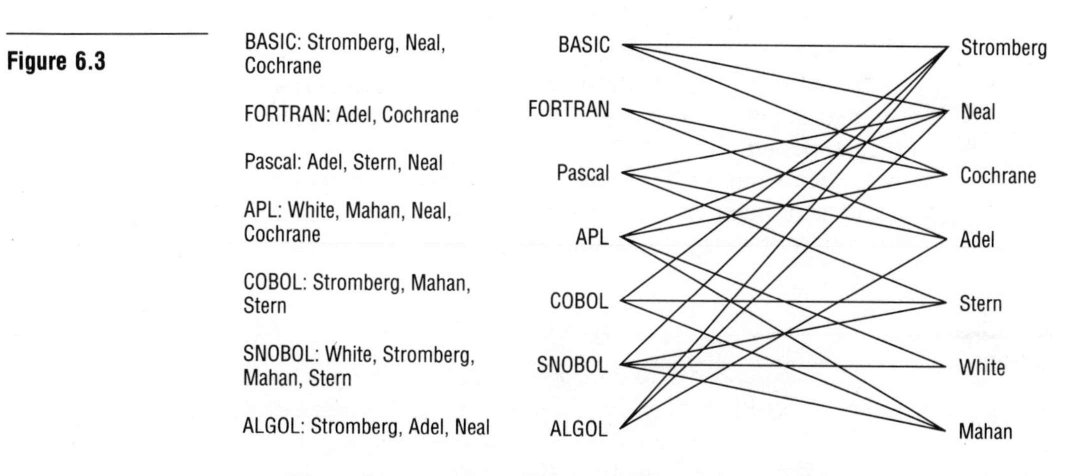

Figure 6.3

How do we solve such a problem? We could take a very crude approach and simply list all possible ways of assigning an instructor to each course. Then for each possibility we could count the number of instructors who are assigned a course they requested. One possibility would be to take the course and instructors in the order they are listed, as in Table 6.1. In this case, only three instructors would get courses they requested, so surely there is a better way.

Table 6.1

Course	Instructor	Requested
BASIC	Stromberg	Yes
FORTRAN	Neal	No
Pascal	Cochrane	No
APL	Adel	No
COBOL	Stern	Yes
SNOBOL	White	Yes
ALGOL	Mahan	No

We might instead list the courses in the same order but rearrange the instructors into the order Stromberg, Adel, Stern, White, Mahan, Neal, and Cochrane. Then five instructors would get courses they requested. Can we do better?

Several questions arise concerning this crude method of attacking the problem: (1) How many arrangements do we have to list? (2) How can we systematically generate all possible arrangements so that we are sure that we have not missed any? We discuss the second question later, but the first question is simply a *counting* problem. Let us begin with the course on BASIC. There are seven instructors who can be assigned to teach the course. We pick one and then consider the FORTRAN course. There are only six instructors left, and we choose one. Then only five instructors are left as possibilities for the Pascal course. We continue in this manner and see that the total number of possible ways of assigning one instructor to each course is $7 \cdot 6 \cdot 5 \cdot 4 \cdot 3 \cdot 2 \cdot 1 = 5{,}040$. This example illustrates a slightly different use of the product rule, different in the sense that the number of possibilities at each stage depends on what has already happened.

In Chapter 3, we defined a permutation to be a one-to-one function of a nonempty set onto itself. The word *permutation* is also used in a different sense, to denote an ordered arrangement of the members of a set.

Definition

Let $n,r \in \mathbb{N}$ with $r \leq n$ and let S be a set with n elements (recall that $I_r = \{m \in \mathbb{N}: m \leq r\}$). An ***r*-permutation** of S is a one-to-one function $\sigma: I_r \to S$. Notice that an r-permutation of S is a one-to-one function whose domain is specified to be I_r and whose codomain is specified to be S. An n-permutation of S is simply called a **permutation** of S. An n-permutation of S is often referred to simply as a permutation when it is clear from context what the domain is intended to be. Whenever $\sigma: I_r \to S$ is an r-permutation, it is convenient to use the notation $\sigma(1)\sigma(2) \cdots \sigma(r)$ for the r-permutation itself.

Do not be frightened by the definition of *permutation*. All the one-to-one function from I_r into S really does is keep track of the order of the members of an r-element subset of S—the image of 1 being the first element, the image of 2 being the second element, and so on. As stated in the definition, the function is usually not mentioned at all. The elements are simply listed in their correct order. Thus, if $n = 4$ and $S = \{a,b,c,d\}$ is a four-element set, then *dba* is a 3-permutation of S. The actual function $\sigma: I_3 \to S$ required by the definition is defined by $\sigma(1) = d$, $\sigma(2) = b$, and $\sigma(3) = a$. Using this notation, we list the twenty-four 3-permutations of the set S:

abc, acb, abd, adb, acd, adc, bac, bca, bad, bda, bcd, bdc, cab, cba, cad, cda, cbd, cdb, dab, dba, dac, dca, dbc, dcb

Notice that the product rule applies when we wish to count 3-permutations. We have a set with four members, so there are four possible ways of choosing $\sigma(1)$. Having chosen $\sigma(1)$, we have three possible ways of choosing $\sigma(2)$, and having chosen $\sigma(1)$ and $\sigma(2)$, we have two possible ways of choosing $\sigma(3)$. So the number of 3-permutations on a set with four members is $4 \cdot 3 \cdot 2 = 24$.

If $n,r \in \mathbb{N}$, S is a finite set with n elements, and $r \leqslant n$, then the same principles used to determine the number of 3-permutations of $\{a,b,c,d\}$ can be used to determine the number of r-permutations of S. We can define an r-permutation of S by letting $\sigma(1)$ be any one of the n elements of S. There remain $n - 1$ choices for $\sigma(2)$, $n - 2$ choices for $\sigma(3)$, and, assuming that for some $p < r$, $\sigma(1),\sigma(2), \ldots ,\sigma(p)$ have been chosen, we may let $\sigma(p + 1)$ be any one of the members of $S - \{\sigma(1),\sigma(2), \ldots ,\sigma(p)\}$. Notice that there are $n - p$ choices for $\sigma(p + 1)$. In particular, for $p = r - 1$ there are $n - (r - 1)$ choices for $\sigma(r - 1 + 1)$; that is, there are $n - r + 1$ choices for $\sigma(r)$. Therefore by the product rule there are $n(n - 1)(n - 2) \cdots (n - r + 1)$ r-permutations of S. Notice that if $r = n$ then this product becomes $n(n - 1)(n - 2) \cdots 1$, which is $n!$.

We let $P(n,r)$ denote the number of r-permutations of a set with n members. We have just obtained a formula for $P(n, r)$, but a quick comparison of this formula with the formula

$$n! = n(n - 1)(n - 2) \cdots (n - r)(n - r - 1) \cdots (3)(2)(1)$$

yields a more compact expression for $P(n,r)$. Since

$$n! = P(n,r)(n - r)(n - r - 1) \cdots (3)(2)(1) = P(n,r)(n - r)!$$

we have $P(n,r) = n!/(n - r)!$. The two formulas for $P(n,r)$, which are fundamental to techniques of counting, are important enough to reemphasize here:

> Let $r,n \in \mathbb{N}$ with $r \leq n$. Then
>
> $$P(n,r) = (n)(n-1)(n-2)\cdots(n-r+1) = \frac{n!}{(n-r)!}$$
>
> where $0!$ is defined to be 1. For convenience, we also define $P(n,0) = 1$.

Example 17

How many different arrangements can be formed on a shelf with space for three books if there are five different books available? Asked another way, how many 3-permutations are there of a set with five members. The answer is $P(5,3) = 5!/2! = 5 \times 4 \times 3 = 60$. ■

Let us return to the problem of assigning instructors to the computer science courses. We have answered our first question; that is, we have found that we would have to list $P(7,7) = 7! = 5{,}040$ arrangements. Clearly we do not want to do this by hand. We use an algorithm to tell a computer how to generate all possible arrangements systematically.

Exercises 6.3

37. Assume that cards from a standard deck are not replaced after drawing.

- **a.** In how many ways can a 2 and then an ace be drawn?
- **b.** In how many ways can a spade and then a heart be drawn?
- **c.** In how many ways can an ace and then a king and then a queen be drawn?
- **d.** In how many ways can two cards be drawn?
- **e.** In how many ways can a 2 and then another 2 be drawn?
- **f.** In how many ways can an ace and then a spade be drawn?

38. Let $S = \{a,b,c,d,e,f,g\}$ and assume that $a,b,c,d,e,f,$ and g are all different.

- **a.** How many permutations are there of S?
- **b.** How many permutations of S end with g?
- **c.** How many permutations of S begin with a and end with g?
- **d.** How many 6-permutations are there of S?
- **e.** How many 5-permutations are there of S?

39. Let $S = \{1,2,3,4,5\}$.

- **a.** List all the 3-permutations of S.
- **b.** List all the 4-permutations of S.

40. **a.** How many permutations of the letters *ABCDEF* contain the word *BAD*? (*Hint:* The letters *BAD* must be kept together in this order.)

b. How many permutations of the letters *ABCDEF* contain the letters *BAD* together in any order?

41. Evaluate **a.** $P(7,4)$ **b.** $P(10,5)$ **c.** $P(12,9)$

42. An investment club has 36 members. In how many ways can the positions of president, vice president, and secretary-treasurer be filled if no person is to hold more than one position?

43. How many different numbers, each consisting of two different digits, can be formed using the digits 2,3,4,7, and 8?

44. Two dice, one red and one white, are rolled. How many different outcomes are there?

45. A baseball manager has decided who his nine starting players are to be. How many different batting orders can he have?

46. A woman has five skirts and seven blouses. How many different outfits consisting of a skirt and a blouse can she select?

47. A Virginia license plate consists of three letters followed by three digits, and the first digit cannot be zero. How many different license plates are possible?

48. A social security number is a nine-digit number. How many social security numbers are possible?

49. A combination lock has five tumblers, and each tumbler can assume ten positions. How many different combinations are possible?

50. The F, H & P Manufacturing Company must choose the design of a container for a liquid vitamin. The container consists of three components: the bottle, the label, and the cap. Several different designs of each are available: five for the bottle, seven for the label, and three for the cap. How many different designs must the company consider?

51. Joe failed English composition and discrete mathematics last semester, so the dean called him in for an explanation. Joe can think of three reasons for failing English composition, and he can think of four reasons for failing discrete mathematics. For each class, he must offer only one excuse. How many different pairs of excuses does he have to choose from?

52. In the game of Scrabble, a player tries to make words from certain letters. Suppose Susan has the letters *A*, *D*, *H*, *O*, *R*, *S*, and *T*. How many different three-letter arrangements are there? Because Susan has an *S*, she will be able to play any word she can form, and there is a 50-point bonus for using all seven letters. How many seven-letter arrangements must Susan consider to be sure she has considered all possibilities?

53. A coin is tossed four times.

a. What is the number of possible outcomes (that is, heads or tails)?

b. How many possible outcomes are there that begin with heads?

c. How many possible outcomes begin and end with heads?

d. How many possible outcomes have two heads and two tails?

e. How many possible outcomes have at least two heads?

f. How many possible outcomes have at least one head and at least one tail?

*54. In draw poker, five cards are dealt and a player may discard no more than three cards. Given a five-card hand, how many sets of cards can be discarded?

6.4 THE KNAPSACK PROBLEM

The knapsack problem is familiar to all backpackers and all travelers to exotic places: When you possess many worldly goods but only one sack, what is to be packed and what is to be left behind?

Example 18

Scientists who are to spend three months in Antarctica face a knapsack problem of major proportions. The following table lists twelve experiments labeled A through L, the space required for each experiment's equipment, and the rating by a group of scientists of the value of each experiment on a scale from 1 to 10 (with 10 being the best score).

Experiment	A	B	C	D	E	F	G	H	I	J	K	L
Space (ft^3)	927	2242	225	1880	1041	364	1277	803	3248	323	656	79
Rating	2	4	4	7	8	6	7	6	8	3	8	5

The scientists agree that they cannot carry more than 7,000 cubic feet of scientific equipment, and under this restriction they decide to choose experiments so that the sum of all their ratings is as large as possible. They simply start down the list from A, observing that the space required to store equipment for experiments $A,B,C,D,E,$ and F is 6,679 cubic feet. They cannot take the equipment for experiment G, since its 1,277 cubic feet puts them over the 7,000-cubic-foot limit. Likewise, they cannot take the equipment for experiments H, I, J, or K. They can still take the equipment for experiment L, however, and this choice brings them up to 6,758 cubic feet. The total rating of the experiments chosen in this way is $2 + 4 + 4 + 7 + 8 + 6 + 5 = 36$.

Can the scientists choose more effectively? It seems likely that they can, since they did not consider the ratings in their selection. Suppose that they now relist the experiments according to ratings and continue as before:

Experiment	E	I	K	D	G	F	H	L	B	C	J	A
Space (ft^3)	1041	3248	656	1880	1277	364	803	79	2242	225	323	927
Rating	8	8	8	7	7	6	6	5	4	4	3	2

Now they can take the equipment for experiments E,I,K, and D, since the space required to store this equipment is 6,825 cubic feet. But they cannot take the equipment for experiments G,F, and H. They can also take the equipment for experiment L, but they cannot take the equipment for B,C,J, and A. The space required is 6,904 cubic feet, and the total rating of the experiments chosen in this way is $8 + 8 + 8 + 7 + 5 = 36$. They have exactly the same rating as before.

After spending hours at another meeting, the scientists decide to list the experiments in order according to the space required, starting with experiment L since it requires the least space:

Experiment	L	C	J	F	K	H	A	E	G	D	B	I
Space (ft^3)	79	225	323	364	656	803	927	1041	1277	1880	2242	3248
Rating	5	4	3	6	8	6	2	8	7	7	4	8

With this ordering, they can take the equipment for experiments L,C,J,F,K,H,A,E, and G, since the space required is 5,695 cubic feet. The rating for these experiments is $5 + 4 + 3 + 6 + 8 + 6 + 2 + 8 + 7 = 49$. They are encouraged by their ability to improve the rating, so they meet again. This time they calculate the ratio of rating points per cubic feet for each experiment and then order the experiments from the highest ratio to the lowest.

Experiment	**Space (ft^3)**	**Rating**	**Ratio**
L	79	5	0.0633
C	225	4	0.0178
F	364	6	0.0165
K	656	8	0.0122
J	323	3	0.0093
E	1041	8	0.0077
H	803	6	0.0075
G	1277	7	0.0055
D	1880	7	0.0037
I	3248	8	0.0025
A	927	2	0.0022
B	2242	4	0.0018

They observe that they can take the equipment for experiments L,C,F,K,J,E,H,G, and D, since the total space required is 6,648 cubic feet. The rating for these experiments is $5 + 4 + 6 + 8 + 3 + 8 + 6 + 7 + 7 = 54$. This selection represents another improvement, and as it happens the rating obtained is the largest possible. But for all their work, the scientists have just been lucky, because for an arbitrary knapsack problem even their last method does not always yield the best possible rating (see Exercise 59). ■

Notice that, like the matching problem of Section 6.3, every knapsack problem is an optimization problem. In the preceding section we found that a matching problem involving n one-to-one matchings yields $P(n,n) = n!$ possible matchings. How many possibilities are there for the knapsack problem of the present section? We have twelve experiments, and each possibility is a collection of experiments chosen from these twelve. The number of possibilities is simply the number of subsets of a set with twelve members, which, by Theorem 4.3, is 2^{12}, or 4,096. We would not want to check this many possibilities by hand, but it would be easy enough for a computer. Suppose, however, that we have sixty experiments rather than twelve. Then the number of possibilities is 2^{60}, or 1,152,921,504,606,846,976. It would take a large computer a tremendous amount of time to check through the possibilities for sixty experiments. Unfortunately, there is no efficient way known to solve an arbitrary knapsack problem.

Exercises 6.4

55. A group of scientists are planning to conduct research in the Antarctic. There has been a problem at the research station, and only 5,000 cubic feet of space is available for storage. The rating and storage space requirement for each of ten possible experiments are given in the following table.

Experiment	A	B	C	D	E	F	G	H	I	J
Space (ft^3)	1384	985	249	1775	39	860	1200	632	1113	930
Rating	6	5	4	8	2	5	7	3	8	6

Use the four methods illustrated in Example 18 to select and rate a group of experiments.

56. Let us return to the problem that our scientists in Example 18 face. Assume that they have sixty possible experiments. To find the optimal set of experiments to conduct, they are faced with checking 2^{60} possibilities. Suppose that they use a computer that checks 10 million possibilities per second. How long does it take this computer to check the 2^{60} possibilities? Is it reasonable for the scientists to wait for the computer to give them an answer? (*Hint:* After 240 days, they will have to eat the sled dogs.)

57. Suppose that, instead of sixty possible experiments, the scientists have twenty-five possible experiments. How many possibilities do they have regarding the set of experiments to consider? If they have a computer that checks 10 million possibilities per second, how long does it take the computer to provide them an answer?

58. Let $S = \{m \in \mathbb{N}\text{: there is an } n \in \mathbb{N} \text{ such that } 1 \leqslant n \leqslant 10 \text{ and } m = 2^n\}$. How many subsets does S have?

59. Make up a knapsack problem involving exactly three items, for which the ratio method (the most successful method in Example 18) does *not* yield the highest rating.

PERMUTATIONS

We defined the concept of permutation as an arrangement or ordering of objects in Section 6.3. We now give a systematic treatment of permutations, and we begin with two problems that appear similar but are very different.

Example 19

Suppose that there are three highways connecting Seattle and San Francisco and five highways connecting San Francisco and Los Angeles. If Mary wants to go from Seattle to San Francisco and then to Los Angeles, how many choices of route does she have?

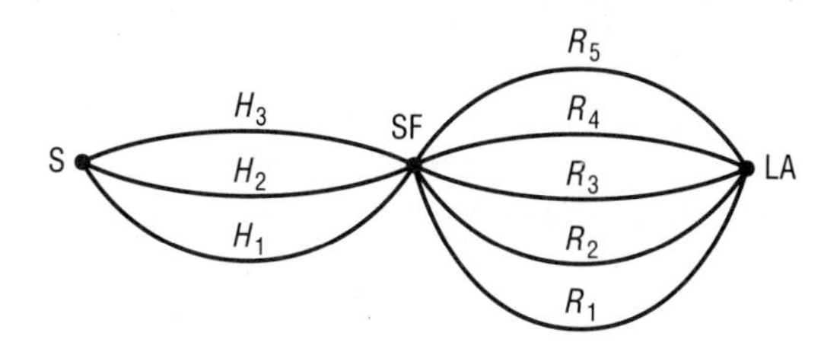

Analysis

We can list all the routes:

H_1R_1, H_1R_2, H_1R_3, H_1R_4, H_1R_5, H_2R_1, H_2R_2, H_2R_3, H_2R_4, H_2R_5, H_3R_1, H_3R_2, H_3R_3, H_3R_4, H_3R_5,

There are 15 choices. Notice that we can obtain the answer by multiplying 3 (the number of highways connecting Seattle and San Francisco) by 5 (the number of highways connecting San Francisco and Los Angeles). The product rule applies, and permutations are not involved. ■

Example 20

It is Christmas and Mary wants to give presents to three friends—Hughes, Peter, and Wayne. She purchases subscriptions to *Newsweek*, *Playboy*, and *Sports Illustrated*. Now she must decide which subscription to give to each friend. How many choices does she have?

Analysis

This is a problem involving permutations. Mary can give any one of the three magazines to Hughes, and for each of these three choices she can give any one of the two remaining magazines to Peter. But having made these two choices, there is only one magazine left to give Wayne. The number of choices is simply the number of 3-permutations on a set with three members, that is, $P(3,3) = 6$. ■

Example 21

How many different numbers are there between 100 and 1,000 each of whose digits is one of 1, 3, 5, 7, 8, 9 and in which no two digits are the same?

Analysis

Here there are so many numbers that it is not feasible to list them. We could have any one of the six digits in the first (hundreds) place. For each of these six choices, we could have any one of the remaining five digits in the second (tens) place. So there are $6 \cdot 5 = 30$ different possibilities for the first two places. For each of these thirty choices, we could have any one of the remaining four digits in the third (units) place. Here the total number of different numbers is $6 \cdot 5 \cdot 4 = 120$. This is simply the number of 3-permutations of a set with six members, $P(6,3)$. Notice that $P(6,3) = 6!/3! = 6 \cdot 5 \cdot 4 = 120$. ■

Example 22

Four men and three women are to occupy a row of seven seats, and men are to occupy the two end seats. In how many different orders can these seven people be seated?

Analysis

There are two end seats, and these seats must be occupied by men. Therefore we have four choices for the first end seat, and after this choice has been made we have three choices for the second end seat. So the total number of ways to occupy the end seats is $4 \cdot 3 = 12$. Notice that this is the number of 2-permutations of a set with four members, $P(4,2)$. The five middle seats can be occupied by any of the remaining five people, so there are five choices for the first seat, four choices for the second seat, and so on. Therefore the total number of ways to occupy the middle seats is $5 \cdot 4 \cdot 3 \cdot 2 \cdot 1 = 120$. This is the number of 5-permutations of a set with five members, $P(5,5)$. Now, for each choice of men to occupy the end seats, we have a choice of people to occupy the middle seats, and the product rule applies. Accordingly, the total number of ways these seven people can be seated with men occupying the end seats is $P(4,2)P(5,5)$

$= (4 \cdot 3)(5 \cdot 4 \cdot 3 \cdot 2 \cdot 1) = 12 \cdot 120 = 1{,}440$. Notice also that, since there are restrictions on the occupants of the end seats, it was easier for us to calculate the number of ways the end seats could be occupied before we calculated the number of ways the middle seats could be occupied. ■

Example 23

How many ways can we seat eight people—Mary, Tom, John, Sue, Bill, Nancy, Lynn, and Peg—in a straight row of chairs if Mary cannot immediately follow Sue?

Analysis

In this problem it is easier to count the number of orderings we do not want and then subtract that number from the total number of ways the people can be ordered. The total number of ways the eight people can be ordered is $P(8,8) = 8!$. To calculate the number of ways the eight people can be ordered *with* Mary immediately following Sue, we reason as follows. Since Mary follows Sue, Sue can occupy any of the first seven seats. But once Sue is seated, there is only one place that Mary can sit. So there are seven ways to seat these two people. Once Mary and Sue have occupied their seats, the remaining six people can occupy any of the remaining six seats, so the total number of ways these six people can be ordered is 6!. Therefore the number of ways the eight people can be ordered with Mary immediately following Sue is $7 \cdot 6! = 7!$. Notice again that, since there are restrictions on two people, it is easier to seat these people first. So the number of ways the eight people can be ordered if Mary cannot follow Sue is $8! - 7! = 40{,}320 - 5{,}040 = 35{,}280$. ■

Example 24

a. How many different eight-digit numbers can be formed if each digit is either a 1 or a 2?

b. How many such numbers contain at least two 1s?

Analysis

a. There are two possibilities for each digit, so the number of different eight-digit numbers is $2^8 = 256$.

b. It is easier to count the number of such numbers that contain fewer than two 1s and then subtract this number from 256. There is exactly one number that does not contain any 1s, 22222222. There are eight that contain exactly one 1; we can put a 1 in each of the eight places. By the sum rule, the number of such numbers that contain at least two 1's is $256 - (1 + 8) = 247$. ■

Example 25

The number of different ways eight people can occupy a row of eight seats is 8!. How many different ways can eight people be arranged in a circle?

Analysis

The answer is *not* 8!, because two circular arrangements are considered to be the same if one can be obtained from the other by a rotation. For example, the two circular arrangements in Figure 6.4 are considered to be the same. So we must determine a way to avoid counting this arrangement twice. If the names of the eight people are Mary, Tom, John, Sue, Bill, Nancy, Lynn, and Peg, we can avoid counting an arrangement twice by agreeing to first put Mary at a position. So Mary occupies a fixed position, and we have seven possibilities for the seat on Mary's immediate right. Then we have six possibilities for the next seat, and so on. So the number of different ways eight people can be arranged in a circle is 7!.

Figure 6.4

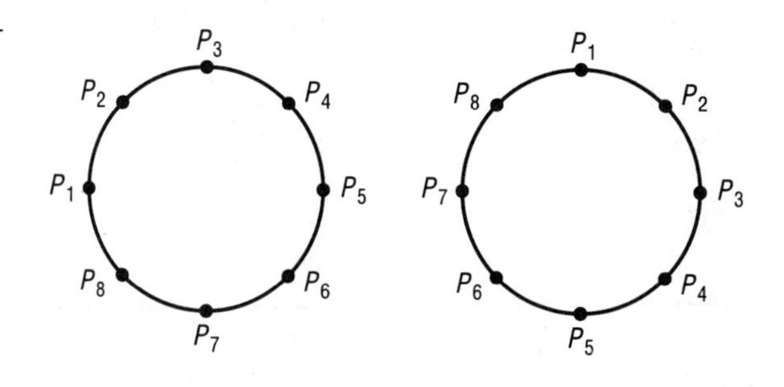

There is an alternative way to arrive at this same answer, one that is in some sense more algebraic. We first calculate that there are 8! different arrangements and then, when we are told that two circular arrangements are considered the same if one can be obtained from the other by a rotation, we divide by the right number to adjust for the change in the rules. Every time we are given an arrangement such as the one on the right-hand side of Figure 6.4, there are really eight arrangements that we consider the same (the first has P_1 at the top, the second has P_2 at the top, and so forth). Therefore each technically different arrangement has been counted eight times, and so we must divide 8! by 8. The result is again 7!. ■

Exercises 6.5

60. Four different cards, marked A, K, Q, and J, are lying on a desk. Two cards are drawn in succession, and the first card is not replaced before the second is drawn. How many different results can there be?

61. Two cards are drawn from a standard deck of 52 cards in succession without replacing the first card.

a. How many ways can this be done if the cards do not make a pair?

b. How many ways can this be done if the cards do make a pair?

c. How many ways are there for the cards to be the same suit?

d. How many ways are there for the cards to be the same color?

62. You throw two dice.

 a. How many ways can you get a 5?

 b. How many ways can you get an 8?

 c. How many ways can you get more than 7?

 d. How many ways can you get an even number?

 e. How many ways can you get doubles?

63. Reconsider Exercise 60, this time with the first card replaced before the second one is drawn.

64. There is space on a shelf for four books, and six different books are available. How many different arrangements are possible?

65. How many ways can four people be arranged in six chairs that are placed in a row?

66. How many ways can seven people—Jack, Diane, Ken, Linda, Dan, Harold, and April—be arranged in a row if Diane must follow Jack and Harold must follow Diane?

67. Using only the digits 1 and 2, how many different ten-digit numbers can be formed that contain at least two 2s?

68. Using only the digits 1, 2, and 3, how many different six-digit numbers can be formed that contain at least two 1s and at least two 2s?

69. How many ways can three numbers (not necessarily distinct) be chosen from the set $\{n \in \mathbb{N}: n \leqslant 9\}$ in such a way that the sum of the numbers is at least 5?

70. Using only the digits 1, 2, 3, 4, 5, 6, and 7, how many five-digit numbers can be formed that satisfy the following conditions?

 a. no additional conditions

 b. at least one 5

 c. at least one 5 and at least one 6

 d. no repeated digits

 e. at least one 5 and no repeated digits

 f. the first and last digits the same

71. A committee consists of five people.

 a. In how many ways can the five be seated around a round table?

 b. How many arrangements are there if we consider the possibility that some but not all the members are absent?

72. At State University there are four freshman English courses and five freshman math courses. In how many ways can a student take exactly two English courses and two math courses?

73. a. Assuming that the number does not begin with 0 and no digit is used

twice, how many different six-digit numbers can be formed with the digits 0, 1, 2, 3, 4, and 5?

b. How many of the numbers in (a) are divisible by 2?

74. Four men and four women are to be seated in eight chairs on the same side of the head table. How many seating arrangements are possible if no two men are to sit next to each other?

75. Four women and three men are to be seated in seven chairs on the same side of the head table. How many seating arrangements are possible if no two women are to sit next to each other?

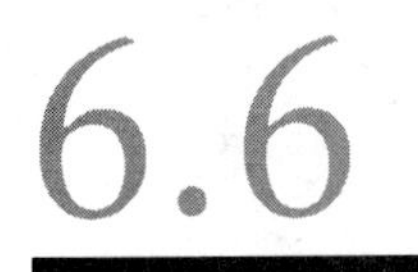

COMBINATIONS

Our study of permutations involved the study of ordered arrangements. We now study finite sets that are not arranged in any way. By definition, sets are not arranged in any way, so actually we are not considering anything new. But there is a shift in our point of view, because we now wish to consider problems of counting, such as determining the number of r-element subsets of a set with n elements.

Definition

Let S be a set with n elements ($n \in \mathbb{N}$), and let $r \in \mathbb{N}$ such that $r \leq n$. A subset of S of cardinality r is called an ***r*-combination** of S.

Definition

The number of r-combinations of a set with n members is denoted by $C(n,r)$ and is called a **binomial coefficient.** It is also convenient to define $C(n,0) = 1$.

Originally, the binomial coefficient $C(n,r)$ was called "the number of combinations of n things taken r at a time." This cumbersome phraseology has been replaced by "n choose r." The symbol $C(n,r)$ is often denoted by $\binom{n}{r}$, which is easy to write on a blackboard but hard to write with a printer. We use $C(n,r)$ throughout this text.

Example 26

We can construct the 3-permutations of the four-element set {1,2,3,4} by selecting the orderings of the 3-combinations of {1,2,3,4}. There are four 3-combinations of {1,2,3,4}, namely, {1,2,3}, {1,2,4}, {1,3,4}, and {2,3,4}, and there are six orderings associated with each 3-combination. For example, the orderings associated with {1,2,3} are 123, 132, 213, 231, 312, and 321. This construction is illustrated in Figure 6.5. ■

Figure 6.5

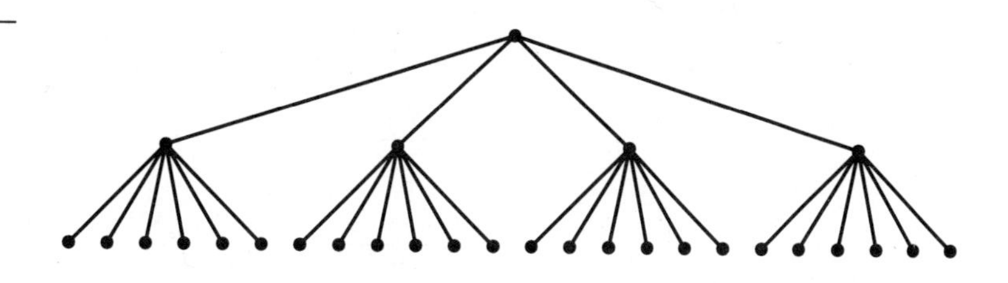

Theorem 6.3

> Let n be a natural number and let r be an integer with $0 \leq r \leq n$. Then $C(n,r) = P(n,r)/r!$.

Proof

Let S be a set with n elements and let V be a subset of S with r elements. The number of r-permutations of V is $r!$. To each of these r-permutations of V there corresponds an r-permutation of S (we simply replace the codomain V with the codomain S). Therefore, corresponding to the set V is a set $P(V)$ of r-permutations of S which has $r!$ members. Let W be a different subset of S with r members and let $P(W)$ be the corresponding set of r-permutations of S. Note that $P(W)$ and $P(V)$ are disjoint. (Why?)

We tally up the r-permutations of S. There are $C(n,r)$ subsets of S having cardinality r and each of these subsets gives rise to $r!$ r-permutations of S. Moreover, no two (different) r-element subsets give rise to the same r-permutation of S. Hence we have *at least* $r!C(n,r)$ r-permutations of S; that is, $r!C(n,r) \leq P(n,r)$.

Could it be that we have failed to count some r-permutation of S? Let $\sigma: I_r \to S$ be an r-permutation of S and let $V = \sigma(I_r)$. Then V is a subset of S with cardinality r and so $\sigma: I_r \to V$ is an r-permutation of a subset of S with cardinality r. So $\sigma: I_r \to S$ was counted. It follows that $P(n,r) = r!C(n,r)$, and since $r! \neq 0$ we may conclude that $C(n,r) = P(n,r)/r!$. □

Since 0! is defined to be 1, the formula in Theorem 6.3 also holds whenever $r = 0$.

Corollary 6.4

> For $r, n \in \mathbb{N}$ with $r \leqslant n$,
>
> $$C(n,r) = C(n,n-r) = \frac{n(n-1)(n-2)\cdots(n-r+1)}{1\cdot 2\cdot 3\cdots r}$$

Proof

We recall that $P(n,r) = n!/(n-r)!$ and so

$$C(n,r) = \frac{n!}{r!(n-r)!} = \frac{n(n-1)(n-2)\cdots(n-r+1)}{r!}$$

$$= \frac{n(n-1)(n-2)\cdots(n-r+1)}{1\cdot 2\cdot 3\cdots r}$$

On the other hand,

$$C(n,n-r) = \frac{P(n,n-r)}{(n-r)!} = \frac{n!}{(n-(n-r))!(n-r)!}$$

$$= \frac{n!}{r!(n-r)!} = C(n,r)$$

□

Before considering the usefulness of Theorem 6.3 and Corollary 6.4, let us pause to consider the difference in the way the theorem and the corollary have been proved. The proof of the corollary is algebraic. Anyone who understands the rules of algebra can establish the corollary mechanically without any thought or understanding. In contrast, the proof of the theorem itself has hardly any algebra to it—none in fact, unless you count our dividing through by $r!$ in the last step. The entire proof is a matter of counting. Such an argument might be called a counting argument, but it is customary to call it a *combinatorial argument*.

The usefulness of Theorem 6.3 is illustrated in the examples that follow. What about the corollary? All you really have to remember about this formula is that "corresponding terms" add to $n + 1$. Suppose, for example, that you wish to calculate $C(n,r)$ when $n = 8$ and $r = 5$. Start with the pairs of natural numbers that add to $n + 1 = 9$, namely, 1 and 8, 2 and 7, 3 and 6, and so forth. Write the smaller numbers in the denominator and the larger numbers in the numerator, and stop when you have written r or $n - r$ in the denominator. The result,

$$\frac{(8)(7)(6)}{(1)(2)(3)} \quad \text{[stop]}$$

is $C(8,5)$.

Example 27

The number of nine-element subsets of a set with twelve members is

$$C(12,9) = \frac{12!}{9!3!} = \frac{12 \cdot 11 \cdot 10}{1 \cdot 2 \cdot 3} = 220 \qquad \blacksquare$$

Example 28

There are 52 cards in an ordinary deck of cards, and a poker hand consists of five cards. How many different poker hands are there?

Analysis

A poker hand is simply a subset of size 5 taken from a set with 52 members, so the answer is

$$C(52,5) = \frac{52!}{5!47!} = \frac{52 \cdot 51 \cdot 50 \cdot 49 \cdot 48}{1 \cdot 2 \cdot 3 \cdot 4 \cdot 5} = 2{,}598{,}960 \qquad \blacksquare$$

Example 29

A bridge hand consists of thirteen cards. The number of different bridge hands is

$$C(52,13) = \frac{52!}{13!39!} = 635{,}013{,}559{,}600 \qquad \blacksquare$$

Example 30

If seven golfers are available, how many different foursomes can be formed?

Analysis

Instead of trying to determine which formula to use, let us attempt to solve the problem on its own merit. We begin the process by selecting a foursome. Since seven golfers are available, we have seven possibilities for the first member of our foursome. Having chosen the first member, we have six possibilities for the second, and so on. Thus we have $7 \cdot 6 \cdot 5 \cdot 4 = 840$ possibilities. However, different orderings of the same four players have been counted as different foursomes; that is, we have counted Tom, Bill, John, and Wayne as being a different foursome than John, Wayne, Bill, and Tom. Since these are not different foursomes, we must eliminate duplications from the total number of possibilities. A given foursome can be ordered in $4! = 24$ different ways, so the 840 possibilities has counted each foursome 24 times. We thus see that there are $840/24 = 35$ possible foursomes. Notice that this is $C(7,4)$. (Incidentally, we think this argument should remind you of the alternative argument given in Example 25.) ■

Example 31

One of the Senate committees consists of six Democrats and three Republicans. How many five-member subcommittees can be formed if each subcommittee is to have at least three Democrats?

Analysis

The number of ways of choosing three Democrats is $C(6,3)$, and the number of ways of choosing two Republicans is $C(3,2)$. Therefore the number of subcommittees that have exactly three Democrats is

$$C(6,3) \cdot C(3,2) = \frac{6!}{3!3!} \cdot \frac{3!}{2!1!} = 20 \cdot 3 = 60$$

The number of ways of choosing four Democrats is $C(6,4)$, and the number of ways of choosing one Republican is $C(3,1)$. Therefore the number of subcommittees that have exactly four Democrats is

$$C(6,4) \cdot C(3,1) = \frac{6!}{4!2!} \cdot \frac{3!}{1!2!} = 15 \cdot 3 = 45$$

The number of ways of choosing five Democrats is $C(6,5) = 6!/(5!1!) = 6$. Hence the number of subcommittees that have at least three Democrats is $60 + 45 + 6 = 111$. ■

Example 32

How many ways can ten different books be divided among Mary, Tom, and Sue so that five are given to Mary, three to Tom, and two to Sue?

Analysis

There are $C(10,5)$ different ways of choosing the five books to give to Mary, and there are $C(5,3)$ different ways of choosing three of the remaining five to give to Tom. Once this has been done, there is only one $[1 = C(2,2)]$ way to choose two of the remaining two books to give to Sue. So the answer to the question is

$$C(10,5) \cdot C(5,3) \cdot C(2,2) = 252 \cdot 10 \cdot 1 = 2{,}520$$

■

Example 33

The number of subsets of size 2 of the set $\{n \in \mathbb{N}: n \leq 10\}$ is $C(10, 2) = 45$. How many of these sets do not contain two consecutive natural numbers?

Analysis

The sets of size 2 that do contain consecutive natural numbers are $\{1,2\}$, $\{2,3\}$, $\{3,4\}$, $\{4,5\}$, $\{5,6\}$, $\{6,7\}$, $\{7,8\}$, $\{8,9\}$, and $\{9,10\}$. So the answer to the question is $45 - 9 = 36$. ■

Example 34

How many routes are there from A to B in the accompanying figure if we can travel only up or to the right?

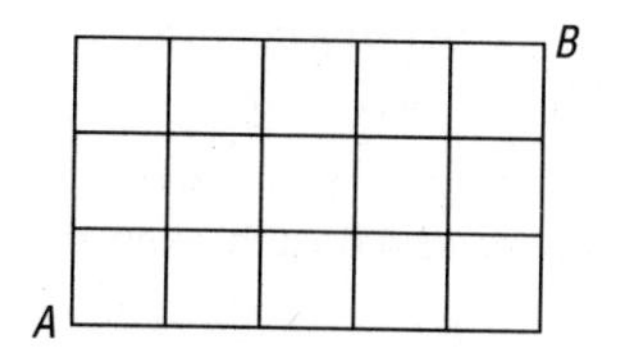

Analysis

Each route consists of eight steps, and we can label each step with an R or a U to indicate whether it is to the right (R) or up (U). Thus the route in the following figure bears the label R, U, R, R, U, R, U, R.

Also, each route consists of five steps to the right and three steps up. Thus each route corresponds to a list of five Rs and three Us, and the number of routes is the number of arrangements of five Rs and three Us. Hence the number of routes is $8!/5!3! = 56$. ■

Exercises 6.6

76. Evaluate the following. **a.** $C(5,3)$ **b.** $C(10,3)$ **c.** $C(15,4)$
77. In how many different ways can we select a committee of five from a group of twelve people?
78. How many different sums of money can be formed by choosing three coins from a penny, a nickel, a dime, a quarter, and a half-dollar?
79. How many different five-card hands can be chosen from seven cards?
80. An honorary organization has ten male and sixteen female members. How many different committees consisting of two males and five females can be formed?
81. From a group of eleven Democrats and nine Republicans, how many different committees of six can be chosen that contain
    a. exactly four Republicans? **b.** at least four Republicans?
    c. at most four Republicans?
82. If $P(n,4) = 15{,}120$, find $C(n,4)$.
83. If $C(n,5) = 462$, find $P(n,5)$.
84. Does there exist a natural number n such that $P(n,4) = 11{,}880$? If so, find n. If not, explain why not.
85. How many different committees of five from a group of fourteen men can be selected if

a. a certain pair of men insist on serving together or not at all?

b. a certain pair of men refuse to serve together?

86. If a baseball league consists of eight teams, how many games are played during the course of a season in which each team plays every other team exactly 22 times?

87. a. How many ways can a student choose ten questions from a thirteen-question exam?

b. How many ways can a student choose ten questions from a thirteen-question exam if eight questions must be chosen from the first ten and two from the last three?

c. How many ways can a student choose ten questions from a thirteen-question exam if at least five questions must be chosen from the first eight and at least three questions from the last five?

88. How many different two-element subsets of $\{n \in \mathbb{N}: n \leq 50\}$ are there such that the sum of the two elements is even?

89. A social committee and a judicial committee are to be formed. It has been decided that the social committee will not have more than nine members and the judicial committee will not have more than seven members. It has also been decided that the two committees will be chosen from a group of fourteen people and each person will serve on exactly one committee. How many different committees can be formed if a group of three people refuse to serve on the judicial committee and a group of four people refuse to serve on the social committee?

90. How many different ways can twelve people be partitioned into three sets, where each set consists of four people?

91. Prove that if $n,r \in \mathbb{N}$ and $r \leqslant n$ then $C(n,r) = (n/r)C(n-1,r-1)$.

92. Prove that if $n \in \mathbb{N}$ then $C(2n,n)$ is even. (*Hint:* For appropriate sets A, consider $A^{\sim}$.)

93. Let $m,n,$ and r be nonnegative integers with $r \leqslant m$ and $r \leqslant n$. Prove that $C(m+n,r) = \Sigma_{i=0}^{r} C(m,r-i)C(n,i)$.

*94. Ten scientists working on a secret project wish to lock documents in a file so that five of the scientists must be present for the file to be opened. What is the smallest number of locks that are needed? What is the smallest number of keys that each scientist must carry?

95. a. How many routes are there from A to B in the following figure if we can travel only up or to the right?

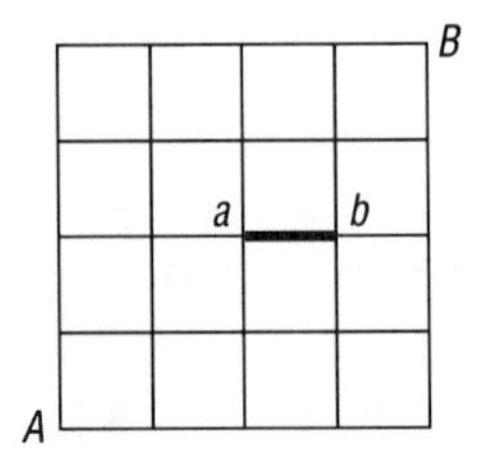

b. How many routes are there from A to B if the step from a to b cannot be traveled? Again assume that we can travel only up or to the right.

6.7 BINOMIAL COEFFICIENTS

In this section we study the binomial coefficients in more detail. We begin by establishing a useful property known as Pascal's formula (Theorem 6.5), named in honor of Blaise Pascal (1623–1662). (The computer language Pascal is also named in his honor, which is why the name of this language has only one capital letter; unlike COBOL and FORTRAN, "Pascal" is not an acronym.)

Theorem 6.5

If $n,r \in \mathbb{N}$ and $r < n$, then $C(n,r) = C(n-1,r) + C(n-1,r-1)$.

Proof

$C(n-1,r) + C(n-1,r-1)$

$$= \frac{(n-1)!}{r!(n-1-r)!} + \frac{(n-1)!}{(r-1)!(n-r)!}$$

$$= (n-1)!\left[\frac{1}{r!(n-1-r)!} + \frac{1}{(r-1)!(n-r)!}\right]$$

$$= (n-1)!\left[\frac{(n-r)}{r!(n-r)!} + \frac{r}{r!(n-r)!}\right]$$

$$= \left[\frac{(n-1)!}{r!(n-r)!}\right][(n-r)+r]$$

$$= \left[\frac{(n-1)!}{r!(n-r)!}\right](n)$$

$$= \frac{n!}{r!(n-r)!}$$

$$= C(n,r)$$

It is instructive to give a combinatorial argument for Theorem 6.5 as well.

Alternative Proof

Let S be a set with n elements and let $x \in S$. We divide the r-element subsets of S into two collections:

$$\mathcal{A} = \{A \subseteq S: n(A) = r \text{ and } x \notin A\}$$

$$\mathcal{B} = \{A \subseteq S: n(A) = r \text{ and } x \in A\}$$

We note that $n(\mathcal{A})$ is just the number of r-element subsets of $S - \{x\}$, namely, $C(n - 1,r)$. Let us consider $\mathcal{B}' = \{A - \{x\}: A \in \mathcal{B}\}$. Note that $\mathcal{B}$ and $\mathcal{B}'$ have the same number of members. Every set in $\mathcal{B}'$ is an $(r - 1)$-element subset of $S - \{x\}$, and in fact every $(r - 1)$-element subset of $S - \{x\}$ belongs to $\mathcal{B}'$. Therefore $n(\mathcal{B}) = n(\mathcal{B}') = C(n - 1,r - 1)$. It follows that $C(n,r) = C(n - 1,r) + C(n - 1,r - 1)$. □

It is just a matter of taste whether one performs the manipulative algebraic proof of Theorem 6.5 or the subsequent thoughtful combinatorial proof. It is possible to prove the next theorem by induction, but the combinatorial proof of this theorem is clearly easier. Notice that the crux of the combinatorial proof is to count the number of subsets of a set in two different ways.

Theorem 6.6 If $n \in \mathbb{N}$, then $\Sigma_{r=0}^{n} C(n,r) = 2^n$.

Proof

By Theorem 4.3, the number of subsets of a set with n members is 2^n. For each $r = 0,1,2, \ldots ,n$, $C(n,r)$ is the number of subsets, consisting of r members, of a set with n elements. Hence $\Sigma_{r=0}^{n} C(n,r)$ is also the number of subsets of a set with n members. □

Our next theorem can be proved combinatorially, but we admit that here the algebraists have the upper hand.

Theorem 6.7 If $n,r \in \mathbb{N}$ and $r \leqslant n$, then $\Sigma_{i=r}^{n} C(i,r) = C(n + 1,r + 1)$.

Proof

The proof is by induction. Let

$$S = \{n \in \mathbb{N}: \Sigma_{i=r}^{n} C(i,r) = C(n + 1,r + 1)\}$$

Since $\Sigma_{i=1}^{1} C(1,1) = C(1,1) = 1 = C(2,2)$, $1 \in S$.

Suppose $n \in S$. Then $\sum_{i=r}^{n} C(i,r) = C(n+1,r+1)$. Therefore

$$\sum_{i=r}^{n+1} C(i,r) = \sum_{i=r}^{n} C(i,r) + C(n+1,r)$$
$$= C(n+1,r+1) + C(n+1,r)$$

By Theorem 6.5, $C(n+1,r+1) + C(n+1,r) = C(n+2,r+1)$. Therefore $\sum_{i=r}^{n+1} C(i,r) = C(n+2,r+1)$, and hence $n+1 \in S$. By the principle of mathematical induction, $S = \mathbb{N}$. □

Numbers of the form $C(n,r)$, where $n \in \mathbb{N}$ and r is an integer such that $0 \leqslant r \leqslant n$, are called *binomial coefficients* because they appear as coefficients in the right-hand side of the following equation, which is called the *binomial formula*.

$$(x+y)^n = \sum_{r=0}^{n} C(n,r)x^{n-r}y^r$$

The proof by induction of Theorem 6.8 is left as Exercise 118. Here we give an informal combinatorial argument. As we have indicated in the introduction to this chapter, we prefer the combinatorial argument because it provides insight into why the theorem is true.

Theorem 6.8

The Binomial Theorem If $x,y \in \mathbb{R}$ and $n \in \mathbb{N}$, then

$$(x+y)^n = \sum_{r=0}^{n} C(n,r)x^{n-r}y^r$$

Informal Argument

We first raise $x + y$ to the third power, but at each stage of multiplication we distinguish the numbers we multiply by from the numbers we multiply at.

$$x_1 + y_1$$
$$\underline{x_2 + y_2}$$
$$x_1x_2 + y_1x_2 + x_1y_2 + y_1y_2$$
$$\underline{x_3 + y_3}$$
$$x_1x_2x_3 + \underline{y_1x_2x_3} + \underline{x_1y_2x_3} + y_1y_2x_3 + \underline{x_1x_2y_3} + y_1x_2y_3 + x_1y_2y_3 + y_1y_2y_3$$

In calculating $(x+y)^3$, all the x's are the same, as are all the y's. Thus the three products that are underlined are all really x^2y. Why should there be exactly three x^2y's? We have a three-element set $\{x_1,x_2,x_3\}$, and each two-element subset of this set gives rise to x^2y. So the coefficient of x^2y

in the expansion of $(x + y)^3$ is $C(3,2)$, where 3 is the power to which $x + y$ is raised and 2 is the power to which x is raised.

We continue our imagined distinction with each multiplication and find the coefficient of x^5y^2 in the expansion of $(x + y)^7$. We have a seven-element set $\{x_1,x_2,x_3,x_4,x_5,x_6,x_7\}$, and each five-element subset of this set gives rise to one on the x^5y^2's. For example, the five-element subset $\{x_1,x_2,x_4,x_6,x_7\}$ gives rise to $x_1x_2y_3x_4y_5x_6x_7$ in which we multiplied at x_1 and multiplied by x_i in stages $i = 2$, 4, 6, and 7 and by y_i in stages $i = 3$ and 5. If at any given stage we did not multiply by x_i, then we must have multiplied by y_i. There are therefore as many x^5y^2 as there are five-element subsets of $\{x_1,x_2,x_3,x_4,x_5,x_6,x_7\}$; that is, there are $C(7,5)x^5y^2$'s in the expansion of $(x + y)^7$. For each natural number n and each r with $r = 0,1,2,\ldots,n$, we have an $x^{n-r}y^r$ in the expansion of $(x + y)^n$, and its coefficient is $C(n,n - r) = C(n,r)$. As a result, $(x + y)^n = \sum_{r=0}^{n} C(n,r)x^{n-r}y^r$. □

In Exercise 119, you are asked to explain what happens if we use y's instead of x's in calculating the coefficient of x^5y^2 in the expansion of $(x + y)^7$.

Definition

Let A and B be algebraic expressions and let n be a natural number. Let $r \in \{1,2,3,\ldots,n + 1\}$. Then $C(n,r - 1)A^{n-r+1}B^{r-1}$ is called the **rth term of the binomial expansion of $(A + B)^n$.** When $n = 2k$, the $(k + 1)$st term is called the **middle term.**

The motivation for this terminology can be seen in Figure 6.6.

Example 35

Use the binomial formula to find each term of the binomial expansion of $(2x + 3y)^5$. Simplify each term.

Analysis

$$\begin{aligned}(2x + 3y)^5 &= C(5,0)(2x)^5 + C(5,1)(2x)^4(3y) + C(5,2)(2x)^3(3y)^2 \\ &\quad + C(5,3)(2x)^2(3y)^3 + C(5,4)(2x)(3y)^4 + C(5,5)(3y)^5 \\ &= (2x)^5 + 5(2x)^4(3y) + 10(2x)^3(3y)^2 + 10(2x)^2(3y)^3 \\ &\quad + 5(2x)(3y)^4 + (3y)^5 \\ &= 32x^5 + 240x^4y + 720x^3y^2 + 1080x^2y^3 + 810xy^4 + 243y^5\end{aligned}$$

■

Example 36

Write the 6th term of the binomial expansion of $(x^2 + 2y)^8$.

Analysis

The 6th term contains $(2y)^5$, so it is

$$C(8,5)(x^2)^3(2y)^5 = \frac{8 \cdot 7 \cdot 6}{3 \cdot 2} x^6(32y^5) = 1792x^6y^5$$

■

If you had been studying algebra a century ago, your text would have illustrated the use of the binomial theorem by examples similar to those we have just given and by exercises such as the ones we are about to give in Exercises 96 through 105. The following theorem illustrates a subtler use of the binomial theorem, one typical of the way the binomial theorem is used in modern discrete mathematics.

Theorem 6.9

Let n be a natural number. Then the following are true:

a. $C(n,0) + C(n,1) + C(n,2) + \cdots + C(n,n) = 2^n$
b. $C(n,0) - C(n,1) + C(n,2) - C(n,3) + \cdots \pm C(n,n) = 0$
c. $C(n,0) + C(n,2) + C(n,4) + \cdots + C(n,2\lfloor n/2 \rfloor)$
$= C(n,1) + C(n,3) + C(n,5) + \cdots + C(n, 2\lfloor (n+1)/2 \rfloor - 1)$
$= 2^{n-1}$

Theorem 6.9a is just Theorem 6.6. In Exercises 107 and 108 you are asked to use the binomial theorem to prove (a) and (b). The proof of (c) does not use the binomial theorem, but it does use (a) and (b) (see Exercise 109). Note that according to Theorem 6.9c the sum of the coefficients of the odd terms equals the sum of the coefficients of the even terms and this common sum is 2^{n-1}.

The binomial coefficients are frequently arranged in the form of a triangle, called *Pascal's triangle* (Figure 6.6). Pascal himself called it the *arithmetic triangle* and discussed its properties in a treatise written in 1653.

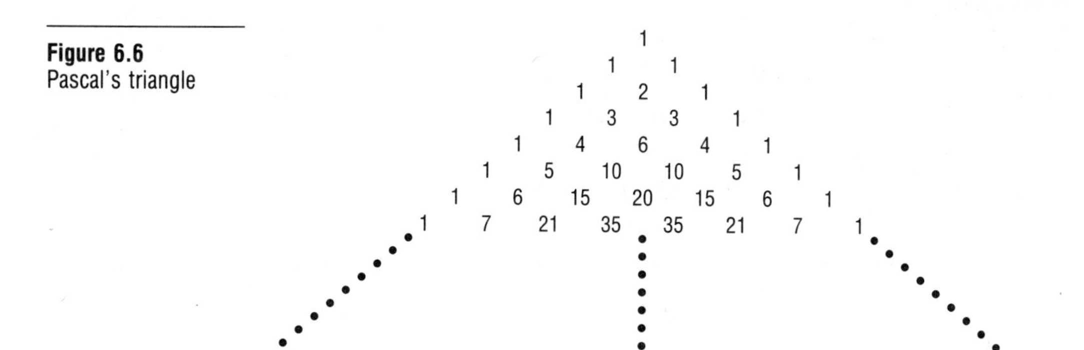

Figure 6.6
Pascal's triangle

One could argue that "arithmetic triangle" is a better name for Pascal's triangle, because the triangle had been discovered by the Chinese at least 400 years before Pascal wrote his treatise.

It is customary to call the top row of Pascal's triangle the 0th row, the next row the 1st row, and so on. With this understanding, the nth row of Pascal's triangle consists of the terms of the expansion of $(1 + 1)^n$. By Pascal's formula (Theorem 6.5), to obtain a particular entry (other than the ends) in this triangle, we simply add the two nearest binomial coefficients in the row immediately above it. Thus the 8th row of Pascal's triangle is

$$\begin{array}{ccccccccccccccccc} & 1 & + & 7 & + & 21 & + & 35 & + & 35 & + & 21 & + & 7 & + & 1 & \\ & & \downarrow & & \downarrow & & \downarrow & & \downarrow & & \downarrow & & \downarrow & & \downarrow & & \\ 1 & & 8 & & 28 & & 56 & & 70 & & 56 & & 28 & & 8 & & 1 \end{array}$$

Pascal's triangle has many interesting properties. We mention one of historical interest. If p is a prime number, then except for the ends, every number in the pth row is divisible by p. The reason for this is that each such number is of the form $C(p,r) = p!/(p - r)!r!$, where $r \neq 0$ and $r \neq p$. So the numerator is $p(p - 1)!$, whereas the prime number p is not to be found in the denominator. Hence the prime number p in the numerator is never divided out. Now the sum of the numbers in the pth row is 2^p. Therefore $2^p = 1 + p(\text{some integer}) + 1$, and it follows that p divides $2^p - 2 = 2(2^{p-1} - 1)$. Fermat, who corresponded and collaborated with Pascal, conjectured that p is a prime number if and only if p divides $2^p - 2$. We have just seen that if p is a prime then indeed p divides $2^p - 2$, and Fermat later showed that if p is prime and a is not a multiple of p then p divides $a^{p-1} - 1$. This result is known as *Fermat's little theorem*. Fermat's conjecture was disproved by Leonhard Euler (pronounced "Oiler," 1707–1783), who showed that 341 divides $2^{341} - 2$ even though $341 = (31)(11)$ is not prime.

Exercises 6.7

96. What is the first term of $(x + 1)^{10}$?

97. What is the third term of $(x - 1)^7$?

98. What is the last term of $(x + y)^8$?

99. What is the middle term of $(2x + y/2)^6$?

100. What is the middle term of $(x + y)^{10}$?

101. Find the binomial expansion of $(x + y)^3$.

102. Use the binomial formula to find each term of the binomial expansion of

a. $(a/2 + b^2/a)^4$ **b.** $(2a - 3b)^6$

103. Write the following in simplified form.

a. the 7th term in the expansion of $(3a^2 - 2b)^{10}$

b. the 4th term in the expansion of $(a + 4b^3)^{16}$

104. Use the binomial formula to compute $(1.01)^{12}$, using only enough terms for a result accurate to three decimal places. (*Hint:* $(1.01)^{12} = (1 + .01)^{12}$.)

105. Use the binomial formula to compute $(.99)^{13}$, using only enough terms for a result accurate to three decimal places.

106. Use induction to prove Theorem 6.6.

107. Let $n \in \mathbb{N}$. Use the binomial theorem to prove that

$$C(n,0) + C(n,1) + C(n,2) + \cdots + C(n,n) = 2^n$$

108. Let $n \in \mathbb{N}$. Use the binomial theorem to prove that

$$C(n,0) - C(n,1) + C(n,2) - C(n,3) + \cdots \mp C(n,n) = 0$$

109. Prove Theorem 6.9c.

110. Prove that $\Sigma_{r=0}^{n} C(2n + 1,r) = 2^{2n}$ for each natural number n.

111. Prove that, for each natural number n, $C(2n,2) = 2C(n,2) + n^2$.

112. Prove that, for each natural number n, $\Sigma_{r=0}^{n} (r + 1)C(n,r) = 2^n + n2^{n-1}$.

113. Prove that it follows from Fermat's little theorem that if p is a prime number then p divides $2^p - 2$.

114. Write Figure 6.6 (through line 7) using appropriate $C(n,r)$ (or $\binom{n}{r}$, if you prefer).

115. Let S be a set with eighteen elements.

a. How many subsets of S have an even number of elements?

b. How many subsets of S have an odd number of elements?

116. Let S be a set with nineteen elements.

a. How many subsets of S have an even number of elements?

b. How many subsets of S have an odd number of elements?

117. Let S be the ordinary English alphabet.

a. How many eleven-element subsets of S do not contain the letter J?

b. How many eleven-element subsets of S contain the letter K?

118. Prove the binomial theorem by induction.

119. Explain informally what would happen if in the proof of Theorem 6.8 you were to keep track of the y's instead of the x's in calculating the coefficient of x^5y^2 in the binomial expansion of $(x + y)^7$.

6.8

ARRANGEMENTS WITH REPETITIONS

In this section, we study arrangements with repetitions. How many different arrangements of the letters in the word *add* are there? It is easy to see that there are only three (the letter *a* can be first, second, or third): *add, dad, dda*. Compare this to the number of different arrangements of the letters of the word *and*: *and, adn, nad, nda, dan, dna*. There are 6 [= $P(3,3)$] arrangements of the letters of the word *and*.

There is another way to arrive at three arrangements of the letters in the word *add*. Suppose the two *d*s are distinguishable; label one of them d_1 and the other d_2. Then the arrangements are ad_1d_2, ad_2d_1, d_1ad_2, d_1d_2a, d_2ad_1, d_2d_1a. Notice that each arrangement in this list gives two arrangements of the preceding list of *add* letters. Thus the number of arrangements of the letters of the word *add* is $P(3,3)/P(2,2)$.

There is yet another way of calculating the number of arrangements of the letters in the word *add*. We have three positions, and we can choose any one of these positions for the letter *a*, so the number of ways we can do this is $C(3,1)$. We then have two positions for the letters *dd*, and there is only one [= $C(2,1)$] way we can do this. Therefore the total number of arrangements is $C(3,1)C(2,1) = 3 \cdot 1 = 3$. Alternatively, we can first note that we have three positions for the letters *dd*, so the number of ways we can place the letters *dd* in these three places is $C(3,2)$. We then have only one position for the letter *a*, so the total number of arrangements is $C(3,2)C(1,1) = 3 \cdot 1 = 3$.

This analysis of the arrangements of the letters of a word illustrates the following theorem.

Theorem 6.10

Let T be a set with n members, and suppose that there is a pairwise disjoint collection $\mathscr{C} = \{T_i: i = 1,2, \ldots, r\}$ of nonempty subsets of T whose union is T. For each $i = 1,2,3, \ldots, r$, let n_i be the number of members of T_i. If we assume that for each i the members of T_i are indistinguishable but that we can distinguish two objects that come from different members of $\mathscr{C}$, then the number of distinguishable permutations of the members of T is

$$C(n,n_1)C(n - n_1,n_2)C(n - (n_1 + n_2),n_3)C(n - (n_1 + n_2 + n_3),n_4) \\ \times \cdots \times C(n - (n_1 + n_2 + n_3 + \cdots + n_{r-1}),n_r) \quad \square$$

As we have said, the arrangements of *add* provide a simple illustration of Theorem 6.10, so before proving this theorem let us apply it to *add*. We present the necessary data in Table 6.2.

Table 6.2

T:	$\{a, d_1, d_2\}$
$\mathscr{C}$:	$\{\{a\}, \{d_1, d_2\}\}$
n, n_1, n_2:	$n = 3\ n_1 = 1\ n_2 = 2$
$C(n, n_1)$:	$C(n, n_1) = C(3,1) = 3$
$C(n - n_1, n_2)$:	$C(n - n_1, n_2) = C(2,2) = 1$

According to Theorem 6.10, there are $3 \cdot 1$ distinguishable permutations of $\{a,d_1,d_2\}$ when we agree that we cannot distinguish d_1 from d_2.

Proof

If we use the notation introduced after the definition of permutation given in Section 6.3 (p. 172), we may think of each permutation of T as an n-letter word where the letters are selected from the alphabet T. Then n_1 letters from T_1 can be located among the n positions in $C(n,n_1)$ ways, leaving $n - n_1$ positions unused. The n_2 letters from T_2 can be located among the $n - n_1$ free positions in $C(n - n_1,n_2)$ ways, leaving $n - (n_1 + n_2)$ positions unused. The n_3 letters from T_3 can be located among the $n - (n_1 + n_2)$ free positions in $C(n - (n_1 + n_2),n_3)$ ways, leaving $n - (n_1 + n_2 + n_3)$ positions unused. When finally we reach the last subset T_r of T, there are $n - (\Sigma_{i=1}^{r-1} n_i)$ free positions and the last n_r letters of T_r can be located among these positions in $C(n - \Sigma_{i=1}^{r-1} n_i,n_r)$ ways. By the product rule, the total number of distinguishable permutations is

$$C(n,n_1)C(n - n_1,n_2) \cdots C(n - \Sigma_{i=1}^{r-1} n_i,n_r) \qquad \square$$

It turns out that the cumbersome product of Theorem 6.10 can be simplified, and the key to the proof of this result is "He robs Peter to pay Paul."

Theorem 6.11

Let $r \in \mathbb{N}$. For each $i \in I_r$, let $n_i \in \mathbb{N}$ and let $n = \Sigma_{i=1}^{r} n_i$. Then

$$C(n,n_1)C(n - n_1,n_2)C(n - (n_1 + n_2),n_3) \times \cdots$$
$$\times\ C(n - \Sigma_{i=1}^{r-1} n_i,n_r) = \frac{n!}{n_1!n_2!n_3! \cdots n_r!}$$

Proof

Recall that, for any integer k with $0 \leq k \leq n$, $C(n,k) = P(n,k)/k! = n!/(n-k)!k!$. Therefore the product may be rewritten as

$$\frac{n!}{n_1!(n-n_1)!} \cdot \frac{(n-n_1)!}{n_2!(n-(n_1+n_2))!} \cdot \frac{(n-(n_1+n_2))!}{n_3!(n-(n_1+n_2+n_3))!} \times \cdots$$

$$\times \frac{(n - \sum_{i=1}^{r-1} n_i)!}{n_r!(n - \sum_{i=1}^{r} n_i)!}$$

$$= \frac{n!}{n_1!n_2!n_3! \cdots n_{r-1}!n_r!(0)!}$$

$$= \frac{n!}{n_1!n_2!n_3! \cdots n_r!} \qquad \text{(since } 0! = 1\text{)}$$

Together Theorems 6.10 and 6.11 provide a quick way to calculate the number of arrangements of letters with repetitions. But it is instructive to work through an example as if we did not know these theorems. Example 37 illustrates the ideas behind the proofs. □

Example 37

Suppose that we have 14 indistinguishable marbles, 22 indistinguishable golf balls, 17 indistinguishable tennis balls, 36 indistinguishable baseballs, and 7 indistinguishable basketballs. Since we have five different kinds of objects, the r in Theorem 6.10 is 5. Also, $n_1 = 14$, $n_2 = 22$, $n_3 = 17$, $n_4 = 36$, $n_5 = 7$, and $n = 14 + 22 + 17 + 36 + 7 = 96$. How many different arrangements of these 96 objects are there?

Analysis

We can place the 14 marbles in 96 possible locations, and the number of ways of doing this is $C(96,14)$. Then we can place the 22 golf balls in $96 - 14 = 82$ locations, and the number of ways of doing this is $C(82,22)$. Next we can place 17 tennis balls in $82 - 22 = 60$ locations, and the number of ways of doing this is $C(60,17)$. Then we can place the 36 baseballs in $60 - 17 = 43$ locations, and the number of ways of doing this is $C(43,36)$. Finally we can place the 7 basketballs in the 7 remaining locations, and the number of ways of doing this is $C(7,7)$. So the number of distinguishable arrangements of the 96 objects is

$$C(96,14)C(82,22)C(60,17)C(43,36)C(7,7)$$

$$= \frac{96!}{14!82!} \cdot \frac{82!}{22!60!} \cdot \frac{60!}{17!43!} \cdot \frac{43!}{7!36!} \cdot 1$$

$$= \frac{96!}{14!22!17!36!7!}$$ ■

Together, Theorems 6.10 and 6.11 constitute an important principle of counting:

> The number of distinguishable permutations of n objects with n_1 indistinguishable objects of type 1, n_2 indistinguishable objects of type 2, . . ., and n_r indistinguishable objects of type r is given by
>
> $$\frac{n!}{n_1!n_2! \cdots n_r!}$$

We now apply this principle to a classic permutation problem, the Mississippi problem. We quote this problem and its solution word for word from Exercise 1, page 443 of *Academic Algebra* (1901), American Book Company.

Example 38

How many permutations may be made with the letters of the word *Mississippi* taken all together? Solution: The number is

$$\frac{\lfloor 11}{\lfloor 4 \ \lfloor 4 \ \lfloor 2} = 34650$$

There are two things of interest here. First, although we now take the notation $n!$ for granted, this notation was introduced in Europe in 1808 by Christian Kramp, and it had still not become standard in the United States by 1901. Second, the Mississippi problem has been worded in *Academic Algebra* with care. Consider the following problem: How many arrangements are there of the set consisting of the letters in the word *Mississippi?* Note that this is a different problem, because the set consisting of the letters in the word *Mississippi* is $\{M,i,s,p\}$, and this set has 4! permutations. ■

Although the Mississippi problem appears in nearly every text that discusses permutations, we concede that it is unimaginable that anyone could ever care that there are 34,650 arrangements. We have in mind a much more important use for Theorems 6.10 and 6.11.

In Section 6.7, we stated and proved the binomial theorem. It is natural to ask whether there is a trinomial theorem, a quadrinomial theorem, and in general an nth-order multinomial theorem. We begin by considering a trinomial theorem. Let a, b, and c be real numbers. By straightforward, but tedious, algebra we find

$$(a + b + c)^1 = a + b + c$$

$$(a + b + c)^2 = (a^2 + b^2 + c^2) + 2(ab + ac + bc)$$

$$(a + b + c)^3 = (a^3 + b^3 + c^3) + 3(ab^2 + ac^2 + bc^2 + a^2b + a^2c + b^2c) + 6abc$$

$$(a + b + c)^4 = (a^4 + b^4 + c^4) + 4(ab^3 + ac^3 + a^3b + bc^3 + a^3c + b^3c) + 6(a^2b^2 + a^2c^2 + b^2c^2) + 12(abc^2 + ab^2c + a^2bc)$$

There is a pattern here. The exponents of each term of $(a + b + c)^n$ add up to n. What is less obvious is that the coefficients are always of the form $n!/n_1!n_2!n_3! \cdots n_j!$, where the n_i's in the denominator are just the exponents of the given term. For $(a + b + c)^4$ there are four types of coefficients: $1 = 4!/4!$, $4 = 4!/1!3!$, $6 = 4!/2!2!$, and $12 = 4!/1!1!2!$. We recognize these coefficients from Theorems 6.10 and 6.11. Why should it be, for example, that there are six a^2b^2's in the expansion of $(a + b + c)^4$? Well, a^2b^2 is a simplification of some one of the following products: *abba*, *aabb*, *abab*, *baba*, *baab*, and *bbaa*. Why should there be exactly six such products? We have looked at this question before. We know that six is the number of distinguishable permutations of the set $\{a_1,a_2,b_1,b_2\}$ if we assume (as is certainly the case) that we cannot distinguish between a_1 and a_2 or between b_1 and b_2.

Notice that the number 3 from the *tri*nomial theorem has never entered into the discussion. We could just as well have been arguing that the coefficient of a^4b^3 in the binomial expansion of $(a + b)^7$ should be $7!/4!3!$ (which it is). After all, $7!/4!3!$ is the number of distinguishable permutations of $\{a_1,a_2,a_3,a_4,b_1,b_2,b_3\}$ when we assume, as is the case, that we cannot distinguish among a_1,a_2,a_3, and a_4 or among b_1,b_2, and b_3. Likewise, we know that the coefficient of $b^3c^4d^5f^2$ in $(a + b + c + d + e + f)^{14}$ is $14!/3!4!5!2!$. So really the binomial and trinomial theorems are just special cases of one general multinomial theorem, which we state informally and without proof.

Theorem 6.12

The Multinomial Theorem To expand $(x_1 + x_2 + x_3 + \cdots + x_m)^n$, write every possible product of the x_i's such that the exponents add to n. In front of each such product write $n!/n_1!n_2! \cdots n_j!$, where the n_i's in the denominator are the exponents of the given product. Sprinkle with $+$ signs as appropriate.

We now consider a different type problem, that of determining the number of choices from a set when members of the set can be chosen more than once.

Example 39

Eight pretty girls have won prizes in a contest at Mother Fletcher's. They may each select a prize, and their choice is limited to a beach ball, a Mother Fletcher's T-shirt, or a beach umbrella. How many different selections are possible?

Analysis

We are selecting eight objects from a set consisting of three different objects. For example, one selection is five beach balls, two T-shirts, and one umbrella. To answer the question, let us think of an order form with heading: beach balls, T-shirts, umbrellas. Each selection will be recorded as a collection of pluses (+). For example, the selection of five beach balls, two T-shirts, and one umbrella is recorded:

beach balls	T-shirts	umbrellas
+ + + + +	+ +	+

To count the number of selections, we can simplify matters by omitting the names of the prizes and separating the prizes by slash marks (/). We simply understand that beach balls are listed first and T-shirts second, so the sample selection is listed as + + + + +/+ +/+. If seven contestants choose beach balls and one chooses a T-shirt, we indicate this selection by + + + + + + +/+/ . Notice that each selection is represented by eight pluses and two slashes and that any selection corresponds uniquely to a pattern of eight pluses and two slashes. Therefore the number of different selections is the number of arrangements of eight pluses and two slashes. Since the number of such arrangements is the number of ways of choosing positions for eight pluses from ten locations, the number of arrangements is $C(10,8) = 45$. Since $C(10,8) = C(10,2)$, the number of arrangements of eight pluses and two slashes can also be thought of as the number of ways of choosing positions for two slashes from ten locations. ■

If, in Example 39, we had twelve contestants and four choices of prizes, the number of different selections would be $C(15,12) = C(15,3) = 455$. We get the 15 because we have twelve pluses (for the twelve contestants) and three slashes (one less than the number of prizes because we are using slashes to separate the prizes). Therefore, if we let n denote the number of prizes and r the number of contestants, then the number of different selections is $C(n + r - 1,r) = C(n + r - 1,n - 1)$.

The argument we have just given illustrates the proof of the following theorem, whose proof we therefore omit.

Theorem 6.13 If repetition is allowed, the number of different ways of choosing r objects from a set containing n different objects is $C(n + r - 1,r)$.

Example 40

A man wishes to purchase a case (24 cans) of soft drinks for his daughter's birthday party. He can choose from Coke, Pepsi, Sprite, Slice, and Dr. Pepper. How many different assortments can he purchase?

Analysis

Since he is choosing 24 objects from a set containing five different objects, the number of different ways of making the selection is $C(5 + 24 - 1,24) = C(28,24) = 20{,}475$. ■

Example 41

Twenty-four Titleist and seventeen Topflite golf balls are to be distributed to eight golfers. If each golfer must receive at least one Titleist and at least one Topflite, how many different distributions are possible?

Analysis

We distribute the Titleists first. Since each golfer must receive at least one Titleist, we begin by giving a Titleist to each golfer. Then we have sixteen Titleists that can be distributed in any way. To decide who will receive each of these sixteen Titleists, we can think of choosing sixteen times with repetition from a set containing the names of the eight golfers. Therefore the number of different distributions of the Titleists is, according to Theorem 6.13, $C(8 + 16 - 1,16) = C(23,16) = 245{,}157$. Now we distribute the Topflite golf balls in the same manner. We give a Topflite to each golfer. Then we have nine Topflites to distribute to eight golfers, so the number of different distributions of the Topflites is $C(8 + 9 - 1,9) = C(16,9) = 11{,}440$. Hence, by the product rule, the number of different distributions of the golf balls is $C(23,16)C(16,9) = 2{,}804{,}596{,}080$. ■

Exercises 6.8

120. Let $T = \{1,2,3,4,5,6,7,8\}$ and let $\mathscr{C} = \{\{1,2\},\{3,5,7,8\},\{4,6\}\}$. Make a list of data like that of Table 6.2 and use it to find the number of distinguishable permutations of T, assuming that we cannot distinguish 1 from 2, 4 from 6, or any of the four members of $\{3,5,7,8\}$.

121. Explain the remark that the key to the proof of Theorem 6.11 is "He robs Peter to pay Paul."

122. Find the coefficient of the term involving $x_1^3x_3^2x_5^5$ in the multinomial expansion of

a. $(x_1 + x_2 + x_3 + x_4 + x_5)^{10}$ **b.** $(x_1 + x_2 + x_3 + x_4 + x_5 + x_6)^{10}$

123. How many different nine-digit numbers can be formed using only the digits 2,3, and 5 and using four 5s, three 2s, and two 3s?

124. How many different eight-digit numbers can be formed using the digits in the following numbers. **a.** 222335555 **b.** 235

125. Three subcommittees of an eleven-member committee are to be appointed. The subcommittees are to contain three, three, and five members, and each member of the committee must be appointed to exactly one of the subcommittees. How many different subcommittee appointments can be made?

126. How many different signals, each consisting of seven pulses, can be formed from four indistinguishable dashes and three indistinguishable dots?

127. A piggy bank contains 84 pennies, 27 nickels, and 13 quarters. Notice that, if we remove five coins from the bank, each possible choice of the five coins results in a unique sum of money. How many different amounts of money can we get by removing five coins? (*Hint:* Does your answer in any way depend on the exact number of coins of each type in the piggy bank?)

128. How many different selections of eight books can be made using only identical mathematics books, identical computer science books, and identical statistics books?

129. In how many different ways can seven identical mathematics books and eleven identical computer science books be distributed among five students?

130. In how many different ways can seven identical mathematics books and eleven identical computer science books be distributed among five students if each student must receive at least one mathematics book and at least one computer science book?

131. In how many different ways can eighteen distinct books be distributed so that Joe receives five, Mary six, and Sally seven?

132. A jogger wishes to jog from the corner of 1st Avenue and 1st Street to the corner of 7th Avenue and 9th Street. In how many ways can this be done if the jogger does not run more than

a. 14 blocks? **b.** 15 blocks? **c.** 16 blocks?

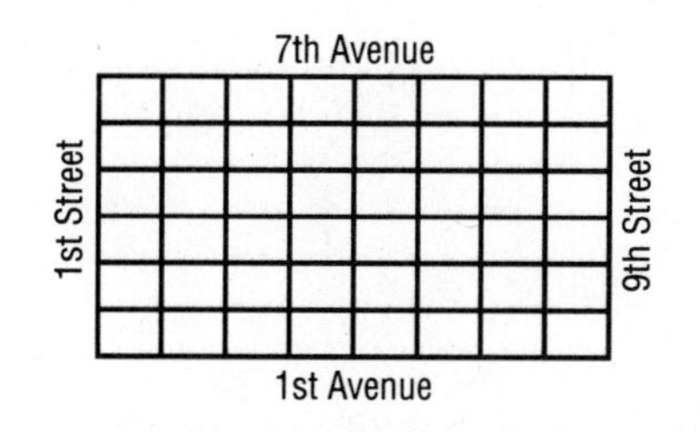

133. Diane made 17 identical ducks, 22 identical rabbits, and 14 identical bears. At Christmas, she plans to give these to nine friends so that each of the nine receives at least one duck, at least one rabbit, and at least one bear. In how many different ways can she do this?

134. **a.** A man wishes to purchase three candy bars from a machine that sells only Hershey bars and Mars bars. How many different purchases of three candy bars can be made? List all possible purchases.

b. How many nonnegative integer-valued solutions does the equation $x_1 + x_2 = 3$ have? List them.

135. How many nonnegative integer-valued solutions does $x_1 + x_2 + x_3 = 11$ have?

136. How many terms are there in the multinomial expansion of $(x_1 + x_2 + x_3 + x_4 + x_5)^{17}$?

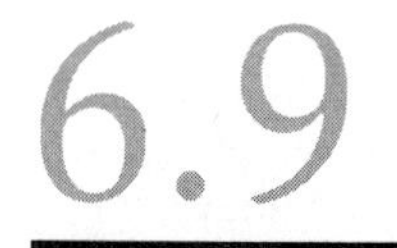

6.9 DISCRETE PROBABILITY

Discrete probability deals with problems of chance in which only discrete sets play a role. It is reasonable to guess that problems of discrete mathematics give rise only to problems of discrete probability, but this is not the case. For example, a typical problem of parallel computing is that two pieces of data cannot be put to use unless they arrive at roughly the same time. Suppose that each of two pieces of data arrive randomly once in every $1/10$-second interval and that if their two arrival times do not differ by more than $1/100$ second the computer can use both pieces of data and continue to the next computation. What is the probability that this happens in a given $1/10$-second interval?

To work this problem, we note that, if the ordered pair (x,y) indicates the pair of arrival times, then the set of all possible pairs of arrival times is $B = [0,1/10] \times [0,1/10]$, whereas the set of all useful arrival times is the subset $A = \{(x,y) \in B: |x - y| < 1/100\}$ (see Figure 6.7).

Figure 6.7

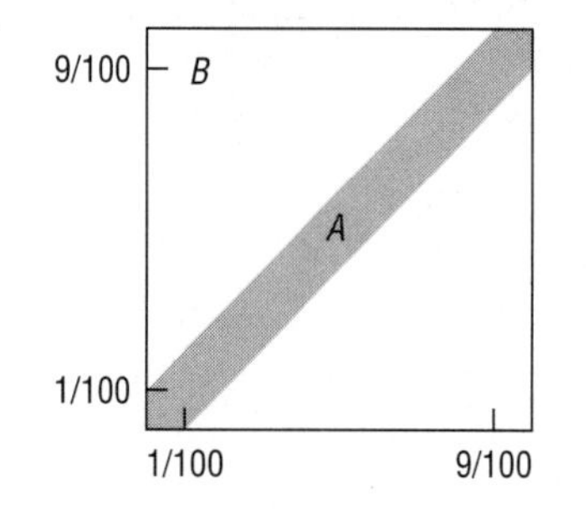

It is intuitive that the probability of success ought to be the ratio of the area of A to that of B, and this intuition is correct. The area of A is $19/10000$ and the area of B is $1/100$ (see Exercise 137). Consequently, the probability of success is $19/100 = .19$.

Since neither A nor B is a discrete subset of the plane, the problem we have just considered is not one of discrete probability, even though it arises in the context of discrete mathematics. For the rest of this section we consider only problems of discrete probability, and, following the historical development of our subject, we pay particular attention to problems of gambling. If you are interested in discrete mathematics only for its applications, do not be put off by our preoccupation with gaming. The usefulness of discrete probability extends to such disparate subjects as genetics, insurance, and queuing theory; but, as the historical development of discrete probability suggests, it is easiest to consider first those problems of probability with which we are most familiar. These problems are the problems of games of chance.

The first text on our subject, *Book on Games of Chance,* was written around 1520 by one of history's more colorful figures, Girolamo Cardano (1501–1576). Cardano was a physician, astrologer, and mathematician, in that order. After his request to practice medicine in Milan was rejected (because of his illegitimate birth), Cardano supported his wife and three children by gambling and casting horoscopes. He played most of the popular games of his day—chess, backgammon, dice games, and an early version of poker called primero. Although he was a knowledgeable gambler, and probably a cheat, Cardano did not always win, and his family spent some time in the Milano poorhouse. His youngest son was jailed for stealing and later exiled; his oldest son was found guilty of murder and executed. As for Cardano himself, his entire life was threaded with controversy. At age 69, he was arrested for heresy and after three months in jail released on the condition that he not reveal the charges against him. We now know that he was charged with casting a horoscope of Christ's life. Cardano's mathematical controversies centered around his publication in *Ars Magna* of what is now known as the Cardano–Tartagalia formula for the solution to an arbitrary cubic equation. Cardano had nothing to do with the discovery of this formula and had learned it from Tartagalia after promising not to publish it. Cardano's *Book on Games of Chance,* which ought to have initiated the study of discrete probability, was not published until 1663, and so Pascal and Fermat are given credit for beginning the study of discrete probability through an exchange of letters in 1654. Thus, ironically, Cardano is credited with a formula he did not discover and not credited with a mathematical theory he did discover.

The French mathematician Pierre Simon de Laplace (1749–1827) defined *finite probability* to be the ratio of the number of favorable outcomes to the total number of outcomes, when it is assumed that any two outcomes are equally likely. There is a circularity built into this definition, because if we do not yet have a notion of finite probability then we do not yet have the notion "equally likely." Furthermore, Laplace's definition rules out the study of certain problems in which outcomes are not equally likely. But for our purposes, Laplace's definition is good enough. Note that counting techniques are necessary because we must know the number of favorable outcomes and the total number of outcomes.

Definition

Any procedure that results in an observable outcome is called an **experiment.**

Flipping two coins, a nickel and a dime, is an example of an experiment. The number of outcomes is four, and these are tabulated here. We see that the probability of two heads is $\frac{1}{4}$ since, in this case, there is only one favorable outcome.

N	H	H	T	T
D	H	T	H	T

Definition

A **sample space** for an experiment is a set consisting of all the possible outcomes, where any two outcomes are equally likely.

The preceding table gives an example of a sample space: $\{HH, HT, TH, TT\}$.

There may be more than one sample space for an experiment. For example, in the experiment of drawing a card from an ordinary deck of cards, we could have a sample space consisting of 52 members; each card in the deck would be a member of the sample space. We could have another sample space {heart, not a heart}. But to use the definition of probability that we have adopted, we must be sure that the various cases in the sample space are equally possible. This is not the case for the {heart, not a heart} space, since there are thirteen hearts and thirty-nine non-hearts, so {heart, not a heart} is not an appropriate sample space. However, {spade, heart, diamond, club} is a perfectly good sample space for our experiments.

Definition

Any subset of a sample space is called an **event.**

Example 42

Flipping three coins, a nickel, a dime, and a quarter, is an example of an experiment. A sample space for the experiment that lists all possible outcomes is $S = \{HHH, HHT, HTH, THH, HTT, THT, TTH, TTT\}$. The following subsets are examples of events:

$$A = \{HHH\}, \qquad B = \{HHH, TTT\}, \qquad C = \{HHT, HTH, THH\}$$ ■

Definition

If E is an event in a nonempty sample space S consisting of equally likely outcomes, then the **probability** of E, denoted by $P(E)$, is defined by

$P(E) =$ (number of members of E)/(number of members of S)

In Example 42, the probability that all three coins will be heads is 1/8, the probability that all three coins will be alike is 1/4, and the probability that exactly two will be heads is 3/8.

Example 43

Three different mathematics books and three different computer science books are to be placed on a shelf in a random order. What is the probability that the mathematics books will be together and the computer science books will be together?

Analysis

We must first determine an appropriate sample space that consists of equally likely outcomes. Since the ordering of the books is the important thing, the set S of all arrangements of the six books is the choice. What is the number of members of S? We have six choices for the first book, five choices for the second, and so on. Thus the number of members of S is $6 \cdot 5 \cdot 4 \cdot 3 \cdot 2 \cdot 1 = 720$. Let E denote the subset of S consisting of those arrangements in which the mathematics books are together and the computer science books are together. What is the number of members of E? The number of ways we can put the mathematics book in the first three spaces and the computer science books in the last three spaces is $3 \cdot 2 \cdot 1 \cdot 3 \cdot 2 \cdot 1 = 36$, and the number of ways we can put the computer science books in the first three spaces and the mathematics books in the last three spaces is also 36. Therefore the number of members of E is 72, and the probability that the mathematics books will be together and the computer science books will be together is $72/720 = 1/10$. Note that $P(E) = [P(3,3) \cdot P(3,3)]/P(6,6)$. ■

Example 44

Suppose that there are three #2 pencils in a box of eighteen pencils. If we choose four pencils at random, what is the probability that we do not choose a #2 pencil?

Analysis

The set S of all choices of four pencils chosen from the box of eighteen pencils is the obvious choice of a sample space. What is the number of members of S? We have eighteen choices for the first pencil, seventeen for the second, sixteen for the third, and fifteen for the fourth. However, the order in which we choose the pencils does not matter, and the number of ways the four pencils can be ordered is $4 \cdot 3 \cdot 2 \cdot 1$. Therefore the number of members of S is $(18 \cdot 17 \cdot 16 \cdot 15)/(4 \cdot 3 \cdot 2 \cdot 1) = 3{,}060$. Note that this is $C(18,4)$. Let E denote the subset of S consisting of all choices of four pencils chosen from the fifteen pencils that are not #2 pencils. The number of members of E is $(15 \cdot 14 \cdot 13 \cdot 12)/(4 \cdot 3 \cdot 2 \cdot 1) = 1{,}365$. Note that this is $C(15,4)$. Therefore the probability that we do not choose a #2 pencil is $1365/3060 = 91/204$. ■

We indicated earlier that gambling played an important part in the early development of probability theory. In this context, what is the meaning of the answer we obtained in Example 44? Suppose that we have a box of eighteen pencils, and that we know that three are #2 pencils. If four are to be chosen at random, would you want to bet that a #2 pencil is chosen? We have seen that the probability of not choosing a #2 pencil is $91/204$. This number is less than $1/2$, so you should bet that a #2 pencil is chosen.

Now suppose that we change the problem so that we are choosing only three pencils rather than four. Would this change your bet? Now the number of members of our sample space S is $(18 \cdot 17 \cdot 16)/(3 \cdot 2 \cdot 1) = 816$, and the number of members of E is $(15 \cdot 14 \cdot 13)/(3 \cdot 2 \cdot 1) = 455$. So the probability that we do not choose a #2 pencil is $455/816$. Since this number is greater than $1/2$, you should bet that a #2 pencil is not chosen.

Example 45

In a single throw of a pair of dice, what is the probability of obtaining a total of 7 or 11?

Analysis

If we throw one die, the natural choice of a sample space is $\{1,2,3,4,5,6\}$, and the natural choice of a sample space S for a pair of dice consists of 36 members. To make visualization easier, suppose that one die is red and the other white. We want E to be the subset of S consisting of all possible combinations that total 7; these are listed in the accompanying table.

Red die	1	2	3	4	5	6	5	6
White die	6	5	4	3	2	1	6	5

The probability of obtaining a total of 7 or 11 in a single throw of a pair of dice is $8/36 = 2/9$. Since this number is considerably less than $1/2$, we would definitely not bet on obtaining a 7 or 11 in a single throw. We examine this situation further. The probability of $2/9$ indicates that, on average, if we bet that a 7 or 11 will be obtained, we will win twice and lose seven times for each nine times that the pair of dice is thrown. But pay heed to the words "on average." If we throw a pair of dice nine times, the result may be significantly different (the probability of winning exactly twice is about .306; see Example 52). If we throw a pair of dice 9,000 times, the total 7 or 11 should appear approximately 2,000 times, though once again it is unlikely to appear exactly 2,000 times. ■

Example 46

Five different mathematics books and three different computer science books are to be placed on a bookshelf with space for eight books. What is the probability that no two computer science books will be next to each other?

Analysis

Let the sample space S be the set of all arrangements of the eight books. Then the number of members of S is $8! = 40{,}320$. Let E denote the subset of S consisting of all arrangements in which no two computer science books are next to each other. What is the number of members of E? There are $5! = 120$ different arrangements of the five mathematics books, and there are $3! = 6$ different arrangements of the three computer science books. Since no two computer science books are to be next to each other, there are six possible slots for the three computer science books. (We can have a computer science book between mathematics books—four slots; and we can have a computer science book on the far left or the far right—two slots.) This gives $C(6,3) = 20$ possibilities. Therefore the total number of members of E is $5! \cdot 3! \cdot C(6,3) = 120 \cdot 6 \cdot 20 = 14{,}400$. The probability that no two computer science books will be next to each other is $14{,}400/40{,}320 = 5/14$. ■

Example 47

Diane has eight ducks and twelve bears that she wants to distribute so that Sibyle gets eight items, Janet seven, and Linda five. If she distributes the items at random, what is the probability that Sibyle will get all eight ducks?

Analysis

Let the sample space S be the set of all distributions of the twenty items in which Sibyle gets eight items, Janet seven, and Linda five. Then the number of members of S is $20!/(8!7!5!) = 99{,}768{,}240$. Let E denote the subset of S consisting of all distributions in which Sibyle gets all eight ducks. Note that the members of E are those distributions in which the twelve bears are distributed so that Janet gets 7 and Linda 5. Therefore the number of members of E is $12!/(7!5!) = 792$, and the probability that Sibyle will get all eight ducks is $792/99{,}768{,}240 = 1/125{,}970$. ■

We now establish a few of the basic principles of finite probability. Our first proposition, Theorem 6.14, contains two evident observations.

Theorem 6.14

Let S be a sample space and let A and B be events.

a. If $A \subseteq B$, then $P(A) \leqslant P(B)$.
b. $P(A) + P(A^{\sim}) = 1$.

Proof

We may assume that S has n members and that A and B have j and k members each.

a. If $A \subseteq B$, then $j \leqslant k$ and so $P(A) = j/n \leqslant k/n = P(B)$.
b. By definition, $P(A) = j/n$ and $P(A^{\sim}) = (n - j)/n$. Thus $P(A) + P(A^{\sim}) = j/n + n/n - j/n = 1$. □

Corollary 6.15

Let S be a sample space. For any event $A \subseteq S$, $P(\varnothing) = 0 \leqslant P(A) \leqslant 1 = P(S)$.

Theorem 6.14b is useful in its own right, but it is just a special case of a theorem that should look familiar.

Theorem 6.16

Let S be a sample space and let A and B be events. Then $P(A \cup B) = P(A) + P(B) - P(A \cap B)$.

Proof

We may assume that S has m members and that A and B have j and k members each. By Theorem 2.7, $n(A \cup B) = n(A) + n(B) - n(A \cap B)$, and so

$$P(A \cup B) = \frac{n(A \cup B)}{m}$$

$$= \frac{n(A) + n(B) - n(A \cap B)}{m}$$

$$= \frac{j}{m} + \frac{k}{m} - \frac{n(A \cap B)}{m}$$

$$= P(A) + P(B) - P(A \cap B) \quad \square$$

Intuitively, two events are independent if the occurrence of one is not affected by the occurrence of the other. Clearly, rolling a 6 and drawing the two of clubs are independent events, but there are other pairs of events whose independence or dependence is not so clear. If two events A and B really are independent, we should expect (by analogy with the product rule) that $P(A \cap B) = P(A)P(B)$. For example, the probability of rolling a 6 and then drawing the two of clubs ought to be $(1/6)(1/52) = 1/312$. We take this observation as our definition.

Definition

Two events A and B are **independent** if $P(A \cap B) = P(A)P(B)$.

We note that it follows from Theorem 6.16 that if A and B are *independent* events then $P(A \cup B) = P(A) + P(B) - P(A)P(B)$. For example, the probability of rolling a 6 or drawing the seven of clubs is $1/6 + 1/52 - 1/312 = 19/104$.

The following theorem is an immediate consequence of the definition of independent events.

Theorem 6.17

If P_1 is the probability that a first event will occur, and after the first event has occurred P_2 is the probability that a second event will occur, then the probability that the two events will occur in succession is P_1P_2.

Example 48

In a single throw of a pair of dice, what is the probability of obtaining a total of 12?

Analysis

To throw a 12, we must throw a 6 with each die. In a single throw of one die, the probability of obtaining a 6 is $1/6$. Therefore the probability of obtaining a 12 is $(1/6)(1/6) = 1/36$. ■

Example 49

Suppose that a card is drawn from an ordinary deck of 52 cards and is not replaced. If a second card is then drawn, what is the probability that both will be aces?

Analysis

The probability that the first card is an ace is $4/52 = 1/13$. If indeed the first card is an ace, then the probability that the second one will be an ace is $3/51 = 1/17$. Therefore the probability that both will be aces is $(1/13)(1/17) = 1/221$. ■

It is a common error to suppose that, since the probability of rolling a 5 with one roll of a die is $1/6$ and since six rolls yield six independent events, the probability of rolling a 5 in six rolls is 1. We can use Theorem 6.14b to determine the correct probability of rolling a 5 in six rolls. Let $A = \{5\}$. Then $A^{\sim} = \{1,2,3,4,6\}$ and $P(A^{\sim}) = 5/6$. Since the six rolls yield six independent events, by Theorem 6.17 the probability of *not* rolling a 5 in six tries is $(5/6)^6$, and by Theorem 6.14b the answer to our problem is $1 - (5/6)^6$. Must we use such a devious approach? Let us try the straightforward argument on a somewhat simpler problem. What is the probability of rolling a 5 in three rolls? Using the formula $P(A \cup B) = P(A) + P(B) - P(A)P(B)$, which only holds for independent events, we see that the probability of rolling a 5 in two tries is $1/6 + 1/6 - 1/36 = 11/36$. Let A be the event consisting of the following eleven pairs:

$$(5,5),(5,1),(5,2),(5,3),(5,4),(5,6),(1,5),(2,5),(3,5),(4,5),(5,6)$$

where an ordered pair stands for rolling the first number and then the second number. Let $B = \{5\}$. Then A and B are independent events and

$$\begin{aligned} P(A \cup B) &= P(A) + P(B) - P(A)P(B) \\ &= \frac{11}{36} + \frac{1}{6} - \frac{11}{6(36)} \\ &= \frac{6(17) - 11}{216} \\ &= \frac{91}{216} \end{aligned}$$

Does this answer jibe with the answer we would obtain using the first method? Yes, $1 - (5/6)^3 = 1 - 125/216 = 91/216$. The lesson is clear: In working problems of gambling you should work deviously.

Definition

Two events are **mutually exclusive** if they are disjoint sets. Equivalently, events A and B are **mutually exclusive** if $P(A \cap B) = 0$.

It is easy to confuse mutually exclusive events and independent events. If events A and B are mutually exclusive, then $P(A \cap B) = 0$. Can A and B also be independent? If so, then $0 = P(A \cap B) = P(A)P(B)$, and so either $P(A) = 0$ or $P(B) = 0$. In other words, unless you are considering one or more impossible events, mutually exclusive events can never be independent. Two disjoint events are mutually exclusive, however, since $P(\varnothing) = 0$.

Theorem 6.18

> If two events are mutually exclusive, with P_1 the probability that the first will occur and P_2 the probability that the second will occur, then the probability that the first or second event will occur is $P_1 + P_2$.

The proof of Theorem 6.18 is left as Exercise 140.

Example 50

The probability of drawing a king or a queen from an ordinary deck of 52 cards in a single draw is $4/52 + 4/52 = 2/13$. ■

Theorem 6.18 is typical of results we have obtained about finite probability. It may be thought of as just a dressed-up version of a theorem about counting, namely Theorem 2.7. No doubt you can recognize our final theorem in spite of its probabilistic finery.

Theorem 6.19

> If P is the probability that an event will occur in a single trial, then the probability that the event will occur exactly r times in n trials is $C(n,r)P^r(1 - P)^{n-r}$.

Proof

We revert to the language of truth tables, writing T if the event occurs and F if it does not. How many ways can we write an n-letter word using altogether r T's and $n - r$ F's? (Each such word corresponds to an n-trial experiment in which we are successful r times and unsuccessful $n - r$ times.) Answer: As anyone from Mississippi should know, there are $n!/r!(n - r)! = C(n,r)$ such words. Pick some one of these $C(n,r)$ different n-letter words, say

$$\underbrace{\text{TTTT}\ldots\text{TT}}_{r\text{ T's}}\underbrace{\text{FFFFFF}\ldots\text{FF}}_{n-r\text{ F's}}$$

What is the probability of getting this particular event in n trials? By Theorem 6.14b and Theorem 6.17, it is $P^r(1 - P)^{n-r}$. Now pick another of the $C(n,r)$ different n-letter words, say

$$\underbrace{\text{FFFFFF}\ldots\text{FF}}_{n\,-\,r\text{ F's}}\underbrace{\text{TTTT}\ldots\text{TT}}_{r\text{ T's}}$$

As before, the probability of getting this event in n trials is $(1 - P)^{n-r}P^r$, but as multiplication is commutative we may write $P^r(1 - P)^{n-r}$ instead. So these two events have the same probability. Are they mutually exclusive? Certainly. If we do the experiment n times, we may get one or the other of these events, or we may not get either of them, but we will not get them both! The upshot is that all $C(n,r)$ events we are considering are mutually exclusive and each has probability $P^r(1 - P)^{n-r}$. By Theorem 6.18, the probability we are looking for, that one of these events occurs, is just the product $C(n,r)P^r(1 - P)^{n-r}$. □

The following example illustrates the proof we have just considered.

Example 51

If a single die is thrown five times, what is the probability that a 1 will occur exactly three times?

Analysis

Let us first examine the probability that a 1 will occur in each of the first three throws and will not occur in the last two throws. The probability of throwing a 1 in the first throw is $1/6$. Therefore, by repeated use of Theorem 6.17, the probability of throwing a 1 in each of the first three throws is $(1/6)(1/6)(1/6) = 1/216$. The probability of not throwing a 1 is $5/6$, so by Theorem 6.17 the probability of not throwing a 1 in each of the last two throws is $(5/6)(5/6) = 25/36$. Therefore the probability that a 1 will occur in each of the first three throws and not occur in the last two throws is $(1/216)(25/36) = 25/7{,}776$. Now this is just like thinking of all the 5-letter words (using T,F) and selecting those words with exactly three T's. The number of ways we can do this is $C(5,3) = 10$. Therefore the probability that a 1 will occur exactly three times is $10 \cdot (25/7{,}776)$. ■

Example 52

A gambler throws a pair of dice nine times. What is the probability that he will throw a 7 or an 11 in exactly two of the nine trials?

Analysis

By Example 45, the probability of throwing a 7 or an 11 in one trial is $2/9$. By Theorem 6.19, the probability of throwing a 7 or an 11 exactly twice in nine trials is

$$C(9,2)\left(\frac{2}{9}\right)^2\left(\frac{7}{9}\right)^7 = (36)\left(\frac{4}{81}\right)\left(\frac{823{,}543}{4{,}782{,}969}\right) \approx .306$$ ■

As you attempt to solve the problems in the following set of exercises, keep in mind that the probability is *always* a real number between 0 and 1, inclusive. If you obtain a probability that is less than 0 or more than 1, then you know that your answer cannot be correct.

Exercises 6.9

137. Find the area of the region labeled A in Figure 6.7.

138. Suppose that a card is drawn from an ordinary deck of 52 cards.

a. What is the probability that it is a heart?

b. What is the probability it is a king or a heart?

139. If a coin is tossed four times, what is the probability that it comes up heads each time?

140. Prove Theorem 6.18. (*Hint:* There is a one-line proof.)

141. If a single die is thrown twice, what is the probability of obtaining a 3 each time?

142. If a single die is thrown three times, what is the probability of obtaining a 3 exactly once?

143. If a single die is thrown five times, what is the probability that the first and last numbers are different?

144. What is the probability of obtaining an "odd coin" (that is, all the coins are heads except one, which is tails, or vice versa) if

a. three coins are tossed? **b.** four coins are tossed?

c. five coins are tossed?

145. A coin is tossed three times. Let E_1 be the event that the first toss lands heads, E_2 be the event that the second toss lands heads, and E_3 be the event that two consecutive tosses land heads but not all three tosses land heads.

a. Find $P(E_1)$, $P(E_2)$, and $P(E_3)$. **b.** Are E_1 and E_2 independent?

c. Are E_1 and E_3 independent? **d.** Are E_2 and E_3 independent?

146. If eight coins are tossed, what is the probability that

a. exactly three will be heads? **b.** at least three will be heads?

147. If a pair of dice is thrown eight times, what is the probability that the total is 7 each time?

148. If a five-digit number is created using the digits 1, 2, 3, 4, 5, and 6 as often as desired, what is the probability that it contains three 2s and two 3s?

149. A difficult problem is assigned to a class, and the probability that Carl will solve it is $3/5$. If the probability that Sue will solve the same problem is $2/3$, what is the probability that at least one of them will solve it? that both will solve it? that neither will solve it?

150. Four different green and four different red books are placed at random on a shelf with space for eight books. What is the probability that they will be placed with colors alternating?

151. Two boxes are sitting on a table. One box contains eight #2 pencils and five #3 pencils. The other box contains six #2 pencils and nine #3 pencils. If Joe draws one pencil, what is the probability that it will be a #2 pencil?

152. A multiple-choice test consists of ten questions. Each question has five possible answers, exactly one of which is correct. If Joe knows absolutely nothing about the material and simply guesses the answer to each question, what is the probability that he will answer at least six correctly?

153. Would the answer in Exercise 152 be different if there were fifty questions on the test and you were asked the probability that Joe would answer at least thirty correctly? Explain. Note that the percentage of questions answered correctly in each case is the same.

154. If five people are chosen from an audience containing ten men and fourteen women, what is the probability that five men are chosen?

155. If the letters of *Mississippi* are randomly arranged, what is the probability that the four *s*'s are adjacent and the four *i*'s are adjacent?

156. If six people (no two of whom are the same age) are chosen at random, what is the probability that they are chosen in order of decreasing age?

157. Suppose that you are dealt the king, ten, seven, and three of hearts and the five of diamonds. If you discard the five of diamonds and draw one card, what is the probability that the card you draw will be a heart?

158. If you are dealt thirteen cards from an ordinary deck of 52 cards, what is the probability that you will have exactly five spades?

159. You have been dealt five cards, four of which are diamonds. You can see twelve other cards which have been dealt to your opponents, and two of these are diamonds. If you are to be dealt two more cards, what is the probability that at least one of them will be a diamond?

160. A poker hand consists of five cards chosen from an ordinary deck of 52 cards. What is the probability that a random poker hand contains exactly one pair? at least one pair?

161. A bridge hand consists of thirteen cards chosen from an ordinary deck of 52 cards. What is the probability that a random bridge hand contains **a.** four aces? **b.** a 5-4-3-1 distribution?

162. A bridge or whist hand containing no card higher than a nine is called a Yarborough, after an Earl of Yarborough who would bet 1,000 to 1 against its occurring. What is the probability of being dealt a Yarborough?

163. In the deal of an ordinary deck of 52 cards, what is the probability that the tenth card is the first ace to be dealt?

***164.** Three men are in jail and they know that one of them is to be executed, but only the jailer knows which one. The first man says to the jailer, "Look, I know that one of the other two is not going to be executed. Why don't you write this man's wife and tell her that her husband is not going to be executed?" The jailer thinks about this overnight and then tells the first man, "I have decided that this wouldn't be fair to you. At the present time, the probability that you will be executed is $1/3$. If I told one of the other's wife that her husband was not going to be executed, then the probability that you will be executed would become 1/2." Is the jailer correct? Explain your answer.

***165.** How many people (none of whom has a birthday on February 29) must there be in a room so that the probability that some two people in the room have the same birthday is greater than 1/2?

Chapter 6 Review Exercises

166. A somewhat disorganized man has 37 brown socks and 17 blue socks in a drawer. How many socks must he choose to be certain that he has two matching socks?

167. Suppose $n \in \mathbb{N}$. In how many ways can a group of n men and n women be arranged in a line if the men and women alternate?

168. Suppose $n \in \mathbb{N}$. In how many ways can n men and $n + 1$ women be arranged in a line if the men and women alternate?

169. **a.** List all three-element subsets of $\{1,2,3,4,5\}$.

b. List all 3-permutations of $\{1,2,3,4,5\}$ corresponding to the three-element subset $\{2,4,5\}$ of I_5.

170. List all the 3-permutations of I_4.

171. How many 4-permutations of I_6 map 1 to 5?

172. Given that $2^{16} = 65{,}536$, evaluate

$$C(16,1) + C(16,3) + C(16,5) + C(16,7) + C(16,9) + C(16,11) + C(16,13) + C(16,15)$$

173. Given that $7! = 5{,}040$, evaluate $P(7,7)$, $P(7,5)$, $C(7,2)$, and $C(7,5)$.

174. How many 4-permutations of I_{100} have 55 in their range?

175. Each of the following sentences is a proposition. Which of them are true?

a. There are more seven-element subsets of a twenty-element set than there are thirteen-element subsets.

b. The coefficient of $a^{35}b^{12}$ in the binomial expansion of $(a + b)^{47}$ is divisible by 47.

c. If a penny is flipped 1,000 times, the probability that the penny lands heads 500 times is 1/2.

176. Professor Simon loves to give F's. If the probability that Jones will make an F is 2/3 and the probability that Smith will make an F is 1/2, what is the probability that at least one of them will make an F?

177. Let r and n be positive integers with $r \leqslant n$.

a. Prove that $C(n + 1,r) = (n + 1)(1/r)C(n,r - 1)$.

b. Use the identity in part (a) and Pascal's formula to show that $C(n,r)$ can be defined inductively by $C(n,0) = 1$ and the recursion relation $C(n,r + 1) = [(n + 1)/(r + 1) - 1]\, C(n,r)$.

c. Use the inductive definition of part (b) to evaluate $C(n,5)$, where it is understood that $n \geqslant 5$.

178. Let $n \in \mathbb{N}$. Prove that $C(n,0)^2 + C(n,1)^2 + \cdots + C(n,n)^2 = C(2n,n)$.

179. John is being interviewed in another city, and his prospective employer, Sam, invites him for dinner. Sam mentions that he has three children and, since his son Sam Jr. is on spring break, all three of them will be having dinner with John and Sam. When John arrives for dinner, he is met at the door by Sam and Sam's daughter Nancy. Assuming that it is equally likely that Sam's third child is a boy or a girl, what is the probability that the next child of Sam's that John sees is a boy?

180. Suppose that k pigeons nest in the ceiling in n nests. Argue that at least one nest contains $\lceil k/n \rceil$ pigeons.

181. A rumor is spread among nine people when someone starts the rumor by telephoning another member of the group. Each person who hears the rumor calls someone else in the group.

a. In how many different ways can a rumor spread in exactly three calls?

b. What is the probability that, if Dan starts the rumor, Dan receives the third call?

182. What is the probability that a four-digit campus telephone number has one or more digits that repeats?

183. Joe, who is a computer operator at a small company, has fourteen programs to process. In how many different orders can he process these programs,

a. if there are no restrictions?

b. if five of the programs have higher priority than the other nine?

c. if four of the programs have top priority and three of the remaining programs are of higher priority than the other seven?

184. Let $\Sigma = \{0,1,2\}$ be an alphabet. By the product rule, for each $n \in \mathbb{N}$ there are 3^n words of length n over Σ. If $x = x_1x_2 \ldots x_n$ is one of these words, we define the *weight* of x, $w(x)$, by $w(x) = x_1 + x_2 + \cdots + x_n$. How many of the words of length 8 over Σ meet the following criteria?

a. three 1s

b. three 1s and two 2s

c. at least six 1s

d. weight 3

e. even weight

f. odd weight

185. A message consisting of ten different symbols and forty spaces between the symbols with at least three spaces between consecutive symbols is to be sent. How many ways can this be done?

186. A word processing software allows filenames consisting of from one to eight characters. Each character may be any one of the 36 alphanumeric characters (26 letters and 10 digits). It is also possible to have a filename extension after the filename; this is accomplished by writing a period after the filename and following the period by one to three additional alphanumeric characters, for example, *LETTER.TXT*.

a. How many filenames are possible if we do not allow extensions?

b. How many filenames with extensions are possible?

c. How many of the filenames in (a) start with AB?

d. How many of the filenames in (b) have a repeated character in the filename extension?

187. The English alphabet consists of five vowels, a, e, i, o, and u, and twenty-one consonants. Prove that any list of twenty-six symbols containing each of the letters of the alphabet must have four consecutive consonants. (For the purpose of this exercise, y and w are always taken to be consonants.)

188. A computer has 50 incoming lines and only 15 ports to accommodate them. The probability that a given line is being used at a given time is $1/4$, and whether a given line is being used is independent of whether other lines are being used. What is the probability that exactly 15 lines are being used at a given time?

189. a. Find the probability that the object of a binary search of a sorted list of 1,023 objects will be found in fewer than 10 trials.

b. Which elements of the list are not found until the last trial?

Chapter

7

BOOLEAN ALGEBRA AND GATE NETWORKS

In 1938, the American Electrical Engineering Society published a paper in which two well-known electrical engineers at General Electric, G. E. Inman and R. N. Thayer, detailed their recent progress in the development of fluorescent light. The article included a photograph of the now familiar fluorescent tube and asserted that "fluorescence makes possible for the first time an efficient practicable low-wattage white light matching daylight in appearance." The authors had announced a major breakthrough in electrical engineering. The left-hand column of the page on which Inman and Thayer's paper began was devoted to the concluding topic (Electric Adder to the Base Two) of a paper by an unknown research assistant, Claude E. Shannon, whom the journal identified as "Enrolled Student AIEE." Shannon's paper, which had no photographs of electrical apparatus, suggested that the study of electrical circuits of on–off switches connected in series or in parallel could be reduced to a mathematical study entirely analogous to the study of propositions in logic.

These propositions of logic were based on the work of the English mathematician George Boole (1815–1864), and we have to believe that Boole would have been amused by the side-by-side positioning of Shannon's paper with that of Inman and Thayer. Whereas Shannon's academic credentials were not yet established, Boole's were nonexistent, for his formal education ended with the third grade. Whereas Shannon's work seemed vaguely too theoretical when placed next to the practical use of fluorescent light, Boole's *Investigation of the Laws of Thought* had never gained acceptance as applied mathematics but instead had been heralded as a classic example of the study of mathematics for its own sake. As Bertrand Russell had put it, "Pure mathematics was discovered by Boole in a work he called *The Laws of Thought.*" But, as indicated in Appendix A, a computer may be thought of as a collection of on–off switches. Thus, as Shannon proposed, Boole's algebra may be used to analyze computer circuits and to simplify circuits in much the same way that one simplifies equations. In all, the laws of algebra as proposed by Boole have become the laws of the digital computer.

7.1 SWITCHING CIRCUITS

We begin this section by considering three fundamental ways that two switches can be arranged: in series, in parallel, and in complement (see Figure 7.1). We caution you to remember that the innards of a digital computer are not actually relay switches, which, being mechanical, are entirely too slow for the job.

Figure 7.1

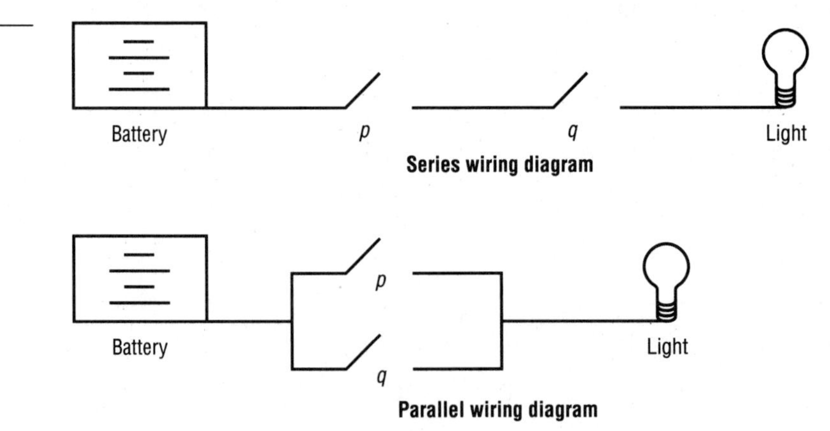

Figure 7.1 (*continued*)

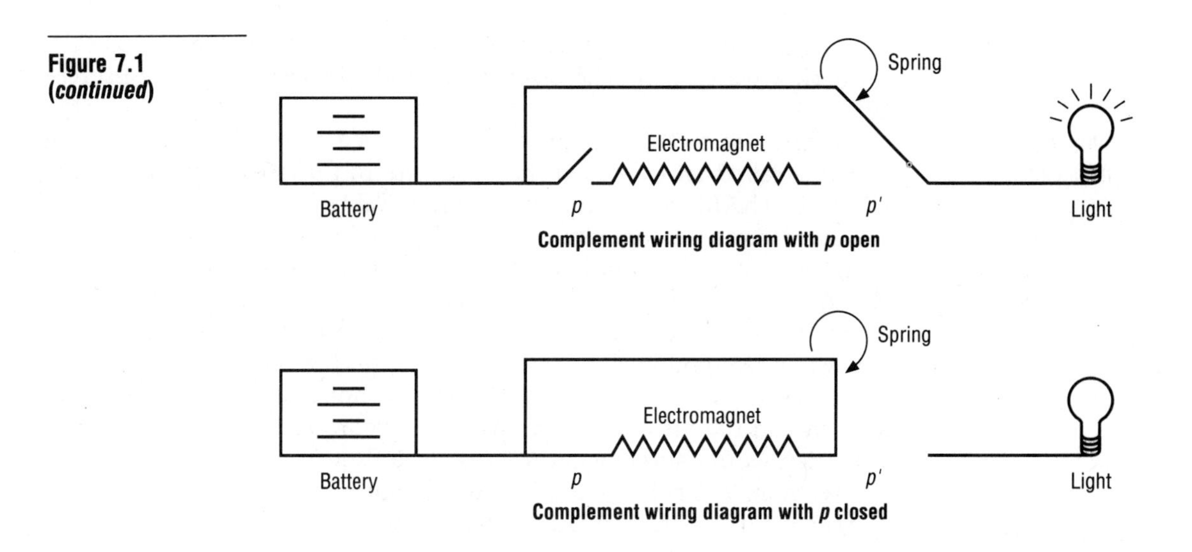

Complement wiring diagram with *p* open

Complement wiring diagram with *p* closed

In computer science, series wiring is often referred to as an *AND-circuit*. The AND-circuit is so named because current flows only when both p *and* q are closed. The proof that current flows only when both p and q are closed is just as easy as it looks; we have only to consider all the possible cases. These are listed in Table 7.1, which shows all possible inputs and the resulting outputs. We see that the current flows only when p and q are closed.

Table 7.1 is somewhat unwieldy. From now on, in such tables, we list all possible inputs with 1 meaning the switch is closed and 0 meaning the switch is open, and the resulting outputs with 1 meaning the light is lit and 0 meaning that it is unlit. Thus Table 7.1 would be written as shown in Table 7.2.

Table 7.1 AND-circuit

p	*q*	Light
closed	closed	lit
closed	open	unlit
open	closed	unlit
open	open	unlit

Table 7.2 AND-circuit

p	*q*	Output
1	1	1
1	0	0
0	1	0
0	0	0

We pause to consider our proof that, when two switches are connected in series, current flows only when both switches are closed. Mathematicians call a proof in which one simply verifies each possible case a *proof by exhaustion*. Electrical engineers and computer scientists sometimes refer to such a proof as a proof by *perfect induction*. Here, our proof by exhaustion works because we have only four immediately obvious cases to check; we are lucky whenever such a proof is reasonable (see Exercise 10).

When two switches p and q are connected in parallel, the resulting circuit is called an *OR-circuit*, because current flows exactly when *p or q* is closed. The four-line proof is shown in Table 7.3.

Table 7.3
OR-circuit

p	q	Output
1	1	1
1	0	1
0	1	1
0	0	0

The word *or* is sometimes used in an exclusive sense, meaning one thing or another but not both. Clearly, current flows when p and q are closed, and so the *or* in OR-circuit takes the nonexclusive sense of the word. Throughout mathematics, discrete or otherwise, the word *or* is always understood in the nonexclusive sense. If a clever lawyer were to ask a mathematician, "Did you or did you not take the knife from the kitchen and hide it under a pillow in the bedroom?" the hapless mathematician would surely answer "Yes" (see Exercise 11).

We come at last to the third circuit of Figure 7.1, the *complement circuit*. It is customary to draw circuit diagrams with all switches in open position, and one often sees the complement switch drawn in this way, even though the switches p and p' can never both be open or both be closed. By drawing first p' and then p in closed position, we have flouted custom in the last two circuit diagrams of Figure 7.1. These two diagrams also indicate the use of an electromagnet. When the switch p is closed, current flows through that tightly wound coil. The force of the resulting magnetic field overcomes the spring holding p' closed, and so the switch p' opens and the light goes out. Note that when switch p is open the spring keeps switch p' closed. Thus p' is always in the opposite state from p. While there is nothing here to prove, it is still comforting to see the little table that catalogues the two cases (Table 7.4).

Table 7.4
Complement circuit

p	Output (p')
0	1
1	0

The switches inside a computer have only one thing in common with the relay switches whose circuit diagrams were given in Figure 7.1: they accomplish the same results as switches wired in series, in parallel, or in complement. It is obvious that we need a general term that does not imply that we are working with the large-scale switches of Figure 7.1.

Definition

An electronic circuit that uses one or more input signals to produce a single output signal is called a **gate**. A gate is entirely determined by the table that lists all possible inputs and each resulting output. This table is called the **truth table** of the gate. A gate in which the output is 1 exactly when at least one of the inputs is 1 is called an OR-gate. A gate in which the output is 1 exactly when every input is 1 is called an AND-gate.

Figure 7.2 gives the truth tables for OR-gates and AND-gates with two inputs, and Figure 7.3 gives the truth tables for OR-gates and AND-gates with three inputs.

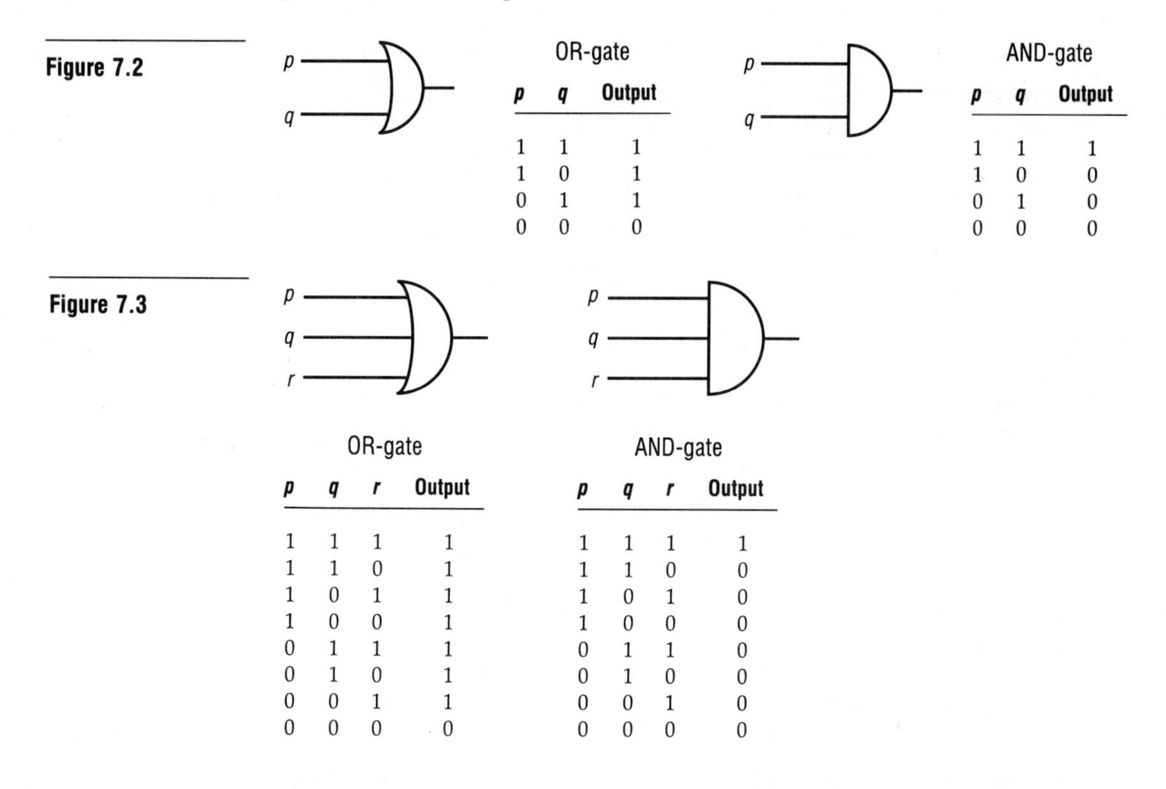

Figure 7.2

OR-gate

p	*q*	Output
1	1	1
1	0	1
0	1	1
0	0	0

AND-gate

p	*q*	Output
1	1	1
1	0	0
0	1	0
0	0	0

Figure 7.3

OR-gate

p	*q*	*r*	Output
1	1	1	1
1	1	0	1
1	0	1	1
1	0	0	1
0	1	1	1
0	1	0	1
0	0	1	1
0	0	0	0

AND-gate

p	*q*	*r*	Output
1	1	1	1
1	1	0	0
1	0	1	0
1	0	0	0
0	1	1	0
0	1	0	0
0	0	1	0
0	0	0	0

By conventions established by the Institute of Electrical and Electronic Engineers, in these diagrams current is always represented as flowing from left to right. The output of any OR-gate is always the maximum of the inputs, and the output of any AND-gate is always the minimum of the inputs.

What about the complement gate? Yes, there is one, but it is called a *NOT-gate*, or sometimes an *inverter*. Its sign is an equilateral triangle. Note that the NOT-gate can have only one input; like any gate whatsoever it can have only one output (see Figure 7.4).

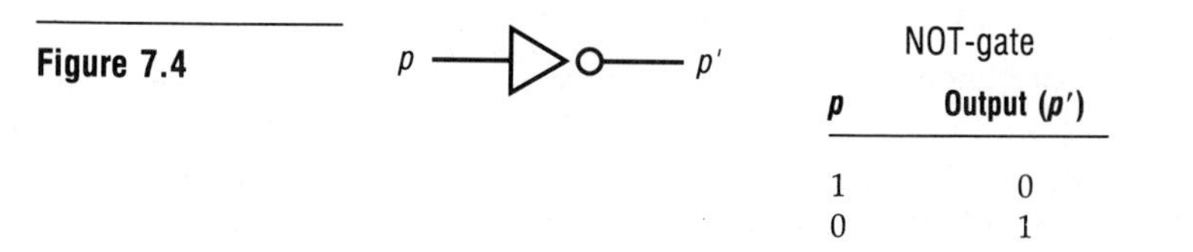

Figure 7.4

NOT-gate

p	Output (p')
1	0
0	1

As is customary, throughout this text we call gates that are made up of combinations of AND-gates, OR-gates, or NOT-gates, *gate networks*, or simply *circuits*.

Exercises 7.1

1. In the simple parallel circuit below, the switches p and q are drawn with p open and q closed.

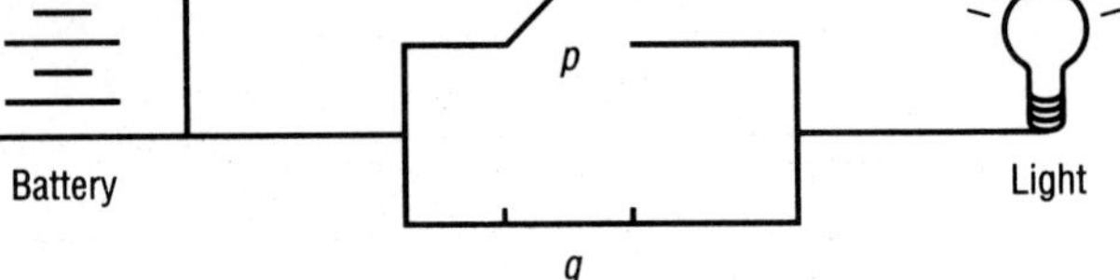

   a. Redraw the circuit above with both p and q open.

   b. Redraw the circuit above with p closed and q open.

   c. Redraw the circuit above with both p and q closed.

   d. Complete the following table to indicate the correct state of the light (lit or unlit) in each of the indicated circumstances.

p	q	Light
closed	closed	
closed	open	lit
open	closed	
open	open	

   e. Give an argument to indicate why parallel circuits pantomime our usual notion of "or."

2. a. Draw the four possible states of the simple two-switch series circuit using the switches p and q.

   b. Complete the following table to indicate the correct state of the light (lit or unlit) in each of the indicated circumstances.

p	q	Light
closed	closed	
closed	open	
open	closed	
open	open	

   c. Give an argument to indicate why series circuits mimic our usual notion of "and."

3. In the following circuit, any two switches labeled with the same letter must both be open or both be closed when a particular state of the circuit is drawn. In the particular state drawn, p is closed and q is open.

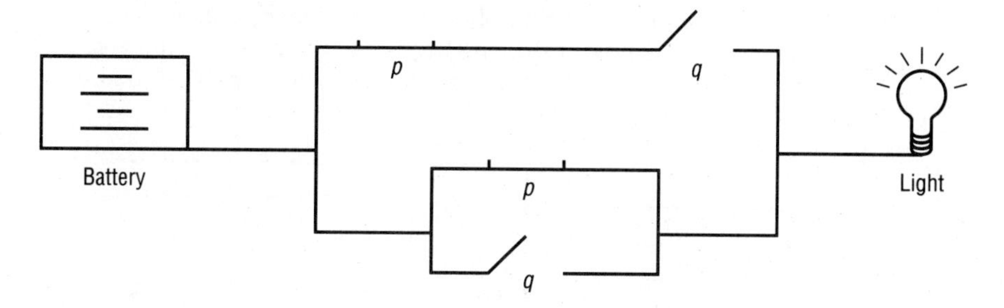

a. Complete the following table to indicate the correct state of the light (lit or unlit) in each of the indicated circumstances.

p	q	Light
closed	closed	
closed	open	
open	closed	
open	open	

b. Using our everyday notions of "and" and "or" together with p's and q's, make a sentence that matches the meaning of this circuit.

4. The English statement "(p and q) or (not (p or q))" may be represented by the following circuit with the understanding that all the switches with the same label must match when any particular state is drawn.

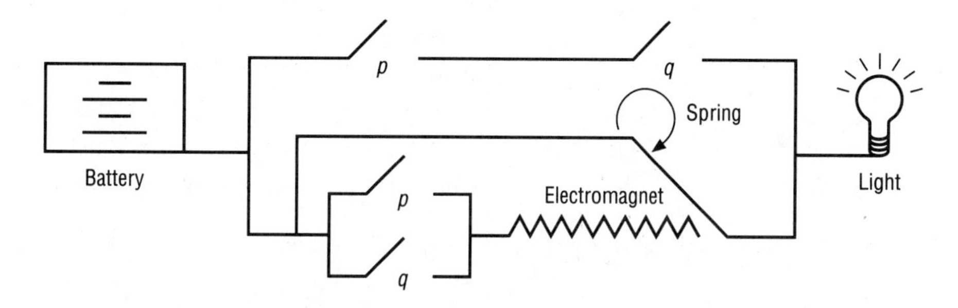

Complete the following table to indicate the correct state of the light (lit or unlit) in each of the indicated circumstances. (*Hint:* Any time you cannot visualize this circuit with one or both switches closed, redraw the circuit with the switches in their actual positions, decide what happens to the electromagnet switch, and then decide what happens to the light.)

p	q	Light
closed	closed	
closed	open	
open	closed	
open	open	

5. **a.** What English statement is represented by the following circuit? (*Hint:* Recall that in Exercise 4 "(p and q) or (not p or q))" was the English statement represented by the diagram in that exercise.)

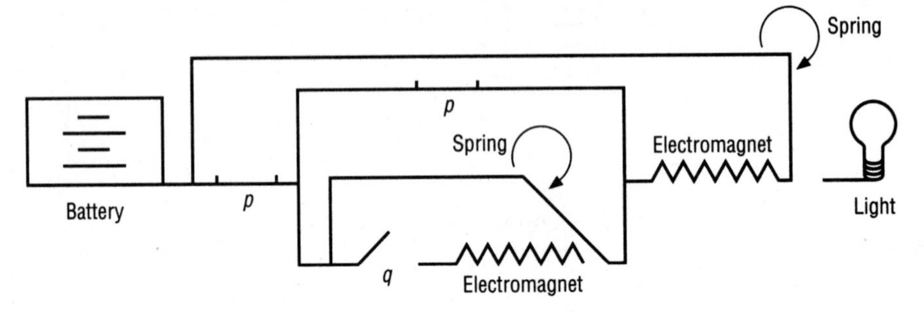

b. What English statement is represented by the following circuit?

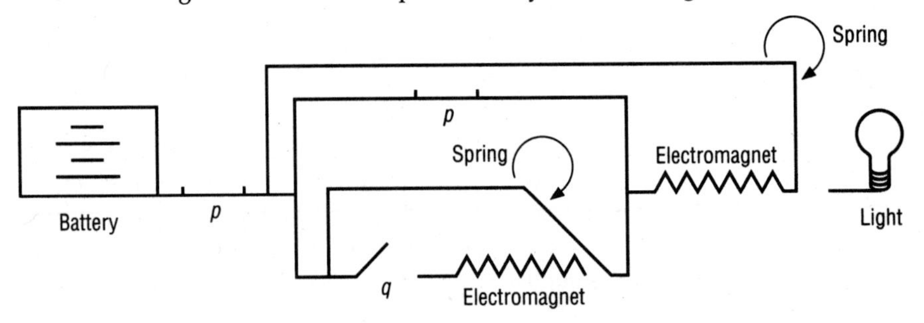

c. Does the English statement represented by a particular circuit depend on the state (open or closed) of the switches?

6. **a.** Draw a circuit that represents the English statement "(not p) and (not q)."

b. Draw a circuit that represents the English statement "not (p and q)."

c. Draw a circuit that represents the English statement "((not p) or q) and ((not q) or p)."

7. **a.** Redraw the following circuit using standard representations of AND-gates, OR-gates, and NOT-gates.

b. Complete the truth table for the given circuit.

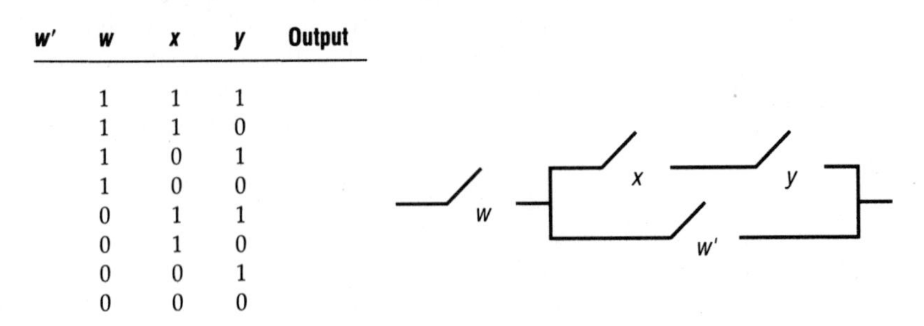

w'	w	x	y	Output
	1	1	1	
	1	1	0	
	1	0	1	
	1	0	0	
	0	1	1	
	0	1	0	
	0	0	1	
	0	0	0	

8. **a.** Redraw the given circuit using standard representations of AND-gates, OR-gates, and NOT-gates.

b. If you were asked to write the truth table of this circuit, which you are *not*, how many rows would your table have?

c. What are the outputs corresponding to the four conditions tabulated here?

s	*v*	*w*	*x*	*y*	*z*
1	1	1	1	1	1
0	0	0	0	0	0
0	1	0	0	1	0
1	0	0	0	0	1

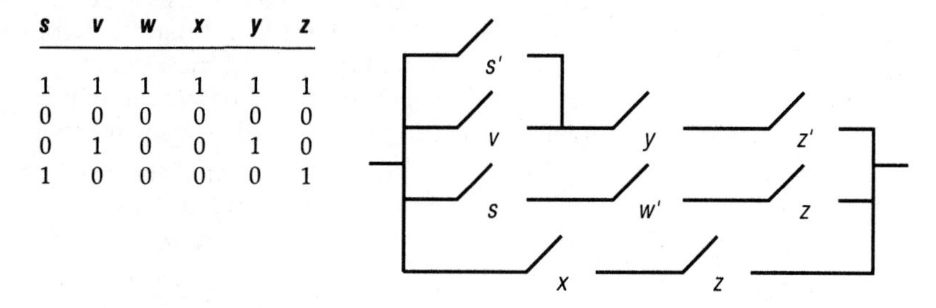

9. Two networks of gates are defined to be *equivalent* if they have the same outputs for all possible choices of inputs.

a. Show that the following networks are equivalent.

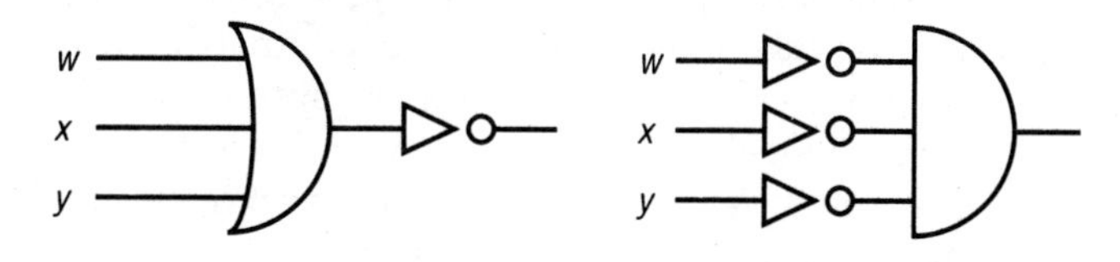

b. Show that the following networks are equivalent and that they do not have the same truth tables.

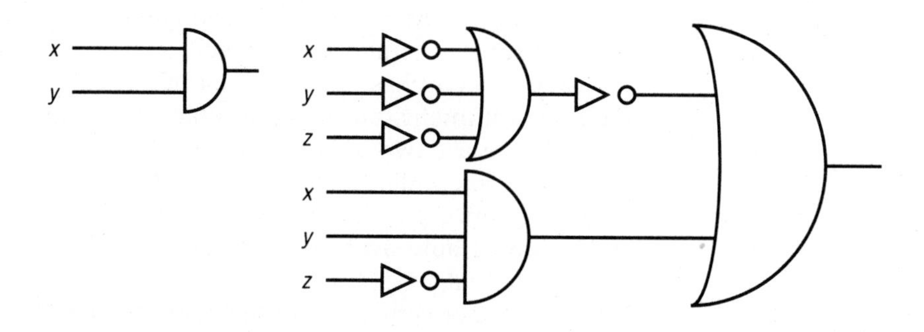

10. A number such as 1, 4, 9, 16, or 25, which is of the form n^2 for some natural number n, is called a *perfect square*.

a. Show that no perfect square between 2 and 10 ends in a 1. (*Hint:* What are the perfect squares between 2 and 10?)

b. Show that no perfect square between 2 and 80 ends in a 1.

c. Show that no perfect square between 2 and 80 million ends in a 3. (*Hint:* Suppose x ends in a 1. What does x^2 end in?)

d. Which, if any, of your proofs is a proof by exhaustion?

11. William Sessions, Director of the FBI, addressed the National Press Club in September 1988 and was asked this final question, which he promised to answer: "As a Texan, would you like to see a Texan president or a Texan vice president?" Bearing in mind that the Republicans were running a Texan for President (Bush) and the Democrats were running a Texan for vice president (Bentsen), and bearing in mind that William Sessions is an intelligent man, guess correctly the answer Sessions gave.

12. Explain why, with the exception of this exercise, you should not expect to find in this text the phrase "and/or," as in the sentence "The patient and/or the attending physician may obtain copies of laboratory reports."

7.2 BOOLEAN ALGEBRAS

The concept of a Boolean algebra was first given by Boole in 1847 in a work called "The Mathematical Analysis of Logic." Boole's algebraic system is much like our familiar system of arithmetic for integers. In a Boolean algebra there is a set B and two ways of combining members of B denoted by $\wedge$ (*meet*) and $\vee$ (*join*). We can think of the set B as corresponding to the set of integers but, in general, B need not even be a set of real numbers. For members a and b of B, $a \wedge b$ ("a meet b") and $a \vee b$ ("a join b") correspond to the product and sum of integers a and b. Just as each integer has a negative, each member b of B has a *complement*, which is noted by b'. Table 7.5 compares the rules of Boolean algebra with the rules of ordinary arithmetic.

Table 7.5 emphasizes the similarities between Boolean algebra and ordinary arithmetic, but it is important to notice several essential differences. There are two distributive laws for a Boolean algebra. According to the first, meet distributes over join (just as multiplication distributes over addition). According to the second, join distributes over meet. Does addition distribute over multiplication; that is, for all integers x, y, and z, is it true that $x + (yz) = (x + y)(x + z)$? Of course not (see Exercise 14). The upshot is that one of the distributive laws for a Boolean algebra is familiar, $a \wedge (b \vee c) = (a \wedge b) \vee (a \wedge c)$, but the other distributive law, $a \vee (b \wedge c) = (a \vee b) \wedge (a \vee c)$, is unfamiliar.

The Boolean complement (b') of an element b does not behave quite like the negative ($-b$) of an integer b. First, $b \vee b' = 1$, whereas for an integer b, $b + (-b) = 0$. Moreover, the property that $b \wedge b' = 0$ has no analogue in ordinary arithmetic.

An important difference between Boolean algebra and ordinary arith-

Table 7.5

Axioms of Boolean algebra	Familiar axioms of arithmetic
a. Commutative laws	
$a \vee b = b \vee a$	$a + b = b + a$
$a \wedge b = b \wedge a$	$ab = ba$
b. associative laws	
$a \vee (b \vee c) = (a \vee b) \vee c$	$a + (b + c) = (a + b) + c$
$a \wedge (b \wedge c) = (a \wedge b) \wedge c$	$a(bc) = (ab)c$
c. distributive laws	
$a \wedge (b \vee c) = (a \wedge b) \vee (a \wedge c)$	$a(b + c) = ab + ac$
$a \vee (b \wedge c) = (a \vee b) \wedge (a \vee c)$	
d. identity laws	
is one element 0 and one element 1 such that $0 \vee b = b \vee 0 = b$ and $1 \wedge b = b \wedge 1 = b$.	There is one number 0 and one number 1 such that $0 + b = b + 0 = b$ and $1 \cdot b = b \cdot 1 = b$.
e. For each element b of B there is exactly one element b' of B, called the complement of b, such that $b \vee b' = b' \vee b = 1$ and $b \wedge b' = b' \wedge b = 0$.	For each integer b there is exactly one integer $-b$, called the negative of b, such that $b + (-b) = 0$.
f. $0 \neq 1$	$0 \neq 1$

metic is that the laws of Boolean algebra come in pairs in which one law in a pair can be obtained from the other by swapping $\vee$ with $\wedge$ and 0 with 1. Assure yourself of this observation; it turns out to be quite useful.

Although we could continue to point out differences between the laws of Boolean algebra and the laws of ordinary arithmetic, the most appropriate way to see the differences is to use the laws of Boolean algebra to establish some results. We begin with elementary facts about the complement. By definition, b' is the complement of b, and b' itself has a complement b'', which, by Theorem 7.1, turns out to be b.

Theorem 7.1

Let b be a member of a Boolean algebra $(B, \vee, \wedge, ')$. Then the complement of b' is b; that is, $b'' = b$.

Proof

By definition, *the* complement of b' is *the* member x of B such that $b' \vee x = x \vee b' = 1$ and $b' \wedge x = x \wedge b' = 0$. Because b itself has these properties and the complement of an element is unique, b is the complement of b'. □

Theorem 7.2

> In any Boolean algebra, $0' = 1$.

Proof

As in the proof of Theorem 7.1, we must show that $0 \vee 1 = 1$ and $0 \wedge 1 = 0$. But the first of these equations is the identity law $0 \vee b = b$ for $b = 1$, and the second is the identity law $b \wedge 1 = b$ for $b = 0$. □

Theorem 7.3

> Let $(B, \vee, \wedge, ')$ be a Boolean algebra and let b be a member of B. Then $b \vee b = b$ and $b \wedge b = b$.

Proof

It is convenient to prove the two parts side by side.

$$\begin{aligned} b \vee b &= (b \vee b) \wedge 1 & \qquad b \wedge b &= (b \wedge b) \vee 0 \\ &= (b \vee b) \wedge (b \vee b') & &= (b \wedge b) \vee (b \wedge b') \\ &= b \vee (b \wedge b') & &= b \wedge (b \vee b') \\ &= b \vee 0 & &= b \wedge 1 \\ &= b & &= b \end{aligned}$$
□

We have not supplied the justification of any of the steps in the proof of Theorem 7.3, but you are asked to justify them in Exercises 17 and 18.

In 1858, Augustus de Morgan (1806–1871) published two theorems which together are known as de Morgan's laws.

Theorem 7.4

> **De Morgan's Laws** Let a and b be elements of a Boolean algebra. Then
>
> $(a \wedge b)' = a' \vee b'$ and
> $(a \vee b)' = a' \wedge b'$

Although the formulas that describe de Morgan's laws are certainly concise, it is also worthwhile stating them in English: The complement of the meet of two elements is the join of their complements, and the complement of the join of two elements is the meet of their complements. The two results are paired in the same manner that the axioms from which they are derived are paired. Thus de Morgan's laws are impossible to get wrong as long as you remember that each law involves both a join and a meet.

Example 1

Let B be a set with only two members, 0 and 1, and let $\vee$, $\wedge$, and $'$ be defined by the following tables. Then $(B, \vee, \wedge, ')$ is a Boolean algebra.

Join

$\vee$	0	1
0	0	1
1	1	1

Meet

$\wedge$	0	1
0	0	0
1	0	1

Complement

	$'$
0	1
1	0

■

We end this section with an example concerning the positive divisors of the number 30. To make use of this example, we need to recall the concepts of greatest common divisor and least common multiple (p. 83).

Example 2

Let B be the positive divisors of 30. For a and b belonging to B, define $a \vee b = \text{lcm}(a,b)$, $a \wedge b = \text{gcd}(a,b)$, and $a' = 30/a$. Then $(B, \vee, \wedge, ')$ is a Boolean algebra whose zero is 1 and whose one is 30. ■

Exercises 7.2

13. As noted in the text, for the real numbers, multiplication distributes over addition; that is, if x, y, and z are real numbers, then $x(y + z) = (xy) + (xz)$.

a. By actual calculation, show that $3 \cdot (4 + 5) = (3 \cdot 4) + (3 \cdot 5)$.

b. By actual calculation, show that $17 \cdot (23 + 4) = (17 \cdot 23) + (17 \cdot 4)$.

c. Doing the arithmetic in your head, find $15 \cdot 112$.

d. Let us find $35 \cdot 112$ using steps that could all be done in your head: (i) find $10 \cdot 112$; (ii) find 3 times the result in (i); (iii) find $1/2$ of the result in (i); (iv) add the results in (ii) and (iii). The result in (iv) is the answer. Explain why this process works.

e. Doing the arithmetic in your head, find $25 \cdot 68$.

14. For our usual number system, to say addition distributes over multiplication would mean that, if x, y, and z are numbers, then $x + (yz) = (x + y)(x + z)$.

a. Find $1 + (2 \cdot 3)$. **b.** Find $(1 + 2) \cdot (1 + 3)$.

c. Does $1 + (2 \cdot 3) = (1 + 2) \cdot (1 + 3)$?

d. Does addition distribute over multiplication?

15. We have seen in Exercise 14 that addition does not distribute over multiplication. In fact, choosing three real numbers x, y, and z at random, it is extremely unlikely that $x + (yz)$ turns out to be $(x + y)(x + z)$.

a. Try it yourself. Choose any three real numbers for x, y, and z. Check as we did in Exercise 14 to see if, for your choices, $x + (yz) = (x + y)(x + z)$.

b. As an interesting aside, can you find numbers a, b, and c such that $a + (bc) = (a + b)(a + c)$?

c. Since it is fairly obvious that $0 + (0 \cdot 0) = (0 + 0)(0 + 0)$, we now ask if there are three nonzero numbers a, b, and c such that $a + (bc) = (a + b)(a + c)$.

d. Show that, if x, y, and z are real numbers such that $x + (yz) = (x + y)(x + z)$, then $x = 0$ or $x + y + z = 1$. (*Hint*: Start with $x + (yz) = (x + y)(x + z)$, multiply it out, cancel like terms, and, assuming x is not 0, divide by x.)

16. Let B be the set of all integers. Let $\vee$ be ordinary addition, $\wedge$ be ordinary multiplication, and for each $x \in B$ define x' to be $-x$. Consider the laws of Boolean algebra in Table 7.5. Determine which of these laws are satisfied by $(B, \vee, \wedge, ')$.

17. Justify each step given in the first part of the proof of Theorem 7.3 by citing an appropriate law of Boolean algebras.

18. Justify each step in the second part of the proof of Theorem 7.3 by citing an appropriate law of Boolean algebras.

19. a. Show that in any Boolean algebra $1' = 0$.

b. Show that in a Boolean algebra if $x' = y$ then $y' = x$.

20. Show that for any element b of a Boolean algebra $1 \vee b = 1$. (*Hint:* Use the fact that $1 = b' \vee b$.)

21. Show that for any element b of a Boolean algebra $0 \wedge b = 0$.

22. Let $(B, \vee, \wedge, ')$ be a Boolean algebra. For all x and y belonging to B, define $x \leqslant y$ provided $x \wedge y = x$.

a. Why is it true that, for each x in B, $x \leqslant y$?

b. Why is it true that if $x \leqslant y$ and $y \leqslant x$ then $x = y$?

c. Show that, if $x \leqslant y$ and $y \leqslant z$, then $x \leqslant z$.

d. What results previously established show that if $x \in B$ then $0 \leqslant x$ and $x \leqslant 1$?

23. Show that no member of a Boolean algebra can be its own complement (*Hint:* Consider $b \vee b$.)

24. For Example 1, complete the following table.

a	**b**	**a ∨ b**	**b ∨ a**	**a ∧ b**	**b ∧ a**
1	1				
1	0				
0	1				
0	0				

If you have completed the table correctly, you have proved that $(B, \vee, \wedge, ')$ satisfies the commutative laws.

25. For Example 1, complete the following table.

a	b	c	$a \vee (b \vee c)$	$(a \vee b) \vee c$	$a \wedge (b \wedge c)$	$(a \wedge b) \wedge c$
1	1	1				
1	1	0				
1	0	1				
1	0	0				
0	1	1				
0	1	0				
0	0	1				
0	0	0				

If you have completed the table correctly, you have proved that $(B, \vee, \wedge, ')$ satisfies the associative laws.

26. Recall that in a Boolean algebra we have defined $a \leq b$ provided that $a \wedge b = a$ (see Exercise 22). Consider the Boolean algebra of Example 2.

a. Is $6 \leq 3$ or is $3 \leq 6$? **b.** Is $6 \leq 5$ or is $5 \leq 6$?

27. In Example 2, verify the associative law $a \vee (b \vee c) = (a \vee b) \vee c$ under the following conditions.

a. $a = 2$, $b = 3$, and $c = 5$ **b.** $a = 30$, $b = 1$, and $c = 6$

28. In Example 2, find each of the following.

a. $15'$ **b.** $5 \vee 15$

c. $6 \wedge 3'$ **d.** $(3 \vee 5) \wedge 6$

e. $3' \wedge 5'$ **f.** $3' \vee 5'$

g. $(5' \wedge 6') \vee 3$

29. In Example 2, verify the distributive law $a \wedge (b \vee c) = (a \wedge b) \vee (a \wedge c)$ under the following conditions.

a. $a = 2$, $b = 3$, and $c = 15$ **b.** $a = 6$, $b = 1$, and $c = 5$

30. Let B be the set of all positive divisors of 15. With $\vee$ and $\wedge$ defined as in Example 2, is $(B, \vee, \wedge, ')$ a Boolean algebra? If not, give an example to show that at least one of Boole's rules fails.

***31.** Repeat Exercise 30 with the following new set of conditions. Let B be the set of all positive divisors of 45. If $(B, \vee, \wedge, ')$ is a Boolean algebra, then $1' = 45$, $3' = 15$, $5' = 9$, $9' = 5$, $15' = 3$, $45' = 1$, and the identity elements 0 and 1 of part (b) of the definition of a Boolean algebra are the natural numbers 1 and 45. Find a property from Table 7.5 that $(B, \vee, \wedge, ')$ does not possess.

32. Let B be the set of all positive divisors of 4. With $\vee$ and $\wedge$ defined as in Example 2, is $(B, \vee, \wedge, ')$ a Boolean algebra? Why?

***33.** There are 100 cells in a certain jail, and the locks are such that turning a key in a locked cell unlocks it and turning a key in an unlocked cell locks it. On the first day, the jailor unlocks all the cells. On the second day, the jailor turns the key in the lock on all even-numbered cells. On the third day, the jailor turns the key in the lock on all cells whose numbers are divisible by 3. Suppose this process continues. Which cells are unlocked at the end of 100 days? (*Hint:* This problem is related to Exercises 31 and 32.)

7.3

CIRCUITS AND GATES

We return to the study of electrical circuits, which has motivated our study of Boolean algebras. We assign each switch a letter and call these letters *circuit variables*. If our circuit is so complex that 26 letters are not enough, we can assign circuit variables x_1, x_2, x_3, and so forth and in this way be sure that we do not exhaust our alphabet. Always, however, we assume that our supply of circuit variables is finite.

In a complicated circuit, it may happen that two switches behave exactly alike without our knowing it, but when we have two or more switches that always switch on and off in concert, we use the same circuit variable to denote each of these switches. If we have two switches that are always at odds, with the second switch off whenever the first one is on, and vice versa, we denote one switch by a variable, say p, and the other by p' (p complement). The variables together with their complements are called *literals*. Using literals, the signs $\wedge$, $\vee$, and $'$, and parentheses, we form *expressions*. Here we rely on your intuition rather than attempt a formal definition of an expression, which is based on Theorem 4.6 and is similar to the definition of a polynomial function given in Example 18 of Chapter 4. Each literal is an expression, and if the preceding signs are put together in a way that appears meaningful, the result is an expression. Thus $)(\vee \wedge x\ p\ q$ is not an expression, but $(x \vee p) \wedge q$ is an expression. Just as each variable has a complement, each expression has a complement, which is also an expression. Thus $(x \vee y)'$ is the complement of the expression $x \vee y$, and $((x \vee y)')'$ is the complement of $(x \vee y)'$.

Corresponding to each expression is a circuit, where $\vee$ means "is connected in parallel" and $\wedge$ means "is connected in series." Also, each circuit has a corresponding expression. For example, $(x \vee p) \wedge q'$ represents the circuit on the left in Figure 7.5 and the circuit on the right is represented by the expression $(x' \wedge y) \vee (y \wedge x')$.

Figure 7.5

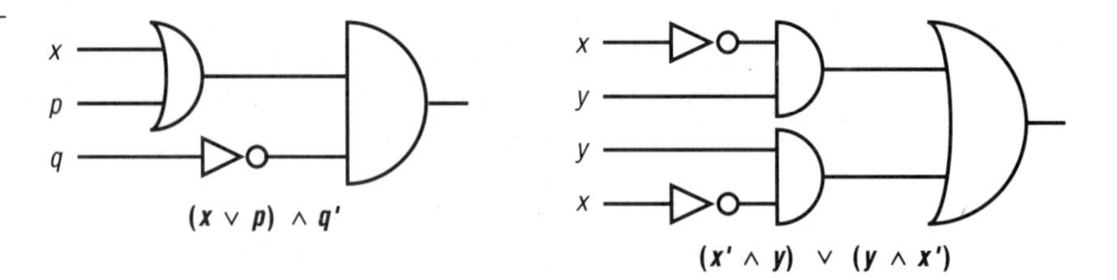

In this section, we say that two expressions are *equivalent* provided that the circuits they represent have the same outputs for all possible choices of inputs, and in this case we say that the circuits themselves are equivalent. (Recall that the notion of equivalent expressions was defined in Chapter 1. This idea is discussed again in the next section.)

The crux of Shannon's 1938 paper is the observation that Boolean algebra can be used to describe and design switching circuits. What makes this observation as useful, say, as fluorescent light? We consider a simple practical problem to see just how helpful a Boolean algebra can be.

Example 3

A company plans to market an electrical device that contains the circuit indicated in Figure 7.6. Because the initial production involves 500,000 devices, the company can save considerably if the circuit can be simplified (reduced to a simpler but equivalent circuit).

Figure 7.6

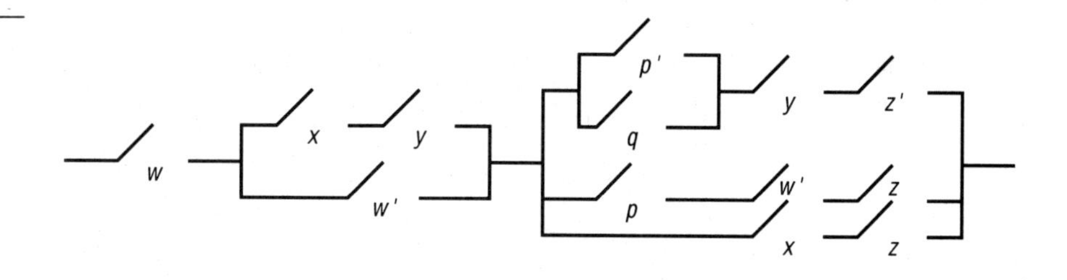

Analysis

We rely on both Boolean algebra and our knowledge of circuitry. Since it is clear that the switches w, x, and y must be closed, we can replace the circuit of Figure 7.6 with that in Figure 7.7.

Figure 7.7

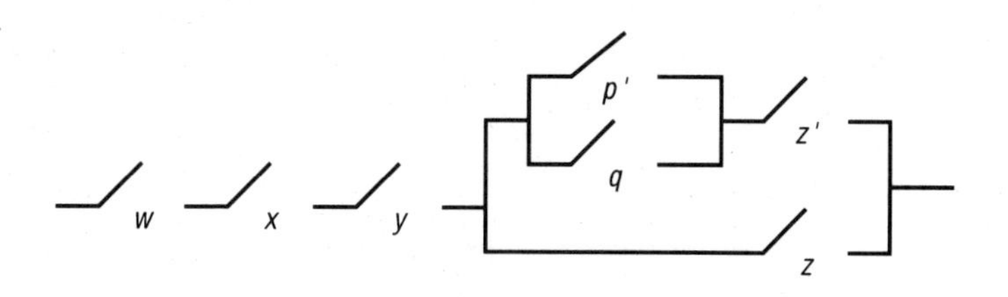

The Boolean expression of the switches wired in parallel is $((p' \vee q) \wedge z') \vee z$, which can be simplified as follows:

$z \vee ((p' \vee q) \wedge z')$	(commutative law)
$(z \vee (p' \vee q)) \wedge (z \vee z')$	(distributive law)
$(z \vee (p' \vee q)) \wedge 1$	(definition of complement)
$z \vee (p' \vee q)$	(identity law)

The simplified circuit is drawn in Figure 7.8.

Figure 7.8

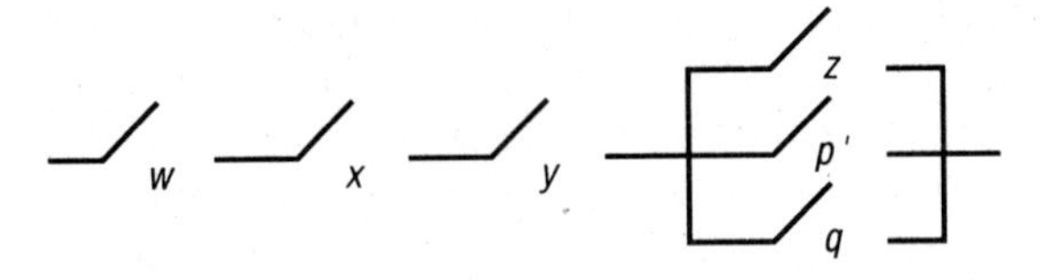

■

The circuit of this problem was given by Shannon to illustrate the usefulness of Boolean algebra in designing circuits. In more complex problems, Boolean algebra provides an important practical approach to simplifying circuits. Although we consider more sophisticated methods of simplification in Chapter 13, these methods also rely on Boolean algebra.

Exercises 7.3

34. Which of the following are expressions?

a. x **b.** x' **c.** $x' \vee y$ **d.** $x \vee (x \vee y')$ **e.** $x \wedge' y$

35. Which of the following pairs of expressions are equivalent?

a. $x, (x')'$ **b.** $((x \vee y) \vee z)', x' \wedge (y' \wedge z')$

c. $(x \wedge x') \wedge (x \vee x), (x \wedge x') \vee (x \vee x)$

36. Make three nonequivalent expressions each of which uses all the following signs exactly as many times as the sign is listed: (,(,),), x, x,y,z, $\vee$,$\vee$,$\wedge$.

37. Draw circuits that are represented by the following expressions.

a. $(x \vee y) \wedge (x \vee z)'$ **b.** $(x' \vee y') \wedge (x' \wedge z')$

c. $(x \vee (y \wedge z)') \vee (x' \vee (y \vee z))$

38. Using standard signs, draw the gates represented by the expressions given in Exercise 37.

39. Write the truth tables of the circuits given in Exercise 37.

40. Write expressions that represent the following three circuits.

a.

x
y
z
y
z

b.

x
y
x
y

c.

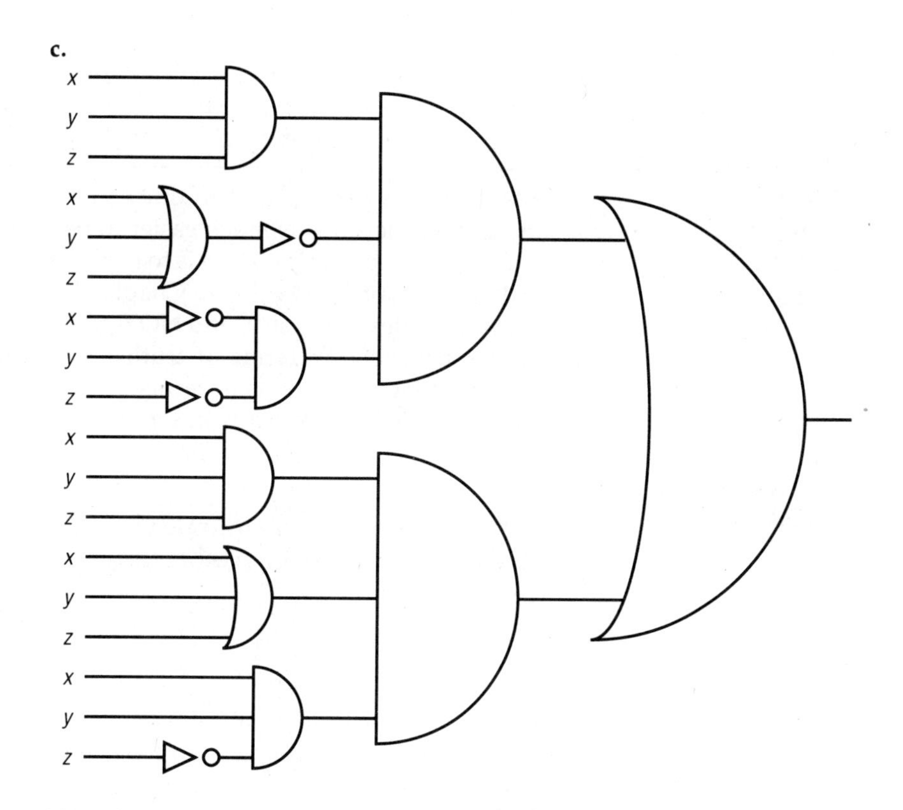

41. Write the complement of each of the following expressions, and in each case simplify the resulting expression. Draw the circuit represented by each simplified complement.

a. $x' \wedge (y \vee z)$ **b.** $x \vee (y' \wedge z)$

c. $(x \wedge y) \vee ((x \wedge z) \wedge (x \vee y))$

42. Simplify the circuits of Exercise 40a and b.

43. Give an example in the Boolean algebra of Example 2 of two nonzero members x and y such that $x \wedge y = 0$.

44. One solution to the equation $X \wedge Z = X \wedge Y$ is $Z = Y$. Show that $Z = (X \wedge Y) \vee X'$ is also a solution. Are Y and $(X \wedge Y) \vee X'$ equivalent expressions?

***45.** Show that, if $X \vee Y = X \wedge Y$, then $X = Y$. (*Hint:* First show that $X \vee Y = X$.)

***46.** Suppose that X, Y, and Z are Boolean expressions satisfying both of the equations $X \wedge Z = X' \vee Y$ and $Y \wedge Z = Y' \vee X'$.

a. Show that it follows that $Y = X \wedge Z$. (*Hint:* First show that $X \wedge Y \wedge Z = (X' \vee Y) \wedge (X' \vee Y') = X'$.)

b. Show that $Y = X \wedge Z$ is *not* a solution to the system of equations because substituting $Y = X \wedge Z$ leads to the conclusion that $1 = 0$. (*Hint:* Show that $Y = X \wedge Y \wedge Z = X \wedge (Y' \vee X') = X \wedge Y'$ and that $Y = 0$.)

BOOLEAN ALGEBRA, LOGIC, AND SETS

Boolean algebra can be used to unify many of the concepts in logic and sets, and frequently it is much easier to be precise if we use Boolean algebra. Our purpose in this section is to explore these ideas. First observe that the definitions of $\wedge$, $\vee$, and $'$ in Chapter 1 are motivated by the truth tables for AND-gates, OR-gates, and NOT-gates. In Chapter 1 we defined equivalent expressions in terms of truth tables, and in Section 7.3 we defined equivalent expressions in terms of circuits they represent. Now we define equivalent expressions in terms of Boolean algebra.

Definition

Two expressions are **equivalent** if it is possible to obtain one of the expressions from the other using only the rules of Boolean algebra.

The proof that this definition and the one given in Chapter 1 lead to the same concept is beyond the scope of our text. We use the preceding definition to illustrate the advantage that Boole's rules can sometimes provide. The following two examples are reconsiderations of Examples 5 and 6 in Chapter 1.

Example 4

Use Boole's rules to show that $X \wedge Y$ and $(X \wedge Y) \vee (W \wedge W')$ are equivalent expressions.

Analysis

$$(X \wedge Y) \vee (W \wedge W') \Leftrightarrow (X \wedge Y) \vee 0$$
$$\Leftrightarrow X \wedge Y$$ ■

Example 5

Use Boole's rules to explain why the expressions $W \vee X$ and $Y \vee X$ are not equivalent.

Analysis

There is no rule of Boolean algebra that allows us to change W into Y. ■

Example 6

Use Boole's rules to prove that $X \rightarrow Y$ and $Y' \rightarrow X'$ are equivalent expressions.

Analysis

Since there is no $\rightarrow$ sign in Boolean algebra, we define $A \rightarrow B$ to be alternative notation for the Boolean expression $A' \vee B$. Then

$$\begin{aligned} X \rightarrow Y &\Leftrightarrow X' \vee Y \\ &\Leftrightarrow Y \vee X' \\ &\Leftrightarrow Y'' \vee X' \\ &\Leftrightarrow Y' \rightarrow X' \end{aligned}$$

■

The following theorem is an immediate consequence of Theorem 2.1.

Theorem 7.5

Let S be a nonempty set. Then $(\mathcal{P}(S), \cup, \cap, \sim)$ is a Boolean algebra.

Proof

We take $\varnothing$ to be 0, S to be 1, and set complement ($\sim$) to be the complement ($'$). □

In Exercise 22 we defined $\leq$ in an arbitrary Boolean algebra by saying $a \leq b$ provided that $a \wedge b = a$. Theorem 7.6, which we state without proof, shows how this $\leq$ translates into set-theoretic language in the Boolean algebra provided by Theorem 7.5.

Theorem 7.6

Let S be a nonempty set and let $(\mathcal{P}(S), \cup, \cap, \sim)$ be the Boolean algebra of subsets of S. For $A,B \in \mathcal{P}(S)$ define $A \leq B$ provided that $A \cap B = A$. Then $A \leq B$ if and only if $A \subseteq B$.

A comparison of Exercise 22 and Theorem 7.6 indicates that it is natural to consider the Boolean algebra of subsets of a set. Moreover, just as we can use the laws of Boolean algebra in place of truth tables to establish a logical equivalence, we can now use those laws in place of an element-chasing argument to establish a set-theoretic equation. Notice that Theorem 7.5 offers an incredibly rich supply of Boolean algebras, one of each nonempty set! In particular, the Boolean algebra of subsets of $\mathbb{N}$ provides us with our first infinite Boolean algebra.

The coincidence of Boolean algebra, sets, and logic is important to the computer scientist because it shows that exactly the same circuits and programs handle all three subjects, and only the interpretation of the user or programmer needs to be changed. The common nature of Boolean algebra, logic, and sets is so important that we summarize their relationships in Table 7.6. In this table, we use the term "equivalence class,"

Table 7.6

Boolean algebra	Logic	Sets
set B of objects	equivalence classes of propositional expressions	$\mathcal{P}(U)$, where U is the universal set
0	c (contradiction)	$\varnothing$ (empty set)
1	t (tautology)	U (universal set)
$\wedge$ (meet)	$\wedge$ (conjunction) "and"	$\cap$ (intersection)
$\vee$ (join)	$\vee$ (disjunction) "or"	$\cup$ (union)
$'$ (complement)	negation (not)	$\sim$ (set complement)

which is discussed at length in Chapter 10. Here, it is intended only as a reminder that in forming a Boolean algebra of propositional expressions we must consider any two logically equivalent expressions to be the same.

In Chapter 2, we studied the concept of an ordered n-tuple. An ordered n-tuple of zeros and ones may be used to describe the Boolean algebra of subsets of n-element set in a form that is palatable for a computer. For example, suppose that we are given the set $U = \{1,2,3,4,5,6,7,8\}$. Each subset A of U is represented by the 8-tuple whose kth term is 1 if $k \in A$ and whose kth term is 0 if $k \notin A$. Thus $(1,0,1,0,1,0,1,0)$ represents $\{x \in U\colon x \text{ is odd}\}$, and $\{x \in U\colon x \text{ is even}\}$ is represented by the 8-tuple $(0,1,0,1,0,1,0,1)$.

Example 7

Let n be a natural number and let B be the set of all zero–one n-tuples. Using the usual rules of Boolean algebra for 0 and 1, we may define the join, meet, and complement of zero–one n-tuples term by term. We illustrate these definitions in the case $n = 8$.

$$(1,1,0,0,1,1,0,1)' = (0,0,1,1,0,0,1,0)$$

$$(1,1,0,0,1,1,0,1) \vee (0,0,0,1,0,0,1,1) = (1,1,0,1,1,1,1,1)$$

$$(1,1,0,0,1,1,0,1) \wedge (0,0,0,1,0,0,1,1) = (0,0,0,0,0,0,0,1)$$ ■

Example 8

Let $U = \{a_1,a_2,a_3,a_4,a_5,a_6,a_7,a_8\}$ be an eight-element indexed set. Then the Boolean equations given in Example 7 have the following set-theoretic interpretations:

$$\{a_1,a_2,a_5,a_6,a_8\}^{\sim} = \{a_3,a_4,a_7\}$$

$$\{a_1,a_2,a_5,a_6,a_8\} \cup \{a_4,a_7,a_8\} = \{a_1,a_2,a_4,a_5,a_6,a_7,a_8\}$$

$$\{a_1,a_2,a_5,a_6,a_8\} \cap \{a_4,a_7,a_8\} = \{a_8\}$$ ■

Note that the only difference between Example 7 and Example 8 is a difference in interpretation.

Exercises 7.4

47. Suppose that you go to the store to buy an AND-gate and sadly discover that the store sells only NOT-gates and OR-gates and has plenty of both.
   a. Can you buy NOT-gates and OR-gates and build a network equivalent to an AND-gate?
   b. How many of each kind of gate must you buy?
   c. Sketch your network.

*48. A NAND-gate for two switches is the gate

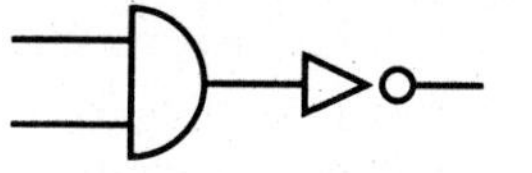

The previous problem is unrealistic because stores sell only NAND-gates. It is easy to make a network equivalent to a NOT-gate from a NAND-gate by joining the two inputs:

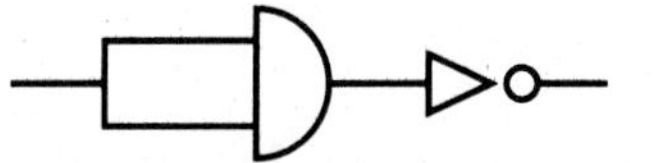

Thus three NAND-gates can be used to make a network equivalent to an OR-gate:

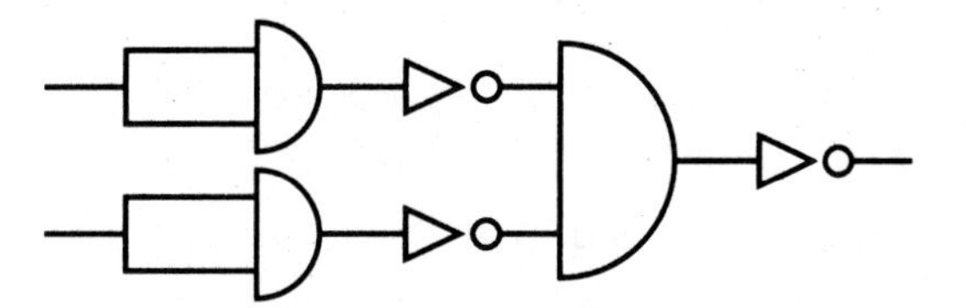

Show how to combine two or more NAND-gates to obtain a network equivalent to

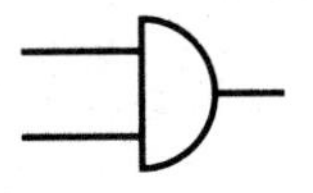

*49. Construct networks for the following truth tables.

a.

p	q	Output
1	1	1
1	0	0
0	1	0
0	0	1

b.

p	q	Output
1	1	0
1	0	1
0	1	1
0	0	0

c.

p	q	Output
1	1	0
1	0	1
0	1	1
0	0	1

d.

p	q	Output
1	1	1
1	0	0
0	1	0
0	0	0

50. Let $x_1, x_2, \ldots, x_8$ be a list and let $U = \{x_i: i \in I_8\}$.
   a. Find the subset A of U represented by (1,0,1,0,1,0,0,1).
   b. Find the subset B of U represented by (0,1,1,0,1,0,1,1).
   c. What 8-tuple represents $A - B$?
   d. What 8-tuple represents $B - A$?
   e. What 8-tuple represents $A \oplus B$?

51. Let I_8 be the universal set and let A be the set represented by (1,0,1,1,0,1,1,1). List all the ordered 8-tuples that represent neighbors of A. (*Neighbor* is defined in Exercise 37 of Chapter 2.)

52. Define $0 \oplus 0$, $0 \oplus 1$, $1 \oplus 0$, and $1 \oplus 1$ in such a way that, if we perform the operation $\oplus$ on two ordered zero–one n-tuples representing sets A and B, the resulting n-tuple represents the symmetric difference $A \oplus B$.

Chapter 7 Review Exercises

53. a. Draw a circuit that represents the English statement "(not p) or (not q)."
   b. Draw a circuit that represents the English statement "not(p or q)."
   c. Draw a circuit that represents the English statement "((not p) and q) or ((not q) and p)."

54. a. Redraw the following circuit using standard representations of AND-gates, OR-gates, and NOT-gates.

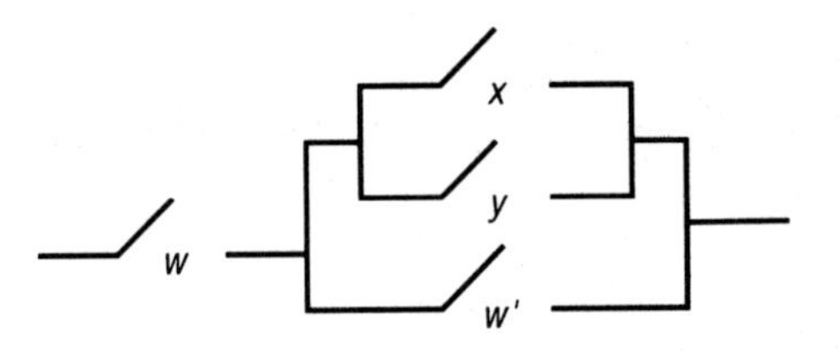

   b. Complete the truth table for the given circuit.

w'	w	x	y	Output
	1	1	1	
	1	1	0	
	1	0	1	
	1	0	0	
	0	1	1	
	0	1	0	
	0	0	1	
	0	0	0	

55. Are the following networks equivalent? Explain your answer.

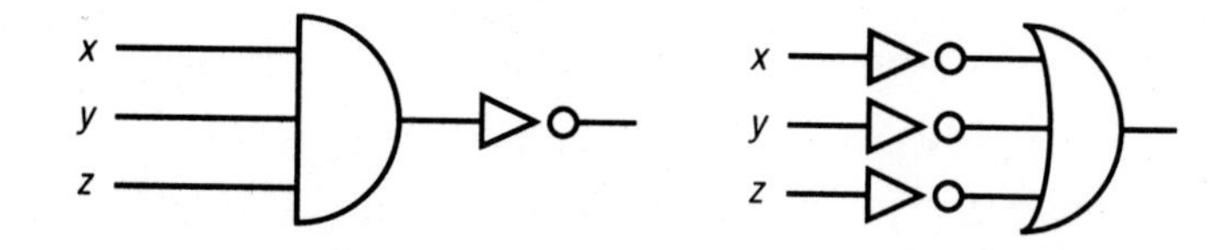

56. Let B be the positive divisors of 210. For a and b belonging to B, define $a \vee b = \text{lcm}(a,b)$, $a \wedge b = \text{gcd}(a,b)$, and $a' = 210/a$. Find each of the following. **a.** $42'$ **b.** $15 \wedge 14$ **c.** $15 \vee 14$ **d.** $6' \wedge 14'$

57. In Example 2, verify the following equalities.

a. $(5 \wedge 6)' = 5' \vee 6'$ **b.** $(15 \wedge 6)' = 15' \vee 6'$

c. $5 \wedge (5 \vee 15) = 5$ **d.** $5 \vee (5 \wedge 15) = 5$

58. Which of the following pairs of expressions are equivalent?

a. $(x \vee y) \wedge (x \wedge z),\ x \vee (y \wedge z)$ **b.** $x \vee (x \wedge y),\ x$

59. Draw circuits that are represented by the following expressions.

a. $(x \vee y) \wedge ((x \wedge y')$ **b.** $(x \vee y) \wedge (x' \vee y')$

60. Write the truth tables of the circuits given in Exercise 59.

61. Construct networks for the following truth tables.

a.

p	q	Output
1	1	1
1	0	1
0	1	0
0	0	0

b.

p	q	Output
1	1	0
1	0	0
0	1	1
0	0	1

62. Using the definition of join, meet, and complement as indicated in Example 7, find each of the following.

a. $(1,0,0,1,1,0)'$ **b.** $(1,0,0,1,1,0) \vee (1,0,1,0,0,1)$

c. $(1,0,0,1,1,0) \wedge (1,0,1,0,0,1)$

Chapter

8

MATRICES

Some words evoke pleasant memories of idleness and play. "Table" is not one of these. Our early introduction to tables is accompanied by homework problems such as "Copy and complete table:"

10	12	?
2	4	?
3	5	15

or "Using the table on page 396, determine whether Rwanda has an adequate supply of manganese." (Adequate for what, we are never told.) Later we are introduced to more sophisticated tables, logarithmic tables and trigonometric tables, or most painful of all a wily combination of the two.

This chapter is about tables, but here we are interested in tables themselves and only secondarily interested in their contents. When seen in this light, tables take on a new life. If they are of the appropriate size, we may add, subtract, or multiply them. As we see in the remainder of this book, tables of one kind or another seem to be able to represent almost everything mathematical, and the concepts we introduce in this chapter are nearly as basic to the study of mathematics as functions.

8.1

MATRIX ARITHMETIC

We can use a table to display the cost of four different personal computers at three different stores (see Table 8.1). In mathematics, such an arrangement of data into rows and columns is called a *matrix*. The rows of a matrix are read horizontally and the columns vertically. We can think of a matrix as a table. In fact, in COBOL *table* is used to denote a matrix. Matrices are used throughout discrete mathematics to express relationships between members of sets. In this section we give a review of matrix arithmetic.

Table 8.1

	Store 1	Store 2	Store 3
PC 1	800	850	795
PC 2	1200	1150	1125
PC 3	1000	1050	1025
PC 4	1400	1200	1300

Definition

A **matrix** is a rectangular display of members of a set. A matrix with m rows and n columns is called an $m \times n$ (read "m by n") matrix. A matrix with the same number of rows as columns is called a **square matrix.**

An example of a 3×4 matrix is

$$A = \begin{bmatrix} 1 & 7 & 3 & 4 \\ 2 & 4 & 6 & 5 \\ 3 & 2 & 1 & 2 \end{bmatrix}$$

It is customary to use the notation

$$A = \begin{bmatrix} a_{11} & a_{12} & \cdots & a_{1n} \\ a_{21} & a_{22} & \cdots & a_{2n} \\ \vdots & \vdots & & \vdots \\ a_{m1} & a_{m2} & \cdots & a_{mn} \end{bmatrix}$$

to denote an $m \times n$ matrix, and we say that the *size of A* is $m \times n$ (again read "m by n"). For each $i = 1,2,\ldots,m$, the *ith row of A* is the $1 \times n$ matrix

$[a_{i1}\ a_{i2}\ .\ .\ .\ a_{in}]$

and for each $j = 1,2, \ldots ,n$, the *jth column of A* is the $m \times 1$ matrix

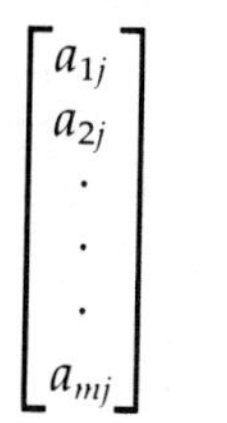

For each $n \in \mathbb{N}$, a $1 \times n$ matrix is called a *row matrix*, and an $n \times 1$ matrix is called a *column matrix*. For each $i = 1,2, \ldots ,m$ and $j = 1,2, \ldots ,n$, the element in the ith row and jth column of A is called the *(i,j)th entry of A*. Each member of the rectangular display is called an *entry*. The shorthand notation $A = [a_{ij}]$ is often used to express the matrix A. This notation indicates that A is the matrix whose (i,j)th entry is a_{ij}. The entries of a matrix are often real numbers, but they may well be other objects such as sets or functions.

Definition

Two $m \times n$ matrices $A = [a_{ij}]$ and $B = [b_{ij}]$ are **equal** if $a_{ij} = b_{ij}$ for each $i = 1,2, \ldots ,m$ and each $j = 1,2, \ldots ,n$.

Example 1

Find real numbers x, y, and z such that

$$\begin{bmatrix} x + y & z \\ y + z & x - z \end{bmatrix} = \begin{bmatrix} 7 & 3 \\ 5 & 2 \end{bmatrix}$$

Analysis

We have two 2×2 matrices, so they are equal if

$$x + y = 7, \qquad z = 3, \qquad y + z = 5, \qquad x - z = 2$$

Solving these equations we obtain

$$z = 3, \qquad y = 2, \qquad x = 5$$

■

There is an alternative way to define a matrix. An $m \times n$ matrix A is a function with domain $I_m \times I_n$, and if $(i,j) \in \text{dom}(A)$, then $A((i,j))$ is denoted a_{ij} and called the (i,j)th entry of A. There are certain advantages to this definition of a matrix. For example, the definition of equality of matrices, which we have just given, becomes the definition of equality of functions and so there is no need to repeat the definition. A second advantage is that the alternative definition is mathematically precise; it does not depend on the reader's good will in agreeing on the meaning of a "rectangular display." For the time being, however, we urge you to stay with the first definition of a matrix, as a rectangular display.

In this chapter, we assume that the entries in our matrices are real numbers, and we describe some operations that can be performed on such matrices.

Definition

If $k \in \mathbb{R}$ and A is a matrix, the **scalar product of k and A** is the matrix obtained from A by multiplying each entry of A by k. The scalar product of k and A is denoted kA.

Example 2

$$\text{Let } A = \begin{bmatrix} 2 & 1 \\ 3 & 5 \\ 7 & 2 \end{bmatrix}$$

Find $3A$.

Analysis

$$3A = 3\begin{bmatrix} 2 & 1 \\ 3 & 5 \\ 7 & 2 \end{bmatrix} = \begin{bmatrix} 3 \cdot 2 & 3 \cdot 1 \\ 3 \cdot 3 & 3 \cdot 5 \\ 3 \cdot 7 & 3 \cdot 2 \end{bmatrix} = \begin{bmatrix} 6 & 3 \\ 9 & 15 \\ 21 & 6 \end{bmatrix}$$ ■

Definition

If $A = [a_{ij}]$ and $B = [b_{ij}]$ are $m \times n$ matrices, the **sum of A and B,** denoted $A + B$, is the $m \times n$ matrix $C = [c_{ij}]$ such that $c_{ij} = a_{ij} + b_{ij}$ for each $i = 1, 2, \ldots, m$ and each $j = 1, 2, \ldots, n$.

Notice that matrices of different sizes cannot be added.

Example 3

$$\text{Let } A = \begin{bmatrix} 2 & -1 \\ 5 & 7 \\ -3 & 4 \end{bmatrix} \text{ and } B = \begin{bmatrix} -3 & 2 \\ 1 & 6 \\ 8 & 2 \end{bmatrix}$$

Find $A + B$.

Analysis

$$A + B = \begin{bmatrix} 2 & -1 \\ 5 & 7 \\ -3 & 4 \end{bmatrix} + \begin{bmatrix} -3 & 2 \\ 1 & 6 \\ 8 & 2 \end{bmatrix} = \begin{bmatrix} 2+(-3) & (-1)+2 \\ 5+1 & 7+6 \\ (-3)+8 & 4+2 \end{bmatrix}$$

$$= \begin{bmatrix} -1 & 1 \\ 6 & 13 \\ 5 & 6 \end{bmatrix}$$ ■

Notice that, if $A = [a_{ij}]$ is an $m \times n$ matrix, then $(-1)A$, or $-A$, is the matrix $B = [b_{ij}]$ such that $b_{ij} = -a_{ij}$ for each $i = 1,2, \ldots ,m$, and $j = 1,2, \ldots ,n$.

Definition

An $m \times n$ matrix in which every entry is 0 is called a **zero matrix** and is denoted by $\mathbf{0}_{m \times n}$. (When there is no question as to the size of the zero matrix, it is customary to omit the subscript $m \times n$.)

Theorem 8.1

Let $A = [a_{ij}]$, $B = [b_{ij}]$, and $C = [c_{ij}]$ be $m \times n$ matrices. If $\mathbf{0}$ is the $m \times n$ zero matrix, then

a. $A + B = B + A$
b. $(A + B) + C = A + (B + C)$
c. $A + \mathbf{0} = A$
d. $A - A = \mathbf{0}$

Proof

a. Both $A + B$ and $B + A$ are $m \times n$ matrices. To show that they are equal it is sufficient to show that, for each $i = 1,2, \ldots ,m$ and each $j = 1,2, \ldots ,n$, the (i,j)th entry of $A + B$ is equal to the (i,j)th entry of $B + A$. But the (i,j)th entry of $A + B$ is $a_{ij} + b_{ij}$ and the (i,j)th entry of $B + A$ is $b_{ij} + a_{ij}$. Since a_{ij} and b_{ij} are real numbers, $a_{ij} + b_{ij} = b_{ij} + a_{ij}$. Therefore $A + B = B + A$.

b. See Exercise 9.

c. We know that $A + \mathbf{0}$ and A are $m \times n$ matrices. Suppose $i,j \in \mathbb{N}$, $1 \leq i \leq m$, and $1 \leq j \leq n$. Then the (i,j)th entry of $A + \mathbf{0}$ is $a_{ij} + 0$ and the (i,j)th entry of A is a_{ij}. Since $a_{ij} + 0 = a_{ij}$, $A + \mathbf{0} = A$.

d. See Exercise 10. □

The four statements of Theorem 8.1 can be put in simple, familiar terms: (a) matrix addition is *commutative;* (b) matrix addition is *associative;* (c) the $m \times n$ zero matrix is an *additive identity;* and (d) $-A$ is an *additive inverse* of A. Notice that $\mathbf{0} + A = A$ because, by (a), $\mathbf{0} + A = A + \mathbf{0}$, and, by (b), $A + \mathbf{0} = A$.

Example 4

In computer graphics there are three primary colors—red, green, and blue—and various intensities of these three colors are used to create all other colors. Nonnegative integers between 0 and 255 are used to denote the intensity of each of the three primary colors. Table 8.2 gives the intensity of each of these three primary colors needed to create six colors that we label c_1, c_2, c_3, c_4, c_5, and c_6.

Table 8.2

	Red	Green	Blue
c_1	45	92	87
c_2	0	255	100
c_3	150	75	99
c_4	0	0	255
c_5	200	50	0
c_6	255	100	200

We have a 6×3 matrix C, and for each $i = 1,2,3,4,5,6$ and $j = 1,2,3$ we can find the (i,j)th entry. For instance, the (2,3)th entry is 100 and the (3,2)th entry is 75. In FORTRAN, the (i,j)th entry of C is denoted by $C(i,j)$. Construct flowcharts for each of the following activities:

a. Calculate and print the percentage of the maximum intensity of each primary color in each created color.

b. Calculate and print the average intensity of the primary colors used to create color c_1.

c. Change the intensity of each of the primary colors by increasing the intensity of the primary color by 100, if possible, and otherwise by increasing the intensity as much as possible.

Analysis

a. First we want to calculate the percentage of the maximum intensity of each primary color in each color that we are creating. Since the maximum intensity of each primary color is 255, we want to divide each entry in C by 255 and then multiply by 100, so we take the scalar product of 100/255 and C. The flowchart in Figure 8.1 performs this calculation.

b. Next we want to calculate the average intensity of the primary colors used to create color c_1. We add the entries in the first row and divide by three. The flowchart in Figure 8.2 performs this calculation.

Figure 8.1

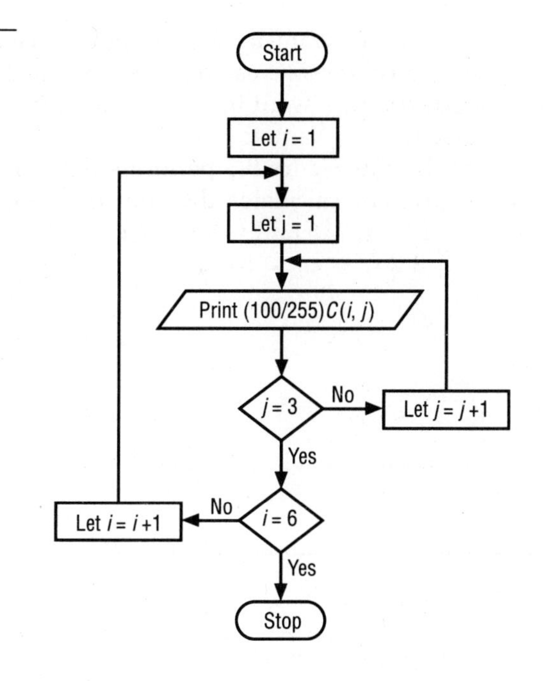

Figure 8.2

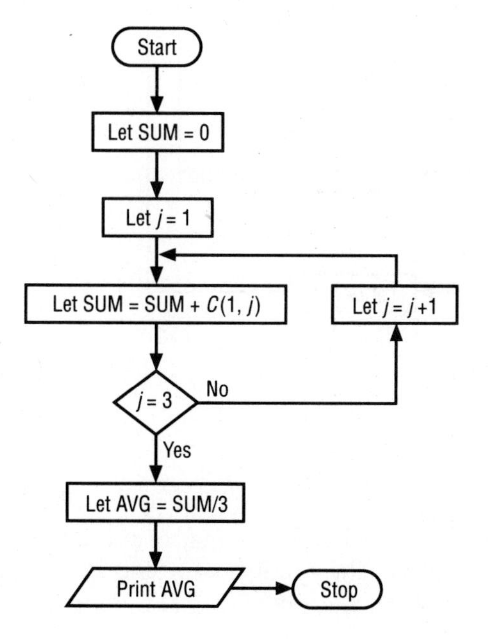

c. Finally, suppose we want to change the intensity of each of the primary colors in each of the six created colors and create six new colors. Suppose, where possible, we want to increase the intensity by 100, and, if this is not possible, we want to increase the intensity as much as possible. The flowchart in Figure 8.3 performs the required calculation. Notice that this flowchart calculates the sum of the matrix $D = [d_{ij}]$, where for each $i = 1,2, \ldots ,6$ and $j = 1,2,3$, $d_{ij} = 100$ if $255 - c_{ij} \geq 100$ and $d_{ij} = 255 - c_{ij}$ if $255 - c_{ij} < 100$.

Figure 8.3

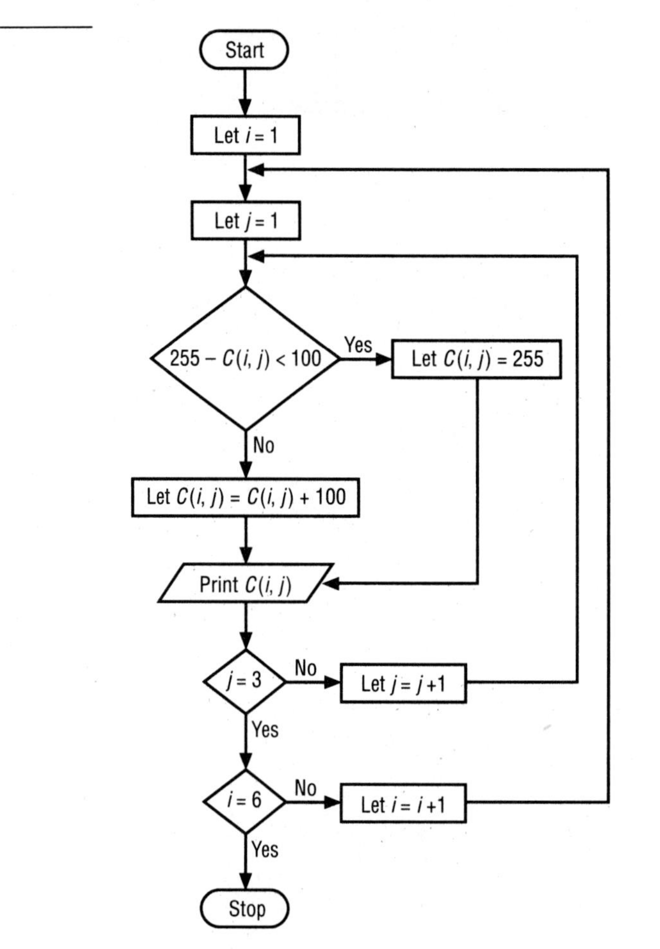

■

Definition

Let $A = [a_{ij}]$ be an $m \times n$ matrix and let $B = [b_{ij}]$ be an $n \times p$ matrix. The **product of A and B,** denoted AB, is the $m \times p$ matrix $[c_{ij}]$, where $c_{ij} = \sum_{k=1}^{n} a_{ik}b_{kj}$.

Notice that the product of A and B is defined only when the number of columns of A is equal to the number of rows of B. The (i,j)th entry of AB is the sum of the products of the corresponding elements from the ith row of A and the jth column of B.

Example 5

Let $A = \begin{bmatrix} 2 & -3 \\ -2 & 4 \\ 5 & -1 \end{bmatrix}$ and $B = \begin{bmatrix} 2 & 3 & 4 & -2 \\ -1 & -3 & 6 & 5 \end{bmatrix}$.

Then A is a 3×2 matrix and B is a 2×4 matrix. Therefore the product AB is defined, and it is a 3×4 matrix. We first calculate the (2,3)th entry c_{23} of the product matrix:

$$\text{row } 2 \rightarrow \begin{bmatrix} 2 & -3 \\ -2 & 4 \\ 5 & -1 \end{bmatrix} \times \begin{bmatrix} 2 & 3 & \overset{\text{column } 3}{\overset{\downarrow}{4}} & -2 \\ -1 & -3 & 6 & 5 \end{bmatrix}$$

$$= \begin{bmatrix} - & - & - & - \\ - & - & c_{23} & - \\ - & - & - & - \end{bmatrix}$$

$$(-2)(4) + 4(6) = 16$$

The remaining entries of the product matrix are calculated in this same way:

$$c_{11} = 2(2) + (-3)(-1) = 7 \qquad c_{12} = 2(3) + (-3)(-3) = 15$$

$$c_{13} = 2(4) + (-3)(6) = -10 \qquad c_{14} = 2(-2) + (-3)(5) = -19$$

$$c_{21} = (-2)(2) + 4(-1) = -8 \qquad c_{22} = (-2)(3) + 4(-3) = -18$$

$$c_{23} = (-2)(4) + 4(6) = 16 \qquad c_{24} = (-2)(-2) + 4(5) = 24$$

$$c_{31} = 5(2) + (-1)(-1) = 11 \qquad c_{32} = 5(3) + (-1)(-3) = 18$$

$$c_{33} = 5(4) + (-1)(6) = 14 \qquad c_{34} = 5(-2) + (-1)(5) = -15$$

So

$$AB = \begin{bmatrix} 7 & 15 & -10 & -19 \\ -8 & -18 & 16 & 24 \\ 11 & 18 & 14 & -15 \end{bmatrix}$$

■

Notice that if A and B are matrices then AB is defined only when the number of columns of A is equal to the number of rows of B and BA is defined only when the number of columns of B is equal to the number of rows of A. Therefore, if A is an $m \times n$ matrix and AB and BA are defined, then B must be an $n \times m$ matrix. If A is an $m \times n$ matrix and B is an $n \times m$ matrix, then AB is an $m \times m$ matrix and BA is an $n \times n$ matrix. Therefore, if $m \neq n$, then AB and BA are not of the same size, so they cannot be equal. As the following example shows, even when A and B are square matrices (so that AB and BA are of the same size) it is not necessarily true that $AB = BA$.

Example 6

Let $A = \begin{bmatrix} 1 & 2 \\ 3 & 1 \end{bmatrix}$ and $B = \begin{bmatrix} 2 & -1 \\ 3 & 2 \end{bmatrix}$

Then

$$AB = \begin{bmatrix} 8 & 3 \\ 9 & -1 \end{bmatrix} \text{ and } BA = \begin{bmatrix} -1 & 3 \\ 9 & 8 \end{bmatrix}$$

So $AB \neq BA$. ■

We have just demonstrated that matrix multiplication is not commutative; however, as the following theorem indicates, matrix multiplication is associative.

Theorem 8.2

Let A be an $m \times n$ matrix, B an $n \times p$ matrix, and C a $p \times q$ matrix. Then $(AB)C = A(BC)$.

Proof

Notice that AB is an $m \times p$ matrix and $(AB)C$ is an $m \times q$ matrix. Also BC is an $n \times q$ matrix and $A(BC)$ is an $m \times q$ matrix. Therefore the indicated products are defined and $(AB)C$ and $A(BC)$ are of the same size. It remains to show that, for each $i = 1,2, \ldots ,m$ and each $j = 1,2, \ldots ,q$, the (i,j)th entry of $(AB)C$ is the same as the (i,j)th entry of $A(BC)$.

Let $A = [a_{ij}]$, $B = [b_{ij}]$, $C = [c_{ij}]$, $AB = [d_{ij}]$, $(AB)C = [e_{ij}]$, $BC = [f_{ij}]$, and $A(BC) = [g_{ij}]$. We must show that, for each i and j, $e_{ij} = g_{ij}$. By the definition of the product we have

$$d_{ik} = \sum_{t=1}^{n} a_{it}b_{tk}, \qquad f_{tj} = \sum_{k=1}^{p} b_{tk}c_{kj}$$

$$e_{ij} = \sum_{k=1}^{p} d_{ik}c_{kj}, \qquad \text{and} \qquad g_{ij} = \sum_{t=1}^{n} a_{it}f_{tj}$$

Therefore

$$e_{ij} = \sum_{k=1}^{p} \left(\sum_{t=1}^{n} a_{it}b_{tk} \right) c_{kj} \qquad \text{and} \qquad g_{ij} = \sum_{t=1}^{n} a_{it} \left(\sum_{k=1}^{p} b_{tk}c_{kj} \right)$$

so by the associative and commutative properties of real numbers, $e_{ij} = g_{ij}$. (The details are tedious, and we choose to omit them. In Exercise 14 you are asked to verify this equality when n and p are 3.) □

As Theorem 8.3 indicates, we also have the distributive laws, but we omit the proof.

Theorem 8.3

If A, B, and C are square matrices of the same size, then $A(B + C) = AB + AC$ and $(A + B)C = AC + BC$.

Theorem 8.4, whose proof we also omit, states some properties of scalar multiplication.

Theorem 8.4

If A and B are square matrices of the same size and α and β are real numbers, then

a. $\alpha(\beta A) = (\alpha\beta)A$
b. $(\alpha + \beta)A = \alpha A + \beta A$
c. $\alpha(A + B) = \alpha A + \alpha B$
d. $\alpha(AB) = (\alpha A)B = A(\alpha B)$

In Figure 8.4 (page 260) we give a flowchart for calculating the product AB of an $m \times n$ matrix A and an $n \times p$ matrix B. If A and B are square $n \times n$ matrices, the number of entries in the product AB is n^2. As illustrated in Figure 8.4, each entry requires n multiplications and $n - 1$ additions. Therefore a total of n^3 multiplications and $n^2(n-1)$ additions are used.

We have established that matrix multiplication is associative. For appropriately sized matrices A, B, and C, however, one of the multiplications $A \times (B \times C)$ and $(A \times B) \times C$ may be more complex than the other, even though the resulting products are equal. Example 7 illustrates this point.

Example 7

Let A be a 10×40 matrix, B a 40×30 matrix, and C a 30×20 matrix. There are two ways to find the product $A \times B \times C$: we can find $(A \times B) \times C$ or $A \times (B \times C)$. The number of entries in $A \times B$ is 300, and each entry requires 40 multiplications (see Figure 8.4). Hence $300 \times 40 = 12{,}000$ multiplications are used. There are 200 entries in $(A \times B) \times C$, and each entry requires 30 multiplications. Hence $200 \times 30 = 6{,}000$ multiplications are used to calculate the product of $A \times B$ and C. Therefore a total of $12{,}000 + 6{,}000 = 18{,}000$ multiplications are used to calculate $(A \times B) \times C$. Using the same reasoning, we find that the number of multiplications required to calculate $A \times (B \times C)$ is $(800 \times 30) + (200 \times 40) = 32{,}000$. It is therefore clear that we want to calculate $(A \times B) \times C$.

Figure 8.4

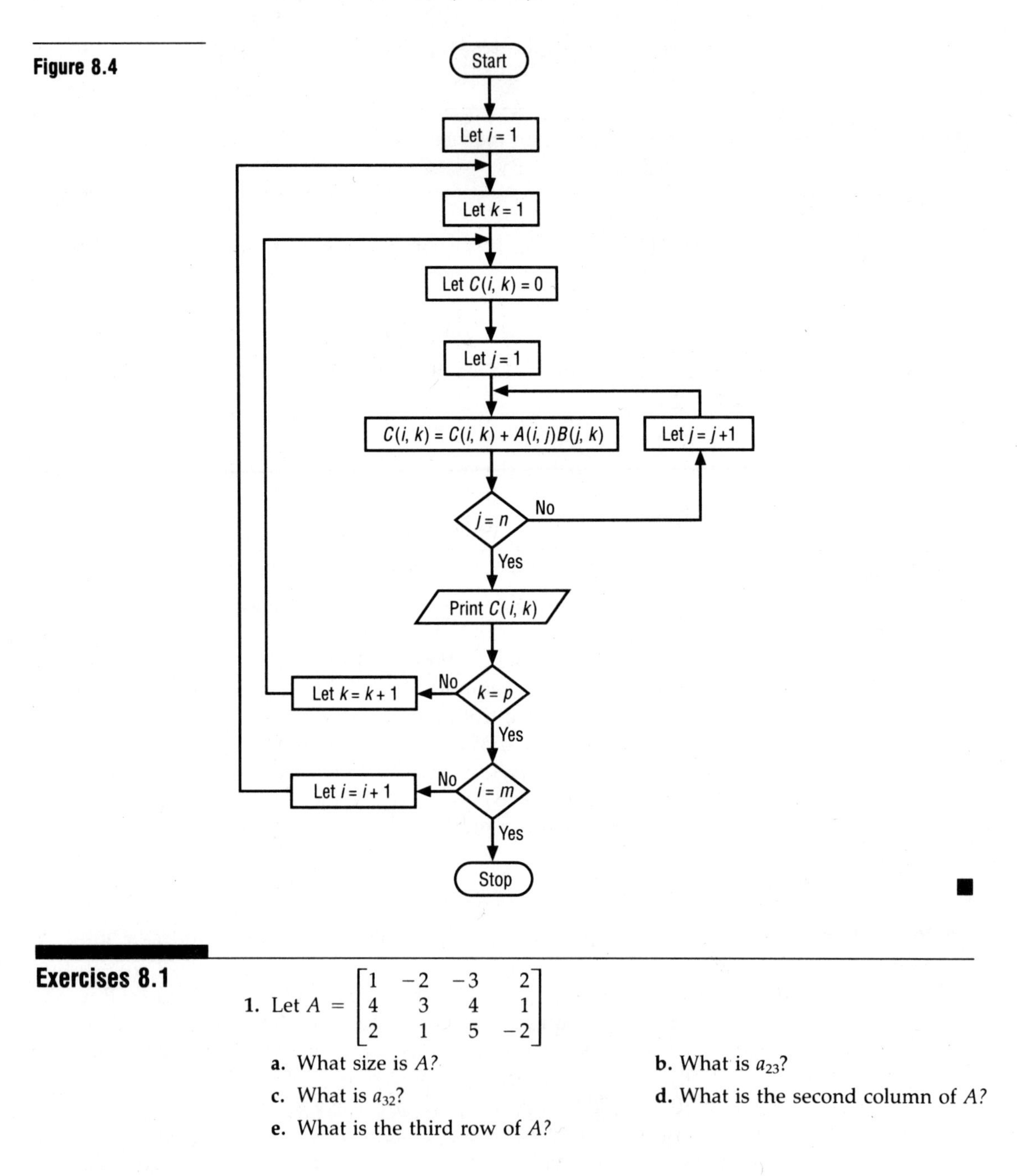

■

Exercises 8.1

1. Let $A = \begin{bmatrix} 1 & -2 & -3 & 2 \\ 4 & 3 & 4 & 1 \\ 2 & 1 & 5 & -2 \end{bmatrix}$

   a. What size is A?
   b. What is a_{23}?
   c. What is a_{32}?
   d. What is the second column of A?
   e. What is the third row of A?

2. Let $A = \begin{bmatrix} 1 & 0 \\ 2 & 1 \end{bmatrix}$ and $B = \begin{bmatrix} 3 & -1 \\ 2 & 1 \end{bmatrix}$. Find

   a. $2A$ **b.** $A + B$ **c.** $B + A$ **d.** AB **e.** BA

3. If $\begin{bmatrix} x & z \\ y & z \end{bmatrix} + \begin{bmatrix} 1 & 4 \\ 2y & z \end{bmatrix} = \begin{bmatrix} 3 & 6 \\ 6 & 4 \end{bmatrix}$, find x, y, and z.

4. If $\begin{bmatrix} 2 & 1 \\ 4 & -1 \end{bmatrix} \times \begin{bmatrix} x \\ y \end{bmatrix} = \begin{bmatrix} 4 \\ 2 \end{bmatrix}$, find x and y.

5. Find x and y such that $\begin{bmatrix} 1 & 2 \\ 2 & 5 \end{bmatrix} \times \begin{bmatrix} x-2 \\ y+1 \end{bmatrix} = \begin{bmatrix} 1 \\ 0 \end{bmatrix}$

6. Let $A = \begin{bmatrix} 1 & -2 & 2 \\ 2 & 4 & 1 \\ 3 & -1 & -3 \end{bmatrix}$ and $B = \begin{bmatrix} 3 & 0 & 1 \\ -2 & 5 & 2 \\ -1 & -3 & 4 \end{bmatrix}$

   a. Find 5A. **b.** Find $A + B$. **c.** Find AB. **d.** Find BA.
   e. Find $2A + 3B$.

7. Let A be a 3×4 matrix, B a 4×4 matrix, and C a 4×5 matrix. Determine which of the following products are defined and state the size of those that are defined.

   a. AB **b.** AC **c.** BA **d.** CA **e.** BC **f.** CB

8. Find a matrix A such that $A \times \begin{bmatrix} 2 & 0 \\ 0 & 3 \end{bmatrix} = \begin{bmatrix} 1 & 0 \\ 0 & 1 \end{bmatrix}$

9. Prove that, if A, B, and C are $m \times n$ matrices, then $(A + B) + C = A + (B + C)$.

10. Let A be an $m \times n$ matrix. Prove that, if $\mathbf{0}$ is the $m \times n$ zero matrix, then $A - A = \mathbf{0}$.

11. Draw a flowchart for finding the sum of two $m \times n$ matrices.

12. Let A be a 30×60 matrix, B a 60×20 matrix, and C a 20×50 matrix. What is the most efficient way to find $A \times B \times C$? Verify your answer by giving the number of real-number multiplications involved in each method.

13. Let A be a 10×80 matrix, B an 80×50 matrix, and C a 50×30 matrix. What is the most efficient way to find $A \times B \times C$? Verify your answer by giving the number of real-number multiplications involved in each method.

14. For each i,j,k, and t, let a_{ij}, b_{ik}, and c_{kt} be real numbers. Prove that

$$\sum_{k=1}^{3} \left(\sum_{t=1}^{3} a_{it} b_{tk} \right) c_{kj} = \sum_{t=1}^{3} a_{it} \left(\sum_{k=1}^{3} b_{tk} c_{kj} \right)$$

8.2

OTHER OPERATIONS ON MATRICES

In this section we introduce some special matrices and define some additional operations on matrices.

Definition

If $A = [a_{ij}]$ is an $n \times n$ matrix, the **main diagonal** of A is the set consisting of $a_{11}, a_{22}, \ldots, a_{nn}$. The matrix A is called a **diagonal matrix** if $a_{ij} = 0$ whenever $i \neq j$.

Example 8

$$\text{If } A = \begin{bmatrix} 1 & 3 & -3 \\ 2 & -2 & -4 \\ -1 & 0 & 4 \end{bmatrix}$$

the main diagonal of A is $\{1, -2, 4\}$. The matrix

$$B = \begin{bmatrix} 2 & 0 & 0 \\ 0 & 1 & 0 \\ 0 & 0 & -3 \end{bmatrix}$$

is a diagonal matrix. ■

Theorem 8.5

> If A and B are diagonal matrices, then the sum $A + B$ and the product AB are diagonal matrices. Furthermore if $A = [a_{ij}]$, $B = [b_{ij}]$, and $AB = [c_{ij}]$ are $n \times n$ matrices, then $c_{ii} = a_{ii}b_{ii}$ for each $i = 1, 2, \ldots, n$.

The proof of Theorem 8.5 is left as Exercise 23.

Definition

The **identity matrix of order *n*** is the $n \times n$ matrix $I_n = [\delta_{ij}]$, where $\delta_{ij} = 1$ if $i = j$ and $\delta_{ij} = 0$ if $i \neq j$.

Note that $\delta: I_n \times I_n \to \{0,1\}$ is a function. For those using the alternative definition of a matrix, this function δ, which is called the *Kroeneker delta function,* is itself the $n \times n$ identity matrix.

The identity matrix of order 2 is the matrix

$$\begin{bmatrix} 1 & 0 \\ 0 & 1 \end{bmatrix}$$

and the identity matrix of order 3 is the matrix

$$\begin{bmatrix} 1 & 0 & 0 \\ 0 & 1 & 0 \\ 0 & 0 & 1 \end{bmatrix}$$

Notice that the identity matrix of order n is an $n \times n$ diagonal matrix $A = [a_{ij}]$ such that $a_{ii} = 1$ for each $i = 1,2, \ldots ,n$.

Theorem 8.6

If A is an $m \times n$ matrix, then $I_m A = AI_n = A$.

Proof

Let $A = [a_{ij}]$, $I_m A = [b_{ij}]$, and $AI_n = [c_{ij}]$. Then, for each $i = 1,2, \ldots ,m$ and $j = 1,2, \ldots ,n$,

$$b_{ij} = \sum_{k=1}^{m} \delta_{ik} a_{kj} = \delta_{ii} a_{ij} = a_{ij}$$

and

$$c_{ij} = \sum_{k=1}^{n} a_{ik} \delta_{kj} = a_{ij} \delta_{jj} = a_{ij}$$

Hence $I_m A = A$ and $AI_n = A$. □

Powers of matrices are defined recursively.

Definition

Let A be an $n \times n$ matrix. We define $A^0 = I_n$, $A^1 = A$, and for each $n > 1$, $A^n = A^{n-1} \times A$.

Theorem 8.7

If A is an $n \times n$ matrix, then $A^m \times A = A \times A^m$ for each natural number m.

Proof

The proof is by induction. Let $S = \{m \in \mathbb{N}: A^m \times A = A \times A^m\}$. Since $A^1 \times A = A \times A = A \times A^1$, $1 \in S$.

Suppose $m \in S$. Then $A^m \times A = A \times A^m$. Now

$$\begin{aligned} A \times A^{m+1} &= A \times (A^m \times A) && \text{(definition)} \\ &= (A \times A^m) \times A && \text{(associativity)} \\ &= (A^m \times A) \times A && (m \in S) \\ &= A^{m+1} \times A && \text{(definition)} \end{aligned}$$

Therefore $m + 1 \in S$. By the principle of mathematical induction, $S = \mathbb{N}$. □

Example 9

If $A = \begin{bmatrix} 1 & 2 & 3 \\ 4 & 2 & 1 \\ 3 & 1 & 2 \end{bmatrix}$, then

$$A^2 = \begin{bmatrix} 18 & 9 & 11 \\ 15 & 13 & 16 \\ 13 & 10 & 14 \end{bmatrix} \quad \text{and} \quad A^3 = \begin{bmatrix} 87 & 65 & 85 \\ 115 & 72 & 90 \\ 95 & 60 & 77 \end{bmatrix}$$ ■

Definition

If $A = [a_{ij}]$ is an $m \times n$ matrix, then the $n \times m$ matrix $[b_{ij}]$, where $b_{ij} = a_{ji}$ for each $i = 1, 2, \ldots, n$ and each $j = 1, 2, \ldots, m$, is called the **transpose** of A and is denoted by A^t.

Notice that the transpose of a matrix A is obtained from A by interchanging the rows and columns of A. Pictorially, the transpose of a matrix is obtained by rotating the matrix 180° around the imaginary line through the main diagonal of the matrix.

Example 10

If $A = \begin{bmatrix} 1 & 2 & 3 \\ 4 & 2 & 1 \\ 3 & 1 & 2 \end{bmatrix}$ then $A^t = \begin{bmatrix} 1 & 4 & 3 \\ 2 & 2 & 1 \\ 3 & 1 & 2 \end{bmatrix}$ ■

Definition

A square matrix is called **symmetric** if $A = A^t$.

Notice that $A = [a_{ij}]$ is symmetric provided that $a_{ij} = a_{ji}$ for each i and j.

Example 11

The matrix $A = \begin{bmatrix} 1 & 2 & 3 \\ 2 & 4 & 5 \\ 3 & 5 & 6 \end{bmatrix}$ is symmetric. ■

Definition

A matrix all of whose entries are either 0 or 1 is called a **zero–one matrix.**

Zero–one matrices are used to represent relations (as we see in Chapter 9) and graphs (as we see in Chapter 11). Relations and graphs are examples of discrete structures, and algorithms using discrete structures are based on Boolean arithmetic with zero–one matrices. Recall that the Boolean operations meet and join operate on pairs of bits:

$$a \wedge b = \begin{cases} 1, & \text{if } a = b = 1 \\ 0, & \text{otherwise} \end{cases} \qquad a \vee B = \begin{cases} 1, & \text{if } a = 1 \text{ or } b = 1 \\ 0, & \text{otherwise} \end{cases}$$

Definition

Let $A = [a_{ij}]$ and $B = [b_{ij}]$ be $m \times n$ zero–one matrices. The **meet** of A and B, denoted by $A \wedge B$, is the zero–one matrix whose (i,j)th entry is $a_{ij} \wedge b_{ij}$. The **join** of A and B, denoted by $A \vee B$, is the zero–one matrix whose (i,j)th entry is $a_{ij} \vee b_{ij}$.

Example 12

Find the meet and join of the matrices

$$A = \begin{bmatrix} 1 & 1 & 0 \\ 0 & 1 & 1 \\ 1 & 0 & 1 \end{bmatrix} \quad \text{and} \quad B = \begin{bmatrix} 1 & 0 & 1 \\ 1 & 1 & 0 \\ 0 & 0 & 0 \end{bmatrix}$$

Analysis

The meet of A and B is

$$A \wedge B = \begin{bmatrix} 1 \wedge 1 & 1 \wedge 0 & 0 \wedge 1 \\ 0 \wedge 1 & 1 \wedge 1 & 1 \wedge 0 \\ 1 \wedge 0 & 0 \wedge 0 & 1 \wedge 0 \end{bmatrix} = \begin{bmatrix} 1 & 0 & 0 \\ 0 & 1 & 0 \\ 0 & 0 & 0 \end{bmatrix}$$

The join of A and B is

$$A \vee B = \begin{bmatrix} 1 \vee 1 & 1 \vee 0 & 0 \vee 1 \\ 0 \vee 1 & 1 \vee 1 & 1 \vee 0 \\ 1 \vee 0 & 0 \vee 0 & 1 \vee 0 \end{bmatrix} = \begin{bmatrix} 1 & 1 & 1 \\ 1 & 1 & 1 \\ 1 & 0 & 1 \end{bmatrix}$$

■

Definition

Let $n \in \mathbb{N}$ and for each $i = 1,2, \ldots ,n$ let a_i be a bit (that is, a_i is 0 or 1). Then $\bigwedge_{i=1}^{n} a_i$ and $\bigvee_{i=1}^{n} a_i$ are defined by the recursion relations

$$\bigwedge_{i=1}^{n} a_i = \left(\bigwedge_{i=1}^{n-1} a_i\right) \wedge a_n \quad \text{and} \quad \bigvee_{i=1}^{n} a_i = \left(\bigvee_{i=1}^{n-1} a_i\right) \vee a_n$$

where $\bigwedge_{i=1}^{1} a_i = a_1 = \bigvee_{i=1}^{1} a_i$.

Definition

If $A = [a_{ij}]$ is an $m \times n$ zero–one matrix and $B = [b_{ij}]$ is an $n \times p$ zero–one matrix, then the **Boolean product** of A and B, denoted by $A \odot$ B, is the $m \times p$ zero–one matrix whose (i,j)th entry is

$$c_{ij} = \bigvee_{k=1}^{n} (a_{ik} \wedge b_{kj})$$

Example 13

Let $A = \begin{bmatrix} 1 & 0 & 1 \\ 0 & 1 & 1 \end{bmatrix}$ and $B = \begin{bmatrix} 1 & 0 \\ 1 & 1 \\ 0 & 0 \end{bmatrix}$.

Then the Boolean product, $A \odot B$, of A and B is

$$A \odot B = \begin{bmatrix} (1 \wedge 1) \vee (0 \wedge 1) \vee (1 \wedge 0) & (1 \wedge 0) \vee (0 \wedge 1) \vee (1 \wedge 0) \\ (0 \wedge 1) \vee (1 \wedge 1) \vee (1 \wedge 0) & (0 \wedge 0) \vee (1 \wedge 1) \vee (1 \wedge 0) \end{bmatrix}$$

$$= \begin{bmatrix} 1 \vee 0 \vee 0 & 0 \vee 0 \vee 0 \\ 0 \vee 1 \vee 0 & 0 \vee 1 \vee 0 \end{bmatrix} = \begin{bmatrix} 1 & 0 \\ 1 & 1 \end{bmatrix}$$

■

Example 14

If A is an $m \times n$ zero–one matrix and B is an $n \times p$ zero–one matrix, what is the number of bit operations used in the flowchart in Figure 8.5 to find $A \odot B$.

Figure 8.5

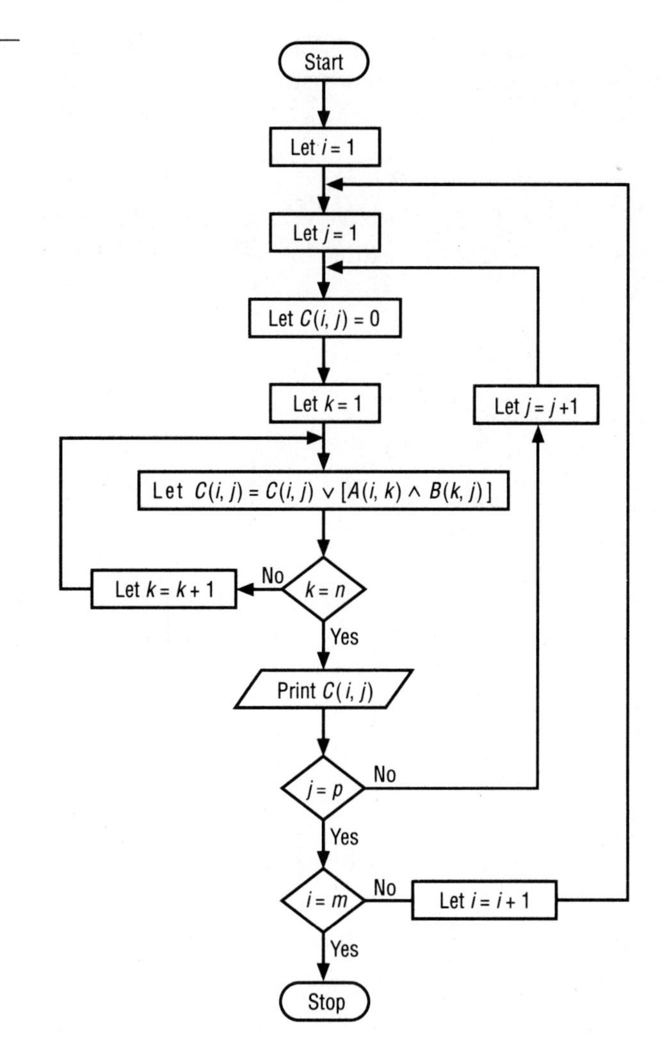

Analysis

The number of entries in $A \odot B$ is mp. We see from the flowchart that a total of n meets and n joins are used to find each entry. Therefore the number of bit operations used to find $A \odot B$ is $2mnp$. The flowchart given in Figure 8.5 is inefficient in that it does not take advantage of the property that $1 \vee x = 1$ for all elements x of a Boolean algebra. If the flowchart is modified to use this property, then $2mnp$ counts the number of bit operations needed in a worst case (see Exercise 37). ■

We can also define the Boolean powers of a square zero–one matrix. These powers are used in our study of the closure of relations and in our study of paths in graphs. The latter is used to model computer networks.

Definition

Let A be an $n \times n$ zero–one matrix. We define $A^{[0]} = I_n$, $A^{[1]} = A$, and for each $m > 1$, $A^{[m]} = A^{[m-1]} \odot A$.

Example 15

Let A = $\begin{bmatrix} 0 & 1 & 1 \\ 0 & 0 & 1 \\ 1 & 0 & 0 \end{bmatrix}$.

Then $A^{[0]} = \begin{bmatrix} 1 & 0 & 0 \\ 0 & 1 & 0 \\ 0 & 0 & 1 \end{bmatrix}$,

$$A^{[1]} = \begin{bmatrix} 0 & 1 & 1 \\ 0 & 0 & 1 \\ 1 & 0 & 0 \end{bmatrix}, A^{[2]} = A^{[1]} \odot A = \begin{bmatrix} 1 & 0 & 1 \\ 1 & 0 & 0 \\ 0 & 1 & 1 \end{bmatrix},$$

$$A^{[3]} = A^{[2]} \odot A = \begin{bmatrix} 1 & 1 & 1 \\ 0 & 1 & 1 \\ 1 & 0 & 1 \end{bmatrix}, A^{[4]} = A^{[3]} \odot A = \begin{bmatrix} 1 & 1 & 1 \\ 1 & 0 & 1 \\ 1 & 1 & 1 \end{bmatrix},$$

and $$A^{[5]} = A^{[4]} \odot A = \begin{bmatrix} 1 & 1 & 1 \\ 1 & 1 & 1 \\ 1 & 1 & 1 \end{bmatrix}$$

Since there is a 1 in each column of A, we can see that $A^{[n]} = A^{[5]}$ for each $n \geq 5$. ■

Exercises 8.2

15. Give an example of a 4×4 diagonal matrix that is not equal to I_4.
16. Find the transpose of $\begin{bmatrix} 0 & 1 & 4 \\ 3 & 0 & 2 \\ 5 & 6 & 0 \end{bmatrix}$.
17. Give an example of a 4×4 symmetric matrix that is different from I_4 and $\mathbf{0}$.
18. Find 3×3 matrices A and B such that $A \neq \mathbf{0}$ and $B \neq \mathbf{0}$ but $AB = \mathbf{0}$.
19. Find a 3×3 matrix A, different from $\mathbf{0}$, such that $A^2 = \mathbf{0}$.
20. Find a matrix A, different from $\mathbf{0}$ and I_n for each n, such that $A^2 = A$.
21. Find a matrix A, different from I_n and $-I_n$ for each n, such that $A^2 = I_p$ for some p.

22. If $A = \begin{bmatrix} 0 & 1 & 2 \\ 1 & 0 & 2 \\ 2 & 1 & 1 \end{bmatrix}$, find A^3.

23. Prove Theorem 8.5.

24. If $A = \begin{bmatrix} 0 & 1 & 1 \\ 1 & 0 & 1 \\ 1 & 1 & 1 \end{bmatrix}$ and $B = \begin{bmatrix} 1 & 0 & 1 \\ 0 & 0 & 0 \\ 0 & 1 & 1 \end{bmatrix}$, find $A \wedge B$ and $A \vee B$.

25. Prove that, if A and B are $m \times n$ zero–one matrices, then $A \vee B = B \vee A$ and $A \wedge B = B \wedge A$.

26. Prove that, if A is a zero–one matrix, then $A \vee A = A$ and $A \wedge A = A$.

27. Prove that, if A, B, and C are $m \times n$ zero–one matrices, then $(A \vee B) \vee C = A \vee (B \vee C)$ and $(A \wedge B) \wedge C = A \wedge (B \wedge C)$.

28. Prove that, if A, B, and C are $m \times n$ zero–one matrices, then $A \vee (B \wedge C) = (A \vee B) \wedge (A \vee C)$ and $A \wedge (B \vee C) = (A \wedge B) \vee (A \wedge C)$.

29. If $A = \begin{bmatrix} 1 & 0 & 1 \\ 1 & 0 & 1 \\ 1 & 1 & 1 \end{bmatrix}$, find $A^{[n]}$ for each natural number n.

30. If $A = \begin{bmatrix} 0 & 1 & 1 \\ 0 & 0 & 0 \\ 1 & 0 & 1 \end{bmatrix}$, find $A^{[n]}$ for each natural number n.

31. If $A = \begin{bmatrix} 0 & 1 & 0 \\ 0 & 0 & 1 \\ 1 & 0 & 0 \end{bmatrix}$, find $A^{[n]}$ for each natural number n.

32. If $A = \begin{bmatrix} 0 & 1 & 1 & 1 \\ 0 & 0 & 1 & 1 \\ 0 & 0 & 0 & 1 \\ 1 & 0 & 0 & 0 \end{bmatrix}$, find $A^{[n]}$ for each natural number n.

33. Prove that, if A is an $n \times n$ zero–one matrix, then $A^{[m]} \odot A = A \odot A^{[m]}$ for each natural number m.

34. Prove that, if A is an $n \times n$ zero–one matrix, then $A \odot I_n = I_n \odot A = A$.

35. Find 3×3 zero–one matrices A and B such that $A \odot B \neq B \odot A$.

36. Prove that, if A is an $m \times n$ zero–one matrix, B is an $n \times p$ zero–one matrix, and C is a $p \times$ q zero–one matrix, then $A \odot (B \odot C) = (A \odot B) \odot C$.

37. **a.** Modify the flowchart of Figure 8.5 to make use of the property that $1 \vee x = 1$ for all $x \in \{0,1\}$.

b. Give an example of a 3×2 matrix A and a 2×4 matrix B such that the calculation of $A \odot B$ using your modified flowchart still requires $2 \cdot 3 \cdot 2 \cdot 4$ bit operations.

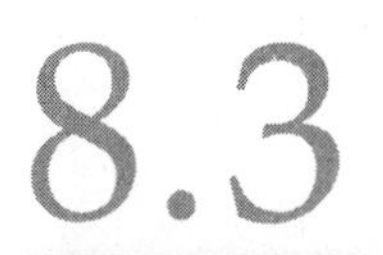

SOLUTIONS OF SYSTEMS OF LINEAR EQUATIONS

Our goal in this section is to provide two methods of solving systems of linear equations. Readers who desire a more complete presentation should consult a linear algebra text. Matrices can be used to express a system of linear equations. For example, the system

$$\begin{aligned} 3x + 2y - z &= 4 \\ x + 3y + 2z &= 7 \\ 2x - y + z &= 5 \end{aligned}$$

can be expressed in terms of matrices:

$$\begin{bmatrix} 3 & 2 & -1 \\ 1 & 3 & 2 \\ 2 & -1 & 1 \end{bmatrix} \begin{bmatrix} x \\ y \\ z \end{bmatrix} = \begin{bmatrix} 4 \\ 7 \\ 5 \end{bmatrix}$$

The matrix on the left is called the *coefficient matrix.* Often the variables x, y, and z are ignored, and the preceding matrix equation is written as an *augmented matrix:*

$$\left[\begin{array}{ccc|c} 3 & 2 & -1 & 4 \\ 1 & 3 & 2 & 7 \\ 2 & -1 & 1 & 5 \end{array}\right]$$

Gaussian elimination [named after Karl Friedrich Gauss (1777–1855)] is a method for solving a system of linear equations by replacing the given system by a system that has the same solution set but is easier to solve. We describe Gaussian elimination in the language of matrix theory. We use row operations to obtain a system of equations that is equivalent to the original system in the sense that the two systems have the same solution. Three elementary row operations, based on allowable operations on systems of equations, are used:

1. Interchange two equations.
2. Multiply an equation by a nonzero scalar.
3. Replace an equation by the sum of it and a multiple of another equation.

1′. Interchange two rows.
2′. Multiply a row by a nonzero scalar.
3′. Replace a row by the sum of it and a multiple of another row.

As we solve a given system of equations, we keep track of our elementary row operations by determining their effect on the system of equations and on the corresponding augmented matrix.

Example 16

Solve, using Gaussian elimination,

$$\begin{aligned} 3x + 2y - z &= 4 \\ x + 3y + 2z &= 7 \\ 2x - y + z &= 5 \end{aligned}$$

Analysis

Elementary row operation	System of equations becomes	Corresponding augmented matrix	Number of operations
1. Make sure 1st coefficient is not 0.	$\begin{aligned} 3x + 2y - z &= 4 \\ x + 3y + 2z &= 7 \\ 2x - y + z &= 5 \end{aligned}$	$\left[\begin{array}{rrr@{\;\vert\;}r} 3 & 2 & -1 & 4 \\ 1 & 3 & 2 & 7 \\ 2 & -1 & 1 & 5 \end{array}\right]$	1
2. Multiply 1st row by $\frac{1}{3}$.	$\begin{aligned} x + \tfrac{2}{3}y - \tfrac{1}{3}z &= \tfrac{4}{3} \\ x + 3y + 2z &= 7 \\ 2x - y + z &= 5 \end{aligned}$	$\left[\begin{array}{rrr@{\;\vert\;}r} 1 & \frac{2}{3} & -\frac{1}{3} & \frac{4}{3} \\ 1 & 3 & 2 & 7 \\ 2 & -1 & 1 & 5 \end{array}\right]$	4
3. Add -1 times 1st row to 2nd row.	$\begin{aligned} x + \tfrac{2}{3}y - \tfrac{1}{3}z &= \tfrac{4}{3} \\ \tfrac{7}{3}y + \tfrac{7}{3}z &= \tfrac{17}{3} \\ 2x - y + z &= 5 \end{aligned}$	$\left[\begin{array}{rrr@{\;\vert\;}r} 1 & \frac{2}{3} & -\frac{1}{3} & \frac{4}{3} \\ 0 & \frac{7}{3} & \frac{7}{3} & \frac{17}{3} \\ 2 & -1 & 1 & 5 \end{array}\right]$	8
4. Add -2 times 1st row to 3rd row.	$\begin{aligned} x + \tfrac{2}{3}y - \tfrac{1}{3}z &= \tfrac{4}{3} \\ \tfrac{7}{3}y + \tfrac{7}{3}z &= \tfrac{17}{3} \\ -\tfrac{7}{3}y + \tfrac{5}{3}z &= \tfrac{7}{3} \end{aligned}$	$\left[\begin{array}{rrr@{\;\vert\;}r} 1 & \frac{2}{3} & -\frac{1}{3} & \frac{4}{3} \\ 0 & \frac{7}{3} & \frac{7}{3} & \frac{17}{3} \\ 0 & -\frac{7}{3} & \frac{5}{3} & \frac{7}{3} \end{array}\right]$	8
5. Make sure 2nd coefficient of 2nd equation is not 0.	$\begin{aligned} x + \tfrac{2}{3}y - \tfrac{1}{3}z &= \tfrac{4}{3} \\ \tfrac{7}{3}y + \tfrac{7}{3}z &= \tfrac{17}{3} \\ -\tfrac{7}{3}y + \tfrac{5}{3}z &= \tfrac{7}{3} \end{aligned}$	$\left[\begin{array}{rrr@{\;\vert\;}r} 1 & \frac{2}{3} & -\frac{1}{3} & \frac{4}{3} \\ 0 & \frac{7}{3} & \frac{7}{3} & \frac{17}{3} \\ 0 & -\frac{7}{3} & \frac{5}{3} & \frac{7}{3} \end{array}\right]$	1
6. Multiply 2nd row by $3/7$.	$\begin{aligned} x + \tfrac{2}{3}y - \tfrac{1}{3}z &= \tfrac{4}{3} \\ y + z &= \tfrac{17}{7} \\ -\tfrac{7}{3}y + \tfrac{5}{3}z &= \tfrac{7}{3} \end{aligned}$	$\left[\begin{array}{rrr@{\;\vert\;}r} 1 & \frac{2}{3} & -\frac{1}{3} & \frac{4}{3} \\ 0 & 1 & 1 & \frac{17}{7} \\ 0 & -\frac{7}{3} & \frac{5}{3} & \frac{7}{3} \end{array}\right]$	3
7. Add $7/3$ times 2nd row to 3rd row.	$\begin{aligned} x + \tfrac{2}{3}y - \tfrac{1}{3}z &= \tfrac{4}{3} \\ y + z &= \tfrac{17}{7} \\ 4z &= 8 \end{aligned}$	$\left[\begin{array}{rrr@{\;\vert\;}r} 1 & \frac{2}{3} & -\frac{1}{3} & \frac{4}{3} \\ 0 & 1 & 1 & \frac{17}{7} \\ 0 & 0 & 4 & 8 \end{array}\right]$	6
8. Multiply 3rd row by $1/4$.	$\begin{aligned} x + \tfrac{2}{3}y - \tfrac{1}{3}z &= \tfrac{4}{3} \\ y + z &= \tfrac{17}{7} \\ z &= 2 \end{aligned}$	$\left[\begin{array}{rrr@{\;\vert\;}r} 1 & \frac{2}{3} & -\frac{1}{3} & \frac{4}{3} \\ 0 & 1 & 1 & \frac{17}{7} \\ 0 & 0 & 1 & 2 \end{array}\right]$	2

We complete the process of solving the given system of equations by *backsolving*. We know that $z = 2$, so the second equation becomes $y + 2 = 17/7$. Hence $y = 3/7$. Then the first equation becomes $x + (2/3)(3/7) - (1/3)2 = 4/3$. Hence $x = 12/7$. ■

Gauss–Jordan elimination (Wilhelm Jordan, 1842–1899, was a German engineer) uses the three elementary row operations to further reduce the given system and avoid backsolving. The goal of the Gauss–Jordan elimination method is to obtain 1s on the main diagonal and 0s as the other entries of the "coefficient" matrix. Stated in the language of matrix algebra, the goal is to replace the coefficient matrix with the identity matrix.

Example 17

Use Gauss–Jordan elimination to solve the system of equations in Example 16.

Analysis

We proceed as in Example 16. In fact, the first eight steps are precisely the same, so we continue from there:

Step	System	Augmented matrix	
9. Add −1 times 3rd row to 2nd row.	$\begin{aligned} x + \tfrac{2}{3}y - \tfrac{1}{3}z &= \tfrac{4}{3} \\ y &= \tfrac{3}{7} \\ z &= 2 \end{aligned}$	$\left[\begin{array}{ccc:c} 1 & \tfrac{2}{3} & -\tfrac{1}{3} & \tfrac{4}{3} \\ 0 & 1 & 0 & \tfrac{3}{7} \\ 0 & 0 & 1 & 2 \end{array}\right]$	4
10. Add 1/3 times 3rd row to 1st row.	$\begin{aligned} x + \tfrac{2}{3}y &= 2 \\ y &= \tfrac{3}{7} \\ z &= 2 \end{aligned}$	$\left[\begin{array}{ccc:c} 1 & \tfrac{2}{3} & 0 & 2 \\ 0 & 1 & 0 & \tfrac{3}{7} \\ 0 & 0 & 1 & 2 \end{array}\right]$	4
11. Add −2/3 times 2nd row to 1st row.	$\begin{aligned} x &= \tfrac{12}{7} \\ y &= \tfrac{3}{7} \\ z &= 2 \end{aligned}$	$\left[\begin{array}{ccc:c} 1 & 0 & 0 & \tfrac{12}{7} \\ 0 & 1 & 0 & \tfrac{3}{7} \\ 0 & 0 & 1 & 2 \end{array}\right]$	4

■

There are four important observations to be made on the basis of Example 17:

1. The augmented matrix is all that is needed to keep track of changes, and solving a system of equations is unpleasant enough without having to drag along the x's, y's, and z's as placeholders.
2. As previously indicated, the goal of the Gauss–Jordan elimination method is to obtain 1s on the main diagonal and 0s as the other entries of the "coefficient" matrix.
3. Every step consists of one of the three elementary row operations listed prior to Example 16.
4. There is a routine to Gauss–Jordan elimination. Make the first coefficient of the top equation a 1, and then eliminate the first coefficient of all the remaining equations. Make the second coefficient of the equation second from the top a 1, and use this equation to eliminate the second coefficient of all the remaining equations. Continue this process. The same routine applies to Gaussian elimination. We simply stop sooner and backsolve.

Comparing an entry to 0, multiplying two numbers, or adding two numbers is an operation. Calculating the number of operations needed at each stage in Example 17 requires knowledge that certain entries in the matrix are 0 and thus that certain operations need not be performed. As indicated, the number of operations that must be performed is 45.

Gaussian elimination and Gauss–Jordan elimination are easy programs to write for a computer. Unfortunately, when the matrices get to be size 8×8 or larger, the large number of multiplications frequently produce numbers outside the limits imposed by the programming language being used. Hence, one of the problems currently receiving much interest in computer science is a method of getting accurate results for large matrices.

Which of the two methods of solving a system of linear equations is more efficient? For large systems, the Gauss–Jordan procedure substantially increases the number of arithmetic operations performed, thus Gaussian elimination and backsolving is more efficient.

The example we have considered consists of three equations in three unknowns. It is not, however, necessary for the number of equations to equal the number of unknowns. We next consider two such examples, but first we introduce some terminology and make a remark.

Definition

An $m \times n$ **system of linear equations** is a system of m linear equations in n unknowns. An $m \times n$ system of linear equations is said to be **consistent** if it has at least one solution and **inconsistent** if it has no solution.

Remark An $m \times n$ system of linear equations has a unique solution, no solution, or an infinite number of solutions.

In the following examples, the row operation we use at each step is indicated at the left of the matrix in abbreviated form. The code we use to describe our row operations must be interpreted with care. For example, $R_1 + R_3$ is shorthand for "add row 1 *to* row 3." Consequently, $R_1 + R_3$ and $R_3 + R_1$ have different meanings. We also write "check" as the abbreviation for "make sure the ith coefficient of the ith equation is not 0."

Example 18

Use Gaussian elimination (and backsolving) to solve the system

$$2x + y - z = 5$$

$$x - y + 2z = 3$$

Analysis

We form the augmented matrix and proceed as in Example 16 (except this time we do not carry along the system of equations).

1. check $\left[\begin{array}{ccc|c} 2 & 1 & -1 & 5 \\ 1 & -1 & 2 & 3 \end{array}\right]$ 2. $\frac{1}{2}R_1$ $\left[\begin{array}{ccc|c} 1 & \frac{1}{2} & -\frac{1}{2} & \frac{5}{2} \\ 1 & -1 & 2 & 3 \end{array}\right]$

3. $-1 \cdot R_1 + R_2$ $\left[\begin{array}{ccc|c} 1 & \frac{1}{2} & -\frac{1}{2} & \frac{5}{2} \\ 0 & -\frac{3}{2} & \frac{5}{2} & \frac{1}{2} \end{array}\right]$ 4. check $\left[\begin{array}{ccc|c} 1 & \frac{1}{2} & -\frac{1}{2} & \frac{5}{2} \\ 0 & -\frac{3}{2} & \frac{5}{2} & \frac{1}{2} \end{array}\right]$

5. $-\frac{2}{3}R_2$ $\left[\begin{array}{ccc|c} 1 & \frac{1}{2} & -\frac{1}{2} & \frac{5}{2} \\ 0 & 1 & -\frac{5}{3} & -\frac{1}{3} \end{array}\right]$

The system of equations corresponding to the matrix after step 5 is

$$\begin{aligned} x + \frac{1}{2}y - \frac{1}{2}z &= \frac{5}{2} \\ y - \frac{5}{3}z &= -\frac{1}{3} \end{aligned}$$

Thus z is an independent variable; that is, for each real number z, we can backsolve to find x and y. In particular,

$$y = \frac{5}{3}z - \frac{1}{3} \quad \text{and}$$

$$x = \frac{1}{2}z + \frac{5}{2} - \frac{1}{2}y = \frac{1}{2}z + \frac{5}{2} - \frac{1}{2}\left(\frac{5}{3}z - \frac{1}{3}\right) = -\frac{1}{3}z + \frac{8}{3}$$

Therefore the system has an infinite number of solutions. ■

Example 19

Use Gaussian elimination to show that the following system is inconsistent.

$$\begin{aligned} 2x - y + z &= 7 \\ x + y + z &= 3 \\ x + 2y - z &= 0 \\ 3x + y &= 5 \end{aligned}$$

Analysis

We form the augmented matrix and proceed as in Example 18.

1. check $\left[\begin{array}{ccc|c} 2 & -1 & 1 & 7 \\ 1 & 1 & 1 & 3 \\ 1 & 2 & -1 & 0 \\ 3 & 1 & 0 & 5 \end{array}\right]$ 2. $\frac{1}{2}R_1$ $\left[\begin{array}{ccc|c} 1 & -\frac{1}{2} & \frac{1}{2} & \frac{7}{2} \\ 1 & 1 & 1 & 3 \\ 1 & 2 & -1 & 0 \\ 3 & 1 & 0 & 5 \end{array}\right]$

3. $-1 \cdot R_1 + R_2$
$$\left[\begin{array}{ccc|c} 1 & -\frac{1}{2} & \frac{1}{2} & \frac{7}{2} \\ 0 & \frac{3}{2} & \frac{1}{2} & -\frac{1}{2} \\ 1 & 2 & -1 & 0 \\ 3 & 1 & 0 & 5 \end{array}\right]$$

4. $-1 \cdot R_1 + R_3$
$$\left[\begin{array}{ccc|c} 1 & -\frac{1}{2} & \frac{1}{2} & \frac{7}{2} \\ 0 & \frac{3}{2} & \frac{1}{2} & -\frac{1}{2} \\ 0 & \frac{5}{2} & -\frac{3}{2} & -\frac{7}{2} \\ 3 & 1 & 0 & 5 \end{array}\right]$$

5. $-3R_1 + R_4$
$$\left[\begin{array}{ccc|c} 1 & -\frac{1}{2} & \frac{1}{2} & \frac{7}{2} \\ 0 & \frac{3}{2} & \frac{1}{2} & -\frac{1}{2} \\ 0 & \frac{5}{2} & -\frac{3}{2} & -\frac{7}{2} \\ 0 & \frac{5}{2} & -\frac{3}{2} & -\frac{11}{2} \end{array}\right]$$

6. check
$$\left[\begin{array}{ccc|c} 1 & -\frac{1}{2} & \frac{1}{2} & \frac{7}{2} \\ 0 & \frac{3}{2} & \frac{1}{2} & -\frac{1}{2} \\ 0 & \frac{5}{2} & -\frac{3}{2} & -\frac{7}{2} \\ 0 & \frac{5}{2} & -\frac{3}{2} & -\frac{11}{2} \end{array}\right]$$

7. $\frac{2}{3}R_2$
$$\left[\begin{array}{ccc|c} 1 & -\frac{1}{2} & \frac{1}{2} & \frac{7}{2} \\ 0 & 1 & \frac{1}{3} & -\frac{1}{3} \\ 0 & \frac{5}{2} & -\frac{3}{2} & -\frac{7}{2} \\ 0 & \frac{5}{2} & -\frac{3}{2} & -\frac{11}{2} \end{array}\right]$$

8. $-\frac{5}{2}R_2 + R_3$
$$\left[\begin{array}{ccc|c} 1 & -\frac{1}{2} & \frac{1}{2} & \frac{7}{2} \\ 0 & 1 & \frac{1}{3} & -\frac{1}{3} \\ 0 & 0 & -\frac{7}{3} & -\frac{8}{3} \\ 0 & \frac{5}{2} & -\frac{3}{2} & -\frac{11}{2} \end{array}\right]$$

9. $-\frac{5}{2}R_2 + R_4$
$$\left[\begin{array}{ccc|c} 1 & -\frac{1}{2} & \frac{1}{2} & \frac{7}{2} \\ 0 & 1 & \frac{1}{3} & -\frac{1}{3} \\ 0 & 0 & -\frac{7}{3} & -\frac{8}{3} \\ 0 & 0 & -\frac{7}{3} & -\frac{14}{3} \end{array}\right]$$

10. check
$$\left[\begin{array}{ccc|c} 1 & -\frac{1}{2} & \frac{1}{2} & \frac{7}{2} \\ 0 & 1 & \frac{1}{3} & -\frac{1}{3} \\ 0 & 0 & -\frac{7}{3} & -\frac{8}{3} \\ 0 & 0 & -\frac{7}{3} & -\frac{14}{3} \end{array}\right]$$

11. $-\frac{3}{7}R_3$
$$\left[\begin{array}{ccc|c} 1 & -\frac{1}{2} & \frac{1}{2} & \frac{7}{2} \\ 0 & 1 & \frac{1}{3} & -\frac{1}{3} \\ 0 & 0 & 1 & \frac{8}{7} \\ 0 & 0 & -\frac{7}{3} & -\frac{14}{3} \end{array}\right]$$

12. $\frac{7}{3}R_3 + R_4$
$$\left[\begin{array}{ccc|c} 1 & -\frac{1}{2} & \frac{1}{2} & \frac{7}{2} \\ 0 & 1 & \frac{1}{3} & -\frac{1}{3} \\ 0 & 0 & 1 & \frac{8}{7} \\ 0 & 0 & 0 & -2 \end{array}\right]$$

The equation associated with the last row of the matrix in step 12 is $0x + 0y + 0z = -2$. There are no x, y, and z that satisfy this equation, so the given system of equations is inconsistent. Notice that in the matrix in step 7 we can add -1 times the third row to the fourth row and obtain the same equation, namely, $0x + 0y + 0z = -2$. Although this step shortens the work, it does not use the method of Gaussian elimination. ■

If in the Gaussian elimination method we obtain a row with a nonzero entry in the "augmented" column and 0s elsewhere, then the system of equations does not have a solution. If in an $n \times n$ linear system of equations we obtain a row all of whose entries are 0s, then the system of equations has an infinite number of solutions.

Exercises 8.3

38. Find the augmented matrix for each of the following systems of linear equations.

a. $\begin{aligned} x + 2y &= 5 \\ 3x - y &= 4 \end{aligned}$

b. $\begin{aligned} 2x - 3y + z &= 5 \\ -x + 2y - z &= 3 \\ 3x - y + 4z &= 7 \end{aligned}$

c. $\begin{aligned} x - 2y &= 1 \\ 3x + 4y &= 3 \\ 2x - y &= 0 \end{aligned}$

d. $\begin{aligned} x + y - 3z &= 2 \\ 2x - 2y + z &= -3 \end{aligned}$

39. Find a system of linear equations corresponding to each of the following augmented matrices.

a. $\left[\begin{array}{ccc|c} 1 & 0 & 2 & 3 \\ 0 & 3 & 0 & 5 \\ 2 & -3 & 4 & -7 \end{array}\right]$

b. $\left[\begin{array}{ccc|c} 0 & 1 & 2 & 3 \\ -2 & 3 & 1 & 2 \\ 1 & -2 & -3 & 5 \end{array}\right]$

c. $\left[\begin{array}{cccc|c} 1 & 0 & 2 & -3 & 5 \\ 2 & 3 & 5 & -4 & 1 \end{array}\right]$

d. $\left[\begin{array}{cc|c} 2 & 1 & 3 \\ -1 & 2 & 2 \\ 3 & 1 & -5 \\ 2 & 3 & -7 \end{array}\right]$

40. Use Gaussian elimination to solve each of the following systems.

a. $\begin{aligned} x + 2y &= 3 \\ 2x - y &= 4 \end{aligned}$

b. $\begin{aligned} 2x + 3y &= 5 \\ x - y &= 6 \end{aligned}$

c. $\begin{aligned} 2x - y + 3z &= 4 \\ x + 2y - 3z &= 5 \\ 3x + 4y - z &= 6 \end{aligned}$

d. $\begin{aligned} x + y + z &= 1 \\ 2x + 3y - z &= 2 \\ 3x - y + 2z &= 3 \end{aligned}$

41. Use the Gauss–Jordan procedure to solve each of the following systems.

a. $\begin{aligned} x + y + 2z &= 9 \\ 2x + 4y - 3z &= 1 \\ 3x + 6y - 5z &= 0 \end{aligned}$

b. $\begin{aligned} x + y + 2z &= 8 \\ 3x - 7y + 4z &= 10 \\ x + 2y - 3z &= -1 \end{aligned}$

c. $\begin{aligned} 2x - y - 3z &= 0 \\ 5x + 2y + 6z &= 0 \end{aligned}$

d. $\begin{aligned} 2x + 3y + 4z + 9w &= 16 \\ 2x - 3z + 7w &= -11 \\ x + y + z + 4w &= 4 \end{aligned}$

42. Solve each of the following systems of linear equations or show that the system is inconsistent.

a. $\begin{aligned} x - 2y &= -5 \\ -3x + 6y &= 15 \end{aligned}$

b. $\begin{aligned} x - 2y &= -5 \\ 2x - 4y &= 7 \end{aligned}$

c. $\begin{aligned} x + z &= 2 \\ -x + y + z &= -1 \\ y + 2z &= 3 \end{aligned}$

d. $\begin{aligned} x + y + z &= 1 \\ -x + 2y - z &= 3 \\ 2x - y + 3z &= 4 \end{aligned}$

INVERSES OF MATRICES

If x is a nonzero real number, we know that there is a real number y (namely, $y = 1/x$) such that $xy = 1$. If A is a nonzero $n \times n$ matrix, is there a matrix B such that $AB = I_n$? It is easy to see that the answer is no. Let

$$A = \begin{bmatrix} 1 & 0 \\ 2 & 0 \end{bmatrix}$$

If

$$B = \begin{bmatrix} a & b \\ c & d \end{bmatrix}$$

is a matrix such that $AB = I_2$, then we see that

$$\begin{aligned} 1 \cdot a + 0 \cdot c &= 1 \qquad & 1 \cdot b + 0 \cdot d &= 0 \\ 2 \cdot a + 0 \cdot c &= 0 \qquad & 2 \cdot b + 0 \cdot d &= 1 \end{aligned}$$

Therefore $a = 1$, $b = 0$, $a = 0$, and $b = 1/2$. This is a contradiction, so no such matrix exists.

Definition

Let A be an $n \times n$ matrix. If there exists an $n \times n$ matrix B such that $AB = BA = I_n$, then A is said to be **invertible** and B is called an **inverse** of A.

Example 20

The matrix

$$B = \begin{bmatrix} 3 & 5 \\ 1 & 2 \end{bmatrix}$$

is an inverse of

$$A = \begin{bmatrix} 2 & -5 \\ -1 & 3 \end{bmatrix}$$

since

$$AB = \begin{bmatrix} 1 & 0 \\ 0 & 1 \end{bmatrix} \quad \text{and} \quad BA = \begin{bmatrix} 1 & 0 \\ 0 & 1 \end{bmatrix}$$ ■

When calculating an inverse of a matrix, it would be helpful to know that an answer obtained is the only inverse of a given matrix. The following theorem tells us that this is the case.

Theorem 8.8 If B and C are inverses of an $n \times n$ matrix A, then $B = C$.

Proof

The proof is a straightforward calculation using previous results.

$$
\begin{aligned}
B &= BI_n && \text{(by Theorem 8.6)} \\
&= B(AC) && \text{(since } C \text{ is an inverse of } A) \\
&= (BA)C && \text{(by Theorem 8.2)} \\
&= I_nC && \text{(since } B \text{ is an inverse of } A) \\
&= C && \text{(by Theorem 8.6)}
\end{aligned}
$$

□

In view of Theorem 8.8, we can speak of *the* inverse of an invertible matrix. If A is invertible, then its inverse is denoted by A^{-1}.

Theorem 8.9 If A and B are invertible $n \times n$ matrices, then AB is invertible and $(AB)^{-1} = B^{-1}A^{-1}$.

Proof

If we can show that $(AB)(B^{-1}A^{-1}) = (B^{-1}A^{-1})(AB) = I_n$, then, by Theorem 8.8, we have shown simultaneously that AB is invertible and $(AB)^{-1} = B^{-1}A^{-1}$. Now

$$
\begin{aligned}
(AB)(B^{-1}A^{-1}) &= A[B(B^{-1}A^{-1})] && \text{(by Theorem 8.2)} \\
&= A[(BB^{-1})A^{-1}] && \text{(by Theorem 8.2)} \\
&= A[I_nA^{-1}] && \text{(since } B^{-1} \text{ is the inverse of } B) \\
&= AA^{-1} && \text{(by Theorem 8.6)} \\
&= I_n && \text{(since } A^{-1} \text{ is the inverse of } A)
\end{aligned}
$$

We leave as Exercise 47 the proof that $(B^{-1}A^{-1})(AB) = I_n$. □

Corollary 8.10

Let $E_1, E_2, \ldots, E_k$ be invertible $n \times n$ matrices. Then

$$(E_1E_2 \cdots E_k)^{-1} = E_k^{-1} E_{k-1}^{-1} \cdots E_1^{-1}$$

The proof of Corollary 8.10 is left as Exercise 48.

Theorem 8.11

If A is an invertible $n \times n$ matrix, then

a. A^{-1} is invertible and $(A^{-1})^{-1} = A$.
b. For each $n \in \mathbb{N}$, A^n is invertible and $(A^n)^{-1} = (A^{-1})^n$.
c. For each nonzero $k \in \mathbb{R}$, kA is invertible and $(kA)^{-1} = (1/k)A^{-1}$.

Proof

a. Since $AA^{-1} = A^{-1}A = I_n$, A^{-1} is invertible and $(A^{-1})^{-1} = A$.
b. The proof is by induction.
Let $S = \{n \in \mathbb{N}: A^n \text{ is invertible and } (A^n)^{-1} = (A^{-1})^n\}$. Since $A^1 = A$, $1 \in S$. Suppose $n \in S$. Then

$$
\begin{aligned}
(A^{n+1})^{-1} &= (A^n \times A)^{-1} && \text{(definition of } A^{n+1}) \\
&= A^{-1} \times (A^n)^{-1} && \text{(by Theorem 8.9)} \\
&= A^{-1} \times (A^{-1})^n && \text{(since } n \in S) \\
&= (A^{-1})^{n+1} && \text{(By Theorem 8.7 and definition of } A^{n+1})
\end{aligned}
$$

c. Once again, it is sufficient to show that
$(kA)[(1/k)A^{-1}] = [(1/k)A^{-1}](kA) = I_n$. Now

$$
\begin{aligned}
(kA)[(1/k)A^{-1}] &= [(1/k)(kA)]A^{-1} && \text{(by Theorem 8.4)} \\
&= (1/k)k(AA^{-1}) && \text{(by Theorem 8.4)} \\
&= AA^{-1} && \text{(since } (1/k)k = 1) \\
&= I_n && \text{(since } A^{-1} \text{ is the inverse of } A)
\end{aligned}
$$

The proof that $[(1/k)A^{-1}](kA) = I_n$ is left as Exercise 49. □

Theorem 8.12

If $A = \begin{bmatrix} a & b \\ c & d \end{bmatrix}$ and $ad - bc \neq 0$, then $A^{-1} = \dfrac{1}{ad - bc}\begin{bmatrix} d & -b \\ -c & a \end{bmatrix}$.

Proof

$$\begin{bmatrix} a & b \\ c & d \end{bmatrix} \begin{bmatrix} \dfrac{d}{ad - bc} & \dfrac{-b}{ad - bc} \\ \dfrac{-c}{ad - bc} & \dfrac{a}{ad - bc} \end{bmatrix} = \begin{bmatrix} \dfrac{ad - bc}{ad - bc} & \dfrac{-ab + ba}{ad - bc} \\ \dfrac{cd - dc}{ad - bc} & \dfrac{-cb + da}{ad - bc} \end{bmatrix} = \begin{bmatrix} 1 & 0 \\ 0 & 1 \end{bmatrix}$$

By a similar calculation, we can show that

$$\begin{bmatrix} \dfrac{d}{ad - bc} & \dfrac{-b}{ad - bc} \\ \dfrac{-c}{ad - bc} & \dfrac{a}{ad - bc} \end{bmatrix} \begin{bmatrix} a & b \\ c & d \end{bmatrix} = \begin{bmatrix} 1 & 0 \\ 0 & 1 \end{bmatrix}$$

□

The inverse of an invertible $n \times n$ matrix can be found by using Gauss–Jordan elimination. Although Gauss–Jordan elimination is a technique for solving systems of equations, our intent is to use this technique to find the inverse of an invertible matrix. It is not even obvious that solving systems of equations and finding inverses of invertible matrices have anything in common. Imagine that we know that the coefficient matrix of Example 16 is invertible and furthermore that we can find its inverse. Then multiplying on the left on both sides of the matrix equation

$$\begin{bmatrix} 3 & 2 & -1 \\ 1 & 3 & 2 \\ 2 & -1 & 1 \end{bmatrix} \begin{bmatrix} x \\ y \\ z \end{bmatrix} = \begin{bmatrix} 4 \\ 7 \\ 5 \end{bmatrix} \quad \text{by} \quad \begin{bmatrix} 3 & 2 & -1 \\ 1 & 3 & 2 \\ 2 & -1 & 1 \end{bmatrix}^{-1}$$

yields

$$I_3 \begin{bmatrix} x \\ y \\ z \end{bmatrix} = \begin{bmatrix} 3 & 2 & -1 \\ 1 & 3 & 2 \\ 2 & -1 & 1 \end{bmatrix}^{-1}_{3 \times 3} \begin{bmatrix} 4 \\ 7 \\ 5 \end{bmatrix}_{3 \times 1}$$

and the resulting 3×1 matrix on the right is

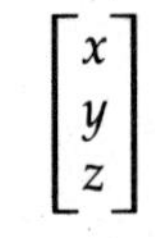

Since knowing the inverse of the coefficient matrix is all that is needed to solve a system of equations, perhaps knowing how to solve a system of equations is all that is needed to find the inverse of that system's coefficient matrix.

Let

$$A = \begin{bmatrix} 3 & 2 & -1 \\ 1 & 3 & 2 \\ 2 & -1 & 1 \end{bmatrix}$$

and suppose we try to find A^{-1}. If we can find invertible matrices $E_1, E_2, \ldots, E_k$ such that $E_k E_{k-1} \cdots E_2 E_1 A = I_3$, then

$$\begin{aligned} A &= E_1^{-1} E_2^{-1} \cdots E_k^{-1} E_k E_{k-1} \cdots E_2 E_1 A \\ &= E_1^{-1} E_2^{-1} \cdots E_k^{-1} I_3 \\ &= E_1^{-1} E_2^{-1} \cdots E_k^{-1} \end{aligned}$$

and by Corollary 8.10,

$$A^{-1} = (E_1^{-1} E_2^{-1} \cdots E_k^{-1})^{-1} = E_k E_{k-1} E_{k-2} \cdots E_2 E_1$$

We have reduced the problem of finding A^{-1} to finding a bunch of invertible matrices $E_1, E_2, \ldots, E_k$ such that $E_k E_{k-1} \cdots E_1 A = I_3$. Our new goal is reminiscent of the goal of Gaussian elimination, which was to use row operations to change A to I_3. Perhaps each of the three types of row operations can be obtained by multiplying on the left by a suitable invertible matrix. Indeed, this is the case, and the matrix needed to perform a given row operation is none other than the matrix obtained when the given row operation is performed on the identity matrix.

Example 21

a. If you switch the first and second rows of I_3, you get

$$\begin{bmatrix} 0 & 1 & 0 \\ 1 & 0 & 0 \\ 0 & 0 & 1 \end{bmatrix}$$

When any 3×3 matrix M is multiplied on the left by

$$\begin{bmatrix} 0 & 1 & 0 \\ 1 & 0 & 0 \\ 0 & 0 & 1 \end{bmatrix}$$

the result is to switch the first and second rows of M.

b. If you multiply the third row of I_3 by a nonzero constant k, you get

$$\begin{bmatrix} 1 & 0 & 0 \\ 0 & 1 & 0 \\ 0 & 0 & k \end{bmatrix}$$

When any 3×3 matrix M is multiplied on the left by

$$\begin{bmatrix} 1 & 0 & 0 \\ 0 & 1 & 0 \\ 0 & 0 & k \end{bmatrix}$$

the result is a multiplication of every entry of the third row of M by k.

c. If you add k times the first row of I_3 to the second row of I_3, the result is

$$\begin{bmatrix} 1 & 0 & 0 \\ k & 1 & 0 \\ 0 & 0 & 1 \end{bmatrix}$$

When any 3×3 matrix M is multiplied on the left side by this matrix, the result is an addition of k times the first row of M to the second row of M. ■

An $n \times n$ matrix of any of the three types illustrated in Example 21 is called an *elementary matrix*. It should be obvious that any elementary matrix is invertible and that in fact its inverse is another elementary matrix of the same type (see Exercise 51).

Example 22

We are at last ready to find A^{-1} when

$$A = \begin{bmatrix} 3 & 2 & -1 \\ 1 & 3 & 2 \\ 2 & -1 & 1 \end{bmatrix}$$

It follows from Example 17 that $E_9E_8 \cdots E_1A = I_3$ when the nine elementary matrices are

$$E_1 = \begin{bmatrix} \frac{1}{3} & 0 & 0 \\ 0 & 1 & 0 \\ 0 & 0 & 1 \end{bmatrix} \quad E_2 = \begin{bmatrix} 1 & 0 & 0 \\ -1 & 1 & 0 \\ 0 & 0 & 1 \end{bmatrix} \quad E_3 = \begin{bmatrix} 1 & 0 & 0 \\ 0 & 1 & 0 \\ -2 & 0 & 1 \end{bmatrix}$$

$$E_4 = \begin{bmatrix} 1 & 0 & 0 \\ 0 & \frac{3}{7} & 0 \\ 0 & 0 & 1 \end{bmatrix} \quad E_5 = \begin{bmatrix} 1 & 0 & 0 \\ 0 & 1 & 0 \\ 0 & \frac{7}{3} & 1 \end{bmatrix} \quad E_6 = \begin{bmatrix} 1 & 0 & 0 \\ 0 & 1 & 0 \\ 0 & 0 & \frac{1}{4} \end{bmatrix}$$

$$E_7 = \begin{bmatrix} 1 & 0 & 0 \\ 0 & 1 & -1 \\ 0 & 0 & 1 \end{bmatrix} \quad E_8 = \begin{bmatrix} 1 & 0 & \frac{1}{3} \\ 0 & 1 & 0 \\ 0 & 0 & 1 \end{bmatrix} \quad E_9 = \begin{bmatrix} 1 & -\frac{2}{3} & 0 \\ 0 & 1 & 0 \\ 0 & 0 & 1 \end{bmatrix}$$

Since $E_9E_8 \cdots E_1A = I_3$ and $E_9, E_8, \ldots, E_1$ are invertible, $A^{-1} = E_9E_8 \cdots E_1$. All that is left to do is to calculate the product (see Exercise 53). ■

The easiest way to calculate the product $E_9E_8 \cdots E_2E_1$ is to realize that E_1 performs the first-row operation of Example 16 on I_3, E_2 performs the second-row operation of Example 16 on E_1, E_3 performs the third-row operation of Example 16 on E_2E_1, and so forth. In other words, to find A^{-1}, we must perform exactly the same row operations on I_3 that we performed on A using Gaussian elimination. If only we had known this fact in advance, we could have done both jobs at once. Our final example illustrates how to find A^{-1} at the same time that we are "row reducing" A to I.

Example 23

As we mentioned, the Gauss–Jordan elimination method can be used to find the inverse of a matrix. We "augment" the matrix by placing an identity matrix on its right. For example, suppose that we want to find the inverse of

$$A = \begin{bmatrix} 1 & 2 & -3 \\ 2 & -1 & 4 \\ -1 & 0 & 5 \end{bmatrix}$$

We form the matrix

$$\left[\begin{array}{rrr|rrr} 1 & 2 & -3 & 1 & 0 & 0 \\ 2 & -1 & 4 & 0 & 1 & 0 \\ -1 & 0 & 5 & 0 & 0 & 1 \end{array}\right]$$

Then, just as before, we perform elementary row operations to obtain the identity matrix I_3 on the left of the dashed line. The row operation we use at each step is indicated on the left side of the matrix in abbreviated form.

$$\begin{array}{r} \\ -2R_1 + R_2 \\ R_1 + R_3 \end{array} \left[\begin{array}{rrr|rrr} 1 & 2 & -3 & 1 & 0 & 0 \\ 0 & -5 & 10 & -2 & 1 & 0 \\ 0 & 2 & 2 & 1 & 0 & 1 \end{array}\right]$$

$$\begin{array}{r} \\ -\frac{1}{5}R_2 \\ \\ \end{array} \left[\begin{array}{rrr|rrr} 1 & 2 & -3 & 1 & 0 & 0 \\ 0 & 1 & -2 & \frac{2}{5} & -\frac{1}{5} & 0 \\ 0 & 2 & 2 & 1 & 0 & 1 \end{array}\right]$$

$$\begin{array}{r} \\ \\ -2R_2 + R_3 \end{array} \left[\begin{array}{rrr|rrr} 1 & 2 & -3 & 1 & 0 & 0 \\ 0 & 1 & -2 & \frac{2}{5} & -\frac{1}{5} & 0 \\ 0 & 0 & 6 & \frac{1}{5} & \frac{2}{5} & 1 \end{array}\right]$$

$$\begin{array}{l} \\ \\ \frac{1}{6}R_3 \end{array} \left[\begin{array}{ccc|ccc} 1 & 2 & -3 & 1 & 0 & 0 \\ 0 & 1 & -2 & \frac{2}{5} & -\frac{1}{5} & 0 \\ 0 & 0 & 1 & \frac{1}{30} & \frac{1}{15} & \frac{1}{6} \end{array}\right]$$

$$\begin{array}{l} 3R_3 + R_1 \\ 2R_3 + R_2 \\ \\ \end{array} \left[\begin{array}{ccc|ccc} 1 & 2 & 0 & \frac{11}{10} & \frac{1}{5} & \frac{1}{2} \\ 0 & 1 & 0 & \frac{7}{15} & -\frac{1}{15} & \frac{1}{3} \\ 0 & 0 & 1 & \frac{1}{30} & \frac{1}{15} & \frac{1}{6} \end{array}\right]$$

$$\begin{array}{l} -2R_2 + R_1 \\ \\ \\ \end{array} \left[\begin{array}{ccc|ccc} 1 & 0 & 0 & \frac{1}{6} & \frac{1}{3} & -\frac{1}{6} \\ 0 & 1 & 0 & \frac{7}{15} & -\frac{1}{15} & \frac{1}{3} \\ 0 & 0 & 1 & \frac{1}{30} & \frac{1}{15} & \frac{1}{6} \end{array}\right]$$

The 3×3 matrix on the right is A^{-1}. We can check our work by finding the products $A \times A^{-1}$ and $A^{-1} \times A$. ■

If with Gaussian elimination we obtain a row of 0s to the left of the dotted line, then the matrix does not have an inverse.

Definition

Two matrices are **equivalent** if one can be obtained from the other by using the three elementary row operations (discussed previously).

We state the following theorem without proof.

Theorem 8.13 An $n \times n$ matrix A is invertible if and only if it is equivalent to I_n. □

Exercises 8.4

43. Find the inverse of each of the following matrices.

a. $\begin{bmatrix} 2 & 1 \\ 3 & 4 \end{bmatrix}$ **b.** $\begin{bmatrix} 2 & 0 \\ 1 & 1 \end{bmatrix}$ **c.** $\begin{bmatrix} 0 & 1 \\ 1 & 0 \end{bmatrix}$

44. Let A be the matrix

$$\begin{bmatrix} 1 & 1 & 0 \\ 0 & 1 & 1 \\ 1 & 0 & 1 \end{bmatrix}$$

Determine whether A is invertible, and if so, find its inverse. (*Hint:* Let

$$X = \begin{bmatrix} a & b & c \\ d & e & f \\ g & h & i \end{bmatrix}$$

and solve $AX = I_3$ by equating corresponding entries on each side.)

45. Let A be an invertible matrix whose inverse is

$$\begin{bmatrix} 1 & 2 \\ 3 & 4 \end{bmatrix}$$

Find the matrix A.

46. Let A be an invertible matrix such that the inverse of $3A$ is

$$\begin{bmatrix} 1 & -2 \\ -4 & 7 \end{bmatrix}$$

Find the matrix A.

47. Prove that, if A and B are invertible $n \times n$ matrices, then $(B^{-1}A^{-1})(AB) = I_n$.

48. Prove Corollary 8.10.

49. Prove that, if A is an invertible $n \times n$ matrix and k is a nonzero member of $\mathbb{R}$, then $[(1/k)A^{-1}](kA) = I_n$.

50. Find the inverse, if it exists, of each of the following matrices.

a. $\begin{bmatrix} 1 & 2 \\ 1 & 3 \end{bmatrix}$ **b.** $\begin{bmatrix} 4 & -12 \\ -2 & 6 \end{bmatrix}$ **c.** $\begin{bmatrix} 1 & 2 \\ 3 & 4 \end{bmatrix}$

d. $\begin{bmatrix} 1 & -1 & 0 \\ 0 & 1 & 2 \\ 2 & 0 & 1 \end{bmatrix}$ **e.** $\begin{bmatrix} 2 & 3 & 4 \\ 5 & 0 & 2 \\ -2 & 1 & 3 \end{bmatrix}$ **f.** $\begin{bmatrix} 1 & 2 & -3 \\ 0 & 1 & 2 \\ -2 & -4 & 6 \end{bmatrix}$

51. By inspection, find the inverses of each of the following 4×4 elementary matrices.

a. $\begin{bmatrix} 1 & 0 & 0 & 0 \\ 0 & 0 & 0 & 1 \\ 0 & 0 & 1 & 0 \\ 0 & 1 & 0 & 0 \end{bmatrix}$ **b.** $\begin{bmatrix} 1 & 0 & 0 & 0 \\ 0 & 1 & 0 & 0 \\ 0 & 0 & -7 & 0 \\ 0 & 0 & 0 & 1 \end{bmatrix}$ **c.** $\begin{bmatrix} 1 & 0 & 0 & 0 \\ 0 & 1 & 0 & -7 \\ 0 & 0 & 1 & 0 \\ 0 & 0 & 0 & 1 \end{bmatrix}$ **d.** $\begin{bmatrix} 1 & 0 & 0 & 0 \\ 0 & 1 & 0 & 0 \\ 0 & 0 & 1 & 0 \\ 1 & 5 & 0 & 1 \end{bmatrix}$

52. By inspection, multiply the following matrix on the left by each of the matrices given in Exercise 51.

$$\begin{bmatrix} 1 & 2 & 3 & 4 \\ 1 & 1 & 1 & 1 \\ 4 & 3 & 2 & 1 \\ 3 & 3 & 3 & 3 \end{bmatrix}$$

53. **a.** Calculate the product $E_9E_8 \cdots E_1$, where $E_9, E_8, \ldots, E_1$ are the elementary matrices of Example 22.

b. Let M be the 3×3 matrix obtained in (a). Verify that

$$M \cdot \begin{bmatrix} 4 \\ 7 \\ 5 \end{bmatrix} = \begin{bmatrix} \frac{12}{7} \\ \frac{3}{7} \\ 2 \end{bmatrix}$$

8.5

DETERMINANTS

In this section we review determinants because they are used in the study of eigenvalues and eigenvectors. We begin by defining the determinants of 2 × 2 and 3 × 3 matrices.

Definition

Let $A = \begin{bmatrix} a_{11} & a_{12} \\ a_{21} & a_{22} \end{bmatrix}$ be a 2 × 2 matrix. The **determinant of** ***A***, denoted $\det(A)$, is the real number $a_{11}a_{22} - a_{21}a_{12}$.

Example 24

Find $\det(A)$ if $A = \begin{bmatrix} 2 & -3 \\ -1 & 5 \end{bmatrix}$

Analysis

$\det(A) = 2 \cdot 5 - (-1)(-3) = 10 - 3 = 7$ ■

Definition

Let $A = \begin{bmatrix} a_{11} & a_{12} & a_{13} \\ a_{21} & a_{22} & a_{23} \\ a_{31} & a_{32} & a_{33} \end{bmatrix}$ be a 3 × 3 matrix. The determinant of A, denoted by $\det(A)$, is the real number

$$a_{11}\det\left(\begin{bmatrix} a_{22} & a_{23} \\ a_{32} & a_{33} \end{bmatrix}\right) - a_{12}\det\left(\begin{bmatrix} a_{21} & a_{23} \\ a_{31} & a_{33} \end{bmatrix}\right) + a_{13}\det\left(\begin{bmatrix} a_{21} & a_{22} \\ a_{31} & a_{32} \end{bmatrix}\right)$$

Example 25

Find $\det(A)$ if $A = \begin{bmatrix} 2 & -1 & 3 \\ 1 & -2 & 3 \\ -3 & 1 & -1 \end{bmatrix}$.

Analysis

$$\begin{aligned}\det(A) &= 2\det\left(\begin{bmatrix} -2 & 3 \\ 1 & -1 \end{bmatrix}\right) - (-1)\det\left(\begin{bmatrix} 1 & 3 \\ -3 & -1 \end{bmatrix}\right) \\ &\quad + 3\det\left(\begin{bmatrix} 1 & -2 \\ -3 & 1 \end{bmatrix}\right) \\ &= 2[(-2)(-1) - 1 \cdot 3] + 1(-1) - (-3)(3) \\ &\quad + 3[1 \cdot 1 - (-2)(-3)] \\ &= 2[2 - 3] + [-1 + 9] + 3[1 - 6] \\ &= 2(-1) + 8 + 3(-5) = -2 + 8 - 15 = -9\end{aligned}$$

■

Note that the three 2×2 matrices that appear in the definition of the determinant of a 3×3 matrix follow a pattern. The first 2×2 matrix can be obtained from A by deleting the first row and the first column. The second 2×2 matrix can be obtained from A by deleting the first row and the second column, and the third 2×2 matrix can be obtained from A by deleting the first row and the third column. The process of deleting rows and columns is the fundamental idea in the definition of the determinant of an $n \times n$ matrix. In Figure 8.6 we describe a device that is useful because it saves us from memorizing the formulas in the definitions of the determinant of a 2×2 and a 3×3 matrix.

Figure 8.6

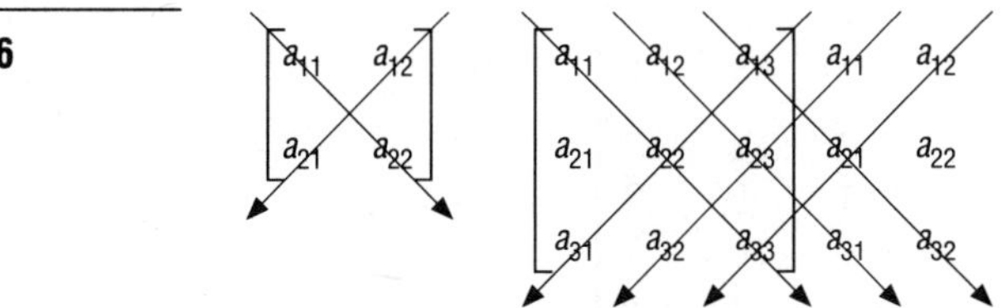

The determinant of a 2×2 matrix is obtained from Figure 8.6 by multiplying the entries on the rightward arrow and subtracting the product of the entries on the leftward arrow. The determinant of a 3×3 matrix is obtained by first copying the first and second columns as shown in Figure 8.6 and then adding the products on the rightward arrows and subtracting the products on the leftward arrows. Alas, neither of the procedures indicated in Figure 8.6 extends to $n \times n$ matrices where $n > 3$.

Definition

Let $n \in \mathbb{N}$ and let A be an $n \times n$ matrix. The $(n - 1) \times (n - 1)$ matrix that is obtained by deleting the rth row and sth column of A is called a **minor matrix of A** and is denoted by M_{rs}.

Example 26

$$\text{Let } A = \begin{bmatrix} 1 & 2 & 3 & 4 & -3 \\ -2 & -1 & 0 & 5 & 2 \\ 6 & -5 & 3 & -6 & 7 \\ -7 & 8 & -4 & 1 & -8 \\ -1 & 9 & 5 & -9 & 6 \end{bmatrix}.$$

List the minor matrices M_{14}, M_{22}, and M_{53} of A.

Analysis

The minor matrix of M_{14} of A is obtained from A by deleting the first row and fourth column of A. The minor matrix M_{22} of A is obtained by deleting the second row and second column, and the minor matrix M_{53} of A is obtained by deleting the fifth row and third column. Therefore

$$M_{14} = \begin{bmatrix} -2 & -1 & 0 & 2 \\ 6 & -5 & 3 & 7 \\ -7 & 8 & -4 & -8 \\ -1 & 9 & 5 & 6 \end{bmatrix}$$

$$M_{22} = \begin{bmatrix} 1 & 3 & 4 & -3 \\ 6 & 3 & -6 & 7 \\ -7 & -4 & 1 & -8 \\ -1 & 5 & -9 & 6 \end{bmatrix}$$

$$M_{53} = \begin{bmatrix} 1 & 2 & 4 & -3 \\ -2 & -1 & 5 & 2 \\ 6 & -5 & -6 & 7 \\ -7 & 8 & 1 & -8 \end{bmatrix}$$

■

Using the concept of minor matrices, we can restate the definition of the determinant of a 3×3 matrix.

Definition

Let A be a 3×3 matrix. The **determinant of A** is

$$a_{11}\det(M_{11}) - a_{12}\det(M_{12}) + a_{13}\det(M_{13})$$

Definition

Let $A = [a_{ij}]$ be an $n \times n$ matrix. The **minor of a_{ij}** is the determinant of M_{ij}. The number $(-1)^{i+j}\det(M_{ij})$ is called the **cofactor of a_{ij}** and is denoted by C_{ij}.

We defined the determinant of a 3×3 matrix to be $a_{11}C_{11} + a_{12}C_{12} + a_{13}C_{13}$. This is known as a *cofactor expansion* that uses the cofactors of entries in the first row. It is natural to ask whether we can find the determinant of a 3×3 matrix with a cofactor expansion that uses the cofactors of entries in the second or third row. The answer is yes. In fact, we can use columns as well as rows. So with a 3×3 matrix there are six cofactor expansions available—three row expansions and three column expansions.

Example 27

Let $A = \begin{bmatrix} 7 & 0 & -3 \\ 2 & 1 & 3 \\ 4 & 0 & 5 \end{bmatrix}$. Find $\det(A)$.

Analysis

We can proceed by using any one of six cofactor expansions. Observe that there are two 0s in the second column. Thus, if we use the cofactor expansion that uses the cofactors of entries in the second column, we need to calculate the determinant of only one 2×2 matrix.

$$\det(A) = 0 \cdot C_{12} + 1 \cdot C_{22} + 0 \cdot C_{32} = C_{22} = (-1)^{2+2}\det(M_{22})$$

$$= \det \begin{bmatrix} 7 & -3 \\ 4 & 5 \end{bmatrix} = 7 \cdot 5 - 4(-3) = 35 + 12 = 47 \qquad \blacksquare$$

In Exercise 55 you are asked to calculate the determinant of the matrix in Example 27 by the cofactor expansion that uses the cofactors of entries in the first row. Based on what we have said, the answer to Exercise 55 is 47.

We next define the determinant of an $n \times n$ matrix when $n \geq 3$. Notice that the determinant is defined recursively.

Definition

Let $A = [a_{ij}]$ be an $n \times n$ matrix ($n \geq 3$). The **determinant of A,** denoted by $\det(A)$, is $\sum_{j=1}^{n} a_{1j}C_{1j}$.

As is the case with 3×3 matrices, the determinant of an $n \times n$ matrix can be found by a cofactor expansion that uses the cofactors of the entries of any row or column.

Example 28

Let $A = \begin{bmatrix} 2 & -1 & 3 & 1 \\ 4 & 0 & -2 & 5 \\ 1 & -3 & 0 & -4 \\ -5 & 6 & 3 & 7 \end{bmatrix}$. Find $\det(A)$.

Analysis

Since A has two 0s, we have to calculate the determinant of only three 3×3 matrices if we use the cofactor expansion that uses the entries of the second row, third row, second column, or third column. In this first example, however, we use the cofactor expansion of the matrix based on the entries in the first row. In Exercise 57 you are asked to calculate the determinant of A by a cofactor expansion that uses the cofactors of the entries of the second column. By definition, $\det(A) = \sum_{j=1}^{4} a_{1j}C_{1j}$. Thus we need to calculate C_{11}, C_{12}, C_{13}, and C_{14}; that is, we need to calculate $\det(M_{11})$, $\det(M_{12})$, $\det(M_{13})$, and $\det(M_{14})$.

$$\begin{aligned}
\det(M_{11}) &= \det\left(\begin{bmatrix} 0 & -2 & 5 \\ -3 & 0 & -4 \\ 6 & 3 & 7 \end{bmatrix}\right) \\
&= 0 \cdot \det\left(\begin{bmatrix} 0 & -4 \\ 3 & 7 \end{bmatrix}\right) - (-2)\det\left(\begin{bmatrix} -3 & -4 \\ 6 & 7 \end{bmatrix}\right) \\
&\qquad + 5 \cdot \det\left(\begin{bmatrix} -3 & 0 \\ 6 & 3 \end{bmatrix}\right) \\
&= 0 + 2[(-3)7 - (-4)6] + 5[(-3)(3) - 0 \cdot 6] \\
&= 2(-21 + 24) + 5(-9) = 6 - 45 = -39
\end{aligned}$$

$$\begin{aligned}
\det(M_{12}) &= \det\left(\begin{bmatrix} 4 & -2 & 5 \\ 1 & 0 & -4 \\ -5 & 3 & 7 \end{bmatrix}\right) \\
&= (-1)^{2+1} \cdot 1 \cdot \det\left(\begin{bmatrix} -2 & 5 \\ 3 & 7 \end{bmatrix}\right) + 0 \\
&\qquad + (-1)^{2+3}(-4)\det\left(\begin{bmatrix} 4 & -2 \\ -5 & 3 \end{bmatrix}\right) \\
&= -[(-2)7 - 3 \cdot 5] + 4[4 \cdot 3 - (-2)(-5)] \\
&= -(-14 - 15) + 4(12 - 10) = 29 + 8 = 37
\end{aligned}$$

$$\begin{aligned}
\det(M_{13}) &= \det\left(\begin{bmatrix} 4 & 0 & 5 \\ 1 & -3 & -4 \\ -5 & 6 & 7 \end{bmatrix}\right) \\
&= (-1)^{1+1}4 \cdot \det\left(\begin{bmatrix} -3 & -4 \\ 6 & 7 \end{bmatrix}\right) + 0 \\
&\qquad + (-1)^{1+3}5 \cdot \det\left(\begin{bmatrix} 1 & -3 \\ -5 & 6 \end{bmatrix}\right) \\
&= 4[(-3)7 - 6(-4)] + 5[1 \cdot 6 - (-5)(-3)] \\
&= 4(-21 + 24) + 5(6 - 15) = 12 - 45 = -33 \\
\det(M_{14}) &= \det\left(\begin{bmatrix} 4 & 0 & -2 \\ 1 & -3 & 0 \\ -5 & 6 & 3 \end{bmatrix}\right) \\
&= (-1)^{1+1}4 \cdot \det\left(\begin{bmatrix} -3 & 0 \\ 6 & 3 \end{bmatrix}\right) + 0 \\
&\qquad + (-1)^{1+3}(-2)\det\left(\begin{bmatrix} 1 & -3 \\ -5 & 6 \end{bmatrix}\right) \\
&= 4[(-3)3 - 6 \cdot 0] - 2[1 \cdot 6 - (-5)(-3)] \\
&= 4(-9) - 2(6 - 15) = -36 + 18 = -18
\end{aligned}$$

Therefore

$$\begin{aligned}
\det(A) &= a_{11}(-1)^{1+1}\det(M_{11}) + a_{12}(-1)^{1+2}\det(M_{12}) \\
&\qquad + a_{13}(-1)^{1+3}\det(M_{13}) + a_{14}(-1)^{1+4}\det(M_{14}) \\
&= 2 \cdot 1 \cdot (-39) + (-1)(-1) \cdot 37 + 3 \cdot 1 \cdot (-33) \\
&\qquad + 1 \cdot (-1)(-18) \\
&= -78 + 37 - 99 + 18 = -122
\end{aligned}$$

■

We conclude this section by stating, without proof, a theorem useful in calculating the determinant. But first we need a definition.

Definition

An $n \times n$ matrix $A = [a_{ij}]$ is **upper triangular** if $a_{ij} = \mathbf{0}$ when $i > j$, and it is **lower triangular** if $a_{ij} = \mathbf{0}$ whenever $i < j$. Then A is **triangular** if it is upper triangular or lower triangular.

Theorem 8.14

Let A be an $n \times n$ matrix.

a. If B is an $n \times n$ matrix, then $\det(AB) = \det(A)\det(B)$.
b. $\det(A^t) = \det(A)$.
c. If B is the matrix obtained from A by interchanging any two rows (or columns), then $\det(B) = -\det(A)$.
d. If A has two identical rows, then $\det(A) = 0$.
e. If B is the matrix obtained from A by multiplying one row (or column) of A by a nonzero scalar c, then $\det(B) = c \cdot \det(A)$.
f. If B is the matrix obtained from A by replacing a row (or column) of A by the sum of it and a multiple of another row (or column), then $\det(B) = \det(A)$.
g. If A contains a row (or column) of 0s, then $\det(A) = 0$.
h. If $A = [a_{ij}]$ and A is triangular, then $\det(A) = a_{11}a_{22} \cdots a_{nn}$.

Exercises 8.5

54. Find the determinant of each of the following matrices.

a. $\begin{bmatrix} 1 & 2 \\ 4 & 3 \end{bmatrix}$ **b.** $\begin{bmatrix} 0 & 5 \\ 3 & 8 \end{bmatrix}$ **c.** $\begin{bmatrix} -2 & -3 \\ 4 & 5 \end{bmatrix}$

55. Find the determinant of the matrix in Example 27 by the cofactor expansion that uses the cofactors of entries in the first row.

56. Find the determinant of each of the following matrices.

a. $\begin{bmatrix} 1 & 2 & 0 \\ 3 & -1 & 2 \\ 4 & 5 & -2 \end{bmatrix}$ **b.** $\begin{bmatrix} 1 & 2 & 3 \\ 0 & 0 & 4 \\ -2 & 5 & 7 \end{bmatrix}$ **c.** $\begin{bmatrix} 1 & 2 & 3 \\ 4 & 3 & 0 \\ -2 & -1 & 0 \end{bmatrix}$

57. Find the determinant of the matrix in Example 28 by the cofactor expansion that uses the cofactors of the entries of the second column.

58. Find the determinant of each of the following matrices.

a. $\begin{bmatrix} 2 & 1 & 3 & -2 \\ 0 & 2 & 1 & 4 \\ 3 & -2 & -3 & 0 \\ 0 & 3 & 0 & -2 \end{bmatrix}$ **b.** $\begin{bmatrix} 3 & 1 & -2 & 2 \\ 0 & 2 & 1 & 4 \\ 1 & -2 & 3 & -1 \\ 0 & 3 & 4 & 5 \end{bmatrix}$

c. $\begin{bmatrix} 2 & -1 & -2 & 1 & 3 \\ 0 & 2 & 1 & 5 & 2 \\ 1 & -3 & 5 & 0 & 1 \\ 4 & -2 & 1 & 2 & -2 \\ 5 & 4 & -3 & 0 & 5 \end{bmatrix}$

8.6

VECTOR SPACES

In the remainder of this chapter, an ordered n-tuple of real numbers $v = (v_1,v_2, \ldots ,v_n)$ is called an *n-dimensional vector*, and it can be written as an $n \times 1$ matrix,

$$v = \begin{bmatrix} v_1 \\ v_2 \\ \cdot \\ \cdot \\ \cdot \\ v_n \end{bmatrix}$$

Each v_i is a *component* of v, and the set of all n-dimensional vectors is called *Euclidean n-space* and is denoted by $\mathbb{R}^n$. A *nonzero vector* is a vector $v = (v_1,v_2, \ldots ,v_n)$ such that $v_i \neq 0$ for some i. In the study of vectors, real numbers are commonly referred to as *scalars*. A 2-dimensional vector $v = (v_1,v_2)$ in the plane can be interpreted as an arrow from the origin to the point (v_1,v_2).

In this section we discuss some properties of $\mathbb{R}^n$. This set of vectors provides an intuitive and natural introduction to the set of abstract vectors in a general vector space. Since we know how to add matrices and how to multiply a matrix by a scalar, we know how to add n-dimensional vectors and how to multiply an n-dimensional vector by a scalar. The geometric interpretation of these concepts in $\mathbb{R}^2$ is illustrated in Figure 8.7.

Figure 8.7

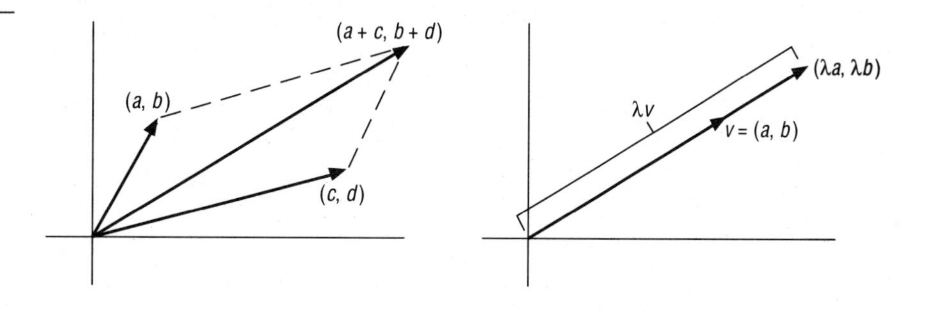

We state several theorems (some without proof) about vectors in $\mathbb{R}^n$ and subsets of $\mathbb{R}^n$.

Theorem 8.15

If u, v, and w are vectors in $\mathbb{R}^n$ and λ and μ are scalars, then the following properties hold:

- **a.** $u + v \in \mathbb{R}^n$
- **b.** $\lambda u \in \mathbb{R}^n$
- **c.** $u + v = v + u$
- **d.** $u + (v + w) = (u + v) + w$
- **e.** There is a **0** vector θ in $\mathbb{R}^n$ such that $v + \theta = v$ for all $v \in \mathbb{R}^n$ (θ is the vector whose components are all 0).
- **f.** For each $v \in \mathbb{R}^n$ there exists $-v \in \mathbb{R}^n$ such that $v + (-v) = \theta$.
- **g.** $\lambda(\mu v) = (\lambda\mu)v$
- **h.** $\lambda(u + v) = \lambda u + \lambda v$
- **i.** $(\lambda + \mu)v = \lambda v + \mu v$
- **j.** $1 \cdot v = v$ for all $v \in \mathbb{R}^n$

An abstract vector space over the set of real numbers (which we do not study) is defined to be a set V for which there is a scalar multiplication and an addition $+ : V \times V \rightarrow V$ satisfying the properties listed in Theorem 8.15, where V replaces the set $\mathbb{R}^n$. A subset V of $\mathbb{R}^n$ that satisfies the properties listed in Theorem 8.15 is called a *subspace* of $\mathbb{R}^n$.

Example 29

Let $V = \{v \in \mathbb{R}^2 : v = (v_1,0)$, where v_1 is a real number$\}$. Then V is a subspace of $\mathbb{R}^2$.

Analysis

We notice that addition of two members of V yields another member of V and that multiplication of a member of V by a scalar yields another member of V. Therefore statements (a) and (b) of Theorem 8.15 are satisfied. Statements (c), (d), (g), (h), (i), and (j) of Theorem 8.15 are automatically satisfied since every member of V is also a member of $\mathbb{R}^2$. To see that (e) is satisfied, we simply observe that $(0,0) \in V$. Finally, if $v \in V$, then $-1 \cdot v = -v$ is a member of V that satisfies (f). ■

Example 29 illustrates the following general theorem.

Theorem 8.16

A subset V of $\mathbb{R}^n$ is a subspace of $\mathbb{R}^n$ if and only if the following conditions hold:

- **a.** The zero vector θ is a member of V.
- **b.** If u and v are members of V, then $u + v$ is a member of V.
- **c.** If $v \in V$ and λ is a scalar, then $\lambda v \in V$.

Example 30

Verify that the subset V of $\mathbb{R}^3$ defined by $V = \{(v_1,v_2,v_3) \in \mathbb{R}^3: v_2 = 3v_1 \text{ and } v_3 = 5v_1\}$ is a subspace of $\mathbb{R}^3$.

Analysis

We show that V is a subspace of $\mathbb{R}^3$ by showing that V satisfies the conditions stated in Theorem 8.16. Since $0 \in \mathbb{R}$, $3 \cdot 0 = 0$, and $5 \cdot 0 = 0$, the zero vector $\theta = (0,0,0)$ is a member of V. If $u = (u_1,3u_1,5u_1)$ and $v = (v_1,3v_1,5v_1)$ are members of V, then $u + v = (u_1 + v_1,3u_1 + 3v_1,5u_1 + 5v_1) = (u_1 + v_1,3(u_1 + v_1),5(u_1 + v_1))$ is a member of V. If $v = (v_1,3v_1,5v_1)$ and $\lambda \in \mathbb{R}$, then $\lambda v = (\lambda v_1,\lambda 3v_1,\lambda 5v_1) = (\lambda v_1,3(\lambda v_1),5(\lambda v_1))$ is a member of V. ■

Example 31

Show that the subset V of $\mathbb{R}^3$ defined by $V = \{(v_1,v_2,v_3) \in \mathbb{R}^3: v_2 = 3v_1 \text{ and } v_3 = v_1 + 5\}$ is not a subspace of $\mathbb{R}^3$.

Analysis

To show that V is not a subspace of $\mathbb{R}^3$, it is sufficient to show that V does not satisfy *one* of the conditions in Theorem 8.16. We show that V does not satisfy statement (a). If $v = (v_1,v_2,v_3)$ is a member of V and $v_1 = 0$, then $v_3 = 5$. Thus the zero vector θ is not a member of V. In Exercise 59 we ask you to show that Theorem 8.16b is not satisfied. ■

Definition

If $v_1,v_2, \ldots ,v_r$ are members of $\mathbb{R}^n$, then a vector u in $\mathbb{R}^n$ is a **linear combination** of $v_1,v_2, \ldots ,v_r$ if there exist scalars $\lambda_1,\lambda_2, \ldots ,\lambda_r$ such that $u = \lambda_1 v_1 + \lambda_2 v_2 + \cdots + \lambda_r v_r$.

Example 32

The vector $(4,-2,3)$ in $\mathbb{R}^3$ is a linear combination of $(2,0,1)$ and $(1,1,0)$ because $(4,-2,3) = 3(2,0,1) + (-2)(1,1,0)$. ■

Theorem 8.17

If $v_1,v_2, \ldots ,v_r$ are members of $\mathbb{R}^n$, then the set V of all linear combinations of $v_1,v_2, \ldots ,v_r$ is a subspace of $\mathbb{R}^n$.

Proof

We show that the three conditions in Theorem 8.16 are satisfied. Since $\theta = 0 \cdot v_1 + 0 \cdot v_2 + \cdots + 0 \cdot v_r$, $\theta \in V$. Suppose that u and v are members of V. Then there exist scalars $\lambda_1,\lambda_2, \ldots ,\lambda_r$, $\mu_1,\mu_2, \ldots ,\mu_r$ such that $u = \lambda_1 v_1 + \lambda_2 v_2 + \cdots + \lambda_r v_r$ and $v = \mu_1 v_1 + \mu_2 v_2 + \cdots + \mu_r v_r$.

Thus $u + v = (\lambda_1 + \mu_1)v_1 + (\lambda_2 + \mu_2)v_2 + \cdots + (\lambda_r + \mu_r)v_r$ and hence $u + v \in V$. Suppose that $v \in V$ and λ is a scalar. Then there exist scalars $\lambda_1, \lambda_2, \ldots, \lambda_r$ such that $v = \lambda_1 v_1 + \lambda_2 v_2 + \cdots + \lambda_r v_r$. So $\lambda v = (\lambda\lambda_1)v_1 + (\lambda\lambda_2)v_2 + \cdots + (\lambda\lambda_r)v_r$ and hence $\lambda v \in V$. □

Definition

If $S = \{v_1, v_2, \ldots, v_r\}$ is a subset of $\mathbb{R}^n$, then the subspace V of $\mathbb{R}^n$ consisting of all linear combinations of $v_1, v_2, \ldots, v_r$ is called the **subspace spanned by *S*.**

Example 33

If $v = (v_1, v_2)$ is a nonzero vector in $\mathbb{R}^2$, then the subspace spanned by $\{v\}$ is the line in the plane that passes through the origin and the point (v_1, v_2). ■

Definition

Let $S = \{v_1, v_2, \ldots, v_r\}$ be a subset of $\mathbb{R}^n$. If there are scalars $\lambda_1, \lambda_2, \ldots, \lambda_r$ such that $\lambda_1 v_1 + \lambda_2 v_2 + \cdots + \lambda_r v_r = 0$ but $\lambda_i \neq 0$ for some i, then S is a **linearly dependent set.** The set S is **linearly independent** if it is not linearly dependent; that is, it is linearly independent if the only scalars for which $\lambda_1 v_1 + \lambda_2 v_2 + \cdots + \lambda_r v_r = \theta$ are the scalars $\lambda_1 = \lambda_2 = \cdots = \lambda_r = 0$.

Example 34

Show that the set {(1,2,)(8,7),(2,1)} is a linearly dependent subset of $\mathbb{R}^2$.

Analysis

Let λ_1, λ_2, and λ_3 be scalars such that $\lambda_1(1,2) + \lambda_2(8,7) + \lambda_3(2,1) = (0,0)$. Then $\lambda_1 + 8\lambda_2 + 2\lambda_3 = 0$ and $2\lambda_1 + 7\lambda_2 + \lambda_3 = 0$. This system of equations is equivalent to the system

$$\lambda_1 + 8\lambda_2 + 2\lambda_3 = 0$$

$$-9\lambda_2 - 3\lambda_3 = 0$$

So $\lambda_2 = -\lambda_3/3$ and $\lambda_1 = 2\lambda_3/3$. Hence $\lambda_3 = 1$, $\lambda_2 = -1/3$, and $\lambda_1 = 2/3$ is a solution of $\lambda_1(1,2) + \lambda_2(8,7) + \lambda_3(2,1) = (0,0)$, and therefore {(1,2)(8,7)(2,1)} is a linearly dependent subset of $\mathbb{R}^2$. ■

Example 35

Show that the set {(3,2,1),(0,1,2),(1,0,2)} is a linearly independent subset of $\mathbb{R}^3$.

Analysis

Let λ_1, λ_2, and λ_3 be scalars such that $\lambda_1(3,2,1) + \lambda_2(0,1,2) + \lambda_3(1,0,2) = (0,0,0)$. Then $3\lambda_1 + \lambda_3 = 0$, $2\lambda_1 + \lambda_2 = 0$, and $\lambda_1 + 2\lambda_2 + 2\lambda_3 = 0$. We use the augmented matrix and Gaussian elimination to solve this system:

$$\left[\begin{array}{ccc|c} 3 & 0 & 1 & 0 \\ 2 & 1 & 0 & 0 \\ 1 & 2 & 2 & 0 \end{array}\right] \xrightarrow{\frac{1}{3}R_1} \left[\begin{array}{ccc|c} 1 & 0 & \frac{1}{3} & 0 \\ 2 & 1 & 0 & 0 \\ 1 & 2 & 2 & 0 \end{array}\right] \xrightarrow{-2R_1 + R_2} \left[\begin{array}{ccc|c} 1 & 0 & \frac{1}{3} & 0 \\ 0 & 1 & -\frac{2}{3} & 0 \\ 1 & 2 & 2 & 0 \end{array}\right]$$

$$\xrightarrow{-R_1 + R_3} \left[\begin{array}{ccc|c} 1 & 0 & \frac{1}{3} & 0 \\ 0 & 1 & -\frac{2}{3} & 0 \\ 0 & 2 & \frac{5}{3} & 0 \end{array}\right] \xrightarrow{-2R_2 + R_3} \left[\begin{array}{ccc|c} 1 & 0 & \frac{1}{3} & 0 \\ 0 & 1 & -\frac{2}{3} & 0 \\ 0 & 0 & 3 & 0 \end{array}\right]$$

$$\xrightarrow{\frac{1}{3}R_3} \left[\begin{array}{ccc|c} 1 & 0 & \frac{1}{3} & 0 \\ 0 & 1 & -\frac{2}{3} & 0 \\ 0 & 0 & 1 & 0 \end{array}\right]$$

Thus $\lambda_3 = 0$, and backsolving we see that $\lambda_2 = 0$ and $\lambda_1 = 0$. Therefore $\{(3,2,1),(0,1,2),(1,0,2)\}$ is a linearly independent subset of $\mathbb{R}^3$. ■

Theorem 8.18

> Let $S = \{v_1, v_2, \ldots, v_r\}$ be a subset of $\mathbb{R}^n$. If $r > n$, then S is linearly dependent.

In the discussion that follows, we want to think of the n-dimensional vector v as being an $n \times 1$ matrix. It is convenient to express an $m \times n$ matrix A in the form $A = [A_1, A_2, \ldots, A_n]$, where, for each $j = 1, 2, \ldots, n$, A_j is the jth column of A. Hence A_j is an $m \times 1$ matrix or a vector.

Example 36

If $A = \begin{bmatrix} 2 & 1 & 3 \\ 4 & 0 & 5 \end{bmatrix}$, then $A = [A_1, A_2, A_3]$, where $A_1 = \begin{bmatrix} 2 \\ 4 \end{bmatrix}$, $A_2 = \begin{bmatrix} 1 \\ 0 \end{bmatrix}$, and $A_3 = \begin{bmatrix} 3 \\ 5 \end{bmatrix}$. ■

Theorem 8.19 Let $A = [A_1, A_2, \ldots, A_n]$ be an $m \times n$ matrix and let $v = (v_1, v_2, \ldots, v_n)$ be an n-dimensional vector. Then the matrix product Av can be expressed as $v_1A_1 + v_2A_2 + \cdots + v_nA_n$.

The following example illustrates Theorem 8.19.

Example 37

Let A, A_1, A_2, and A_3 be as defined in Example 36 and let $v = (v_1, v_2, v_3)$. Then

$$\begin{aligned} Av &= \begin{bmatrix} 2 & 1 & 3 \\ 4 & 0 & 5 \end{bmatrix} \begin{bmatrix} v_1 \\ v_2 \\ v_3 \end{bmatrix} \\ &= \begin{bmatrix} 2v_1 + v_2 + 3v_3 \\ 4v_1 + 0v_2 + 5v_3 \end{bmatrix} = v_1 \begin{bmatrix} 2 \\ 4 \end{bmatrix} + v_2 \begin{bmatrix} 1 \\ 0 \end{bmatrix} + v_3 \begin{bmatrix} 3 \\ 5 \end{bmatrix} \\ &= v_1A_1 + v_2A_2 + v_3A_3 \end{aligned}$$

■

Notice that if $A = [A_1, A_2, \ldots, A_n]$ is an $n \times n$ matrix and v and b are n-dimensional vectors then, by Theorem 8.19, the matrix equation $Av = b$ can be expressed as $v_1A_1 + v_2A_2 + \cdots + v_nA_n = b$.

Example 38

Solve

$$v_1 \begin{bmatrix} 1 \\ 2 \\ 3 \end{bmatrix} + v_2 \begin{bmatrix} -1 \\ -2 \\ 3 \end{bmatrix} + v_3 \begin{bmatrix} 3 \\ 3 \\ 2 \end{bmatrix} = \begin{bmatrix} -13 \\ -16 \\ 1 \end{bmatrix}$$

Analysis

The given equation is equivalent to the matrix equation $Av = b$, where

$$A = \begin{bmatrix} 1 & -1 & 3 \\ 2 & -2 & 3 \\ 3 & 3 & 2 \end{bmatrix}, \quad v = \begin{bmatrix} v_1 \\ v_2 \\ v_3 \end{bmatrix}, \quad \text{and } b = \begin{bmatrix} -13 \\ -16 \\ 1 \end{bmatrix}$$

Thus we can set up the augmented matrix and use Gaussian elimination and backsolving to solve the given system.

$$\left[\begin{array}{ccc|c} 1 & -1 & 3 & -13 \\ 2 & -2 & 3 & -16 \\ 3 & 3 & 2 & 1 \end{array}\right] \underset{-2R_1 + R_2}{} \left[\begin{array}{ccc|c} 1 & -1 & 3 & -13 \\ 0 & 0 & -3 & 10 \\ 3 & 3 & 2 & 1 \end{array}\right] \underset{-3R_1 + R_3}{} \left[\begin{array}{ccc|c} 1 & -1 & 3 & -13 \\ 0 & 0 & -3 & 10 \\ 0 & 6 & -7 & 40 \end{array}\right]$$

$$\begin{array}{l}\text{interchange} \\ R_2 \,\&\, R_3\end{array} \left[\begin{array}{ccc|c} 1 & -1 & 3 & -13 \\ 0 & 6 & -7 & 40 \\ 0 & 0 & -3 & 10 \end{array}\right] \xrightarrow{\frac{1}{6}R_2} \left[\begin{array}{ccc|c} 1 & -1 & 3 & -13 \\ 0 & 1 & -\frac{7}{6} & \frac{20}{3} \\ 0 & 0 & -3 & 10 \end{array}\right] \xrightarrow{-\frac{1}{3}R_3} \left[\begin{array}{ccc|c} 1 & -1 & 3 & -13 \\ 0 & 1 & -\frac{7}{6} & \frac{20}{3} \\ 0 & 0 & 1 & -\frac{10}{3} \end{array}\right]$$

We see that $v_3 = -10/3$, and backsolving we find that $v_2 = 25/9$ and $v_1 = -2/9$. ■

The previous example illustrates that the system of linear equations

$$\begin{aligned} a_{11}x_1 + a_{12}x_2 + \cdots + a_{1n}x_n &= b_1 \\ a_{21}x_1 + a_{22}x_2 + \cdots + a_{2n}x_n &= b_2 \\ \vdots \qquad \vdots \qquad\quad & \quad \vdots \qquad \vdots \\ a_{m1}x_1 + a_{m2}x_2 + \cdots + a_{mn}x_n &= b_m \end{aligned}$$

is consistent if and only if $b = (b_1, b_2, \ldots, b_m)$ is a linear combination of the columns

$$\begin{bmatrix} a_{11} \\ a_{21} \\ \vdots \\ a_{m1} \end{bmatrix} \begin{bmatrix} a_{12} \\ a_{22} \\ \vdots \\ a_{m2} \end{bmatrix} \cdots \begin{bmatrix} a_{1n} \\ a_{2n} \\ \vdots \\ a_{mn} \end{bmatrix}$$

Example 39

Let $A_1 = \begin{bmatrix} 0 \\ -2 \\ 1 \end{bmatrix}$, $A_2 = \begin{bmatrix} 2 \\ 2 \\ -1 \end{bmatrix}$, $A_3 = \begin{bmatrix} 1 \\ 1 \\ 1 \end{bmatrix}$, and $b = \begin{bmatrix} 3 \\ 4 \\ 1 \end{bmatrix}$.

Write b as a linear combination of A_1, A_2, and A_3.

Analysis

If $A = [A_1, A_2, A_3]$, then

$$A = \begin{bmatrix} 0 & 2 & 1 \\ -2 & 2 & 1 \\ 1 & -1 & 1 \end{bmatrix}$$

and expressing b as a linear combination of A_1, A_2, and A_3 is equivalent to solving the linear system represented by the matrix equation $Av = b$.

The augmented matrix for this equation is

$$\left[\begin{array}{rrr|r} 0 & 2 & 1 & 3 \\ -2 & 2 & 1 & 4 \\ 1 & -1 & 1 & 1 \end{array}\right]$$

and we can use Gaussian elimination and backsolving to solve this system. The solution is $v_1 = -1/2$, $v_2 = 1/2$, and $v_3 = 2$. Therefore $b = -A_1/2 + A_2/2 + 2A_3$. ■

Definition

An $n \times n$ matrix A is **nonsingular** if the matrix equation $Av = \theta$ has the unique solution $v = \theta$. The matrix A is **singular** if it is not nonsingular.

Example 40

Determine whether the matrix

$$A = \begin{bmatrix} 1 & -2 & 4 \\ 0 & -1 & 5 \\ 2 & 3 & -4 \end{bmatrix}$$

is nonsingular.

Analysis

We answer this question by solving the system $Av = \theta$.

$$\left[\begin{array}{rrr|r} 1 & -2 & 4 & 0 \\ 0 & -1 & 5 & 0 \\ 2 & 3 & -4 & 0 \end{array}\right] \quad \underset{-2R_1 + R_3}{} \left[\begin{array}{rrr|r} 1 & -2 & 4 & 0 \\ 0 & -1 & 5 & 0 \\ 0 & 7 & -12 & 0 \end{array}\right] \quad \underset{-1 \cdot R_2}{} \left[\begin{array}{rrr|r} 1 & -2 & 4 & 0 \\ 0 & 1 & -5 & 0 \\ 0 & 7 & -12 & 0 \end{array}\right]$$

$$\underset{-7R_2 + R_3}{} \left[\begin{array}{rrr|r} 1 & -2 & 4 & 0 \\ 0 & 1 & -5 & 0 \\ 0 & 0 & 23 & 0 \end{array}\right]$$

We see that the only solution is $v = (0,0,0)$. Therefore A is nonsingular. ■

Theorem 8.20 states, without proof, several characterizations of nonsingular matrices.

Theorem 8.20

Let $A = [A_1, A_2, \ldots, A_n]$ be an $n \times n$ matrix. Then the following statements are equivalent:

a. A is nonsingular.
b. $\{A_1, A_2, \ldots, A_n\}$ is a linearly independent set of vectors.
c. For each n-dimensional vector v, the equation $Av = b$ has a unique solution.
d. $\det(A) \neq 0$.
e. A is invertible.

Example 41

Find a 3×3 singular matrix.

Analysis

By Theorem 8.20, we know that a 3×3 matrix $A = [A_1, A_2, A_3]$ is singular if and only if $\{A_1, A_2, A_3\}$ is a linearly dependent set of vectors. Therefore we can find a 3×3 singular matrix A by letting A_1 and A_2 be arbitrary 3-dimensional vectors and then taking A_3 to be any 3-dimensional vector that can be expressed as a linear combination of A_1 and A_2. Let $A_1 = (0,1,2)$, $A_2 = (2,1,0)$, and $A_3 = 2A_1 + (-3)A_2 = (0,2,4) + (-6,-3,0) = (-6,-1,4)$. Since we did not prove Theorem 8.20, we solve the equation $Av = \theta$ in order to see that A is singular:

$$\begin{bmatrix} 0 & 2 & -6 & 0 \\ 1 & 1 & -1 & 0 \\ 2 & 0 & 4 & 0 \end{bmatrix} \xrightarrow{\text{interchange } R_1 \,\&\, R_2} \left[\begin{array}{ccc|c} 1 & 1 & -1 & 0 \\ 0 & 2 & -6 & 0 \\ 2 & 0 & 4 & 0 \end{array}\right] \xrightarrow{-2R_1 + R_3} \left[\begin{array}{ccc|c} 1 & 1 & -1 & 0 \\ 0 & 2 & -6 & 0 \\ 0 & -2 & 6 & 0 \end{array}\right]$$

$$\xrightarrow{\frac{1}{2}R_2} \left[\begin{array}{ccc|c} 1 & 1 & -1 & 0 \\ 0 & 1 & -3 & 0 \\ 0 & -2 & 6 & 0 \end{array}\right] \xrightarrow{2R_2 + R_3} \left[\begin{array}{ccc|c} 1 & 1 & -1 & 0 \\ 0 & 1 & -3 & 0 \\ 0 & 0 & 0 & 0 \end{array}\right]$$

Since we have a row of 0s, we know that the equation $Av = \theta$ has an infinite number of solutions. In particular, v_3 can be any real number, $v_2 = 3v_3$, and $v_1 = -2v_3$. Thus, for example, $(-2,3,1)$ is a nonzero solution of $Av = \theta$. Hence A is singular. ■

Definition

Let V be a subspace of $\mathbb{R}^n$ and let $S = \{v_1, v_2, \ldots, v_r\}$ be a subset of V. We say that S **spans** V if every vector in V can be expressed as a linear combination of vectors in S.

Example 42

Let $S = \{v_1,v_2,v_3\}$, where

$$v_1 = \begin{bmatrix}1\\1\\1\end{bmatrix}, \quad v_2 = \begin{bmatrix}1\\2\\2\end{bmatrix}, \quad \text{and} \quad v_3 = \begin{bmatrix}1\\2\\3\end{bmatrix}$$

Determine whether S spans $\mathbb{R}^3$.

Analysis

We must determine whether an arbitrary vector $b = (b_1,b_2,b_3)$ in $\mathbb{R}^3$ can be expressed as a linear combination of v_1, v_2, and v_3; that is, we must determine whether the equation $x_1v_1 + x_2v_2 + x_3v_3 = b$ has a solution. This equation is equivalent to the matrix equation $Ax = b$, where $A = [v_1,v_2,v_3]$ and $x = (x_1,x_2,x_3)$. The augmented matrix is

$$\left[\begin{array}{ccc|c} 1 & 1 & 1 & b_1 \\ 1 & 2 & 2 & b_2 \\ 1 & 2 & 3 & b_3 \end{array}\right]$$

and this matrix is equivalent to

$$\left[\begin{array}{ccc|c} 1 & 1 & 1 & b_1 \\ 0 & 1 & 1 & b_2 - b_1 \\ 0 & 0 & 1 & b_3 - b_2 \end{array}\right]$$

Backsolving yields $x_3 = b_3 - b_2$, $x_2 = 2b_2 - b_1 - b_3$, and $x_1 = 2b_1 - b_2$. Therefore S spans $\mathbb{R}^3$. ■

Definition

Let V be a subspace of $\mathbb{R}^n$ such that $V \neq \{\theta\}$. A **basis** for V is a linearly independent set that spans V.

The concept of basis is not meaningful for $\{\theta\}$, because $\{\theta\}$ contains only the zero vector θ, and, although $\{\theta\}$ spans $\{\theta\}$, $\{\theta\}$ is linearly dependent.

Example 43

The unit vectors $e_1 = (1,0,0)$, $e_2 = (0,1,0)$, and $e_3 = (0,0,1)$ form a basis for $\mathbb{R}^3$.

Analysis

If $b = (b_1,b_2,b_3)$ is any vector in $\mathbb{R}^3$, then $b = b_1e_1 + b_2e_2 + b_3e_3$. Therefore $\{e_1,e_2,e_3\}$ spans $\mathbb{R}^3$. If $A = [e_1,e_2,e_3]$, then the only solution of the matrix equation $Av = \theta$ is θ. This is easy to see because the augmented matrix is

$$\left[\begin{array}{ccc|c} 1 & 0 & 0 & 0 \\ 0 & 1 & 0 & 0 \\ 0 & 0 & 1 & 0 \end{array}\right]$$

Therefore $\{e_1, e_2, e_3\}$ is a basis for $\mathbb{R}^3$. ■

We state, without proof, the following theorem.

Theorem 8.21

If $S = \{v_1, v_2, \ldots, v_r\}$ is a basis for a subspace V of $\mathbb{R}^n$, then every basis for V contains r vectors.

Definition

The **dimension** of a subspace V of $\mathbb{R}^n$ is the number of vectors in a basis for V.

By Example 43, the dimension of $\mathbb{R}^3$ is 3.

Definition

Let A be an $n \times n$ matrix. The **null space** of A, denoted by $N(A)$, is defined to be $\{v \in \mathbb{R}^n: Av = \theta\}$.

Example 44

Describe the null space of the matrix $A = \begin{bmatrix} 1 & -2 & 1 \\ 2 & -3 & 5 \\ 1 & 0 & 7 \end{bmatrix}$.

Analysis

The null space of A is determined by solving the matrix equation $Av = \theta$. The augmented matrix is

$$\left[\begin{array}{ccc|c} 1 & -2 & 1 & 0 \\ 2 & -3 & 5 & 0 \\ 1 & 0 & 7 & 0 \end{array}\right]$$

and this matrix is equivalent to

$$\left[\begin{array}{rrr|r} 1 & -2 & 1 & 0 \\ 0 & 1 & 3 & 0 \\ 0 & 0 & 0 & 0 \end{array}\right]$$

Thus a vector $v = (v_1,v_2,v_3)$ is in the null space of A if and only if $v_2 = -3v_3$ and $v_1 = -7v_3$. Therefore N(A) consists of all vectors that can be written in the form $(-7v_3,-3v_3,v_3)$. ■

Theorem 8.22

If A is an $m \times n$ matrix, then N(A) is a subspace of $\mathbb{R}^n$.

Definition

Let A be an $m \times n$ matrix. The range of A, denoted by R(A), is defined to be $\{u \in \mathbb{R}^m\colon u = Av \text{ for some } v \in \mathbb{R}^n\}$.

Example 45

Describe the range of the matrix in Example 44.

Analysis

Let b be a vector in $\mathbb{R}^3$. Then b is in R(A) if and only if the matrix equation $Av = b$ has a solution. The augmented matrix is

$$\left[\begin{array}{rrr|l} 1 & -2 & 1 & b_1 \\ 2 & -3 & 5 & b_2 \\ 1 & 0 & 7 & b_3 \end{array}\right]$$

and this matrix is equivalent to

$$\left[\begin{array}{rrr|l} 1 & -2 & 1 & b_1 \\ 0 & 1 & 3 & b_2 - 2b_1 \\ 0 & 0 & 0 & b_3 - 2b_2 + 3b_1 \end{array}\right]$$

Hence $b = (b_1,b_2,b_3)$ is in R(A) if and only if $b_3 - 2b_2 + 3b_1 = 0$. Thus R(A) consists of all vectors that can be written in the form $(b_1,b_2,2b_2 - 3b_1)$. ■

Definition

Let $A = [A_1,A_2, \ldots ,A_n]$ be an $m \times n$ matrix. The **column space** of A is the subspace of $\mathbb{R}^m$ spanned by $\{A_1,A_2, \ldots ,A_n\}$. The rows $B_1,B_2, \ldots ,B_m$ of A can be regarded as n-dimensional vectors and the **row space** of A is the subspace of $\mathbb{R}^n$ spanned by $\{B_1,B_2, \ldots ,B_m\}$.

Theorem 8.23 Let A be an $m \times n$ matrix. The column space of A is the range of A.

Definition

Let A be an $m \times n$ matrix. The dimension of the null space of A is called the **nullity** of A, and the dimension of the range of A is called the **rank** of A.

Example 46

Find the dimension, nullity, and rank of the row space of the matrix

$$A = \begin{bmatrix} 1 & 1 & 1 & 3 \\ -2 & -1 & 0 & -8 \\ 0 & 1 & 2 & 2 \end{bmatrix}$$

Analysis

The matrix A is equivalent to the matrix

$$B = \begin{bmatrix} 1 & 1 & 1 & 3 \\ 0 & 1 & 2 & -2 \\ 0 & 0 & 0 & 4 \end{bmatrix}$$

Since the rows of B (considered as vectors) are a linearly independent set (if one of them could be expressed in terms of the other two, then one row would consist of 0s), the three rows form a basis for the row space of A. Therefore the dimension of the row space of A is 3.

Since the matrix equation $Av = \theta$ is equivalent to $Bv = \theta$, the null space of A can be found by backsolving $Bv = \theta$. Thus $4v_4 = 0$, or $v_4 = 0$, $v_2 + 2v_3 = 0$, and $v_1 + v_2 + v_3 = 0$. Therefore $v_2 = -2v_3$ and $v_1 = -v_2 - v_3 = -(-2v_3) - v_3 = v_3$. Therefore the null space of A consists of all vectors that can be written in the form $(v_3, -2v_3, v_3, 0)$. Thus the nullity of A is 1, since the vector $(1, -2, 1, 0)$ forms a basis for $N(A)$.

Since the range of A is the column space, we can find the range of A by finding the row space of

$$A^t = \begin{bmatrix} 1 & -2 & 0 \\ 1 & -1 & 1 \\ 1 & 0 & 2 \\ 3 & -8 & 2 \end{bmatrix}$$

The matrix A^t is equivalent to the matrix

$$C = \begin{bmatrix} 1 & -2 & 0 \\ 0 & 1 & 1 \\ 0 & 0 & 4 \\ 0 & 0 & 0 \end{bmatrix}$$

The three nonzero rows (considered as vectors) of C form a basis for the row space of A^t. Therefore the rank of A is 3. ■

Exercises 8.6

59. Let V be the subset of $\mathbb{R}^3$ defined by $V = \{(v_1,v_2,v_3) \in \mathbb{R}^3\colon v_2 = 3v_1 \text{ and } v_3 = v_1 + 5\}$. Show that there exist two vectors u and v in V such that $u + v \notin V$.

60. Verify that the subset V of $\mathbb{R}^3$ defined by $V = \{(v_1,v_2,v_3) \in \mathbb{R}^3\colon v_2 = -2v_1 \text{ and } v_3 = 3v_2\}$ is a subspace of $\mathbb{R}^3$.

61. Show that the vector $(2,-1,0)$ is a linear combination of $(1,1,1)$ and $(1,2,0)$.

62. Show that the vector $(0,1,1)$ is not a linear combination of $(1,1,1)$ and $(1,2,0)$.

63. Which of the following vectors are linear combinations of $(1,2,0)$ and $(0,3,2)$? Prove your answers.

a. $(3,3,-2)$ **b.** $(1,0,0)$ **c.** $(-1,1,-2)$ **d.** $(1,1,1)$

64. Which of the following sets are linearly independent subsets of $\mathbb{R}^3$? Prove your answers.

a. $\{(1,2,0),(8,7,0),(2,1,0)\}$ **b.** $\{(1,1,0),(0,1,1),(1,0,1)\}$

c. $\{(1,2,3),(2,3,4),(3,4,5)\}$ **d.** $\{(1,0,2),(0,1,0),(1,1,2)\}$

65. Solve the following systems.

a. $$v_1\begin{bmatrix} 1 \\ 1 \\ 0 \end{bmatrix} + v_2\begin{bmatrix} 0 \\ 1 \\ 2 \end{bmatrix} + v_3\begin{bmatrix} 2 \\ 3 \\ 4 \end{bmatrix} = \begin{bmatrix} -2 \\ -5 \\ -10 \end{bmatrix}$$

b. $$v_1\begin{bmatrix} 1 \\ 1 \\ 1 \end{bmatrix} + v_2\begin{bmatrix} 1 \\ 2 \\ 2 \end{bmatrix} + v_3\begin{bmatrix} 1 \\ 3 \\ 3 \end{bmatrix} = \begin{bmatrix} 1 \\ 4 \\ 4 \end{bmatrix}$$

c. $$v_1\begin{bmatrix} 1 \\ -1 \\ 2 \end{bmatrix} + v_2\begin{bmatrix} 2 \\ 1 \\ -1 \end{bmatrix} + v_3\begin{bmatrix} 5 \\ 0 \\ -1 \end{bmatrix} = \begin{bmatrix} 1 \\ 1 \\ 1 \end{bmatrix}$$

66. Let $A_1 = \begin{bmatrix} 0 \\ 2 \\ 1 \end{bmatrix}$, $A_2 = \begin{bmatrix} 2 \\ 1 \\ 4 \end{bmatrix}$, and $A_3 = \begin{bmatrix} 1 \\ 1 \\ 1 \end{bmatrix}$

Write each of the following vectors as a linear combination of A_1, A_2, and A_3.

a. $\begin{bmatrix} 0 \\ 1 \\ 4 \end{bmatrix}$ **b.** $\begin{bmatrix} 0 \\ 2 \\ 4 \end{bmatrix}$ **c.** $\begin{bmatrix} 4 \\ 1 \\ 6 \end{bmatrix}$ **d.** $\begin{bmatrix} 0 \\ 3 \\ 2 \end{bmatrix}$ **e.** $\begin{bmatrix} 2 \\ 1 \\ 3 \end{bmatrix}$

67. Determine whether each of the following matrices is nonsingular.

a. $\begin{bmatrix} 3 & 2 & 1 \\ 2 & -1 & -1 \\ 1 & 4 & 0 \end{bmatrix}$ **b.** $\begin{bmatrix} 1 & 1 & -1 \\ 2 & 1 & 1 \\ -1 & 1 & -5 \end{bmatrix}$ **c.** $\begin{bmatrix} 1 & 7 & -2 \\ 3 & 0 & 4 \\ 2 & -1 & -3 \end{bmatrix}$

68. **a.** Find a 2×2 singular matrix.

b. Find a 4×4 singular matrix.

69. Determine whether each of the following sets spans $\mathbb{R}^3$.

a. $S = \left\{ \begin{bmatrix} 1 \\ 1 \\ 1 \end{bmatrix}, \begin{bmatrix} 1 \\ 2 \\ 2 \end{bmatrix}, \begin{bmatrix} 1 \\ 5 \\ 6 \end{bmatrix} \right\}$ **b.** $S = \left\{ \begin{bmatrix} 0 \\ 2 \\ 1 \end{bmatrix}, \begin{bmatrix} 3 \\ 1 \\ 4 \end{bmatrix}, \begin{bmatrix} -6 \\ 6 \\ -4 \end{bmatrix} \right\}$

c. $S = \left\{ \begin{bmatrix} 1 \\ -1 \\ 2 \end{bmatrix}, \begin{bmatrix} 2 \\ 1 \\ -1 \end{bmatrix}, \begin{bmatrix} 0 \\ 1 \\ 1 \end{bmatrix} \right\}$

70. Describe the null space and the range of each of the following matrices.

a. $\begin{bmatrix} 1 & 2 \\ 3 & 2 \end{bmatrix}$ **b.** $\begin{bmatrix} 1 & 2 \\ 2 & 4 \end{bmatrix}$ **c.** $\begin{bmatrix} 1 & 2 & 0 \\ -1 & 1 & 1 \\ 2 & -1 & 1 \end{bmatrix}$ **d.** $\begin{bmatrix} 0 & 2 & 1 \\ 3 & 1 & 4 \\ -6 & 6 & -4 \end{bmatrix}$

71. Find the nullity, rank, and dimension of the row space of each of the following matrices.

a. $\begin{bmatrix} 1 & -1 & 1 \\ 2 & 0 & 5 \\ 1 & 3 & 7 \end{bmatrix}$ **b.** $\begin{bmatrix} 0 & 1 & 2 \\ -1 & 3 & -2 \\ -3 & 4 & -4 \end{bmatrix}$ **c.** $\begin{bmatrix} 2 & 1 & 0 & 1 \\ 2 & 2 & 4 & 4 \\ 5 & 2 & -2 & 1 \end{bmatrix}$

8.7

EIGENVALUES AND EIGENVECTORS

In this section we introduce the concepts of eigenvalues and eigenvectors. These important concepts are used in many branches of mathematics. One way in which they are used in discrete mathematics is in finding the solutions of optimization problems.

Definition

> Let A be an $n \times n$ matrix. An n-dimensional nonzero vector v is an **eigenvector** of A if there is a scalar λ such that $Av = \lambda v$. The scalar λ is called an **eigenvalue** of A and v is an **eigenvector corresponding to λ.** Note that the zero vector is not an eigenvector.

Note that the product Av is the product of two matrices, and the result is an $n \times 1$ matrix. The product λv is the *scalar product* obtained by multiplying the n-dimensional vector $v = (v_1, v_2, \ldots, v_n)$ by the real number λ. Thus $\lambda v = (\lambda v_1, \lambda v_2, \ldots, \lambda v_n)$. But remember, an n-dimensional vector can be written as an $n \times 1$ matrix.

Example 47

Let $A = \begin{bmatrix} 9 & 0 \\ 3 & -2 \end{bmatrix}$. Find a scalar λ and a 2-dimensional nonzero vector v such that $Av = \lambda v$.

Analysis

Let $v = (v_1, v_2)$. Then

$$Av = \begin{bmatrix} 9v_1 \\ 3v_1 - 2v_2 \end{bmatrix}$$

and $\lambda v = (\lambda v_1, \lambda v_2)$. So $9v_1 = \lambda v_1$, and $3v_1 - 2v_2 = \lambda v_2$. Therefore $\lambda = 9$ and $3v_1 - 2v_2 = 9v_2$. Hence $3v_1 = 11v_2$, or $v_1 = 11v_2/3$. Hence $\lambda = 9$ is a scalar and $v = (11,3)$ is a nonzero vector such that $Av = \lambda v$. ■

The geometric interpretation of eigenvalues and eigenvectors in the plane is that, if λ is an eigenvalue of A corresponding to v, then since $Av = \lambda v$ multiplication by A dilates v if $\lambda > 1$, contracts v if $0 < \lambda < 1$, and reverses the direction if $\lambda < 0$ (see Figure 8.8).

Figure 8.8

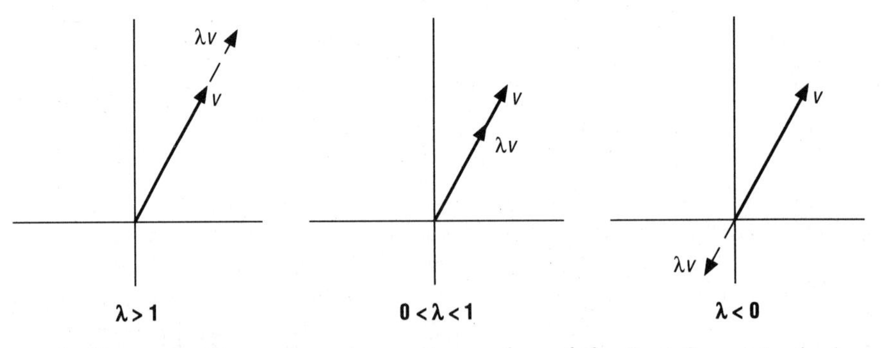

In Example 47, we found one eigenvalue of the 2×2 matrix A. An

$n \times n$ matrix A may, however, have more than one eigenvalue. To find the eigenvalues of A, we write $Av = \lambda v$ as $Av = \lambda Iv$, where I is the $n \times n$ identity matrix, or equivalently as $(\lambda I - A)v = \theta$, where $v \neq \theta$. The scalar λ is an eigenvalue of A if and only if the equation $(\lambda I - A)v = \theta$ has a nonzero solution. By Theorem 8.20, the latter is true if and only if $\det(\lambda I - A) = 0$.

Definition

The equation $\det(\lambda I - A) = 0$ is called the **characteristic equation** of A. The scalars λ that satisfy this equation are the eigenvalues of A. When $\det(\lambda I - A)$ is expanded, it is a polynomial in λ, and this polynomial is called the **characteristic polynomial** of A.

It can be shown that if A is an $n \times n$ matrix then the characteristic polynomial of A has degree n and the coefficient of λ^n is 1. Therefore there are real numbers $c_1, c_2, \ldots, c_n$ such that $\det(\lambda I - A) = \lambda^n + c_1\lambda^{n-1} + \cdots + c_n$.

Example 48

Find the eigenvalues of the matrix $A = \begin{bmatrix} 1 & 2 \\ -1 & 4 \end{bmatrix}$.

Analysis

Since

$$\lambda I - A = \lambda \begin{bmatrix} 1 & 0 \\ 0 & 1 \end{bmatrix} - \begin{bmatrix} 1 & 2 \\ -1 & 4 \end{bmatrix} = \begin{bmatrix} \lambda - 1 & -2 \\ 1 & \lambda - 4 \end{bmatrix}$$

the characteristic polynomial of A is

$$\det\left(\begin{bmatrix} \lambda - 1 & -2 \\ 1 & \lambda - 4 \end{bmatrix}\right) = \lambda^2 - 5\lambda + 6$$

Thus the characteristic equation of A is $\lambda^2 - 5\lambda + 6 = 0$. The solutions of this equation are $\lambda = 2$ and $\lambda = 3$, and hence the eigenvalues of A are 2 and 3. ■

Once we have found an eigenvalue λ of A, we can find an eigenvector v of A by solving the equation $(\lambda I - A)v = \theta$.

Example 49

For the matrix A in Example 48 find the eigenvectors corresponding to $\lambda = 2$ and $\lambda = 3$.

Analysis

The eigenvectors corresponding to $\lambda = 2$ are the nonzero vectors v such that $(2I - A)\,v = \theta$. Since

$$2I - A = \begin{bmatrix} 2 & 0 \\ 0 & 2 \end{bmatrix} - \begin{bmatrix} 1 & 2 \\ -1 & 4 \end{bmatrix} = \begin{bmatrix} 1 & -2 \\ 1 & -2 \end{bmatrix}$$

we need to solve the equation

$$\begin{bmatrix} 1 & -2 \\ 1 & -2 \end{bmatrix} \begin{bmatrix} v_1 \\ v_2 \end{bmatrix} = \theta$$

The solution is given by $v_1 = 2v_2$. Thus all nonzero solutions v of $(2I - A)v = \theta$ are of the form

$$v = \begin{bmatrix} 2v_2 \\ v_2 \end{bmatrix}$$

The eigenvectors corresponding to $\lambda = 3$ are the nonzero vectors v such that $(3I - A)v = \theta$. The solutions of this equation are found by solving the equation

$$\begin{bmatrix} 2 & -2 \\ 1 & -1 \end{bmatrix} \begin{bmatrix} v_1 \\ v_2 \end{bmatrix} = \theta$$

The solution is given by $v_1 = v_2$. Thus all nonzero solutions v of $(3I - A)v = \theta$ are of the form

$$v = \begin{bmatrix} v_2 \\ v_2 \end{bmatrix}$$

We list in summary form the eigenvalues and the corresponding eigenvectors of A:

eigenvalue $\lambda = 2$ eigenvectors $v = a\begin{bmatrix} 2 \\ 1 \end{bmatrix}$, $a \neq 0$

eigenvalue $\lambda = 3$ eigenvectors $v = a\begin{bmatrix} 1 \\ 1 \end{bmatrix}$, $a \neq 0$ ■

Theorem 8.24

If A is an $n \times n$ matrix, then the following statements are equivalent:

a. λ is an eigenvalue of A.
b. The equation $(\lambda I - A)v = \theta$ has nonzero solutions.
c. There is a nonzero vector v in $\mathbb{R}^n$ such that $Av = \lambda v$.
d. λ is a solution of the characteristic equation of A.

Example 50

Find the eigenvalues of the matrix $A = \begin{bmatrix} -2 & 0 & 1 \\ -6 & -2 & 0 \\ 19 & 5 & -4 \end{bmatrix}$.

Analysis

As in Example 48, the characteristic polynomial of A is

$$\det(\lambda I - A) = \det\left(\begin{bmatrix} \lambda+2 & 0 & -1 \\ 6 & \lambda+2 & 0 \\ -19 & -5 & \lambda+4 \end{bmatrix}\right) = \lambda^3 + 8\lambda^2 + \lambda + 8$$

Thus the eigenvalues of A must satisfy the characteristic equation of A: $\lambda^3 + 8\lambda^2 + \lambda + 8 = 0$. This is a cubic equation, and to solve it we begin by searching for integer solutions. A fact that is useful is that all integer solutions (if there are any) of a polynomial equation with integral coefficients and leading coefficient 1 must be divisors of the constant term. Therefore any integer solution of this characteristic equation must be a divisor of 8, so the possibilities are ± 1, ± 2, ± 4, and ± 8. We can immediately see that no positive integer is a solution, because any positive value of λ yields a positive number. We begin checking the negative possibilities:

$$(-1)^3 + 8(-1)^2 + (-1) + 8 = -1 + 8 - 1 + 8 = 14$$
$$(-2)^3 + 8(-2)^2 + (-2) + 8 = -8 + 32 - 2 + 8 = 30$$
$$(-4)^3 + 8(-4)^2 + (-4) + 8 = -64 + 128 - 4 + 8 = 68$$
$$(-8)^3 + 8(-8)^2 + (-8) + 8 = 0$$

We divide $\lambda^3 + 8\lambda^2 + \lambda + 8$ by $\lambda - (-8)$ and obtain $\lambda^3 + 8\lambda^2 + \lambda + 8 = (\lambda + 8)(\lambda^2 + 1)$. Thus we want to solve the equation $(\lambda + 8)(\lambda^2 + 1) = 0$. Since we are dealing with real numbers and $\lambda^2 + 1 = 0$ does not have a real solution, $\lambda + 8 = 0$ and hence $\lambda = -8$ is the only eigenvalue of A. ■

The computer can be used to find or approximate the roots of a polynomial equation. Thus the examples and exercises in this section are usually constructed so that the characteristic equation has some integer solutions.

Some of the theorems previously stated can be used to prove the following theorem. You are asked to prove this theorem in Exercises 80, 81, and 82.

Theorem 8.25

> Let $A = [a_{ij}]$ be an $n \times n$ matrix. Then
>
> **a.** A and A^t have the same eigenvalues.
> **b.** A is singular if and only if $\lambda = 0$ is an eigenvalue of A.
> **c.** If A is triangular, then the eigenvalues of A are $a_{11}, a_{22}, \ldots, a_{nn}$.

The proof of Theorem 8.26 is left as Exercises 83, 84, and 85. It tells us the eigenvalues of certain matrices associated with a matrix A if we know the eigenvalues of A.

Theorem 8.26

> Let λ be an eigenvalue of the $n \times n$ matrix A. Then
>
> **a.** For each $n \in \mathbb{N}$, λ^n is an eigenvalue of A^n.
> **b.** If A is nonsingular, then $1/\lambda$ is an eigenvalue of A^{-1}.
> **c.** If a is a scalar, then $\lambda + a$ is an eigenvalue of $aI + A$.

Example 51

Let A be the matrix in Example 50. Find the eigenvalues of A^3, A^{-1}, and $5I + A$.

Analysis

We use Theorem 8.26. The only eigenvalue of A is -8. Thus the eigenvalue of A^3 is $(-8)^3 = -512$. The eigenvalue of A^{-1} is $-1/8$, and the eigenvalue of $5I + A$ is $-8 + 5 = -3$. ■

Solving a polynomial equation is a difficult task because there are no formulas for the roots of polynomials of degree 5 or higher. In general, the best we can do is find a good approximation. However, once we have found the eigenvalues, we can find the eigenvectors in a finite number of steps.

As Example 49 illustrates, the nonzero solutions of $(\lambda I - A)v = \theta$ are the eigenvectors of A that correspond to the eigenvalue λ. Thus we can use Gaussian elimination to find the eigenvectors that correspond to a given eigenvalue.

Example 52

Find the eigenvectors that correspond to the eigenvalues of the matrix A in Example 50.

Analysis

The only eigenvalue of A is -8. The eigenvectors that correspond to -8 are the nonzero solutions of the equation $-8I - A = \theta$. This matrix equation is represented by the augmented matrix

$$\left[\begin{array}{rrr|r} -6 & 0 & -1 & 0 \\ 6 & -6 & 0 & 0 \\ -19 & -5 & -4 & 0 \end{array}\right]$$

The solutions of this system are $v_2 = v_1$ and $v_3 = -6v_1$. Thus the eigenvectors of A that correspond to the eigenvalue $\lambda = -8$ are of the form

$$\begin{bmatrix} v_1 \\ v_1 \\ -6v_1 \end{bmatrix}$$

that is, the eigenvectors are of the form

$$a\begin{bmatrix} 1 \\ 1 \\ -6 \end{bmatrix}$$

where $a \neq 0$. ■

Definition

Let A be an $n \times n$ matrix. If λ is an eigenvalue of A, then the eigenvectors of A corresponding to λ are the nonzero vectors in the null space of $\lambda I - A$. The null space of $\lambda I - A$ is called the **eigenspace** of λ and is denoted E_λ. The **geometric multiplicity** of λ is the dimension of E_λ.

If A is the matrix in Examples 50 and 52, then the eigenspace of -8, E_{-8}, is all vectors of the form

$$a\begin{bmatrix} 1 \\ 1 \\ -6 \end{bmatrix}$$

The geometric multiplicity of $\lambda = -8$ is 1.

We conclude this chapter by stating, without proof, the following theorem.

Theorem 8.27 Let A be an $n \times n$ matrix, and let $\lambda_1,\lambda_2, \ldots ,\lambda_r$ be the distinct eigenvalues of A. For each $i = 1,2, \ldots ,r$, let v_i be an eigenvector of A corresponding to λ_i. Then $\{v_1,v_2, \ldots ,v_r\}$ is a linearly independent set.

Exercises 8.7

72. Find the eigenvalues of each of the following matrices.

a. $\begin{bmatrix} 5 & 1 \\ 4 & 8 \end{bmatrix}$ **b.** $\begin{bmatrix} 2 & -1 \\ -1 & 2 \end{bmatrix}$ **c.** $\begin{bmatrix} 4 & 3 \\ 1 & 2 \end{bmatrix}$ **d.** $\begin{bmatrix} 3 & 2 \\ 2 & 0 \end{bmatrix}$

73. For each matrix in Exercise 72, find the eigenvectors corresponding to each eigenvalue.

74. Write the characteristic polynomial of each of the following matrices.

a. $\begin{bmatrix} 0 & 0 & 2 \\ 0 & 1 & 0 \\ 2 & 0 & 0 \end{bmatrix}$ **b.** $\begin{bmatrix} 5 & 1 & -7 \\ 0 & 1 & 1 \\ 1 & 0 & 0 \end{bmatrix}$ **c.** $\begin{bmatrix} 2 & 0 & 0 \\ 4 & 1 & 1 \\ 4 & -1 & 3 \end{bmatrix}$

75. Find the eigenvalues of each of the matrices in Exercise 74.

76. For each matrix in Exercise 74, find the eigenvectors corresponding to each eigenvalue.

77. Let A be the matrix in Exercise 74b. Find the eigenvalues of A^5, A^{-1}, and $3I + A$.

78. Find the eigenvalues of each of the following matrices.

a. $\begin{bmatrix} -7 & 8 & 32 \\ 4 & -3 & -16 \\ -3 & 3 & 13 \end{bmatrix}$ **b.** $\begin{bmatrix} 2 & -2 & 0 \\ 0 & 4 & 0 \\ 2 & 5 & 1 \end{bmatrix}$ **c.** $\begin{bmatrix} 1 & 1 & 1 \\ 0 & 3 & 3 \\ -2 & 1 & 1 \end{bmatrix}$

79. For each matrix in Exercise 78, find the eigenspace and the geometric multiplicity of each eigenvalue.

80. Let A be an $n \times n$ matrix. Prove that A and A^t have the same eigenvalues.

81. Let A be an $n \times n$ matrix. Prove that A is singular if and only if $\lambda = 0$ is an eigenvalue of A.

82. Let $A = [a_{ij}]$ be an $n \times n$ triangular matrix. Prove that the eigenvalues of A are $a_{11},a_{22}, \ldots ,a_{nn}$.

83. Prove by induction that, if λ is an eigenvalue of the $n \times n$ matrix A, then λ^n is an eigenvalue of A^n for each $n \in \mathbb{N}$.

84. Let A be an $n \times n$ nonsingular matrix. Prove that, if λ is an eigenvalue of A, then $1/\lambda$ is an eigenvalue of A^{-1}.

85. Let A be an $n \times n$ matrix, let λ be an eigenvalue of A, and let a be a scalar. Prove that $\lambda + a$ is an eigenvalue of $aI + A$.

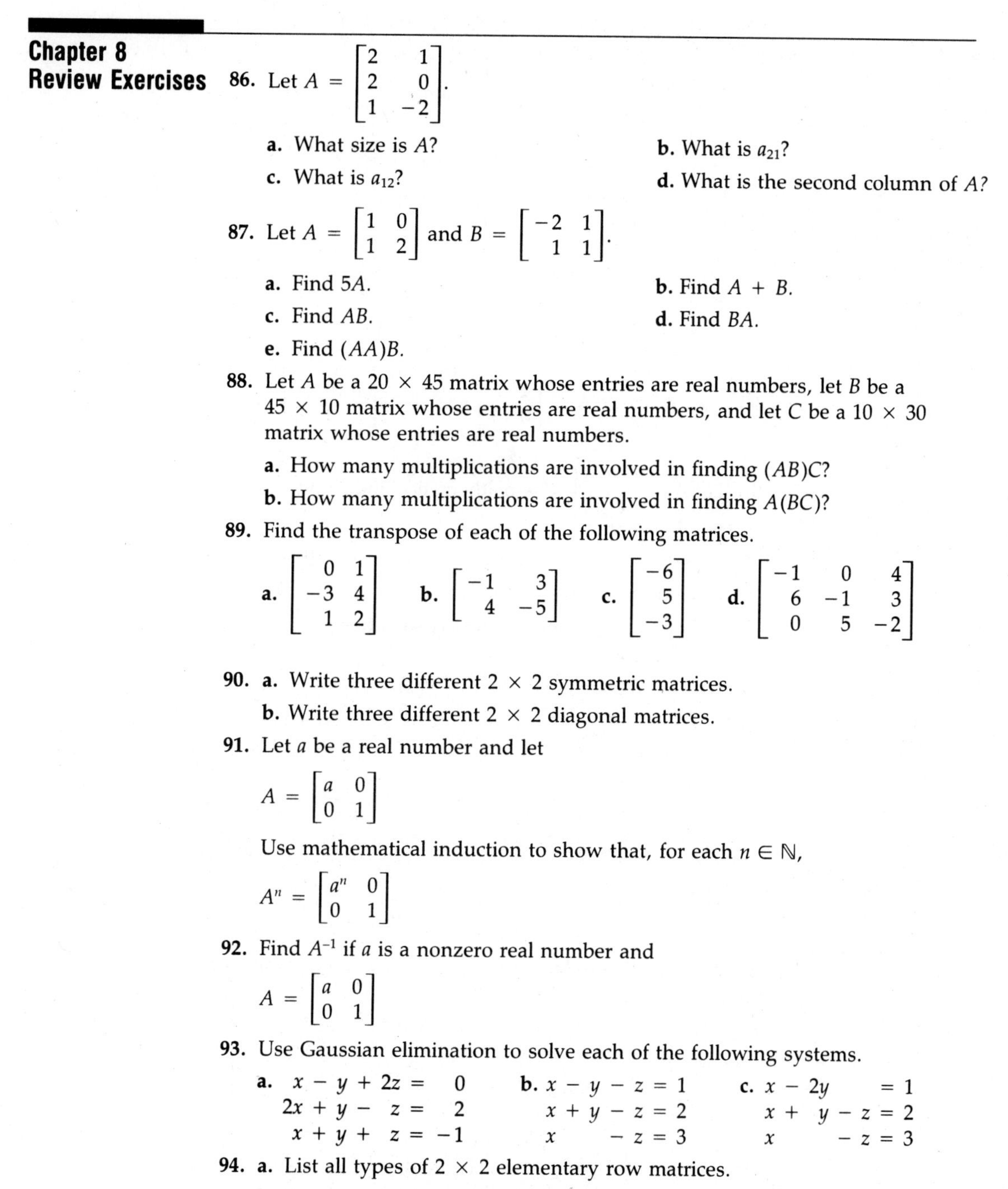

Chapter 8 Review Exercises

86. Let $A = \begin{bmatrix} 2 & 1 \\ 2 & 0 \\ 1 & -2 \end{bmatrix}$.

a. What size is A?

b. What is a_{21}?

c. What is a_{12}?

d. What is the second column of A?

87. Let $A = \begin{bmatrix} 1 & 0 \\ 1 & 2 \end{bmatrix}$ and $B = \begin{bmatrix} -2 & 1 \\ 1 & 1 \end{bmatrix}$.

a. Find $5A$.

b. Find $A + B$.

c. Find AB.

d. Find BA.

e. Find $(AA)B$.

88. Let A be a 20×45 matrix whose entries are real numbers, let B be a 45×10 matrix whose entries are real numbers, and let C be a 10×30 matrix whose entries are real numbers.

a. How many multiplications are involved in finding $(AB)C$?

b. How many multiplications are involved in finding $A(BC)$?

89. Find the transpose of each of the following matrices.

a. $\begin{bmatrix} 0 & 1 \\ -3 & 4 \\ 1 & 2 \end{bmatrix}$ **b.** $\begin{bmatrix} -1 & 3 \\ 4 & -5 \end{bmatrix}$ **c.** $\begin{bmatrix} -6 \\ 5 \\ -3 \end{bmatrix}$ **d.** $\begin{bmatrix} -1 & 0 & 4 \\ 6 & -1 & 3 \\ 0 & 5 & -2 \end{bmatrix}$

90. **a.** Write three different 2×2 symmetric matrices.

b. Write three different 2×2 diagonal matrices.

91. Let a be a real number and let

$$A = \begin{bmatrix} a & 0 \\ 0 & 1 \end{bmatrix}$$

Use mathematical induction to show that, for each $n \in \mathbb{N}$,

$$A^n = \begin{bmatrix} a^n & 0 \\ 0 & 1 \end{bmatrix}$$

92. Find A^{-1} if a is a nonzero real number and

$$A = \begin{bmatrix} a & 0 \\ 0 & 1 \end{bmatrix}$$

93. Use Gaussian elimination to solve each of the following systems.

a. $\begin{aligned} x - y + 2z &= 0 \\ 2x + y - z &= 2 \\ x + y + z &= -1 \end{aligned}$ **b.** $\begin{aligned} x - y - z &= 1 \\ x + y - z &= 2 \\ x - z &= 3 \end{aligned}$ **c.** $\begin{aligned} x - 2y &= 1 \\ x + y - z &= 2 \\ x - z &= 3 \end{aligned}$

94. **a.** List all types of 2×2 elementary row matrices.

b. For each matrix in (a), tell what elementary row operation on the 2×2 identity matrix produces that matrix.

95. Find the meet and join of each of the following pairs of zero–one matrices.

a. $A = \begin{bmatrix} 1 & 0 \\ 1 & 1 \end{bmatrix}$ and $B = \begin{bmatrix} 1 & 1 \\ 0 & 0 \end{bmatrix}$

b. $A = \begin{bmatrix} 1 & 1 \\ 0 & 1 \\ 1 & 0 \end{bmatrix}$ and $B = \begin{bmatrix} 0 & 1 \\ 0 & 1 \\ 1 & 1 \end{bmatrix}$

c. $A = \begin{bmatrix} 1 & 1 & 0 \\ 0 & 1 & 1 \\ 1 & 1 & 1 \end{bmatrix}$ and $B = \begin{bmatrix} 1 & 0 & 1 \\ 0 & 1 & 0 \\ 1 & 1 & 0 \end{bmatrix}$

96. Find $A \odot B$ if $A = \begin{bmatrix} 1 & 1 \\ 0 & 1 \end{bmatrix}$ and $B = \begin{bmatrix} 1 & 0 & 1 \\ 1 & 1 & 0 \end{bmatrix}$

97. Let $A = \begin{bmatrix} 0 & 0 \\ 0 & 1 \end{bmatrix}$. Use mathematical induction to show that if n is a natural number then $A^{[n]} = A$.

***98.** **a.** How many 2×2 zero–one matrices are there?

b. Find a zero–one matrix A and a natural number n such that $A^{[n]} \neq A$.

c. Find all zero–one matrices A such that if n is a natural number then $A^{[n]} = A$.

99. Use the Gauss-Jordan method to solve each of the following systems.

a. $\begin{aligned} x + y &= 7 \\ x - y &= 3 \\ x - 2y &= 1 \end{aligned}$

b. $\begin{aligned} x + y - 5z &= 0 \\ x - y + 3z &= 4 \end{aligned}$

c. $\begin{aligned} x + 2y + 3z &= 5 \\ 2x + 5y + 3z &= 3 \\ x + 8z &= 17 \end{aligned}$

d. $\begin{aligned} x + 2y + 2z &= -1 \\ x + 3y + z &= 4 \\ x + 3y + 2z &= 3 \end{aligned}$

e. $\begin{aligned} 2x + y + z &= 7 \\ 3x + 2y + z &= -3 \\ y + z &= 5 \end{aligned}$

100. Find the determinant of each of the following matrices.

a. $\begin{bmatrix} 1 & 2 & -3 \\ -2 & -1 & 2 \\ 3 & 1 & 1 \end{bmatrix}$ **b.** $\begin{bmatrix} 3 & -1 & 3 \\ 2 & 5 & -3 \\ 5 & 4 & -1 \end{bmatrix}$ **c.** $\begin{bmatrix} 1 & 3 & 2 & -1 \\ 1 & 1 & 0 & 2 \\ 2 & -1 & 0 & 3 \\ -2 & -1 & 1 & 2 \end{bmatrix}$

101. Let $v_1 = \begin{bmatrix} 1 \\ 2 \\ 1 \end{bmatrix}$ and $v_2 = \begin{bmatrix} 0 \\ 1 \\ 1 \end{bmatrix}$ Find a vector v_3 such that $\{v_1, v_2, v_3\}$ spans $\mathbb{R}^3$.

102. Let $v_1 = \begin{bmatrix} 1 \\ 2 \\ 3 \end{bmatrix}$ and $v_2 = \begin{bmatrix} 1 \\ 0 \\ 2 \end{bmatrix}$ and let V be the subspace of $\mathbb{R}^3$ spanned by $\{v_1, v_2\}$. Find two distinct nonzero vectors u_1 and u_2 in V such that $u_i \neq v_j$ for any i and j. Is it true that $\{u_1, u_2\}$ spans V (no matter which vectors u_1 and u_2 you have found)? Explain your answer.

103. Determine whether each of the following sets spans $\mathbb{R}^3$.

a. $S = \left\{ \begin{bmatrix} 1 \\ 0 \\ 1 \end{bmatrix}, \begin{bmatrix} 4 \\ 3 \\ 1 \end{bmatrix}, \begin{bmatrix} 2 \\ 2 \\ 0 \end{bmatrix} \right\}$ **b.** $S = \left\{ \begin{bmatrix} 0 \\ 1 \\ 2 \end{bmatrix}, \begin{bmatrix} 1 \\ 2 \\ 3 \end{bmatrix}, \begin{bmatrix} 2 \\ 3 \\ 4 \end{bmatrix} \right\}$

104. Describe the null space and the range of each of the following matrices, and in each case find the nullity, rank, and dimension of the row spaces of each of the matrices.

a. $\begin{bmatrix} 2 & -1 & 4 \\ 5 & 1 & -2 \\ 3 & -3 & -4 \end{bmatrix}$ **b.** $\begin{bmatrix} 2 & 2 & 1 \\ 2 & -1 & -2 \\ 4 & 1 & -1 \end{bmatrix}$ **c.** $\begin{bmatrix} 1 & 4 & -5 \\ 2 & 3 & -1 \\ 3 & 7 & 2 \end{bmatrix}$

105. Find the eigenvalues of each of the following matrices.

a. $\begin{bmatrix} 2 & -2 \\ -1 & 3 \end{bmatrix}$ **b.** $\begin{bmatrix} -5 & 5 & 5 \\ 1 & -4 & -2 \\ -3 & 5 & 3 \end{bmatrix}$ **c.** $\begin{bmatrix} 1 & -1 & -1 & -1 \\ -1 & 1 & -1 & -1 \\ -1 & -1 & 1 & -1 \\ -1 & -1 & -1 & 1 \end{bmatrix}$

106. For each matrix in Exercise 105, find the eigenvectors corresponding to each eigenvalue.

107. For each matrix in Exercise 105, find the eigenspace and the geometric multiplicity of each eigenvalue.

108. Let A be the matrix in Exercise 105c. Find the eigenvalues of A^7, A^{-1}, and $7I + A$.

Chapter

9

RELATIONS

Let X and Y be sets. Recall from Chapter 3 that the graph $G(f)$ of a function $f\colon X \to Y$ is a subset of $X \times Y$ with the property that each member of X is the first term of exactly one member of $G(f)$. The graph of a function describes a "relation" between X and Y. In particular, a member x of X and a member y of Y are related if $f(x) = y$. Because of the restriction that each member of X is the first term of exactly one member of $G(f)$, the graph associated with a function $f\colon X \to Y$ is not sufficient to describe many important relations. For example, let X be the set whose members are the last names of all the students in a certain discrete mathematics class and let $Y = \{A,B,C,D,F\}$. Then the subset R of $X \times Y$ consisting of all ordered pairs (x,y) such that y is the grade that student x receives in the course is not necessarily a graph of a function $R\colon X \to Y$, because if there are two students whose last names are Smith and one of them receives a B in the course while the other receives a D, then we have two distinct ordered pairs with the same first term belonging to R. Nor is there a graph of a function that describes the relation *less than* on the set of natural numbers. We state this observation formally. If $S = \{(m,n) \in \mathbb{N} \times \mathbb{N}: m < n\}$, then S is not the graph of a function $S: \mathbb{N} \to \mathbb{N}$, because (2,3) and (2,4) are two distinct members of S with the same first term.

These two examples are subsets of Cartesian products. Our purpose in this chapter is to study such subsets.

9.1

BASIC CONCEPTS

We begin by defining the concept of a relation. As we have indicated, a relation is a generalization of the graph of a function. We might have taken a different approach and begun with relations and then narrowed our perspective by considering functions to be relations of a special kind.

Relations have important applications in a number of disciplines. In computer science they are used in the study of state diagrams and databases, and in mathematics they are used in topology and algebra.

Example 1

Suppose we are given a set of names: {Emma, Sibyle, Janet, Linda, Diane, Scott, John, Angie, Tera, Tom, Hubert, Johnny, Carolyn, Kathy, Sherry, Kim}. The set has 16 members and 2^{16} subsets, but there is not much else we can say about it. If, however, we know that Emma is the mother of Sibyle and Hubert, that Sibyle is the mother of Janet, Linda, and Diane, that Hubert is the father of Johnny, Carolyn, and Kathy, that Janet is the mother of Scott and John, that Linda is the mother of Angie, Tera, and Tom, and that Johnny is the father of Sherry and Kim, then we know how the names are related to each other, and we can draw the family tree (Figure 9.1).

Figure 9.1

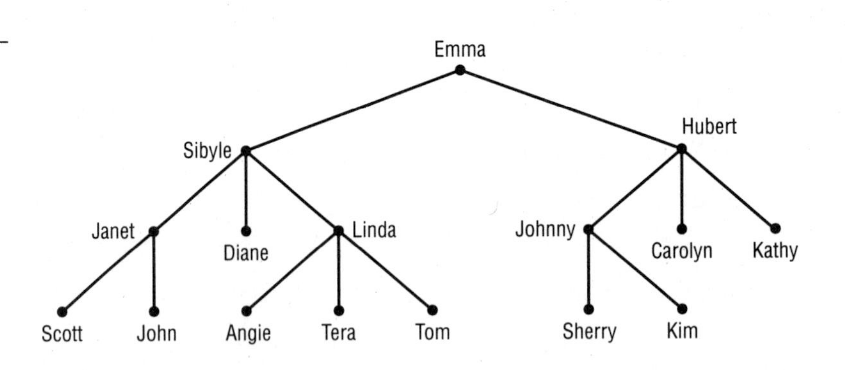

Furthermore, if R denotes the phrase "is a parent of," then we have

Emma R Sibyle	Hubert R Carolyn	Linda R Tom
Sibyle R Diane	Janet R John	Emma R Hubert

Sibyle R Linda	Johnny R Sherry	Janet R Scott
Hubert R Kathy	Sibyle R Janet	Linda R Tera
Linda R Angie	Hubert R Johnny	Johhny R Kim

■

Definition

A **relation** is a set of ordered pairs. If R is a relation, the set of first terms of all the ordered pairs of R is called the **domain** of R and the set of all second terms of the ordered pairs of R is called the **range** of R. We denote the domain of R by Dom(R) and the range of R by Rng(R).

Example 2

Let $R = \{(-2,3),(-1,1/2),(0,4),(-1,6),(2,4),(3,3)\}$. Then R is a relation. The domain of R is $\{-2,-1,0,2,3\}$ and the range of R is $\{1/2,3,4,6\}$. ■

Definition

Let A and B be sets. A **relation** (sometimes called a **binary relation**) **from A to B** is a subset of $A \times B$. A **relation on A** is a relation from A to A.

Let A and B be sets and let R be a relation from A to B. Then each member of R is an ordered pair. Thus R is a relation, $\text{Dom}(R) \subseteq A$, and $\text{Rng}(R) \subseteq B$.

Example 3

Let $A = \{1,3,5,7\}$ and $B = \{1,2,6,8,9\}$. Then $R = \{(1,1),(1,6),(5,1),(5,8),(7,2)\}$ is a relation from A to B, $\text{Dom}(R) = \{1,5,7\}$, and $\text{Rng}(R) = \{1,2,6,8\}$. Also $S = \{(1,3),(3,3),(7,7)\}$ is a relation on A, $\text{Dom}(S) = \{1,3,7\}$, and $\text{Rng}(S) = \{3,7\}$. ■

Example 4

A certain cable television company offers 28 channels and has 5,287 customers. Let A be the set whose members are these 5,287 customers, and let B be the set whose members are the 28 channels. Let R be the set whose members are those ordered pairs (a,b), where a is a customer of this cable television company who watched channel b for at least 30 minutes last week. Then R is a relation from A to B, Dom(R) is the subset of A consisting of those customers who watched at least one channel for at least 30 minutes last week, and Rng(R) is the subset of B consisting of those channels that were watched for at least 30 minutes last week by at least one customer. ■

Example 5

Suppose that a program contains a set of procedures and we say that procedure A is related to procedure B if A calls B. Then *calls* is a relation on the set of procedures in the program. We can use a diagram to illustrate calls by drawing procedure A above procedure B and connecting them with a line if A calls B. For example, the diagram in Figure 9.2 indicates that A calls B and C, that B calls D, that C calls E, F, and G, that D and E call H, and that F and G call I. ■

Figure 9.2

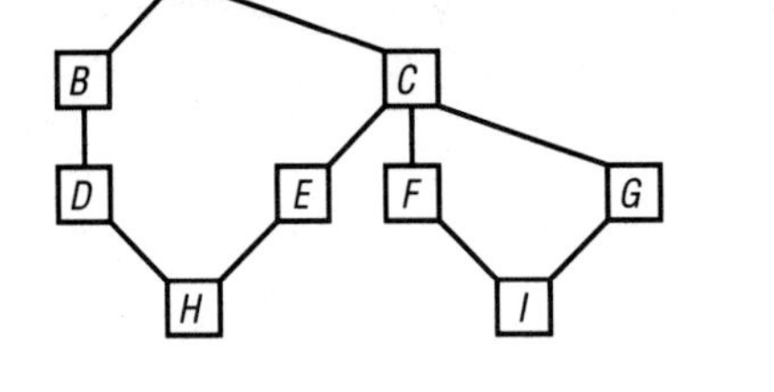

If R is a relation and $(a,b) \in R$, then we say that *a is related to b by R*, and we write aRb. We also use the notation $a \not R b$ to mean that $(a,b) \notin R$.

Let R be a relation of a set A. If a and b are members of A, then $(a,b) \in R$ or $(a,b) \notin R$. We can define a function $T: A \times A \rightarrow \{\text{true,false}\}$ by $T(a,b) = \text{true}$ if $(a,b) \in R$ and $T(a,b) = \text{false}$ if $(a,b) \notin R$; that is, the value of the function T at (a,b) is true if a is related to b by R, and the value of the function T at (a,b) is false if a is not related to b.

Example 6

Let $A = \{1,2\}$ and $B = \{3,4\}$. List the relations from A to B.

Analysis

Every subset of $A \times B$ is a relation from A to B, so we simply list the subsets of $A \times B$. Since $A \times B$ has four members, the number of relations from A to B is $2^4 = 16$ (see Theorem 4.3). These relations are

$\varnothing$	$\{(1,3)\}$	$\{(1,4)\}$
$\{(2,3)\}$	$\{(2,4)\}$	$\{(1,3),(1,4)\}$
$\{(2,3),(2,4)\}$	$\{(1,3),(2,3)\}$	$\{(1,4),(2,4)\}$
$\{(1,3),(2,4)\}$	$\{(1,4),(2,3)\}$	$\{(1,3),(1,4),(2,3)\}$
$\{(1,3),(1,4),(2,4)\}$	$\{(1,3),(2,3),(2,4)\}$	$\{(1,4),(2,3),(2,4)\}$
$A \times B$		

■

Example 7

There are many important relations on $\mathbb{R}$. One is *equality;* we define this relation R on $\mathbb{R}$ by saying that R is $\{(a,b) \in \mathbb{R} \times \mathbb{R}: a \text{ is } b\}$. Then, for example, $(3,3) \in R$ but $(3,4) \notin R$; that is, 3 is related to 3 but 3 is not related to 4. Another relation on $\mathbb{R}$ is *less than;* we define this relation S

on $\mathbb{R}$ by saying that $S = \{(a,b) \in \mathbb{R} \times \mathbb{R}: b - a \text{ is positive}\}$. Then, for example, $(3,4) \in S$ but $(3,3) \notin S$ and $(4,3) \notin S$. Using the alternative notation, we would write $3S4$ but $3\not{S}3$ and $4\not{S}3$. The relation we have been calling S is usually indicated by $<$, and so we would normally write $3 < 4$ but $3 \not< 3$ and $4 \not< 3$. ■

Example 8

It is possible to consider relations that relate pairs of numbers; that is, we may consider relations on the plane ($= \mathbb{R} \times \mathbb{R}$). We can define a relation R on $\mathbb{R} \times \mathbb{R}$ by saying that $(a,b)R(c,d)$ provided that $a < c$. Then $(2,8)R(3,6)$, $(1,7)R(2,7)$, and $(3,4)R(4,5)$, but $(3,6)\not{R}(2,8)$, $(2,7)\not{R}(1,7)$, and $(4,5)\not{R}(3,4)$. The geometric interpretation of R is that (a,b) is related to (c,d) if and only if the point (a,b) lies to the left of the line $x = c$. We can also define a relation S on $\mathbb{R} \times \mathbb{R}$ by saying that $(a,b)S(c,d)$ provided that $b < d$. Then $(3,4)S(4,5)$, $(3,6)S(2,8)$, and $(1,7)S(1,9)$, but $(4,5)\not{S}(3,4)$, $(2,8)\not{S}(3,6)$, and $(1,9)\not{S}(1,7)$. The geometric interpretation of S is that (a,b) is related to (c,d) if and only if the point (a,b) lies below the line $y = d$. We can define a third relation T on $\mathbb{R} \times \mathbb{R}$ by saying that $(a,b)T(c,d)$ provided that $a < c$ and $b < d$. In Exercise 8 we ask for examples of points that are related with respect to T and of points that are not related with respect to T, and we ask for a geometric interpretation of points that are related with respect to T. ■

Example 9

Let $X = \{1,2,3,4,5\}$ and let R be the relation on $\mathcal{P}(X)$ defined by $(A,B) \in R$ provided that $A \subseteq B$. Then $(\{2,3\},\{2,3,4\}) \in R$, $(\{1\},\{1,5\}) \in R$, $(\varnothing,\{2\}) \in R$, and $(\{3,4\},\{3,4\}) \in R$, whereas $(\{1,2\},\{1,3,5\}) \notin R$ and $(\{1,3,5\},\{3,5\}) \notin R$. ■

Definition

> If R is a relation, then the **inverse** of R, denoted by R^{-1}, is the set $\{(x,y): (y,x) \in R\}$.

Example 10

Let $R = \{(1,1),(1,6),(5,1),(5,8),(7,2)\}$. Find R^{-1}.

Analysis

$R^{-1} = \{(1,1),(6,1),(1,5),(8,5),(2,7)\}$. ■

Example 11

Let $R = \{(m,n) \in \mathbb{N} \times \mathbb{N}: m \text{ divides } n\}$. Then R is a relation on $\mathbb{N}$, and we can determine whether an ordered pair (a,b) of integers belongs to R by determining whether a divides b. Thus $(7,91) \in R$ but $(4,78) \notin R$. Describe R^{-1} and give an example of a member of $\mathbb{N} \times \mathbb{N}$

that is a member of R^{-1} and an example of a member of $\mathbb{N} \times \mathbb{N}$ that is not a member of R^{-1}.

Analysis

$R^{-1} = \{(m,n) \in \mathbb{N} \times \mathbb{N}: (n,m) \in R\}$. Therefore $R^{-1} = \{(m,n) \in \mathbb{N} \times \mathbb{N}: n \text{ divides } m\}$. Thus $(9,3) \in R^{-1}$ but $(7,2) \notin R^{-1}$. ■

If $f: A \to B$ is an invertible function, then the inverse of the relation $G(f)$ is the graph of the inverse function $f^{-1}: B \to A$.

If R and S are relations from a set A to a set B, then R and S are subsets of $A \times B$. Therefore we can form $R \cup S$, $R \cap S$, $R - S$, $S - R$, and $R \oplus S$.

Example 12

If R is the relation given in Example 11 and $S = R^{-1}$, describe $R \cup S$, $R \cap S$, $R - S$, $S - R$, and $R \oplus S$. Also, give a member of each of these five sets as well as a member of $\mathbb{N} \times \mathbb{N}$ that is not a member of the set.

Analysis

$R \cup S = \{(m,n) \in \mathbb{N} \times \mathbb{N}: m \text{ divides } n \text{ or } n \text{ divides } m\}$

$R \cap S = \{(m,n) \in \mathbb{N} \times \mathbb{N}: m = n\}$

$R - S = \{(m,n) \in \mathbb{N} \times \mathbb{N}: m \text{ divides } n \text{ but } n \text{ does not divide } m\}$

$S - R = \{(m,n) \in \mathbb{N} \times \mathbb{N}: n \text{ divides } m \text{ but } m \text{ does not divide } n\}$

$R \oplus S = \{(m,n) \in \mathbb{N} \times \mathbb{N}: m \text{ divides } n \text{ or } n \text{ divides } m \text{ but } m \neq n\}$

$(2,4) \in R \cup S$ but $(3,5) \notin R \cup S$

$(1,1) \in R \cap S$ but $(1,2) \notin R \cap S$

$(2,4) \in R - S$ but $(4,2) \notin R - S$

$(4,2) \in S - R$ but $(2,4) \notin S - R$

$(2,4) \in R \oplus S$ but $(2,2) \notin R \oplus S$ ■

If R is a relation and x belongs to the domain of R, the $R[x]$ is defined to be $\{y \in \text{Rng}(R): (x,y) \in R\}$. When the relation R is the only relation under discussion, then $R[x]$ is often abbreviated $[x]$. Note that $y \in R[x]$ means that $(x,y) \in R$. We can summarize the three methods commonly used to indicate that an object a is related to an object b by a relation R as follows:

If A and B are sets and R is a relation from A to B, then

$$\left.\begin{array}{l}(a,b) \in R \\ b \in R[a] \\ aRb\end{array}\right\} \begin{array}{l}a \text{ is an element in the domain of } R \\ \text{that is related to the element} \\ b \text{ in the range of } R.\end{array}$$

Note that, if R is a relation from A to B and x belongs to the domain of R, then $R \subseteq A \times B$ and $R[x] \subseteq B$. In other words, a member of R is an ordered pair, whereas a member of $R[x]$ is the second term of some member of R. If R is the relation in Example 3, then $R[1] = \{1,6\}$, $R[5] = \{1,8\}$, and $R[7] = \{2\}$.

Recall that if R is a function then the image of x under R is denoted by $R(x)$. Thus it is customary not to use the notation $R[x]$ when R is a function. In this case, however, $R[x]$ is $\{R(x)\}$; that is, $R[x]$ is the set whose only member is the image of x under R.

Example 13

The family tree in Figure 9.1 represents the relation R, where R is the phrase *is a parent of*. Notice that

$$R[\text{Emma}] = \{\text{Sibyle,Hubert}\}, \qquad R[\text{Sibyle}] = \{\text{Janet,Diane,Linda}\}$$

$$R[\text{Hubert}] = \{\text{Johnny,Carolyn,Kathy}\}, \qquad R[\text{Janet}] = \{\text{Scott,John}\}$$

$$R[\text{Linda}] = \{\text{Angie,Tera,Tom}\}, \qquad R[\text{Johnny}] = \{\text{Sherry,Kim}\},$$

$$\begin{aligned} R[\text{Diane}] &= R[\text{Carolyn}] = R[\text{Kathy}] = R[\text{Scott}] = R[\text{John}] \\ &= R[\text{Angie}] = R[\text{Tera}] = R[\text{Tom}] = R[\text{Sherry}] \\ &= R[\text{Kim}] = \varnothing \end{aligned}$$

In other words, for each member x of the family tree, $R[x]$ denotes the children of x. ■

Example 14

If S is the relation *calls* of Example 5, then $S[A] = \{B,C\}$, $S[B] = \{D\}$, $S[C] = \{E,F,G\}$, $S[D] = S[E] = \{H\}$, $S[F] = S[G] = \{I\}$, and $S[H] = S[I] = \varnothing$. If x is a procedure, then $S[x]$ denotes the set of procedures that are called by x. ■

Exercises 9.1

1. Give an example of each of the following.
   a. a relation on $A = \{1,2,3,4,5,6,7\}$ whose domain is A
   b. a relation on $A = \{1,2,3,4,5,6,7\}$ whose domain is not A but whose range is A

c. a relation from $A = \{1,2,3,4\}$ to $B = \{5,6,7\}$ whose domain is A and whose range is B

d. a relation from $A = \{1,2,3,4\}$ to $B = \{5,6,7\}$ whose domain is A but whose range is not B

e. a relation on $\mathbb{N}$

f. a relation from $\mathbb{Z}$ to $\mathbb{N}$

g. a relation on $\mathbb{N}$ whose domain is $\mathbb{N}$ but whose range is not $\mathbb{N}$

h. a relation on $\mathbb{N}$ whose domain and range are $\mathbb{N}$

i. a relation from $\mathbb{Z}$ to $\mathbb{N}$ whose domain is $\mathbb{Z}$ and whose range is $\mathbb{N}$

j. a relation from $\mathbb{Z}$ to $\mathbb{N}$ whose domain is $\mathbb{Z}$ but whose range is not $\mathbb{N}$

2. Let $R = \{(1,3),(0,3),(1,4),(-1,3),(-2,5),(0,5)\}$.
   a. Find the domain of R.
   b. Find the range of R.
   c. Is R a relation on $\{-2,-1,0,1,2,3,4,5\}$?
   d. Is R a relation from $\{-5,-4,-3,-2,-1,0,1\}$ to $\{2,3,4,5,6\}$?
3. a. List the relations from $\{1\}$ to $\{2,3\}$.
   b. How many relations are there from $\{1,2,3\}$ to $\{3,4\}$?
4. Define a relation R on $\mathbb{N}$ by saying that $R = \{(m,n) \in \mathbb{N} \times \mathbb{N}: n \leq 10, m < n, \text{ and } m \text{ divides } n\}$. List the ordered pairs that belong to R.
5. Define a relation R on $\mathbb{R}$ by saying that $R = \{(m,n) \in \mathbb{R} \times \mathbb{R}: -n < m < n\}$.
   a. Is $(-1, 4) \in R$? b. Is $(1, -1) \in R$?
   c. List five ordered pairs that belong to R.
   d. List five members of $\mathbb{R} \times \mathbb{R}$ that do not belong to R.
6. Let $R = \{(0,1),(0,2)(2,3),(2,5),(2,0),(0,8),(4,3),(2,6),(4,7)\}$.
   a. What is the domain of R?
   b. For which numbers x is $R[x]$ defined?
   c. For those numbers x such that $R[x]$ is defined, find $R[x]$.
   d. Find R^{-1}.
   e. For which numbers x is $R^{-1}[x]$ defined?
   f. What is the range of R?
   g. For those numbers x such that $R^{-1}[x]$ is defined, find $R^{-1}[x]$.
7. Let $R = \{(0,1),(0,2),(2,3),(2,5),(4,0),(4,2)\}$ and $S = \{(0,5),(1,3),(1,4),(2,3),(2,6),(3,3),(3,5),(4,7)\}$.
   a. Find $(R \cup S)\,[2]$. b. Find $(R \cup S)^{-1}[2]$. c. Find $(R \cap S)\,[2]$.
8. a. In Example 8, give three pairs of points such that the first member of each pair is related to the second member with respect to T.
   b. In Example 8, give three pairs of points such that the first member of each pair is not related to the second member with respect to T.
   c. In Example 8, give a geometric interpretation of points that are related with respect to T.

9. Let $m,n \in \mathbb{N}$, let X be a set with m members, and let Y be a set with n members. How many distinct relations are there from X to Y? Explain.

10. Let $p \in \mathbb{N}$ and let X be a set with p members. Let $R = \{(A,B) \in \mathcal{P}(X) \times \mathcal{P}(X): A \cap B = \varnothing\}$.

    a. Find the number of members of R when $p = 3$.

    b. Find a formula for the number of members of R for arbitrary p.

9.2 DIRECTED GRAPHS

In Section 3.3 we discussed the graph of a function. We can graph a relation in the same manner, and we give two examples to illustrate this concept.

Example 15

Let $R = \{(x,y) \in \mathbb{R} \times \mathbb{R}: |y| \leq 5 \text{ and } -|y| < x < |y|\}$. Then R is a relation whose graph is given in Figure 9.3.

Figure 9.3

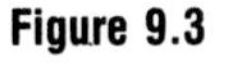

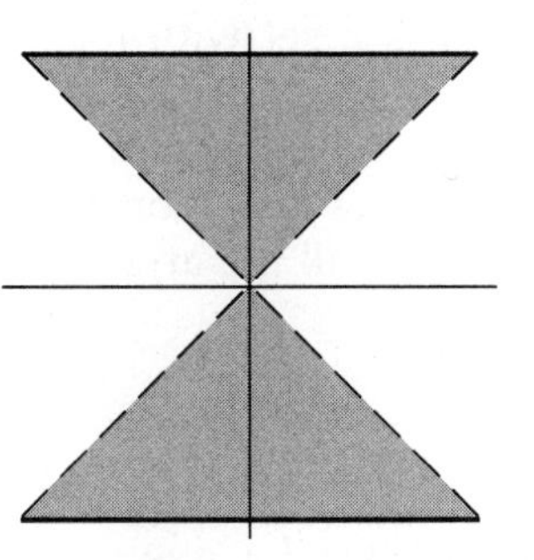

■

Example 16

Let $R = \{(1,1)(2,2),(4,4),(5,1),(1,3),(2,5),(4,3)\}$. Then R is a finite relation. We can circle members of R when it is considered a subset of the plane, as in Figure 9.4. ■

Figure 9.4

(5, 5)

(1, 1)

The graphs in Examples 15 and 16 are called *Cartesian graphs* because they are subsets of a Cartesian product. A *directed graph* (or *digraph*) is another way to graph a relation. If $A = \{a,b,c,d,e\}$ is a set with five elements and R is a relation on A, we choose five points to represent a, b, c, d, and e, and for each $(x,y) \in R$ we draw an arrow starting at x and ending at y. If $(x,x) \in R$, we simply draw a loop that begins and ends at x.

Example 17

Suppose that $A = \{a,b,c,d,e\}$ is a set with five elements and that $R = \{(a,b),(b,b),(b,d),(c,e),(d,b),(d,e)\}$ is a relation on A. A directed graph of R is given in Figure 9.5.

Figure 9.5

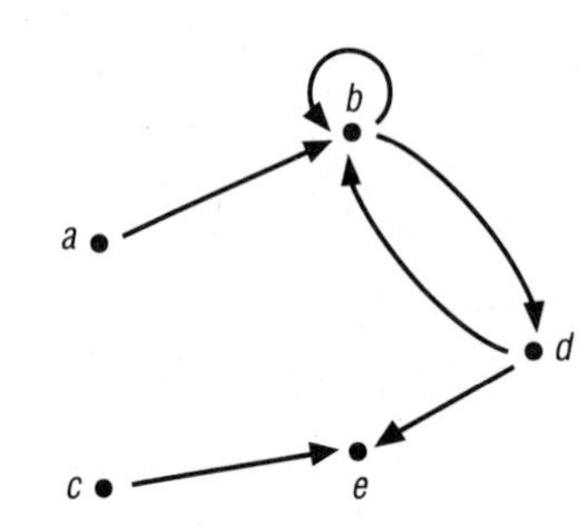

■

In a directed graph, the points are called *vertices* and the arrows are called *directed edges.* If R is a relation on a nonempty set A, then there is a directed graph that represents R. If, however, A is a large finite set or an infinite set, we may not be able to draw the directed graph that represents R. Nevertheless, we can still define the directed graph of the relation R to be the ordered pair (A,E), where A is the set of vertices and E is the set of directed edges to which the directed edge $\vec{xy}$ belongs if and only if $(x,y) \in R$.

Example 18

Suppose that $A = \{a,b,c,d\}$ is a set with four members and that $R = \{(a,a), (a,c), (b,a), (b,c), (c,b), (c,c), (d,a)\}$ is a relation on A. Then a directed graph (A,E) that represents R is given in Figure 9.6a.

If we have the directed graph of a relation R, then we can draw the directed graph of R^{-1} by reversing all the arrows, as in Figure 9.6b.

Figure 9.6

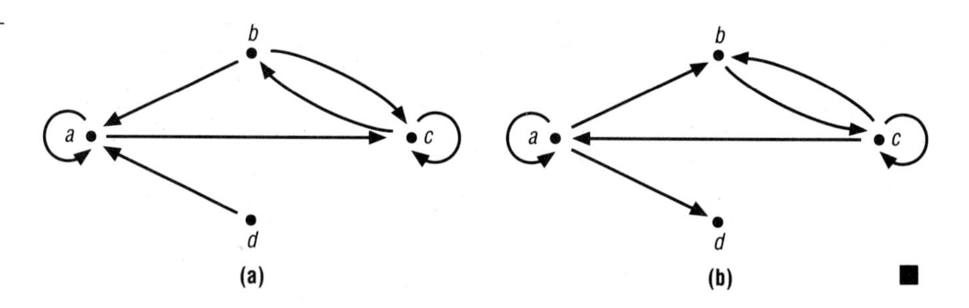

■

We have indicated how a directed graph can represent a given relation, but it is not necessary to have a relation in mind in order to consider a directed graph. We can simply start with some points and arrows, as in Figure 9.7. Of course there is a relation R represented by such a directed graph. In this case the relation R is

$$\{(a,a),(a,b),(a,c),(c,b),(c,c),(c,d),(d,e),(e,d),(e,a)\}.$$

Each time we have a relation R, there is a directed graph that represents R, and each time we have a directed graph (A,E), there is a relation

Figure 9.7

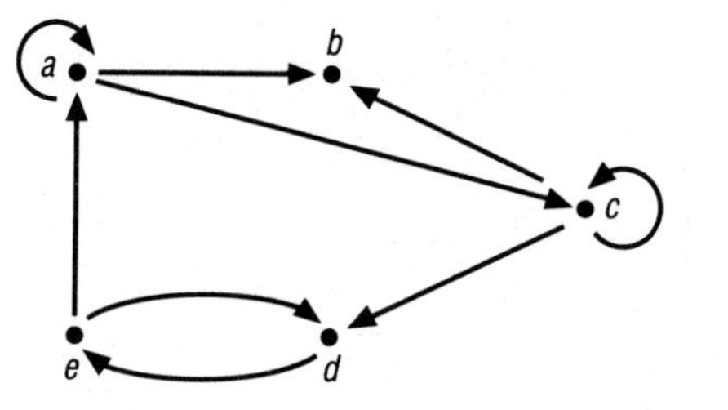

represented by the directed graph. Furthermore, if we start with a relation, draw the directed graph that represents it, and then find a relation that represents the directed graph, we get back our original relation. Also, if we start with a directed graph, obtain the relation that represents it, and then draw the directed graph that represents the relation, we get back our original directed graph.

Example 19

Let $A = \{1,2,3,4,5,6\}$ and let R be the relation on A defined by $R = \{(m,n) \in A \times A: m \text{ divides } n \text{ and } m \neq n\}$. Then the directed graph of this relation is given in Figure 9.8.

Figure 9.8

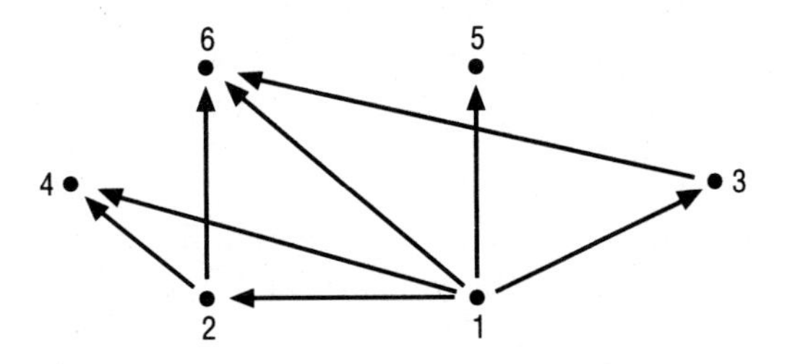

■

Recall from Chapter 3 that a permutation on a set A is a one-to-one function from A onto A. As we have observed, the graph of any function is a relation. Thus, if $\alpha: A \to A$ is a permutation, then α has a directed graph and a Cartesian graph. The next example shows that, in determining the composition of two permutations, directed graphs are far more useful than Cartesian graphs.

Example 20

Let $A = \{1,2,3,4\}$, $\alpha = \{(1,2),(2,4),(3,3),(4,1)\}$, and $\beta = \{(1,4),(2,3),(3,2),(4,1)\}$. Draw the directed graphs of the permutations α and β, and using these graphs draw the directed graphs of α^{-1}, $\alpha \circ \beta$, and $\alpha^{-1} \circ \beta$.

Analysis

We start by drawing the directed graphs of the permutations α, α^{-1}, and β, as in Figure 9.9.

Figure 9.9

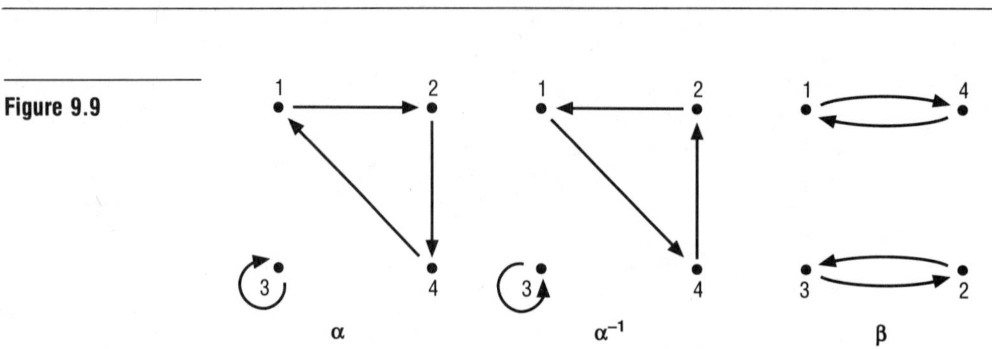

Using the directed graphs of α and β, we can obtain the directed graph of $\alpha \circ \beta$. Imagine that the vertices 1, 2, 3, and 4 are the subway stops in a city served by two subway lines α and β, and imagine that to encourage greater use of these subway systems the transit authority has issued special tickets that entitle a citizen to ride the β-line one stop and then ride the α-line one stop. The directed graph of $\alpha \circ \beta$ (Figure 9.10) has an arrow from vertex x to vertex y provided that the special ticket is valid for travel from x to y. For example, a ticketholder is allowed to travel the β-line from 3 to 2 and then travel the α-line from 2 to 4. Since the ticket allows travel from 3 to 4, there is a directed edge from 3 to 4 in the directed graph of $\alpha \circ \beta$. Notice that a ticketholder is not allowed to exit the subway system until he has traveled first one stop on the β-line and then traveled one stop on the α-line. He cannot use the special ticket for travel from 3 to 2, because he is required to follow his one-stop trip on the β-line by a one-stop trip on the α-line. Since the α-line has a nonstop tour of the city that begins and ends at 3, the ticketholder can, however, travel from 2 to 3, if he is not in a hurry. In a similar way, the directed graph of $\alpha^{-1} \circ \beta$ (Figure 9.10) can be determined from the directed graphs of α^{-1} and β.

Figure 9.10

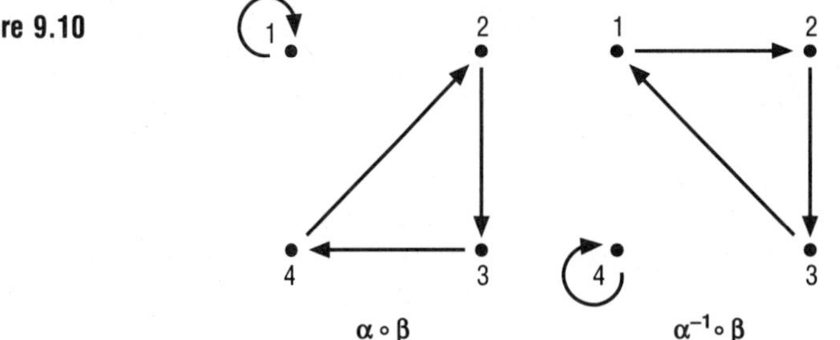

■

The description of α and β as sets of ordered pairs is awkward, and the use of directed graphs suggests a more compact description known as *cycle notation*. Since a permutation is a bijection, in the directed graph of a permutation exactly one arrow leaves each vertex and exactly one arrow arrives at each vertex. Thus the directed graph of a permutation splits into one or more closed figures. In Example 20, α consists of a circle and a triangle, and β consists of two ovals (see Figure 9.9). In cycle notation, α is described as (124)(3), meaning that under α, 1 maps to 2, 2 maps to 4, 4 maps to 1, and 3 maps to itself. In cycle notation, β is described as (14)(23), meaning that under β, 1 maps to 4, 4 maps to 1, 2 maps to 3, and 3 maps to 2. Notice that α can also be written as (241)(3) or as (412)(3) and β can be written as (41)(32). But $\gamma = (142)(3)$ is different from α, since γ maps 1 to 4 whereas α maps 1 to 2. Of course, if you draw the directed graph of γ and compare it with the directed graph of α, it is obvious that $\gamma \neq \alpha$. In cycle notation it is customary to omit singleton cycles. With this convention, α can be written as (124) and γ can be written as (142). If this convention were applied to the identity permutation (1)(2)(3)(4), the permutation would vanish without a trace. Instead, (1) is often used to denote the identity permutation on a finite set.

The last example of this section should serve as a warning that the directed graphs of permutations are too well behaved to be considered typical.

Example 21

Let $A = \{1,2,3\}$ and let R be the relation on $\mathcal{P}(A)$ defined by $R = \{(B,C) \in \mathcal{P}(A) \times \mathcal{P}(A)$: B is a proper subset of $C\}$. Then the directed graph of R is given in Figure 9.11.

Figure 9.11

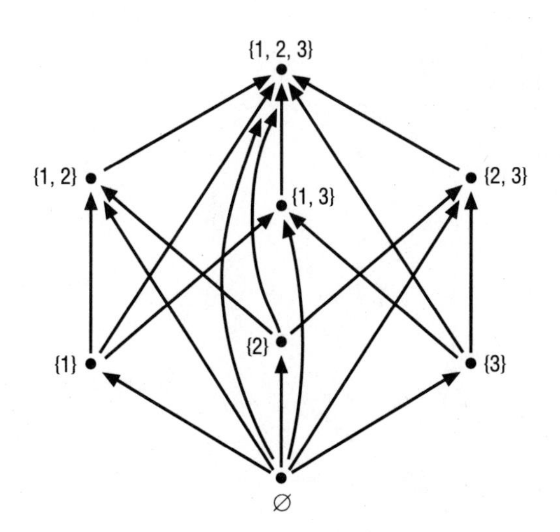

■

Exercises 9.2

11. Let $A = \{1,2,3,4,5\}$ and suppose that R is a relation on A such that the accompanying directed graph represents R. List the members of R.

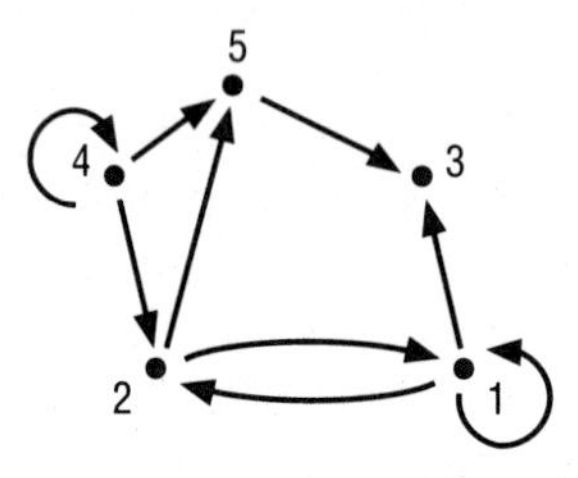

12. Let $A = \{1,2,3,4,5\}$. Draw the directed graph of each of the following relations on A.
    - a. $R = \{(a,b) \in A \times A: b = a^2\}$
    - b. $S = \{(a,b) \in A \times A: b^2 = a\}$
    - c. $T = \{(a,b) \in A \times A: a < b\}$
    - d. $U = \{(a,b) \in A \times A: a \neq b\}$
    - e. $E = \{(a,b) \in A \times A: a + b \text{ is even}\}$
    - f. $O = \{(a,b) \in A \times A: a + b \text{ is odd}\}$
13. Let $A = \{1,2,3,4,5\}$ and let R be the relation on A defined by $R = \{(a,b) \in A \times A: a \text{ divides } b + 1\}$. Draw the directed graph that represents R.
14. Let $A = \{1,2,3,4,5\}$ and let R be the relation on A defined by $R = \{(a,b) \in A \times A: a^2 + b = 6\}$. Draw the directed graph that represents R.
15. Let $A = \{1,2,3,4,5\}$ and let R be the relation on A defined by $R = \{(a,b) \in A \times A: a^2 + b^2 = 51\}$. Draw the directed graph that represents R.
16. Draw the directed graphs of the following permutations on $\{1,2,3,4,5\}$.
    - a. (12345)
    - b. (1234)
    - c. (241)(35)
    - d. (25)(143)
    - e. (15)(23)
    - f. (24)
17. Write in cycle notation the permutations on $\{1,2,3,4,5,6\}$ whose directed graphs are as follows.

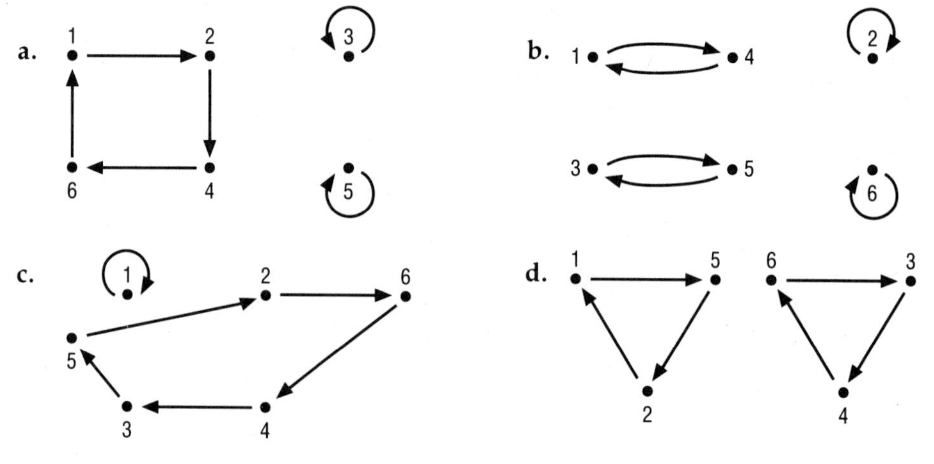

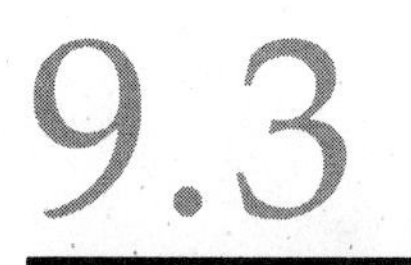

COMPOSITION OF RELATIONS

In Section 3.2 we studied the composition of functions. Recall that if $g: A \to B$ and $f: B \to C$ then f composite g is the function defined by $f \circ g(a) = f(g(a))$ for each $a \in A$. In this section, we generalize this concept to relations.

Definition

If A, B, and C are sets, $S \subseteq A \times B$ and $R \subseteq B \times C$, then **R composite S** is the relation $R \circ S$ defined by

$$R \circ S = \{(a,c) \in A \times C: \text{there exists } b \in B \text{ such that } (a,b) \in S \text{ and } (b,c) \in R\}$$

The relation $R \circ S$ is a relation from A to C. In other words, the domain of $R \circ S$ is a subset of A and the range of $R \circ S$ is a subset of C. Let us take an ordered pair (a,c) that belongs to $A \times C$ and ask whether (a,c) belongs to $R \circ S$ (see Figure 9.12). If we can find an element b in B such that $(a,b) \in S$ and $(b,c) \in R$, then $(a,c) \in R \circ S$. If no such element exists, then $(a,c) \notin R \circ S$. Let us consider some examples in order to clarify this concept.

Figure 9.12

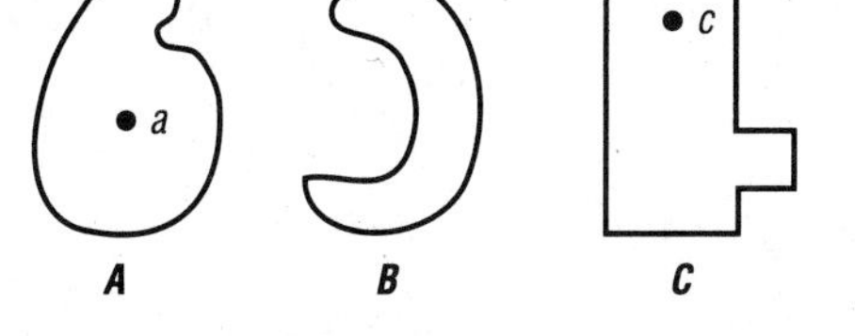

Example 22

Let $A = \{2,3,8,9\}$, $B = \{4,6,18\}$, and $C = \{1,4,7,9\}$. Let S be the relation *divides* from A to B, and let R be the relation *less than or equal to* from B to C. Then $S = \{(2,4),(2,6),(2,18),(3,6),(3,18),(9,18)\}$ and $R = \{(4,4),(4,7),(4,9),(6,7),(6,9)\}$. Figure 9.13 illustrates the composition $R \circ S$; certain elements of A go through elements in B to elements of C.

2 divides 4 and $4 \leq 4$, so $(2,4) \in R \circ S$

2 divides 4 and $4 \leq 7$, so $(2,7) \in R \circ S$

2 divides 4 and $4 \leq 9$, so $(2,9) \in R \circ S$

2 divides 6 and $6 \leq 7$, so $(2,7) \in R \circ S$

2 divides 6 and $6 \leq 9$, so $(2,9) \in R \circ S$

3 divides 6 and $6 \leq 7$, so $(3,7) \in R \circ S$

3 divides 6 and $6 \leq 9$, so $(3,9) \in R \circ S$

Figure 9.13

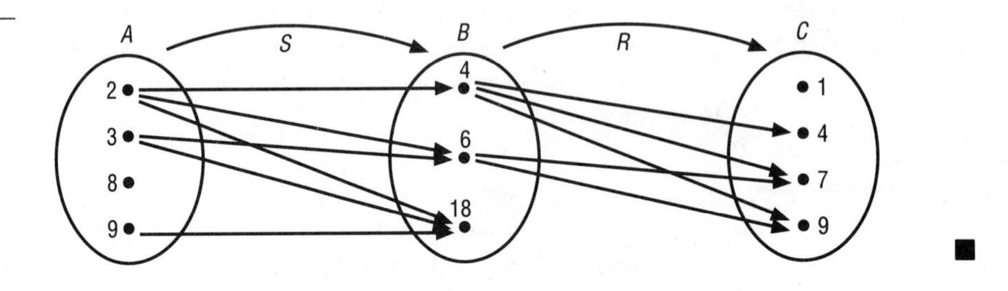

■

Example 23

Let $A = \{1,2\}$, $B = \{3,4\}$, $C = \{5,6\}$, $S = \{(1,3),(1,4)\}$, and $R = \{(3,5),(4,5)\}$. Which elements of $A \times C$ belong to $R \circ S$?

Analysis

First we observe that $A \times C = \{(1,5),(1,6),(2,5),(2,6)\}$. Is $(1,5) \in R \circ S$; that is, is there a member b of B such that $(1,b) \in S$ and $(b,5) \in R$? By looking at the definition of R and S, we see that there are two such b's, namely, 3 and 4. We only need one, so $(1,5)$ is a member of $R \circ S$. Is $(1,6) \in R \circ S$; that is, is there a member b of B such that $(1,b) \in S$ and $(b,6) \in R$? Since 6 is not the second term of any member of R, the answer is no. Therefore $(1,6)$ is not a member of $R \circ S$. Is $(2,5) \in R \circ S$; that is, is there a member b of B such that $(2,b) \in S$ and $(b,5) \in R$? Since 2 is not the first term of any member of S, the answer is no. Therefore $(2,5)$ is not a member of $R \circ S$. We also see that $(2,6)$ is not a member of $R \circ S$, and hence $R \circ S = \{(1,5)\}$. ■

Example 24

Let $A = \{-2,-1,1\}$, $B = \{0,2,3\}$, $C = \{-3,4,-4\}$, $S = \{(-2,0),(-2,3),(-1,3)\}$, and $R = \{(0,4),(2,4),(2,-4),(0,-3)\}$. Which elements of $A \times C$ belong to $R \circ S$?

Analysis

There are nine members of $A \times C$, and once again we could examine each of these members. As you can see, this method is going to become tedious if the sets are very large. So let us find a shortcut. First observe that 1 is not the first term of any member of S, so 1 cannot be the first term of any member of $R \circ S$. Therefore we have eliminated three of the nine members of $A \times C$. Now consider -2. It is the first term of two members of S, namely, $(-2,0)$ and $(-2,3)$. We observe that 3 is not the first term of any member of R, so we can disregard $(-2,3)$. Now 0 is the first term of two members, $(0,4)$ and $(0,-3)$, of R. Therefore $(-2,4)$ and $(-2,-3)$ are members of $R \circ S$. Finally, consider -1. It is the first term of one member, $(-1,3)$, of S. However, as we previously observed, 3 is not the first term of any member of R, so we can disregard $(-1,3)$. Hence $R \circ S = \{(-2,4),(-2,-3)\}$. ■

Example 25

Suppose that both R and S are the *is a parent of* relation of Example 1. Then $R \circ S$ is the grandparent–grandchild relation; that is, $(x,y) \in R \circ S$ if and only if x is a grandparent of y. Thus, in Example 1, $R \circ S$ is the set whose members are (Emma,Janet), (Emma,Diane), (Emma,Linda), (Emma,Johnny), (Emma,Carolyn), (Emma,Kathy), (Sibyle,Scott), (Sibyle,John), (Sibyle,Angie), (Sibyle,Tera), (Sibyle,Tom), (Hubert,Sherry), and (Hubert,Kim). This relation $R \circ S$ is indicated in Figure 9.14.

Figure 9.14

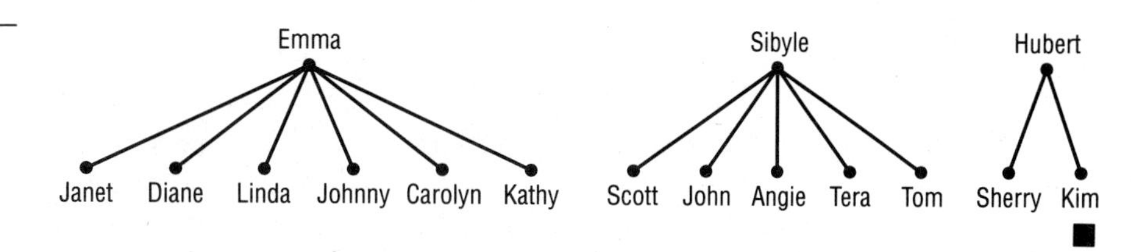

■

Example 26

Suppose that both R and S are the *calls* relation of Example 5. Then $R \circ S = \{(A,D),(A,E),(A,F),(A,G),(B,H),(C,H),(C,I)\}$. This relation $R \circ S$ is indicated in Figure 9.15. ■

Figure 9.15

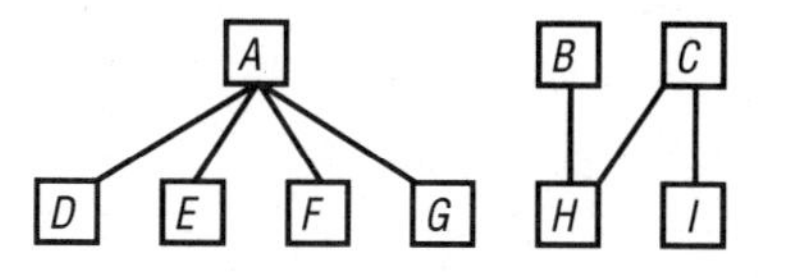

■

We know that composition of permutations is not commutative (Theorem 3.10). Therefore, since permutations are functions and graphs of functions are relations, the composition of relations is not commutative. By Theorem 3.9, composition of functions is associative. As the following theorem indicates, composition of relations is also associative.

Theorem 9.1

> Let A, B, C, and D be sets. Let $T \subseteq A \times B$, $S \subseteq B \times C$, and $R \subseteq C \times D$. Then $R \circ (S \circ T) = (R \circ S) \circ T$.

Proof

Since $R \circ (S \circ T)$ and $(R \circ S) \circ T$ are sets, to prove that they are equal we need to show that each is a subset of the other. Let

$(x,y) \in R \circ (S \circ T)$. Then there exists $z \in C$ such that $(x,z) \in S \circ T$ and $(z,y) \in R$. Since $(x,z) \in S \circ T$, there exists $w \in B$ such that $(x,w) \in T$ and $(w,z) \in S$. Since $(w,z) \in S$ and $(z,y) \in R$, $(w,y) \in R \circ S$. Since $(x,w) \in T$ and $(w,y) \in R \circ S$, $(x,y) \in (R \circ S) \circ T$. Therefore $R \circ (S \circ T) \subseteq (R \circ S) \circ T$. We leave as Exercise 23 the proof that $(R \circ S) \circ T \subseteq R \circ (S \circ T)$. □

By Theorem 3.11, if f and g are permutations on a nonempty set S, then $(f \circ g)^{-1}$ is $g^{-1} \circ f^{-1}$. The following theorem generalizes this result to relations.

Theorem 9.2

If A, B, and C are the sets, $S \subseteq A \times B$, and $R \subseteq B \times C$, then $(R \circ S)^{-1} = S^{-1} \circ R^{-1}$.

Proof

Once again, $(R \circ S)^{-1}$ and $S^{-1} \circ R^{-1}$ are sets, so to prove that they are equal we need to show that each is a subset of the other. Let $(x,y) \in (R \circ S)^{-1}$. Then $(y,x) \in R \circ S$ and hence there exists $z \in B$ such that $(y,z) \in S$ and $(z,x) \in R$. Since $(y,z) \in S$, $(z,y) \in S^{-1}$. Since $(z,x) \in R$, $(x,z) \in R^{-1}$. Since $(x,z) \in R^{-1}$ and $(z,y) \in S^{-1}$, $(x,y) \in S^{-1} \circ R^{-1}$. Therefore $(R \circ S)^{-1} \subseteq S^{-1} \circ R^{-1}$. We leave as Exercise 24 the proof that $S^{-1} \circ R^{-1} \subseteq (R \circ S)^{-1}$. □

Theorem 9.3

Let A and B be sets, and let R be a relation from A to B. Then the following statements hold:

a. $R \circ i_A = i_B \circ R = R$.
b. $i_B \subseteq R \circ R^{-1}$ whenever the range of R is B.
c. $i_A \subseteq R^{-1} \circ R$ whenever the domain of R is A.

We leave the proof of Theorem 9.3 to Exercises 26 and 27, and in Exercise 20 we give an example to illustrate Theorem 9.3.

The powers of a relation R can be defined recursively from the definition of the composition of two relations. The remainder of this section is devoted to the study of R^n.

Definition

Let R be a relation on a set A. For each $n \in \mathbb{N}$, define R^n recursively by the initial value $R^1 = R$ and the recursion relation $R^{n+1} = R^n \circ R$.

Example 27

Let $R = \{(1,1),(3,1),(2,3),(4,2)\}$. Find R^2, R^3, and R^4.

Analysis

Since $R^2 = R \circ R$, $R^2 = \{(1,1),(2,1),(4,3),(3,1)\}$

Since $R^3 = R^2 \circ R$, $R^3 = \{(1,1),(3,1),(2,1),(4,1)\}$

Since $R^4 = R^3 \circ R$, $R^4 = \{(1,1),(3,1),(2,1),(4,1)\}$

Notice that $R^4 = R^3$. In Exercise 28 we ask for a proof that $R^n = R^3$ for each $n > 3$. ■

Example 28

Let $R = \{(1,4),(3,1),(3,2),(4,3)\}$. Find R^2, R^3, and R^4.

Analysis

$$R^2 = R \circ R = \{(1,3),(3,4),(4,1),(4,2)\}$$

$$R^3 = R^2 \circ R = \{(1,1),(1,2),(3,3),(4,4)\}$$

$$R^4 = R^3 \circ R = \{(1,4),(3,1),(3,2),(4,3)\} = R$$ ■

Intuitively, we can see that, for each $n \in \mathbb{N}$, R^n is going to be one of R, R^2, or R^3. To prove this, however, we need some properties of R^n.

Theorem 9.4 If R is a relation on a set A and $m,n \in \mathbb{N}$, then $R^m \circ R^n = R^{m+n}$.

Proof

The proof is by induction. Let

$$S = \{n \in \mathbb{N}\text{: if } m \in \mathbb{N} \text{ then } R^m \circ R^n = R^{m+n}\}$$

By definition, $R^m \circ R^1 = R^m \circ R = R^{m+1}$ for each $m \in \mathbb{N}$. Therefore $1 \in S$.

Suppose $n \in S$. Then $R^m \circ R^n = R^{m+n}$ for each $m \in \mathbb{N}$. So

$$\begin{aligned} R^m \circ R^{n+1} &= R^m \circ (R^n \circ R) && \text{(by definition)} \\ &= (R^m \circ R^n) \circ R && \text{(by Theorem 9.1)} \\ &= R^{m+n} \circ R && \text{(since } n \in S\text{)} \\ &= R^{m+n+1} && \text{(by definition)} \end{aligned}$$

Therefore $n + 1 \in S$, and by the principle of mathematical induction $S = \mathbb{N}$. □

Corollary 9.5

> If R is a relation and $m,n \in \mathbb{N}$, then $R^m \circ R^n = R^n \circ R^m$.

Proof

By Theorem 9.4, $R^m \circ R^n = R^{m+n}$ and $R^n \circ R^m = R^{n+m}$. Since $m + n = n + m$, the proof is complete. □

Theorem 9.6

> If R is a relation on a set A and $m,n \in \mathbb{N}$, then $(R^m)^n = R^{mn}$.

Proof

Again the proof is by induction. Let

$$S = \{n \in \mathbb{N}: \text{if } m \in \mathbb{N} \text{ then } (R^m)^n = R^{mn}\}$$

By definition, $(R^m)^1 = R^m = R^{m \cdot 1}$ for all $m \in \mathbb{N}$, so $1 \in S$.

Suppose $n \in S$. Then $(R^m)^n = R^{mn}$ for all $m \in \mathbb{N}$. Hence

$$\begin{aligned}
(R^m)^{n+1} &= (R^m)^n \circ R^m && \text{(by definition)}\\
&= R^{mn} \circ R^m && \text{(since } n \in S\text{)}\\
&= R^{mn+m} && \text{(by Theorem 9.4)}\\
&= R^{m(n+1)}
\end{aligned}$$

Therefore $n + 1 \in S$, and by the principle of mathematical induction $S = \mathbb{N}$. □

We now return to Example 28 to illustrate how we can compute R^n for any $n \in \mathbb{N}$, For example,

$$\begin{aligned}
R^{1000} &= (R^4)^{250}\\
&= R^{250} && \text{(since } R^4 = R\text{)}\\
&= (R^4)^{62} \circ R^2 && \text{(by Theorems 9.6 and 9.4)}\\
&= R^{62} \circ R^2 && \text{(since } R^4 = R\text{)}\\
&= [(R^4)^{15} \circ R^2] \circ R^2 && \text{(by Theorems 9.6 and 9.4)}\\
&= R^{15} \circ R^4 && \text{(by Theorems 9.6, 9.1, and 9.4 since } R^4 = R\text{)}\\
&= R^{15} \circ R && \text{(since } R^4 = R\text{)}\\
&= R^{16} && \text{(by Theorem 9.4)}\\
&= (R^4)^4 && \text{(by Theorem 9.6)}\\
&= R^4 && \text{(since } R^4 = R\text{)}\\
&= R && \text{(since } R^4 = R\text{)}
\end{aligned}$$

Exercises 9.3

18. Let $R = \{(1,0),(3,-2),(4,1),(-1,2)\}$ and $S = \{(2,-1),(1,7),(1,4),(0,3)\}$. Find the following.

a. R^{-1} **b.** S^{-1} **c.** $R \circ S$ **d.** $S \circ R$

e. $R^{-1} \circ S^{-1}$ **f.** $S^{-1} \circ R^{-1}$ **g.** $(R \circ S)^{-1}$ **h.** $(S \circ R)^{-1}$

19. Let $R = \{(x,y) \in \mathbb{R} \times \mathbb{R}: y = 2x^2 + 1\}$ and $S = \{(x,y) \in \mathbb{R} \times \mathbb{R}: y^2 = x + 3\}$. Find the following.

a. R^{-1} **b.** S^{-1} **c.** $R \circ S$ **d.** $S \circ R$

20. Let $A = \{1,2,3\}$, $B = \{4,5,6,7\}$, and $R = \{(1,5),(1,7),(2,6),(2,7),(3,4),(3,6)\}$. Find the following and compare your results with Theorem 9.3.

a. $R \circ i_A$ **b.** $i_B \circ R$ **c.** $R \circ R^{-1}$ **d.** $R^{-1} \circ R$

21. Let $R = \{(1,3),(1,2),(2,1),(2,2),(3,2),(3,3)\}$. Find the following.

a. R^2 **b.** R^3 **c.** R^5 **d.** R^8

22. Let $R = \{(1,1),(1,2),(2,3),(2,4),(3,4),(4,1),(4,5),(5,2)\}$. Find the following.

a. R^2 **b.** R^3 **c.** R^5 **d.** R^{15}

23. Let A, B, C, and D be sets and let $T \subseteq A \times B$, $S \subseteq B \times C$, and $R \subseteq C \times D$. Prove that $(R \circ S) \circ T \subseteq R \circ (S \circ T)$.

24. Let A, B, and C be sets and let $S \subseteq A \times B$ and $R \subseteq B \times C$. Prove that $S^{-1} \circ R^{-1} \subseteq (R \circ S)^{-1}$.

25. Let A, B, and C be sets and let $S \subseteq A \times B$, $T \subseteq A \times B$, and $R \subseteq B \times C$.

a. Prove that $R \circ (S \cup T) = (R \circ S) \cup (R \circ T)$.

b. Prove that $R \circ (S \cap T) \subseteq (R \circ S) \cap (R \circ T)$.

c. Give an example to show that $R \circ (S \cap T) \neq (R \circ S) \cap (R \circ T)$.

26. Let A and B be sets and let R be a relation from A to B. Prove the following.

a. $R \circ i_A = R$ **b.** $i_B \circ R = R$

27. Let A and B be sets and let R be a relation from A to B.

a. Prove that if A is the domain of R then $i_A \subseteq R^{-1} \circ R$.

b. Prove that if B is the range of R then $i_B \subseteq R \circ R^{-1}$.

28. With respect to the relation in Example 27, prove by induction that $R^n = R^3$ for each $n > 3$.

29. Let R and S be relations from A to B such that $R \subseteq S$, and let T be a relation from B to C. Prove that $T \circ R \subseteq T \circ S$.

9.4

USING MATRICES TO REPRESENT RELATIONS

There are many ways to represent a relation R from one finite set to another. As we have seen, we can simply list the ordered pairs that belong to R, we can draw the Cartesian graph of R, or we can draw the directed graph of R. In this section we discuss using a zero–one matrix to represent a relation on a finite set.

Let $A = \{a_1, a_2, \ldots, a_n\}$ be a set with exactly n elements and let R be a relation on A. We associate an $n \times n$ matrix M_R with the relation R by agreeing that $m_{ij} = 1$ if $(a_i, a_j) \in R$ and $m_{ij} = 0$ otherwise. For example, if $A = \{a_1, a_2, a_3, a_4\}$ and R is the relation whose directed graph is given in Figure 9.16a, then the associated matrix M_R is as shown in Figure 9.16b.

Figure 9.16

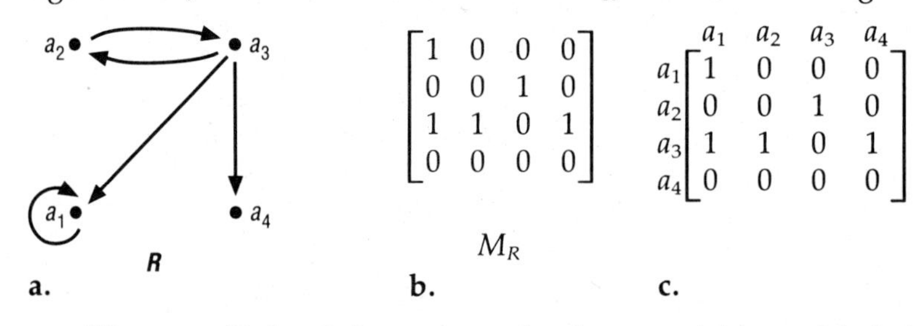

We may think of the rows and columns of M_R as labeled by the elements of A, as in Figure 9.16c. Thus the row labeled a_3 indicates that in the directed graph of R there is an arrow from a_3 to a_1, from a_3 to a_2, from a_3 to a_4, but not from a_3 to a_3. The column labeled a_3 indicates that the only arrow going to a_3 comes from a_2. In principle, there is some ambiguity in labeling the rows and columns of M_R by the elements of A. For example, if in Figure 9.16c $a_3 = 2$, then the row labeled by 2 is row 3 and not row 2. In practice, it is easy to distinguish between the rth row and the row labeled r. Every time we are given an n-element set A, there are $n!$ different ways of labeling the elements of A and so for most relations R there is more than one way to associate a zero–one matrix M_R.

It is easy to generalize the idea of a matrix associated with a relation on a nonempty set A to a matrix associated with a relation from A to B (where A and B are two different nonempty finite sets). As before, we represent R by the zero–one matrix $M_R = [m_{ij}]$, where $m_{ij} = 1$ if $(a_i, b_j) \in R$ and $m_{ij} = 0$ if $(a_i, b_j) \notin R$.

In this section, the order of the members of a set are always as indicated inside the set brackets; that is, if $A = \{1,4,3,2\}$ and $B = \{7,5,6\}$, then the rows of the matrix of a relation from A to B are labeled in order by 1,4,3, and 2, and the columns by 7,5, and 6. Likewise, the labels of the rows and columns of a matrix determine the order in which these "labels" are listed inside the set brackets.

Example 29

Let $A = \{1,2,3,4\}$, let $R = \{(m,n) \in A \times A: m < n\}$, and let $S = \{(m,n) \in A \times A: m = n$, or m and n are squares$\}$. Find M_R and M_S.

Analysis

$$M_R = \begin{array}{c} \\ 1 \\ 2 \\ 3 \\ 4 \end{array}\begin{array}{c} \begin{array}{cccc} 1 & 2 & 3 & 4 \end{array} \\ \begin{bmatrix} 0 & 1 & 1 & 1 \\ 0 & 0 & 1 & 1 \\ 0 & 0 & 0 & 1 \\ 0 & 0 & 0 & 0 \end{bmatrix} \end{array} \qquad M_S = \begin{array}{c} \\ 1 \\ 2 \\ 3 \\ 4 \end{array}\begin{array}{c} \begin{array}{cccc} 1 & 2 & 3 & 4 \end{array} \\ \begin{bmatrix} 1 & 0 & 0 & 1 \\ 0 & 1 & 0 & 0 \\ 0 & 0 & 1 & 0 \\ 1 & 0 & 0 & 1 \end{bmatrix} \end{array}$$

■

Each zero–one matrix represents a relation from one finite set to another. We simply label the rows and columns, let A be the set $\{a_1,a_2, \ldots ,a_m\}$ consisting of the row labels, let B be the set $\{b_1,b_2, \ldots ,b_n\}$ consisting of the column labels, and let R be the relation from A to B defined by $(a_i,b_j) \in R$ if and only if the (i,j)th entry is 1.

Example 30

Label the rows and columns of the following zero-one matrix as indicated. Find the relation that this matrix illustrates.

$$\begin{array}{c} \\ a \\ b \\ c \end{array}\begin{array}{c} \begin{array}{cccc} d & e & f & g \end{array} \\ \begin{bmatrix} 1 & 0 & 1 & 0 \\ 0 & 1 & 1 & 0 \\ 0 & 0 & 1 & 0 \end{bmatrix} \end{array}$$

Analysis

This matrix represents the relation

$\{(a,d),(a,f),(b,e),(b,f),(c,f)\}$ ■

Example 31

Find the matrix of the relation

$R = \{(1,c),(1,d),(2,b),(3,a),(3,c), (4,a)\}$

from $A = \{1,2,3,4\}$ to $B = \{a,b,c,d\}$.

Analysis

We label the rows 1,2,3,4 and the columns a,b,c,d as indicated here:

$$\begin{array}{c} \\ 1 \\ 2 \\ 3 \\ 4 \end{array}\begin{array}{c} \begin{array}{cccc} a & b & c & d \end{array} \\ \begin{bmatrix} & & & \\ & & & \\ & & & \\ & & & \end{bmatrix} \end{array}$$

We then set the six entries corresponding to the members of R to 1 and the remaining entries to 0 and obtain the matrix M_R:

$$\begin{array}{c} \\ 1 \\ 2 \\ 3 \\ 4 \end{array} \begin{array}{c} \begin{array}{cccc} a & b & c & d \end{array} \\ \begin{bmatrix} 0 & 0 & 1 & 1 \\ 0 & 1 & 0 & 0 \\ 1 & 0 & 1 & 0 \\ 1 & 0 & 0 & 0 \end{bmatrix} \end{array}$$

■

Example 32

If $A = \{a_1,a_2,a_3\}$ and $B = \{b_1,b_2,b_3,b_4,b_5\}$, which ordered pairs belong to the relation R represented by the matrix

$$M_R = \begin{bmatrix} 1 & 0 & 1 & 0 & 1 \\ 0 & 1 & 0 & 1 & 0 \\ 1 & 1 & 0 & 1 & 1 \end{bmatrix}$$

Analysis

Since $(a_i,b_j) \in R$ if and only if $m_{ij} = 1$,

$$R = \{(a_1,b_1),(a_1,b_3), (a_1,b_5),(a_2,b_2), (a_2,b_4),(a_3,b_1), (a_3,b_2),(a_3,b_4), (a_3,b_5)\}$$

■

The Boolean operations meet and join can be used to find the matrices that represent the intersection and union of two relations.

Theorem 9.7

> If R and S are relations on a set A, then $M_{R\cap S} = M_R \wedge M_S$ and $M_{R\cup S} = M_R \vee M_S$.

Proof

The (i,j)th entry of the matrix $M_{R\cap S}$, representing the intersection of R and S, is 1 provided that the (i,j)th entry of both M_R and M_S is 1, and it is 0 otherwise. But this is precisely the (i,j)th entry of $M_R \wedge M_S$.

The (i,j)th entry of the matrix $M_{R\cup S}$, representing the union of R and S, is 1 provided that the (i,j)th entry of M_R or M_S is 1, and it is 0 otherwise. But this is precisely the (i,j)th entry of $M_R \vee M_S$. □

Example 33

Let $A = \{1,2,3,\}$, $R = \{(1,1),(1,3),(2,3),(3,2)\}$, and $S = \{(1,1),(1,3),(2,1),(2,2),(3,3)\}$. Find M_R and M_S and use Theorem 9.7 to find $M_{R\cap S}$ and $M_{R\cup S}$.

Analysis

$$M_R = \begin{bmatrix} 1 & 0 & 1 \\ 0 & 0 & 1 \\ 0 & 1 & 0 \end{bmatrix} \quad \text{and} \quad M_S = \begin{bmatrix} 1 & 0 & 1 \\ 1 & 1 & 0 \\ 0 & 0 & 1 \end{bmatrix}$$

So

$$M_{R \cap S} = M_R \wedge M_S = \begin{bmatrix} 1 & 0 & 1 \\ 0 & 0 & 0 \\ 0 & 0 & 0 \end{bmatrix} \quad \text{and}$$

$$M_{R \cup S} = M_R \vee M_S = \begin{bmatrix} 1 & 0 & 1 \\ 1 & 1 & 1 \\ 0 & 1 & 1 \end{bmatrix}$$

Of course $R \cap S = \{(1,1),(1,3)\}$ and $R \cup S = \{(1,1),(1,3),(2,1),(2,2),(2,3),(3,2),(3.3)\}$. ■

As stated in the following theorem, the matrix for the composition of two relations is the Boolean product of the two matrices.

Theorem 9.8

> Suppose $A = \{a_1,a_2, \ldots, a_m\}$, $B = \{b_1, b_2, \ldots, b_n\}$, $C = \{c_1,c_2, \ldots, c_p\}$, R is a relation from A to B, and S is a relation from B to C. Then $M_{S \circ R} = M_R \odot M_S$.

Proof

Let $M_R = [r_{ij}]$, $M_S = [s_{ij}]$, and $M_{S \circ R} = [t_{ij}]$. Then $(a_i,c_k) \in S \circ R$ if and only if there exists $b_j \in B$ such that $(a_i,b_j) \in R$ and $(b_j,c_k) \in S$. Therefore $t_{ik} = 1$ if and only if there exists j such that $r_{ij} = s_{jk} = 1$. So t_{ik} is the (i,k)th entry of the Boolean product $M_R \odot M_S$. Hence $M_{S \circ R} = M_R \odot M_S$. □

Example 34

Let $A = \{a_1,a_2,a_3\}$, $B = \{b_1,b_2,b_3\}$, $C = \{c_1,c_2,c_3\}$, $R = \{(a_1,b_1),(a_1,b_3),(a_2,b_2),(a_2,b_3)\}$, and $S = \{(b_1,c_2),(b_2,c_1),(b_3,c_1),(b_3,c_3)\}$. Find $M_R \odot M_S$ and $M_{S \circ R}$.

Analysis

$S \circ R = \{(a_1,c_2),(a_1,c_1),(a_1,c_3),(a_2,c_1),(a_2,c_3)\}$,

$$M_R = \begin{bmatrix} 1 & 0 & 1 \\ 0 & 1 & 1 \\ 0 & 0 & 0 \end{bmatrix}, \quad M_S = \begin{bmatrix} 0 & 1 & 0 \\ 1 & 0 & 0 \\ 1 & 0 & 1 \end{bmatrix}, \quad M_R \odot M_S = \begin{bmatrix} 1 & 1 & 1 \\ 1 & 0 & 1 \\ 0 & 0 & 0 \end{bmatrix},$$

$$\text{and } M_{S \circ R} = \begin{bmatrix} 1 & 1 & 1 \\ 1 & 0 & 1 \\ 0 & 0 & 0 \end{bmatrix}$$

■

The following observation, though simple, is often useful: If A, B, and C are finite sets, R is a relation from A to B, and S is a relation from B to C, then the matrix $M_{S \circ R}$ is the matrix obtained from the ordinary product $M_R \times M_S$ by changing all nonzero entries to 1s.

A simple induction proof can be used to establish the following theorem. We leave this as Exercise 37.

Theorem 9.9 If R is a relation on a finite set A, then $M_{R^n} = (M_R)^{[n]}$.

Exercises 9.4

Throughout these exercises we continue the convention of ordering the elements of A or B as they appear between set brackets when considering a relation from A to B or on A.

30. Find the matrix that represents each of the following relations on $\{1,2,3,4\}$.

a. $\{(1,1),(1,4),(2,3)\}$ **b.** $\{(1,1),(1,2),(1,3),(1,4),(4,4)\}$

c. $\{(1,2),(2,3),(3,4),(4,1)\}$ **d.** $\{(1,1),(1,3),(2,2),(2,4),(3,3),(4,4)\}$

31. List the ordered pairs that belong to the relations on $\{1,2,3\}$ that correspond to the following matrices.

a. $\begin{bmatrix} 1 & 1 & 0 \\ 0 & 1 & 1 \\ 0 & 0 & 1 \end{bmatrix}$ **b.** $\begin{bmatrix} 0 & 1 & 0 \\ 1 & 0 & 1 \\ 1 & 1 & 0 \end{bmatrix}$ **c.** $\begin{bmatrix} 0 & 0 & 1 \\ 0 & 1 & 0 \\ 1 & 0 & 0 \end{bmatrix}$

32. Explain how we can quickly determine whether a relation R is the graph of a function by examining the matrix M_R that represents R.

33. Suppose that we are given the matrix that represents a relation R from a finite set A to a finite set B. Explain how we can find the matrix that represents R^{-1}.

34. Let R be the relation represented by the matrix $M_R = \begin{bmatrix} 1 & 0 & 1 \\ 0 & 1 & 1 \\ 1 & 1 & 0 \end{bmatrix}$

Find the matrix that represents the following.

a. R^{-1} **b.** R^2 **c.** R^3 **d.** R^4

35. Let R and S be relations on a set A represented by the matrices

$$M_R = \begin{bmatrix} 1 & 1 & 0 \\ 1 & 1 & 1 \\ 1 & 0 & 0 \end{bmatrix} \quad \text{and} \quad M_S = \begin{bmatrix} 0 & 1 & 0 \\ 0 & 1 & 1 \\ 1 & 1 & 0 \end{bmatrix}$$

Find the matrices that represent the following.

a. $R \cap S$ **b.** $R \cup S$ **c.** $R \circ S$ **d.** $S \circ R$

36. Let M_f be the matrix that represents the graph of a function f from a finite set A into a finite set B.

a. What conditions must M_f satisfy in order for f to be one to one?

b. What conditions must M_f satisfy in order for f to map A onto B?

37. Prove that if R is a relation on a finite set A then $M_{R^n} = (M_R)^{[n]}$.

38. Let $A = \{1,2,3,4,5\}$ and let R be the relation whose directed graph is as shown here. Let M_R be the associated matrix. Draw the directed graph of the relation obtained from the transpose of M_R.

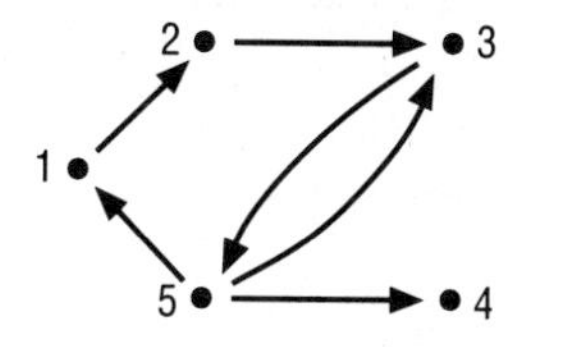

Chapter 9 Review Exercises

39. Let $R = \{(m,n) \in \mathbb{N} \times \mathbb{N}: 1/m < 1/n\}$.

a. List five members of R.

b. List five ordered pairs of natural numbers which are not members of R.

40. Let $A = \{1,2,3,4\}$. Draw the directed graph of each of the following relations on A.

a. $R = \{(1,1),(1,2)\}$

b. $S = \{(1,1),(2,2),(3,3),(4,4),(1,2),(2,1)\}$

c. $T = \{(1,2),(2,3),(3,4),(4,1),(2,2),(1,3)\}$

d. $U = \{(1,2),(1,3),(1,4),(1,1),(2,4)\}$

41. Let $A = \{1,2,3,4\}$. Write each of the following permutations on A in cycle notation.

a. $P_1 = \{(1,1),(2,3),(3,4),(4,2)\}$ **b.** $P_2 = \{(2,2),(4,4),(1,3),(3,1)\}$

c. $P_3 = \{(1,4),(4,1),(2,3),(3,2)\}$

42. Let $A = \{1,2,3,4\}$. Write each of the following permutations on A in ordered-pair notation.

a. (12)(34) **b.** (123) **c.** (2134) **d.** (34)

43. Let $A = \{1,2,3,4,5\}$. Draw the directed graphs of each of the following permutations on A.

a. (2135) **b.** (12)(34) **c.** (123)(45) **d.** (35)

44. Write in cycle notation the permutations on $\{1,2,3,4,5\}$ whose directed graphs are shown here.

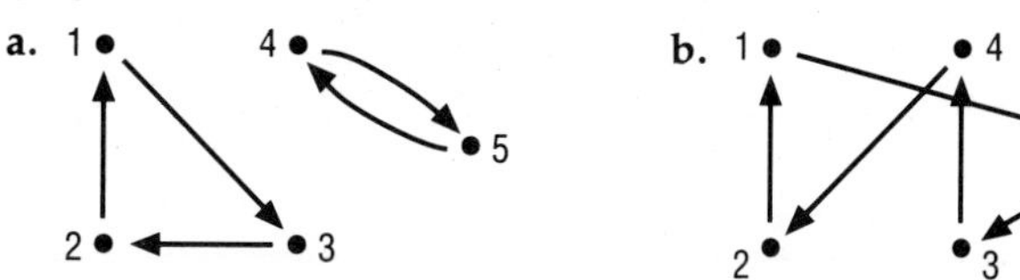

45. Let $A = \{1,2,3,4\}$. The following directed graph represents the restriction of a permutation P on A to $\{1,2,3\}$.

   a. Explain why the directed graph of P can have at most one arrow that starts at 4.

   b. Explain why the directed graph of P can have at most one arrow that ends at 3.

   c. Draw the directed graph of P.

46. Let $A = \{1,2,3,4,5\}$, and let $\alpha = (12)(345)$ and $\beta = (234)$ be permutations on A.

   a. Draw the directed graphs of α and β.

   b. Draw the directed graphs of α^{-1} and β^{-1}.

   c. Draw the directed graph of $\alpha^{-1} \circ \beta$.

   d. Draw the directed graph of $(\alpha \circ \beta) \circ \alpha^{-1}$.

   e. Write the permutation $\alpha^{-1} \circ \beta$ in cycle notation.

   f. Write the permutation $(\alpha \circ \beta) \circ \alpha^{-1}$ in cycle notation.

47. Let $R = \{(1,2),(3,4),(2,3),(3,1)\}$ and let $S = \{(1,1),(2,4),(4,1),(3,2)\}$. Find each of the following.

   a. R^{-1} b. S^{-1} c. $R \circ S^{-1}$ d. $(R \circ S)^{-1}$ e. $R^{-1} \circ S \circ R$

48. Let $A = \{4,3,2,1\}$. Using the order of the elements of A as they appear in A, find the matrix that represents each of the following relations.

   a. $\{(2,3),(1,4)\}$ b. $\{(2,3),(2,1),(1,4),(1,2)\}$

   c. $\{(1,1),(1,4),(2,2),(2,4),(3,3),(3,4),(4,4)\}$

   d. $\{(1,1),(3,3),(4,4)\}$

49. Let $A = \{3,2,1\}$. Assuming that the following matrices represent relations on A and using the order of the elements of A as they appear in A, write each of the indicated relations in ordered-pair format.

   a. $\begin{bmatrix} 1 & 0 & 0 \\ 0 & 1 & 0 \\ 0 & 1 & 1 \end{bmatrix}$ b. $\begin{bmatrix} 0 & 1 & 0 \\ 0 & 0 & 1 \\ 0 & 1 & 0 \end{bmatrix}$ c. $\begin{bmatrix} 0 & 1 & 1 \\ 1 & 0 & 1 \\ 1 & 1 & 0 \end{bmatrix}$

Chapter

10

EQUIVALENCE RELATIONS AND ORDER RELATIONS

In this chapter, we study two types of relations that arise in mathematics and its applications. One type consists of those relations that express the idea that items are equivalent, and the other type consists of those relations that express the concept that items are arranged in some order. The real numbers, for example, are ordered by the relation consisting of all ordered pairs (x,y) for which $y - x$ is positive. The idea that relations can establish an ordering of items is important in discrete mathematics because, for example, algorithms execute instructions in a specific order.

10.1

PROPERTIES OF RELATIONS

We begin by studying some important properties that a relation on A can satisfy. The property we study first is the property that every member of A is related to itself. For example, if R is the relation on $\mathbb{N}$ consisting of all ordered pairs (m,n) such that m divides n, then every natural number is related to itself. In other words, if $n \in \mathbb{N}$, then nRn, or $(n,n) \in R$.

Definition

If R is a relation on a set A, then R is **reflexive on A** provided that, for each $x \in A$, $(x,x) \in R$.

Notice that a relation R on a set A is reflexive on A provided that for each $x \in A$ the directed graph of R has a loop that begins and ends at x.

Example 1

Which directed graph in Figure 10.1 is the directed graph of a relation that is reflexive on $\{1,2,3,4\}$?

Figure 10.1

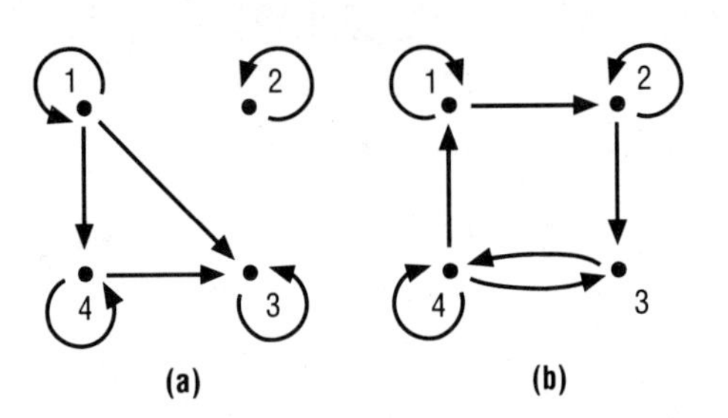

Analysis

Directed graph (a) is the directed graph of a relation that is reflexive on $\{1,2,3,4\}$ because for each $i = 1,2,3,4$ there is a loop that begins and ends at i. Directed graph (b) is the directed graph of a relation that is not reflexive on $\{1,2,3,4\}$ because there is no loop that begins and ends at 3. ■

Example 2

Which of the following relations are reflexive on $\{1,2,3,4\}$?

$$R_1 = \{(1,1),(1,2),(2,3),(2,4),(3,3),(3,4),(4,1),(4,4)\}$$
$$R_2 = \{(1,1),(1,4),(2,2),(2,4),(3,3),(3,4),(4,4)\}$$

$R_3 = \{(1,2),(2,1),(3,4),(4,3),(1,1),(2,2),(3,3)\}$
$R_4 = \varnothing$
$R_5 = \{(1,1),(2,2),(3,3),(4,4)\}$
$R_6 = \{(3,2),(4,4),(2,1),(1,1),(4,2),(3,3),(3,1),(2,2)\}$

Analysis

The relations R_2, R_5, and R_6 are reflexive on $\{1,2,3,4\}$ since (1,1), (2,2), (3,3), and (4,4) are members of each. The other relations are not reflexive on $\{1,2,3,4\}$ because they do not contain all these ordered pairs. In particular, $(2,2) \notin R_1$, $(4,4) \notin R_3$, and none of the ordered pairs belongs to R_4. ■

Notice that reflexivity is defined in terms of a relation R and a set A. The relation $R = \{(x,y) \in \mathbb{R} \times \mathbb{R}: xy \text{ is positive}\}$ is a relation both on $\mathbb{R} - \{0\}$ and on $\mathbb{R}$, but it is reflexive on $\mathbb{R} - \{0\}$ and not reflexive on $\mathbb{R}$.

There are relations with the property that, if x is related to y, then y is related to x. For example, let R be the relation represented by the directed graph in Figure 10.2. Notice that if there is an arrow starting at x and ending at y then there is also an arrow starting at y and ending at x. Thus yRx whenever xRy, or $(y,x) \in \mathbb{R}$ whenever $(x,y) \in \mathbb{R}$.

Figure 10.2

Definition

> A relation R is **symmetric** provided that if $(x,y) \in R$ then $(y,x) \in R$.

The relations R_3, R_4, and R_5 of Example 2 are symmetric since $(b,a) \in R_i$, $i = 3,4,5$, whenever $(a,b) \in R_i$. The other relations are not symmetric. In particular, R_1 is not symmetric because, for example, $(2,3) \in R_1$ but $(3,2) \notin R_1$. Also, R_2 is not symmetric because, for example, $(1,4) \in R_2$ but $(4,1) \notin R_2$. We see in a similar way that R_6 is not symmetric since $(3,2) \in R_6$ but $(2,3) \notin R_6$.

Note that symmetry is an internal property of a relation in the sense that we can determine whether a relation is symmetric without considering anything other than the given relation. The test to see that a given relation R is symmetric requires checking *every* member (a,b) of R to see if we can interchange the first and second terms. Thus, if *there is* a member (a,b) of R such that $(b,a) \notin R$, then R is not symmetric.

Example 3

Which of the following relations are symmetric?

$R_1 = \{(m,n) \in \mathbb{Z} \times \mathbb{Z}: m \text{ is } n\}$
$R_2 = \{(m,n) \in \mathbb{Z} \times \mathbb{Z}: m \geqslant n\}$
$R_3 = \{(m,n) \in \mathbb{Z} \times \mathbb{Z}: m = n \text{ or } m = -n\}$
$R_4 = \{(m,n) \in \mathbb{Z} \times \mathbb{Z}: n = m + 1\}$
$R_5 = \{(m,n) \in \mathbb{Z} \times \mathbb{Z}: m < n\}$
$R_6 = \{(m,n) \in \mathbb{Z} \times \mathbb{Z}: m + n \leqslant 0\}$
$R_7 = \varnothing$

Analysis

The relation R_1 is symmetric because if $(m,n) \in R_1$ then m is n, so $(n,m) \in R_1$. The relation R_3 is symmetric because if $(m,n) \in R_3$ then $m = n$ or $m = -n$. Thus $n = m$ or $n = -m$, so $(n,m) \in R_3$. Likewise, the relation R_6 is symmetric since if $(m,n) \in R_6$ then $m + n \leqslant 0$. Hence $n + m \leqslant 0$, so $(n,m) \in R_6$. Finally, R_7 is symmetric since, in the conditional proposition, if $(m,n) \in R_7$ then $(n,m) \in R_7$, the hypothesis is false. The other relations are not symmetric. In particular, R_2 is not symmetric since $(4,3) \in R_2$ but $(3,4) \notin R_2$. Likewise, R_4 and R_5 are not symmetric since $(1,2) \in R_i$ for each $i = 4,5$, but $(2,1) \notin R_i$ for any $i = 4,5$. ■

There are relations with the property that, if a is related to b and b is related to c, then a is related to c. For example, if $R = \{(m,n) \in \mathbb{N} \times \mathbb{N}: m < n\}$, then $(m,p) \in R$ whenever $(m,n) \in R$ and $(n,p) \in R$. Notice that we are simply saying that, if m,n, and p are natural numbers, $m < n$, and $n < p$, then $m < p$.

Definition

A relation R is **transitive** provided that if $(a,b) \in R$ and $(b,c) \in R$ then $(a,c) \in R$.

Once again we can see this concept by drawing the directed graph of a relation. The relation is transitive provided that, if there is an arrow from a to b and an arrow from b to c, then there is also an arrow from a to c.

Example 4

Which directed graph in Figure 10.3 is the directed graph of a relation that is transitive?

Figure 10.3

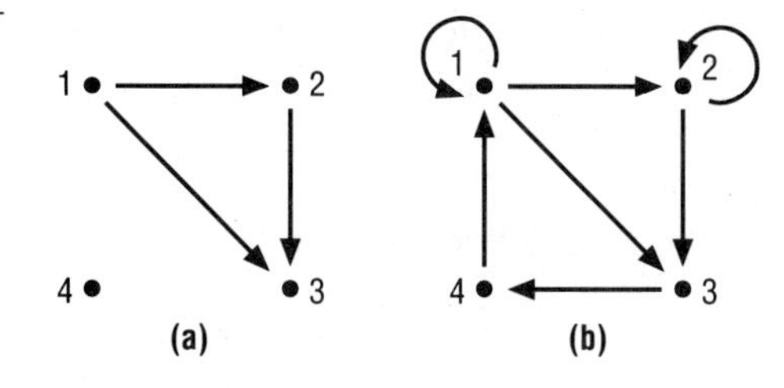

Analysis

Directed graph (a) is the directed graph of a relation that is transitive. The only vertex with an arrow that ends at it and an arrow that begins at it is 2; that is, there is an arrow that begins at 1 and ends at 2, and there is an arrow that begins at 2 and ends at 3. But there is also an arrow that begins at 1 and ends at 3. Directed graph (b) is the directed graph of a relation that is not transitive because there is an arrow that begins at 4 and ends at 1 and an arrow that begins at 1 and ends at 2, but there is no arrow that begins at 4 and ends at 2. ■

Like symmetry, transitivity is an internal property of a relation; that is, we can determine whether a relation is transitive without considering anything other than the given relation. Suppose that (a,b) and (b,c) are members of a relation R. If $a = b$, then $(a,c) = (b,c)$, so $(a,c) \in R$. Likewise, if $b = c$, then $(a,c) = (a,b)$, so $(a,c) \in R$. Therefore, in checking to see if a given relation is transitive, we need to consider only those members (a,b) and (b,c) of R where $a \neq b$ and $b \neq c$. However, the test to see if a given relation is transitive requires checking *every* pair $(a,b),(b,c)$, where $a \neq b$ and $b \neq c$, of members of R with a "middleman" b to see if the middleman can be eliminated. Thus $R = \{(1,2),(2,3),(2,4),(3,4),(1,3)\}$ is not a transitive relation because $\{(1,3),(3,4)\} \subseteq R$ but $(1,4) \notin R$. Hence R is not transitive, since we have found two ordered pairs $(1,3),(3,4)$ in R with a middleman 3 but the pair $(1,4)$ that we obtain by "eliminating" 3 is not a member of R.

The relations R_2, R_4, and R_5 of Example 2 are transitive. For each $i = 2,4,5$, we can show that R_i is transitive by verifying that if (a,b) and (b,c) belong to R_i then so does (a,c).

Example 5

Which of the following relations are transitive?

$R_1 = \{(1,2),(3,1),(3,2),(4,1),(4,2),(4,3)\}$
$R_2 = \{(1,1),(1,3),(1,4),(2,2),(3,1),(3,3),(4,1),(4,4)\}$
$R_3 = \{(1,1),(1,3),(2,2),(3,1),(3,4),(4,1),(4,4)\}$
$R_4 = \{(1,1),(1,3),(2,2),(3,1),(3,3),(3,4),(4,1),(4,4)\}$
$R_5 = \{(1,1),(1,2),(1,3),(1,4),(2,2),(2,3),(2,4),(3,3),(3,4)\}$

$R_6 = \{(1,1),(1,3),(3,1),(3,4)\}$
$R_7 = \{(1,2)\}$

Analysis

The relation R_1 is transitive because

$\{(3,1),(1,2)\} \subseteq R_1$ and $(3,2) \in R_1$
$\{(4,1),(1,2)\} \subseteq R_1$ and $(4,2) \in R_1$
$\{(4,3),(3,1)\} \subseteq R_1$ and $(4,1) \in R_1$
$\{(4,3),(3,2)\} \subseteq R_1$ and $(4,2) \in R_1$

Likewise, the relation R_5 is transitive because

$\{(1,2),(2,3)\} \subseteq R_5$ and $(1,3) \in R_5$
$\{(1,2),(2,4)\} \subseteq R_5$ and $(1,4) \in R_5$
$\{(1,3),(3,4)\} \subseteq R_5$ and $(1,4) \in R_5$
$\{(2,3),(3,4)\} \subseteq R_5$ and $(2,4) \in R_5$

Finally, the relation R_7 is transitive because there is no pair $(a,b),(c,d)$ of members of R_7 such that $b = c$. The other relations are not transitive. In particular,

$\{(3,1),(1,4)\} \subseteq R_2$ but $(3,4) \notin R_2$
$\{(3,1),(1,3)\} \subseteq R_3$ but $(3,3) \notin R_3$
$\{(1,3),(3,4)\} \subseteq R_4$ but $(1,4) \notin R_4$
$\{(3,1),(1,3)\} \subseteq R_6$ but $(3,3) \notin R_6$ ■

Example 6

Here is an interesting political example of a nontransitive relation. Suppose that a committee with three members is considering an issue with three alternatives P, Q, and R. The first member prefers P to Q and Q to R, the second member prefers Q to R and R to P, and the third prefers R to P and P to Q. If the three alternatives are voted on at the same time, each of P, Q, and R receives one vote and no decision is made. Suppose instead that a vote is taken on two alternatives and then the committee chooses between the preliminary winner and the third alternative. If P and Q are voted on first, P wins this vote. Then R wins the vote between P and R, so alternative R is chosen. If P and R are voted on first, R wins. Then Q wins the vote between Q and R, so alternative Q is chosen. If Q and R are voted on first, Q wins. Then P wins the vote between P and Q, so alternative P is chosen. The relation $>$ defined by $X > Y$ provided that X gets a majority vote when pitted against Y is not a transitive relation because $P > Q$ and $Q > R$ but $P \not> R$. ■

Let us summarize the use of a directed graph of a relation R on a set A to determine whether R is reflexive on A, symmetric, and transitive:

a. For R to be reflexive on A, it is necessary and sufficient that there is a loop at each point of A.
b. For R to be symmetric, it is necessary and sufficient that whenever there is an arrow from x to y there is also an arrow from y to x.
c. For R to be transitive, it is necessary and sufficient that whenever one can travel from one point to another in two steps there is always a corresponding shortcut going from the first point to the second point in one step.

These characterizations of reflexivity, symmetry, and transitivity can also be formulated in terms of a relation's associated matrix (see Exercise 5). In Figure 10.4, (a) is a directed graph of a relation that is reflexive on $\{1,2,3,4\}$, symmetric, and transitive; (b) is a directed graph of a relation that is reflexive on $\{1,2,3,4\}$ and transitive but not symmetric; (c) is a directed graph of a relation that is symmetric and transitive but not reflexive on $\{1\}$; and (d) is a directed graph of a relation that is reflexive on $\{1,2,3\}$ and symmetric but not transitive.

Figure 10.4

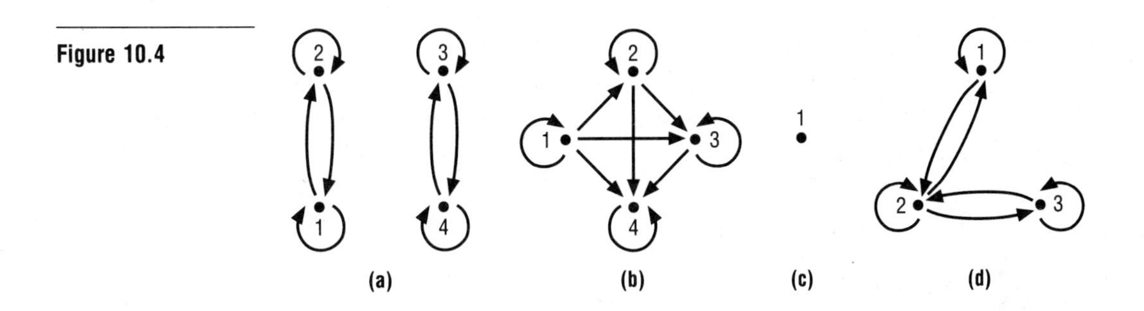

Example 7

Let $R = \{(x,y) \in \mathbb{R} \times \mathbb{R}: |x| + |y| = 1\}$. Then R is a symmetric relation that is not reflexive on the closed interval $[-1,1]$. Furthermore, this relation is not transitive because $(1,0) \in R$ and $(0,1) \in R$ but $(1,1) \notin R$. ■

The following theorem relates reflexivity and transitivity to composition.

Theorem 10.1

Let R be a relation on a set A. The following statements hold:

a. R is reflexive on A if and only if $i_A \subseteq R$.
b. R is transitive if and only if $R \circ R \subseteq R$.
c. R is reflexive on A and transitive if and only if $i_A \subseteq R = R \circ R$.

Proof

a. Suppose R is reflexive on A. Let $(a,a) \in i_A$. Since R is reflexive on A, $(a,a) \in R$. Therefore $i_A \subseteq R$. Suppose $i_A \subseteq R$. Let $a \in A$. Then $(a,a) \in i_A$ and, since $i_A \subseteq R$, $(a,a) \in R$. Therefore R is reflexive on A.

b. Suppose R is transitive. Let $(a,c) \in R \circ R$. Then by the definition of composition there exists b such that $(a,b) \in R$ and $(b,c) \in R$. Since R is transitive, $(a,c) \in R$. Hence $R \circ R \subseteq R$. Suppose $R \circ R \subseteq R$. Let (a,b) and (b,c) be members of R. By the definition of composition, $(a,c) \in R \circ R$. Since $R \circ R \subseteq R$, $(a,c) \in R$. Therefore R is transitive.

c. We leave the proof of part (c) as Exercise 6. □

According to Theorem 10.1b, if R is transitive, then $R^2 \subseteq R$. Theorem 10.2 gives a stronger result.

Theorem 10.2 If R is a transitive relation, then $R^n \subseteq R$ for each $n \in \mathbb{N}$.

Proof

We use mathematical induction to prove this theorem. Let $S = \{n \in \mathbb{N}: R^n \subseteq R\}$. Since $R^1 = R$, $1 \in S$. Suppose $n \in S$. Then $R^n \subseteq R$. To show that $n+1 \in S$, we must show that $R^{n+1} \subseteq R$. Let $(a,c) \in R^{n+1}$. Since $R^{n+1} = R^n \circ R$, there exists an element b such that $(a,b) \in R$ and $(b,c) \in R^n$. Since $R^n \subseteq R$, $(b,c) \in R$. Since R is transitive and (a,b) and (b,c) belong to R, $(a,c) \in R$. Therefore $R^{n+1} \subseteq R$ and $n+1 \in S$. By the principle of mathematical induction, $S = \mathbb{N}$. □

Exercises 10.1

1. Let $A = \{1,2,3,4\}$. Which of the following relations on A is reflexive on A? is symmetric? is transitive?

   a. $R_1 = \{(1,3),(3,1)\}$

   b. $R_2 = \{(1,2),(2,3),(3,4)\}$

   c. $R_3 = \{(1,1),(2,2,),(3,3),(4,4)\}$

   d. $R_4 = \{(1,1),(1,2),(2,2),(3,3),(4,4)\}$

   e. $R_5 = \{(1,1),(2,2),(3,3),(3,4),(4,3),(4,4)\}$

   f. $R_6 = \{(2,1),(2,4),(3,1),(3,4),(4,1),(4,2)\}$

   g. $R_7 = \{(1,1),(1,2),(1,3),(2,1),(2,2),(2,3)\}$

2. Which of the following relations on $\mathbb{R}$ is reflexive on $\mathbb{R}$? is symmetric? is transitive?

   a. $R_1 = \{(x,y) \in \mathbb{R} \times \mathbb{R}: xy = 0\}$

   b. $R_2 = \{(x,y) \in \mathbb{R} \times \mathbb{R}: |x - y| < 5\}$

   c. $R_3 = \{(x,y) \in \mathbb{R} \times \mathbb{R}: xy \neq 0\}$

   d. $R_4 = \{(x,y) \in \mathbb{R} \times \mathbb{R}: x \geqslant y\}$

   e. $R_5 = \{(x,y) \in \mathbb{R} \times \mathbb{R}: x^2 + y^2 = 1\}$

3. Which of the following relations on $\mathbb{Z}$ is reflexive on $\mathbb{Z}$? is symmetric? is transitive?
   a. $R_1 = \{(m,n) \in \mathbb{Z} \times \mathbb{Z}: m \neq n\}$
   b. $R_2 = \{(m,n) \in \mathbb{Z} \times \mathbb{Z}: m \text{ divides } n\}$
   c. $R_3 = \{(m,n) \in \mathbb{Z} \times \mathbb{Z}: m + n \text{ is even}\}$
   d. $R_4 = \{(m,n) \in \mathbb{Z} \times \mathbb{Z}: m + n \text{ is odd}\}$
   e. $R_5 = \{(m,n) \in \mathbb{Z} \times \mathbb{Z}: mn \geqslant 1\}$
   f. $R_6 = \{(m,n) \in \mathbb{Z} \times \mathbb{Z}: m = n + 1 \text{ or } m = n - 1\}$
4. Let $A = \{1,2,3,4\}$. Construct a directed graph such that the relation associated with the directed graph meets the following conditions.
   a. reflexive on A but neither symmetric nor transitive
   b. symmetric but not reflexive on A and not transitive
   c. transitive but not reflexive on A and not symmetric
   d. not reflexive on A and neither symmetric nor transitive
5. Let R be a relation on a nonempty set A and as usual let M_R denote the associated zero–one matrix. Complete each of the following statements meaningfully and correctly.
   a. The relation R is reflexive on A if and only if each entry ________ of M_R is a ________.
   b. The relation R is symmetric if and only if $M_R =$ ________.
   c. The relation R is transitive if and only if, whenever the (i,j)th entry of ________ is a 1, the (i,j)th entry of ________ is also a 1.
6. Let R be a relation on a set A. Prove that R is reflexive on A and transitive if and only if $i_A \subseteq R = R \circ R$ (*Hint*: Use Theorem 10.1a and b.)
7. Show that a relation R is symmetric if and only if $R = R^{-1}$.
8. Let R and S be relations that are reflexive on a set X. Show that $R \cap S$ and $R \cup S$ are reflexive on X.
9. Let R and S be symmetric relations. Show that $R \cap S$ and $R \cup S$ are symmetric.
10. Let R and S be transitive relations. Show that $R \cap S$ is transitive.
11. Find two transitive relations R and S such that $R \cup S$ is not transitive.

10.2 EQUIVALENCE RELATIONS

Often in mathematics, as in real life, it is convenient to think of two things that are different as being essentially the same. Consider the triangles ABC and DEF in Figure 10.5. Certainly as sets these two triangles are not equal; indeed they are disjoint. But they are congruent triangles, from which it follows that they share geometric properties such as area, length of hypotenuse, and measures of corresponding angles.

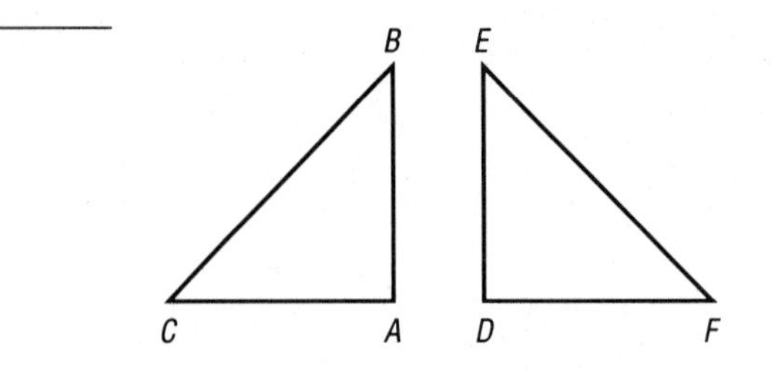

Figure 10.5

It is also possible to think of properties that the two triangles do not share. For example, the triangle on the right cannot be fitted exactly onto $\triangle ABC$ without lifting it out of the plane and flipping it over. Still, to anyone working in plane geometry these two triangles, being congruent, have no essential differences. The definition of congruence is somewhat involved, but we believe you have a notion of it and can agree that congruence shares with equality the following properties:

1. If A is a subset of the plane, then A is congruent to A (reflexivity on $\mathcal{P}(\mathbb{R} \times \mathbb{R})$).
2. If A and B are subsets of the plane and A is congruent to B, then B is congruent to A (symmetry).
3. If A, B, and C are subsets of the plane, A is congruent to B, and B is congruent to C, then A is congruent to C (transitivity).

Turning from geometry to number theory, we might agree that two integers are essentially the same provided that their difference is a multiple of 2. Let us write $m \simeq n$ provided that $m - n$ is a multiple of 2. We again have the following properties:

1. For each integer n, $n \simeq n$ (reflexivity on $\mathbb{Z}$).
2. If $m \simeq n$, then $n \simeq m$ (symmetry).
3. If $m \simeq n$ and $n \simeq p$, then $m \simeq p$ (transitivity).

If m and n are even integers, then their difference is even, so $m \simeq n$. Likewise, if m and n are odd integers, then their difference is even, so $m \simeq n$. If, however, m and n are integers, with one even and the other odd, then $m \not\simeq n$. Therefore the relation $\simeq$ splits the integers into two disjoint sets, the set consisting of all even integers and the set consisting of all odd integers. Recall that, if R is a relation on a set A and a is a member of the domain of R, then $R[a] = \{y \in A\colon aRy\}$. Also recall that we have agreed to write $[a]$ in place of $R[a]$ whenever this contraction causes no confusion. Therefore [0] is a name for the set of all even integers and [1] is a name for the set of all odd integers. Furthermore, $\mathbb{Z} = [0] \cup [1]$ and $[0] \cap [1] = \varnothing$.

Definition

A relation R on a set A is an **equivalence relation on A** provided that R is reflexive on A, symmetric, and transitive.

We have seen that the relation $\simeq$ defined on the set of integers is an equivalence relation on $\mathbb{Z}$. Also, the directed graph (a) in Figure 10.4 represents an equivalence relation on $\{1,2,3,4\}$.

Example 8

If $R = \{(m,n) \in \mathbb{Z} \times \mathbb{Z}: m = n \text{ or } m = -n\}$, then R is an equivalence relation on $\mathbb{Z}$.

Analysis

If $n \in \mathbb{Z}$, then $n = n$, so $(n,n) \in R$. Therefore R is reflexive on $\mathbb{Z}$. Suppose $(m,n) \in R$. If $m = n$ then $n = m$, and if $m = -n$ then $n = -m$. Therefore $(n,m) \in R$ and R is symmetric. Suppose $(m,n) \in R$ and $(n,p) \in R$. There are four cases to consider:

$$\begin{aligned}
m &= n \text{ and } n = p; &&\text{then } m = p \\
m &= n \text{ and } n = -p; &&\text{then } m = -p \\
m &= -n \text{ and } n = p; &&\text{then } m = -p \\
m &= -n \text{ and } n = -p; &&\text{then } m = p
\end{aligned}$$

In each case, $(m,p) \in R$. Therefore R is transitive. With respect to the relation R, $[0] = \{0\}$ and if $n \neq 0$ then $[n]$ is a set with two members, n and $-n$. ■

Definition

Let R be an equivalence relation on a set A. If $(a,b) \in R$, then a and b are said to be **equivalent.** (Since an equivalence relation is symmetric, this definition makes sense.) If $a \in A$, the set $R[a]$ (which consists of all elements that are equivalent to a) is called the **equivalence class of a.** If $b \in R[a]$, then b is called a **representative** of the equivalence class $R[a]$. We denote the set of all equivalence classes by A/R.

As we observed earlier, the directed graph (a) in Figure 10.4 represents an equivalence relation on $\{1,2,3,4\}$. There are two distinct equivalence classes; one of them is $R[1]$ and the other is $R[3]$. Thus the set $\{1,2,3,4\}/R$ of equivalence classes is $\{R[1],R[3]\}$.

Since we usually deal with only one equivalence relation at a time, we write $[a]$ as an abbreviation for $R[a]$. Notice that if R is an equivalence relation on A then the following hold:

1. For each $a \in A$, $a \in [a]$ (reflexivity on A).
2. If $b \in [a]$, then $a \in [b]$ (symmetry).
3. If $b \in [a]$ and $c \in [b]$, then $c \in [a]$ (transitivity).

Example 9

Let $R = \{(m,n) \in \mathbb{Z} \times \mathbb{Z}: m - n \text{ is divisible by } 3\}$. We prove that R is an equivalence relation on $\mathbb{Z}$.

Analysis

For each $n \in \mathbb{Z}$, $n - n = 0$ is divisible by 3. Therefore $(n,n) \in R$ and R is reflexive on $\mathbb{Z}$. Suppose $(m,n) \in R$. Then there is an integer q such that $m - n = 3q$. Hence

$$n - m = -(m - n) = -3q = 3(-q)$$

Therefore $(n,m) \in R$ and R is symmetric. Suppose $(m,n) \in R$ and $(n,p) \in R$. Then there are integers q_1 and q_2 such that $m - n = 3q_1$ and $n - p = 3q_2$. Hence

$$m - p = (m - n) + (n - p) = 3q_1 + 3q_2 = 3(q_1 + q_2)$$

Therefore $(m,p) \in R$ and R is transitive. Thus R is an equivalence relation on $\mathbb{Z}$. The equivalence class $R[0]$ is the set of all multiples of 3, the equivalence class $R[1]$ is the set of all integers that can be written in the form $3n + 1$, where n is an integer, and $R[2]$ is the set of all integers that can be written in the form $3n + 2$, where n is an integer. So $\mathbb{Z}/R = \{[0],[1],[2]\}$. ■

Example 10

Let $A = \{1,2,3,4\}$. For $X,Y \in \mathcal{P}(A)$, define $X \sim Y$ provided that X and Y have the same number of members. Evidently $\sim$ is an equivalence relation on $\mathcal{P}(A)$. List the equivalence classes of $\mathcal{P}(A)$ relative to $\sim$.

Analysis

There are five equivalence classes:

$[\varnothing] = \{\varnothing\}$
$[\{1\}] = \{\{1\},\{2\},\{3\},\{4\}\}$
$[\{1,2\}] = \{\{1,2\},\{1,3\},\{1,4\},\{2,3\},\{2,4\},\{3, 4\}\}$
$[\{1,2,3\}] = \{\{1,2,3\},\{1,2,4\},\{1,3,4\},\{2,3,4\}\}$
$[A] = \{A\}$

Thus $\mathcal{P}(A)/\sim = \{[\varnothing],[\{1\}],[\{1,2\}],[\{1,2,3\}],[A]\}$. ■

Notice that an equivalence class may have more than one name. For instance, in Example 10, $[\{2\}]$ is the same as $[\{1\}]$ However, $[\{2\}]$ is called the equivalence class of the set whose only member is 2, and $[\{1\}]$ is called the equivalence class of the set whose only member is 1.

Observe also in Example 10 that $\mathcal{P}(A)$ has 16 members, $[\varnothing]$ has 1 member, $[\{1\}]$ has 4 members, $[\{1,2\}]$ has 6 members, $[\{1,2,3\}]$ has 4 members, and $[A]$ has 1 member. Also

$$\mathcal{P}(A) = [\varnothing] \cup [\{1\}] \cup [\{1,2\}] \cup [\{1,2,3\}] \cup [A]$$

and finally, if $X,Y \in \mathcal{P}(A)$, then $[X] = [Y]$ or $[X] \cap [Y] = \varnothing$.

We end this section by proving that if R is an equivalence relation on A then the equivalence classes of two members of A are either equal or disjoint.

Theorem 10.3

If R is an equivalence relation on a set A and $x,y \in A$, then the following statements are equivalent:

a. xRy
b. $[x] = [y]$
c. $[x] \cap [y] \neq \varnothing$

Proof

We establish the equivalence of these three statements by proving that (a) implies (b), that (b) implies (c), and that (c) implies (a).

(a) implies (b): Suppose xRy. We prove that $[x] = [y]$ by showing that $[x] \subseteq [y]$ and $[y] \subseteq [x]$. Let $z \in [x]$. Then xRz. Since xRy and R is symmetric, yRx. Since R is transitive, yRx, and xRz, it follows that yRz. Therefore $z \in [y]$ and $[x] \subseteq [y]$. The proof that $[y] \subseteq [x]$ is similar, and it is left as Exercise 19.

(b) implies (c): Suppose $[x] = [y]$. Since R is reflexive on A, $x \in [x]$. Since $[x] \subseteq [y]$, $x \in [y]$. Hence $x \in [x] \cap [y]$ and $[x] \cap [y] \neq \varnothing$.

(c) implies (a): Suppose $[x] \cap [y] \neq \varnothing$. Then there exists $z \in [x] \cap [y]$, so $z \in [x]$ and $z \in [y]$. Therefore xRz and yRz. Since yRz and R is symmetric, zRy. Since R is transitive, xRz, and zRy, it follows that xRy. □

Exercises 10.2

12. Which of the relations in Exercise 1 are equivalence relations on $\{1,2,3,4\}$? For each such R_i list the equivalence classes.

13. Let A denote the set of all functions from $\mathbb{N}$ to $\mathbb{N}$, and consider the following relations on A.

a. $R_1 = \{(f,g) \in A \times A: f(5) = g(5)\}$

b. $R_2 = \{(f,g) \in A \times A: f(n) - g(n) = 1 \text{ for each } n \in \mathbb{N}\}$

c. $R_3 = \{(f,g) \in A \times A: |f(n) - g(n)| = 1 \text{ for each } n \in \mathbb{N}\}$

d. $R_4 = \{(f,g) \in A \times A: f(5) = g(5) \text{ or } f(7) = g(7)\}$

e. $R_5 = \{(f,g) \in A \times A: f(n) = g(n) \text{ for all } n \in \mathbb{N} \text{ such that } n > 10\}$

Which of these relations is an equivalence relation on A? For each $i = 1,2,3,4,5$ such that R_i is not an equivalence relation on A, determine the properties of an equivalence relation that R_i does not possess.

14. Let A and B be nonempty sets, let $f: A \to B$ be a function, and let $R = \{(a,b) \in A \times A: f(a) = f(b)\}$.

a. Prove that R is an equivalence relation on A.

b. Describe the equivalence classes of R.

15. Let $R = \{(x,y) \in \mathbb{R} \times \mathbb{R}: x^2 - y^2 = 0\}$.

a. Prove that R is an equivalence relation on $\mathbb{R}$.

b. List all members of $[5]$.

16. Let $R = \{(x,y) \in \mathbb{R} \times \mathbb{R}: y - x \text{ is an integer}\}$.

a. Prove that R is an equivalence relation on $\mathbb{R}$.

b. List four members of $[\sqrt{2}]$.

c. Which real numbers belong to $[-3]$?

17. Let $A_1 = \{1\}$, $A_2 = \{2,3\}$, $A_3 = \{4,5,6\}$, and $X = \bigcup_{i=1}^{3} A_i$.

a. List the members of X.

b. Let $R = \{(x,y) \in X \times X: \text{there is an } i \ (i = 1,2,3) \text{ such that } x \text{ and } y \text{ belong to } A_i\}$. Prove that R is an equivalence relation on X.

c. List all the members of $[5]$.

d. List all the members of X/R.

18. A student says, "I don't understand why you include 'reflexive on the set' in the definition of an equivalence relation. I can prove that any relation R on a set A that is symmetric and transitive is reflexive on A. My proof is as follows: Let $a \in A$. Since R is symmetric, (a,b) and (b,a) both belong to R. Since R is transitive, $(a,b) \in R$, and $(b,a) \in R$, it follows that $(a,a) \in R$. Therefore R is reflexive on A." What is wrong with the student's proof?

19. Prove that if R is an equivalence relation on a set A and xRy then $[y] \subseteq [x]$.

20. Let $\mathcal{M}_n$ denote the set of all $n \times n$ matrices with real entries. Define a relation $\sim$ on $\mathcal{M}_n$ by $A \sim B$ provided that there are invertible matrices P and Q such that $B = PAQ$. Prove that $\sim$ is an equivalence relation on $\mathcal{M}_n$.

21. Let X be a set and define a relation $\sim$ on $\mathcal{P}(X)$ by $A \sim B$ provided that $A \oplus B$ is a finite set.

a. Prove that $\sim$ is an equivalence relation on $\mathcal{P}(X)$.

b. Describe the members of $[\varnothing]$.

c. Describe the members of $[X]$.

10.3

CONGRUENCE

Congruence modulo n, where $n \in \mathbb{N}$ and $n > 1$, is such an important equivalence relation that we devote a section to its study. We use it extensively in Appendix F in our discussion of operations in a computer. We have seen one example, namely

$$\{(m,n) \in \mathbb{Z} \times \mathbb{Z}: m - n \text{ is divisible by } 2\}$$

The idea was first exploited by Karl Friedrich Gauss.

Definition

Let $n \in \mathbb{N}$. If $a,b \in \mathbb{Z}$, we say that ***a* is congruent to *b* modulo *n*,** written $a \equiv b \pmod{n}$, provided that $a - b$ is divisible by n. This is commonly said "a is congruent to b mod n" rather than "a is congruent to b modulo n."

Theorem 10.4

If $n \in \mathbb{N}$, congruence modulo n is an equivalence relation on the set of integers.

Proof

Let $a \in \mathbb{Z}$. Since $a - a = 0$, $a - a$ is divisible by n. Therefore $a \equiv a \pmod{n}$ and congruence modulo n is reflexive on $\mathbb{Z}$. Suppose $a \equiv b \pmod{n}$. Then $a - b$ is divisible by n. Therefore $b - a = -(a - b)$ is divisible by n and $b \equiv a \pmod{n}$. Hence congruence modulo n is symmetric. Suppose $a \equiv b \pmod{n}$ and $b \equiv c \pmod{n}$. Then there are integers q_1 and q_2 such that $a - b = nq_1$ and $b - c = nq_2$. Therefore

$$a - c = (a - b) + (b - c) = nq_1 + nq_2 = n(q_1 + q_2)$$

Hence $a \equiv c \pmod{n}$, and congruence modulo n is transitive. Therefore congruence modulo n is an equivalence relation on $\mathbb{Z}$. □

This concept is of no interest if $n = 1$, because any two integers are congruent modulo 1, so we only have one congruence class.

We have already seen that if $n = 2$ there are two congruence classes—one is the set of all even integers and the other is the set of all odd integers. If $n = 3$, there are three congruence classes:

$$\{\ldots,-6,-3,0,3,6,\ldots\}$$

$$\{\ldots,-5,-2,1,4,7,\ldots\}$$

$$\{\ldots,-7,-4,-1,2,5,\ldots\}$$

In general, for each $n \in \mathbb{N}$, there are n congruence classes. We use the division algorithm for integers (Theorem 3.4) to prove this:

Theorem 10.5

> If $n \in \mathbb{N}$, then each integer p is congruent, modulo n, to precisely one of the integers $0,1,2,\ldots,n-1$. In particular, $p \equiv r \pmod{n}$, where r is the remainder when p is divided by n.

Proof

Let $n \in \mathbb{N}$, and let a be an integer. By the division algorithm for integers, there are unique integers q and r such that $a = nq + r$ and $0 \leqslant r < n$. Therefore $a - r = nq$, and hence n divides $a - r$. So a is congruent, modulo n, to r, and r is one of the integers $0,1,2,\ldots,n-1$.

To complete the proof of the theorem, we need to show that, if a is congruent, modulo n, to s, and s is one of the integers $0,1,2,\ldots,n-1$, then $r = s$. Suppose $s \in \mathbb{Z}$, $a \equiv s \pmod{n}$, and $0 \leqslant s < n$. Then there is an integer t such that $a - s = nt$. Hence $a = nt + s$, and $r = s$ by the uniqueness of r in the division algorithm for integers. □

Example 11

In many automobiles, the odometer turns over at 100,000. Thus, if the odometer on a 1973 Chevrolet shows a reading of 43,218, we do not know whether the actual mileage is 43,218, 143,218, 243,218, or even more. The odometer tells us the true mileage modulo 100,000. ■

Example 12

One example of an application of congruence modulo n to error-correcting codes is in the publishing industry. Each book is assigned a ten-digit code number called an International Standard Book Number (ISBN). For example, the ISBN for *Foundations of Higher Mathematics* by Fletcher and Patty is 0-87150-164-3. This number is an aid in computerizing inventories, filling orders, and billing. The digits are divided into certain group codes, but the last digit (3, in this example) is a check digit. This digit permits publishers to detect an incorrect ISBN. The check digit has eleven possible values: 0, 1, 2, 3, 4, 5, 6, 7, 8, 9, X (the X represents the number 10). The check digit is chosen as follows. The first nine digits of the ISBN are multiplied by 10, 9, 8, 7, 6, 5, 4, 3, and 2, respectively. Then the resulting nine numbers are added to obtain a number p, and the check digit q is chosen so that $p + q \equiv 0 \pmod{11}$. In the case of *Foundations of Higher Mathematics*, the number p is

$$10 \cdot 0 + 9 \cdot 8 + 8 \cdot 7 + 7 \cdot 1 + 6 \cdot 5 + 5 \cdot 0 + 4 \cdot 1 + 3 \cdot 6 + 2 \cdot 4 = 0 + 72 + 56 + 7 + 30 + 0 + 4 + 18 + 8 = 195$$

Since division of 195 by 11 yields a quotient of 17 with a remainder of 8, 3 is the number between 0 and 10 such that $p + 3 \equiv 0 \pmod{11}$. ■

Example 13

In this example we discuss congruence modulo 7. The seven equivalence classes may be written [0],[1],[2],[3],[4],[5],[6]. Given an integer n, Theorem 10.5 tells us that $n \in [r]$, where r is the remainder when n is divided by 7. For instance, since $128 = 7 \cdot 18 + 2$, $128 \in [2]$, and since $216 = 7 \cdot 30 + 6$, $216 \in [6]$. ■

Theorem 10.6

If $n \in \mathbb{N}$ and $a,b \in \mathbb{Z}$, then a is congruent to b modulo n if and only if a and b have the same remainder when divided by n.

Proof

Suppose that a is congruent to b modulo n. Then there is an integer p such that $a - b = pn$. By Theorem 3.4, there are unique integers q_1, r_1, q_2, and r_2 such that $a = nq_1 + r_1$, $b = nq_2 + r_2$, and for each $i = 1,2$, $0 \leqslant r_i \leqslant n - 1$. By Theorem 10.5, $a \equiv r_1 \pmod{n}$ and $b \equiv r_2 \pmod{n}$. Since $-(n - 1) \leqslant r_1 - r_2 \leqslant n - 1$ and $r_1 \equiv r_2 \pmod{n}$, $r_1 = r_2$.

Now suppose a and b have the same remainder r when divided by n. By Theorem 10.5, $a \equiv r \pmod{n}$ and $b \equiv r \pmod{n}$. Therefore $a \equiv b \pmod{n}$. □

The following theorem shows that congruence mod n is preserved under sums and products.

Theorem 10.7

Let $n \in \mathbb{N}$ and let $a, b, c, d, \in \mathbb{Z}$. If $a \equiv b \pmod{n}$ and $c \equiv d \pmod{n}$, then

a. $a + c \equiv (b + d) \pmod{n}$
b. $ac \equiv (bd) \pmod{n}$

Proof

We prove (b) and leave the proof of (a) as Exercise 35. Suppose $a \equiv b \pmod{n}$ and $c \equiv d \pmod{n}$. Then there exist integers q_1 and q_2 such that $a - b = nq_1$ and $c - d = nq_2$. Hence

$$ac - bd = ac - ad + ad - bd = a(c - d) + d(a - b)$$
$$= anq_2 + dnq_1 = n(aq_2 + dq_1)$$

Therefore $ac \equiv (bd) \pmod{n}$. □

A common error in dealing with equivalence classes is attempting to define a function on a set of equivalence classes in such a way that the function depends on the members of the equivalence classes. We illustrate this with an example involving fractions.

Example 14

Let $\mathbb{F}$ be the set of all fractions. Can we define a function f mapping $\mathbb{F}$ into $\mathbb{Z}$ by the rule $f(m/n) = m + n$?

Analysis

Since $1/3 = 2/6$, for f to define a function we need $f(1/3) = f(2/6)$. However, $f(1/3) = 1 + 3 = 4$, and $f(2/6) = 2 + 6 = 8$. The problem is that $1/3$ is not really equal to $2/6$. Instead, they are equivalent under the equivalence relation $\sim$ (see Exercise 36) on $\mathbb{F}$ defined by $a/b \sim c/d$ provided that $ad = bc$. The set $\mathbb{Q}$ of rational numbers is really the set of all such equivalence classes, and a function on $\mathbb{Q}$ is well defined provided that it does not depend on the choice of the members of the equivalence classes. Thus we cannot define a function f mapping $\mathbb{Q}$ into $\mathbb{Z}$ by the rule $f([m/n]) = m + n$ because the definition depends on the choice of the member of $[m/n]$. We say that f *is not well defined* (but we really mean that it does not define a function). ■

We give two examples of attempts to define functions on sets of congruence classes. But first we introduce some notation. For each $n \in \mathbb{N}$, let $\mathbb{Z}_n$ denote the set of congruence classes of $\mathbb{Z}$ with respect to congruence modulo n, and for each $m \in \mathbb{Z}$, let $[m]_n$ denote the congruence class of $\mathbb{Z}$ with respect to congruence modulo n that contains m. Then, for example,

$$\mathbb{Z}_3 = \{[0]_3, [1]_3, [2]_3\}$$

$$\mathbb{Z}_5 = \{[0]_5, [1]_5, [2]_5, [3]_5, [4]_5\} \quad \text{and}$$

$$\mathbb{Z}_6 = \{[0]_6, [1]_6, [2]_6, [3]_6, [4]_6, [5]_6\}$$

Example 15

Can we define a function f mapping $\mathbb{Z}_3$ into $\mathbb{Z}_5$ by the rule $f([n]_3) = [n^2]_5$?

Analysis

Since $[2]_3 = [5]_3$, for f to define a function we need $f([2]_3) = f([5]_3)$. However, $f([2]_3) = [2^2]_5 = [4]_5$, $f([5]_3) = [5^2]_5 = [25]_5$, and $[4]_5 \neq [25]_5$. So we cannot define a function in this manner. Again we say that f *is not well defined.* ■

Example 16

Can we define a function f mapping $\mathbb{Z}_4$ into $\mathbb{Z}_8$ by the rule $f([n]_4) = [n^2]_8$?

Analysis

This time we prove that we can define such a function. We need to prove that, if $[m]_4 = [n]_4$, then $f([m]_4) = f([n]_4)$. Since $f([m]_4) = [m^2]_8$ and $f([n]_4) = [n^2]_8$, we need to prove that $[m^2]_8 = [n^2]_8$. In short, we need to prove that if $m \equiv n \pmod 4$ then $m^2 \equiv n^2 \pmod 8$.

Suppose $m \equiv n \pmod 4$. Then there is an integer p such that $m - n = 4p$. Thus $m = n + 4p$, and $m^2 = (n + 4p)^2 = n^2 + 8np + 16p^2$. Therefore $m^2 - n^2 = 8(np + 2p^2)$ and hence $m^2 \equiv n^2 \pmod 8$. In this case we say that f *is well defined.* ■

We conclude this section with a discussion of addition and multiplication in $\mathbb{Z}_n$. This material is used in Appendix F.

Definition

Let n be a natural number greater than 1. For $[a],[b] \in \mathbb{Z}_n$, define $[a] + [b] = [a + b]$ and $[a][b] = [ab]$.

Warning The definitions of addition and multiplication of congruence classes in $\mathbb{Z}_n$ appear deceptively simple, but there is a catch. Each congruence class has infinitely many aliases. For example, in $\mathbb{Z}_5$, $[9] = [14] = [-1]$ and $[3] = [13] = [-2]$. According to our definition, the sum of these two congruence classes is $[12]$ and $[17]$ and $[-3]$ and $[22]$ and $[7]$ and $[2]$. How do we know that all the answers are the same? The following theorem, which answers this question, is an immediate consequence of Theorem 10.7.

Theorem 10.8

Let n be a natural number greater than 1, and let a, b, c, and d be integers. If $[a] = [c]$ and $[b] = [d]$, then $[a + b] = [c + d]$ and $[ab] = [cd]$.

The proof of Theorem 10.8 is left as Exercise 43.

The sum and product of congruence classes in $\mathbb{Z}_n$ remain the same no matter what names of the congruence classes are used.

Example 17

Using only the names $[0],[1],[2],[3]$, and $[4]$, write addition and multiplication tables for $\mathbb{Z}_5$.

+	[0]	[1]	[2]	[3]	[4]
[0]	[0]	[1]	[2]	[3]	[4]
[1]	[1]	[2]	[3]	[4]	[0]
[2]	[2]	[3]	[4]	[0]	[1]
[3]	[3]	[4]	[0]	[1]	[2]
[4]	[4]	[0]	[1]	[2]	[3]

	[0]	[1]	[2]	[3]	[4]
[0]	[0]	[0]	[0]	[0]	[0]
[1]	[0]	[1]	[2]	[3]	[4]
[2]	[0]	[2]	[4]	[1]	[3]
[3]	[0]	[3]	[1]	[4]	[2]
[4]	[0]	[4]	[3]	[2]	[1]

■

We state, without proof, the following theorem:

Theorem 10.9 Let n be a natural number greater than 1, and let $[a],[b],[c] \in \mathbb{Z}_n$. Then

a. $([a] + [b]) + [c] = [a] + ([b] + [c])$ and $([a][b])[c] = [a]([b][c])$
b. $[a] + [b] = [b] + [a]$ and $[a][b] = [b][a]$
c. $[0]$ is the unique member of $\mathbb{Z}_n$ such that $[a] + [0] = [a]$, and $[1]$ is the unique member of $\mathbb{Z}_n$ such that $[a][1] = [a]$
d. $[1] \neq [0]$
e. $[a]([b] + [c]) = ([a][b]) + ([a][c])$
f. There is a unique member $[x]$ of $\mathbb{Z}_n$ such that $[a] + [x] = [0]$

Exercises 10.3

22. Which of the following are true?

a. $4 \equiv 2 \pmod 2$ **b.** $2 \equiv -1 \pmod 3$
c. $13 \equiv 1 \pmod{12}$ **d.** $2 \equiv 7 \pmod 6$
e. $29 \equiv 512 \pmod 2$ **f.** $12062 \equiv 15981 \pmod 3$

23. Find all members x of $\{0,1,2,3,4\}$ such that $2x + 1 \equiv 4 \pmod 5$.

24. Find all members x of $\{0,1,2,3,4,5,6\}$ such that $3x + 1 \equiv 0 \pmod 7$.

25. Let $S = \{1, 2, 3, 4, 5, 6, 7, 8, 9, 10, 11, 12\}$. Find all members x of S such that $2x + 3 \equiv 7 \pmod{12}$.

26. **a.** Describe the set consisting of all integers that are congruent to 6 modulo 2.

b. Describe the set consisting of all integers that are congruent to 1 modulo 2.

27. **a.** Describe the set consisting of all integers that are congruent to 1 modulo 3.

b. Show that this set is different from the set of odd integers.

28. With respect to congruence modulo n, describe the equivalence class [4] when **a.** $n = 2$ **b.** $n = 3$ **c.** $n = 5$ **d.** $n = 6$ **e.** $n = 8$

29. Give a description of each of the equivalence classes with respect to congruence modulo **a.** 4 **b.** 5 **c.** 6 **d.** 8 **e.** 9

30. With respect to congruence modulo 9, indicate which congruence class, [0],[1],[2],[3],[4],[5],[6],[7],[8], contains each of the following numbers.

    a. 354 **b.** 427 **c.** 187 **d.** 529 **e.** 684

31. Find the check digit for the ISBN whose first nine digits are 0-03-012813.

32. Either show that we can, or give an example to illustrate that we cannot, define a function f mapping $\mathbb{Z}_3$ into $\mathbb{Z}_6$ by the rule $f([n]_3) = [n^2]_6$.

33. Either show that we can, or give an example to illustrate that we cannot, define a function f mapping $\mathbb{Z}_6$ into $\mathbb{Z}_{12}$ by the rule $f([n]_6) = [n^2]_{12}$.

34. Either show that we can, or give an example to illustrate that we cannot, define a function f mapping $\mathbb{Z}_3$ into $\mathbb{Z}_5$ by the rule $f([n]_3) = [2n + 7]_5$.

35. Let $n \in \mathbb{N}$ and $a,b,c,d \in \mathbb{Z}$. Prove that, if $a \equiv b \pmod{n}$ and $c \equiv d \pmod{n}$, then $a + c \equiv (b + d) \pmod{n}$.

36. Let $\mathbb{F}$ denote the set of all fractions, and define a relation $\sim$ on $\mathbb{F}$ by $a/b \sim c/d$ provided that $ad = bc$. Prove that $\sim$ is an equivalence relation on $\mathbb{F}$.

37. Find a natural number n and integers a and b such that $a^2 \equiv b^2 \pmod{n}$ but $a \not\equiv b \pmod{n}$.

38. Prove by induction that, if $n \in \mathbb{N}$, $a,b \in \mathbb{Z}$, and $a \equiv b \pmod{n}$, then $a^m \equiv b^m \pmod{n}$ for each natural number n.

39. Prove that, if $n \in \mathbb{N}$ and $a,b \in \mathbb{Z}$, there is an integer c such that $a + c \equiv b \pmod{n}$.

40. Prove by induction that $10^n \equiv 1 \pmod{9}$ for each natural number n.

41. Write the addition table for $\mathbb{Z}_6$.

42. Write the multiplication table for $\mathbb{Z}_6$.

43. Prove Theorem 10.8.

10.4 PARTITIONS

We begin by restating the definition of a partition of a set X, which we first encountered in Section 2.5.

Definition

A **partition of a set** X is a collection of pairwise disjoint nonempty subsets of X whose union is X.

We need to make sure that we understand fully the meaning of each term used in this definition. We start with a set X. Each member of a partition of X is a nonempty subset of X, and each member of X belongs to some member of the partition. Furthermore if A and B are distinct members of the partition, then $A \cap B = \varnothing$. A partition of a set X into nine pairwise disjoint nonempty subsets is shown in Figure 10.6.

Figure 10.6

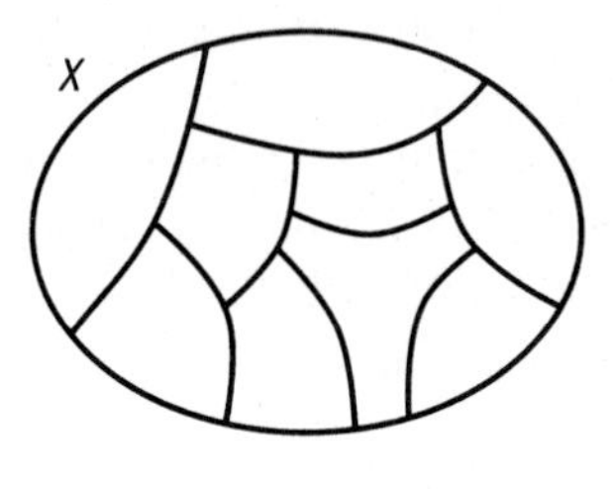

Example 18

Let $X = \{n \in \mathbb{N}: n \leqslant 10\}$. If $\mathcal{A} = \{\{1\},\{2,3,4\},\{5\},\{6,7\},\{8,9,10\}\}$, then $\mathcal{A}$ is a partition of X because

a. each member of $\mathcal{A}$ is a nonempty subset of X;
b. each member of X belongs to some member of $\mathcal{A}$; and
c. if A and B are distinct members of $\mathcal{A}$, then $A \cap B = \varnothing$.

However, $\mathcal{B} = \{\varnothing,\{1\},\{2,3,4\},\{5\},\{6,7\},\{8,9,10\}\}$ is not a partition of X because $\varnothing \in \mathcal{B}$. Likewise, $\mathcal{C} = \{\{1\},\{2,3,4\},\{5\},\{6,7,8,9\},\{10,11\}\}$ is not a partition of X because $\{10,11\}$ is not a subset of X. Also, $\mathcal{D} = \{\{1\},\{2,3,4\},\{6,7\},\{8,9,10\}\}$ is not a partition of X because $5 \in X$ but 5 is not a member of any member of $\mathcal{D}$. Finally, $\mathcal{E} = \{\{1\},\{2,3,4\},\{5\},\{6,7,8\},\{8,9,10\}\}$ is not a partition of X because $\{6,7,8\}$ and $\{8,9,10\}$ are distinct members of X but $\{6,7,8\} \cap [8,9,10\} \neq \varnothing$. ■

It is often convenient to use the concept of an indexed family of sets (see Section 2.3) to describe a partition.

Definition

Let X be a set and let $\mathcal{A} = \{A_i: i \in \Lambda\}$ be a collection of nonempty subsets of X. Then $\mathcal{A}$ is a **partition of X** provided that

a. $\cup\{A_i: i \in \Lambda\} = X$
b. if $A_i \neq A_j$, then $A_i \cap A_j = \varnothing$

We can also write this definition as follows:

Definition

Let X be a set and let $\mathcal{A} = \{A_i : i \in \Lambda\}$ be a collection of subsets of X. Then $\mathcal{A}$ is a **partition of X** provided that

a. For each $i \in \Lambda$, $A_i \subseteq X$
b. For each $i \in \Lambda$, $A_i \neq \emptyset$
c. If $x \in X$, then there exists $i \in \Lambda$ such that $x \in A_i$
d. If $x \in A_i \cap A_j$, then $A_i = A_j$

Example 19

For each $n \in \mathbb{N}$, let $A_n = \{2n - 1, 2n\}$. Then for each $n \in \mathbb{N}$, A_n is a nonempty subset of $\mathbb{N}$. Show that $\mathcal{A} = \{A_n : n \in \mathbb{N}\}$ is a partition of $\mathbb{N}$.

Analysis

It is sufficient to prove two statements:

a. If $m \in \mathbb{N}$, then there exists $n \in \mathbb{N}$ such that $m \in A_n$.
b. If $m \neq n$, then $A_m \cap A_n = \emptyset$. ■

Proof

a. Let $m \in \mathbb{N}$. If m is even, then $m \in A_{m/2}$. If m is odd, then $m \in A_{(m+1)/2}$.

b. Suppose $A_m \cap A_n \neq \emptyset$. Then there exists $p \in A_m \cap A_n$. Since $p \in A_m$, $p = 2m - 1$ or $p = 2m$. Since $p \in A_n$, $p = 2n - 1$ or $p = 2n$. Since $2m - 1$ and $2n - 1$ are odd, and $2m$ and $2n$ are even, there are two possibilities: either $p = 2m - 1$ and $p = 2n - 1$, or $p = 2m$ and $p = 2n$. In either case, $m = n$. □

Example 20

Show that, with respect to congruence modulo 3 on $\mathbb{Z}$, $\mathcal{A} = \{[0],[1],[2]\}$ is a partition of $\mathbb{Z}$.

Analysis

We have already seen that $[0]$ is the subset of $\mathbb{Z}$ consisting of all multiples of 3, $[1]$ is the subset of $\mathbb{Z}$ consisting of all integers of the form $3n + 1$, where n is an integer, and $[2]$ is the subset of $\mathbb{Z}$ consisting of all integers of the form $3n + 2$, where n is an integer. Therefore it is evident that $\mathcal{A}$ is a collection of nonempty subsets of $\mathbb{Z}$, $\mathbb{Z} = [0] \cup [1] \cup [2]$, and if $x \in [a] \cap [b]$ then $[a] = [b]$. Therefore $\mathcal{A}$ is a partition of $\mathbb{Z}$. ■

Example 20 gives a special case of a general result, namely, that if R is an equivalence relation on a nonempty set X then the collection of all equivalence classes is a partition of X.

Theorem 10.10

> If R is an equivalence relation on a nonempty set X, then X/R is a partition of X.

Proof

Let $x \in X$. Since R is reflexive on X, $(x,x) \in R$. Therefore $x \in [x]$. Since $[x] \in X/R$, each member of X belongs to some member of X/R. It also follows that each member of X/R is a nonempty subset of X. If $[x]$ and $[y]$ are distinct members of X/R, then, by Theorem 10.3, $[x] \cap [y] = \varnothing$. Therefore X/R is a partition of X. □

We give an example to show how we can use a partition of a nonempty set X to define an equivalence relation on X.

Example 21

Let X and $\mathscr{A}$ be defined as in Example 18. Let $R = \{(m,n) \in X \times X$: m and n belong to the same member of $\mathscr{A}\}$. Show that R is an equivalence relation on X.

Analysis

If $n \in X$, it is certainly true that n and n belong to the same member of $\mathscr{A}$. Hence $(n,n) \in R$ and R is reflexive on X. If $(m,n) \in R$, then m and n belong to the same member of $\mathscr{A}$. Hence $(n,m) \in R$ and R is symmetric. Suppose $(m,n),(n,p) \in R$. Then m and n belong to the same member A of $\mathscr{A}$, and n and p belong to the same member B of $\mathscr{A}$. Since $n \in A \cap B$, by Theorem 10.3, $A = B$. Therefore m and p belong to the same member of $\mathscr{A}$ and $(m,p) \in R$. Therefore R is transitive, and we have proved that R is an equivalence relation on X. ■

Notice that in Example 21

$$R[1] = \{1\}$$

$$R[2] = R[3] = R[4] = \{2,3,4\}$$

$$R[5] = \{5\}$$

$$R[6] = R[7] = \{6,7\} \quad \text{and}$$

$$R[8] = R[9] = R[10] = \{8,9,10\}$$

So $X/R = \{R[1],R[2],R[5],R[6],R[8]\} = \mathscr{A}$. By Theorem 10.10, X/R is a partition of X, and we have just seen that it is the same as the partition $\mathscr{A}$ that we started with.

Example 21 provides us with a model of how to prove in general that we can use a partition of a nonempty set X to define an equivalence relation on X. As the following theorem indicates, the resulting collection of equivalence classes is the partition that we started with.

Theorem 10.11

> Let $\mathscr{A}$ be a partition of a nonempty set X, and let $R = \{(x,y) \in X \times X$: x and y belong to the same member of $\mathscr{A}\}$. Then R is an equivalence relation on X. Furthermore $X/R = \mathscr{A}$.

Proof

As we have indicated previously, the proof is modeled on the pattern provided in Example 21. Let $a \in X$. Then a and a belong to the same member of $\mathscr{A}$, so $(a,a) \in R$. Therefore R is reflexive on X.

Let $(a,b) \in R$. Then a and b belong to the same member of $\mathscr{A}$, so $(b,a) \in R$. Therefore R is symmetric.

Let $(a,b),(b,c) \in R$. Then there are members A and B of $\mathscr{A}$ such that a and b belong to A and b and c belong to B. Since $b \in A \cap B$, by Theorem 10.3, $A = B$. Therefore a and c belong to the same member of $\mathscr{A}$, so $(a,c) \in R$. Hence R is transitive. Therefore R is an equivalence relation on X.

We need to prove that $X/R = \mathscr{A}$. Since both X/R and $\mathscr{A}$ are sets, we need to prove that each is a subset of the other. Let $A \in \mathscr{A}$. Since $\mathscr{A}$ is a partition, $A \neq \varnothing$. Therefore there exists $x \in A$. Now $R[x] \in X/R$, and to show that $A \in X/R$ it is sufficient to prove that $A = R[x]$. Once again, both A and $R[x]$ are sets, so we need to show that each is a subset of the other. Let $y \in A$. Then x and y belong to the same member A of $\mathscr{A}$, so $(x,y) \in R$, or $y \in R[x]$. We have proved that $A \subseteq R[x]$. Let $y \in R[x]$. Then $(x,y) \in R$, so x and y belong to the same member of $\mathscr{A}$. Since $x \in A$, $y \in A$. Hence $R[x] \subseteq A$ and $A = R[x]$. We have proved that $\mathscr{A} \subseteq X/R$.

Now let $A \in X/R$. Since X/R is the set of equivalence classes, A is an equivalence class. Therefore there exists $x \in X$ such that $A = R[x]$. Let B be the member of $\mathscr{A}$ such that $x \in B$. To prove that $A \in \mathscr{A}$, it is sufficient to show that $R[x] = B$. Let $y \in R[x]$. Then x and y belong to the same member of $\mathscr{A}$. Since $x \in B$, $y \in B$. Therefore $R[x] \subseteq B$. Now let $y \in B$. Then x and y belong to the same member of $\mathscr{A}$, so $(x,y) \in R$. Hence $y \in R[x]$ and $B \subseteq R[x]$. Therefore $B = R[x]$. □

Example 22

Let $R = \{(m,n) \in \mathbb{Z} \times \mathbb{Z}$: $m \equiv n \pmod 5\}$. Then, by Theorem 10.4, R is an equivalence relation on $\mathbb{Z}$. The set $\mathbb{Z}/R$ of equivalence classes is $\{[0],[1],[2],[3],[4]\}$. By Theorem 10.10, $\mathbb{Z}/R$ is a partition of $\mathbb{Z}$. Let $S = \{(m,n) \in \mathbb{Z} \times \mathbb{Z}$: m and n belong to the same member of $\mathbb{Z}/R\}$. Then, by Theorem 10.11, S is an equivalence relation on $\mathbb{Z}$. But since m and n

belong to the same member of $\mathbb{Z}/R$ if and only if they are congruent mod 5, S is the same as R.

This example illustrates Theorem 10.12. ■

Theorem 10.12 Let R be an equivalence relation on a nonempty set X, and let S be the equivalence relation (use Theorem 10.11 to see that it is an equivalence relation) defined by $S = \{(x,y) \in X \times X: x \text{ and } y \text{ belong to the same member of } X/R\}$. Then $R = S$.

Proof

Let $(x,y) \in R$. Then $y \in R[x]$. Since $x \in R[x]$, x and y belong to the same member of X/R. Therefore $(x,y) \in S$ and $R \subseteq S$.

Let $(x,y) \in S$. Then x and y belong to the same member of X/R, so there exists $z \in X$ such that $x,y \in R[z]$. Then $(z,x) \in R$ and $(z,y) \in R$. Since R is symmetric and $(z,x) \in R$, $(x,z) \in R$. Since R is transitive, $(x,z) \in R$, and $(z,y) \in R$, it follows that $(x,y) \in R$. Therefore $S \subseteq R$ and $R = S$. □

Example 23 The directed graph in Figure 10.7 represents an equivalence relation R on $X = \{1,2,3,4,5,6,7,8\}$. The equivalence classes are [1],[2],[4], and [7], so $X/R = \{\{1\},\{2,3\},\{4,5,6\},\{7,8\}\}$. Notice that X/R is a partition of X. ■

Figure 10.7

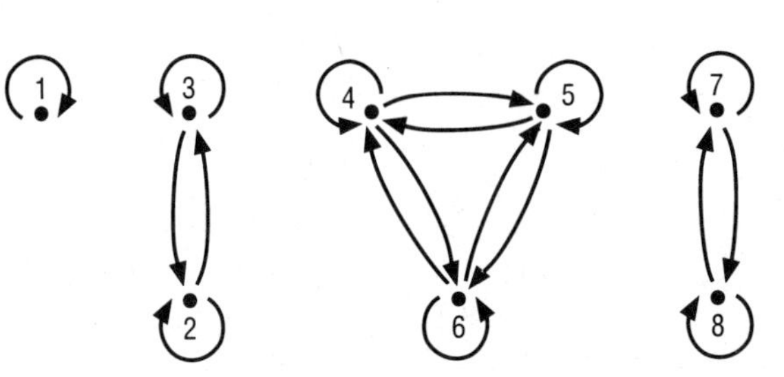

■

Example 24 If $X = \{1,2,3,4,5,6,7,8,9\}$ and $\mathcal{A} = \{\{1,2\},\{3,4,5\},\{6,7,8,9\}\}$, then $\mathcal{A}$ is a partition of X. If $R = \{(x,y) \in X \times X: x \text{ and } y \text{ belong to the same member of } \mathcal{A}\}$, then the directed graph in Figure 10.8 represents the equivalence relation R. Notice that

$$R = \{(1,1),(1,2),(2,1),(2,2),(3,3),(3,4),(3,5),(4,3),(4,4),(4,5),(5,3),(5,4),(5,5),(6,6),(6,7),(6,8),(6,9),(7,6),(7,7),(7,8),(7,9),(8,6),(8,7),(8,8),(8,9),(9,6),(9,7),(9,8),(9,9)\}$$

Figure 10.8

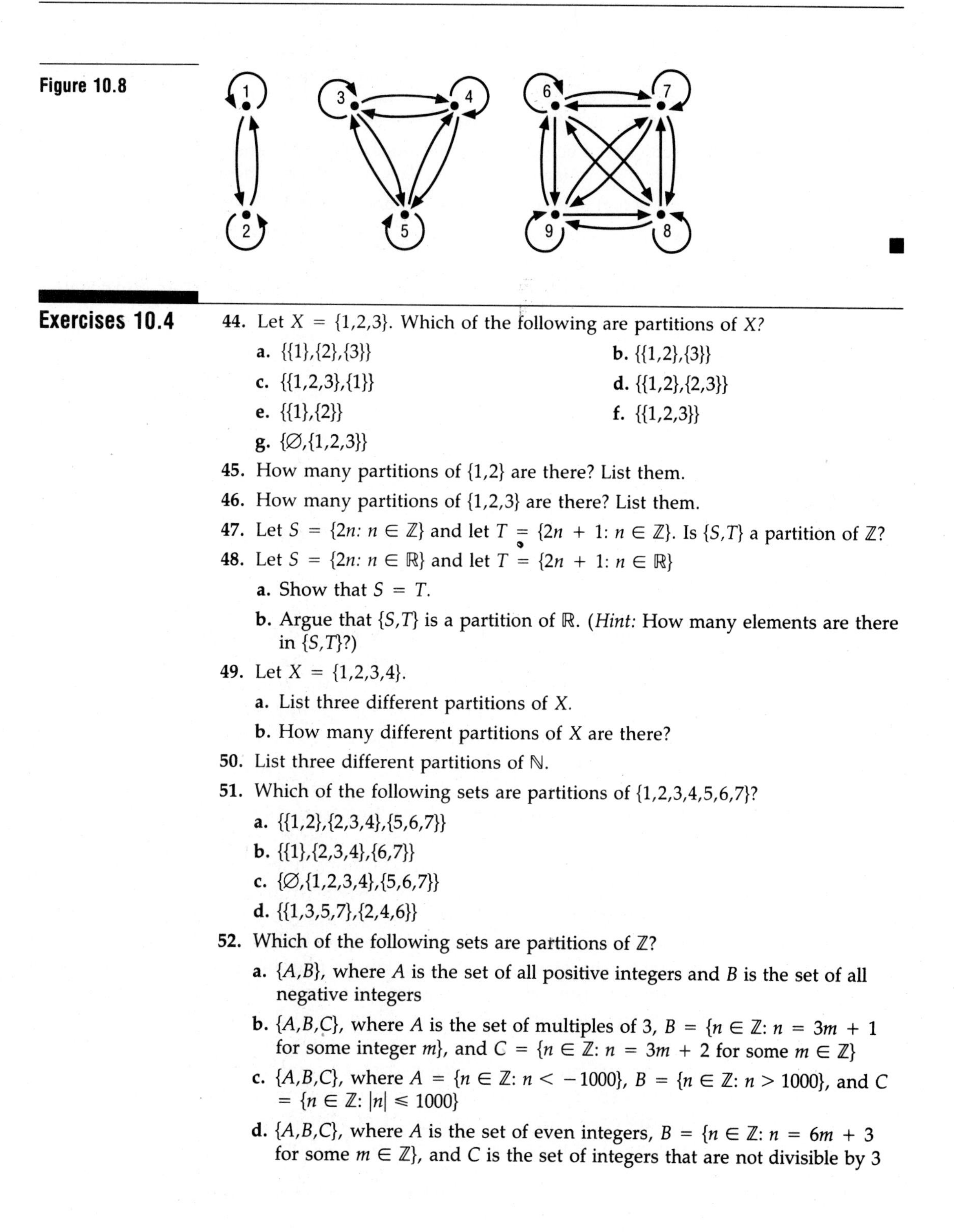

■

Exercises 10.4

44. Let $X = \{1,2,3\}$. Which of the following are partitions of X?

   a. $\{\{1\},\{2\},\{3\}\}$
   b. $\{\{1,2\},\{3\}\}$
   c. $\{\{1,2,3\},\{1\}\}$
   d. $\{\{1,2\},\{2,3\}\}$
   e. $\{\{1\},\{2\}\}$
   f. $\{\{1,2,3\}\}$
   g. $\{\varnothing,\{1,2,3\}\}$

45. How many partitions of $\{1,2\}$ are there? List them.

46. How many partitions of $\{1,2,3\}$ are there? List them.

47. Let $S = \{2n: n \in \mathbb{Z}\}$ and let $T = \{2n + 1: n \in \mathbb{Z}\}$. Is $\{S,T\}$ a partition of $\mathbb{Z}$?

48. Let $S = \{2n: n \in \mathbb{R}\}$ and let $T = \{2n + 1: n \in \mathbb{R}\}$

   a. Show that $S = T$.

   b. Argue that $\{S,T\}$ is a partition of $\mathbb{R}$. (*Hint:* How many elements are there in $\{S,T\}$?)

49. Let $X = \{1,2,3,4\}$.

   a. List three different partitions of X.

   b. How many different partitions of X are there?

50. List three different partitions of $\mathbb{N}$.

51. Which of the following sets are partitions of $\{1,2,3,4,5,6,7\}$?

   a. $\{\{1,2\},\{2,3,4\},\{5,6,7\}\}$
   b. $\{\{1\},\{2,3,4\},\{6,7\}\}$
   c. $\{\varnothing,\{1,2,3,4\},\{5,6,7\}\}$
   d. $\{\{1,3,5,7\},\{2,4,6\}\}$

52. Which of the following sets are partitions of $\mathbb{Z}$?

   a. $\{A,B\}$, where A is the set of all positive integers and B is the set of all negative integers

   b. $\{A,B,C\}$, where A is the set of multiples of 3, $B = \{n \in \mathbb{Z}: n = 3m + 1$ for some integer $m\}$, and $C = \{n \in \mathbb{Z}: n = 3m + 2$ for some $m \in \mathbb{Z}\}$

   c. $\{A,B,C\}$, where $A = \{n \in \mathbb{Z}: n < -1000\}$, $B = \{n \in \mathbb{Z}: n > 1000\}$, and $C = \{n \in \mathbb{Z}: |n| \leq 1000\}$

   d. $\{A,B,C\}$, where A is the set of even integers, $B = \{n \in \mathbb{Z}: n = 6m + 3$ for some $m \in \mathbb{Z}\}$, and C is the set of integers that are not divisible by 3

53. Let $R = \{(m,n) \in \mathbb{Z} \times \mathbb{Z}: m \equiv n \pmod 6\}$ and let $S = \{(m,n) \in \mathbb{Z} \times \mathbb{Z}: m \equiv n \pmod 3\}$. Prove that every member of $\mathbb{Z}/R$ is a subset of some member of $\mathbb{Z}/S$. *Do not use Exercise 54.*

*54. Let R and S be equivalence relations on a set A. Prove that $R \subseteq S$ if and only if every member of A/R is a subset of some member of A/S.

55. Let $\mathcal{A}$ and $\mathcal{B}$ be partitions of a nonempty set X.

  a. Let $\mathcal{P} = \{A \cap B: A \in \mathcal{A} \text{ and } B \in \mathcal{B}\}$. Is $\mathcal{P}$ a partition of X? Prove your answer.

  b. Let $\mathcal{Q} = \{A \cup B: A \in \mathcal{A} \text{ and } B \in \mathcal{B}\}$. Is $\mathcal{Q}$ a partition of X? Prove your answer.

10.5 CLOSURES OF RELATIONS

Bell Laboratories is a major employer of mathematicians, and discrete mathematics plays an important role in the telephone industry. We begin this section with an oversimplified and unrealistic example. A military communications network has data centers in Denver, Washington, Atlanta, Seattle, Minneapolis, Little Rock, and Chicago. There are direct one-way telephone lines from Denver to Washington, from Denver to Atlanta, from Denver to Minneapolis, from Washington to Seattle, from Washington to Minneapolis, from Atlanta to Minneapolis, from Atlanta to Little Rock, from Atlanta to Chicago, from Chicago to Washington, from Little Rock to Minneapolis, and from Seattle to Chicago.

Let $R = \{(C,D)$: there is a direct telephone line from C to $D\}$. How can we tell if there is a telephone line (possibly indirect) from one city to another? The relation R cannot be used to make this determination because, for example, there is a telephone line from Denver to Chicago; one such line is composed of two lines (from Denver to Atlanta and from Atlanta to Chicago), another is composed of three links (from Denver to Washington, from Washington to Seattle, and from Seattle to Chicago). There is, however, no direct telephone line from Denver to Chicago, so the relation R is not transitive. One of the problems we discuss in this section is the problem of finding all pairs of cities that have a link.

We have previously discussed three properties of a relation—reflexivity on a set, symmetry, and transitivity—and there are other properties to be discussed later. At this point, it is evident that we can discuss the following abstract property P, which a relation might or might not satisfy.

Definition

Let P be a property that a relation might satisfy, and let R be a relation on a set A. If there is a relation S on A such that

a. S has property P,
b. $R \subseteq S$, and
c. if T is a relation such that $R \subseteq T$ and T has property P, then $S \subseteq T$,

then S is called **the closure of R with respect to P.**

Intuitively, S is the smallest relation with property P that contains R. Note that the word "the" in the phrase "the closure of R with respect to P" assures that no relation can have two different closures with respect to the given property P. It is your job to ascertain that this assurance is correct.

We next show how to find the closures of relations with respect to reflexivity on a set, symmetry, and transitivity. It is convenient to use the terminology *reflexive closure of R on A* rather than the more cumbersome closure of R with respect to reflexivity on A. Also, we use the terminology *symmetric closure of R* and *transitive closure of R.*

Example 25

Let $A = \{1,2,3,4\}$ and let $R = \{(1,2),(2,1),(2,2),(3,2)\}$. Find the reflexive closure of R on A.

Analysis

R is not reflexive on A, but we can produce a reflexive relation on A that contains R by adding (1,1), (3,3), and (4,4) to R. So $S = \{(1,1),(1,2),(2,1),(2,2),(3,2),(3,3),(4,4)\}$ is the reflexive closure ot R on A. ■

This example illustrates Theorem 10.13. The proof, which is simple, is left as Exercise 56.

Theorem 10.13

If R is a relation on a set A, the reflexive closure of R is $R \cup \{(a,a): a \in A\}$.

Example 26

If $R = \{(m,n) \in \mathbb{N} \times \mathbb{N}: m > n\}$, the reflexive closure of R on $\mathbb{N}$ is $R \cup \{(m,n) \in \mathbb{N} \times \mathbb{N}: m = n\}$ or $\{(m,n) \in \mathbb{N} \times \mathbb{N}: m \geqslant n\}$. ■

Example 27

If $R = \{(1,2),(2,1),(2,2),(3,2)\}$, find the symmetric closure of R.

Analysis

R is not symmetric. We can, however, produce a symmetric relation from R by adding $(2,3)$ to R. So $S = \{(1,2),(2,1),(2,2),(2,3),(3,2)\}$ is the symmetric closure of R. ■

This example illustrates Theorem 10.14.

Theorem 10.14 If R is a relation on a set A, the symmetric closure of R is $R \cup \{(b,a) \in A \times A: (a,b) \in R\}$.

Proof

Let $S = R \cup \{(b,a) \in A \times A: (a,b) \in R\}$, and suppose $(x,y) \in S$. Then $(x,y) \in R$ or $(x,y) \in \{(b,a) \in A \times A: (a,b) \in R\}$. If $(x,y) \in R$, then $(y,x) \in \{(b,a) \in A \times A: (a,b) \in R\}$. If $(x,y) \in \{(b,a) \in A \times A: (a,b) \in R\}$, then $(y,x) \in R$. In either case, $(y,x) \in S$ and S is symmetric.

It is evident that $R \subseteq S$.

Let T be a symmetric relation such that $R \subseteq T$. We must show that $S \subseteq T$. Let $(x,y) \in S$. If $(x,y) \in R$, then $(x,y) \in T$ since $R \subseteq T$. If $(x,y) \in \{(b,a) \in A \times A: (a,b) \in R\}$, then $(y,x) \in R$. Since $R \subseteq T$, $(y,x) \in T$. Since T is symmetric, $(x,y) \in T$. Therefore S is the symmetric closure of R. □

Corollary 10.15 Let R be a relation on a set A. Then the symmetric closure of R is $R \cup R^{-1}$.

The proof of Corollary 10.15 is left as Exercise 57.

Example 28

Let $R = \{(m,n) \in \mathbb{N} \times \mathbb{N}: m > n\}$. The symmetric closure of R is $R \cup \{(m,n) \in \mathbb{N} \times \mathbb{N}: (n,m) \in R\}$ or, $\{(m,n) \in \mathbb{N} \times \mathbb{N}: m > n \text{ or } m < n\} = \{(m,n) \in \mathbb{N} \times \mathbb{N}: m \neq n\}$. ■

Example 29

Let $A = \{1,2,3,4\}$ and $R = \{(1,4),(3,1),(3,2),(4,3)\}$. Then R is a relation on A, but R is not transitive since $(1,4),(4,3) \in R$ but $(1,3) \notin R$. Can we obtain the transitive closure of R by adding all pairs of the form (a,c), where (a,b) and (b,c) are in R for some b, to R?

Analysis

If we add all such pairs to R, we obtain the relation $S = \{(1,3),(1,4),(3,1),(3,2),(3,4),(4,1),(4,2),(4,3)\}$. But S is not transitive, since $(1,3),(3,1) \in S$ and $(1,1) \notin S$. Therefore S cannot be the transitive closure of R. ■

Example 29 illustrates that finding the transitive closure of a relation is more complicated than finding the reflexive closure or the symmetric closure. We can use Example 29 to illustrate that the transitive closure of a relation can be found by adding pairs and then repeating the process until no new pairs are needed. If we add all pairs of the form (a,c), where (a,b) and (b,c) are in S for some b, to S, we obtain the relation

$$T = \{(1,1),(1,2),(1,3),(1,4),(3,1),(3,2),(3,3),(3,4),(4,1),(4,2),(4,3),(4,4)\}$$

and we can argue that T is the transitive closure of R. It is straightforward to check that T is a transitive relation that contains R. Furthermore, we have just seen that any transitive relation that contains R contains S and that any transitive relation that contains S contains T. Therefore T is the transitive closure of S.

In Section 9.2 we discussed the directed graph of a relation. We now introduce some graph-theoretic terminology. Let R be a relation on a set A. The members of A are called *vertices* and the members of R are called *directed edges*. In this context we denote the directed edge (a,b) by $\langle a,b\rangle$ because secretly we are thinking of an arrow drawn on the page rather than the ordered pair. Let a and b be vertices of a directed graph (A,R). If there exists a finite sequence of one or more directed edges $\langle x_0,x_1\rangle,\langle x_1,x_2\rangle,\langle x_2,x_3\rangle, \ldots, \langle x_{n-1},x_n\rangle$ in (A,R) such that $x_0 = a$ and $x_n = b$, then this finite sequence is a *path* from a to b in (A,R), and it is said to be of *length n*. A path that begins and ends at a is called a *circuit*, or *cycle*. Intuitively, a path is a route that does not use any one-way streets in the wrong direction, and a circuit is such a route that starts and ends at the same place.

In this definition we have sacrificed precision for clarity. Recall from Section 4.1 that a finite sequence is a function whose domain is $I_n = \{1,2,3, \ldots ,n\}$ for some natural number n. Thus, in the definition of a path, we really mean: "If there exists a finite sequence f such that $f(i) = \langle x_{i-1},x_i\rangle$ for each $i = 1,2,3, \ldots ,n$."

Example 30

The following are paths from 1 to 2 in the directed graph given in Figure 10.9:

$\langle 1,2\rangle$

$\langle 1,5\rangle,\langle 5,4\rangle,\langle 4,2\rangle$

$\langle 1,6\rangle,\langle 6,1\rangle,\langle 1,2\rangle$

$\langle 1,6\rangle,\langle 6,5\rangle,\langle 5,4\rangle,\langle 4,2\rangle$

$\langle 1,6\rangle,\langle 6,5\rangle,\langle 5,4\rangle,\langle 4,2\rangle,\langle 2,1\rangle,\langle 1,2\rangle$

The path $\langle 1,5\rangle,\langle 5,4\rangle,\langle 4,2\rangle,\langle 2,1\rangle$ is an example of a cycle.

Figure 10.9

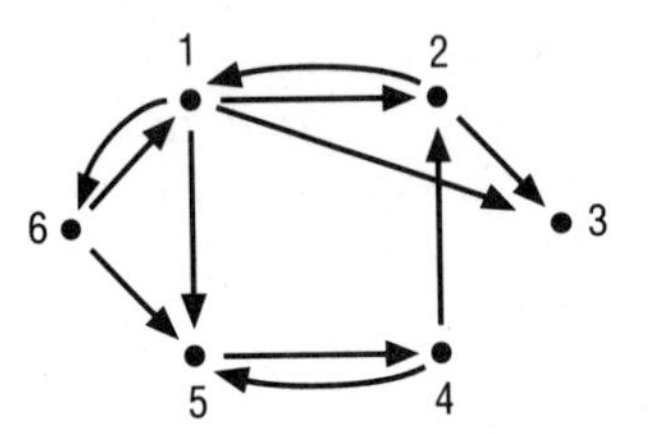

■

Some mathematicians permit paths of length zero. We assume, however, that all paths have positive length. Others insist that each vertex have a directed edge to it or from it, so that every vertex belongs to at least one path. Although we do not formally make this requirement, most of the directed graphs we consider meet this condition.

We observed in Section 9.2 that there is a relation associated with each directed graph and a directed graph associated with each relation. We define the concept of a path in a relation in the obvious manner. Again, we sacrifice precision for clarity.

Definition

Let R be a relation on a set A and let $a,b \in A$. If there exists a finite sequence of ordered pairs $(x_0,x_1),(x_1,x_2),(x_2,x_3), \ldots,$ (x_{n-1},x_n) in R such that $x_0 = a$ and $x_n = b$, then this finite sequence is a **path** in R from a to b, and it is said to be of **length** $\boldsymbol{n}$. A path that begins and ends at a is called a **cycle,** or **circuit.**

Theorem 10.16 Let R be a relation on a set A, let $a,b \in A$, and let $m \in \mathbb{N}$. Then there is a path in R of length m from a to b if and only if $(a,b) \in R^m$.

Proof

Let $S = \{n \in \mathbb{N}$: if $c,d \in A$ and there is a path in R of length n from c to d, then $(c,d) \in R^n\}$. If there is a path of length 1 in R from c to d, then by definition $(c,d) \in R$. Therefore, since $R^1 = R$, $1 \in S$.

Suppose $n \in S$, let $c,d \in A$, and assume that there is a path of length $n + 1$ in R from c to d. Then there is a member e of A such that there is a path of length n from c to e and a pair of length 1 from e to d. Since $n \in S$, $(c,e) \in R^n$; since there is a path of length 1 from e to d, $(e,d) \in R$. Hence $(c,d) \in R \circ R^n$. By Corollary 9.5, $R \circ R^n = R^n \circ R$, and by definition $R^n \circ R = R^{n+1}$. Therefore $n + 1 \in S$ and, by the principle of mathematical induction, $S = \mathbb{N}$. We have thus proved that if there is a path in R of length m from a to b then $(a,b) \in R^m$.

Now let $S = \{n \in \mathbb{N}$: if $(c,d) \in R^n$, then there is a path in R of length n from c to $d\}$. If $(c,d) \in R^1 = R$, then by definition there is a path in R of length 1 from c to d. Therefore $1 \in S$.

Suppose $n \in S$, and assume $(c,d) \in R^{n+1}$. Since $R^{n+1} = R^n \circ R = R \circ R^n$, there exists $e \in A$ such that $(c,e) \in R^n$ and $(e,d) \in R$. Since $n \in S$, there is a path in R of length n from c to e. Since $(e,d) \in R$, by definition there is a path in R of length 1 from e to d. Thus there is a path in R of length $n + 1$ from c to d, and hence $n + 1 \in S$. By the principle of mathematical induction, $S = \mathbb{N}$, and we have proved that if $(a,b) \in R^m$ then there is a path in R of length m from a to b. □

Definition

Let R be a relation on a set A. The relation $R^* = \{(a,b) \in A \times A$: there is a path in R from a to $b\}$ is called the **connectivity relation on A with respect to R.**

Example 31

Let $A = \{1,2,3,4,5,6\}$, and let $R = \{(1,2),(1,5),(1,6),(2,3),(2,4),(4,3),(4,5),(5,4),(5,6),(6,1),(6,2)\}$. Find R^*.

Analysis

The relation R^* is $\{(1,1),(1,2),(1,3),(1,4),(1,5),(1,6),(2,1),(2,2),(2,3),(2,4),(2,5),(2,6),(4,1),(4,2),(4,3),(4,4),(4,5),(4,6),(5,1),(5,2),(5,3),(5,4),(5,5),(5,6),(6,1),(6,2),(6,3),(6,4),(6,5),(6,6)\}$. Notice that R^* is transitive. ■

Theorem 10.17 If R is a relation on a set A, then $R^* = \bigcup_{n=1}^{\infty} R^n$.

Proof

By Theorem 10.16, for each $n \in \mathbb{N}$, $R^n = \{(a,b) \in A \times A$: there is a path in R of length n from a to $b\}$. It follows that $R^* = \bigcup_{n=1}^{\infty} R^n$. □

Lemma 10.18

If R, S, T, and U are relations such that $R \subseteq T$ and $S \subseteq U$, then $R \circ S \subseteq T \circ U$.

Proof

Let $(a,c) \in R \circ S$. Then there exists b such that $(a,b) \in S$ and $(b,c) \in R$. Since $S \subseteq U$ and $R \subseteq T$, $(a,b) \in U$ and $(b,c) \in T$. Therefore $(a,c) \in T \circ U$. □

Theorem 10.19

If R is a relation on a set A, the transitive closure of R is R^*.

Proof

First we prove that R^* is transitive. Let $(a,b),(b,c) \in R^*$. Then, by Theorem 10.17, there exist $m,n \in \mathbb{N}$ such that $(a,b) \in R^n$ and $(b,c) \in R^m$. Thus, by the definition of composition, $(a,c) \in R^m \circ R^n$. By Theorem 9.4, $R^m \circ R^n = R^{m+n}$. Therefore $(a,c) \in R^{m+n}$. By Theorem 10.17, $(a,c) \in R^*$ and R^* is transitive.

Since $R = R^1$, $R \subseteq R^*$ by Theorem 10.17.

Now let T be a transitive relation such that $R \subseteq T$. We prove that $R^* \subseteq T$ by showing that $R^n \subseteq T$ for each $n \in \mathbb{N}$. Let $S = \{n \in \mathbb{N}: R^n \subseteq T\}$. Since $R^1 = R \subseteq T$, $1 \in S$. Suppose $n \in S$. Then $R^n \subseteq T$. Now $R^{n+1} = R^n \circ R$, and by Lemma 10.18 $R^n \circ R \subseteq T \circ T$. Since T is transitive, $T \circ T \subseteq T$ by Theorem 10.1b. It follows that $n + 1 \in S$. By the principle of mathematical induction, $S = \mathbb{N}$.

Therefore, R^* is the transitive closure of R. □

In the remainder of this section we are concerned with the problem of computing the transitive closure of a relation, and this is optional material. In Section 9.4 we introduced the concept of using zero–one matrices to represent relations, and here we show how to compute the zero–one matrix of the transitive closure of a relation on a finite set. But first we prove that, if A is a finite set with n members and R is a relation on A, then $R^* = \bigcup_{i=1}^{n} R^i$.

Theorem 10.20

Let $p \in \mathbb{N}$, let $A = \{m \in \mathbb{N}: m \leq p\}$, and let R be a relation on A. Then $R^* = \bigcup_{i=1}^{p} R^i$. Furthermore, if $a,b \in A$ and $a \neq b$, then there is a path in R from a to b whose length does not exceed $p - 1$.

Proof

By Theorem 10.17, $\bigcup_{i=1}^{p} R^i \subseteq R^*$. Thus we need to prove that $R^* \subseteq \bigcup_{i=1}^{p} R^i$. Suppose $(a,b) \in R^*$ and $a \neq b$. Then there is a path in R from a to b, so $S = \{n \in \mathbb{N}$: there is a path in R of length n from a to $b\}$ is a nonempty subset of $\mathbb{N}$. By the least-natural-number principle, S has a smallest member m. Let $\langle x_0,x_1\rangle,\langle x_1,x_2\rangle, \ldots ,\langle x_{m-1},x_m\rangle$ be a path in R of length m from a to b. We need to show that $m \leq p - 1$. Suppose $m > p - 1$. Then, by the pigeonhole principle, at least two of the $m + 1$ vertices $x_0,x_1,x_2, \ldots ,x_m$ are equal. Since $x_0 \neq x_m$, there exists i with $(0 \leq i \leq m - 1)$ such that $x_i = x_j$ for some $j = 1,2, \ldots ,m$ $(j \neq i)$. We may assume without loss of generality that $i<j$. Then $\langle x_0,x_1\rangle,\langle x_1,x_2\rangle, \ldots ,\langle x_{i-1},x_i\rangle,$ $\langle x_j,x_{j+1}\rangle, \langle x_{j+1},x_{j+2}\rangle, \ldots ,\langle x_{m-1},x_m\rangle$ is a path in A from a to b whose length is less than m. This is a contradiction, so $m \leq p - 1$. Therefore $(a,b) \in \bigcup_{i=1}^{p-1} R^i$.

Now suppose that there is an $a \in A$ such that $(a,a) \in R^*$. Then there is a path in R from a to a, so $S = \{n \in \mathbb{N}$: there is a path in R of length n from a to $a\}$ is a nonempty subset of $\mathbb{N}$. By the least-natural-number principle, S has a smallest member m. Let $\langle x_0,x_1\rangle,\langle x_1,x_2\rangle,\langle x_2,x_3\rangle, \ldots ,\langle x_{m-1},x_m\rangle$ be a path in R of length m from a to a. Let i be the largest integer such that $x_i \neq x_m$. By the preceding proof, there is a path in R from a to x_i whose length does not equal $p - 1$. Therefore there is a path in R from a to a whose length does not exceed p. Hence $(a,a) \in \bigcup_{i=1}^{p} R^i$. □

We have previously defined the meet and join of two $m \times n$ matrices. Recursion relations can be used to extend these definitions to the meet and join of any finite number of $m \times n$ zero–one matrices.

Definition

For each $i \in \mathbb{N}$, let A_i be an $m \times n$ zero–one matrix. Then for each $n \in \mathbb{N}$, $\bigwedge_{i=1}^{n} A_i$ and $\bigvee_{i=1}^{n} A_i$ are defined by the recursion relations

$\bigwedge_{i=1}^{n} A_i = \bigwedge_{i=1}^{n-1} A_i \wedge A_n$ and

$\bigvee_{i=1}^{n} A_i = \bigvee_{i=1}^{n-1} A_i \vee A_n$, where $\bigwedge_{i=1}^{1} A_i = \bigvee_{i=1}^{1} A_i = A_1$

We have the following generalization of Theorem 9.7. Recall that M_R denotes the associated zero–one matrix of a relation R.

Theorem 10.21

Let A be a nonempty finite set, let $n \in \mathbb{N}$ and, for each $i = 1,2, \ldots ,n$ let R_i be a relation on A. If $S = \bigcup_{i=1}^{n} R_i$ and $T = \bigcap_{i=1}^{n} R_i$, then $M_S = \bigvee_{i=1}^{n} M_{R_i}$ and $M_T = \bigwedge_{i=1}^{n} M_{R_i}$.

The proof of Theorem 10.21 is essentially the same as the proof of Theorem 9.7 and is left as Exercise 73.

Theorem 10.22 tells us that the zero–one matrix of the transitive closure of a relation on a set with p members is the join of the first p powers of M_R.

Theorem 10.22

Let $p \in \mathbb{N}$, let $A = \{m \in \mathbb{N}: m \leq p\}$, and let R be a relation on A. Then $M_{R^*} = \bigvee_{i=1}^{p} (M_R)^{[i]}$.

Proof

By Theorem 10.20, $R^* = \bigcup_{i=1}^{p} R^i$. So by Theorem 10.21, $M_{R^*} = \bigvee_{i=1}^{p} M_{R^i}$. Finally, by Theorem 9.9, $\bigvee_{i=1}^{p} M_{R^i} = \bigvee_{i=1}^{p} (M_R)^{[i]}$. □

Example 32

Find the transitive closure of the relation $R = \{(1,1),(1,2),(2,2),(3,1),(3,3)\}$ on $\{1,2,3\}$.

Analysis

By Theorem 10.22, the zero–one matrix of the transitive closure of R is $M_R \vee (M_R)^{[2]} \vee (M_R)^{[3]}$. Now

$$M_R = \begin{bmatrix} 1 & 1 & 0 \\ 0 & 1 & 0 \\ 1 & 0 & 1 \end{bmatrix} \quad (M_R)^{[2]} = M_R \odot M_R = \begin{bmatrix} 1 & 1 & 0 \\ 0 & 1 & 0 \\ 1 & 1 & 1 \end{bmatrix}$$

$$(M_R)^{[3]} = (M_R)^{[2]} \odot M_R = \begin{bmatrix} 1 & 1 & 0 \\ 0 & 1 & 0 \\ 1 & 1 & 1 \end{bmatrix}$$

Therefore,

$$M_R \vee (M_R)^{[2]} \vee (M_R)^{[3]} = \begin{bmatrix} 1 & 1 & 0 \\ 0 & 1 & 0 \\ 1 & 1 & 1 \end{bmatrix}$$

So the transitive closure of R is $\{(1,1),(1,2),(2,2),(3,1),(3,2),(3,3)\}$. ■

Example 33

The flowchart in Figure 10.10 is used to find the zero–one matrix of the transitive closure of a relation R on a set with n members ($n \geq 2$). What is the number of bit operations used in the flowchart?

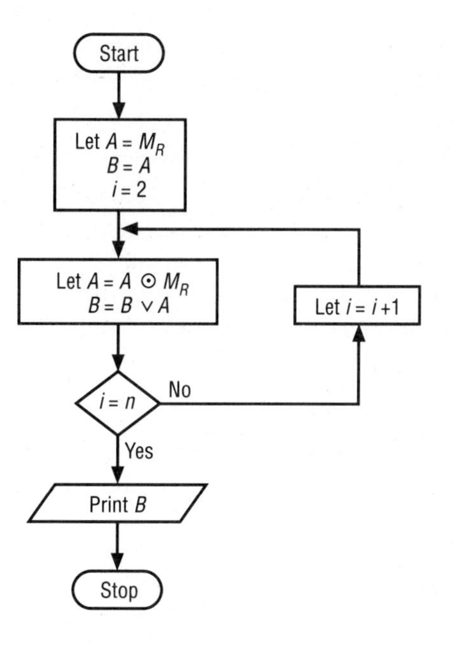

Figure 10.10

Analysis

To find each Boolean product, we use n^3 bit operations. To find the Boolean powers, we must find $n - 1$ Boolean products, so the number of bit operations used to find the Boolean powers is $n^3(n - 1)$. To find the zero–one matrix of the transitive closure of R, we must find $n - 1$ joins of zero–one matrices. To find each of these joins, we use n^2 bit operations. Thus the number of bit operations used to find the zero–one matrix of the transitive closure of R is

$$n^3(n - 1) + n^2(n - 1) = n^4 - n^2$$

There is a more efficient method, known as Warshall's algorithm, for computing the transitive closure of a relation on n members. Warshall's algorithm uses $2n^3$ bit operations. The difference in the number of bit operations used by the two methods is tabulated here:

n	2	3	4	5	6	7	8	9	10
$n^4 - n^2$	12	72	240	600	1260	2352	4032	6480	9900
$2n^3$	16	54	128	250	432	686	1024	1458	2000

■

Exercises 10.5

56. Prove that if R is a relation on a set A then the reflexive closure of R is $R \cup \{(a,a): a \in A\}$.

57. Prove that the symmetric closure of a relation R is $R \cup R^{-1}$.

58. Let $R = \{(1,2),(2,2),(2,3),(3,1),(3,3),(4,1)\}$ be a relation on $\{1,2,3,4\}$.

a. Find the reflexive closure of R. **b.** Find the symmetric closure of R.

59. Let $R = \{(1,2),(2,2),(2,3),(3,1),(3,3),(4,1)\}$ be a relation on $\{1,2,3,4,5\}$.

a. Find the reflexive closure of R. **b.** Find the symmetric closure of R.

60. Let $R = \{(m,n) \in \mathbb{N} \times \mathbb{N}: m \neq n\}$ be a relation on $\mathbb{N}$. What is the reflexive closure of R?

61. Each of the following directed graphs represents a relation on $\{1,2,3,4\}$. For each case, draw the directed graph that represents the reflexive closure of the relation.

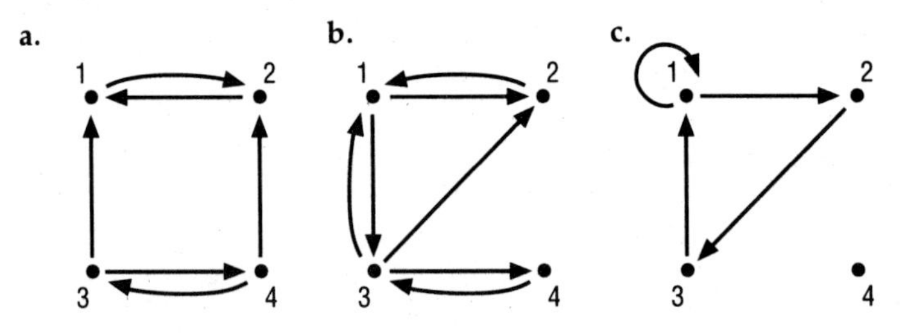

62. If $R = \{(m,n) \in \mathbb{N} \times \mathbb{N}: m \text{ divides } n\}$, what is the symmetric closure of R?

63. If $R = \{(x,y) \in \mathbb{R} \times \mathbb{R}: x < y\}$ is a relation on $\mathbb{R}$, find

a. the symmetric closure of the reflexive closure of R

b. the reflexive closure of the symmetric closure of R

64. Each of the following directed graphs represents a relation on $\{1,2,3,4\}$. For each case, draw a directed graph that represents the symmetric closure of the relation.

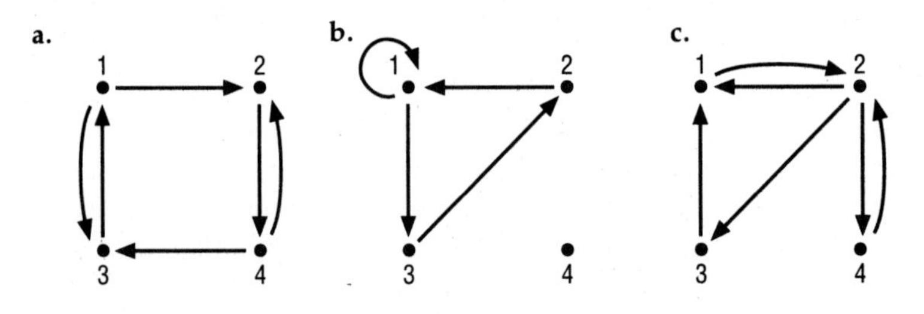

65. Let $p \in \mathbb{N}$, let $A = \{m \in \mathbb{N}: m \leq p\}$, and let R be a relation on A.

a. Prove that the zero–one matrix that represents the reflexive closure of R is $M_R \vee I_p$.

b. Prove that the zero–one matrix that represents the symmetric closure of R is $M_R \vee (M_R)^t$.

c. What is the zero–one matrix that represents the smallest relation containing R that is both reflexive on A and symmetric?

66. If $R = \{(1,2),(2,1),(2,5),(3,4),(4,2),(5,1),(5,3),(5,4)\}$ is a relation on $\{1,2,3,4,5\}$, find R^*.

67. Find the transitive closure of the relation $R = \{(1,2),(2,3),(3,2)\}$ on $\{1,2,3\}$.

68. Find the transitive closure of the relation $R = \{(1,4),(2,1),(2,3),(3,1),(3,4),(4,3)\}$ on $\{1,2,3,4\}$.

69. A relation R on $\{1,2,3,4\}$ is represented by the following matrix. Find the transitive closure of R.

$$\begin{bmatrix} 1 & 0 & 1 & 0 \\ 0 & 1 & 0 & 0 \\ 1 & 0 & 1 & 1 \\ 0 & 1 & 0 & 0 \end{bmatrix}$$

70. Find the smallest relation containing the relation $R = \{(1,3),(1,4),(2,2),(4,1)\}$ on $\{1,2,3,4\}$ that is

a. reflexive on $\{1,2,3,4\}$ **b.** symmetric **c.** transitive

d. reflexive on $\{1,2,3,4\}$ and symmetric **e.** symmetric and transitive

f. reflexive on $\{1,2,3,4\}$ and transitive

g. an equivalence relation on $\{1,2,3,4\}$

71. Prove that if R is reflexive on a set A then R^* is reflexive on A.

72. Prove that if R is symmetric then R^* is symmetric.

73. Prove Theorem 10.21.

10.6 PARTIAL ORDERS

Suppose that we have an *order relation* on a set X. It may be possible to compare some members of X with others, but there may be some pairs of members of X that are not comparable with respect to this order relation.

Example 34

Let $X = \{n \in \mathbb{N}: n < 100\}$ and $R = \{(m,n) \in X \times X: m \text{ divides } n\}$. Then R is a relation on X, 5 and 60 are comparable since $(5,60) \in R$, but 7 and 83 are not comparable since $(7,83) \notin R$. ■

Order relations, such as that in Example 34, which allow the possibility of incomparable elements, are called *partial orders*. We need a precise definition of this term, but first we define the concept of antisymmetry.

Definition

A relation R on a set A is **antisymmetric** provided that whenever $(a,b) \in R$ and $(b,a) \in R$ then $a = b$.

An equivalent way of defining antisymmetry is to say that a relation R on a set A is antisymmetric provided that, if $(a,b) \in R$ and $a \neq b$, then $(b,a) \notin R$.

Observe that the usual order, $\leq$, on $\mathbb{R}$ is antisymmetric because, if $x,y \in R$, $x \leq y$, and $y \leq x$, then $x = y$. Also notice that if R is an equivalence relation on A and R is antisymmetric then each equivalence class $R[x]$ consists of exactly one member of A, namely, x. Why? See Exercise 74. Finally observe that the terms *symmetric* and *antisymmetric* are not negatives of each other, since the relation $=$ on $\mathbb{R}$ is a symmetric relation that is also antisymmetric.

If R is a relation on a finite set A, we can easily see whether R is antisymmetric by drawing its directed graph. The relation R is antisymmetric provided that, if x and y are distinct members of A and there is an arrow from x to y, then there is no arrow from y to x. Thus if in the directed graph of R there is a pair x and y of distinct points such that there is an arrow from x to y and an arrow from y to x, then R cannot be antisymmetric. The relation represented by the directed graph in Figure 10.11a is antisymmetric, whereas the relation represented by the directed graph in Figure 10.11b is not antisymmetric.

Figure 10.11

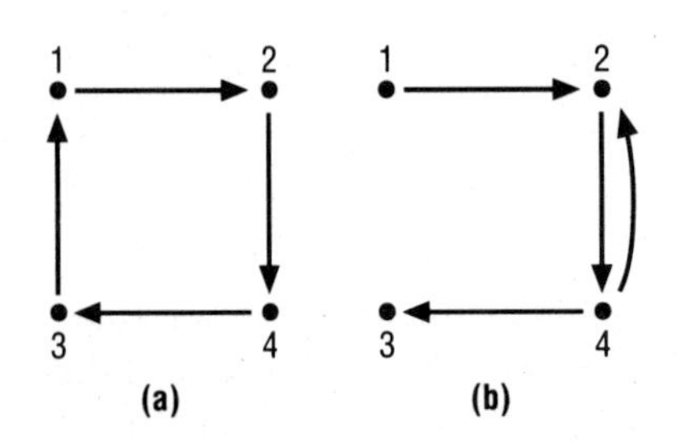

Now let us return to the concept of ordering a set. The properties that characterize relations used to order sets are reflexivity on the set, antisymmetry, and transitivity.

Definition

A relation $\leq$ on a set A is a **partial ordering of A** if it is reflexive on A, antisymmetric, and transitive. A **partially ordered set,** or **poset,** is an ordered pair $(A,\leq)$, where A is a set and $\leq$ is a partial ordering of A. If a and b are members of A, $a \leq b$, and $a \neq b$, we write $a < b$.

As with other relations, it is sometimes convenient to denote a partial order by R, S, or some other letter. But with this alternative notation we no longer have the concise notation $a < b$ to mean a is related to b and $a \neq b$, so we must write aRb and $a \neq b$.

Example 35

The usual order $\leqslant$ on $\mathbb{N}$ is a partial ordering of $\mathbb{N}$; that is, $(\mathbb{N},\leqslant)$ is a poset.

Analysis

Let $x \in \mathbb{N}$. Since $x = x$, $x \leqslant x$. Hence $\leqslant$ is reflexive on $\mathbb{N}$. We have already observed that $\leqslant$ is antisymmetric. If $x,y,z \in \mathbb{N}$, $x \leqslant y$ and $y \leqslant z$, then $x \leqslant z$. Thus $\leqslant$ is transitive. ■

Example 36

Each of the three directed graphs in Figure 10.12 represents a relation that is a partial ordering of $\{1,2,3,4\}$.

Figure 10.12

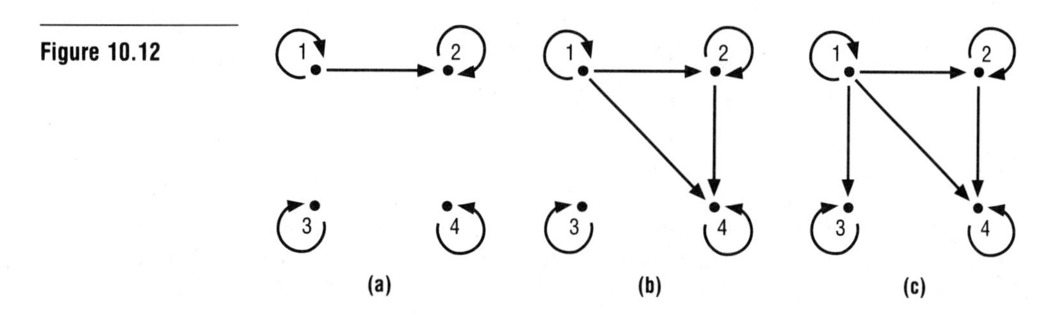

Observe that there are no nontrivial circuits (paths that begin at one vertex, go to another vertex, and then eventually return to the beginning vertex) in these directed graphs. ■

Example 37

Define a relation $\leqslant$ on $\mathbb{N}$ by $a \leqslant b$ provided that a divides b. Then $(\mathbb{N},\leqslant)$ is a poset.

Analysis

Since each natural number divides itself, $\leqslant$ is reflexive on $\mathbb{N}$. If $m,n \in \mathbb{N}$, m divides n, and n divides m, then $m = n$. Therefore $\leqslant$ is antisymmetric. If $m,n,p \in \mathbb{N}$, m divides n, and n divides p, then m divides p. Thus $\leqslant$ is transitive. ■

Example 38

Let A be a set. Then $(\mathscr{P}(A),\subseteq)$ is a poset.

Analysis

Since each set is a subset of itself, $\subseteq$ is reflexive on $\mathcal{P}(A)$. If $X,Y \in \mathcal{P}(A)$, $X \subseteq Y$, and $Y \subseteq X$, then $X = Y$. Hence $\subseteq$ is antisymmetric. If $X,Y,Z \in \mathcal{P}(A)$, $X \subseteq Y$, and $Y \subseteq Z$, then $X \subseteq Z$. Thus $\subseteq$ is transitive. ■

For many partially ordered sets $(A,\leq)$, it is possible to find $x,y \in A$ such that neither $x \leq y$ nor $y \leq x$ (see Example 34 and Exercise 79).

Definition

Let $(A,\leq)$ be a poset and let $a,b \in A$. Then a and b are **comparable** provided that $a \leq b$ or $b \leq a$, and a and b are **incomparable** provided that $a \nleq b$ and $b \nleq a$.

Definition

If $(A,\leq)$ is a poset and each pair of members of A is comparable, then $\leq$ is a **linear order,** or **total order,** and $(A,\leq)$ is a **linearly ordered set,** or **totally ordered set,** or **chain.**

The poset in Example 35 is a linearly ordered set, whereas the posets in Examples 36, 37, and 38 are not linearly ordered sets (see Exercise 80).

The words in a dictionary are listed in alphabetical, or lexicographic, order. We can use this concept to define a partial order on the Cartesian product of two posets.

Definition

Let (A,R) and (B,S) be posets. The **lexicographic order** $\leq$ on $A \times B$ is defined by $(a_1,b_1) \leq (a_2,b_2)$ provided that $(a_1,a_2) \in R$ and if $a_1 = a_2$ then $(b_1,b_2) \in S$.

In Exercise 83, we ask you to prove that the lexicographic order on the Cartesian product of two posets is a partial order on their product.

Example 39

Let $\leq$ be the usual order on $\mathbb{N}$, and let $\lesssim$ denote the lexicographic order on $\mathbb{N} \times \mathbb{N}$ with respect to $\leq$ on $\mathbb{N}$. Is $(4,7) \lesssim (5,9)$? Is $(4,9) \lesssim (5,7)$? Is $(4,7) \lesssim (4,9)$?

Analysis

Since $4 \leq 5$, $(4,7) \lesssim (5,9)$ and $(4,9) \lesssim (5,7)$. Also, $(4,7) \lesssim (4,9)$ since $4 = 4$ and $7 \leq 9$. ■

In Example 39, the set $\{(m,n) \in \mathbb{N} \times \mathbb{N}: m < 4, \text{ or } m = 4 \text{ and } n \leq 7\}$ is the subset of $\mathbb{N} \times \mathbb{N}$ consisting of all ordered pairs (m,n) such that $(m,n) \lesssim (4,7)$.

Directed graphs can be used to represent partially ordered sets, but even simple partial orders have messy directed graphs. The German mathematician Helmut Hasse invented a more economical way to indicate a finite poset, which in his honor is called a *Hasse diagram*. Like a Fibonacci number, a Hasse diagram is usually defined by giving a recipe for finding it, and we admit that the formal definition we give here need not be memorized. It is, nonetheless, a useful exercise to construct a Hasse diagram of a simple poset from the definition (see Exercise 76).

Definition

Let (A,R) be a finite poset and let S be the smallest relation such that $S^* = R$. (Technically, S is the intersection of all relations on A whose transitive closure is R.) Let $T = S - \{(a,a): a \in A\}$ and let H be the symmetric closure of T. The directed graph of H, thought of as a diagram rather than as a relation, is the **Hasse diagram** of (A,R) provided that it has been drawn so that the line segment (a,b) is drawn with a below b exactly when $(a,b) \in R$.

Nobody, not even Hasse himself, uses the definition of a Hasse diagram, because there is a simple routine for drawing the Hasse diagram of a finite poset (A,R). Start with the directed graph of R. Remove all the loops that must be present because R is reflexive on A. Next remove all edges that must be present because of the transitivity of R and arrange each edge so that the arrow points upward. Finally, replace all arrows by ordinary line segments.

Example 40

Draw the Hasse diagram of $(A,\leq)$, where $A = \{1,2,4,5,8,10\}$ and $m \leq n$ provided that m divides n.

We draw the directed graph of $\leq$, as shown in Figure 10.13a. Then we remove all the loops that must be present because $\leq$ is reflexive on A and obtain the graph shown in Figure 10.13b. Next we remove all edges that must be present because of the transitivity of $\leq$ and obtain the graph shown in Figure 10.13c. Finally, we arrange each edge so that the arrow points upward and remove the arrows to obtain the Hasse diagram of $(A,\leq)$ shown in Figure 10.13d.

Figure 10.13

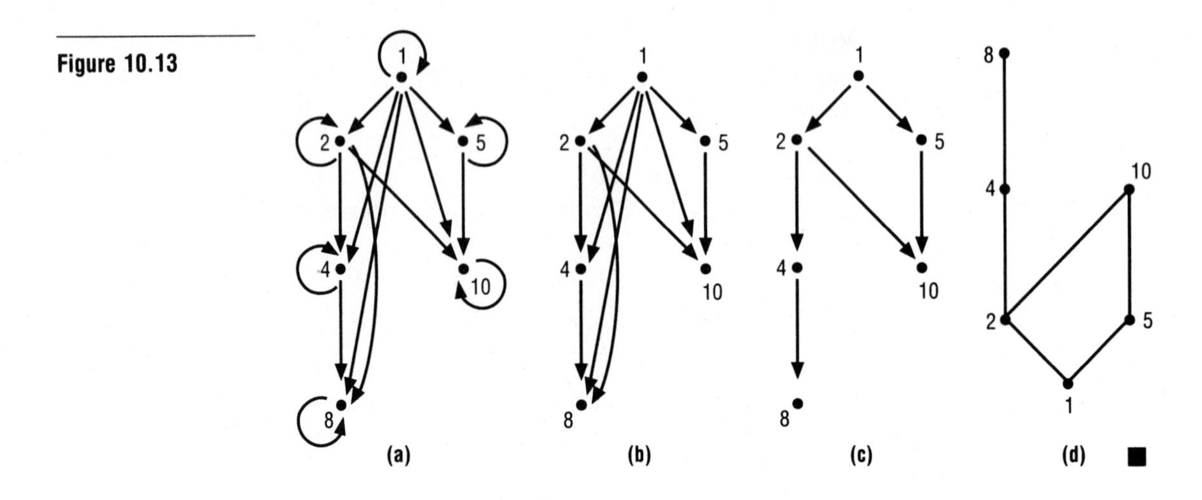

■

Example 41

If $A = \{1,2,3\}$, the Hasse diagram of $(\mathcal{P}(A),\subseteq)$ is shown in Figure 10.14.

Figure 10.14

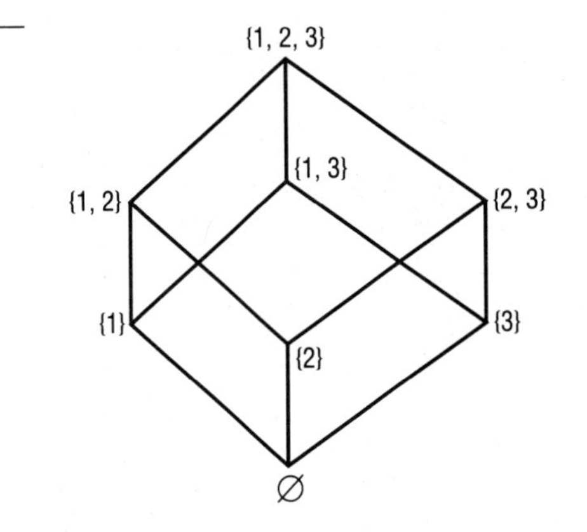

■

Example 42

Let Σ be an alphabet. Define a relation $\leqslant$ on Σ^* by $w_1 \leqslant w_2$ provided that there exists $w \in \Sigma^*$ such that $w_2 = w_1 w$. Then $(\Sigma^*,\leqslant)$ is a poset. If $\Sigma = \{0,1\}$, part of the Hasse diagram for $(\Sigma^*,\leqslant)$ is shown in Figure 10.15.

Figure 10.15

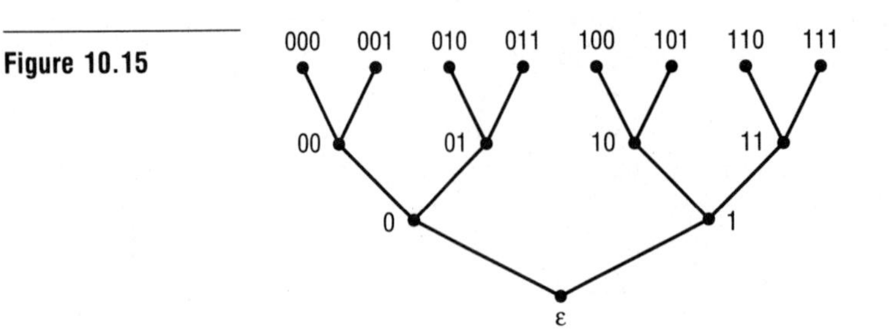

■

The poset whose Hasse diagram is shown in Figure 10.16a is a linearly ordered set, whereas that in Figure 10.16b is not a linearly ordered set.

Figure 10.16

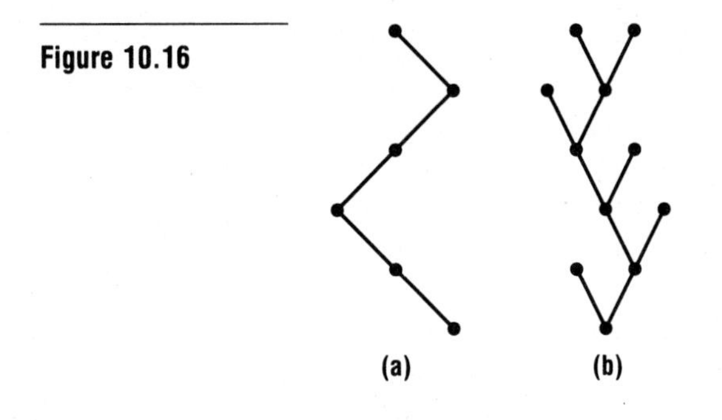

Definition

Let $(A,\leqslant)$ be a poset. A member a of A is **maximal in** $(A,\leqslant)$ provided that there is no $b \in A$, different from a, such that $a \leqslant b$, and a member a of A is **minimal in** $(A,\leqslant)$ provided that there is no $b \in A$, different from a, such that $b \leqslant a$.

The maximal and minimal elements in $(A,\leqslant)$ are the top and bottom elements in the Hasse diagram of $(A,\leqslant)$. For example, 8 and 10 are the maximal elements and 1 is the minimal element of the poset in Example 40. In Example 41, $\{1,2,3\}$ is the maximal element and $\varnothing$ is the minimal element.

Definition

Let $(A,\leqslant)$ be a poset. A member a of A is the **greatest element** of $(A,\leqslant)$ if $b \leqslant a$ for all $b \in A$, and a member a of A is the **least element** of $(A,\leqslant)$ if $a \leqslant b$ for all $b \in A$.

In Example 41, $\varnothing$ is the least element of $(\mathcal{P}(A),\subseteq)$ and $\{1,2,3\}$ is the greatest element. It is easy to see that a poset cannot have two greatest elements or two least elements (see Exercise 87), but a greatest or least element of a poset may not exist. For example, the poset in Example 40 has no greatest element. On the other hand, every finite poset has at least one maximal element and one minimal element and may have more than one maximal element and more than one minimal element.

Example 43

Consider the posets whose Hasse diagrams are given in Figure 10.17.

Figure 10.17

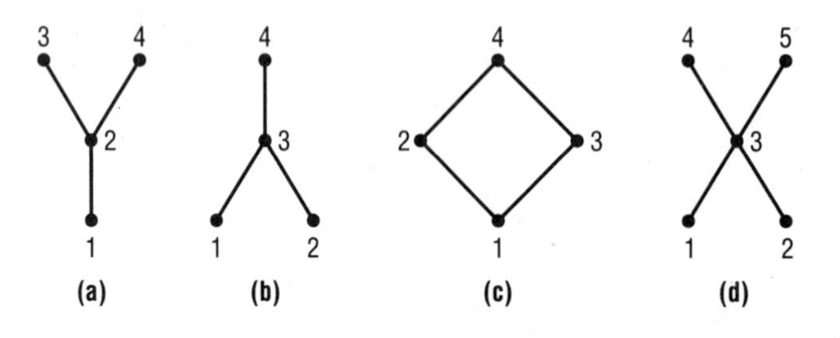

The least element of the poset represented by (a) is 1, and this poset does not have a greatest element. The greatest element of the poset represented by (b) is 4, and this poset does not have a least element. The greatest element of the poset represented by (c) is 4, and the least element of this poset is 1. The poset represented by (d) has neither a least element nor a greatest element. ■

Definition

Let $(A,\leqslant)$ be a poset and let $B \subseteq A$. If $a \in A$ such that $b \leqslant a$ for all $b \in B$, then a is an **upper bound** of B. If $a \in A$ such that $a \leqslant b$ for all $b \in B$, then a is a **lower bound** of B.

Example 44

Consider the poset whose Hasse diagram is given in Figure 10.18. The upper bounds of $\{1,2,3\}$ are 5, 7, and 9, and $\{1,2,3\}$ does not have a lower bound. The upper bounds of $\{1,3,4\}$ are 5, 7, and 9, and 1 is a lower bound of $\{1,3,4\}$. The upper bound of $\{6,7,8\}$ is 9, and the lower bounds of $\{6,7,8\}$ are 1, 2, 4, and 6. The upper bounds of $\{5,6\}$ are 7 and 9, and the lower bounds of $\{5,6\}$ are 1, 2, and 4.

Figure 10.18

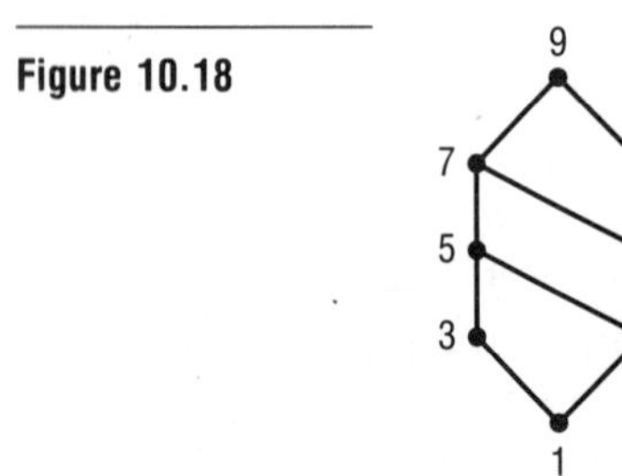

■

Definition

Let $(A,\leqslant)$ be a poset and let $B \subseteq A$. An element a of A is a **least upper bound** of B if a is an upper bound of B and $a \leqslant b$ for every upper bound b of B different from a. An element a of A is a **greatest lower bound** of B if a is a lower bound of B and $b \leqslant a$ for every lower bound b of B different from a.

In Example 44, the least upper bound of $\{5,6\}$ is 7, and $\{1,2\}$ does not have a (greatest) lower bound. Also in Example 44, the greatest lower bound of $\{3,4\}$ is 1.

If the greatest lower bound and least upper bound of a subset of a poset exist, then they are unique (see Exercise 121).

Definition

A poset $(A,\leqslant)$ is **well ordered** provided that it is linearly ordered and every nonempty subset of A has a least element.

The integers with the usual order are not a well-ordered set since the negative integers do not have a least element. However, the natural numbers with the usual order are a well-ordered set. In Exercise 90, we ask you to prove that $\mathbb{N} \times \mathbb{N}$ with the lexicographic order with respect to the usual order on $\mathbb{N}$ is a well-ordered set.

Many sorting techniques are used in computer science. Topological sorting is a method of constructing a linear ordering S from a partial ordering R in such a way that $(a,b) \in S$ whenever $(a,b) \in R$. The need for constructing such a linear ordering arises in a variety of situations. You may be asked to perform some job, let us say to assemble a bicycle, for example. You learn, probably the hard way, that some tasks must be performed before others. In this way a partial order is defined on the set of tasks needed to assemble the bicycle. If you succeed in assembling the bicycle, you will have performed a finite sequence of tasks one at a time, and this obviously defines a linear order. So the problem of figuring out how to assemble a bicycle is the problem of determining a linear order consistent with a given partial order.

Far more difficult problems of assembly arise in subjects connected to computer science such as networking and linguistics. Although we do not present the theory of topological sorting, we give an example of how it works, and we note that it is of potential use whenever we have a problem involving a partial ordering. Since we are presumably going to use a computer to do our sorting, we naturally assume that our set A is a finite set. A partial ordering on a finite set can be illustrated by a diagram in which the members of the set are represented by boxes and the relation is represented by arrows between the boxes. A partial ordering of the set {1,2,3,4,5,6,7,8,9} is given in Figure 10.19. We might think of these nine boxes as the nine tasks needed to assemble a bicycle. It is clear from Figure 10.19 that our problem is to rearrange the boxes into a line so that all arrows point toward the right. In Figure 10.20 we give the ordering relation of the members of {1,2,3,4,5,6,7,8,9} after topological sorting according to the partial ordering given in Figure 10.19.

Figure 10.19

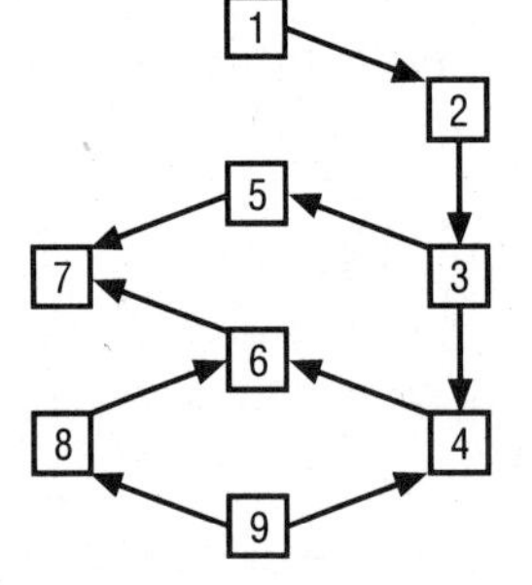

The linear ordering on the diagram in Figure 10.20 is constructed by simply adding enough arrows (all pointing to the right) so that a linearly ordered set is obtained. Thus, one way to assemble the bike is to perform the tasks in order 1, 2, 3, 5, 9, 8, 4, 6, 7. The interesting thing is that such a rearrangement is possible for every partial ordering.

Figure 10.20

1 → 2 → 3 → 5   9 → 8   4 → 6 → 7

Example 45

Let $A = \{1,2,4,5,8,10\}$ and $R = \{(m,n) \in A \times A\colon m \text{ divides } n\}$. We construct a linear ordering S on A such that $(a,b) \in S$ whenever $(a,b) \in R$ as follows:

1. We select a minimal element of (A,R). This must be 1 since it is the only minimal element.
2. We select a minimal element of $(A - \{1\},R_1)$, where $R_1 = \{(m,n) \in (A - \{1\}) \times (A - \{1\})\colon m \text{ divides } n\}$. There are two minimal elements of $(A - \{1\},R_1)$, namely, 2 and 5. We select one of them. Suppose we select 5.
3. We select a minimal element of $(A - \{1,5\},R_2)$, where $R_2 = \{(m,n) \in (A - \{1,5\}) \times (A - \{1,5\})\colon m \text{ divides } n\}$. The only minimal element is 2.

4. We select a minimal element of $(A - \{1,2,5\},R_3)$, where $R_3 = \{(m,n) \in (A - \{1,2,5\}) \times (A - \{1,2,5\})$: m divides $n\}$. There are two minimal elements of $(A - \{1,2,5\},R_3)$, namely, 4 and 10. Suppose we select 10.
5. We select a minimal element of $(A - \{1,2,5,10\},R_4)$, where $R_4 = \{(m,n) \in (A - \{1,2,5,10\}) \times (A - \{1,2,5,10\})$: m divides $n\}$. The only minimal element is 4.
6. The only element left is 8, so we have a linear ordering:

This example illustrates topological sorting. ■

Example 46

Suppose that the installation of a computer requires the completion of seven tasks, and some of these tasks cannot be started until others are completed. We can define a partial order $\leq$ on the set X of tasks by defining $P \leq Q$ provided that task Q cannot be started until task P has been completed. The Hasse diagram for this poset $(X,\leq)$ is given in Figure 10.21a. Use topological sorting to find an order in which the tasks can be performed.

Figure 10.21

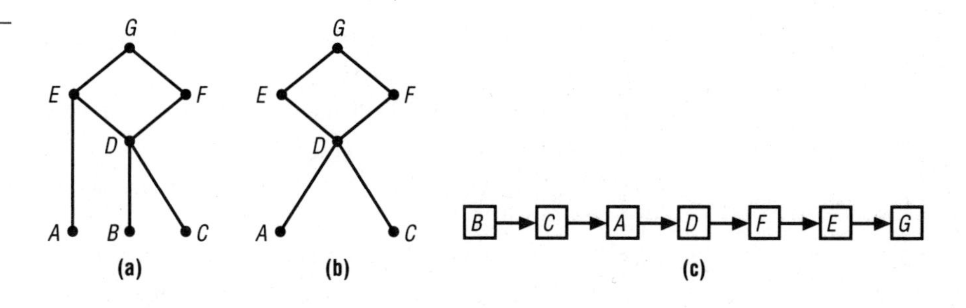

Analysis

We select a minimal element B and delete B from the diagram (as in Figure 10.21b). Then we select a minimal element C. We continue this process and finally obtain the linear ordering shown in Figure 10.21c. ■

Exercises 10.6

74. Let R be an equivalence relation on a set A and suppose that R is also antisymmetric. Prove that, for each $x \in A$, $R[x] = \{x\}$.
75. Which of the following are posets?
    a. $(\mathbb{Z},R_1)$, where $R_1 = \{(m,n) \in \mathbb{Z} \times \mathbb{Z}: m \geqslant n\}$
    b. $(\mathbb{Z},R_2)$, where $R_2 = \{(m,n) \in \mathbb{Z} \times \mathbb{Z}: m = n\}$

76. Determine which of the relations represented by the following directed graphs are partial orders. For the relation R whose directed graph is given in (c), draw the directed graphs of the relations S, T, and H as given in the definition of a Hasse diagram and draw the Hasse diagram of the relation R. Draw the Hasse diagram of all other partial orders whose directed graphs appear in this exercise using either the definition of a Hasse diagram or the recipe for finding it.

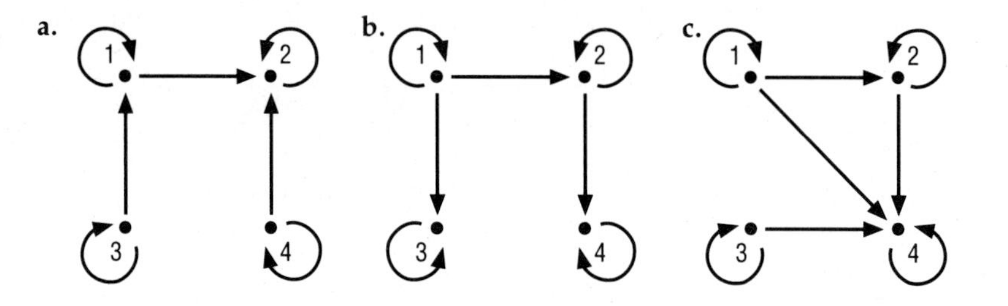

77. Determine which of the relations represented by the following zero–one matrices are partial orders. Draw the Hasse diagram of those relations that are partial orders.

a. $\begin{bmatrix} 1 & 1 & 0 \\ 0 & 1 & 0 \\ 1 & 0 & 1 \end{bmatrix}$ **b.** $\begin{bmatrix} 1 & 1 & 0 \\ 0 & 1 & 0 \\ 0 & 0 & 1 \end{bmatrix}$ **c.** $\begin{bmatrix} 1 & 0 & 1 \\ 0 & 1 & 0 \\ 0 & 0 & 1 \end{bmatrix}$

78. Show that if (A,R) is a poset then (A,R^{-1}) is also a poset. The poset (A,R^{-1}) is called the *dual* of (A,R).

79. Find two incomparable elements in each of the following posets.

a. $(\mathcal{P}(A),\subseteq)$, where $A = \{1,2,3,4\}$

b. $(\{1,2,3,4\},R)$, where R is the relation represented by the directed graph in Figure 10.12c.

c. the poset whose Hasse diagram is given in Figure 10.17a.

80. **a.** Explain why the relation represented by the directed graph in Figure 10.12c is not a linear order.

b. Explain why the poset in Example 37 is not a linearly ordered set.

c. Let A be a set with at least two members. Explain why $(\mathcal{P}(A),\subseteq)$ is not a linearly ordered set.

81. Let A be the set of all natural numbers that divide 30, that is, $A = \{1,2,3,5,6,10,15,30\}$, and let $R = \{(m,n) \in A \times A: m \text{ divides } n\}$. Then (A,R) is a poset.

a. Find all lower bounds of $\{6,10\}$.

b. Find the greatest lower bound of $\{6,10\}$.

c. Find all upper bounds of $\{6,10\}$.

d. Find the least upper bound of $\{6,10\}$.

e. Draw the Hasse diagram of (A,R).

82. Draw the Hasse diagram of the poset $(\mathcal{P}(\{1,2,3,4\}),\subseteq)$.

83. Let (A,R) and (B,S) be posets. Prove that the lexicographic order on $A \times B$ is a partial order on $A \times B$.

84. For each of the given Hasse diagrams representing the poset (A,R), list the ordered pairs that belong to R.

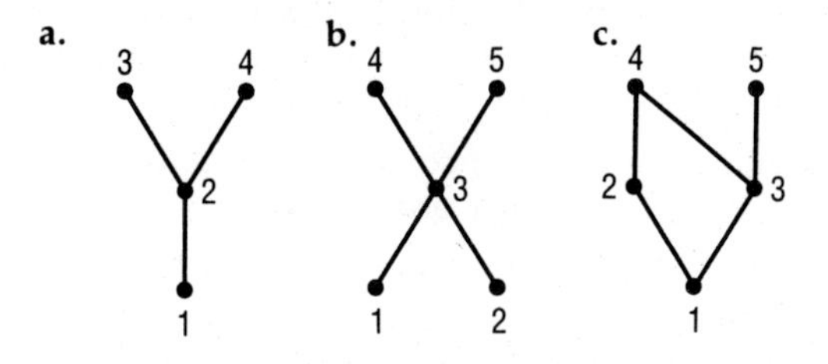

85. Let $(A,\leq)$ be the poset with the following Hasse diagram.

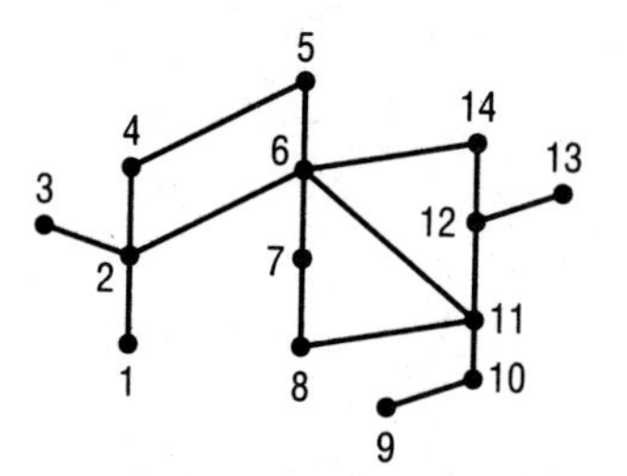

a. Find the maximal elements.

b. Find the minimal elements.

c. Is there a least element? If so, what is it?

d. Is there a greatest element? If so, what is it?

e. Find all upper bounds of $\{1,8,9\}$.

f. If it exists, find the least upper bound of $\{1,8,9\}$.

g. Find all lower bounds of $\{4,11\}$.

h. If it exists, find the greatest lower bound of $\{4,11\}$.

i. Find all lower bounds of $\{6,12\}$.

j. If it exists, find the greatest lower bound of $\{6,12\}$.

86. Give an example of a poset fitting each description.

a. has a minimal element but no maximal element

b. has a maximal element but no minimal element

c. has neither a maximal nor a minimal element

87. Let $(A,\leq)$ be a poset.

a. Show that, if a and b are greatest elements of $(A,\leq)$, then $a = b$.

b. Show that, if a and b are least elements of $(A,\leq)$, then $a = b$.

88. Let $A = \{1,2,3,6,8,12,24,36,48\}$ and let $R = \{(m,n) \in A \times A\colon m \text{ divides } n\}$. Find a linear order S on A such that $(a,b) \in S$ whenever $(a,b) \in R$.

89. Let R be the relation on $A = \{n \in \mathbb{N}: n \leq 14\}$ given in Exercise 85. Find a linear order S on A such that $(a,b) \in S$ whenever $(a,b) \in R$.

90. Prove that $\mathbb{N} \times \mathbb{N}$ with the lexicographic order with respect to the usual order on $\mathbb{N}$ is a well-ordered set.

91. Show that every finite poset has a minimal element. (*Hint:* Use induction.)

92. Let Σ be an alphabet. Define a relation $\leq$ on Σ^* by $w_1 \leq w_2$ provided that there exist $w,w' \in \Sigma^*$ such that $w_2 = ww_1w'$. Is $(\Sigma^*,\leq)$ a poset? Prove your answer.

93. Let $(A_1,\leq_1)$ and $(A_2,\leq_2)$ be disjoint linearly ordered sets. Describe a way to make $A_1 \cup A_2$ a linearly ordered set.

***94.** For sets A and B we denote by B^A the set of all functions $f: A \to B$.

a. Suppose $(B,\leq_B)$ is a poset and define a relation $\leq$ on B^A by $f \leq g$ provided that $f(a) \leq_B g(a)$ for all $a \in A$. Show that $(B^A,\leq)$ is a poset.

b. Find properties for $(B,\leq_B)$ and A such that $(B^A,\leq)$ is a linearly ordered set.

c. Let $A = \{2,3,4\}$ and $B = \{0,1\}$. With respect to the usual order $\leq_B$, $(B,\leq_B)$ is a poset. Draw the Hasse diagram for $(B^A,\leq)$.

10.7 LISTS AND ARRAYS

Lists are a part of our everyday lives. We make grocery lists, shopping lists, Christmas card lists, and lists of telephone numbers of our friends. In fact, we make a list anytime we are given too much information to keep track of in our heads. The same principle applies when a computer is involved. For instance, suppose that we have a set S of numbers and we want to make a new set T of numbers generated from S by multiplying each member of S by 2. The set T is easy to describe as an indexed set: $T = \{2s: s \in S\}$. To give the computer a set of instructions on how to make the set T, given the set S, is a different story. If the set S has only three elements, we might very well call them A, B, and C and tell the computer to find $2 \cdot A$, $2 \cdot B$, and $2 \cdot C$ and put these numbers in T. But if there are 300 numbers in S, we would probably call them something like $N_1, N_2, \ldots, N_{300}$. There are two reasons for labeling the numbers this way. One obvious reason is that our alphabet is not large enough to provide a symbol for each number. The less obvious reason is that there is an advantage to using subscripts to tell a computer what to do. With them and just a few instructions, the computer can be instructed to perform hundreds of steps. The instructions in Figure 10.22 are one way to tell the computer to find T. As you can see, it takes only six steps to instruct the computer to perform hundreds of steps.

We have previously defined both *function* and *list*, but in working with lists and arrays in computer programming it is useful to extend both

Figure 10.22

Doubling the Numbers in a Set

Let S be the set of numbers $\{N_1, N_2, \ldots, N_{300}\}$.
Step 1. Let $i = 1$.
Step 2. Let $M_i = 2 \cdot N_i$.
Step 3. Let $i = i + 1$.
Step 4. If $i = 301$, then go to step 6.
Step 5. Go to step 2.
Step 6. Stop.
Then $T = \{M_1, M_2, \ldots, M_{300}\}$ is the required set.

of these concepts. In this section we agree that a function f not only has a codomain that contains its range but also a set of conceivable inputs, called the *input set,* that contains its domain (see Figure 10.23).

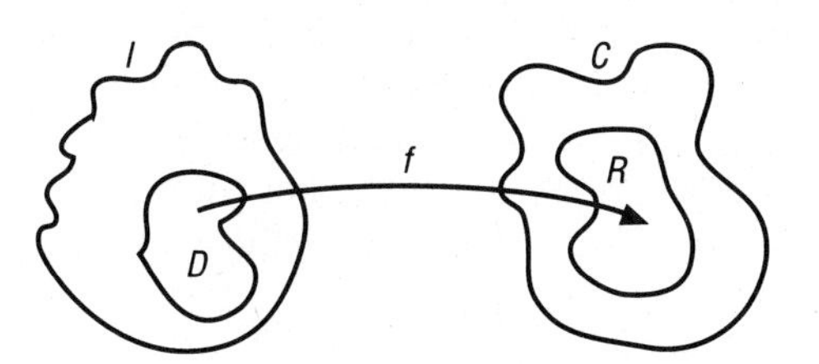

Figure 10.23 A function with domain D, range R, input set I, and codomain C

Definition

Let $(A,\leqslant)$ be a linearly ordered set with first element a. A **linear list on** $(A,\leqslant)$ is a function L whose input set is A and whose domain is a finite subset of A such that (a) a is in the domain of L, and (b) if p is in the domain of L and q is an element of A such that $q < p$, then q is in the domain of L.

At first glance this definition seems formidable, but it is really quite simple. A linear list is a function whose input set is a linearly ordered set with a first member. The domain of a linear list must contain the first member of the input set and, once an element of the input set gets in the domain of a linear list, all the preceding elements of the linearly ordered set must also be there. Notice that the codomain of the linear list can be anything, although, in theory, like any function a linear list must have some specified set for its codomain. In practice the codomain is often supplied as an afterthought, because it is the input set that you have to be careful about.

Example 47

The set $(\mathbb{N},\leqslant)$ is a linearly ordered set with first element 1. Let $L = \{(1,3),(3,\text{dog}),(2,-4)\}$. Then L is a linear list on $(\mathbb{N},\leqslant)$.

Analysis

The domain of L is $\{1,3,2\}$, and its input set is $\mathbb{N}$. Therefore 1 is in the domain of L, and we must check part (b) of the previous definition. There are no natural numbers less than 1, so there is nothing to check for 1. Since 3 is in the domain of L, 1 and 2 must be in the domain of L, and they are. Since 2 is in the domain of L, 1 must be in the domain of L, and it is. Thus L is a linear list on $(\mathbb{N},\leqslant)$. ■

Example 48

The set $(\mathbb{N},\leqslant)$ is a linearly ordered set with first element 1. Let $J = \{(1,3),(2,0),(0,4)\}$, $K = \{(2,3),(3,17)\}$, and $L = \{(2,5),(4,2),(1,-3)\}$. Then J, K, and L are not linear lists on $(\mathbb{N},\leqslant)$.

Analysis

The domain of J is not a subset of the input set, and the first element of the input set does not belong to K. Although the domain of L is a subset of the input set and contains its least element 1, 4 is in the domain of L, and 3 is a member of the input set that is less than 4 and yet 3 is not in the domain L. Thus L does not satisfy the definition and so is not a linear list on $(\mathbb{N},\leqslant)$. ■

Example 49

Let A = {SUNDAY, MONDAY, TUESDAY, WEDNESDAY, THURSDAY, FRIDAY, SATURDAY} and let $\leqslant$ be the usual (linear) ordering of A. Then $(A,\leqslant)$ is a linearly ordered set with first term SUNDAY. Let L = {(MONDAY,8),(TUESDAY,8),(SUNDAY,0)}. Then L is a linear list on $(A,\leqslant)$.

Analysis

The domain of L is {MONDAY,TUESDAY,SUNDAY}. Since SUNDAY is in the domain of L, L satisfies part (a) of the definition of a linear list. Since MONDAY is in the domain of L, we must see if SUNDAY is in the domain of L, and it is. Since TUESDAY is in the domain of L, we need to see if SUNDAY and MONDAY are in the domain of L, and they are. Thus L satisfies part (b) of the definition and is a linear list on $(A,\leqslant)$. ■

In classical mathematics, a one-dimensional array was defined to be a function whose domain is either $\mathbb{N}$ or an initial segment of $\mathbb{N}$, a two-dimensional array was defined to be a function whose domain is $S_1 \times S_2$, where for each $i = 1,2$, S_i is either $\mathbb{N}$ or an initial segment of $\mathbb{N}$, and in general an n-dimensional array was defined to be a function whose domain is $S_1 \times S_2 \times \cdots \times S_n$, where for each $i = 1,2,\ldots,n$, S_i is either $\mathbb{N}$ or an initial segment of $\mathbb{N}$. (Notice that, if there are integers m and n such that the domain of a two-dimensional array is $I_m \times I_n$, then the classical definition of a two-dimensional array coincides with the alternative definition of a matrix given in Chapter 8.) In recent years, with the invention of the computer and high-level programming languages, the word *array* has taken on a different, though similar, meaning. The easiest way to think of an array is to think of it as a linear-list variable. By a *variable* in a computer (or a computer program), we mean a storage location in the computer. The size of a particular variable's storage location depends on what type of information the variable is representing. The value of a variable depends on what is currently in its storage location. Any time a variable is used in calculations, the computer program gets the value of the variable from its storage location and uses the value in its calculations. An array is a contiguous collection of storage locations in a computer. This collection of storage locations (or variables) can be accessed using the single name of the array. Most high-level programming languages have some method for using arrays inside a program. The programmer must provide the high-level language with an array's name, the linearly ordered set which is the input set of the array, and the type of data that is allowed to be inside each storage location—namely, the codomain of the array.

The customary notation for specifying the codomain of an array is "type ______". Thus "type integer" means that the codomain is the set of integers, or more precisely whatever set the programming language and computer will accept for integers. For example, in the programming language Pascal, the lines

```
var
  a: array [1 . . 4] of real;
```

mean that one of the variables in the program has name a, that this is an array variable, that the domain of a is $\{1,2,3,4\}$, and that the codomain of a is the set of all real numbers. Thus, in this program, you can talk about $a[1]$, $a[2]$, $a[3]$, and $a[4]$, and each of these is a storage location that can contain the computer's version of a real number. It means even more than this. Inside the program, if i is a variable of type integer and i is either 1, 2, 3, or 4, then $a[i]$ is a permissible variable name. Thus in one step you can tell the computer how to perform certain operations. Then you can make the computer repeat the operations by changing the value of the variable i and looping through the operations again. It should be noted that the computer has to do the operations over and over again, so the array does not save the computer any time. It is the programmer who profits from the use of an array, since the programmer has only to write the operations once and to provide a looping mechanism.

To review:

a. A *linear list* is a function whose domain is a finite linearly ordered set.

b. An *array* is a data type in a high-level programming language that acts as a linear-list variable.

It should be understood that when an array variable is declared in a Pascal program no values are actually entered into the storage locations. The values must be entered while the program is running, and there is no requirement that all the storage locations provided for in the array be filled. If convenient, a programmer might well choose to leave some of the locations blank. Suppose, for example, that you are going to write a program to sort a list of real numbers and you know that there will be no more than 200 numbers in your list. You must provide an array variable with domain the natural numbers from 1 to 200. But a user may wish to input only 35 real numbers to be sorted. In this case, the last 165 locations in your array would remain blank.

To deal with higher dimensional arrays such as matrices, Pascal allows the codomain of an array to be a previously defined array data type. For example, the following lines in a Pascal program

```
type
  a: array [1 . . 4] of real;

var
  b: array [1 . . 3] of a;
```

define b to be a two-dimensional array of real numbers. Thus, the array variable b reserves room for a complete array of size a for each of the elements in the domain of b, giving b twelve storage locations that can hold the computer's version of a real number:

$(b[1])[1]$	$(b[1])[2]$	$(b[1])[3]$	$(b[1])[4]$
$(b[2])[1]$	$(b[2])[2]$	$(b[2])[3]$	$(b[2])[4]$
$(b[3])[1]$	$(b[3])[2]$	$(b[3])[3]$	$(b[3])[4]$

Once again, this idea of a two-dimensional array is for the convenience of the programmer—not the computer. The computer stores the numbers in order as if they were a usual linear list. But the programmer is allowed to think of the variables as if they were in two-dimensional space. This can be useful when sending information to the two-dimensional screen of a modern-day computer because it makes it easier to keep track of the various symbols on the screen.

We have glossed over how one might enter elements into the storage locations provided by an array, as this is more properly a subject in a beginning programming course. We briefly mention two such methods in an intuitive way.

Let M be a consecutive collection of n storage locations in a computer. Then M is said to operate like a stack provided that an entry or deletion can only be made at one end of the storage locations. Also, M is said to operate like a queue provided that an entry can only be made at one end of the storage locations and a deletion can only be made at the other end of the storage locations.

A picture of a stack in the real world is a plate-dispensing container at a cafeteria. This is a spring-loaded device that holds plates. The plates get a natural linear ordering from their position in the stack (see Figure 10.24). Plate 1 comes first, then Plate 2, and so on. If you want to take a plate off the stack, you must take Plate 1 first. To get to Plate 3, you must take Plate 1 and Plate 2 before Plate 3 becomes available.

Figure 10.24

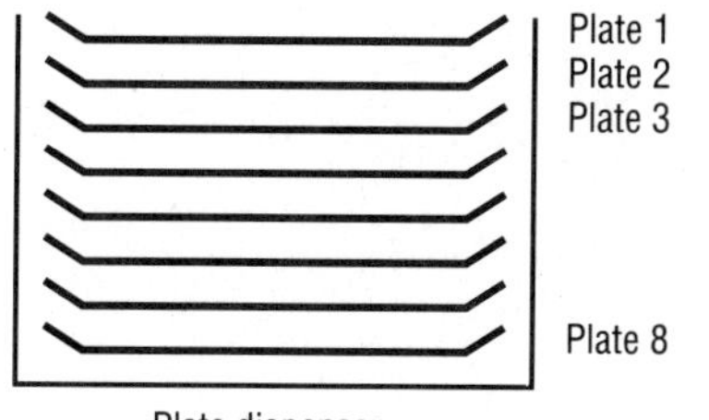

A real world example of a queue can also be found in a cafeteria, since a line of people waiting to be served is a queue. The ordering comes from the natural ordering starting at the front of the line. You delete from this line at the front after the customer pays the cashier. You add to the line at the back, since it is considered rude to cut into line.

Exercises 10.7

95. Decide whether each of the following is a linear list on $(\mathbb{N},\leqslant)$.

a. $\{(3,2)\}$ **b.** $\{(1,0)\}$

c. $\{(-1,2),(0,3),(1,5)\}$ **d.** $\{(4,-1),(3,2),(2,-3),(1,4)\}$

96. Explain why you can never have a linear list on $(\mathbb{Z},<)$.

97. Suppose that a and b are linear lists of real numbers on $(\mathbb{N},\leqslant)$. Suppose that 25 is the largest natural number in the domain of a and that 25 is also the largest natural number in the domain of b.

a. What is the domain of a? **b.** What is the codomain of a?

c. Is 15 in the domain of b? **d.** Is 30 in the domain of a?

e. Write a process in pseudocode that exchanges the values of a and b; that is, for each natural number n between 1 and 25, $a[n]$ is to become $b[n]$ while $b[n]$ is to become $a[n]$. (*Hint:* If you let $a[n] = b[n]$ then, when it comes time to let $b[n] = a[n]$, $a[n]$ has already disappeared.)

98. Suppose that today is Tuesday. What day of the week will it be exactly 1,000 days from today?

99. What is the minimum number of calendars you can print so that one of your calendars is suitable for any year?

100. Suppose that you like the pictures on the calendar you are using so much that you do not want to throw it away. How many years do you have to wait before you are assured that the calendar will again match the year? (*Hint:* The answers to Exercise 99 and this exercise are different.)

101. Give an example of a real world stack other than the one given in the text.

102. Give an example of a real world queue other than the one given in the text.

Chapter 10 Review Exercises

103. Let $A = \{1,2,3,4,5\}$. Which of the following relations on A are reflexive on A? symmetric? transitive?

a. $R_1 = \{(1,2),(2,1),(1,1),(2,2)\}$

b. $R_2 = \{(1,1),(2,2),(3,3),(4,4),(5,5)\}$

c. $R_3 = \{(1,2),(2,3),(3,4),(4,5),(5,1)\}$

d. $R_4 = \{(1,2),(2,1),(3,4),(4,3)\}$

104. Let $A = \{1,2,3\}$.

a. Find a relation on A that is symmetric but not reflexive on A and not transitive.

b. Find a relation on A that is transitive but not reflexive on A and not symmetric.

c. Find a relation on A that is not reflexive on A, not symmetric, and not transitive.

d. Find a relation on A that is reflexive on A and transitive but not symmetric.

105. Which of the following relations on the indicated sets are reflexive on the set? symmetric? transitive?

a. is a brother of, boys

b. is a brother of, people

c. is a mother of, girls

d. is an ancestor of, people

e. is a sibling of, people

106. Let $R = \{(x,y) \in \mathbb{R} \times \mathbb{R}: y - x \text{ is a rational number}\}$.

a. Prove that R is an equivalence relation on $\mathbb{R}$.

b. Describe the set $[4/3]$.

c. Describe the set $[2]$.

d. Find an equivalence class different from $[2]$.

107. Let $R = \{(x,y) \in \mathbb{N} \times \mathbb{N}: x \equiv y \pmod 2\}$. Describe the equivalence classes of $\mathbb{N}$ with respect to R.

108. Assume that it is now 2:00 P.M.

a. What time is it after 24 hours have passed?

b. What time is it after 200 hours have passed?

c. What time is it after 2,000 hours have passed?

d. Describe an easy way to answer questions of this type.

109. Assume that today is Friday.

a. What day of the week is it 10 days from now?

b. What day of the week is it 100 days from now?

c. What day of the week was it 100 days ago?

d. What day of the week is it 6,000 days from now?

e. Describe an easy way to answer questions of this type.

***110.** Use mathematical induction to show that if n is a natural number then $2^{2n+1} + 1 \equiv 0 \pmod 3$.

111. Let $X = \{1,2,3,4,5\}$.

a. List all two-element partitions of X.

b. List all three-element partitions of X.

c. List all four-element partitions of X.

d. How many different partitions of X are there?

112. Let $X = \{1,2,3,4\}$ and $R = \{(1,2),(2,1),(3,4),(4,3)\}$. Let $S_1 = \{x \in X: (x,1) \in R\}$, $S_2 = \{x \in X: (x,2) \in R\}$, $S_3 = \{x \in X: (x,3) \in R\}$, and $S_4 = \{x \in X: (x,4) \in R\}$.

a. Is $\{S_1,S_2\}$ a partition of X? Explain.

b. Is $\{S_1,S_2,S_3,S_4\}$ a partition of X? Explain.

113. Let $A = \{1,2,3,4,5\}$ and $R = \{(1,2),(2,3)\}$.

a. Find the reflexive closure of R.

b. Find the symmetric closure of R.

c. Find the transitive closure of R.

d. Find a relation S on A such that $R \subseteq S$, S is an equivalence relation on A, and if T is a proper subset of S that contains R then T is not an equivalence relation on A.

114. Let $A = \{1,2,3,4\}$. The following directed graph represents a relation R on A.

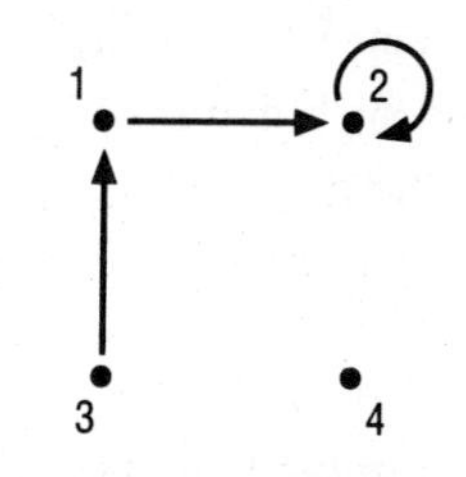

Draw the directed graph that represents each of the following.

a. the symmetric closure of R

b. the reflexive closure of R

c. the transitive closure of R

d. the smallest equivalence relation on A that contains R

115. Let $A = \{1,2,3\}$ and $R = \{(1,2),(2,3),(3,3),(1,3)\}$. Find R^*.

116. Which of the following are posets?

a. $(\mathbb{Z},R_3)$, where $R_3 = \{(m,n) \in \mathbb{Z} \times \mathbb{Z}: m \neq n\}$

b. $(\mathbb{Z},R_4)$, where $R_4 = \{(m,n) \in \mathbb{Z} \times \mathbb{Z}: m \text{ does not divide } n\}$

117. Determine which of the relations represented by the following directed graphs are partial orders. Draw the Hasse diagram of all partial orders whose directed graphs appear in this exercise using either the definition of a Hasse diagram or the recipe for finding it.

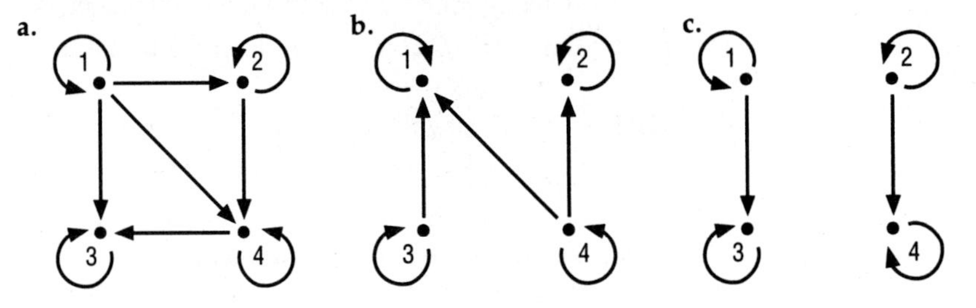

118. Determine which of the relations represented by the following zero–one matrices are partial orders. Draw the Hasse diagram of those relations that are partial orders.

a. $\begin{bmatrix} 1 & 0 & 1 & 0 \\ 0 & 1 & 0 & 1 \\ 1 & 0 & 1 & 0 \\ 0 & 1 & 0 & 1 \end{bmatrix}$ **b.** $\begin{bmatrix} 1 & 0 & 0 & 1 \\ 0 & 1 & 1 & 0 \\ 0 & 1 & 1 & 0 \\ 1 & 0 & 1 & 1 \end{bmatrix}$

c. $\begin{bmatrix} 1 & 0 & 1 & 1 \\ 0 & 1 & 1 & 1 \\ 1 & 1 & 1 & 0 \\ 1 & 1 & 1 & 1 \end{bmatrix}$

119. Let $A = \{1,2,3,4\}$ and let R be the lexicographic order on $A \times A$ with respect to the usual $\leqslant$.

a. Find all pairs $(m,n) \in A \times A$ such that $(m,n)R(3,2)$.

b. Find all pairs $(m,n) \in A \times A$ such that $(1,3)R(m,n)$.

c. Draw the Hasse diagram of the poset $(A \times A,R)$.

120. Let $(A,\leqslant)$ be a poset in which every nonempty set has a least member. Prove that $\leqslant$ is a well-order.

121. Let $(A,\leqslant)$ be a poset and let $B \subseteq A$.

a. Prove that if a and b are least upper bounds of B then $a = b$.

b. Prove that if a and b are greatest lower bounds of B then $a = b$.

122. Let $(A,\leqslant)$ be the linearly ordered set in Example 49.

a. Is WEDNESDAY $\leqslant$ SUNDAY? **b.** Is WEDNESDAY $\leqslant$ FRIDAY?

c. Is FRIDAY $\leqslant$ TUESDAY?

d. Suppose that M is a linear list on A and THURSDAY is in the domain of M. Does TUESDAY have to be in the domain of M? Does FRIDAY have to be in the domain of M?

123. Suppose that today is Wednesday. What day of the week will it be 5,000 days from today?

124. Let $n,p \in \mathbb{N}$, let A be a set with n elements, and let R be an equivalence relation on A. Prove that if R has p elements then $p - n$ is even.

125. Suppose that $n \in \mathbb{N}$ and that A is a set with n members.

a. How many relations on A are reflexive on A?

b. How many relations on A are symmetric?

126. Let R be a relation that is reflexive on a finite set A with n elements. Prove that $M_{R^*} = (M_R)^{[n]}$.

Chapter

11

GRAPH THEORY

Hitherto most of the problems we have addressed in this text have been algebraic as opposed to geometric, and, with the exception of the use of directed graphs to describe relations, our basic tools have been tools of algebra rather than of geometry. Indeed, it is not even apparent that there should be such a thing as discrete geometry, because the usual notions of ray, line, and line segment do not involve discrete sets. But there is such a thing, and in this chapter we consider one of its branches, graph theory.

Although Leonhard Euler initiated the study of graph theory in 1736 with a problem of no practical significance, today graph theory has important applications in such diverse areas as computer science, chemistry, physics, economics, biology, psychology, engineering, and urban and highway planning. The graphs with which we are concerned are similar to road maps in that they represent connections or relations between various points of interest.

11.1

SIMPLE GRAPHS

Intuitively, a *graph* is a finite set of points, called *vertices,* together with a finite set of lines, called *edges,* that join some or all of these points (see Figure 11.1). Some mathematicians, however, use the word *graph* to mean what we call a *simple graph,* and it is important to study both graphs and simple graphs. We begin with the definition of a simple graph, which, as the terminology suggests, is the easier of the two concepts to understand. As we see later in this chapter, every simple graph can be thought of as being a graph, so we sometimes refer to a simple graph as a graph.

Figure 11.1

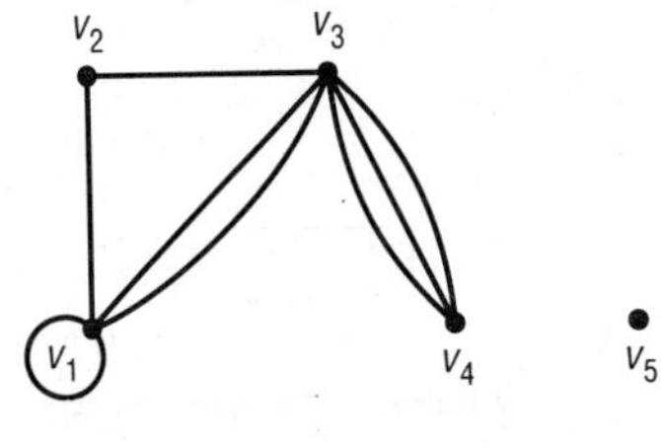

Definition

A **simple graph** G is an ordered pair (V,E), where V is a finite nonempty set whose members are called **vertices** and E is a set of two-element subsets of V whose members are called **edges.**

Example 1

Let $V = \{1,2,3,4\}$ and $E = \{\{1,2\},\{2,3\},\{3,4\},\{2,4\},\{1,4\}\}$. Then $G = (V,E)$ is a simple graph.

An appealing feature of graph theory is the geometric aspect of the subject. It is often useful to draw a picture in the plane of a graph, where each member of V is represented by a point and each member of E is represented by a line segment. We say that such a picture represents the graph G. Figure 11.2, for instance, represents the simple graph G given in this example.

Figure 11.2

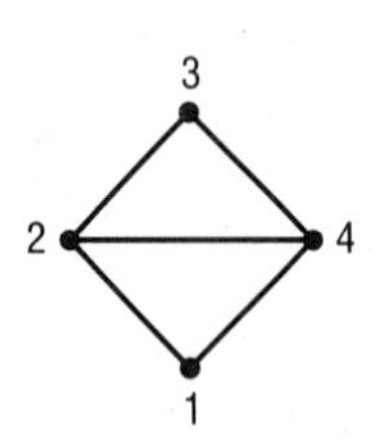

■

Note that in a simple graph (a) the set E of edges may be empty; (b) the set E must be disjoint from V; and (c) each member of E consists of two distinct members of V.

Example 2

Let $V = \{v_1,v_2,v_3,v_4,v_5\}$ and $E = \{\{v_1,v_2\}, \{v_1,v_3\}, \{v_2,v_4\}, \{v_2,v_3\}, \{v_2,v_5\}, \{v_3,v_4\}, \{v_4,v_5\}\}$. Then $G = (V,E)$ is a simple graph and is represented by Figure 11.3.

Figure 11.3

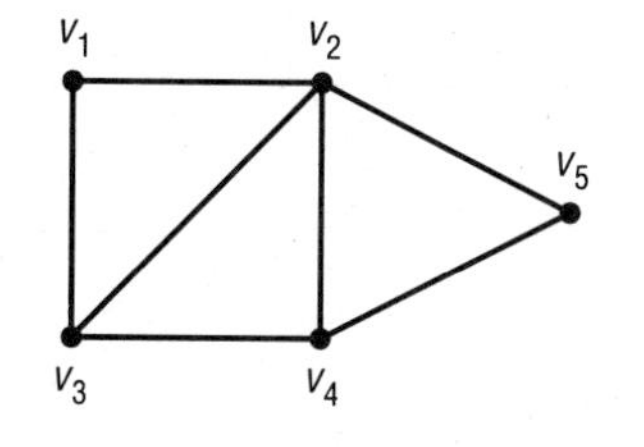

■

Recall that we introduced directed graphs in Section 9.2. Before we continue our study of graphs, let us digress and compare simple graphs with directed graphs. Suppose that $G = (V,E)$ is a simple graph. For each member $\{u,v\}$ of E we define an ordered pair (u,v) by arbitrarily choosing one of u and v to be the first term of (u,v). It is easy to see that, if R is the set consisting of all such ordered pairs, then (V,R) is a directed graph.

Example 3

Let $V = \{v_1,v_2,v_3,v_4\}$ and $E = \{\{v_1,v_2\},\{v_1,v_4\},\{v_3,v_2\}, \{v_3,v_4\}\}$. Then $G = (V,E)$ is a simple graph. How do we associate with G a directed graph according to the preceding discussion?

Analysis

For each member of E we define an ordered pair, and we let R be the set of all such ordered pairs. For example, if $R = \{(v_1,v_2),(v_4,v_1),(v_3,v_2),(v_4,v_3)\}$, then (V,R) is a directed graph. We can also obtain a different directed graph by choosing a different member of some member of E to be the first term of the ordered pair. If $S = \{(v_1,v_2),(v_1,v_4),(v_2,v_3),(v_4,v_3)\}$, then (V,S) is a directed graph. For some member e of E, we can also define two ordered pairs associated with e. Thus, if $T = \{(v_1,v_2),(v_2,v_1),(v_4,v_1), (v_2,v_3),(v_3,v_4),(v_4,v_3)\}$, then (V,T) is a directed graph. ■

Suppose that we have a directed graph $G = (V,R)$. Can we obtain a simple graph from G by choosing an edge e for each member (a,b) of R?

Why not? It is possible that there is a vertex v of V such that the ordered pair (v,v) belongs to R. But if $G' = (V,E)$ is a simple graph, then no member of E can be equal to $\{v,v\} = \{v\}$.

Example 4

Let (V,R) be the directed graph given in Figure 11.4. Then $V = \{a,b,c,d\}$ and $R = \{(a,a),(a,b),(a,c),(b,a),(c,b),(c,c)\}$. But if $E = \{\{a\},\{a,b\},\{a,c\},\{b,c\},\{c\}\}$, then (V,E) is not a simple graph.

Figure 11.4

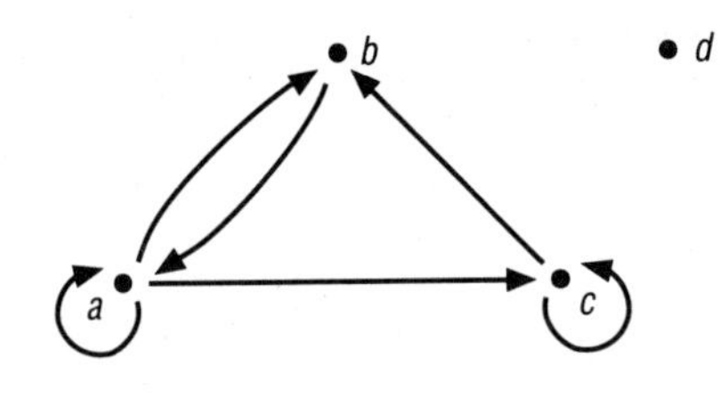

In the remainder of this section we introduce some basic terminology and consider some special simple graphs. ■

Definition

> Let $G = (V,E)$ be a simple graph. Two vertices u and v of V are **adjacent** provided that $\{u,v\} \in E$. Let $e \in E$ and $u,v \in V$. Then e is said to **connect** u and v provided that $e = \{u,v\}$, and e is said to be **incident with** u, or u is said to be an **endpoint** of e, provided that $u \in e$.

Example 5

In Figure 11.3, which pairs of vertices are adjacent?

Analysis

The following pairs of vertices are adjacent:

v_1 and v_2 v_1 and v_3 v_2 and v_3 v_2 and v_4

v_2 and v_5 v_3 and v_4 v_4 and v_5 ■

Definition

> Let $G = (V,E)$ be a simple graph. If $v \in V$, the **degree** of v, denoted $\deg(v)$, is the number of edges that are incident with v.

Example 6

Let $G = (V,E)$ be the simple graph represented by Figure 11.5. What is the degree of each member of V?

Figure 11.5

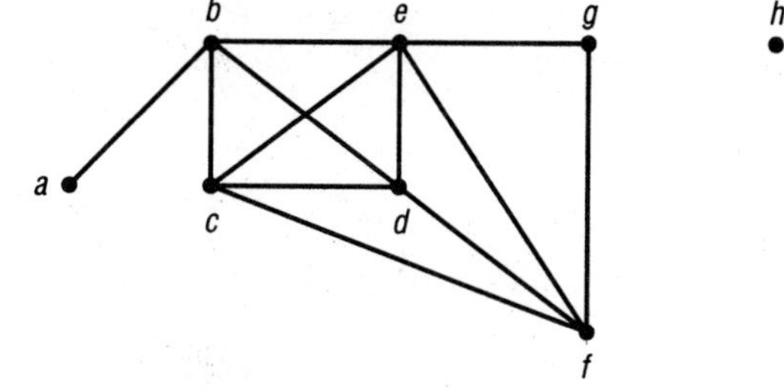

Analysis

$\deg(a) = 1$, $\deg(b) = \deg(c) = \deg(d) = \deg(f) = 4$, $\deg(e) = 5$, $\deg(g) = 2$, and $\deg(h) = 0$. ■

Notice that in Figure 11.5, the graph is drawn so that the line segment that represents the edge $\{b,d\}$ crosses the line segment that represents the edge $\{c,e\}$. But remember that edge $\{b,d\}$ is really the set whose members are b and d. You might think of the place where these edges cross as being an overpass without a cloverleaf. Sometimes, as is the case here, an edge can be represented by a curved line segment, and the graph drawn so that line segments representing edges do not cross except at vertices.

Definition

A simple graph $G = (V,E)$ is **complete** provided that any two members of V are adjacent. If $G = (V,E)$ is a complete graph and the number of members of V is n, then G is called a **complete graph on n vertices.** It is customary to denote a complete graph on p vertices by K_p.

The drawings in Figure 11.6 represent K_p for each $p = 1,2,3,4,5$.

Figure 11.6

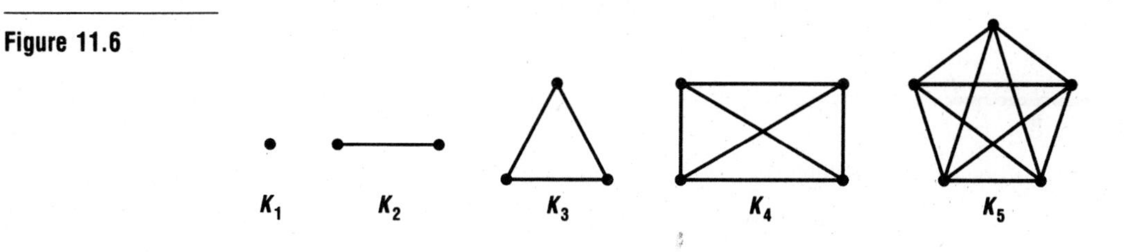

Definition

If $r \in \mathbb{N} \cup \{0\}$, and each vertex of the simple graph $G = (V,E)$ has degree r, then G is said to be ***r*-regular.** A simple graph $G = (V,E)$ is said to be **regular** if there is a member r of $\mathbb{N} \cup \{0\}$ such that G is r-regular.

Definition

If $G = (V,E)$ is a simple graph, the number of members of V is referred to as the **order** of G.

Example 7

Draw the regular graphs of order 4.

Analysis

The drawings in Figure 11.7 represent the regular graphs of order 4.

Figure 11.7

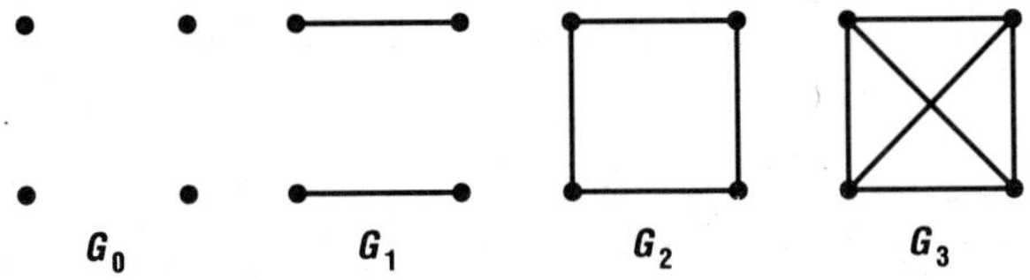

G_0 G_1 G_2 G_3

Notice that G_3 is K_4. ■

Definition

A simple graph $G = (V,E)$ is said to be **bipartite** if (a) there are nonempty disjoint subsets V_1 and V_2 of V such that $V = V_1 \cup V_2$, and (b) u and v are adjacent vertices if and only if one of them belongs to V_1 and the other to V_2. The partition $\{V_1,V_2\}$ of V is called a **bipartition** of G. A **complete bipartite graph** is a bipartite graph $G = (V,E)$ with bipartition $\{V_1,V_2\}$. A complete bipartite graph with partition $\{V_1,V_2\}$, where m is the number of members of V_1 and n is the number of members of V_2, is denoted by $K_{m,n}$.

Example 8

Draw $K_{1,1}$, $K_{1,2}$, $K_{2,2}$, $K_{2,3}$, $K_{3,3}$, and $K_{2,4}$.

Analysis

The drawings in Figure 11.8 represent $K_{1,1}$, $K_{1,2}$, $K_{2,2}$, $K_{2,3}$, $K_{3,3}$, and $K_{2,4}$.

Figure 11.8

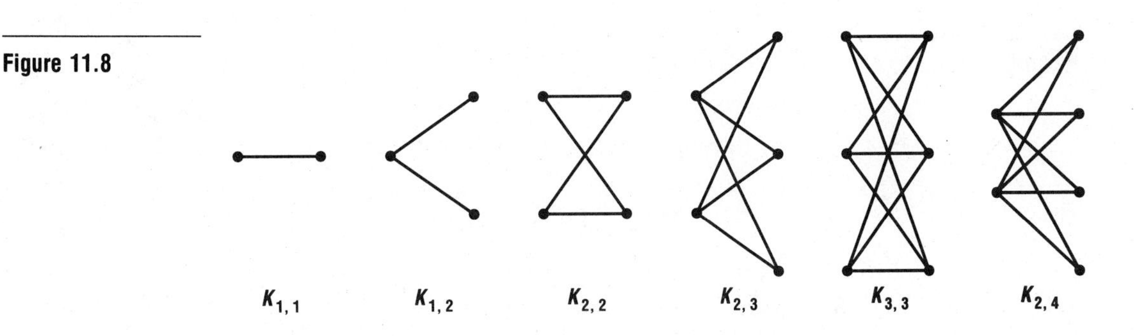

In each case, the vertices on the left represent the members of V_1 and the vertices on the right represent the members of V_2. ■

Example 9

Are the simple graphs G and H represented by the drawings in Figure 11.9 bipartite?

Figure 11.9

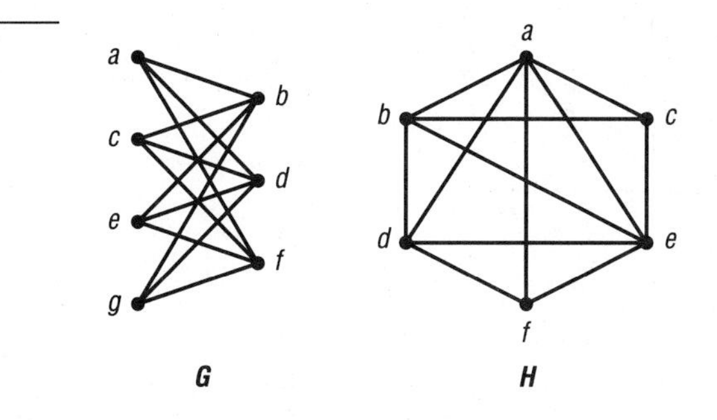

Analysis

The simple graph G is bipartite since $\{\{a,c,e,g\},\{b,d,f\}\}$ is a bipartition. The simple graph H is not bipartite, and the proof is by contradiction. Suppose $\{V_1,V_2\}$ is a bipartition. We may assume, without loss of generality, that $a \in V_1$. Then, since a is adjacent to each of the other vertices, V_2 must be $\{b,c,d,e,f\}$. However, b and c are adjacent, so we have a contradiction. ■

Definition

Let $n \in \mathbb{N}$ such that $n \geqslant 3$. Let $V_n = \{v_1,v_2, \ldots ,v_n\}$ be a set with n members, let $e_n = \{v_n,v_1\}$, for each $i = 1,2, \ldots ,n-1$, let $e_i = \{v_i,v_{i+1}\}$, and let $E_n = \{e_1,e_2, \ldots ,e_n\}$. Then $C_n = (V_n,E_n)$ is a simple graph, called a **cycle of length *n*** or an ***n*-cycle.** We say that a simple graph G is a **cycle** if there is a natural number $n \geqslant 3$ such that G is an n-cycle.

Example 10

Draw the cycles C_3, C_4, C_5, and C_6.

Analysis

The drawings in Figure 11.10 represent C_3, C_4, C_5, and C_6.

Figure 11.10

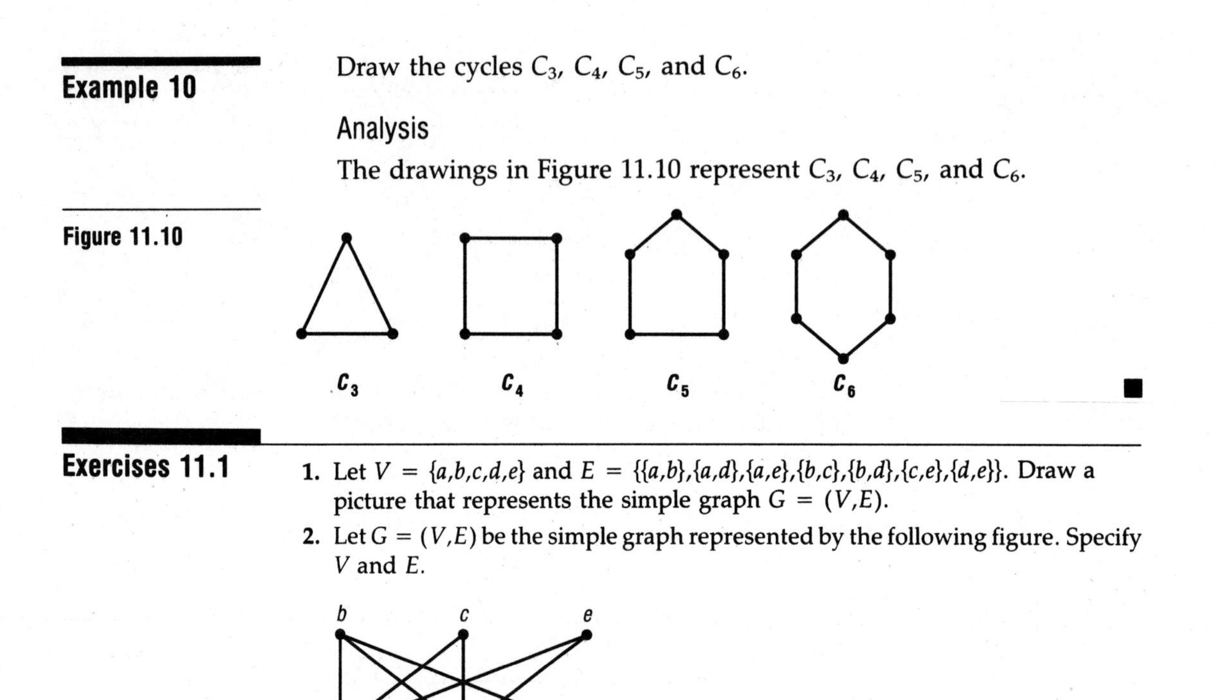

■

Exercises 11.1

1. Let $V = \{a,b,c,d,e\}$ and $E = \{\{a,b\},\{a,d\},\{a,e\},\{b,c\},\{b,d\},\{c,e\},\{d,e\}\}$. Draw a picture that represents the simple graph $G = (V,E)$.
2. Let $G = (V,E)$ be the simple graph represented by the following figure. Specify V and E.

3. Find the vertices that are adjacent to vertex a in each of the following simple graphs.

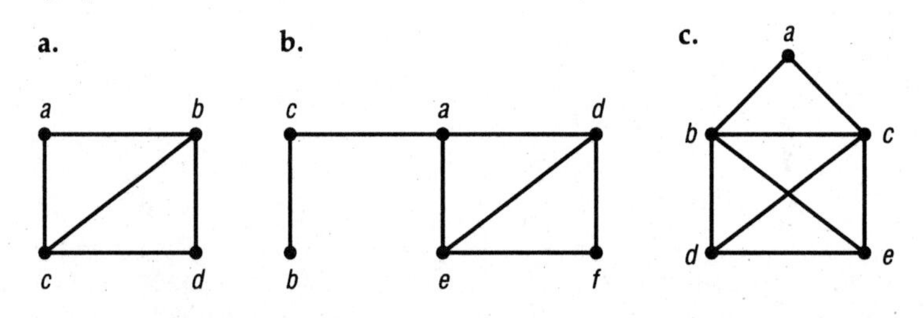

4. Find the degree of each of the vertices of the simple graph drawn in Exercise 3b.
5. Draw a figure that represents the complete graph on six vertices.
6. Draw a figure that represents $K_{3,4}$.
7. How would you partition the vertices of each of the following simple graphs to verify that each is a bipartite graph?

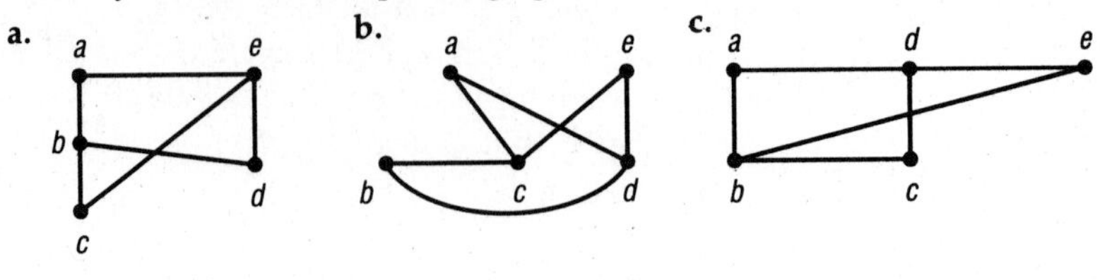

8. Which of the following simple graphs are bipartite? Justify your answers.

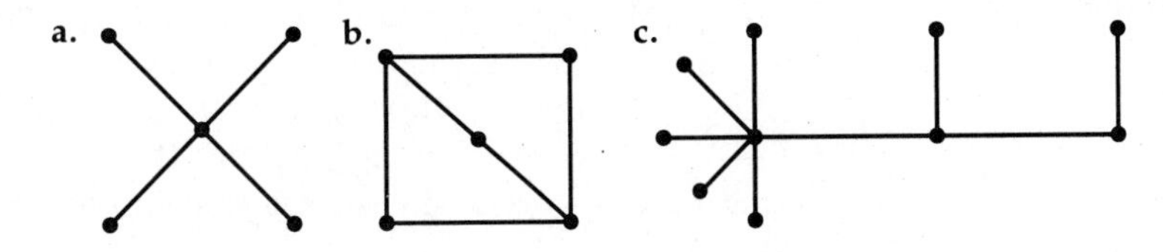

9. Draw a simple graph with five vertices and ten edges.
10. How many vertices and how many edges do each of the following simple graphs have? Assume that $m,n \in \mathbb{N}$ and $n \geqslant 3$.

    a. C_n b. K_n c. $K_{m,n}$

11. Let $n \in \mathbb{N}$ such that $n \geqslant 3$. Let $V_n = \{v_1,v_2, \ldots ,v_n,v_{n+1}\}$ be a set with $n + 1$ members. Let $e_n = \{v_n,v_1\}$, and for each $i = 1,2, \ldots ,n - 1$ let $e_i = \{v_i,v_{i+1}\}$. Also, for each $i = n + 1,n + 2, \ldots ,2n$ let $e_i = \{v_{i-n},v_{n+1}\}$. Finally, let $E_n = \{e_1,e_2, \ldots ,e_{2n}\}$. Then $W_n = (V_n,E_n)$ is a simple graph. Draw diagrams that represent W_3, W_4, and W_5. (*Hint:* W_n is called a *wheel*, and for each $i = n + 1,n + 2, \ldots ,2n$, e_i is called a *spoke* of the wheel.)
12. Show that for each natural number $n \geqslant 2$ the number of simple graphs with n vertices is $2^{C(n,2)}$.

11.2 GRAPHS

Figure 11.11a does not represent a simple graph because there are two lines joining points a and c. In like manner, Figure 11.11b does not represent a simple graph because there is a line that starts at a and ends at a. Yet in discrete mathematics there is a need to study such figures. For example, there are computer networks in which there is more than one telephone line from one computer to another and in which there is a telephone line from a computer to itself. Consequently, we generalize the concept of a simple graph. In doing so we must introduce some additional terminology because, for example, we cannot represent each of the two distinct lines from a to c in Figure 11.11a by $\{a,c\}$. We are therefore led to use a function in our definition of a graph.

Figure 11.11

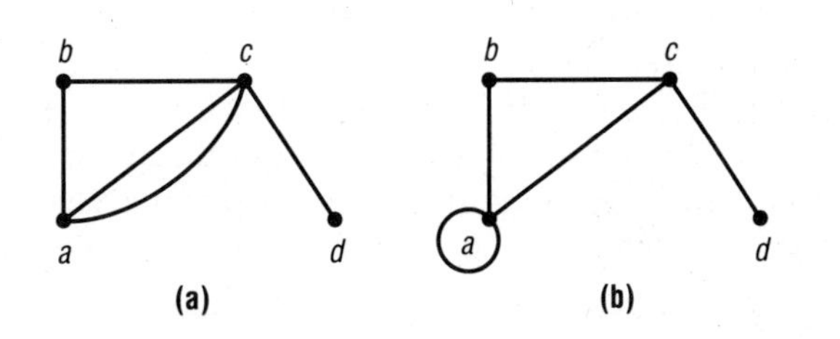

Definition

A **graph** G is an ordered triple (V,E,f), where V is a finite nonempty set whose members are called **vertices,** E is a finite set, disjoint from V, whose members are called **edges,** and $f: E \to \{\{u,v\}: u,v \in V\}$ is a function. If $v \in V$ and $e \in E$, then v is said to be a **vertex** of e, or e is said **to have vertex** v, provided that $v \in f(e)$.

This definition is abstract, but if we keep a picture in mind it is easy to see what we are doing. Consider Figure 11.12a. We have a labeling of the vertices, so we can specify the set V of vertices. In particular, $V = \{a,b,c,d\}$. To specify the set E of edges, we label the lines as shown in Figure 11.12b. Then $E = \{e_1,e_2,e_3,e_4,e_5,e_6\}$.

Figure 11.12

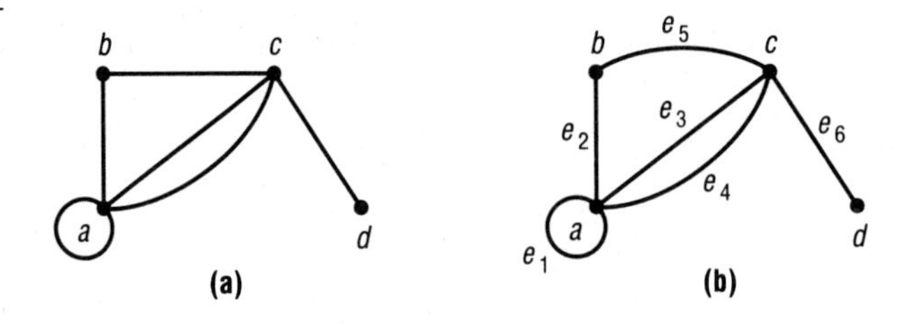

By considering only the sets V and E, we cannot tell whether a given vertex belongs to a specified edge—but to know what graph we are talking about without looking at pictures we need this information. This is the purpose of the function f. It specifies which vertices belong to a given edge. We can, for example, look at Figure 11.12b and define the function $f: E \to \{\{u,v\}: u,v \in V\}$. Since a is the only vertex that belongs to the edge e_1, $f(e_1) = \{a\}$. Since a and b are the vertices of e_2, $f(e_2) = \{a,b\}$. In like manner, $f(e_3) = \{a,c\}$, $f(e_4) = \{a,c\}$, $f(e_5) = \{b,c\}$, and $f(e_6) = \{c,d\}$. Therefore the graph G pictured in Figure 11.12b is the graph $G = (V,E,f)$, where $V = \{a,b,c,d\}$, $E = \{e_1,e_2,e_3,e_4,e_5,e_6\}$, and $f: E \to \{\{u,v\}: u,v \in V\}$ is the function defined by $f(e_1) = \{a\}$, $f(e_2) = \{a,b\}$, $f(e_3) = \{a,c\}$, $f(e_4) = \{a,c\}$, $f(e_5) = \{b,c\}$, and $f(e_6) = \{c,d\}$.

Notice that the codomain of the function $f: E \to \{\{u,v\}: u,v \in V\}$ given in the definition of a graph is the set whose members are two-member subsets of V and one-member subsets of V. The graph $G = (V,E,f)$ shown in Figure 11.12b has two properties that prevent it from being a simple graph: $f(e_3) = f(e_4)$, and $f(e_1) = \{a\}$.

In the introduction to Section 11.1, we promised that every simple graph could be thought of as being a graph. Since a simple graph is an ordered pair and a graph is an ordered triple, it is clear that simple graphs are *not* graphs. However, if $G = (V,E)$ is a simple graph, then we can define a function $f: E \to \{\{u,v\}: u,v \in V\}$ so that (V,E,f) is a graph. Let $e \in E$. Then there are two unique members u and v of V such that $e = \{u,v\}$. Define $f(e) = \{u,v\}$. It is this function that we are thinking of as being the third term of the ordered triple when we refer to a simple graph as a graph.

Definition

Let $G = (V,E,f)$ be a graph. If e_1 and e_2 are members of E, and $f(e_1) = f(e_2)$, then e_1 and e_2 are called **multiple,** or **parallel, edges.** If e is an edge in E and v is a vertex in V such that $f(e) = \{v\}$, then e is called a **loop at** v, or simply a **loop.**

Example 11

Let $G = (V,E,f)$ be the graph shown in Figure 11.13. Specify V, E, and f. Which edges are multiple edges and which edges are loops?

Figure 11.13

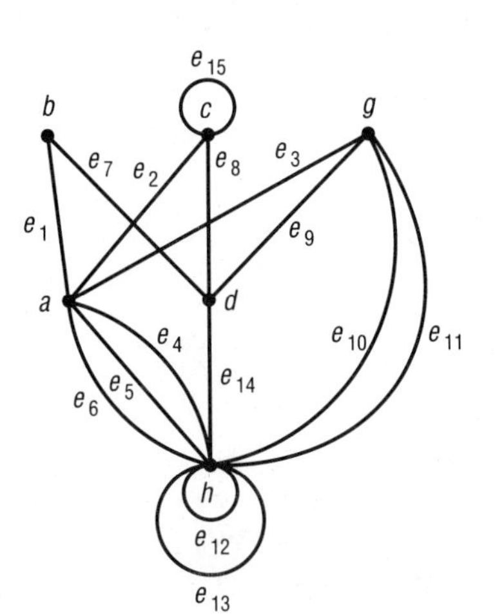

Analysis

The vertices are a, b, c, d, g, and h, so $V = \{a,b,c,d,g,h\}$. The edges are e_1, e_2, e_3, e_4, e_5, e_6, e_7, e_8, e_9, e_{10}, e_{11}, e_{12}, e_{13}, e_{14}, and e_{15}, so

$$E = \{e_1,e_2,e_3,e_4,e_5,e_6,e_7,e_8,e_9,e_{10},e_{11},e_{12},e_{13},e_{14},e_{15}\}$$

We write the definition of the function $f: E \to \{\{u,v\}: u,v \in V\}$ by looking at the picture and determining which vertices belong to which edges. So this function is defined by $f(e_1) = \{a,b\}$, $f(e_2) = \{a,c\}$, $f(e_3) = \{a,g\}$, $f(e_4) = \{a,h\}$, $f(e_5) = \{a,h\}$, $f(e_6) = \{a,h\}$, $f(e_7) = \{b,d\}$, $f(e_8) = \{c,d\}$, $f(e_9) = \{d,g\}$, $f(e_{10}) = \{g,h\}$, $f(e_{11}) = \{g,h\}$, $f(e_{12}) = \{h\}$, $f(e_{13}) = \{h\}$, $f(e_{14}) = \{d,h\}$, and $f(e_{15}) = \{c\}$. Edges e_4, e_5, and e_6 are multiple edges, and edges e_{10} and e_{11}

are multiple edges. Edges e_{12} and e_{13} are loops at h, and edge e_{15} is a loop at c. ■

Notice that a graph $G = (V,E,f)$ can be thought of as being a simple graph provided that G has no multiple edges and no loops. If G has no multiple edges and no loops, then we can identify the edge e with $f(e)$; that is, we can think of them as being one and the same.

Example 12

Let $V = \{a,b,c,d\}$ and $E = \{e_1,e_2,e_3,e_4,e_5,e_6,e_7,e_8,e_9,e_{10}\}$. Define $f: E \rightarrow \{\{u,v\}: u,v \in V\}$ by $f(e_1) = \{a\}$, $f(e_2) = \{a,b\}$, $f(e_3) = \{a,b\}$, $f(e_4) = \{b,c\}$, $f(e_5) = \{c\}$, $f(e_6) = \{c,d\}$, $f(e_7) = \{c,d\}$, $f(e_8) = \{c,d\}$, $f(e_9) = \{d\}$, $f(e_{10}) = \{d\}$. Draw the graph $G = (V,E,f)$ and label the vertices and edges.

Analysis

The set V of vertices consists of four points, so we label these four points and then draw the ten edges joining these vertices according to the definition of f as shown in Figure 11.14.

Figure 11.14

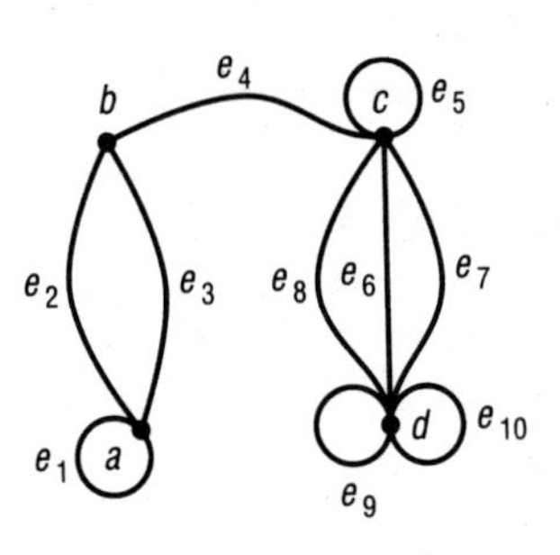

■

We next give an example of a famous problem and show how a graph can be used to model this problem.

Example 13

In the town of Königsberg in the eighteenth century, there were seven bridges that crossed the Pregel River. These bridges connected two islands A and B in the river with each other and connected the islands to the banks of the river as shown in Figure 11.15a.

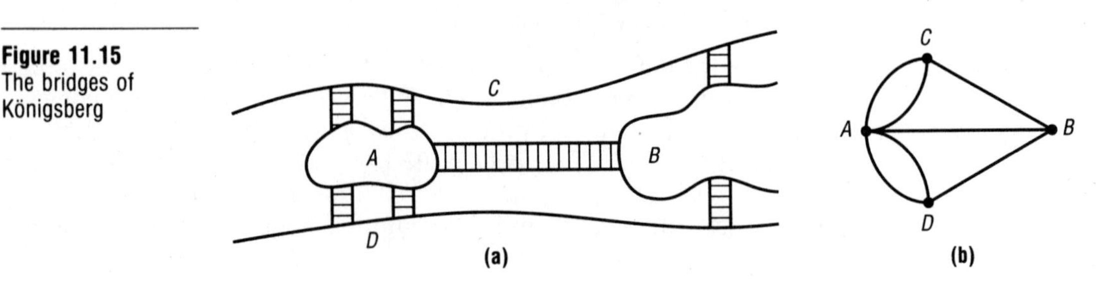

Figure 11.15 The bridges of Königsberg

The townspeople liked to take walks through town across these bridges, but they could not figure out a way to start at some point in town, walk across each bridge just once, and return to the starting point. They brought the problem to the attention of the Swiss mathematician Leonhard Euler, who solved the problem by using a graph. He represented each of the four land areas $A,B,C,$ and D by vertices and each of the bridges by edges, obtaining the graph represented by Figure 11.15b. Euler observed that if such a walk were possible it could be represented by a sequence of eight letters, where each letter was one of A, B, C, or D and consecutive letters meant that the edge (or bridge) joining these two vertices (land areas) had been traversed. Since five edges (or bridges) led to vertex A (or land area A), the letter A had to appear in the sequence three times—twice to denote an entrance to and exit from vertex A and once to indicate an exit from A or entrance to A. Likewise, each of the letters B, C, and D had to appear in the sequence two times. Thus he saw that nine letters had to appear in the sequence, so he knew that such a walk was impossible. ■

Some concepts that we have defined for simple graphs have obvious analogues for graphs.

Definition

Let $G = (V,E,f)$ be a graph. Vertices u and v (not necessarily distinct) of V are **adjacent** provided that there is an edge e of E such that $f(e) = \{u,v\}$. If u and v are vertices (not necessarily distinct) of V, and e is a member of E such that $f(e) = \{u,v\}$, then e is said to **connect** u and v. Let e be an edge and let u be a vertex. Then e is **incident with** u and u is an **endpoint** of e if $u \in f(e)$. If $v \in V$, the **degree** of V, denoted $\deg(v)$, is the number of edges that are incident with v, where each loop is counted as two edges.

Example 14

Let $G = (V,E,f)$ be the graph shown in Figure 11.14. What is the degree of each member of V?

Analysis

The degree of a is 4 because e_1, e_2, and e_3 are incident with a and e_1 is a loop. The degree of b is 3 because e_2, e_3, and e_4 are incident with b and none of them is a loop. The degree of c is 6 since e_4, e_5, e_6, e_7, and e_8 are incident with c and e_5 is a loop. Finally, the degree of d is 7 since e_6, e_7, e_8, e_9, and e_{10} are incident with d and e_9 and e_{10} are loops. ■

Definition

Let $G = (V,E,f)$ be a graph. A vertex v of V is said to be **isolated** if $\deg(v) = 0$ and is said to be **pendant** if $\deg(v) = 1$. If $n \in \mathbb{N} \cup \{0\}$, a vertex of degree n is said to be ***n*-valent.** A vertex is said to be **odd** or **even** depending on whether its degree is odd or even.

Example 15

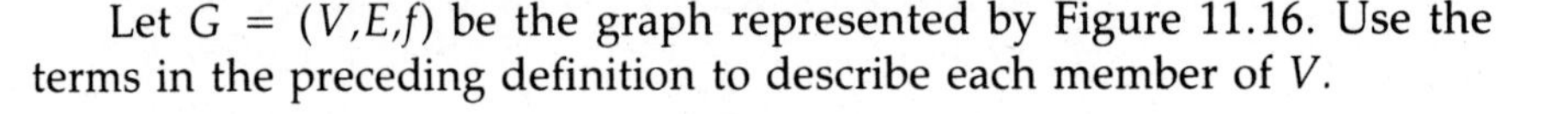

Let $G = (V,E,f)$ be the graph represented by Figure 11.16. Use the terms in the preceding definition to describe each member of V.

Figure 11.16

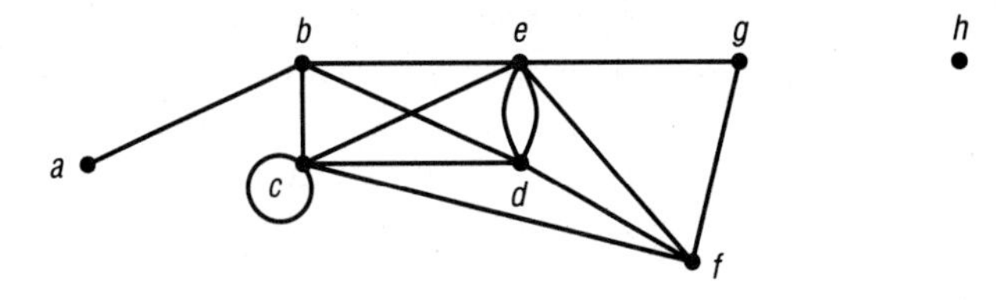

Analysis

Vertex h is isolated; it is not adjacent to any vertex, and no edge is incident with it. Vertex a is pendant; it is adjacent to one vertex, b, and one edge, e, such that $f(e) = \{a,b\}$ is incident with it. Vertex g is 2-valent, vertices b and f are 4-valent, vertex d is 5-valent, and vertices c and e are 6-valent. ■

Theorem 11.1

> If $G = (V,E,f)$ is a graph, where $V = \{v_1,v_2, \ldots ,v_p\}$ and $E = \{e_1,e_2, \ldots ,e_q\}$, then $\Sigma_{i=1}^{p} \deg(v_i) = 2q$.

Proof

This result follows immediately because each edge that is not a loop has two vertices, and if e is a loop at v then e adds 2 to the degree of v. □

Example 16

Let $G = (V,E,f)$ be a graph. If V has twelve members and the degree of each member of V is 4, how many members does E have?

Analysis

Let $V = \{v_1,v_2, \ldots v_{12}\}$. Then, by Theorem 11.1, if q is the number of members of E, $2q = \Sigma_{i=1}^{12} \deg(v_i) = 12 \cdot 4 = 48$. Hence $q = 24$. ■

Theorem 11.2

If $G = (V,E,f)$ is a graph, then the number of vertices of odd degree is even.

Proof

Let $G = (V,E,f)$ be a graph, let $v_1, v_2, \ldots, v_k$ be the vertices of odd degree, let $v_{k+1}, v_{k+2}, \ldots, v_p$ be the vertices of even degree, and suppose that q is the number of members of E. By Theorem 11.1,

$$2q = \sum_{i=1}^{p} \deg(v_i) = \sum_{i=1}^{k} \deg(v_i) + \sum_{i=k+1}^{p} \deg(v_i)$$

Since $\Sigma_{i=k+1}^{p} \deg(v_i)$ is even, $\Sigma_{i=1}^{k} \deg(v_i)$ is even and therefore k is even. □

As with simple graphs, the number of members of V is referred to as the *order* of the graph $G = (V,E,f)$.

Definition

A graph $H = (W,F,g)$ is a **subgraph** of a graph $G = (V,E,f)$ provided that $W \subseteq V$, $F \subseteq E$, and $g = f|W$. A **spanning subgraph** of a graph $G = (V,E,f)$ is a subgraph $H = (W,F,g)$ such that $W = V$.

Let $G = (V,E,f)$ be a graph and suppose that we (a) remove each member of E that is a loop, and (b) for each pair u,v of distinct members of V, remove all but one edge that connects u and v. Then we obtain a simple graph $H = (W,F,g)$ that is a spanning subgraph of G, called an **underlying simple graph** of G.

Example 17

Let $G = (V,E,f)$ be the graph represented in Figure 11.17. Find an underlying simple graph of G.

Figure 11.17

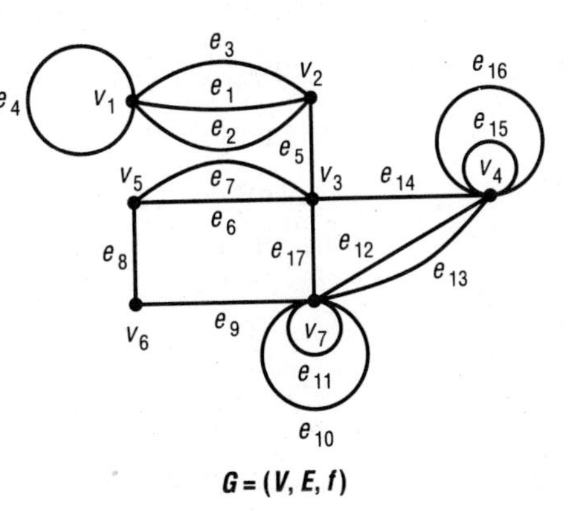

G = (V, E, f)

Analysis

Let $F = \{e_1,e_5,e_6,e_8,e_9,e_{12},e_{14}\}$ and define $g: F \rightarrow \{\{u,v\}: u,v \in V\}$ by $g(e_1) = \{v_1,v_2\}$, $g(e_5) = \{v_2,v_3\}$, $g(e_6) = \{v_3,v_5\}$, $g(e_8) = \{v_5,v_6\}$, $g(e_9) = \{v_6,v_7\}$, $g(e_{12}) = \{v_7,v_4\}$, and $g(e_{14}) = \{v_4,v_3\}$. Then $H = (V,F,g)$ is an underlying simple graph of G, and it is represented in Figure 11.18.

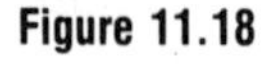

Figure 11.18

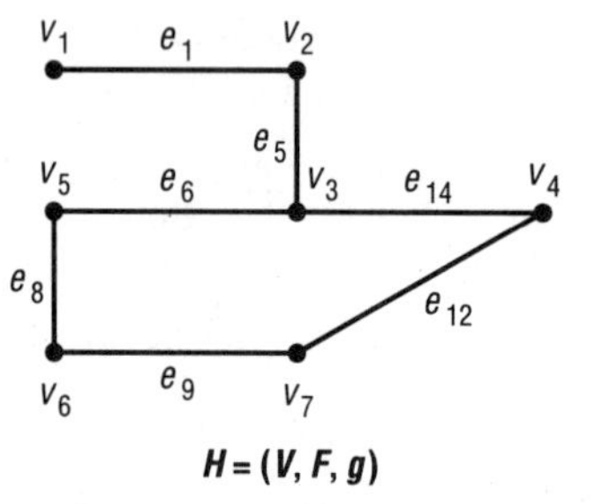

■

Definition

A graph $G = (V,E,f)$ is said to be **empty** provided that $E = \varnothing$.

Notice that if $G = (V,E,f)$ is an empty graph then each vertex is isolated.

Definition

Let $G = (V,E,f)$ be a graph and let $G_1 = (V_1,E_1,f_1)$ and $G_2 = (V_2,E_2,f_2)$ be subgraphs of G. We say that G_1 and G_2 are **disjoint** if they have no vertex in common, and we say that G_1 and G_2 are **edge-disjoint** if they have no edge in common. The **union** $G_1 \cup G_2$ of G_1 and G_2 is the subgraph $(V_1 \cup V_2, E_1 \cup E_2, f_1 \cup f_2)$ of G. If G_1 and G_2 have at least one vertex in common, the **intersection** $G_1 \cap G_2$ is the subgraph $(V_1 \cap V_2, E_1 \cap E_2, f_1 \cap f_2)$ of G.

It is possible that $G_1 \cap G_2$ is an empty graph; that is, it may not have any edges.

Definition

Let $G = (V,E,f)$ be a graph without loops. Then G is said to be **bipartite** if (a) there are nonempty disjoint subsets V_1 and V_2 of V such that $V = V_1 \cup V_2$, and (b) u and v are adjacent vertices if and only if one of them belongs to V_1 and the other to V_2. The partition $\{V_1,V_2\}$ of V is called a **bipartition** of G.

Example 18

Let $G = (V,E,f)$ be the graph shown in Figure 11.19. Show that G is bipartite.

Figure 11.19

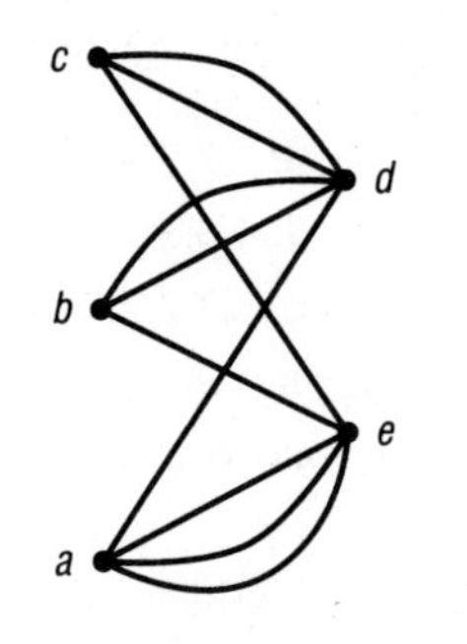

Analysis

If $V_1 = \{a,b,c\}$ and $V_2 = \{d,e\}$, then $\{V_1,V_2\}$ is a bipartition of G. The fact that G has multiple edges is of no consequence. The bipartition of the underlying simple graph of G is a bipartition of G. ■

Exercises 11.2

13. Find the number of vertices and the degree of each vertex in each of the following graphs. Identify the simple graphs and the graphs without loops.

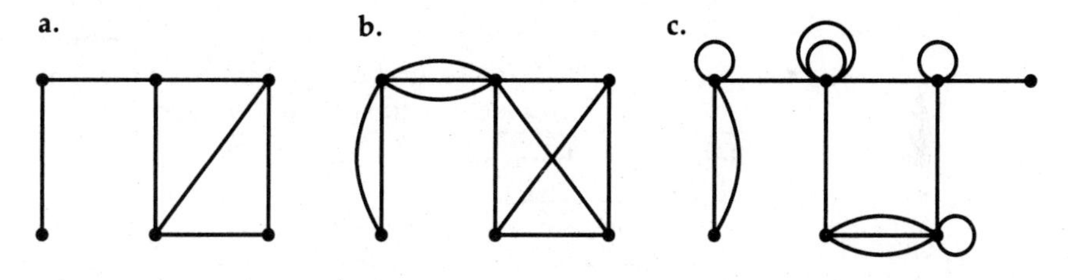

14. Find the number of edges and the sum of the degrees of the vertices of each graph in Exercise 13.
15. Explain why, for each of the following simple graphs, it is not possible to begin a walk at vertex a, cross each edge exactly once, and end the walk at vertex a.

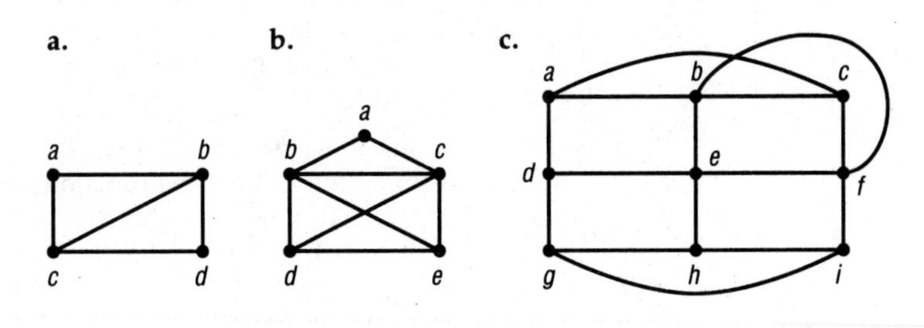

16. Let $G = (V,E,f)$ be a graph such that the number of members of V is five, and suppose that the degrees of these five vertices are 4, 4, 3, 2, 1. How many edges does G have?

17. Does there exist a simple graph with five vertices such that the degrees of these five vertices are as follows?

    a. 3,3,2,2,2 **b.** 3,3,3,3,2

    c. 3,3,3,2,2 **d.** 4,3,3,2,2

    e. 4,4,3,2,1 **f.** 5,3,2,2,2

    In each case, if it exists, draw such a graph.

18. How many subgraphs does each of the following graphs have?

    a. K_3 **b.** G_2 **c.** $K_{2,2}$

19. Find the underlying simple graph of the graph shown here.

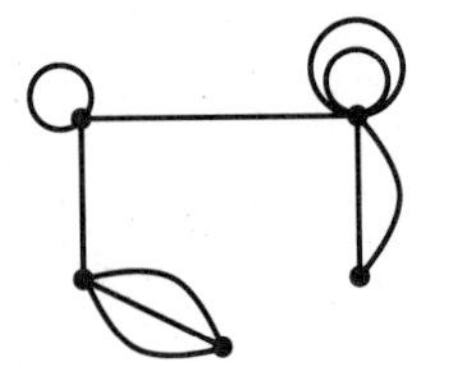

20. Let $V = \{a,b,c,d\}$ and $E = \{e_1,e_2,e_3,e_4,e_5,e_6,e_7,e_8,e_9\}$, and define $f: E \to \{\{u,v\}: u,v \in V\}$ by $f(e_1) = \{a,b\}$, $f(e_2) = \{a,c\}$, $f(e_3) = \{a,d\}$, $f(e_4) = \{b\}$, $f(e_5) = \{b,d\}$, $f(e_6) = \{c\}$, $f(e_7) = \{c,d\}$, $f(e_8) = \{c,d\}$, and $f(e_9) = \{c,d\}$. Draw a figure that represents the graph $G = (V,E,f)$.

21. Let $G = (V,E,f)$ be the graph represented by the figure shown here. Label the vertices and edges and specify V, E, and f.

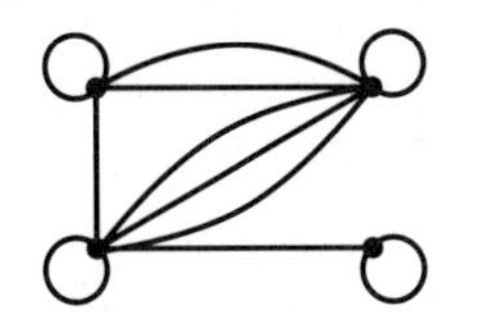

22. Let $V = \{a,b,c\}$ and $E = \{\{a,b\},\{a,c\},\{b,c\}\}$.

    a. Define a function $f: E \to \{\{u,v\}: u,v \in V\}$ such that the graph (V,E,f) is the simple graph (V,E).

    b. Define a set F and a function $g: F \to \{\{u,v\}: u,v \in V\}$ such that the graph (V,F,g) is *not* the simple graph (V,E) but the simple graph (V,E) is the underlying simple graph of (V,F,g).

23. Let $G = (V,E,f)$, where $V = \{v_1,v_2,v_3,v_4,v_5,v_6\}$, $E = \{e_1,e_2,e_3,e_4,e_5,e_6,e_7,e_8,e_9,e_{10},e_{11}\}$, and $f: E \to \{\{u,v\}: u,v \in V\}$ is the function defined by $f(e_1) = \{v_1,v_2\}$, $f(e_2) = \{v_1,v_3\}$, $f(e_3) = \{v_1,v_6\}$, $f(e_4) = \{v_2,v_3\}$, $f(e_5) = \{v_2,v_4\}$, $f(e_6) = \{v_2,v_5\}$, $f(e_7) = \{v_3,v_4\}$, $f(e_8) = \{v_4,v_5\}$, $f(e_9) = \{v_4,v_6\}$, $f(e_{10}) = \{v_5,v_6\}$, and $f(e_{11}) = \{v_2,v_6\}$. Let $G_1 = (V_1,E_1,f_1)$, where $V_1 = \{v_1,v_2,v_3,v_4,v_5,v_6\}$, $E_1 = \{e_1,e_3,e_4,e_7,e_8,e_{10}\}$, and $f_1 = f|E_1$, and let $G_2 = (V_2,E_2,f_2)$, where $V_2 =$

$\{v_2,v_4,v_6\}$, $E_2 = \{e_5,e_9,e_{11}\}$, and $f_2 = f|E_2$. Then G_1 and G_2 are subgraphs of G. Find $G_1 \cup G_2$ and $G_1 \cap G_2$. (*Hint:* Draw figures that represent G_1 and G_2.)

11.3 ADJACENCY AND INCIDENCE MATRICES

We have already seen in Section 9.4 that there is a matrix associated with every finite directed graph. In this section we discuss two matrices that we associate with graphs. We begin by constructing a list of vertices that are adjacent to each vertex of a graph.

Example 19

Let $V = \{a,b,c,d\}$ and $E = \{e_1,e_2,e_3,e_4,e_5,e_6,e_7,e_8,e_9\}$, and define $f\colon E \rightarrow \{\{u,v\}: u,v \in V\}$ by $f(e_1) = \{a,b\}$, $f(e_2) = \{a,b\}$, $f(e_3) = \{b,c\}$, $f(e_4) = \{b,c\}$, $f(e_5) = \{c,d\}$, $f(e_6) = \{a,d\}$, $f(e_7) = \{a,c\}$, $f(e_8) = \{d\}$, and $f(e_9) = \{d\}$. Then $G = (V,E,f)$ is a graph. Construct a list of vertices of G that are adjacent to each vertex of G.

Analysis

We begin by drawing a picture of G (Figure 11.20). (As we indicated in the previous section, the concept of a graph is not difficult if one works with the picture that represents the graph rather than the abstract idea

Figure 11.20

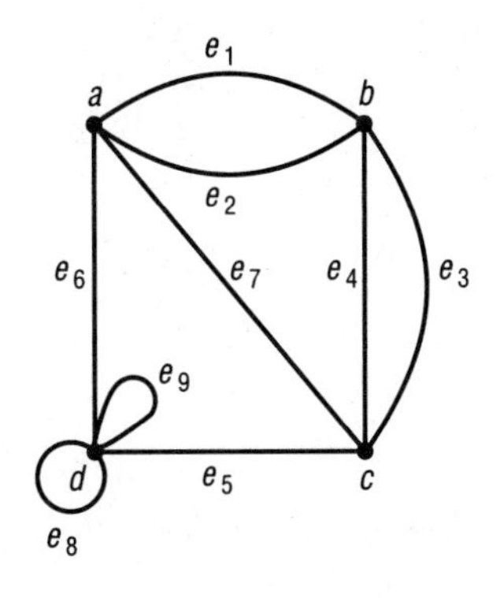

of a list of vertices and edges and a function that maps the edges into certain subsets of the set of vertices.) Next, in Table 11.1 we list each vertex along with its adjacent vertices:

Table 11.1

Vertex	Adjacent vertices
a	b, c, d
b	a, c
c	a, b, d
d	a, c, d

A moment's thought should show that this list does not provide enough information for one to draw the graph of G. To draw the graph, we need to know not only which vertices are adjacent to each vertex, but also the number of edges joining any two vertices. Thus, if we agree to list a vertex n times if there are n edges joining it to a given vertex, our complete table of vertex adjacency will provide enough information for us to recreate the graph. Our complete table (Table 11.2) looks like this:

Table 11.2

Vertex	Adjacent vertices
a	b, b, c, d
b	a, a, c, c
c	a, b, b, d
d	a, c, d, d

If $G = (V,E,f)$ is a graph and n is the order of G, then there is an $n \times n$ matrix associated with G that provides the information given in such a table. ■

Definition

Let $G = (V,E,f)$ and suppose that the order of G is n. If $V = \{v_1,v_2, \ldots ,v_n\}$, the $n \times n$ matrix $A(G) = [a_{ij}]$, where a_{ij} is the number of edges incident with v_i and v_j, is called the **adjacency matrix of G.**

Example 20

What is the adjacency matrix of the graph $G = (V,E,f)$ in Example 19?

Analysis

We use the members of V to label the rows and columns of $A(G)$ and find the entries of $A(G)$ by looking at Figure 11.20 (or Table 11.2).

$$A(G) = \begin{array}{c} \\ a \\ b \\ c \\ d \end{array} \begin{array}{c} \begin{array}{cccc} a & b & c & d \end{array} \\ \begin{bmatrix} 0 & 2 & 1 & 1 \\ 2 & 0 & 2 & 0 \\ 1 & 2 & 0 & 1 \\ 1 & 0 & 1 & 2 \end{bmatrix} \end{array}$$

■

Since the adjacency matrix of a graph G provides a complete description of G, adjacency matrices are commonly used to store graphs in a computer.

Theorem 11.3

> If $G = (V,E,f)$ is a simple graph of order n, then $A(G) = [a_{ij}]$ is a zero–one matrix, and $a_{ii} = 0$ for each $i = 1,2, \ldots ,n$.

Proof

Let G be a simple graph and let u and v be any two vertices of G. Then the number of edges incident with u and v is either 0 or 1. Therefore the adjacency matrix of G is a zero–one matrix.

Since a simple graph does not have any loops, no vertex is adjacent to itself. Therefore $a_{ii} = 0$ for each $i = 1,2, \ldots ,n$. □

Theorem 11.4

> If $G = (V,E,f)$ is a graph, then $A(G)$ is symmetric.

Proof

Let v_i and v_j be members of V. If v_i and v_j are not adjacent, then $a_{ij} = 0 = a_{ji}$. If v_i and v_j are adjacent, then both a_{ij} and a_{ji} denote the number of edges that connect v_i and v_j. Therefore $a_{ij} = a_{ji}$. □

Example 21

Draw the graph whose adjacency matrix is

$$\begin{array}{c} \\ a \\ b \\ c \\ d \\ e \end{array} \begin{array}{c} \begin{array}{ccccc} a & b & c & d & e \end{array} \\ \begin{bmatrix} 1 & 0 & 2 & 0 & 0 \\ 0 & 0 & 1 & 1 & 3 \\ 2 & 1 & 0 & 2 & 2 \\ 0 & 1 & 2 & 2 & 0 \\ 0 & 3 & 2 & 0 & 0 \end{bmatrix} \end{array}$$

Analysis

The row and column labels of the adjacency matrix of a graph are the vertices of the graph. We therefore join points a, b, c, d, and e according to the entries in the adjacency matrix and obtain the graph in Figure 11.21.

Figure 11.21

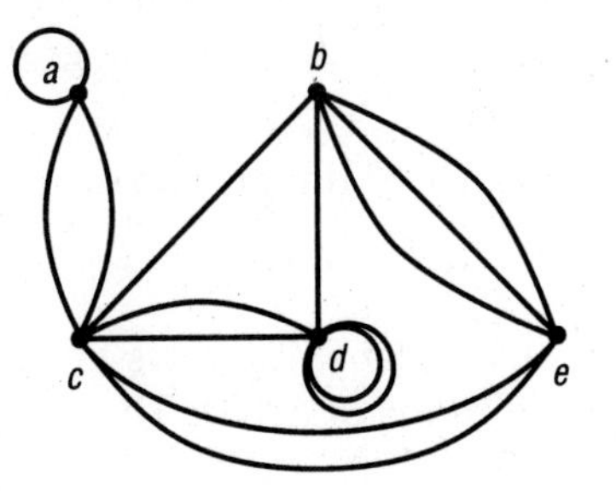

■

Theorem 11.5

Let $G = (V,E,f)$ be a graph and let $v \in V$. The sum of the entries in the column of $A(G)$ labeled v is the number of edges incident with v.

Proof

Each edge incident with v contributes 1 to the column of $A(G)$ labeled v. Therefore the sum of the entries in the column labeled v is the number of edges incident with v. □

Corollary 11.6

Let $G = (V,E,f)$ be a graph and let $v \in V$. The sum of the entries in the column of $A(G)$ labeled v is the degree of v minus the number of loops at v.

Proof

In calculating the degree of v, each loop is counted twice. □

Let $G = (V,E,f)$ be a graph with m vertices and n edges. Besides the $n \times n$ adjacency matrix $A(G)$, there is an $m \times n$ matrix associated with G that provides another way of specifying the graph G.

Definition

Let $G = (V,E,f)$ be a graph, let m be the order of G, and let n be the number of edges of G. If $V = \{v_1,v_2, \ldots ,v_m\}$ and $E = \{e_1,e_2, \ldots ,e_n\}$, then the $m \times n$ matrix $M(G) = [m_{ij}]$, where m_{ij} is the number of times (0, 1, or 2) that e_j is incident with v_i, is called the **incidence matrix of G.**

In the definition of an incidence matrix, if e_j is a loop at v_i, then we say that the number of times that e_j is incident with v_i is 2. If e_j is incident with v_i but is not a loop at v_i, then we say that the number of times that e_j is incident with v_i is 1, and if e_j is not incident with v_i, then we say that the number of times that e_j is incident with v_i is 0. This convention is necessary for $M(G)$ to completely describe the graph G.

Example 22

What is the incidence matrix of the graph shown in Figure 11.20?

Analysis

We use the members of V to label the rows of $M(G)$ and we use the members of E to label the columns of $M(G)$. Then we find the entries of $M(G)$ by looking at Figure 11.20

$$M(G) = \begin{array}{c} \\ a \\ b \\ c \\ d \end{array} \begin{array}{c} \begin{array}{ccccccccc} e_1 & e_2 & e_3 & e_4 & e_5 & e_6 & e_7 & e_8 & e_9 \end{array} \\ \begin{bmatrix} 1 & 1 & 0 & 0 & 0 & 1 & 1 & 0 & 0 \\ 1 & 1 & 1 & 1 & 0 & 0 & 0 & 0 & 0 \\ 0 & 0 & 1 & 1 & 1 & 0 & 1 & 0 & 0 \\ 0 & 0 & 0 & 0 & 1 & 1 & 0 & 2 & 2 \end{bmatrix} \end{array}$$

■

Example 23

Draw the graph whose incidence matrix is

$$\begin{array}{c} \\ a \\ b \\ c \\ d \\ g \end{array} \begin{array}{c} \begin{array}{cccccccc} e_1 & e_2 & e_3 & e_4 & e_5 & e_6 & e_7 & e_8 \end{array} \\ \begin{bmatrix} 2 & 1 & 0 & 0 & 0 & 0 & 1 & 1 \\ 0 & 1 & 1 & 1 & 0 & 0 & 0 & 0 \\ 0 & 0 & 0 & 1 & 2 & 2 & 1 & 0 \\ 0 & 0 & 1 & 0 & 0 & 0 & 0 & 0 \\ 0 & 0 & 0 & 0 & 0 & 0 & 0 & 1 \end{bmatrix} \end{array}$$

Analysis

The row labels of the incidence matrix of a graph are the vertices of a graph. We join points a, b, c, d, and g according to the entries in the incidence matrix and obtain the graph in Figure 11.22.

Figure 11.22

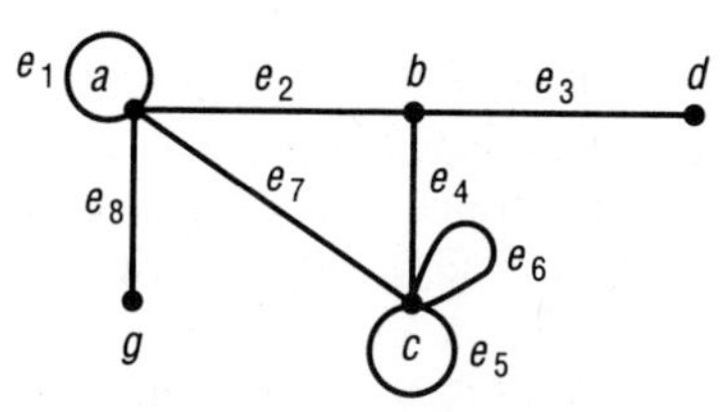

■

The adjacency matrix of a graph is usually smaller than the incidence matrix of the graph, which is why the adjacency matrix is commonly used to store graphs in a computer.

The following theorem is an immediate consequence of the definition of an incidence matrix.

Theorem 11.7

> If $G = (V,E,f)$ is a graph without loops, then $M(G)$ is a zero–one matrix.

The following theorem is an analogue of Theorem 11.5.

Theorem 11.8

> Let $G = (V,E,f)$ be a graph. The sum of the entries in each column of $M(G)$ is 2.

Proof

Let $e \in E$. Then either there exists $v \in V$ such that $f(e) = \{v\}$ or there exist distinct vertices u and v of V such that $f(e) = \{u,v\}$. If there exists a vertex v such that $f(e) = \{v\}$, then the entry in the row labeled v and column labeled e is 2 and all the other entries in this column are 0. If there exist distinct vertices u and v of V such that $f(e) = \{u,v\}$, then the entry in the row labeled u and the column labeled e is 1, the entry in the row labeled v and column labeled e is 1, and all the other entries in the column labeled e are 0. In either case, the sum of the entries in the column labeled e is 2. Since e is an arbitrary member of E, we have proved the theorem. □

Exercises 11.3

24. Let $G = (V,E,f)$ be the graph that is represented by the following picture.

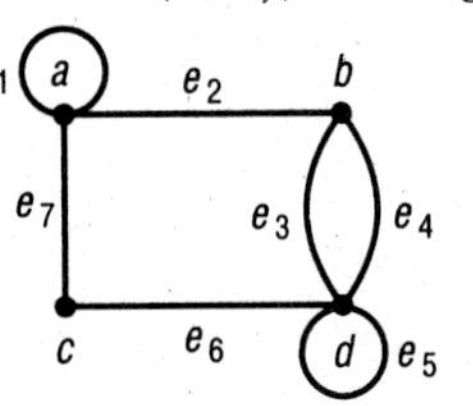

a. Find the adjacency matrix of G. **b.** Find the incidence matrix of G.

25. Let $V = \{a,b,c,d\}$, let $E = \{e_1,e_2,e_3,e_4,e_5,e_6,e_7,e_8,e_9\}$, and let $f\colon E \to \{\{u,v\}: u,v \in V\}$ be the function defined by $f(e_1) = \{a,b\}$, $f(e_2) = \{a,b\}$, $f(e_3) = \{b\}$, $f(e_4) = \{b,c\}$, $f(e_5) = \{c\}$, $f(e_6) = \{c\}$, $f(e_7) = \{c,d\}$, $f(e_8) = \{c,d\}$, and $f(e_9) = \{d,a\}$.

a. Find the adjacency matrix of G. **b.** Find the incidence matrix of G.

26. Draw a picture of the graph that is represented by each of the following adjacency matrices.

a.
$$\begin{array}{c} \\ a \\ b \\ c \end{array}\begin{array}{c} \begin{array}{ccc} a & b & c \end{array} \\ \begin{bmatrix} 1 & 1 & 1 \\ 1 & 0 & 1 \\ 1 & 1 & 0 \end{bmatrix} \end{array}$$

b.
$$\begin{array}{c} \\ a \\ b \\ c \\ d \end{array}\begin{array}{c} \begin{array}{cccc} a & b & c & d \end{array} \\ \begin{bmatrix} 0 & 2 & 0 & 1 \\ 2 & 3 & 1 & 0 \\ 0 & 1 & 1 & 1 \\ 1 & 0 & 1 & 0 \end{bmatrix} \end{array}$$

27. Draw a picture of the graph whose incidence matrix is

$$\begin{array}{c} \\ a \\ b \\ c \\ d \\ g \end{array}\begin{array}{c} \begin{array}{ccccccc} e_1 & e_2 & e_3 & e_4 & e_5 & e_6 & e_7 \end{array} \\ \begin{bmatrix} 1 & 1 & 0 & 0 & 0 & 0 & 1 \\ 1 & 1 & 2 & 1 & 0 & 0 & 0 \\ 0 & 0 & 0 & 1 & 1 & 0 & 0 \\ 0 & 0 & 0 & 0 & 1 & 1 & 0 \\ 0 & 0 & 0 & 0 & 0 & 1 & 1 \end{bmatrix} \end{array}$$

28. Represent each of the following graphs with adjacency matrices.

a. K_3 **b.** K_4 **c.** $K_{3,2}$ **d.** C_5

29. Represent each of the following graphs with incidence matrices.

a. K_4 **b.** K_5 **c.** $K_{3,4}$ **d.** C_6

30. Let $G = (V,E,f)$ be a graph. Find the sum of the entries in each column of $M(G)$.

31. Let $G = (V,E,f)$ be the graph represented here.

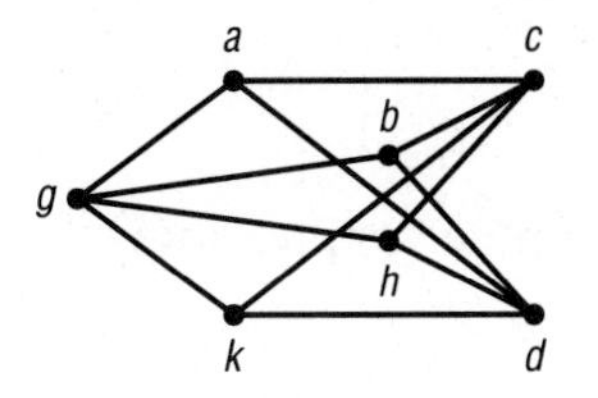

a. Show that G is a bipartite graph.

b. It is often convenient to write a matrix $A = [a_{ij}]$ in the form

$$\left[\begin{array}{c|c} 0 & B \\ \hline C & 0 \end{array}\right]$$

to indicate that all the entries in the upper left corner and the lower right corner are 0s. Show that the vertices of G can be ordered so that the adjacency matrix of G has the form

$$\left[\begin{array}{c|c} 0 & B \\ \hline B^t & 0 \end{array}\right]$$

(In other words, choose an ordering of the vertices of G so that the adjacency matrix of G is in this form.)

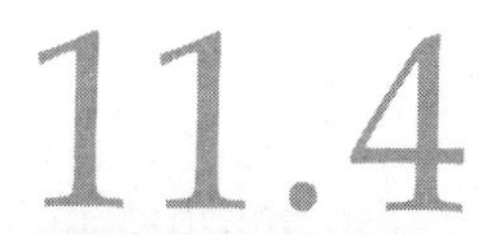

11.4 ISOMORPHIC GRAPHS

There is more than one way to draw a given graph. Are the graphs in Figure 11.23 the same?

Figure 11.23

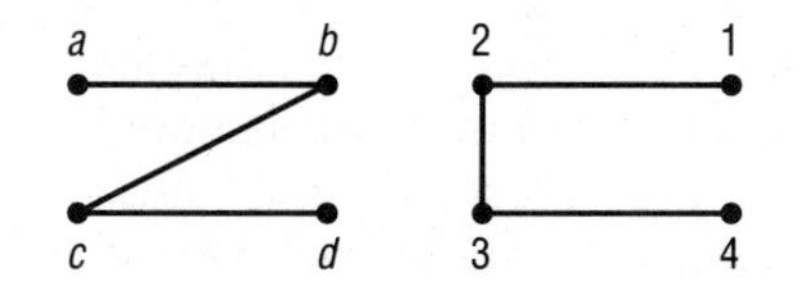

Although as sets these two graphs cannot be equal unless $\{a,b,c,d\} = \{1,2,3,4\}$, it seems reasonable to believe that they are essentially the same in that they share the same geometric properties. The graphs have been drawn and labeled differently, but these differences are inconsequential. We begin, then, with the definition of an isomorphism between two simple graphs.

Definition

Let $G = (V_1,E_1)$ and $H = (V_2,E_2)$ be simple graphs. Then G and H are **isomorphic** provided that there is a bijection θ: $V_1 \rightarrow V_2$ such that $\{u,v\} \in E_1$ if and only if $\{\theta(u),\theta(v)\} \in E_2$. A bijection θ: $V_1 \rightarrow V_2$ such that $\{u,v\} \in E_1$ if and only if $\{\theta(u), \theta(v)\} \in E_2$ is called an **isomorphism** of G and H.

Example 24

Show that the two simple graphs shown in Figure 11.23 are isomorphic.

Analysis

One simple graph, call it G, is (V_1,E_1), where $V_1 = \{a,b,c,d\}$ and $E_1 = \{\{a,b\},\{b,c\},\{c,d\}\}$. The other one, call it H, is (V_2,E_2), where $V_2 = \{1,2,3,4\}$ and $E_2 = \{\{1,2\},\{2,3\},\{3,4\}\}$. To show that G is isomorphic to H, it is sufficient to define a bijection θ: $V_1 \rightarrow V_2$ such that $\{u,v\} \in E_1$ if and only if $\{\theta(u),\theta(v)\} \in E_2$. The pictures in Figure 11.23 of these two simple graphs give us an indication of how to do this. Since each of the pairs a and b, b and c, and c and d are adjacent in G, the pairs $\theta(a)$ and $\theta(b)$, $\theta(b)$ and $\theta(c)$, and $\theta(c)$ and $\theta(d)$ must be adjacent in H. We define the bijection θ: $V_1 \rightarrow V_2$ by $\theta(a) = 1$, $\theta(b) = 2$, $\theta(c) = 3$, and $\theta(d) = 4$. We see that $\{a,b\} \in E_1$ and $\{\theta(a),\theta(b)\} = \{1,2\} \in E_2$, $\{b,c\} \in E_1$ and $\{\theta(b),\theta(c)\} = \{2,3\} \in E_2$, and $\{c,d\} \in E_1$ and $\{\theta(c),\theta(d)\} = \{3,4\} \in E_2$. Therefore θ is an isomorphism. ■

Suppose that $G = (V_1,E_1)$ and $H = (V_2,E_2)$ are simple graphs. If G and H have different orders, it is clear that they are not isomorphic. But suppose their orders are the same. Even then it is often difficult to determine whether they are isomorphic. In particular, if n is the common order, then the number of possible bijections θ: $V_1 \rightarrow V_2$ is $n!$. If n is very large, it is impractical (if not impossible) to examine each such bijection to see whether it preserves adjacency and nonadjacency.

To show that two simple graphs are isomorphic, it is sufficient to find one bijection that preserves adjacency and nonadjacency. Without additional information, to show that two simple graphs are not isomorphic it may be necessary to examine each bijection and show that it does not preserve adjacency or nonadjacency. Fortunately, there are some properties of simple graphs that are preserved under isomorphism, and we can often use these properties to show that two simple graphs are not isomorphic. We do not attempt to define what we mean by a *property* of a graph here. Instead, after defining *invariant*, we give some examples.

Definition

> A property P of a simple graph $G = (V_1,E_1)$ is called an **invariant** with respect to isomorphism of simple graphs provided that, if G has property P and $H = (V_2,E_2)$ is a simple graph isomorphic to G, then H must also have property P.

As we have already observed, the order of a simple graph is an invariant with respect to isomorphism of simple graphs. If $G = (V_1,E_1)$ and $H = (V_2,E_2)$ are isomorphic simple graphs, then the isomorphism $\theta: V_1 \to V_2$ of G and H establishes a one-to-one correspondence between E_1 and E_2. Therefore the number of edges of a simple graph is also an invariant with respect to isomorphism of simple graphs. As the following theorem indicates, we can also establish that the degrees of vertices in isomorphic simple graphs must be the same.

Theorem 11.9

> Suppose that $\theta: V_1 \to V_2$ is an isomorphism between the simple graphs $G = (V_1,E_1)$ and $H = (V_2,E_2)$. If $u \in V_1$, $\deg(u) = n$, and $v = \theta(u)$, then $\deg(v) = n$.

Proof

Let $u \in V_1$, $n = \deg(u)$, and $v = \theta(u)$. Since $\deg(u) = n$, there are n vertices $u_1, u_2, \ldots, u_n$ in V_1 that are adjacent to u. Since θ is one-to-one and preserves adjacency, $\theta(u_1), \theta(u_2), \ldots, \theta(u_n)$ are distinct vertices in V_2 that are adjacent to $\theta(u)$. Therefore $\deg[\theta(u)] \geq n$. Since θ is a surjection and preserves nonadjacency, no vertex of V_2 other than $\theta(u_1)$, $\theta(u_2), \ldots, \theta(u_n)$ can be adjacent to $\theta(u)$. Hence $\deg[\theta(u)] = n$. □

Example 25

Show that the simple graphs pictured in Figure 11.24 are not isomorphic.

Figure 11.24

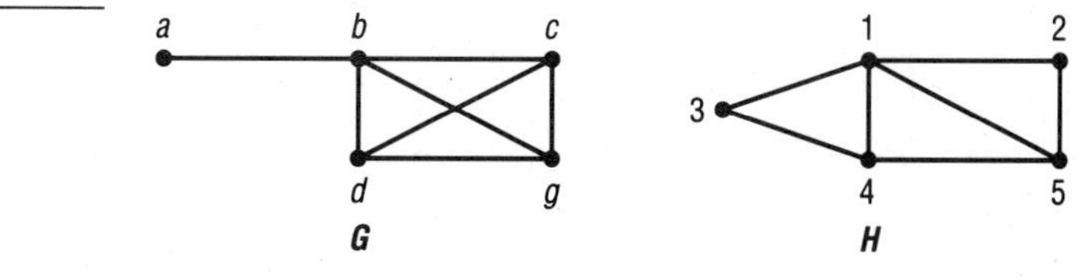

Analysis

Graphs G and H have the same number of vertices and the same number of edges. However, since $\deg(a) = 1$ and no vertex of H has degree 1, by Theorem 11.9, G and H are not isomorphic. ■

Example 26

Determine whether the simple graphs shown in Figure 11.25 are isomorphic.

Figure 11.25

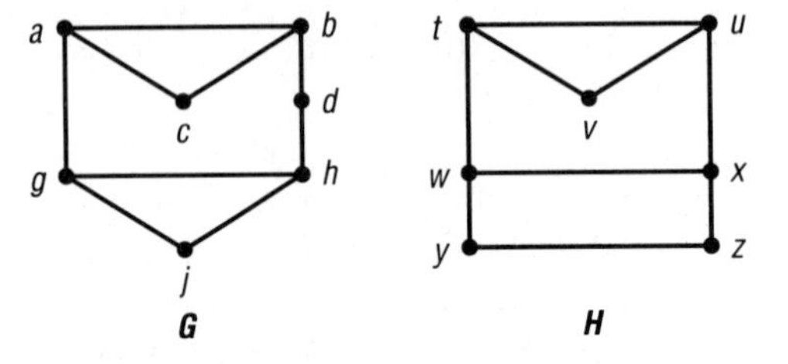

Analysis

The simple graphs G and H each have seven vertices and nine edges. Each graph also has four vertices of degree 3 and three vertices of degree 2. Thus it is conceivable that the two simple graphs are isomorphic. Let $V_1 = \{a,b,c,d,g,h,j\}$ and $V_2 = \{t,u,v,w,x,y,z\}$, and suppose that $\theta: V_2 \rightarrow V_1$ is an isomorphism between the simple graphs H and G. Since $\deg(y) = 2$, $\theta(y)$ must be one of $c,d,$ or j since these are the vertices of degree 2 in H. Since $\deg(z)$ is also 2, $\theta(z)$ must also be one of c, d, or j. However, y and z are adjacent, while no two of c, d, and j are adjacent. Therefore $\theta: V_2 \rightarrow V_1$ cannot be an isomorphism between the two simple graphs.

■

We turn now to the concept of isomorphic graphs. We begin by giving an example to show that the concept of isomorphism between simple graphs is not sufficient to define isomorphism between graphs.

Example 27

Let $G = (V_1,E_1,f)$ be the graph and $H = (V_2,E_2)$ the simple graph shown in Figure 11.26. Can we define a bijection $\theta: V_1 \rightarrow V_2$ that preserves adjacency and nonadjacency? Do we want to say that G and H are isomorphic?

Figure 11.26

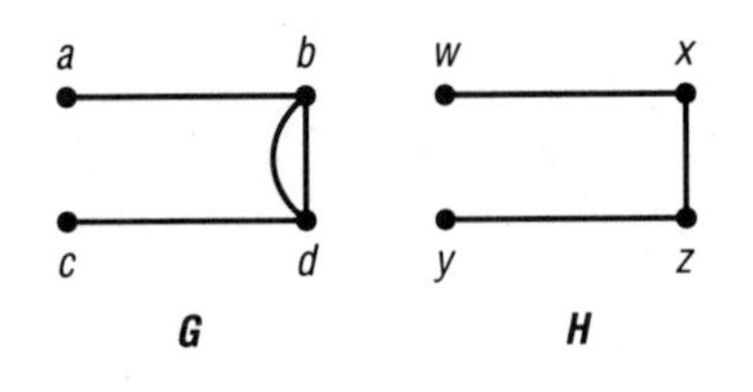

Analysis

The function $\theta: V_1 \rightarrow V_2$ defined by $\theta(a) = w$, $\theta(b) = x$, $\theta(c) = y$, and $\theta(d) = z$ is a bijection, and θ preserves adjacency and nonadjacency.

Therefore, under the definition of isomorphism between two simple graphs, $\theta: V_1 \to V_2$ is an isomorphism. We do not, however, want to say that G and H are isomorphic, because H is a simple graph and G is a graph with multiple edges. Also, G has four edges while H has only three. ■

The definition of isomorphism between graphs involves two functions.

Definition

Two graphs $G = (V_1,E_1,f_1)$ and $H = (V_2,E_2,f_2)$ are **isomorphic** provided that there are bijections $\theta: V_1 \to V_2$ and $\phi: E_1 \to E_2$ such that, for each $e \in E_1$, $f_1(e) = \{u,v\}$ if and only if $f_2(\phi(e)) = \{\theta(u),\theta(v)\}$. The pair (θ,ϕ) of functions is called an **isomorphism.**

Example 28

Let $G = (V_1,E_1,f_1)$ and $H = (V_2,E_2,f_2)$ be the graphs shown in Figure 11.27. Show that G and H are isomorphic.

Figure 11.27

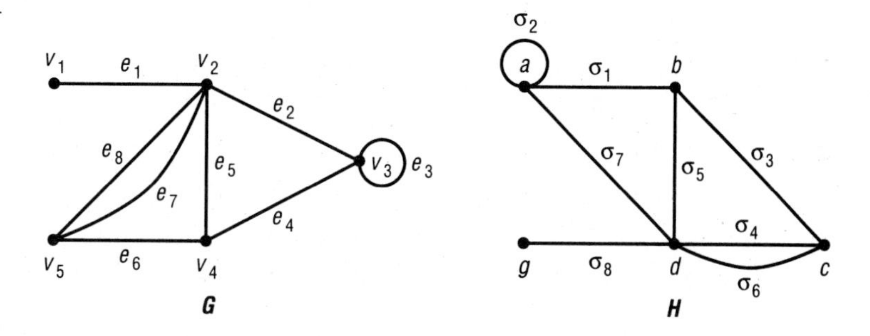

Analysis

Define $\theta: V_1 \to V_2$ by

$$\theta(v_1) = g, \quad \theta(v_2) = d, \quad \theta(v_3) = a, \quad \theta(v_4) = b, \quad \text{and} \quad \theta(v_5) = c$$

and define $\phi: E_1 \to E_2$ by

$$\phi(e_1) = \sigma_8, \quad \phi(e_2) = \sigma_7, \quad \phi(e_3) = \sigma_2, \quad \phi(e_4) = \sigma_1,$$

$$\phi(e_5) = \sigma_5, \quad \phi(e_6) = \sigma_3, \quad \phi(e_7) = \sigma_4, \quad \phi(e_8) = \sigma_6$$

Then θ and ϕ are bijections. To complete the proof it is sufficient to show that, for each $e \in E_1$, $f_1(e) = \{u,v\}$ if and only if $f_2(\phi(e)) = \{\theta(u),\theta(v)\}$.

We calculate:

$$f_1(e_1) = \{v_1,v_2\} \text{ and } f_2(\phi(e_1)) = f_2(\sigma_8) = \{d,g\} = \{\theta(v_2),\theta(v_1)\}$$

$$f_1(e_2) = \{v_2,v_3\} \text{ and } f_2(\phi(e_2)) = f_2(\sigma_7) = \{a,d\} = \{\theta(v_3),\theta(v_2)\}$$

$$f_1(e_3) = \{v_3\} \text{ and } f_2(\phi(e_3)) = f_2(\sigma_2) = \{a\} = \{\theta(v_3)\}$$

$$f_1(e_4) = \{v_3,v_4\} \text{ and } f_2(\phi(e_4)) = f_2(\sigma_1) = \{a,b\} = \{\theta(v_3),\theta(v_4)\}$$

$$f_1(e_5) = \{v_2,v_4\} \text{ and } f_2(\phi(e_5)) = f_2(\sigma_5) = \{b,d\} = \{\theta(v_4),\theta(v_2)\}$$

$$f_1(e_6) = \{v_4,v_5\} \text{ and } f_2(\phi(e_6)) = f_2(\sigma_3) = \{b,c\} = \{\theta(v_4),\theta(v_5)\}$$

$$f_1(e_7) = \{v_2,v_5\} \text{ and } f_2(\phi(e_7)) = f_2(\sigma_4) = \{c,d\} = \{\theta(v_5),\theta(v_2)\}$$

$$f_1(e_8) = \{v_2,v_5\} \text{ and } f_2(\phi(e_8)) = f_2(\sigma_6) = \{c,d\} = \{\theta(v_5),\theta(v_2)\}$$

■

If $G = (V_1,E_1)$ and $H = (V_2,E_2)$ are simple graphs, then (as we previously indicated) we may also think of G and H as graphs. The following theorem states that the two concepts of isomorphism are equivalent.

Theorem 11.10 Let $G = (V_1,E_1)$ and $H = (V_2,E_2)$ be two simple graphs. Then G and H, considered as simple graphs, are isomorphic if and only if G and H, considered as graphs, are isomorphic.

Proof

As we indicated in Section 11.2, the functions $f_1: E_1 \rightarrow \{\{u,v\}: u,v \in V_1\}$ and $f_2: E_2 \rightarrow \{\{u,v\}: u,v \in V_2\}$ such that $G = (V_1,E_1,f_1)$ and $H = (V_2,E_2,f_2)$ are the identity functions.

Suppose that G and H are isomorphic as simple graphs. Then there is a bijection $\theta: V_1 \rightarrow V_2$ such that $\{u,v\} \in E_1$ if and only if $\{\theta(u),\theta(v)\} \in E_2$. Define $\phi: E_1 \rightarrow E_2$ by $\phi(\{u,v\}) = \{\theta(u),\theta(v)\}$. Then ϕ is a bijection. Let $\{u,v\} \in E_1$. Then $f_1(\{u,v\}) = \{u,v\}$ and $f_2(\phi(\{u,v\})) = f_2(\{\theta(u),\theta(v)\}) = \{\theta(u),\theta(v)\}$. Therefore G and H are isomorphic as graphs.

Suppose that G and H are isomorphic as graphs. Then there are bijections $\theta: V_1 \rightarrow V_2$ and $\phi: E_1 \rightarrow E_2$ such that $\{u,v\} \in E_1$ if and only if $\phi(\{u,v\}) = \{\theta(u),\theta(v)\}$. Therefore $\{u,v\} \in E_1$ if and only if $\{\theta(u),\theta(v)\} \in E_2$. Hence G and H are isomorphic as simple graphs. □

We conclude this section with the following theorem, which we state without proof.

Theorem 11.11

Suppose that $G = (V_1,E_1,f_1)$ and $H = (V_2,E_2,f_2)$ are graphs and that (θ,ϕ) is an isomorphism from G to H. Then the following statements are true:

a. V_1 and V_2 have the same number of members.
b. E_1 and E_2 have the same number of members.
c. u and v are adjacent in G if and only if $\theta(u)$ and $\theta(v)$ are adjacent in H.
d. If $v \in V_1$ and $n = \deg(v)$, then $\deg(\theta(v)) = n$.

Exercises 11.4

32. Determine which of the following pairs of simple graphs are isomorphic.

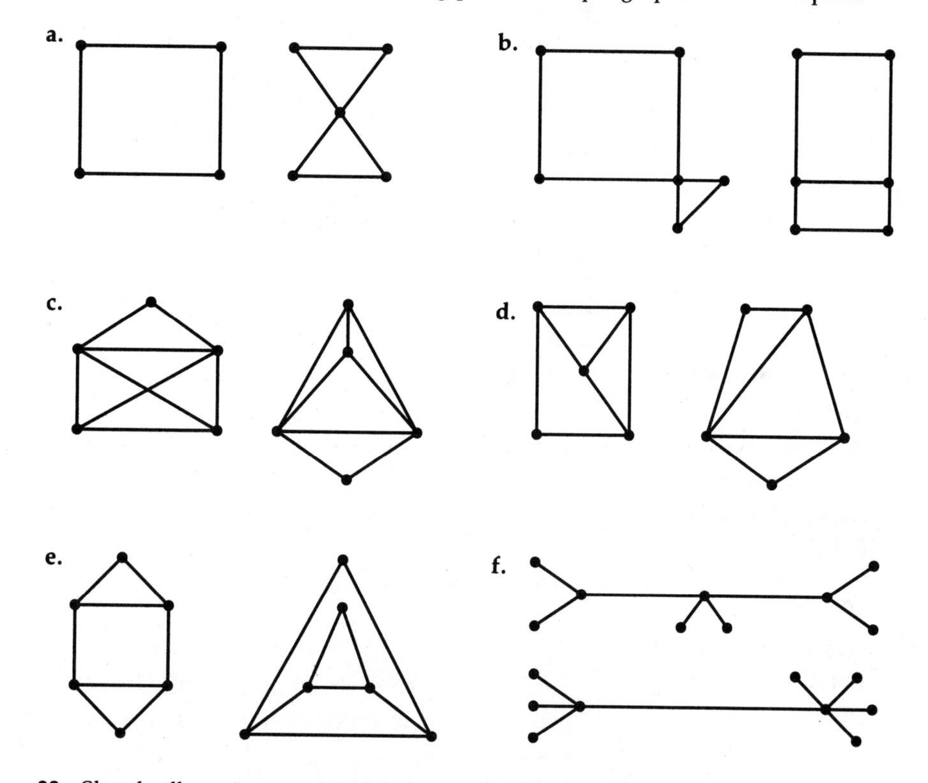

33. Sketch all nonisomorphic simple graphs that have four vertices.
34. Sketch all nonisomorphic simple graphs that have five vertices and three edges.
35. Sketch all nonisomorphic simple graphs that have five vertices and four edges.
36. Show that isomorphism is an equivalence relation on the set of simple graphs.

37. Show that two simple graphs are isomorphic if and only if there is a permutation of the vertices of one of them such that they have the same adjacency matrix.

38. Show that the following two graphs are isomorphic

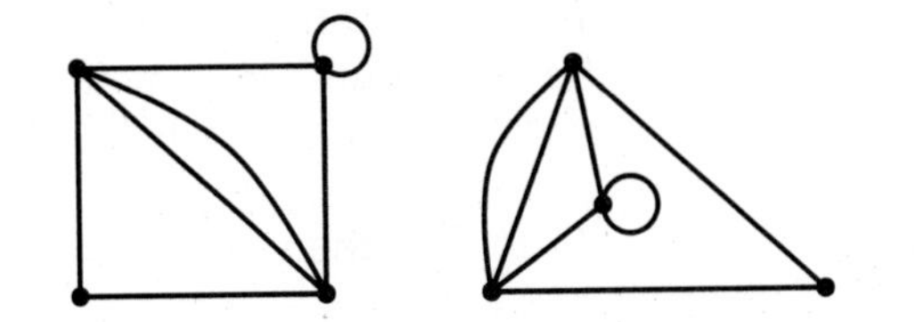

39. Give an example of two graphs with loops and multiple edges that have the same number of vertices and the same number of edges but are not isomorphic.

40. Show that the following graph is isomorphic to the graph of Figure 11.39 (see p. 457).

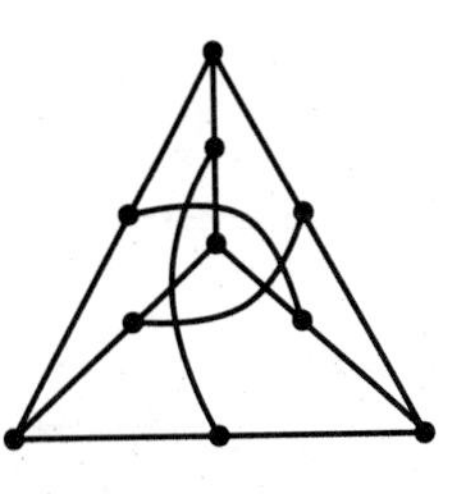

11.5

CONNECTIVITY OF GRAPHS

Bell Laboratories is a major contributor to graph theory. One of their concerns is the number of ways one vertex is connected to another through a sequence of edges and vertices.

Definition

Let u and v be vertices of a graph $G = (V,E,f)$. A **walk of length n from u to v** is a sequence whose terms are alternately vertices and edges, $w = ue_1v_1e_2v_2 \ldots v_{n-1}e_nv$, such that $u \in f(e_1)$, $v \in f(e_n)$, and for each $i = 1,2, \ldots, n-1$, $v_i \in f(e_i) \cap f(e_{i+1})$. We say that w is a **(u,v)-walk.**

To simplify the notation for a walk, we may omit the vertices from the sequence specified in the definition. In this case, we denote a walk of length n by $e_1e_2 \dots e_n$, or by $ue_1e_2 \dots e_nv$ if we want to denote a walk from u to v.

Example 29

Find a walk of length 8, a walk of length 6, and a walk of length 4 from u to v in the graph shown in Figure 11.28.

Figure 11.28

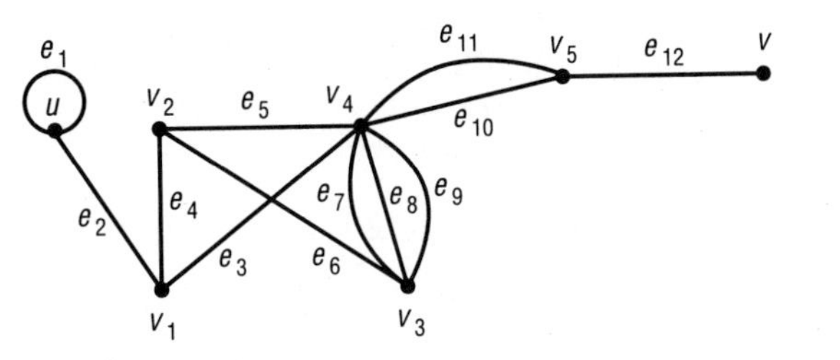

Analysis

There are many walks from u to v, so we have many choices. The walk $ue_1e_2e_3e_9e_6e_5e_{10}e_{12}v$ is a walk of length 8, the walk $ue_2e_4e_6e_8e_{10}e_{12}v$ is a walk of length 6, and the walk $ue_2e_3e_{11}e_{12}v$ is a walk of length 4 from u to v. ■

In a simple graph, a walk $ue_1v_1e_2v_2 \dots v_{n-1}e_nv$ is determined by the sequence $uv_1v_2 \dots v_{n-1}v$ of its vertices. It is not necessary to specify edge e_i, because it is $\{u,v_1\}$, $\{v_{i-1},v_i\}$, or $\{v_{n-1},v\}$. Even when the graph contains multiple edges, a walk can be denoted by its vertex sequence when it is not necessary to distinguish between multiple edges.

Definition

A walk whose edges are distinct is called a **trail.** We use the notation (u,v)-trail to denote a trail from u to v.

The three walks given in Example 29 are trails. However, the walk $ue_2,e_3e_8e_6e_4e_3e_{10}e_{12}v$ in the graph shown in Figure 11.28 is not a trail, since e_3 occurs twice.

Definition

A trail whose vertices are distinct is called a **path.** We use the notation (u,v)-path to denote a path from u to v.

The walks $uv_1v_2v_3v_4v_5v$ and $uv_1v_4v_5v$ in Example 29 are paths. However, the walk $uuv_1v_4v_3v_2v_4v_5v$ is not a path, because the vertices u and v_4 occur twice.

Definition

Vertices u and v of a graph G are **connected** if $u = v$ or if there is a (u,v)-path in G, and the graph G is **connected** if each pair of vertices of G is connected.

Theorem 11.12 Let $G = (V,E,f)$ be a graph. The relation $R = \{(u,v) \in V \times V$: u and v are connected$\}$ is an equivalence relation on V.

The proof of Theorem 11.12 is left as Exercise 50.

Definition

Let $G = (V,E,f)$ be a graph. The equivalence classes of V with respect to the equivalence relation defined in Theorem 11.12 form a partition $\{V_1, V_2, \ldots, V_n\}$ of V (Theorem 10.10). For each $i = 1,2, \ldots, n$, let $E_i = \{e \in E: f(e) \subseteq V_i\}$. Then for each $i = 1,2, \ldots, n$, $G_i = (V_i, E_i, f|E_i)$ is a subgraph of G, called a **component** of G.

Notice that each component of a graph is a connected graph, and if G has one component then G is connected.

It has been said that a picture is worth a thousand words, and this is certainly true regarding connectivity of a graph. We can look at a picture of a graph to determine whether it is connected, and if it is not we can identify its components.

Example 30 Identify the components of the graph $G = (V,E,f)$ shown in Figure 11.29.

Figure 11.29

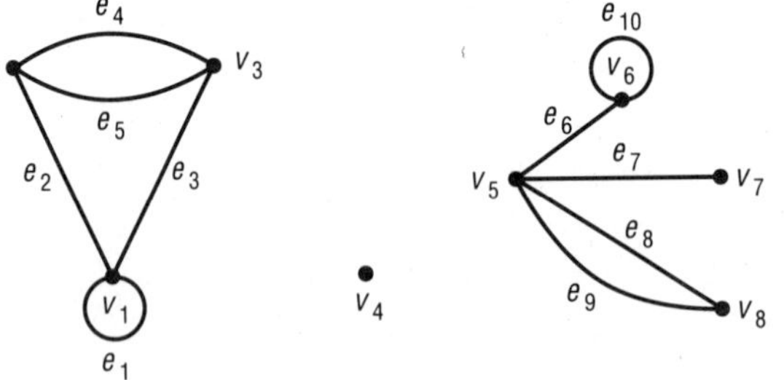

Analysis

We can easily see that G has three components, and we can identify them:

a. $G_1 = (V_1,E_1,f_1)$, where $V_1 = \{v_1,v_2,v_3\}$, $E_1 = \{e_1,e_2,e_3,e_4,e_5\}$, and f_1: $E_1 \to \{\{u,v\}: u,v \in V_1\}$ is the function defined by $f_1(e_1) = \{v_1\}$, $f_1(e_2) = \{v_1,v_2\}$, $f_1(e_3) = \{v_1,v_3\}$, $f_1(e_4) = \{v_2,v_3\}$, and $f_1(e_5) = \{v_2,v_3\}$.

b. $G_2 = (V_2,E_2)$, where $V_2 = \{v_4\}$ and $E_2 = \varnothing$.

c. $G_3 = (V_3,E_3,f_3)$, where $V_3 = \{v_5,v_6,v_7,v_8\}$, $E_3 = \{e_6,e_7,e_8,e_9,e_{10}\}$, and f_3: $E_3 \to \{\{u,v\}: u,v \in V_3\}$ is the function defined by $f_3(e_6) = \{v_5,v_6\}$, $f_3(e_7) = \{v_5,v_7\}$, $f_3(e_8) = \{v_5,v_8\}$, $f_3(e_9) = \{v_5,v_8\}$, and $f_3(e_{10}) = \{v_6\}$.

The subgraphs G_1, G_2, and G_3 of G are labeled in Figure 11.30.

Figure 11.30

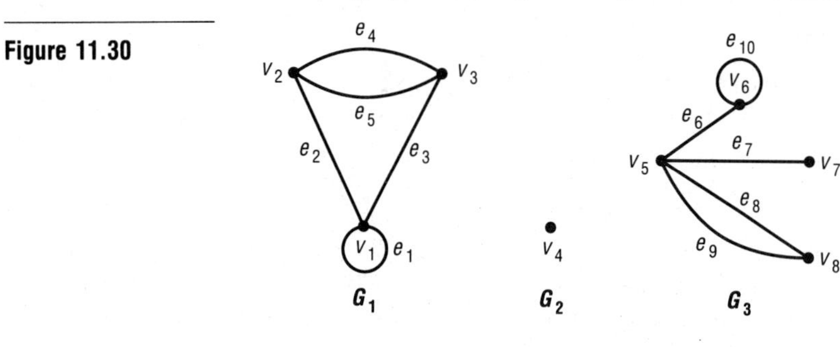

■

Definition

A walk w in a graph that begins and ends at the same vertex u is said to be **closed,** and the vertex u is called the **origin** of w.

Definition

A closed trail $T = ue_1v_1e_2v_2 \ldots v_{n-1}e_nu$ in a graph such that $u \neq v_i$ for any i and $v_i \neq v_j$ whenever $i \neq j$ is called a **cycle,** or **circuit,** in G.

Theorem 11.13 If u and v are distinct vertices of a graph $G = (V,E,f)$ and there is a (u,v)-walk in G, then there is a (u,v)-path in G.

Proof

Let u and v be distinct vertices of a graph $G = (V,E,f)$, and let $T = \{n \in \mathbb{N}$: there is a (u,v)-walk in G of length $n\}$. By the least-natural-number principle, T has a smallest member p. Let $w = ue_1v_1e_2v_2 \ldots v_{p-1}e_pv$ be a (u,v)-walk in G of length p. We claim that w is a (u,v)-path in G. To see this, suppose that it is not. Then $v_i = v_j$ for some i and j where $i < j$. Then $w = ue_1v_1e_2v_2 \ldots v_{i-1}e_iv_je_{j+1} \ldots v_{p-1}e_pv$ is a (u,v)-walk in G of length q, where $q < p$, and we have a contradiction. □

If G is a graph, we let $\omega(G)$ denote the number of components of G. Also, if e is an edge of a graph G, we let $G - \{e\}$ denote the subgraph of G that is obtained by deleting e. The function $f: E \to \{\{u,v\}: u,v \in V\}$ is replaced by the function $f|(E - \{e\}): (E - \{e\}) \to \{\{u,v\}: u,v \in V\}$.

Theorem 11.14

> If $G = (V,E)$ is a simple graph and $e \in E$, then $\omega(G) \leq \omega(G - \{e\}) \leq \omega(G) + 1$.

Proof

Since adding an edge to a graph cannot increase the number of components, $\omega(G) \leq \omega(G - \{e\})$. Let H be any subgraph of G, and suppose that we add an edge e to H. Now $e = \{u,v\}$, where u and v are vertices of H. If there is a (u,v)-path in G, then adding e does not change the number of components of H. If there is no (u,v)-path in H, then u is a member of one component H_1 of H and v is a member of a different component H_2 of H. So $H_1 \cup H_2 \cup \{e\}$ is now one component of H, and hence H has one less component. Therefore $\omega(G - \{e\}) \leq \omega(G) + 1$. □

Theorem 11.15

> Let $G = (V,E,f)$ be a graph, and let $A = [a_{ij}]$ be the adjacency matrix of G with respect to the ordering $v_1,v_2, \ldots ,v_r$ of the vertices. The number of different walks of length n ($n \in \mathbb{N}$) from v_i to v_j is the (i,j)th entry of A^n.

Proof

We use the principle of mathematical induction. Let $S = \{n \in \mathbb{N}$: for each $i,j = 1,2, \ldots ,n$, the number of different walks of length n from v_i to v_j is the (i,j)th entry of $A^n\}$.

Since, for each $i,j = 1,2, \ldots ,n$, the (i,j)th entry of A is the number of edges incident with v_i and v_j, $1 \in S$.

Suppose $n \in S$. For each $i,j = 1,2, \ldots ,n$, the number of different walks of length n from v_i to v_j is the (i,j)th entry of $A^n = [b_{ij}]$. Since $A^{n+1} = A^n \times A$, the (i,j)th entry of A^{n+1} is $b_{i1}a_{1j} + b_{i2}a_{2j} + \cdots + b_{in}a_{nj}$. Since $n \in S$, the number of different walks of length n from v_i to v_k is b_{ik}. A walk of length $n + 1$ from v_i to v_j is made up of a path of length n from v_i to a vertex v_k (where v_k is adjacent to v_j) and an edge e such that $f(e) = \{v_k,v_j\}$. Thus the number of different walks of length $n + 1$ from v_i to v_j is the product of the number of walks of length n from v_i to v_k (where v_k is adjacent to v_i) and the number of edges e such that $f(e) = \{v_i,v_k\}$. Thus, given a vertex v_k, the number of different walks of length $n + 1$

from v_i to v_j that include v_k is $b_{ik}a_{kj}$. Therefore the number of different walks of length $n + 1$ from v_i to v_j is the sum over all v_k, and hence it is $\sum_{k=1}^{n} b_{ik}a_{kj}$. Therefore $n + 1 \in S$. By the principle of mathematical induction, $S = \mathbb{N}$. □

Example 31

How many different walks of length 4 are there from a to d in the graph $G = (V,E,f)$ in Figure 11.31? How many different walks of length 4 are there from c to g?

Figure 11.31

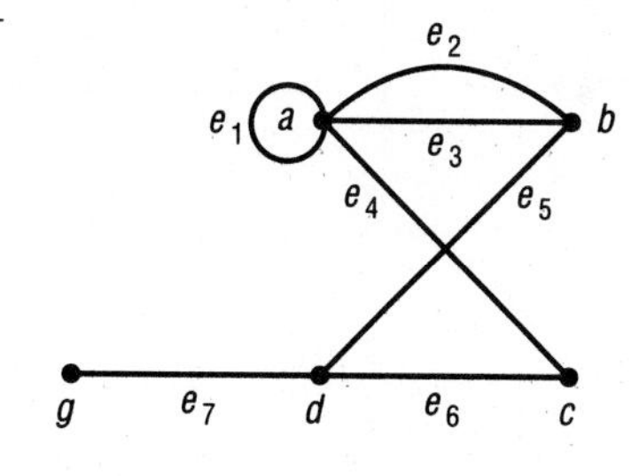

Analysis

The adjacency matrix A is

$$\begin{array}{c} \\ a \\ b \\ c \\ d \\ g \end{array}\begin{array}{c} \begin{array}{ccccc} a & b & c & d & g \end{array} \\ \begin{bmatrix} 1 & 2 & 1 & 0 & 0 \\ 2 & 0 & 0 & 1 & 0 \\ 1 & 0 & 0 & 1 & 0 \\ 0 & 1 & 1 & 0 & 1 \\ 0 & 0 & 0 & 1 & 0 \end{bmatrix} \end{array}$$

By Theorem 11.15, the number of different walks of length 4 from a to d is the (1,4)th entry of A^4. Since

$$A^4 = \begin{bmatrix} 50 & 25 & 14 & 27 & 3 \\ 25 & 39 & 24 & 6 & 9 \\ 14 & 24 & 15 & 3 & 6 \\ 27 & 6 & 3 & 18 & 0 \\ 3 & 9 & 6 & 0 & 3 \end{bmatrix}$$

there are 27 walks of length 4 from a to d and 6 walks of length 4 from c to g. The list of 6 walks of length 4 from c to g is

$ce_4ae_3be_5de_7g$ $\quad$ $ce_4ae_2be_5de_7g$

$ce_4ae_4ce_6de_7g$ $\quad$ $ce_6de_5be_5de_7g$

$ce_6de_6ce_6de_7g$ $\quad$ $ce_6de_7ge_7de_7g$ ■

Exercises 11.5

41. **a.** For each graph, find a walk from a to g that is not a trail.

b. For each graph, find a trial from a to g that is not a path.

c. For each graph, find a path from a to g.

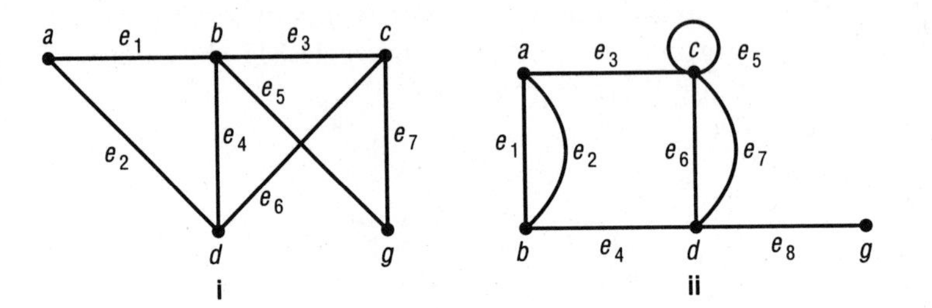

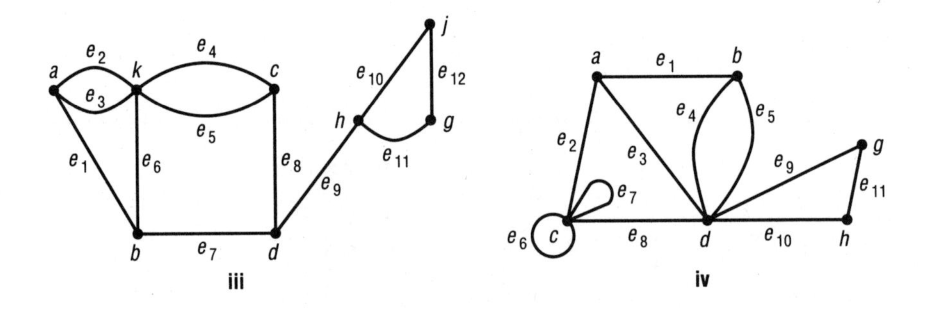

42. Does each of the following form a walk in the illustrated graph? Which walks are trails? Which walks are paths? Which walks are cycles? What is the length of each walk?

a. $ae_3ge_{13}ke_{16}j$ **b.** $ae_1be_6he_{13}ge_3a$ **c.** $ae_4je_{10}ce_9ge_{14}j$

d. $ae_{13}ge_9ce_2ae_4j$ **e.** $ae_3ge_{14}je_{16}ke_{12}de_{12}k$ **f.** $ae_1be_6he_{13}ge_3ae_2ce_9g$

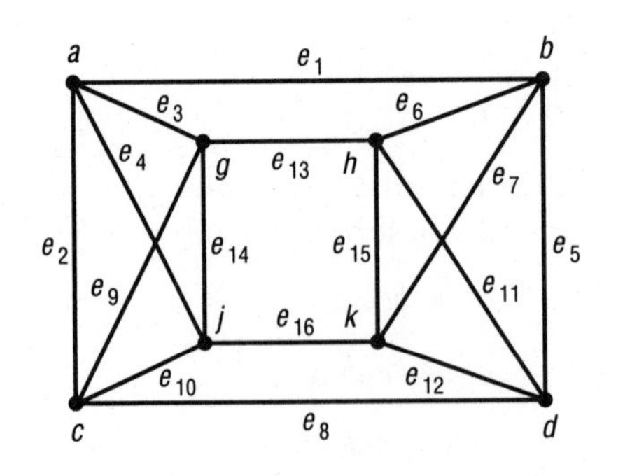

43. Which of the following graphs are connected?

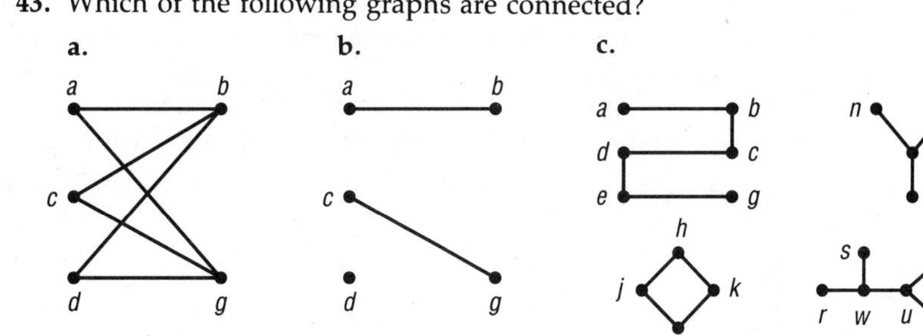

44. How many components does each of the graphs in Exercise 43 have?

45. Find the number of paths between a and d in graph (a) in Exercise 43 of the following lengths.

a. 2 **b.** 3 **c.** 4 **d.** 5

46. Explain how Theorem 11.15 can be used to find a shortest walk from a vertex u to a vertex v in a graph $G = (V,E,f)$.

47. Use Theorem 11.15 to find the length of a shortest path from a to d in the following graph.

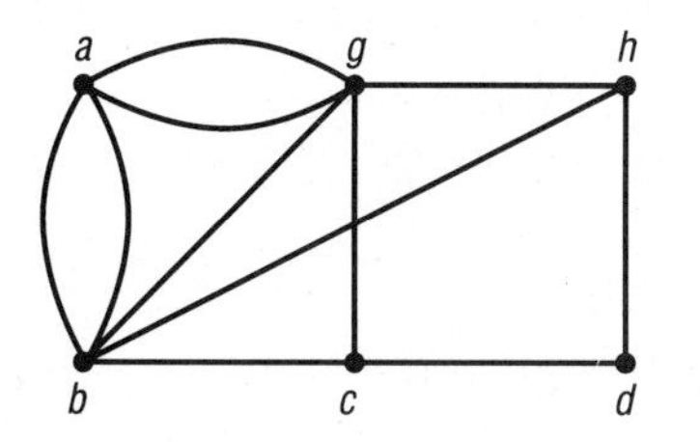

48. Let $G = (V,E,f)$ be a graph with four components and suppose that the vertices of G are ordered such that the vertices of each component are listed successively. Describe the adjacency matrix of G.

49. There are four houses and three utilities. Representing each house and each utility by a vertex, draw a graph that shows each house connected to each utility. Identify the graph you have drawn by its standard name.

50. Let $G = (V,E,f)$ be a graph. Show that the relation $R = \{(u,v) \in V \times V:$ u and v are connected$\}$ is an equivalence relation on V.

51. Prove or give a counterexample to the following proposition: If a graph has exactly two vertices u and v of odd degree then there is a path from u to v.

52. Prove that a graph $G = (V,E,f)$ is connected if and only if for each partition of V into two sets V_1 and V_2 there is an edge e, a vertex v_1 of V_1, and a vertex v_2 of V_2 such that $f(e) = \{v_1,v_2\}$.

53. Let G be a simple graph with k components. Prove that, if ε is the number of edges of G and ν is the number of vertices of G, then $\varepsilon \geqslant \nu - k$.

54. A regional telephone company serves five towns. During severe ice storms it is not uncommon to lose two transmission lines. Draw a connected graph of order 5 (whose vertices represent towns and whose edges represent transmission lines) having the minimum number of edges such that with the removal of any two edges the graph is still connected.
55. Describe an algorithm for determining the number of components of a simple graph.

11.6

EULER TOURS

Recall (Example 13) that Euler settled the problem of the people of Königsberg by showing that it was not possible for them to start at some point in town, walk across each bridge just once, and return to their starting point. In this section we study the general concept of traversing a graph by traveling along each edge exactly once.

Definition

Let $G = (V,E,f)$ be a graph. A trail that contains every edge of G is called an **Euler trail** in G. A closed walk that contains every edge of G is called a **tour** in G, and a closed Euler trail is called an **Euler tour.** A graph is **eulerian** if it contains an Euler tour.

Observe that each edge of the graph $G = (V,E,f)$ occurs exactly once in an Euler trail.

Example 32

Which of the graphs in Figure 11.32 are eulerian?

Figure 11.32

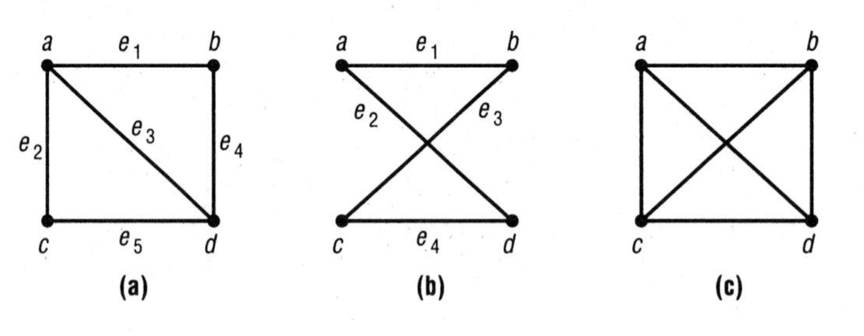

Analysis

Graph (a) contains an Euler trail: $ae_1e_4e_5e_2e_3d$. It is not, however, eulerian, as we establish shortly using Corollary 11.17. Graph (b) contains an Euler tour: $ae_1e_3e_4e_2a$; hence it is eulerian. Graph (c) does not even contain an Euler trail; we prove this after presenting Theorem 11.16. ■

Theorem 11.16

Let $G = (V,E,f)$ be a graph. There is an Euler trail in G if and only if (a) G is connected, and (b) all vertices are of even degree or there are exactly two vertices that are of odd degree.

Since the proof of Theorem 11.16 is rather lengthy, we do not present it here, but we do illustrate part of it in Example 33.

Corollary 11.17

A graph $G = (V,E,f)$ is eulerian if and only if it is connected and all its vertices are of even degree.

Example 32 Continued

To prove that graph (a) in Figure 11.32 is not eulerian, we observe that $\deg(a) = 3$ and use Corollary 11.17. To prove that graph (c) in Figure 11.32 does not contain an Euler trail, we have only to observe that this graph has four vertices of order 3. ■

In the next example, we construct an Euler tour in a connected graph in which all the vertices are of even degree. This illustrates the proof of part of Theorem 11.16.

Example 33

Construct an Euler tour in the graph G shown in Figure 11.33.

Figure 11.33

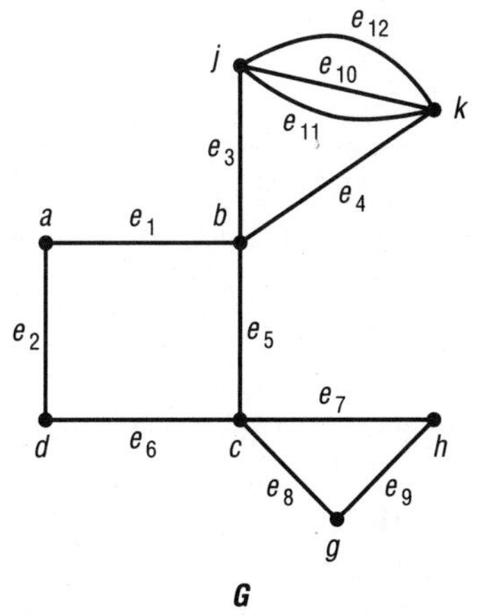

Analysis

We construct an Euler tour in G that begins and ends at a. We choose e_1 as the first edge in the tour and begin by constructing the cycle $C_1 = ae_1be_5ce_6de_2a$. Notice there is no edge on which we can exit a, so the first stage of the construction terminates. Next we form a subgraph H_1 of G by deleting all the edges of G that are in C_1 and all the vertices that are not incident with any of the remaining edges. Observe that H_1 is not connected.

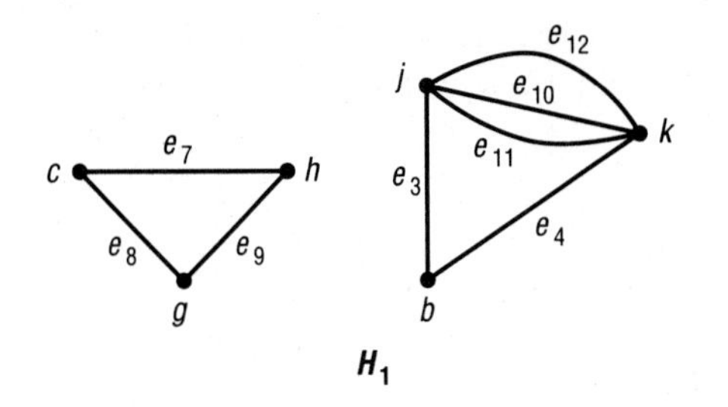

The second stage of the construction of the Euler tour begins by choosing a vertex of H_1 that is in the cycle C_1. In this example we have two choices, b or c. We choose b and construct the cycle $C_2 = be_3je_{12}ke_{11}je_{10}ke_4b$. Then we splice C_1 and C_2 to obtain the cycle $C_3 = ae_1be_3je_{12}ke_{11}je_{10}ke_4be_5ce_6de_2a$. Now we form a subgraph H_2 of G by deleting all the edges of G that are in C_3 and all the vertices that are not incident with any of the remaining edges.

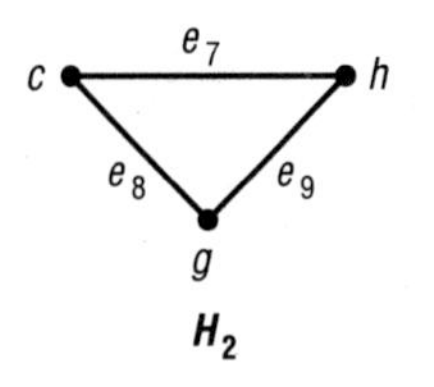

The third stage of the construction of the Euler tour begins with the construction of a cycle C_4 in H_2 that begins and ends at c (c is the only vertex of H_2 that is in the cycle C_3). Suppose that we construct the cycle $C_4 = ce_7he_9ge_8c$. Then we splice C_3 and C_4 to obtain the Euler tour: $ae_1be_3je_{12}ke_{11}je_{10}ke_4be_5ce_7\ he_9ge_8ce_6de_2a$. ■

There are algorithms for constructing Euler trails and Euler tours. One such algorithm is known as Fleury's algorithm. In using Fleury's algorithm, we assume that the graph $G = (V,E,f)$ is connected, that the order of G is at least two, and that all vertices are of even degree or else there are exactly two vertices that are of odd degree. We also use the term *cut edge;* an edge e of a connected graph $G = (V,E,f)$ is a *cut edge* of

G if the graph $(V,E - \{e\},f|(E - \{e\}))$ is not connected. The following result is also used.

Theorem 11.18

> An edge e of a graph $G = (V,E,f)$ is a cut edge of G if and only if e is not contained in any cycle in G.

Proof

The first part of the proof is by contradiction. Suppose that there is a cut edge e of G that is contained in a cycle C. Since $G - \{e\}$ is not connected, there exist vertices u and v of G that are connected in G but not in $G - \{e\}$. Thus there is a (u,v)-path P in G, and e must belong to P. Suppose that $e = \{a,b\}$ and a precedes b in P. Then a section of P forms a (u,a)-path in G and a section of P forms a (b,v)-path in P. Since e is a member of the cycle C, $C - e$ is an (a,b)-path. Hence u and v are connected in $G - \{e\}$. This is a contradiction. We have proved that, if e is a cut edge of G, then e is not contained in any cycle in G.

Now suppose that e is not a cut edge of G. If e is a loop, then the loop is a cycle. Suppose that e is not a loop and let $f(e) = \{u,v\}$, where $u \neq v$. Since e is not a cut edge of G, $G - \{e\}$ is connected. Therefore there is a path P in $G - \{e\}$ from u to v. Then $P \cup \{e\}$ is a cycle in G, and e is contained in this cycle. Thus we have proved that, if e is not contained in any cycle in G, then e is a cut edge of G. □

Fleury's Algorithm

1. Choose a vertex u_0 of odd degree if there is one. Otherwise choose an arbitrary vertex u_0, set $W_0 = u_0$, $V_0 = V$, $E_0 = E$, and $i = 0$.
2. If there is no edge in E_i incident with u_i, stop.
3. If there is exactly one edge in E_i incident with u_i, let e_{i+1} denote this edge, let u_{i+1} denote the "other" vertex that is incident with e_{i+1} (if e_{i+1} is a loop at u_i, then $u_{i+1} = u_i$; otherwise $u_{i+1} \neq u_i$), let $W_{i+1} = W_i e_{i+1} u_{i+1}$, $E_{i+1} = E_i - \{e_{i+1}\}$, $V_{i+1} = V_i - \{u_i\}$, replace i by $i + 1$, and go to step 2.
4. If there is more than one edge in E_i incident with u_i, choose an edge e_{i+1} that is not a cut edge of the graph $(V_i,E_i,f|E_i)$, let u_{i+1} denote the "other" vertex that is incident with e_{i+1}, let $W_{i+1} = W_i e_{i+1} u_{i+1}$, $E_{i+1} = E_i - \{e_{i+1}\}$, $V_{i+1} = V_i$, replace i by $i + 1$, and go to step 2.

We give an example to illustrate Fleury's algorithm.

Example 34

Use Fleury's algorithm to construct an Euler trail in the graph shown in Figure 11.34.

Figure 11.34

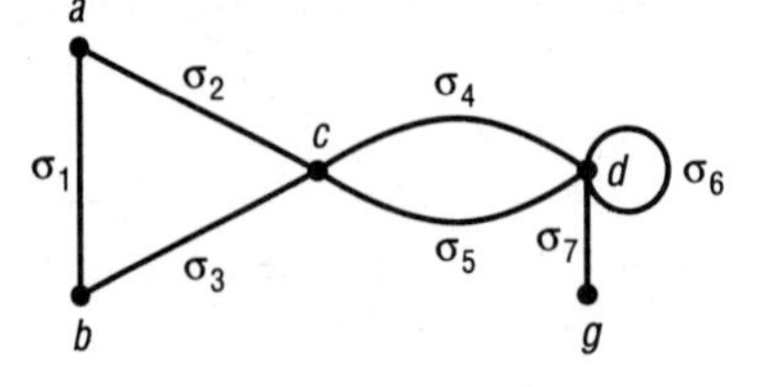

Analysis

We suggest that you draw a picture of the graph that remains each time we remove an edge or an edge and a vertex. We proceed with Fleury's algorithm as follows:

1. There are two vertices d and g of odd degree. Let $u_0 = d$, set $W_0 = u_0$, $V_0 = V$, $E_0 = E$, and $i = 0$.
4. There are four edges σ_4, σ_5, σ_6 and σ_7 in E_0 incident with u_0, and σ_7 is a cut edge. Let $e_1 = \sigma_4$. Then $u_1 = c$, $W_1 = u_0e_1u_1$, $E_1 = E_0 - \{e_1\}$, $V_1 = V_0$, and i becomes 1.
4. There are three edges σ_2, σ_3, and σ_5 in E_1 incident with u_1, and σ_5 is a cut edge. Let $e_2 = \sigma_2$. Then $u_2 = a$, $W_2 = u_0e_1u_1e_2u_2$, $E_2 = E_1 - \{e_2\}$, $V_2 = V_1$, and i becomes 2.
3. There is one edge σ_1 in E_2 incident with u_2. Let $e_3 = \sigma_1$. Then $u_3 = b$, $W_3 = u_0e_1u_1e_2u_2e_3u_3$, $E_3 = E_2 - \{e_3\}$, $V_3 = V_2 - \{u_2\}$, and i becomes 3.
3. There is one edge σ_3 in E_3 incident with u_3. Let $e_4 = \sigma_3$. Then $u_4 = u_1$, $W_4 = u_0e_1u_1e_2u_2e_3u_3,e_4u_4$, $E_4 = E_3 - \{e_4\}$, $V_4 = V_3 - \{u_3\}$, and i becomes 4.
3. There is one edge σ_5 in E_4 incident with u_4. Let $e_5 = \sigma_5$. Then $u_5 = u_0$, $W_5 = u_0e_1u_1e_2u_2e_3u_3,e_4u_4e_5u_5$, $E_5 = E_4 - \{e_5\}$, $V_5 = V_4 - \{u_4\}$, and i becomes 5.
4. There are two edges σ_6 and σ_7 in E_5 incident with u_5, and σ_7 is a cut edge. Let $e_6 = \sigma_6$. Then $u_6 = u_5$, $W_6 = u_0e_1u_1e_2u_2e_3u_3e_4u_4e_5\ u_5e_6u_6$, $E_6 = E_5 - \{e_6\}$, $V_6 = V_5$, and i becomes 6.
3. There is one edge σ_7 in E_6 incident with u_6. Let $e_7 = \sigma_7$. Then $u_7 = g$, $W_7 = u_0e_1u_1e_2u_2e_3u_3e_4u_4e_5u_5e_6u_6e_7u_7$, $E_7 = E_6 - \{e_7\}$, $V_7 = V_6 - \{u_6\}$, and i becomes 7.
2. There are no edges in E_7, so we stop.

Notice that W_7 is an Euler trail. ■

It is more difficult to establish the complexity of Fleury's algorithm than some of the other algorithms. As a matter of information, if n denotes the number of vertices of the graph and ε denotes the number of edges, then the complexity is $O(\varepsilon n^2)$. Since in a simple graph $\varepsilon \leqslant n(n-1)/2$, the complexity of Fleury's algorithm when applied to a simple graph is $O(n^4)$.

Exercises 11.6

56. Determine whether each of the following graphs is eulerian. Construct an Euler tour when one exists.

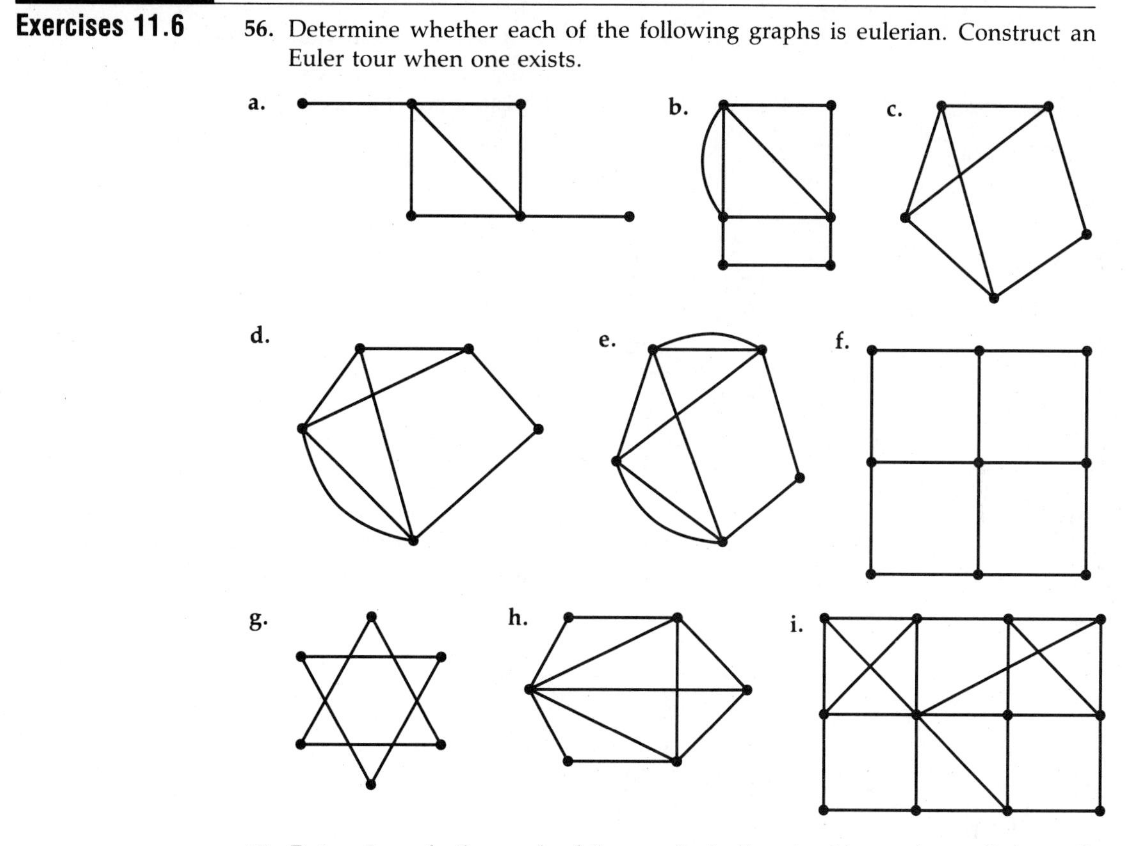

57. Determine whether each of the graphs in Exercise 56 contains an Euler trail. Construct such a trail when one exists.

58. What additional bridges are needed in Königsberg (see Example 13) so that the graph associated with this new set of bridges is eulerian?

59. Is it possible to start at some point in the town illustrated here, walk across each bridge exactly once, and return to the starting point?

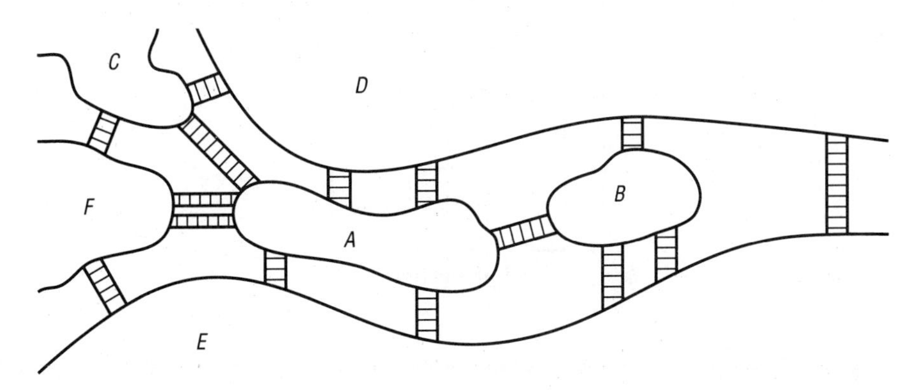

60. For which values of n is C_n eulerian? Explain your answer.

61. For which values of n is K_n eulerian? Explain your answer.

62. If possible, draw an eulerian graph with an even number of edges and an odd number of vertices. If it is not possible, explain why.

63. **a.** Use Fleury's algorithm to find an Euler trail in the following graph.

b. Sketch the intermediate graphs obtained as edges or edges and vertices are deleted.

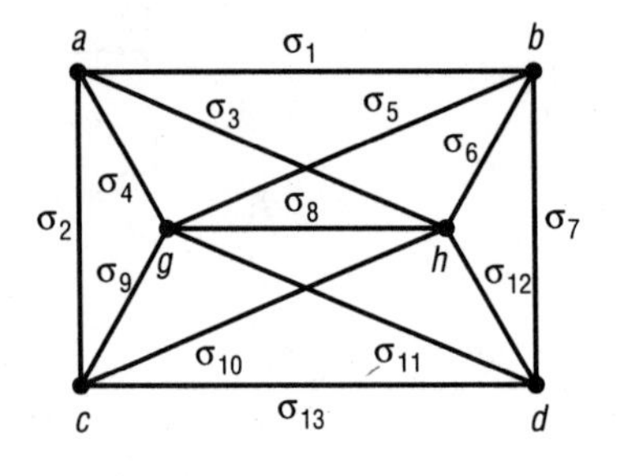

64. **a.** Use Fleury's algorithm to find an Euler tour in the following graph.

b. Sketch the intermediate graphs obtained as edges or edges and vertices are deleted.

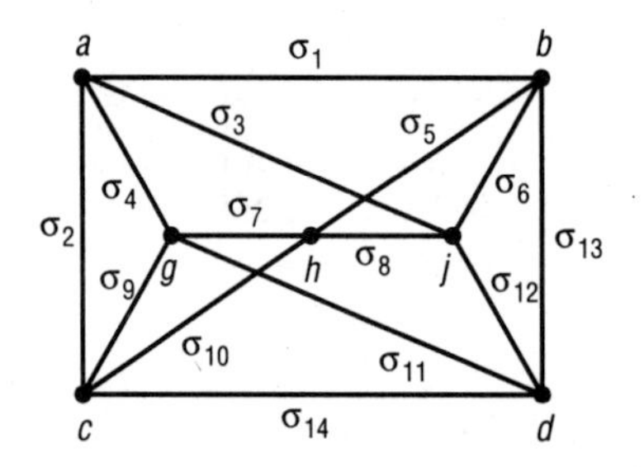

65. A city highway inspector is responsible for the streets shown on the following map. Is it possible for the inspector to start at the office, travel each street exactly once, and return to the office? Explain your answer.

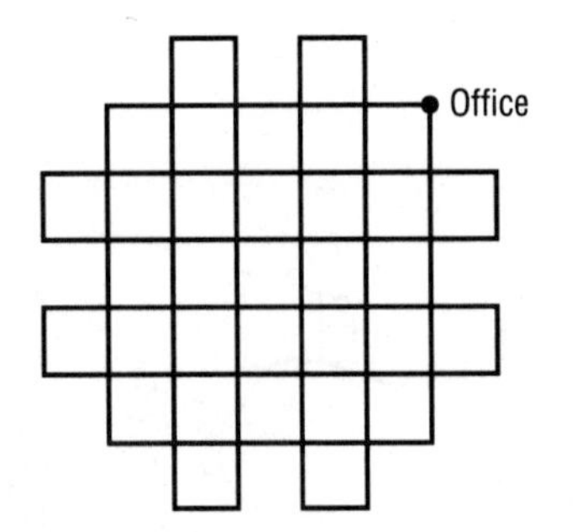

11.7

HAMILTON CYCLES

In the previous section we determined which graphs can be traversed by crossing each edge exactly once. In this section we consider the more difficult task of determining which graphs can be traversed by crossing each vertex exactly once. This problem has grown out of a puzzle invented by the Irish mathematician Sir William Hamilton (1805–1865). From the perspective of graph theory, Hamilton's puzzle was to find a cycle that contained every vertex of the dodecahedron graph, shown in Figure 11.35. This graph is the graph of the edges of the dodecahedron (the unique twelve-sided regular three-dimensional polyhedron on which calendars are sometimes printed).

Figure 11.35

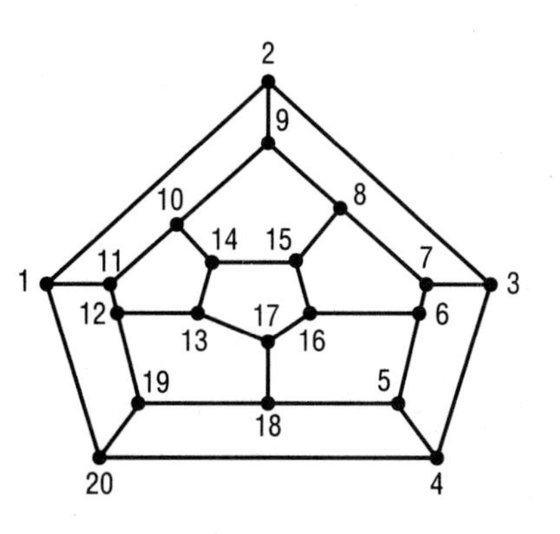

Definition

A **Hamilton path** in a graph G is a path that contains every vertex of G. A **Hamilton cycle** in G is a cycle that contains every vertex of G. A **hamiltonian graph** is a graph that contains a Hamilton cycle.

In contrast to Euler's characterization of eulerian graphs as those connected graphs with exactly two vertices of even degree, one of the primary unsolved problems of graph theory is to give a nontrivial characterization of hamiltonian graphs. There are, however, many conditions that are known to guarantee that a graph is hamiltonian, and we discuss a few of them. But first we consider some examples. We begin by providing a solution to Hamilton's puzzle. If we number the vertices as indicated in Figure 11.35, then the natural order of the vertices provides a Hamilton cycle.

Example 35

We can provide a Hamilton cycle of the graph shown in Figure 11.36 by numbering the vertices as indicated.

Figure 11.36

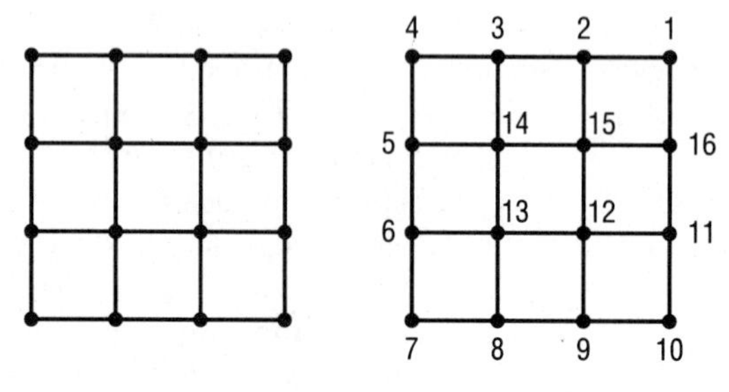

■

Example 36

Let G be the graph shown in Figure 11.37. Show that G is not hamiltonian. Find a Hamilton path in G.

Figure 11.37

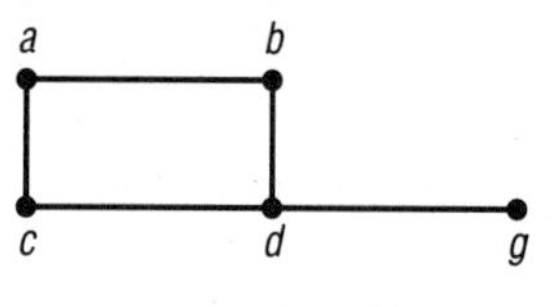

Analysis

Edge $\{d,g\}$ occurs twice in every cycle in G that contains each vertex. Therefore there cannot be a cycle in G that contains each vertex. Hence G is not hamiltonian. The path *cabdg* is a Hamilton path in G. ■

Example 36 illustrates that a graph with a vertex of degree 1 is not hamiltonian. But, in general, showing that a graph is not hamiltonian is difficult. There are a few properties that every Hamilton cycle must possess. These properties can provide useful rules to construct a Hamilton cycle, and they can often be used to show that a graph is not hamiltonian. Four of these properties are as follows:

1. If a vertex has degree 2, then both edges incident with it must be included in every Hamilton cycle, for otherwise the vertex is not visited at all.
2. If a vertex has degree greater than 2, only two edges incident with it can be included in any Hamilton cycle, for otherwise the vertex is visited more than once.
3. Hamilton cycles cannot contain any subcycles.
4. Once a vertex, together with two edges incident with it, has been included in a Hamilton cycle, all other edges incident with the vertex can be deleted from the graph.

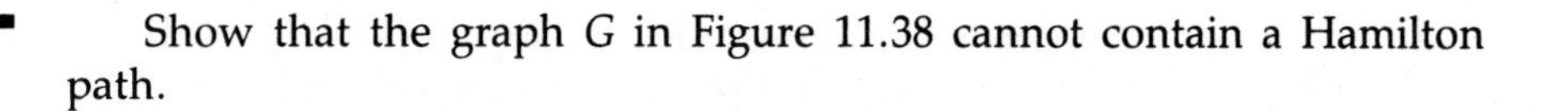

Example 37

Show that the graph G in Figure 11.38 cannot contain a Hamilton path.

Figure 11.38

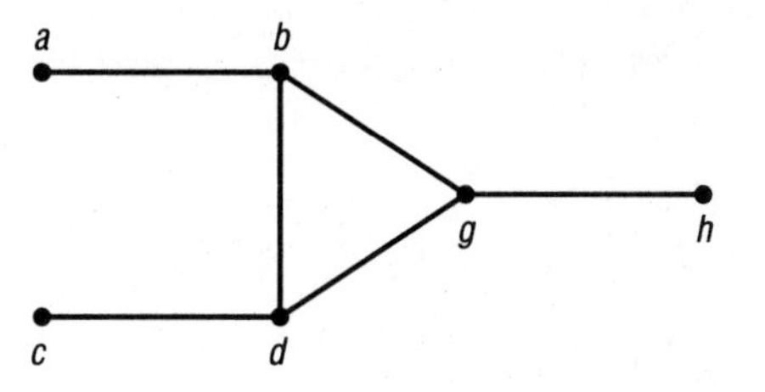

Analysis

At least one of the edges $\{a,b\}$, $\{c,d\}$, and $\{g,h\}$ must occur twice in any Hamilton path in G. Therefore there cannot be a Hamilton path in G. ■

Example 38

Use the properties that we have listed to show that the graph in Figure 11.39 (known as Petersen's graph) is not hamiltonian.

Figure 11.39

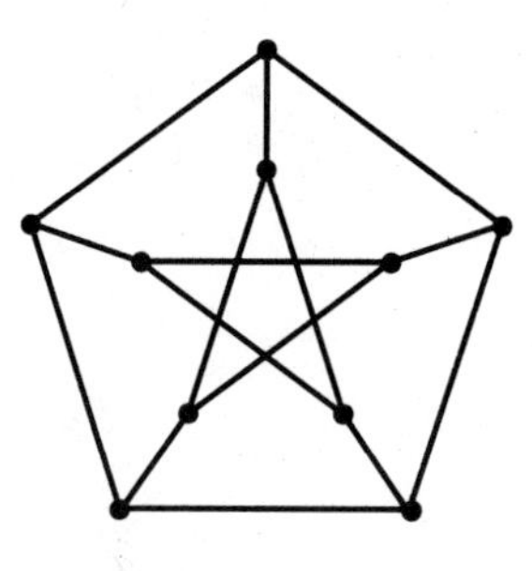

Analysis

In an attempt to simplify the problem, we number the edges rather than the vertices (Figure 11.40). Then we attempt to construct a Hamilton

Figure 11.40

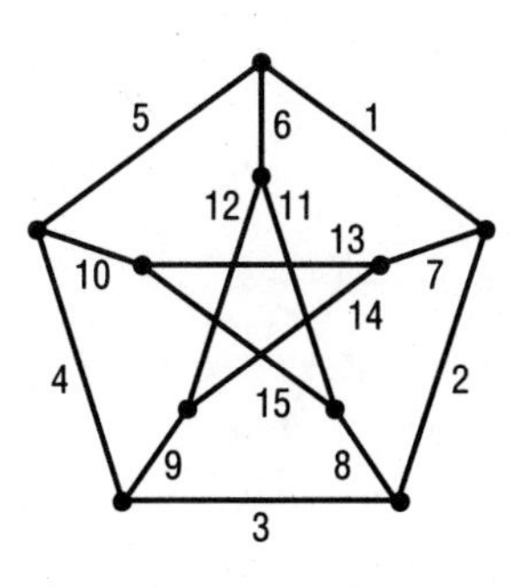

cycle and reach a contradiction. Suppose that the graph has a Hamilton cycle C. Not all the edges numbered 1 through 5 can belong to C, for otherwise C would contain a subcycle. Since the graph is symmetric, we may assume that C does not contain the edge numbered 5. Then we know that edges 1, 6, 4, and 10 must be part of C (see Figure 11.41). Edges 2

Figure 11.41

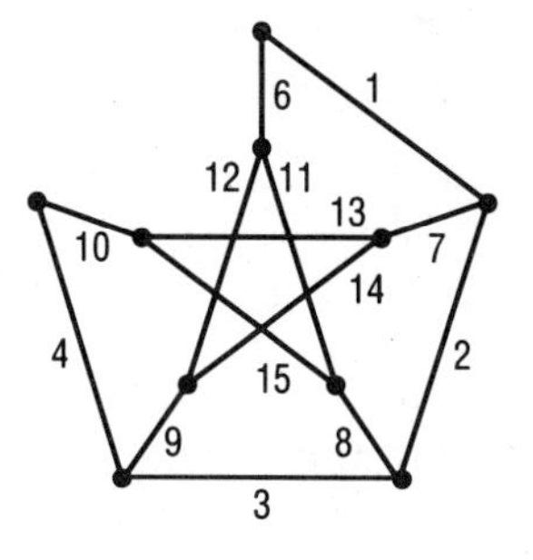

or 3 (or both) must be included in C, and again, since the graph is symmetric, we may assume that edge 2 is in C. Then we know that edge 7 must be excluded from C (see Figure 11.42). We see that edges 13 and 14 must be included in C. Thus 15 must be excluded (see Figure 11.43a). We see that edge 8 must be included, so edge 3 must be excluded and edge 9 must be included. Finally, edge 12 must be excluded and edge 11 in-

Figure 11.42

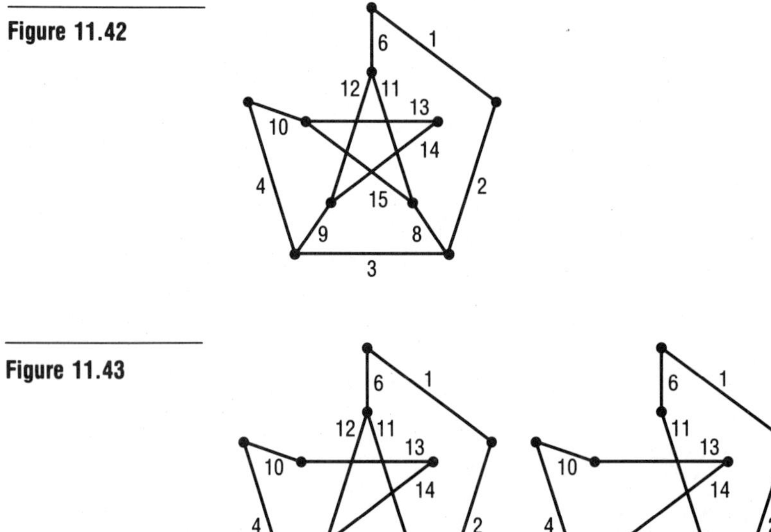

Figure 11.43

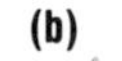

cluded. We are left with two subcycles (see Figure 11.43b), so a Hamilton cycle cannot exist. ■

We now state and illustrate two theorems that are helpful in deciding whether a graph is hamiltonian. We assume that whenever we remove a vertex from a graph we also remove all edges that are incident with the vertex. Theorem 11.19 gives a sufficient condition for a graph not to be hamiltonian.

Theorem 11.19 Let G be a graph and let $p \in \mathbb{N}$. If the removal of p vertices from G results in a graph with more than p components, then G is not hamiltonian.

We use this result in the following example.

Example 39

Show that the graphs in Figure 11.44 are not hamiltonian.

Figure 11.44

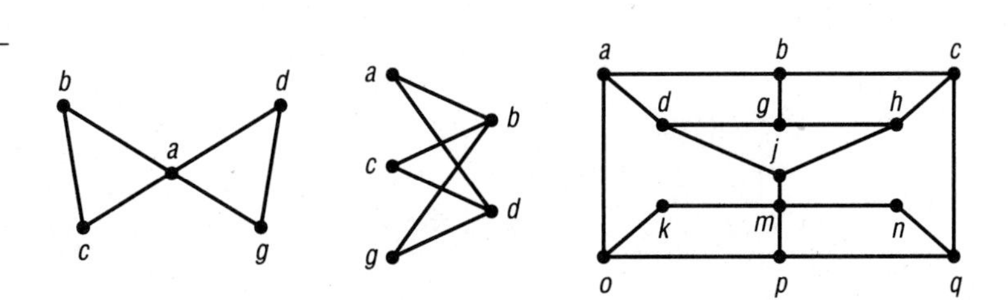

Analysis

Removal of one vertex a in Figure 11.44a results in a graph with two components, as shown in Figure 11.45a. Removal of two vertices b and d in Figure 11.44b results in a graph with three components, as shown in Figure 11.45b. Removal of the six vertices b, d, h, m, o, and q in Figure 11.44c results in a graph with seven components, as shown in Figure

Figure 11.45

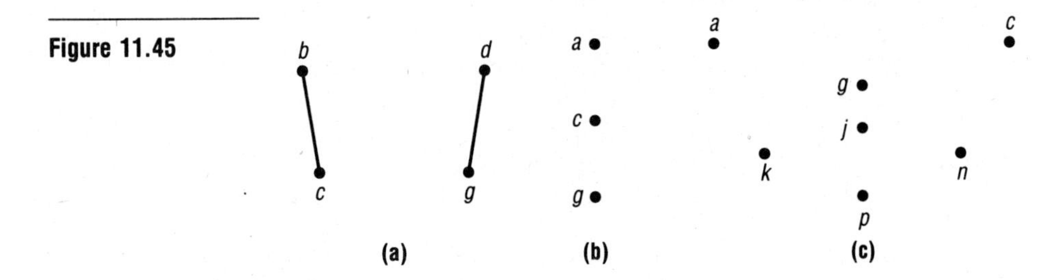

11.45c. Therefore, by Theorem 11.19, the graphs in Figure 11.44 are not hamiltonian. ■

Theorem 11.20 gives a sufficient condition for a graph to be hamiltonian.

Theorem 11.20

> Let G be a simple graph of order $n \geqslant 3$. If, for each vertex v of G, $\deg(v) \geqslant n/2$, then G is hamiltonian.

We use this result in the following example.

Example 40

Show that the graph G in Figure 11.46 is hamiltonian.

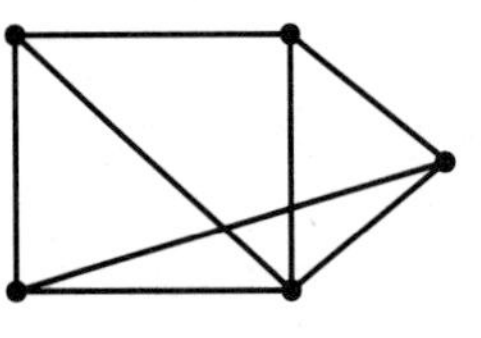

Figure 11.46

Analysis

The order of G is 5, and each vertex has degree greater than or equal to 3. Since $3 \geqslant 5/2$, by Theorem 11.20, G is hamiltonian. ■

Example 41

A class at State College consists of 19 students who sit at a circular table. The instructor wants each student to sit next to two different classmates each day. For how many days can they do this?

Analysis

The graph K_{19} has 19 vertices and $C(19,2) = 171$ edges. Since a Hamilton cycle in K_{19} has 19 edges, we can have at most $9 = 171/19$ Hamilton cycles with no two having an edge in common.

We can construct these nine Hamilton cycles geometrically. Label the center of a circle v_1 and label 18 equally spaced points on the circle, as in Figure 11.47. Then the collection of edges $\{v_1,v_2\}$, $\{v_2,v_3\}$, $\{v_3,v_4\}$, . . . ,$\{v_{18},v_{19}\}$, $\{v_{19},v_1\}$ are the edges of a Hamilton cycle in K_{19}. Keep the circle fixed and rotate the interior in a clockwise direction through the angle $2\pi/18$. We obtain a Hamilton cycle in K_{19} whose edges are $\{v_1,v_3\}$, $\{v_3,v_5\}$, $\{v_5,v_2\}$, $\{v_2,v_7\}$, . . . ,$\{v_{19},v_{16}\}$, $\{v_{16},v_{18}\}$, $\{v_{18},v_1\}$. In general, for each $k = 2,3, \ldots ,8$, a rotation of the inside of the circle in a clockwise direction through the angle $2k\pi/18$ results in a Hamilton cycle in K_{19} whose edges

Figure 11.47

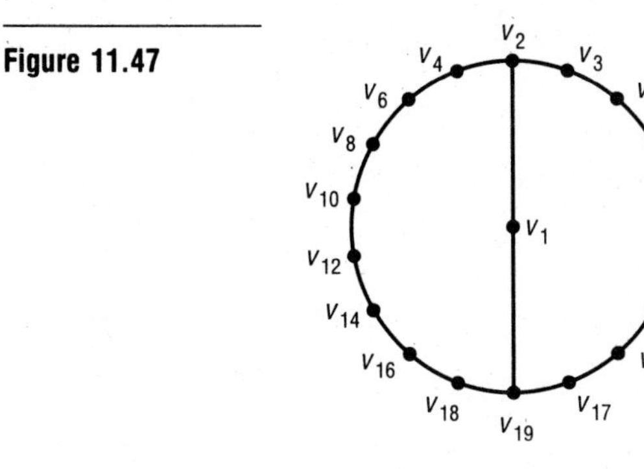

are all different from the edges of the Hamilton cycles previously constructed. Therefore the 19 students can attend class for 9 days and sit next to two different classmates each day. ■

Exercises 11.7

66. Which of the following graphs are hamiltonian? For those that are, find a Hamilton cycle. For those that are not, explain why a Hamilton cycle does not exist.

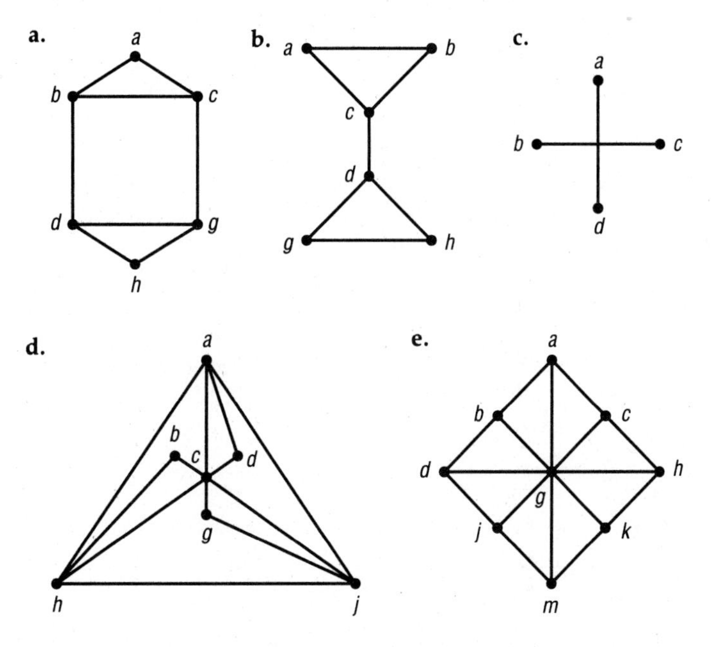

67. Which graphs in Exercise 66 that are not hamiltonian contain a Hamilton path? In each such case find a Hamilton path.

68. Explain why K_n is hamiltonian for each natural number $n \geqslant 3$.

69. Give an example of a graph that is eulerian but not hamiltonian.
70. Give an example of a graph that is hamiltonian but not eulerian.
71. Show that, for each natural number $n \geq 2$, $K_{n,n}$ is hamiltonian.
72. Show that, if m and n are natural numbers and $m \neq n$, then $K_{m,n}$ is not hamiltonian.
73. Show that the following graph is not hamiltonian.

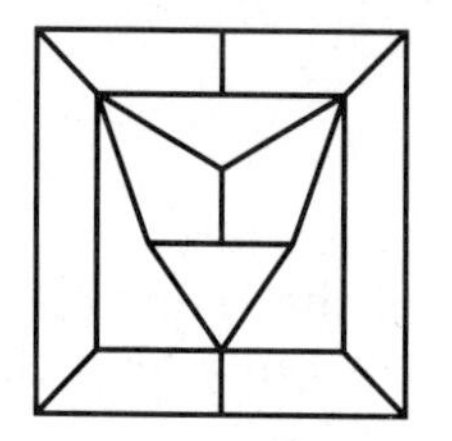

74. Let $G = (V,E)$ be a simple graph, and let $F = \{\{u,v\}: u$ and v are distinct vertices of V and $\{u,v\} \notin E\}$. The *complement* of G is the simple graph (V,F). Prove that, if the simple graph G is r-regular ($r \in \mathbb{N}$) and has at least $2r + 2$ vertices, then the complement of G is hamiltonian. (*Hint:* Use Theorem 11.20.)
75. There are m men and w women at a fraternity party at the end of the school year, and some of the pairs (one man, one woman) dated during the school year. This situation can be represented as a graph in which the vertices represent people at the party and two vertices u and v are adjacent if and only if u and v dated during the school year. Argue that if the graph is hamiltonian then $m = w$.
76. Let G be a simple graph of order n and suppose that the degree of each vertex of G is at least as large as $n/2$. Explain why G is connected. (*Note:* We did not give a proof of Theorem 11.20. For G to be hamiltonian it must be connected, and we are asking you to prove this part of Theorem 11.20.)
77. Let G be a connected simple graph and suppose that G contains a vertex v whose removal leaves a graph that is not connected. Show that G is not hamiltonian. (*Note:* This is a special case of Theorem 11.19. Since we did not give a proof of Theorem 11.19, we are asking you to prove this special case.)

11.8 SHORTEST-PATH ALGORITHMS

Often in applications of graph theory a number is associated with each edge. For example, if a graph represents a highway system, then the number assigned to an edge could be its length. In a communications graph, the numbers assigned to the edges could represent the construction or maintenance costs of the various communication links.

Definition

A **weighted graph** is a graph in which a number $w(e)$, called its **weight,** is assigned to each edge e. If H is a subgraph of a weighted graph, the **weight $w(H)$** of H is the sum of the weights of the edges in H.

We begin with two examples of weighted graphs.

Example 42

In the manufacture of an airplane, it is necessary to drill holes in sheets of metal. A drill press under the control of a computer can be used to bore the holes. The time required to move the drill press between holes should be minimized. A weighted graph G can be used as a model. The vertices of the graph correspond to the holes in the metal, and each pair of vertices is connected by an edge. The weight of each edge is the time required to move the drill press. The problem is to find a subgraph H of G that contains all the vertices of G such that the weight of H is minimized. ■

Connecting each pair of vertices with an edge, as suggested in Example 42, is a time-consuming way of finding a minimum length path that visits each vertex exactly once. Unfortunately, no one knows a general solution. The problem of finding a Hamilton cycle in a graph is related to the Traveling Salesman problem, which we state in the next example.

Example 43

The Traveling Salesman problem is to find a shortest route so that the salesman can start and end at the same city and visit each city exactly once. Again, a weighted graph can be used as a model. The vertices of the graph correspond to cities, the edges correspond to highways between cities, and the weights of the edges correspond to the lengths of the highways. The problem is to find a minimum length Hamilton cycle in G. ■

Since a path in a weighted graph G is a subgraph of G, we have defined the weight of a path in G. There are many optimization problems concerned with finding a subgraph of a certain type with minimum or maximum weight. One such problem is the *shortest-path problem*. Given a railway system connecting various towns, for example, we may want to find a shortest route between two given towns. In this case, we must find a weighted path of minimum weight connecting two given vertices, where the weights represent distances between towns that have a direct link. In general, when we are talking about a shortest path in a weighted graph, we mean the path of minimum weight.

Example 44

Find a (u,v)-path of minimum weight in the weighted graph shown in Figure 11.48.

Figure 11.48

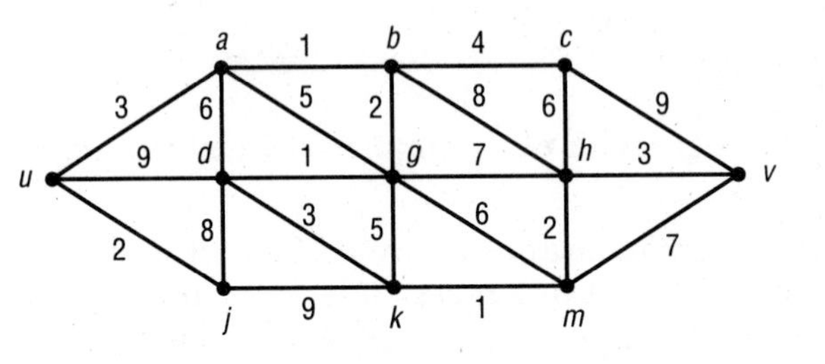

Analysis

At this stage we do not have a formal procedure for finding such a (u,v)-path. Without the use of a computer it is not feasible to consider all (u,v)-paths. Instead, we simply find by inspection a (u,v)-path that appears to be of minimum weight. Such a path is *uabgdkmhv*, and its weight is $3 + 1 + 2 + 1 + 3 + 1 + 2 + 3 = 16$. We recommend that you examine some other (u,v)-paths and find their weights. ■

There are several different algorithms that find a path of minimum weight from one vertex to another. The first algorithm that we describe was discovered by the Dutch mathematician E. W. Dijkstra in 1959. For convenience, we refer to the weight of a path in a weighted graph as its *length*, and we refer to the minimum weight of a (u,v)-path as the *distance*, denoted by $d(u,v)$, between u and v. If $u = v$, we agree that $d(u,v) = 0$. Since it is sufficient to deal with the shortest-path problem for simple graphs, in the remainder of this section we assume that $G = (V,E)$ is a simple graph. We also assume that the weights that have been assigned to the edges are positive real numbers. Furthermore, we assume that the order of G is n ($n \in \mathbb{N}$) and that G is connected in order to assure the existence of a (u,v)-path for each pair u and v of vertices of G. We adopt the convention that, if $u,v \in V$ and $\{u,v\} \notin E$, then $w(\{u,v\}) = \infty$ (if we are using a computer, we replace ∞ by a sufficiently large number).

Dijkstra's algorithm does not find a shortest path from a vertex u_0 to all other vertices of G; it determines only the distances from u_0 to all the other vertices. So before we discuss Dijkstra's algorithm we describe a procedure that permits us to find a shortest path from a vertex to all other vertices of a graph. If S is a proper subset of V and $u_0 \in S$, we let $S' = V - S$. If $P = u_0 \ldots ab$ is a shortest path from u_0 to S', then $a \in S$ and the (u_0,a) section of P is a shortest (u_0,a)-path. Therefore

$$d(u_0,b) = d(u_0,a) + w(\{a,b\}) \tag{11.1}$$

If $u_0 = a$, then $d(u_0,a) = 0$ and hence $d(u_0,b) = w(\{a,b\})$. The basis of Dijkstra's algorithm is the formula for the distance, denoted by $d(u_0,S')$, from u_0 to S':

$$d(u_0,S') = \min\{d(u_0,v) + w(\{v,w\}): v \in S \text{ and } w \in S'\}$$

Starting with the set $S_0 = \{u_0\}$, we construct an increasing sequence $S_0,S_1, \ldots ,S_{n-1}$ of subsets of V so that, at the end of step i, a shortest path from u_0 to each vertex of S_i is known.

Step 1: We find a vertex u_1 nearest to u_0 by computing $d(u_0,S_0')$ and choosing a vertex $u_1 \in S_0'$ such that $d(u_0,u_1) = d(u_0,S_0')$:

$$\begin{aligned} d(u_0,S_0') &= \min\{d(u_0,a) + w(\{a,b\}): a \in S_0 \text{ and } b \in S'\} \\ &= \min\{w(\{u_0,b\}): b \in S'\} \end{aligned}$$

Let $S_1 = \{u_0,u_1\}$ and let P_1 denote the path $\{u_0,u_1\}$.

Step 2: We compute $d(u_0,S_1')$:

$$d(u_0,S_1') = \min\{d(u_0,a) + w(\{a,b\}): a \in S_1 \text{ and } b \in S_1'\}$$

Then we select a vertex $u_2 \in S_1'$ such that $d(u_0,u_2) = d(u_0,S_1')$. By equation 11.1, $d(u_0,u_2) = d(u_0,u_i) + w(\{u_i,u_2\})$ for some $i = 0,1$. Choose such an i, let $S_2 = \{u_0,u_1,u_2\}$, and let P_2 denote the path obtained by adjoining the edge $\{u_i,u_2\}$ to the path P_1.

Step $k + 1$, (where $2 < k < n$): We have the sequence $S_0,S_1, \ldots ,S_k$ of subsets of V and the sequence $P_0,P_1, \ldots ,P_k$ of connected subgraphs without cycles, where for each $i = 0,1, \ldots ,k$, $S_i = \{u_0,u_1, \ldots ,u_i\}$ and P_i contains a shortest path from u_0 to u_i. We compute $d(u_0,S_k')$:

$$d(u_0,S_k') = \min\{d(u_0,a) + w(\{a,b\}): a \in S_k \text{ and } b \in S_k'\}$$

We select a vertex $u_{k+1} \in S_k'$ such that $d(u_0,u_{k+1}) = d(u_0,S_k')$. By equation 11.1, $d(u_0,u_{k+1}) = d(u_0,u_i) + w(\{u_i, u_{k+1}\})$ for some $i = 0,1, \ldots ,k$. Choose such an i, let $S_{k+1} = \{u_0,u_1, \ldots ,u_k,u_{k+1}\}$, and let P_{k+1} denote the path obtained by adjoining the edge $\{u_i,u_{k+1}\}$ to the path P_k.

Dijkstra's algorithm is a refinement of the procedure we have just described. But before discussing and presenting this refinement, we illustrate the procedure with an example.

Example 45

Find a shortest path between u_0 and v in the weighted graph shown in Figure 11.49.

Analysis

We suggest that you relabel the vertices and darken the paths as you proceed through this analysis.

Let $S_0 = \{u_0\}$. Then $d(u_0,S_0') = \min\{w(\{u_0,u): u \in S_0'\}$. Since

Figure 11.49

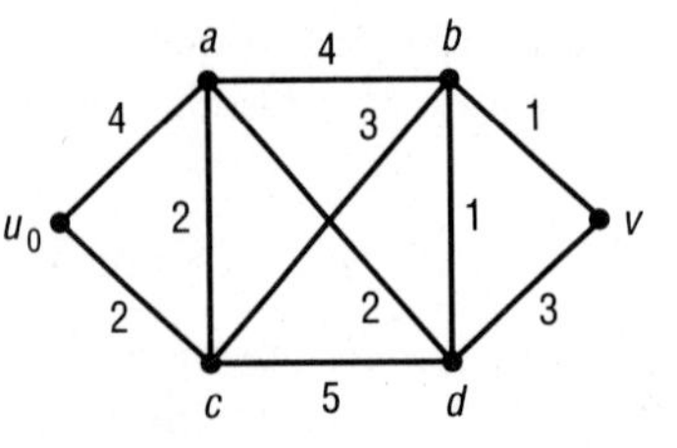

$\min\{w(\{u_0,u\}): u \in S_0'\} = \min\{4,2\}$, $d(u_0,S_0') = 2$. Thus $u_1 = c$, $S_1 = \{u_0,u_1\}$, and P_1 is the path consisting of the edge $\{u_0,u_1\}$.

Now

$$\begin{aligned} d(u_0,S_1') &= \min\{d(u_0,u) + w(\{u,s\}): u \in S_1 \text{ and } s \in S_1'\} \\ &= \min\{w(\{u_0,a\}), d(u_0,u_1) + w(\{u_1,a\}), d(u_0,u_1) + w(\{u_1,b\}), \\ &\quad d(u_0,u_1) + w(\{u_1,d\})\} \\ &= \min\{4,4,5,7\} \\ &= 4 \end{aligned}$$

Thus $u_2 = a$, $S_2 = \{u_0,u_1,u_2\}$, and we have a choice for the path P_2. We can choose the path consisting of the edges $\{u_0,u_1\}$ and $\{u_1,u_2\}$ or the path consisting of the edges $\{u_0,u_1\}$ and $\{u_0,u_2\}$. We choose the path consisting of $\{u_0,u_1\}$ and $\{u_1,u_2\}$.

Now

$$d(u_0,S_2') = \min\{d(u_0,u) + w(\{u,s\}): u \in S_2 \text{ and } s \in S_2'\}$$

so $d(u_0,S_2')$ is the minimum of $d(u_0,u_1) + w(\{u_1,b\})$, $d(u_0,u_1) + w(\{u_1,d\})$, $d(u_0,u_2) + w(\{u_2,b\})$, and $d(u_0,u_2) + w(\{u_2,d\})$. Therefore $d(u_0,S_2') = \min\{2 + 3, 2 + 5, 4 + 4, 4 + 2\} = 5$. Thus $u_3 = b$, $S_3 = \{u_0,u_1,u_2,u_3\}$, and P_3 is the subgraph consisting of the edges $\{u_0,u_1\}$, $\{u_1,u_2\}$, and $\{u_1,u_3\}$.

Now

$$d(u_0,S_3') = \min\{d(u_0,u) + w(\{u,s\}): u \in S_3 \text{ and } s \in S_3'\}$$

so $d(u_0,S_3')$ is the minimum of $d(u_0,u_1) + w(\{u_1,d\})$, $d(u_0,u_2) + w(\{u_2,d\})$, $d(u_0,u_3) + w(\{u_3,d\})$, and $d(u_0,u_3) + w(\{u_3,v\})$. Therefore $d(u_0,S_3') = \min\{2 + 5, 4 + 2, 5 + 1, 5 + 1\} = 6$. Hence we have a choice for u_4. We can choose d or v. Since we are interested in finding a shortest path between u_0 and v, we choose u_4 to be v. Then $P_4 = \{u_0,u_1\}\ \{u_1,u_2\}\ \{u_1,u_3\}\ \{u_3,v\}$ contains a shortest path from u_0 to v. Its weight is 8.

As we mentioned earlier, the procedure we are describing is designed to find a shortest path from u_0 to all other vertices of the graph. We can accomplish this in one more step. Let $S_4 = \{u_0,u_1,u_2,u_3,u_4\}$, and we have already chosen P_4.

Now $d(u_0,S_4')$ is the minimum of $d(u_0,u_1) + w(\{u_1,d\})$, $d(u_0,u_2) + w(\{u_2,d\})$,

$d(u_0,u_3) + w(\{u_3,d\})$, and $d(u_0,u_4) + w(\{u_4,d\})$. Therefore $d(u_0,S_4') = \min\{2 + 5, 4 + 2, 5 + 1, 6 + 3\} = 6$. Therefore $u_5 = d$ and $\{u_0,u_1\}\{u_1,u_3\}\{u_3,d\}$ is a shortest path from u_0 to d. Note that if we adjoin $\{u_3,d\}$ to P_4 we have a subgraph P_5 of the graph in Figure 11.49 that contains this shortest path from u_0 to d. Notice that P_5 is a connected graph without cycles. Such a graph is called a *tree*, and trees are studied in the next chapter. ■

Let us return to Dijkstra's algorithm. Dijkstra's algorithm does not find a shortest path from u_0 to all the other vertices. It determines only the distances from u_0 to all the other vertices. As mentioned earlier, Dijkstra's algorithm is a refinement of the preceding procedure. The motivation for the refinement is that if the minimum of equation 11.1 is computed from scratch at each step of the procedure, then many comparisons must be made repeatedly. To avoid such repetitions and to retain information from one step to another, we adapt a labeling procedure. For each vertex a of G, we use $L(a)$ to denote an upper bound on $d(u_0,a)$. To be specific, initially $L(u_0) = 0$, $S_0 = \{u_0\}$, and $L(a) = \infty$ for $a \neq u_0$ (once again, if we are using a computer we replace ∞ by a sufficiently large number). At each step, for each vertex a of G, $L(a)$ is the length of a shortest path from u_0 to a that has been found at that step.

In step 1, for each vertex $a \in S_0'$, we replace $L(a)$ by $w(\{u_0,a\})$ (so $L(a) = \infty$ if a is not adjacent to u_0, and $L(a)$ is a positive real number if a is adjacent to u_0). Thus we choose a vertex u_1 in S_0' such that $L(u_1) = \min\{L(a)\colon a \in S_0'\}$. Finally we let $S_1 = S_0 \cup \{u_1\}$.

In step 2, for each vertex $a \in S_1'$, we replace $L(a)$ by $\min\{L(u_1) + w(\{u_1,a\}), L(a)\}$. Then we choose a vertex $u_2 \in S_1'$ such that $L(u_2) = \min\{L(a)\colon a \in S_1'\}$, and we let $S_2 = S_1 \cup \{u_2\}$.

In general, in step $k + 1$, $(2 < k < n)$, for each vertex $a \in S_k'$, we replace $L(a)$ by

$$\min\{L(a), L(u_1) + w(\{u_1,a\}), L(u_2) + w(\{u_2,a\}), \ldots, L(u_k) + w(\{u_k,a\})\}$$

Then we choose a vertex $u_{k+1} \in S_k'$ such that $L(u_{k+1}) = \min\{L(a)\colon a \in S_k'\}$ and let $S_{k+1} = S_k \cup \{u_{k+1}\}$.

Dijkstra's Algorithm

1. Set $L(u_0) = 0$, $L(a) = \infty$ if $a \neq u_0$, $S_0 = \{u_0\}$, and $i = 0$.
2. For each $a \in S_i'$, replace $L(a)$ by $\min\{L(a), L(u_j) + w(\{u_j,a\})\colon j = 0,1, \ldots, i\}$.
3. Choose a vertex $u_{i+1} \in S_i'$ such that $L(u_{i+1}) = \min\{L(a)\colon a \in S_i'\}$.
4. Set $S_{i+1} = S_i \cup \{u_{i+1}\}$.
5. If $i = n - 2$, stop.
6. Replace i by $i + 1$.
7. Go to step 2.

Example 46

Use Dijkstra's algorithm to find the length of a shortest path from u_0 to all the other vertices in the weighted graph in Figure 11.50.

Figure 11.50

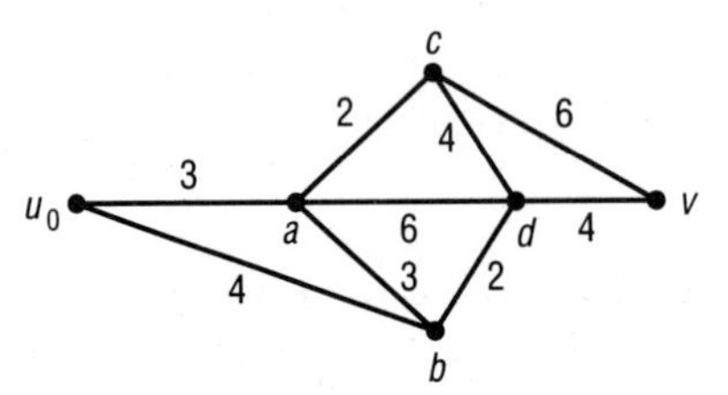

Analysis

1. Set $L(u_0) = 0$, $L(g) = \infty$ if $g \neq u_0$, $S_0 = \{u_0\}$, and $i = 0$.
2. $L(u_0) = 0$, $L(a) = 3$, $L(b) = 4$, and $L(g) = \infty$ if $g \notin \{u_0, a, b\}$.
3. $u_1 = a$.
4. $S_1 = S_0 \cup \{u_1\}$.
6. Change i to 1.
2. $L(u_0) = 0$, $L(a) = 3$, $L(b) = 4$, $L(c) = 5$, $L(d) = 9$, and $L(v) = \infty$.
3. $u_2 = b$.
4. $S_2 = S_1 \cup \{u_2\}$.
6. Change i to 2.
2. $L(u_0) = 0$, $L(a) = 3$, $L(b) = 4$, $L(c) = 5$, $L(d) = 6$, and $L(v) = \infty$.
3. $u_3 = c$.
4. $S_3 = S_2 \cup \{u_3\}$.
6. Change i to 3.
2. $L(u_0) = 0$, $L(a) = 3$, $L(b) = 4$, $L(c) = 5$, $L(d) = 6$, and $L(v) = 11$.
3. $u_4 = d$.
4. $S_4 = S_3 \cup \{u_4\}$.
6. Change i to 4.
2. $L(u_0) = 0$, $L(a) = 3$, $L(b) = 4$, $L(c) = 5$, $L(d) = 6$, and $L(v) = 10$.
3. $u_5 = v$.
4. $S_5 = S_4 \cup \{v\}$.
5. Stop.

Dijkstra's algorithm involves no more than $n - 1$ iterations. At each iteration, steps 2 and 3 each involve no more than $n - 1$ computations. Therefore the total number of operations is $(n - 1)[(n - 1) + (n - 1)] = 2(n - 1)^2$. Hence the complexity of Dijkstra's algorithm is $O(n^2)$.

Warshall's algorithm also finds the length of a shortest path between any two vertices of a graph. We first present Warshall's algorithm for finding the length of a "shortest path" from one vertex to another in a directed graph. Then we describe the modification necessary to find the length of a shortest path between vertices in a simple graph.

Definition

A **weighted directed graph** is a graph in which a number $w(\overrightarrow{uv})$, called its **weight**, is assigned to each directed edge $\overrightarrow{uv}$.

Definition

A **directed walk** in a directed graph is a finite nonempty sequence (there is at least one directed edge in the sequence) $W = v_0e_1v_1e_2v_2 \ldots e_nv_n$, whose terms are alternately vertices and directed edges, such that, for each $i = 1,2, \ldots ,n$, e_i is a directed edge from v_{i-1} to v_i. The **weight** of a directed walk in a directed graph is the sum of the weights of the directed edges in the directed walk.

If u and v are vertices of a directed graph, we denote by $w(u,v)$ the minimum weight of a directed walk from u to v. We agree that, if there is no directed walk from u to v, then $w(u,v) = \infty$. Warshall's algorithm for finding $w(u,v)$ in a directed graph works with an $n \times n$ matrix, where n is the number of vertices of the directed graph. The rows and columns are labeled by the vertices $v_1,v_2, \ldots ,v_n$ of the directed graph. The algorithm begins with the matrix $W_0 = [w_{ij}]$, whose (i,j)th entry is the weight assigned to the directed edge from v_i to v_j if there is a directed edge from v_i to v_j and whose (i,j)th entry is ∞ if there is no directed edge from v_i to v_j (as before, if we are using a computer, we replace ∞ by a sufficiently large number). Warshall's algorithm produces n matrices $W_1,W_2, \ldots ,W_n$, and the (i,j)th entry of W_n is the minimum weight of a directed walk from v_i to v_j. (The (i,j)th entry of W_n is ∞ if and only if there is no directed walk from v_i to v_j.)

Warshall's Algorithm

1. Set $k = 1$.
2. Set $i = 1$.
3. For each $j = 1,2, \ldots ,n$, if $w_{ij} > w_{ik} + w_{kj}$, replace w_{ij} by $w_{ik} + w_{kj}$.
4. If $i < n$, replace i by $i + 1$ and go to step 3.
5. Let $W_k = [w_{ij}]$.
6. If $k < n$, replace k by $k + 1$ and go to step 2.
7. Stop.

We illustrate Warshall's algorithm in the following example.

Example 47

Use Warshall's algorithm to find the minimum weight of a directed walk from each vertex of the weighted directed graph shown in Figure 11.51 to each vertex.

Figure 11.51

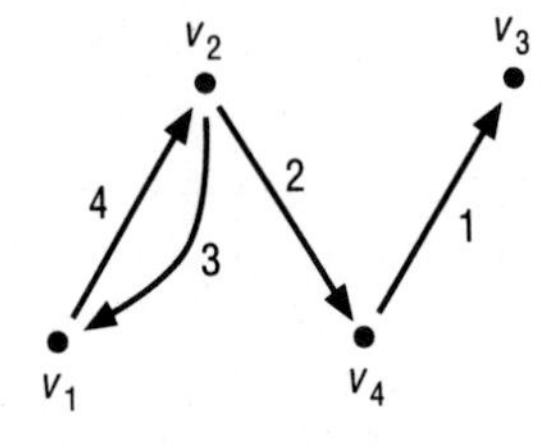

Analysis

The matrix W_0 is the matrix

$$\begin{array}{c} \\ v_1 \\ v_2 \\ v_3 \\ v_4 \end{array} \begin{array}{c} \begin{array}{cccc} v_1 & v_2 & v_3 & v_4 \end{array} \\ \begin{bmatrix} \infty & 4 & \infty & \infty \\ 3 & \infty & \infty & 2 \\ \infty & \infty & \infty & \infty \\ \infty & \infty & 1 & \infty \end{bmatrix} \end{array}$$

We take k to be 1 and produce the matrix W_1 by comparing, for each $i = 1,2, \ldots, n$ and each $j = 1,2, \ldots, n$, w_{ij} and $w_{ik} + w_{kj}$ and replacing w_{ij} by $w_{ik} + w_{kj}$ whenever $w_{ij} > w_{ik} + w_{kj}$. It is a lengthy process to write all the steps necessary to find $W_1, W_2, \ldots, W_n$, but it is not difficult to write a program to find these matrices. We write the steps necessary to find W_1 and then simply list W_2, W_3, and W_4. In Exercise 82, we ask you to supply the missing details.

1. Let $k = 1$.
2. Let $i = 1$.
3. Let $j = 1$. Is $w_{11} > w_{11} + w_{11}$? No.
Let $j = 2$. Is $w_{12} > w_{11} + w_{12}$? No.
Let $j = 3$. Is $w_{13} > w_{11} + w_{13}$? No.
Let $j = 4$. Is $w_{14} > w_{11} + w_{14}$? No.
4. Let $i = 2$.
3. Let $j = 1$. Is $w_{21} > w_{21} + w_{11}$? No.
Let $j = 2$. Is $w_{22} > w_{21} + w_{12}$? Yes.
Replace w_{22} by $w_{21} + w_{12} = 3 + 4 = 7$.
Let $j = 3$. Is $w_{23} > w_{21} + w_{13}$? No.
Let $j = 4$. Is $w_{24} > w_{21} + w_{14}$? No.
4. Let $i = 3$.
3. Let $j = 1$. Is $w_{31} > w_{31} + w_{11}$? No.
Let $j = 2$. Is $w_{32} > w_{31} + w_{12}$? No.
Let $j = 3$. Is $w_{33} > w_{31} + w_{13}$? No.
Let $j = 4$. Is $w_{34} > w_{31} + w_{14}$? No.

4. Let $i = 4$.

3. Let $j = 1$. Is $w_{41} > w_{41} + w_{11}$? No.
Let $j = 2$. Is $w_{42} > w_{41} + w_{12}$. No.
Let $j = 3$. Is $w_{43} > w_{41} + w_{13}$? No.
Let $j = 4$. Is $w_{44} > w_{41} + w_{14}$? No.

5. The matrix W_1 is the matrix

$$\begin{array}{c} \\ v_1 \\ v_2 \\ v_3 \\ v_4 \end{array} \begin{array}{c} \begin{array}{cccc} v_2 & v_2 & v_3 & v_4 \end{array} \\ \left[\begin{array}{cccc} \infty & 4 & \infty & \infty \\ 3 & 7 & \infty & 2 \\ \infty & \infty & \infty & \infty \\ \infty & \infty & 1 & \infty \end{array}\right] \end{array}$$

6. Let $k = 2$ and go to step 2.

We go through a similar process to find that W_2 is the matrix

$$\begin{array}{c} \\ v_1 \\ v_2 \\ v_3 \\ v_4 \end{array} \begin{array}{c} \begin{array}{cccc} v_1 & v_2 & v_3 & v_4 \end{array} \\ \left[\begin{array}{cccc} 7 & 4 & \infty & 6 \\ 3 & 7 & \infty & 2 \\ \infty & \infty & \infty & \infty \\ \infty & \infty & 1 & \infty \end{array}\right] \end{array}$$

Now we let $k = 3$, go through a similar process, and find that $W_3 = W_2$. Finally we let $k = 4$ and repeat the process to find that W_4 is the matrix

$$\begin{array}{c} \\ v_1 \\ v_2 \\ v_3 \\ v_4 \end{array} \begin{array}{c} \begin{array}{cccc} v_1 & v_2 & v_3 & v_4 \end{array} \\ \left[\begin{array}{cccc} 7 & 4 & 7 & 6 \\ 3 & 7 & 3 & 2 \\ \infty & \infty & \infty & \infty \\ \infty & \infty & 1 & \infty \end{array}\right] \end{array}$$

Let us compare W_4 with some of the walks of minimum weight in our graph (Figure 11.51). Observe that to obtain a walk from v_1 to v_1 we must start at v_1, go to v_2, and then back to v_1. The weight of this walk is 7, and 7 is the (1,1)th entry of W_4.

We can easily observe that the minimum weight of a walk from v_1 to v_3 is $4 + 2 + 1 = 7$, and 7 is the (1,3)th entry of W_4.

Observe that there is no walk that begins at v_3, and indeed all the entries in the third row of W_4 are ∞. ■

In Warshall's algorithm, step 3 is executed once for each possible choice of i, j, and k. Hence step 3 is executed n^3 times. Therefore the complexity of Warshall's algorithm is $O(n^3)$.

Warshall's algorithm can be applied to simple graphs by replacing each edge $\{u,v\}$ by two directed edges $\vec{uv}$ and $\vec{vu}$. If the edge $\{u,v\}$ has weight $w(\{u,v\})$, assign the two directed edges the same weight. If the simple graph does not have weights, assign 1 to all edges. A simple graph does not have loops, and loops are irrelevant in the search for a shortest path, so it is convenient to take the (i,i)th entry of W_0 to be 0 for each $i = 1,2, \ldots, n$.

Example 48

Use Warshall's algorithm to find the minimum weight of a path from each vertex of the weighted graph shown in Figure 11.52 to each vertex.

Figure 11.52

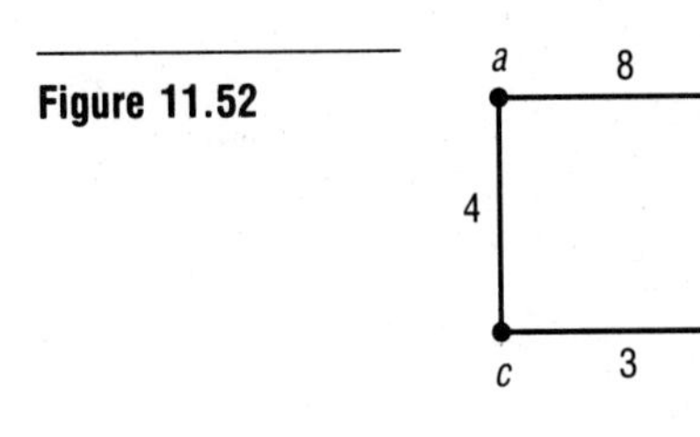

Analysis

The process is the same as in Example 47. The matrices, W_0, W_1, W_2, W_3, and W_4 are as follows:

$$W_0 = \begin{array}{c} \\ a \\ b \\ c \\ d \end{array} \begin{array}{c} \begin{array}{cccc} a & b & c & d \end{array} \\ \begin{bmatrix} 0 & 8 & 4 & \infty \\ 8 & 0 & \infty & 2 \\ 4 & \infty & 0 & 3 \\ \infty & 2 & 3 & 0 \end{bmatrix} \end{array} \qquad W_1 = \begin{array}{c} \\ a \\ b \\ c \\ d \end{array} \begin{array}{c} \begin{array}{cccc} a & b & c & d \end{array} \\ \begin{bmatrix} 0 & 8 & 4 & \infty \\ 8 & 0 & 12 & 2 \\ 4 & 12 & 0 & 3 \\ \infty & 2 & 3 & 0 \end{bmatrix} \end{array}$$

$$W_2 = \begin{array}{c} \\ a \\ b \\ c \\ d \end{array} \begin{array}{c} \begin{array}{cccc} a & b & c & d \end{array} \\ \begin{bmatrix} 0 & 8 & 4 & 10 \\ 8 & 0 & 12 & 2 \\ 4 & 12 & 0 & 3 \\ 10 & 2 & 3 & 0 \end{bmatrix} \end{array}$$

$$W_3 = \begin{array}{c} \\ a \\ b \\ c \\ d \end{array} \begin{array}{c} \begin{array}{cccc} a & b & c & d \end{array} \\ \begin{bmatrix} 0 & 8 & 4 & 7 \\ 8 & 0 & 12 & 2 \\ 4 & 12 & 0 & 3 \\ 7 & 2 & 3 & 0 \end{bmatrix} \end{array} \qquad W_4 = \begin{array}{c} \\ a \\ b \\ c \\ d \end{array} \begin{array}{c} \begin{array}{cccc} a & b & c & d \end{array} \\ \begin{bmatrix} 0 & 8 & 4 & 7 \\ 8 & 0 & 5 & 2 \\ 4 & 5 & 0 & 3 \\ 7 & 2 & 3 & 0 \end{bmatrix} \end{array}$$

The zeros on the diagonal of W_4 are not meaningful. ■

Another algorithm for finding a shortest path between vertices in a simple graph is the *breadth-first search algorithm*. This algorithm finds a shortest path and the distance from a vertex u to a vertex v ($u \neq v$). In the algorithm, L denotes a set of labeled vertices, U denotes a set of unlabeled vertices, and the *predecessor* of a vertex w is the vertex in L that is used to label w. We summarize the algorithm as follows:

The Breadth-First Search Algorithm

1. Assign the label 0 to the vertex u (that is, let $k = 0$) and let $L = \{u\}$.
2. If $v \in L$, stop. The label $k + 1$ is the distance from u to v. A shortest path is obtained by using, in reverse order, the vertex v and the vertices that have been chosen as predecessors.
3. Let k be the largest label that has been used, and find the unlabeled vertices U in G that are adjacent to vertices in L that bear the label k. If there are no such vertices, go to 4. Assign the label $k + 1$ to the vertices in U and put them in L. For each vertex p with label $k + 1$, choose a vertex q with label k and call q the predecessor of p. Go to 2.
4. There is no path in G from u to v. Stop.

Example 49

Use the breadth-first search algorithm to find a shortest path from u to v in the simple graph in Figure 11.53.

Figure 11.53

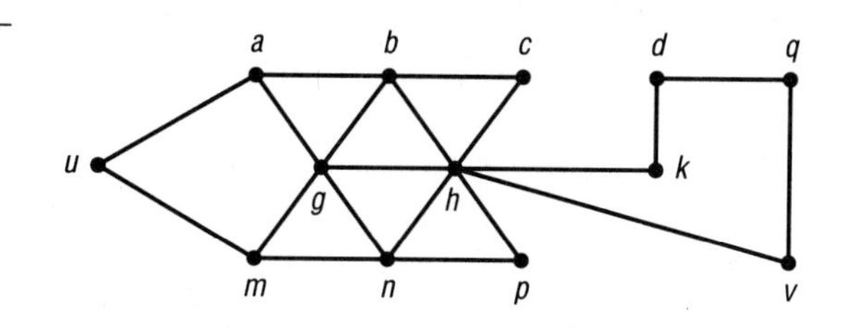

Analysis

1. Assign the label 0 to vertex u and let $L = \{u\}$.
2. $v \notin L$.
3. There are two unlabeled vertices a and m that are adjacent to u. Assign the label 1 to these vertices and let $L = \{u,a,m\}$. Vertex u is the predecessor of both a and m.
2. $v \notin L$.
3. There are three unlabeled vertices b, g, and n that are adjacent to vertices in L that bear the label 1. Assign the label 2 to these three vertices and let $L = \{u,a,m,b,g,n\}$. Vertex a is the predecessor of b, and vertex m is the predecessor of n. Choose a as the predecessor of g.
2. $v \notin L$.
3. There are three unlabeled vertices c, h, and p that are adjacent to vertices in L that bear the label 2. Assign the label 3 to these three vertices and let $L = \{u,a,m,b,g,n,c,h,p\}$. Vertex b is the predecessor of c, and vertex n is the predecessor of p. Choose g as the predecessor of h.

2. $v \notin L$.
3. There are two unlabeled vertices k and v that are adjacent to vertices in L that bear the label 3. Assign the label 4 to these two vertices and let $L = \{u,a,m,b,g,n,c,h,p,k,v\}$. Vertex h is the predecessor of both k and v.
2. $v \in L$. Stop. The distance from u to v is 4, and a shortest path from u to v is obtained by reversing the order of $vhgau$. ■

The complexity of the breadth-first search algorithm is discussed in Chapter 12.

Exercises 11.8

78. Use Dijkstra's algorithm to find the length of a shortest path from u_0 to all the other vertices in each of the following weighted graphs.

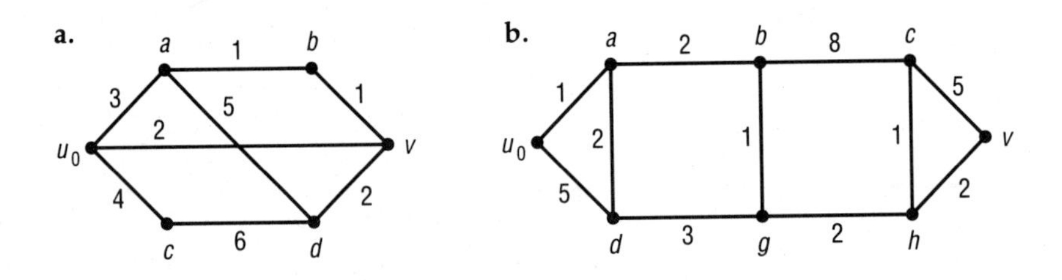

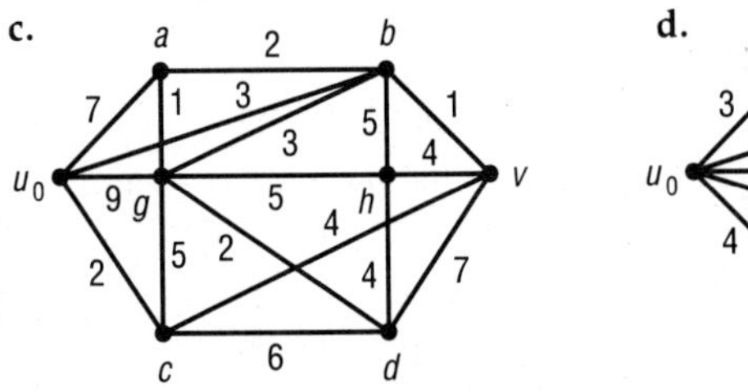

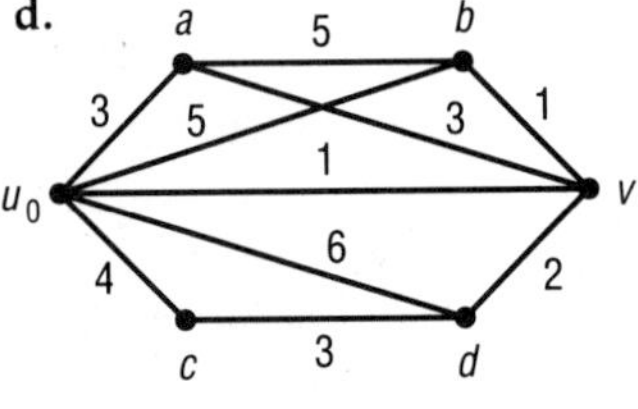

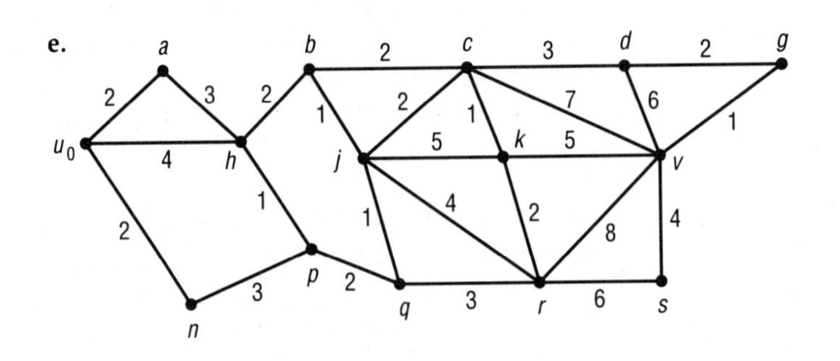

f.

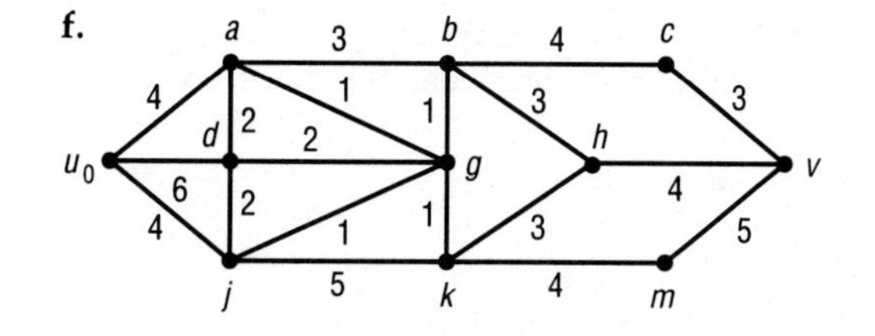

79. Find a shortest path between u_0 and v in each of the weighted graphs in Exercise 78.

80. Use Dijkstra's algorithm to find the shortest route between any two cities in the following communication network.

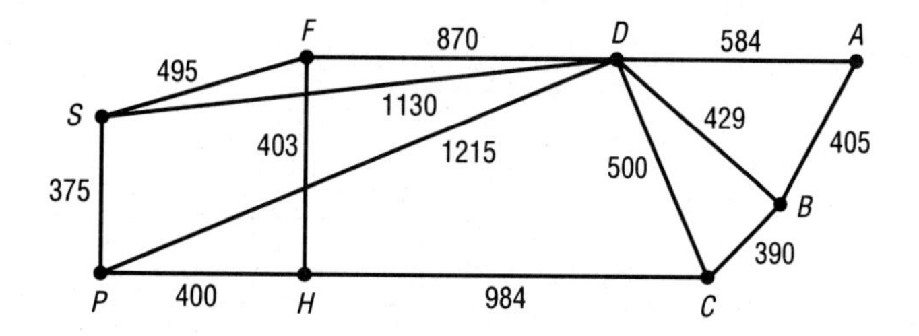

81. A company has branches in cities A, B, C, Q, P, and S. The fare for a direct flight between two cities is given in the following matrix (∞ is used to indicate that there is no direct flight).

	A	B	C	Q	P	S
A	0	500	∞	400	250	100
B	500	0	150	200	∞	250
C	∞	150	0	100	200	∞
Q	400	200	100	0	100	250
P	250	∞	200	100	0	550
S	100	250	∞	250	550	0

The company wants to prepare a table of cheapest routes. Draw a weighted graph that models this matrix and use Dijkstra's algorithm to prepare this table.

82. Supply the missing details needed to find W_2, W_3, and W_4 in Example 47.

83. Use Warshall's algorithm to find the minimum weight of a directed walk from each vertex of the following weighted directed graph to each vertex.

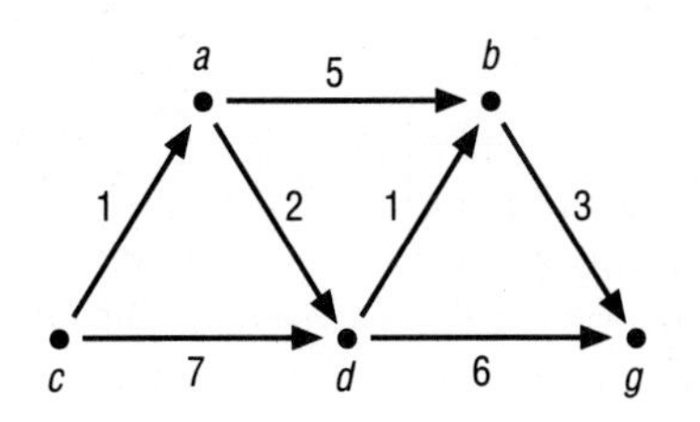

84. Use Warshall's algorithm to find the minimum weight of a path from each vertex of the following weighted graph to each vertex.

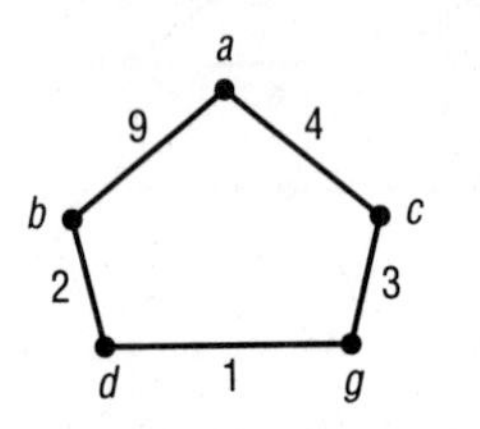

85. Use the breadth-first search algorithm to find a shortest path and the distance from the vertex u to the vertex v in each of the following simple graphs.

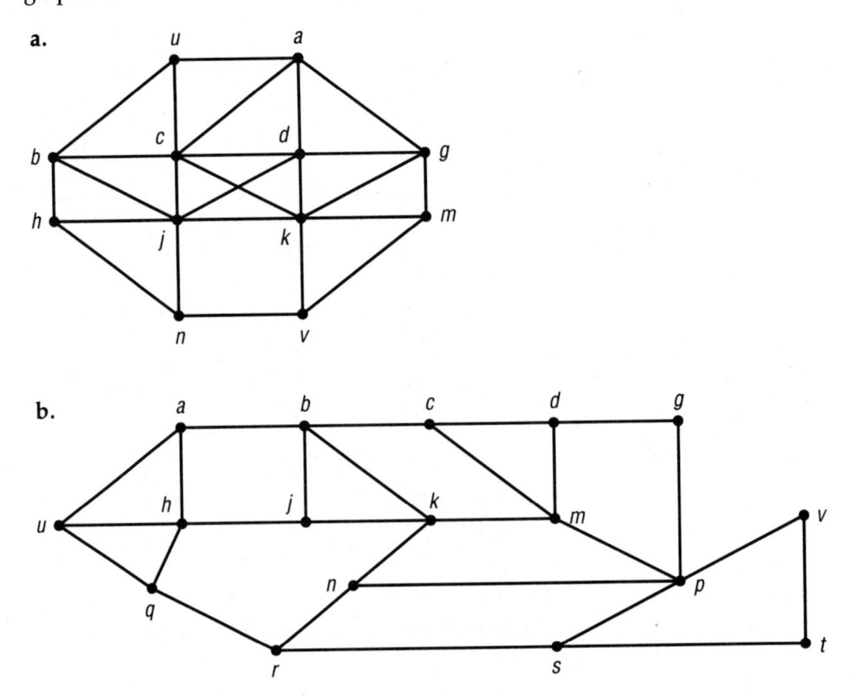

11.9 PLANAR GRAPHS

We have been drawing pictures of graphs where each vertex is represented by a point and each edge is represented by a line segment. Of course, any given graph can be drawn in different ways. For example, the three pictures in Figure 11.54 represent the same graph.

Figure 11.54

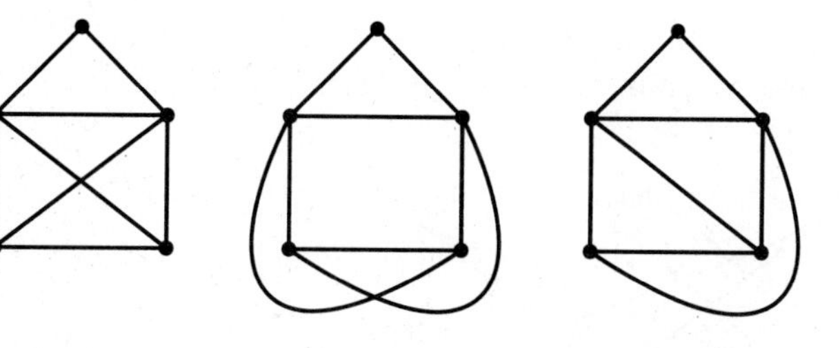

Definition

A graph is said to be **embeddable in the plane,** or **planar,** if it can be drawn in the plane so that its edges intersect only at its vertices. A **plane graph** is a planar embedding of a planar graph.

We are content in this section to adopt a naive point of view in our study of planar graphs, and so we are content to deal with the fundamental connections between graph theory and topology in an intuitive way. For our purposes, an informal discussion of a theorem from topology known as the *Jordan curve theorem* suffices. Informally, we define a *Jordan curve* to be the union of the edges (each edge is now considered to be a line in the plane) in a cycle of a planar graph. For example, the dark outlined area in Figure 11.55 constitutes a Jordan curve. (The Jordan curve and the corresponding Jordan curve theorem are named in honor of the analyst C. Jordan, who should not be confused with W. Jordan of Gauss–Jordan elimination.)

Figure 11.55

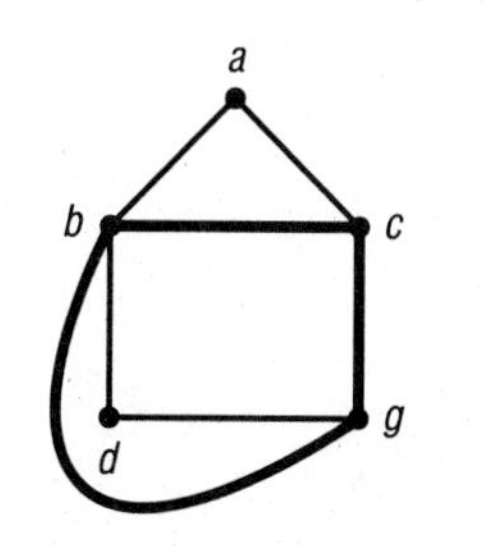

According to the Jordan curve theorem, the complement of a Jordan curve J in the plane is partitioned into two disjoint sets called the *interior* and *exterior* of J. For example, in Figure 11.55 the region *inside* $\{b,c\} \cup \{c,g\} \cup \{g,b\}$ is the interior of a Jordan curve and the region *outside* $\{b,c\} \cup \{c,g\} \cup \{g,b\}$ is the exterior of a Jordan curve.

A famous puzzle in graph theory is the problem of three houses and three utilities: Is it possible to connect each house to each utility without crossing any lines? (See Figure 11.56.) In Example 50 we use the Jordan curve theorem to solve this puzzle.

Figure 11.56

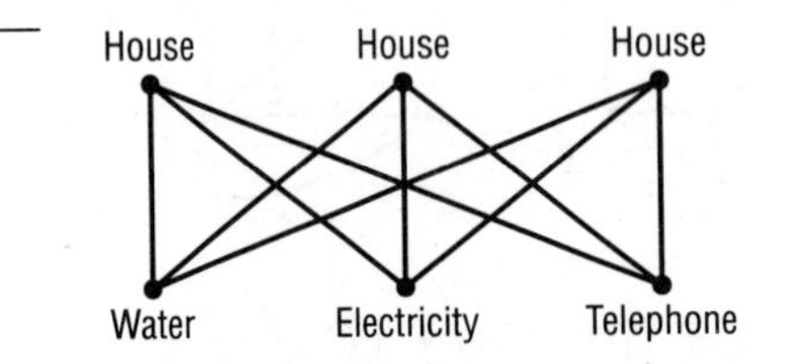

Example 50

Show that $K_{3,3}$ is not planar.

Analysis

The argument is by contradiction. Suppose $K_{3,3}$ is planar. Label the vertices of $K_{3,3}$ as indicated in Figure 11.57. The union of the four edges $\{a,w\}$, $\{w,b\}$, $\{b,u\}$, and $\{u,a\}$ is a Jordan curve J. By the Jordan curve theorem, the complement of J in the plane is the union of two disjoint sets—the interior of J (denoted int(J)) and the exterior of J (denoted ext(J)). Vertex c is either in int(J) or in ext(J), so we consider two cases.

Figure 11.57

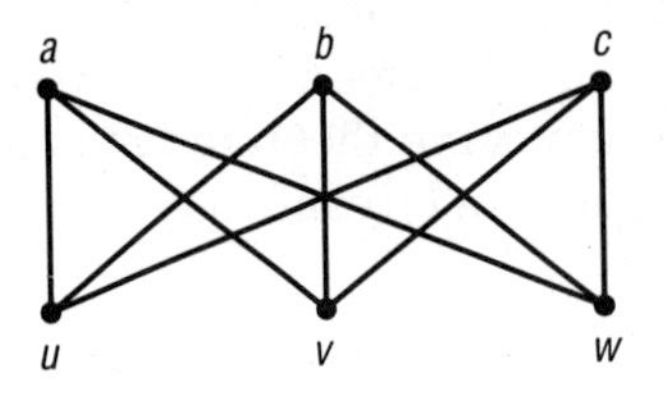

Case 1: Suppose $c \in$ int(J). Then the edges $\{u,c\}$ and $\{w,c\}$ divide int(J) into two regions, as shown here:

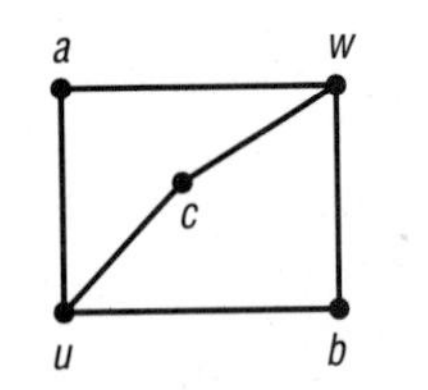

Where do we place vertex v? If v is in the interior of the Jordan curve $J_1 = \{a,w\} \cup \{w,c\} \cup \{c,u\} \cup \{u,a\}$, then by the Jordan curve theorem v cannot be joined to b without crossing J_1. If v is in the interior of the Jordan curve $J_2 = \{w,b\} \cup \{b,u\} \cup \{u,c\} \cup \{c,w\}$, then v cannot be joined to a without crossing J_2. If v is on the outside of the Jordan curve $J_3 = \{a,w\} \cup \{w,b\} \cup \{b,u\} \cup \{u,a\}$, then v cannot be joined to c without crossing J_3.

Case 2: Suppose $c \in \text{ext}(J)$. The argument is essentially the same as for case 1, and it is left as Exercise 90.

As we see in this example, it is impossible to connect the three houses and three utilities without crossing two lines. ■

As illustrated in Figure 11.58, there is more than one way to draw a planar graph in the plane. Euler showed, however, that if P_1 and P_2 are planar representations of a graph then the complements of P_1 and P_2 consist of the same number of regions. Observe that the complement of each planar representation of the graph in Figure 11.58 consists of three regions, an outside region and two inside regions labeled I_1 and I_2. He did this by finding a relationship between the number of vertices, the number of edges, and the number of regions in the complement of a plane graph. Each region in the complement of a plane graph is called a *face* of the graph. Notice that each planar graph has an "outside" face, so that, for example, the planar graph of Figure 11.58 has three faces.

Figure 11.58

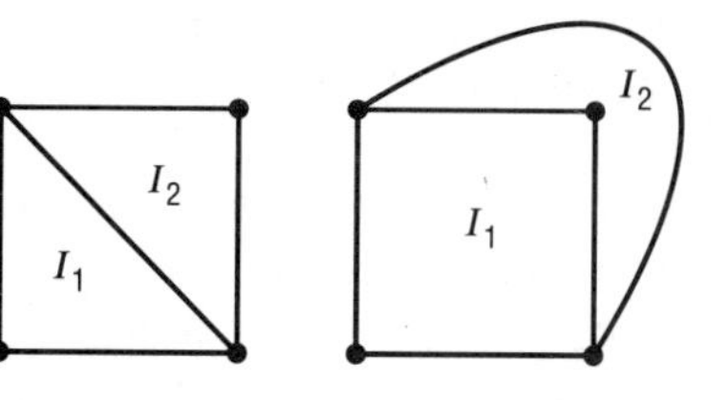

Let G be a connected planar graph. If f is a face of G, we denote by $b(f)$ the *boundary* of f. If $b(f)$ does not have any cut edges of G, it is a cycle of G. If $b(f)$ does have cut edges of G, it can be regarded as a closed walk in which each cut edge is traversed twice. In the connected planar graph shown in Figure 11.59, $b(f_3) = be_3de_7he_8ge_6b$ and $b(f_1) = ae_1be_3de_4ce_5ce_4de_2a$.

A face f of G is said to be *incident* with the vertices and edges in $b(f)$. One face of G is incident with each edge, and two faces of G are incident with each edge that is not a cut edge. We say that an edge *separates* the faces incident with it. In Figure 11.59, edge e_5 separates f_1 and f_2, and

Figure 11.59

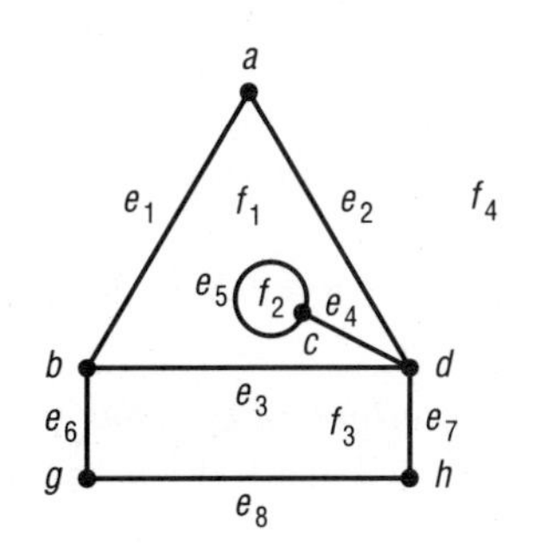

edge e_4 separates f_1. The degree, denoted by $\deg_G(f)$, or $\deg(f)$, of a face f is the number of edges in $b(f)$, that is, the number of edges with which it is incident, where each cut edge is counted twice. In Figure 11.59, $\deg(f_3) = 4$ and $\deg(f_1) = 6$.

It is customary to denote the number of vertices of a planar graph by ν, the number of edges by ε, and the number of faces by ϕ. If more than one graph is being discussed, it is customary to let $\nu(G)$, $\varepsilon(G)$, and $\phi(G)$ denote the number of vertices, the number of edges, and the number of faces of G, respectively. Euler's formula is given by Theorem 11.21. In the proof of Theorem 11.21, we use a result that we prove in Chapter 12 (see Corollary 12.3): If G is a tree, then $\varepsilon = \nu - 1$ (or $\nu - \varepsilon = 1$).

Theorem 11.21 If G is a connected planar graph, then $\nu - \varepsilon + \phi = 2$.

Proof

The proof is by induction on the number of faces of G. Let $S = \{n \in \mathbb{N}$: if G is a connected planar graph with n faces, then $\nu - \varepsilon + \phi = 2\}$.

If G is a connected planar graph with one face, then each edge of G is a cut edge of G. Hence G is a tree and therefore (by Corollary 12.3) $\nu - \varepsilon = 1$. Since $\phi = 1$, $\nu - \varepsilon + \phi = 2$. Hence $1 \in S$.

Suppose $n \in S$, and let G be a connected planar graph with $n + 1$ faces. Since $n + 1 \geqslant 2$, G contains an edge e that is not a cut edge. Let G' be the graph obtained from G by removing e. The number of faces of G' is n since the two faces of G separated by e combine to form one face of G'. Since $n \in S$, $\nu(G') - \varepsilon(G') + \phi(G') = 2$. Since $\nu(G) = \nu(G')$, $\varepsilon(G') = \varepsilon(G) - 1$, and $\phi(G') = \phi(G) - 1$, $\nu(G) - [\varepsilon(G) - 1] + [\phi(G) - 1] = 2$. Hence $\nu(G) - \varepsilon(G) + \phi(G) = 2$, and hence $n + 1 \in S$. By the principle of mathematical induction, $S = \mathbb{N}$. □

Example 51 Let G be a connected planar graph with five vertices of degree 3, six vertices of degree 4, and seven vertices of degree 5. What is the number of faces of G?

Analysis

Let $v_1, v_2, \ldots, v_{18}$ denote the vertices of G. Then

$$\sum_{i=1}^{18} \deg(v_i) = 5 \cdot 3 + 6 \cdot 4 + 7 \cdot 5 = 15 + 24 + 35 = 74$$

By Theorem 11.1, the number of edges of G is $74/2 = 37$. Therefore, by Theorem 11.21, $\phi = 2 - \nu + \varepsilon = 2 - 18 + 37 = 21$. ■

Definition

Let G be a planar graph. There is another graph G^*, called the **dual of G,** that is defined as follows: Associated with each face f of G there is a vertex f^* of G^*, associated with each edge e of G there is an edge e^* of G^*, and two vertices f_1^* and f_2^* are joined by the edge e^* in G^* if and only if the associated faces f_1 and f_2 are separated by the edge e in G.

Example 52

Find the dual G^* of the graph G shown in Figure 11.60.

Figure 11.60

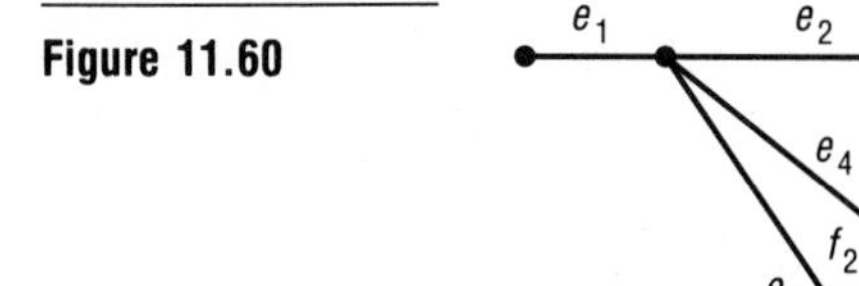

Analysis

For the five faces and nine edges in G, we have five vertices, f_1^*, f_2^*, f_3^*, f_4^*, and f_5^*, and nine edges, e_1^*, e_2^*, e_3^*, e_4^*, e_5^*, e_6^*, e_7^*, e_8^*, and e_9^* of G^* (see graph G^* in Figure 11.61).

Figure 11.61

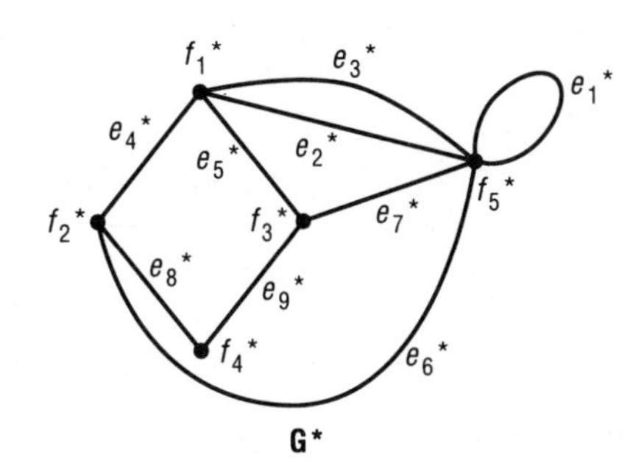

■

It can be shown that the dual G^* of a planar graph G is planar. In fact, the natural way to embed G^* in the plane is to place each vertex f^* of G^* in the face f of G and draw the edge e^* of G^* so that it crosses e exactly once and does not cross any other vertex.

It can also be shown that the dual G^{**} of G^* is isomorphic to G. Isomorphic plane graphs may, however, have nonisomorphic duals.

Example 53

Show that the plane graphs in Figure 11.62 are isomorphic but that their duals are not isomorphic.

Figure 11.62

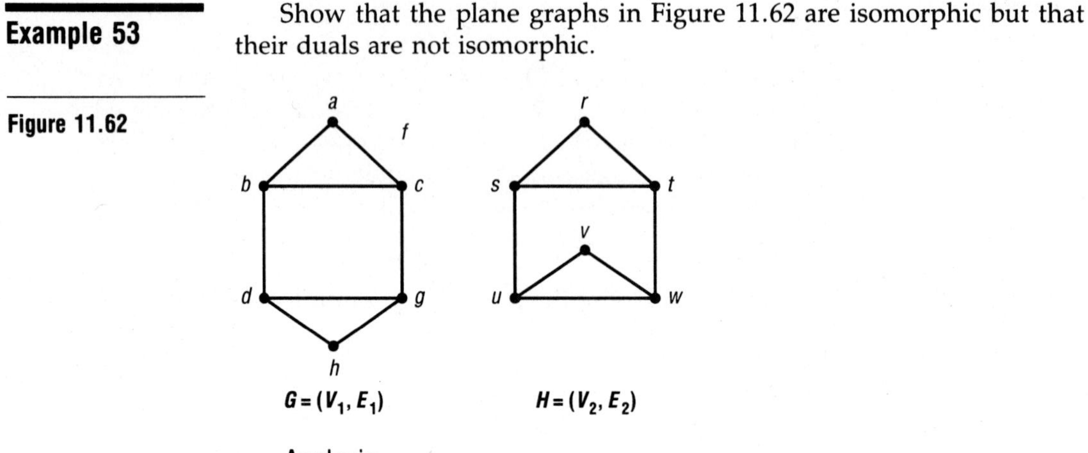

Analysis

The bijection $\theta: V_1 \to V_2$ defined by $\theta(a) = r$, $\theta(b) = s$, $\theta(c) = t$, $\theta(d) = u$,, $\theta(g) = w$, and $\theta(h) = v$ is an isomorphism of G and H. However, face f of G has degree 6, whereas there is no face of H of degree 6. Therefore G^* has a vertex of degree 6, whereas H^* does not have a vertex of degree 6. Hence G^* is not isomorphic to H^*. ■

As a direct consequence of the definition of the dual G^* of a plane graph G, we have the following formulas:

$$\nu(G^*) = \phi(G)$$
$$\varepsilon(G^*) = \varepsilon(G) \qquad (11.2)$$
$$\deg(f^*) = \deg(f) \text{ for all faces } f \text{ in } G$$

Theorem 11.22

> If G is a plane graph, then $\Sigma\{\deg(f): f \text{ is a face of } G\} = 2\varepsilon$.

Proof

Let G^* be the dual of G. Then

$$\begin{aligned}\Sigma\{\deg(f): f \text{ is a face of } G\} &= \Sigma\{\deg(f^*): f^* \in V(G^*)\} && \text{(by equation 11.2)}\\ &= 2\varepsilon(G^*) && \text{(by Theorem 11.1)}\\ &= 2\varepsilon(G) && \text{(by equation 11.2)}\end{aligned}$$ □

Theorem 11.23

> If G is a connected planar simple graph and $\varepsilon \geq 3$, then for each face f of G, $\deg(f) \geq 3$.

Proof

Let G be a connected planar simple graph such that $\varepsilon \geqslant 3$. If G has only one face f, then the boundary of f contains all the edges of G. Thus, in this case, $\deg(f) \geqslant 3$. If G has more than one face, then the boundary of each face contains a cycle. Cycles of length 2 can occur only if the graph has multiple edges, and cycles of length 1 are loops. Since G is a simple graph, the boundary of each face contains at least three edges. □

We use Euler's formula and Theorems 11.22 and 11.23 to prove the following theorem.

Theorem 11.24

> If G is a connected planar simple graph and $\nu \geqslant 3$, then $\varepsilon \leqslant 3\nu - 6$.

Proof

Let G be a connected planar simple graph such that $\nu \geqslant 3$. First observe that $\varepsilon \neq 1$. If $\varepsilon = 2$, then $\varepsilon \leqslant 3 \cdot 3 - 6 \leqslant 3\nu - 6$. Suppose $\varepsilon \geqslant 3$. Then, by Theorem 11.23, for each face f of G, $\deg(f) \geqslant 3$. Hence $\Sigma\{\deg(f)\colon f$ is a face of $G\} \geqslant 3\phi$. So, by Theorem 11.22, $2\varepsilon \geqslant 3\phi$. Thus, by Theorem 11.21, $\nu - \varepsilon + 2\varepsilon/3 \geqslant 2$. Therefore $\nu - 2 \geqslant \varepsilon/3$, or $\varepsilon \leqslant 3\nu - 6$. □

We can use Theorem 11.24 to exhibit a connected simple graph that is nonplanar and does not contain $K_{3,3}$.

Example 54

Show that K_5 is nonplanar.

Analysis

The connected simple graph K_5 has five vertices and ten edges. Thus $3\nu - 6 = 9$ and $\varepsilon = 10$. Hence $\varepsilon > 3\nu - 6$, so, by Theorem 11.24, K_5 is nonplanar. ■

Theorem 11.25

> If G is a connected planar simple graph, then G has a vertex of degree less than or equal to 5.

Proof

Let G be a connected planar simple graph. If ν is 1 or 2, then it is clear that G has a vertex of degree less than or equal to 5. Suppose $\nu \geqslant 3$ and let $v_1, v_2, \ldots, v_\nu$ denote the vertices of G. Then, by Theorem

11.1, $\Sigma_{i=1}^{\nu} \deg(v_i) = 2\varepsilon$, and, by Theorem 11.24, $2\varepsilon \leqslant 6\nu - 12$. If $\delta = \min\{\deg(v_i): i = 1,2, \ldots, \nu\}$, then $\delta\nu \leqslant \Sigma_{i=1}^{\nu} \deg(v_i)$. Therefore $\delta\nu \leqslant 6\nu - 12$, and hence $\delta \leqslant 5$. □

Theorem 11.26

> If G is a connected planar simple graph with no cycles of length 3 and $\nu \geqslant 3$, then $\varepsilon \leqslant 2\nu - 4$.

Proof

Let G be a connected planar simple graph with no cycles of length 3. First observe that $\varepsilon \neq 1$. If $\varepsilon = 2$, then $\nu = 3$ since there are no cycles of length 2. If $\varepsilon = 3$, then $\nu = 4$ since there are no cycles of length 3. In these two cases, $\varepsilon \leqslant 2\nu - 4$. Suppose $\varepsilon \geqslant 4$. Since G has no cycles of length 3, for each face f of G, $\deg(f) \geqslant 4$. Hence $\Sigma\{\deg(f): f \text{ is a face of } G\} \geqslant 4\phi$. So, by Theorem 11.22, $2\varepsilon \geqslant 4\phi$, or $\varepsilon \geqslant 2\phi$. Thus, by Theorem 11.21, $\nu - \varepsilon + \varepsilon/2 \geqslant 2$. Therefore $\nu - 2 \geqslant \varepsilon/2$, or $\varepsilon \leqslant 2\nu - 4$. □

Definition

> Let $G = (V,E)$ be a simple graph. The procedure of removing an edge $\{u,v\}$ from G and adding a vertex w ($w \notin V$) together with the edges $\{u,w\}$ and $\{w,v\}$ is called an **elementary subdivision** of G. Two simple graphs $G_1 = (V_1,E_1)$ and $G_2 = (V_2,E_2)$ are **homeomorphic** if there is a simple graph H such that G_1 can be obtained from H by a finite sequence of elementary subdivisions and G_2 can be obtained from H by a finite sequence of elementary subdivisions.

Example 55

Show that the graphs G_1 and G_2 in Figure 11.63 are homeomorphic.

Figure 11.63

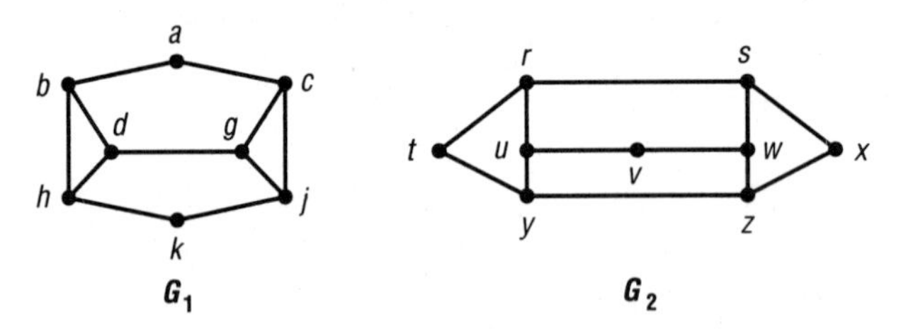

Analysis

Let H be the graph shown in Figure 11.64.

We obtain G_1 from H by the following sequence of elementary subdivisions:

1. Delete edge {1,2} from H and add vertex a and edges $\{1,a\}$ and $\{2,a\}$.
2. Delete edge {5,6} from H and add vertex k and edges $\{5,k\}$ and $\{6,k\}$.
3. Rename the vertices to obtain G_1.

Figure 11.64

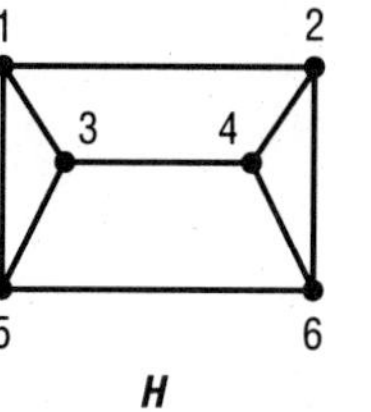

We obtain G_2 from H by the following sequence of elementary subdivisions:

1. Delete edge {3,4} from H and add vertex v and edges $\{3,v\}$ and $\{4,v\}$.
2. Delete edge {1,5} from H and add vertex t and edges $\{1,t\}$ and $\{5,t\}$.
3. Delete edge {2,6} from H and add vertex x and edges $\{2,x\}$ and $\{6,x\}$.
4. Rename the vertices to obtain G_2. ■

We have already noted that the problem of determining which graphs are hamiltonian is unsolved, and the previous two theorems make it appear that the problem of determining which graphs are planar might be even more untoward. Surprisingly, there is an elegant characterization of planar graphs, which was established by the Polish mathematician Kasimir Kuratowski in 1930. If a graph is nonplanar, one cannot improve the situation by adding edges or vertices to it. With this observation in mind, you may be able to guess Kuratowski's result. Even so, Kuratowski's theorem is one of the great surprises of mathematics, because it tells us in discrete terms about a continuous geometric surface, the plane (see Exercise 103).

Theorem 11.27

Kuratowski's Theorem A graph is planar if and only if it does not contain a subgraph that is homeomorphic to K_5 or $K_{3,3}$. □

We give two examples that illustrate the use of this theorem.

Example 56

Show that K_6 is nonplanar.

Analysis

K_5 is the subgraph of K_6 obtained by deleting one vertex and all edges that are incident with this vertex. ■

Example 57

Show that Petersen's graph (Figure 11.65) is nonplanar.

Figure 11.65

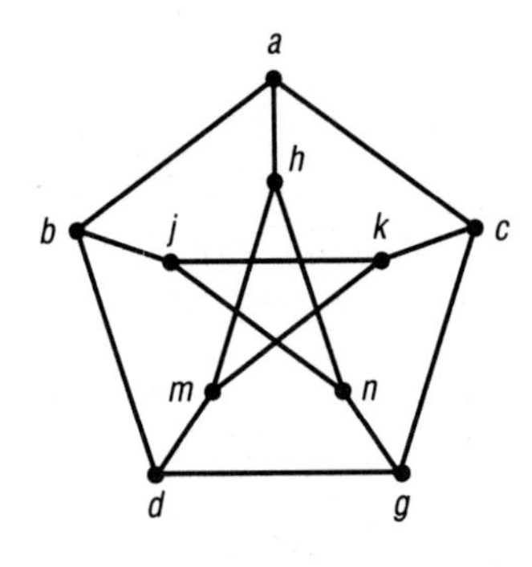

Analysis

Let H_1 be the subgraph of Petersen's graph obtained by deleting edges $\{b,d\}$ and $\{c,k\}$ (see Figure 11.66). Let H_2 be the second graph shown in Figure 11.66. By taking $\{a,m,n\}$ to be one set of vertices and $\{g,h,j\}$ to be

Figure 11.66

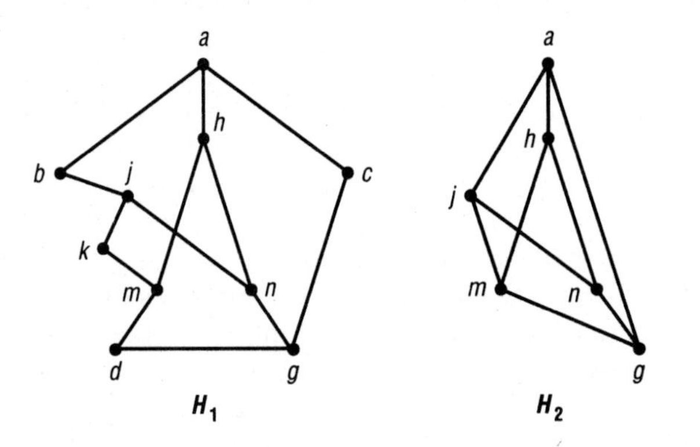

the other set, we see that H_2 is $K_{3,3}$. Furthermore, we can obtain H_1 from H_2 by the following sequence of elementary subdivisions.

1. Delete edge $\{a,g\}$ from H_2 and add vertex c and edges $\{a,c\}$ and $\{c,g\}$.
2. Delete edge $\{g,m\}$ from H_2 and add vertex d and edges $\{g,d\}$ and $\{d,m\}$.

3. Delete edge $\{m,j\}$ from H_2 and add vertex k and edges $\{m,k\}$ and $\{k,j\}$.
4. Delete edge $\{j,a\}$ from H_2 and add vertex b and edges $\{j,b\}$ and $\{b,a\}$.

Therefore H_1 is homeomorphic to $K_{3,3}$. ■

Exercises 11.9

86. Draw planar graphs that represent
   a. K_4
   b. $K_{2,4}$
87. Is $K_{2,5}$ planar? Either draw a planar representation of it or show that it is not planar.
88. Draw planar representations of the following planar graphs.
   a. b. c.

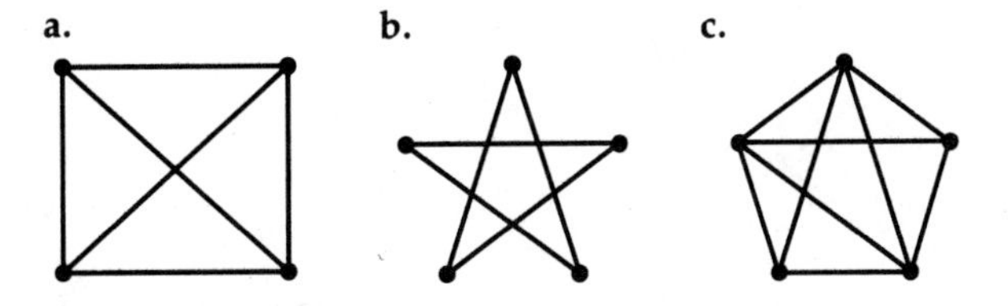

89. Verify Euler's formula for the planar graphs in Exercise 88.
90. Complete the argument in Example 50 to show that $K_{3,3}$ is nonplanar.
91. Give an argument similar to that in Example 50 to show that K_5 is nonplanar.
92. Construct a connected planar simple graph G with at least four edges which has a face whose degree is 3.
93. Use Theorem 11.26 to show that $K_{3,3}$ is nonplanar.
94. Let G be a connected planar simple graph with six vertices each of degree 3. How many faces does G have?
95. Let G be a connected planar simple graph with seven vertices each of degree 4. How many faces does G have?
96. Is there a connected planar graph with four faces, each of which has a boundary consisting of three edges? If so, draw such a graph. If not, explain why.
97. Is there a connected planar graph with ten faces, each of which has a boundary consisting of three edges? If so, draw such a graph. If not, explain why.
98. Show that the following two graphs are homeomorphic.

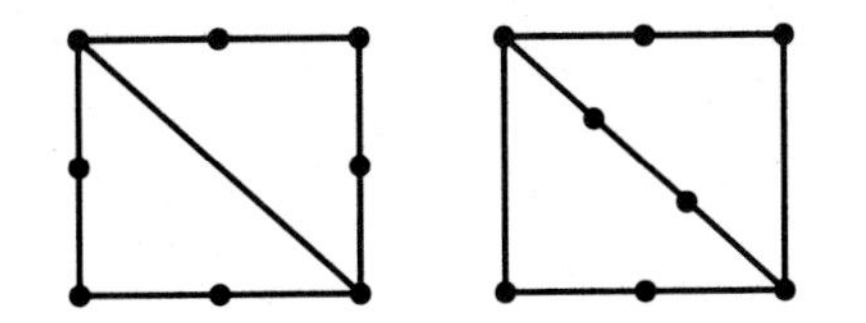

99. Find a simple graph with the smallest possible number of vertices that is homeomorphic to the following graph.

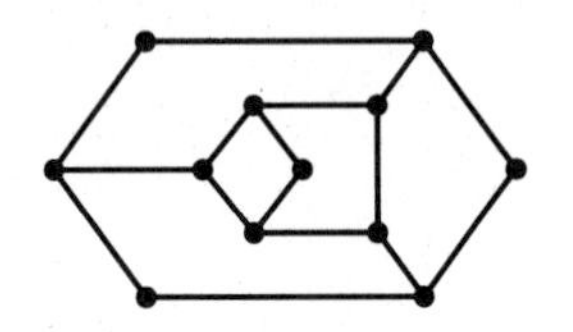

100. **a.** Are two homeomorphic simple graphs necessarily isomorphic?

b. Are two isomorphic simple graphs necessarily homeomorphic?

Explain your answers.

101. Show that the following planar graph is self-dual, that is, that the graph and its dual are isomorphic.

102. The graphs of the edges of the five platonic solids are illustrated here. Draw the duals of each of these graphs and identify each dual as the graph of the edges of some one of five platonic solids. (*Hint:* Theorem 11.21 is helpful.)

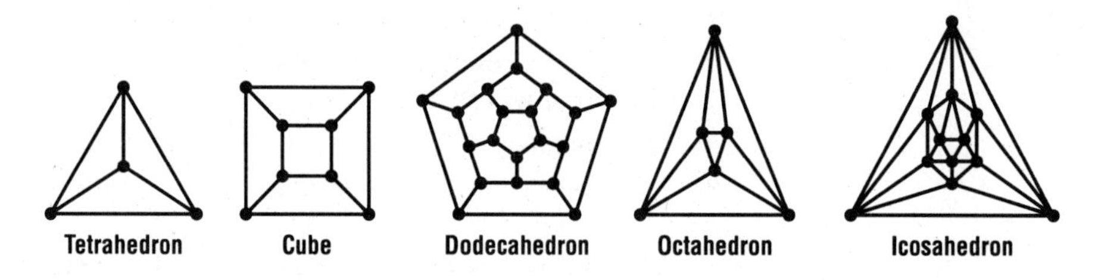

103. **a.** Sketch a torus (the surface of a doughnut) and indicate how to draw $K_{3,3}$ on it so that its edges intersect only at vertices.

b. Sketch a torus and indicate how to draw K_5 on it so that its edges intersect only at vertices.

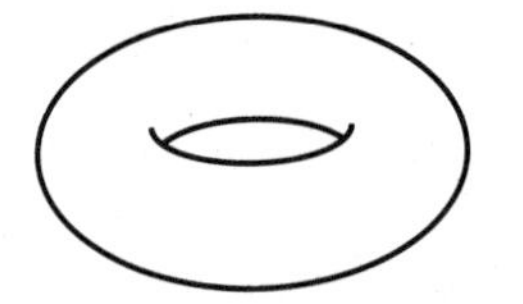

11.10

GRAPH COLORING

The four-color problem is one of the famous problems of mathematics. The problem is to decide how many colors suffice to color any map so that any two regions of the map with a common border are colored differently. In formulating the problem, we require two interpretations: first, by a common border we mean a common edge; second, by a region we mean a connected set (that is, each region is a face of a graph).

Apparently the problem was first raised in 1840 by Auguste Möbius, but around 1850 the brothers Francis and Frederick Guthrie conjectured that four colors were sufficient, and they communicated their conjecture to Augustus de Morgan. It was de Morgan, and later Arthur Cayley, who popularized the problem. Thus the problem began as a mathematical curiousity of no practical import. In 1976, in an article in *Scientific American,* Kenneth Appel and Wolfgang Haken proved the conjecture. Their proof used a computer to give a case-by-case analysis of over 1,400 cases.

We give some simple illustrations of the map-coloring problem. First consider circles divided into regions by radii (Figure 11.67).

Figure 11.67

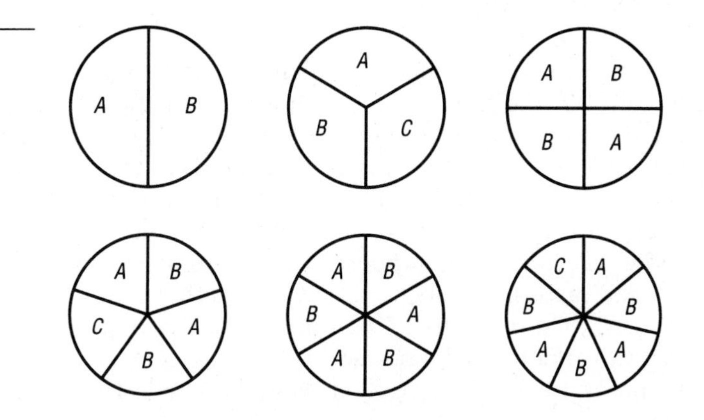

It is clear that if we have an even number of regions then two colors are sufficient, whereas if we have an odd number of regions then three colors are needed. In addition, in each case the outside face requires one more color.

Now consider circles with an inner circle surrounded by regions that have been divided by radii (Figure 11.68).

In this case, if there are an even number of regions surrounding the inner circle then three colors are sufficient, whereas if there are an odd

Figure 11.68

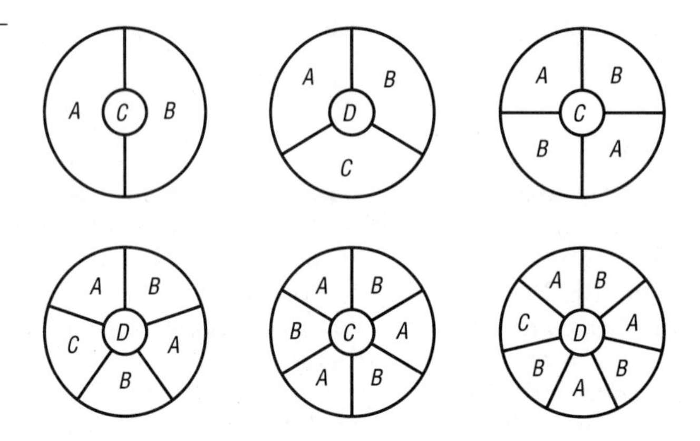

number of regions surrounding the inner circle then four colors are necessary.

Each map in the plane can be represented by a simple plane graph G, and the problem of coloring the regions of the map is equivalent to the problem of coloring the vertices of G^* so that no two adjacent vertices have the same color. Thus we can study the map-coloring problem by studying the problem of coloring the vertices of a graph. However, before we begin the systematic study, we give an example of a different but related type of problem.

Example 58

The Fletcher Chemical Company needs to ship a variety of chemicals by rail. Mr. Fletcher has no desire to be sued, and a violent explosion could occur if certain chemicals mix in the event of an accident. Thus Mr. Fletcher does not want two such chemicals in the same railroad car, and yet he does not want to use any more railroad cars than are necessary. Suppose that there are seven chemical products C_1, C_2, C_3, C_4, C_5, C_6, and C_7. Furthermore, suppose that C_1 cannot be transported in the same railroad car as C_2, C_3, C_4, or C_7; that C_2 cannot be transported in the same car as C_4, C_5, or C_7; that C_3 cannot be transported in the same car as C_6; that C_4 cannot be transported in the same car as C_5; and that C_6 and C_7 cannot be transported together. Find the smallest number of railroad cars that can be used.

Analysis

We draw a graph (Figure 11.69) whose vertices represent the chemicals and whose edges join chemicals that cannot be transported in the same car. We know that C_1 can be in car 1. Since C_2 is adjacent to C_1, C_2 must be in a different car, say, car 2. Since C_3 is adjacent to C_1, it cannot be in car 1, but C_3 is not adjacent to C_2, so it can be in car 2. Now C_4 is adjacent to both C_1 and C_2, so it cannot be in car 1 or car 2. Thus we need

Figure 11.69

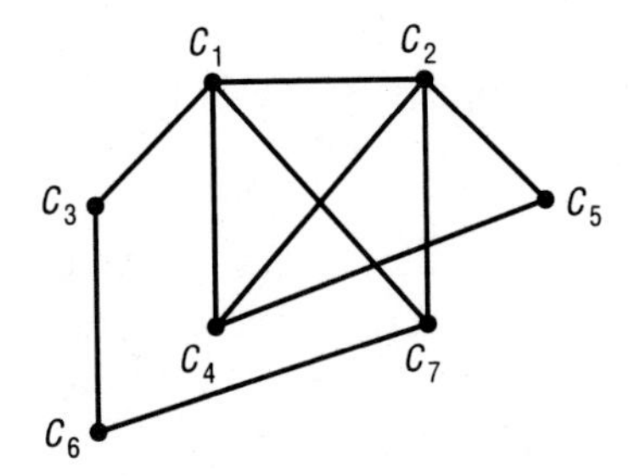

car 3 for C_4. Since C_5 is adjacent to C_2 and C_4, it cannot be in car 2 or car 3. However, C_5 is not adjacent to C_1, so it can be in car 1. Since C_6 is adjacent to C_3, it cannot be in car 2. However, C_6 can be in either car 1 or car 3. Since we have one more chemical, we consider these two possibilities. First, suppose we put C_6 in car 3. Then, since C_7 is adjacent to C_1, C_2, and C_6, we need a fourth car for C_7. However, if we put C_6 in car 1 instead, then we can put C_7 in car 3. Thus three cars are needed to ship the seven chemicals, and Figure 11.70 shows a labeling for shipping the chemicals safely.

Figure 11.70

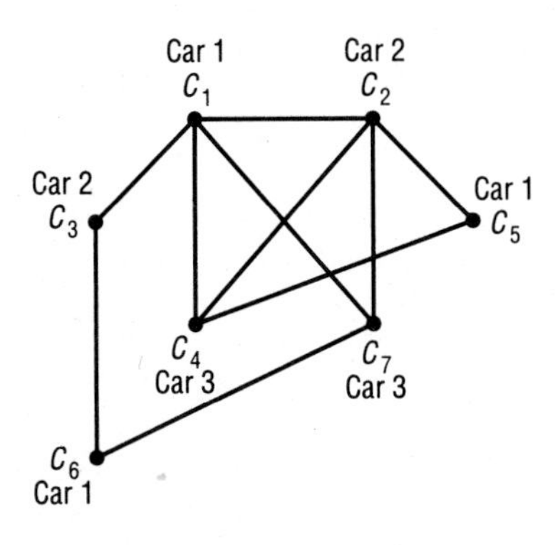

■

Definition

A **coloring** of a loopless graph G is an assignment of a color to each vertex of G such that no two adjacent vertices have the same color. The **chromatic number** $\chi(G)$ of G is the least number of colors needed for a coloring of G.

The chromatic number of a loopless graph and the chromatic number of its underlying simple graph are the same. Thus we restrict ourselves to simple graphs in the remainder of this section.

Theorem 11.28

The chromatic number of a simple graph G is 1 if and only if it is empty.

Proof

If G does not have any edges, then clearly one color is sufficient. If G has an edge e, then the two vertices that e connects must have different colors. □

Theorem 11.29

The chromatic number of a simple graph G is 2 if and only if G is bipartite.

Proof

If $\{V_1, V_2\}$ is a partition of the vertices of a bipartite graph G, then we can use one color for all the members of V_1 and another color for the members of V_2. On the other hand, if we have a coloring of a simple graph G that uses only two colors, then we can form a partition of the vertices by letting V_1 be all the vertices that are colored with one color and letting V_2 be all the vertices that are colored the other color. In this case $\{V_1, V_2\}$ is a partition of V, so G is bipartite. □

Corollary 11.30

If m and n are natural numbers, then the chromatic number of $K_{m,n}$ is 2.

Proof

Since $K_{m,n}$ is bipartite, the desired result follows immediately from Theorem 11.29. □

Theorem 11.31

If $n \in \mathbb{N}$, the chromatic number of K_n is n.

Proof

Let u and v be any two vertices of K_n. Then u and v are adjacent, so they must have different colors. Therefore no two vertices of K_n can have the same color. □

Theorem 11.32

If G is a connected simple graph, then $\chi(G) = 2$ if and only if G does not contain any cycle of odd length.

Proof

Let G be a simple graph and suppose $\chi(G) = 2$. If C is a cycle in G, then as we traverse C the colors alternate and hence the length of C is even.

Now suppose that G is a connected simple graph that does not contain any cycles of odd length. Choose an arbitrary vertex v and color it red. Then color all the vertices adjacent to v green. Next color red all the vertices adjacent to the vertices that have been colored green. Continue this process until all vertices are colored. Since all cycles are of even length, no vertex is assigned two colors. Hence $\chi(G) = 2$. □

As a brief indication of how graph coloring has applications in computing, we consider the concepts of interfering and noninterfering variables. We say that two variables are *noninterfering* if at no time during the execution of the program are both variables needed and at no time are both needed in memory for the execution of some subsequent step. Otherwise the two variables are *interfering*.

Example 59

Suppose that A, B, and C are variables that represent real numbers. A program is designed to find D, where $D = AB + C$. Then A and B are interfering variables in this program. Likewise, A and C are interfering variables, as are B and C. However, A and D, B and D, and C and D are three pairs of noninterfering variables.

Given a program and a list of variables used in the program, we can construct an *interference graph*. In such a graph, each variable is represented by a vertex and two vertices are joined by an edge if they represent interfering variables. The graph shown here is the interference graph of this example.

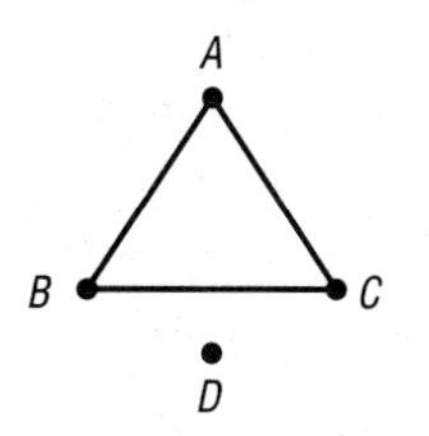

If two vertices are joined by an edge in the interference graph, then the two variables represented by these vertices must be assigned different storage locations. Consequently, the chromatic number of the interference graph is the number of storage locations that are needed for the program. ■

Theorem 11.33

> Let G be a simple graph and let $\alpha = \max\{\deg(v): v \text{ is a vertex of } G\}$. Then $\chi(G) \leq \alpha + 1$.

Proof

We randomly select an ordering $v_1, v_2, \ldots, v_\nu$ of the vertices of G. We use one color C_1 for v_1. If v_2 is not adjacent to v_1, we also use C_1 to color v_2. If v_2 is adjacent to v_1, we select a second color C_2 for v_2. In general, for each $i = 3,4, \ldots, \nu$, v_i is adjacent to at most α vertices (that have been colored), so we select a different color (than those used for the vertices adjacent to v_i) for v_i and so use at most $\alpha + 1$ colors. □

The Welsh and Powell algorithm is used for coloring the vertices of a graph. We summarize this algorithm as follows:

The Welsh and Powell Algorithm

1. Label the vertices $v_1, v_2, \ldots, v_\nu$ of G such that $\deg(v_1) \geqslant \deg(v_2) \geqslant \cdots \geqslant \deg(v_\nu)$.
2. Assign a color C_1 to v_1.
3. Go through the list of vertices in order, assigning the color C_1 to every vertex not adjacent to v_1.
4. Let i be the smallest natural number such that v_i has not been colored. Assign an unused color C to v_i and go through the list of uncolored vertices in order, assigning the color C to every such vertex not adjacent to any other vertex with color C.
5. If all vertices have been colored, stop.
6. Go to step 4.

Example 60

There are eight courses: Math 1534, Math 1215, CS 1705, ENG 1105, MATH 1216, CS 1706, PHY 2305, and HIST 1115. There are no students taking both MATH 1215 and MATH 1216, both CS 1705 and CS 1706, both MATH 1534 and PHY 2305, both CS 1705 and PHY 2305, both MATH 1215 and PHY 2305, and both MATH 1534 and CS 1706. There are, however, students in all the other possible pairs of courses. Schedule final exams so that no student has a conflict (that is, two exams at the same time).

Analysis

We construct a graph model (Figure 11.71). Since there are eight courses, there are eight final exams to be scheduled. We let each final exam (or course) be represented by a vertex, and we join two vertices with an edge if there are students taking both courses.

We use the Welsh and Powell algorithm to color this graph.

1. Let $v_1 = 1105$, $v_2 = 1115$, $v_3 = 1216$, $v_4 = 1534$, $v_5 = 1705$, $v_6 = 1706$, $v_7 = 1215$, and $v_8 = 2305$.

Figure 11.71

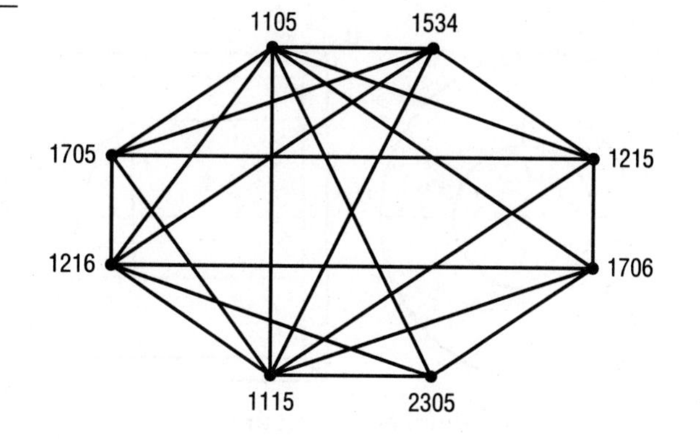

2. Color v_1 red.
3. The other vertices are all adjacent to v_1.
4. Color v_2 green.
6. Go to 4.
4. Color v_3 yellow. Also color v_7 yellow.
6. Go to 4.
4. Color v_4 blue. Also color v_6 and v_8 blue.
6. Go to 4.
4. Color v_5 orange.
5. Stop.

Since the chromatic number of the graph is 5, five time slots are needed to schedule the final exams. We can see which exams can be scheduled at the same time by the vertices that have the same color. Thus MATH 1215 and MATH 1216 can be scheduled at the same time, and MATH 1534, CS 1706, and PHY 2305 can be scheduled at the same time. ■

Exercises 11.10

104. Find the number of colors needed to color each of the following maps so that no two regions with a common border have the same color.

a.

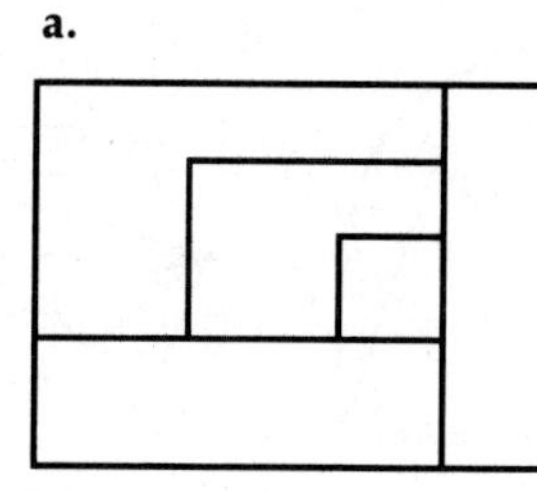

b.

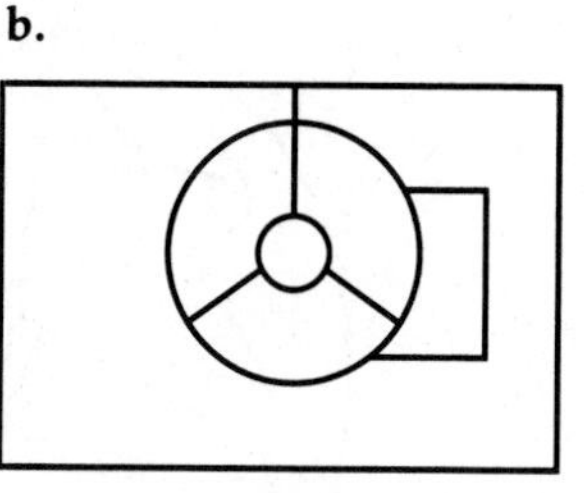

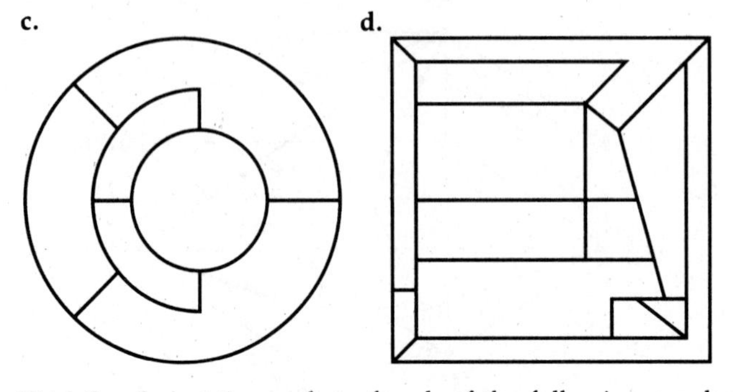

105. Find the chromatic number of each of the following graphs.

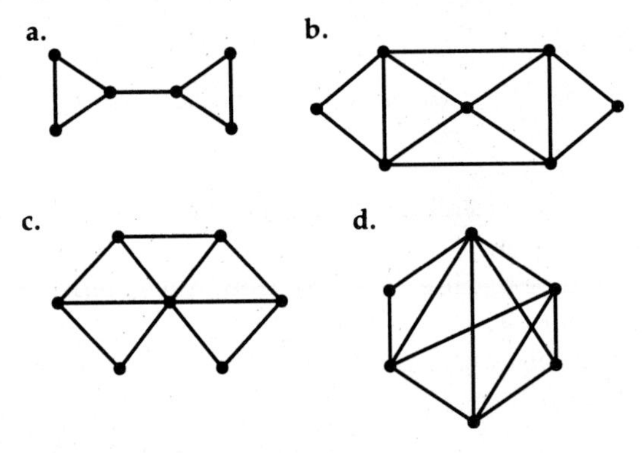

106. What is the chromatic number of C_n?

107. What is the chromatic number of W_n? (The definition of W_n can be found in Exercise 11.)

108. Use the Welsh and Powell algorithm to color the vertices of the following graphs.

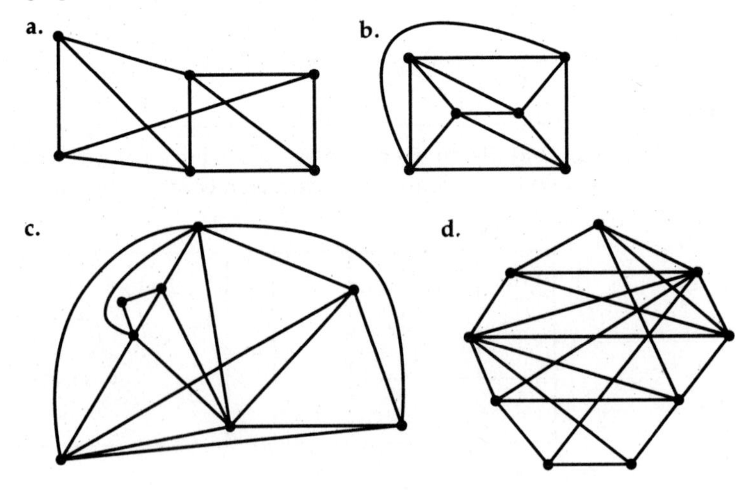

109. Give an example of a simple graph $G = (V,E)$ such that $\chi(G) = \max\{\deg(v): v \in V\} + 1$.

110. Give an example of a simple graph $G = (V,E)$ such that $\nu \geq 12$, $\deg(v) \geq 3$ for each $v \in V$, and $\chi(G) = 2$.

111. A student organization has seven committees, C_1, C_2, C_3, C_4, C_5, C_6, C_7, which meet once a week. Suppose the membership of each committee is as follows:

C_1 = {Lagoyda, Milewski, Mierke, Hadley} C_2 = {Chen, Adrien, Milewski}

C_3 = {Lagoyda, Patty, Wohlford} C_4 = {Mierke, Adrien, Hoyle}

C_5 = {Karabinus, Hadley, Wohlford} C_6 = {Chen, Patty, Kinter}

C_7 = {Hadley, Adrien}

Use graph coloring to give a schedule of the meetings so that no person has two meetings at the same time.

112. What is the complexity of the Welsh and Powell algorithm? Explain your answer.

113. During the execution of a computer program, fourteen stacks, $S_1, S_2, \ldots, S_{14}$, are to be constructed and stacks i and j are to be used at the same time if and only if $i \equiv j \pmod 3$ or $i \equiv j \pmod 4$. A location can store only one stack at a time. Use a graph to model this situation and find the minimum number of locations that are needed during the execution of this program.

114. Ten different tropical fishes, $F_1, F_2, \ldots, F_{10}$, are to be shipped. The problem is that F_1 feeds on F_4 and F_7; F_2 feeds on F_3, F_5, and F_7; F_4 feeds on F_7 and F_{10}; F_6 feeds on F_1 and F_3; F_8 feeds on F_3 and F_4; F_9 feeds on F_2, F_3, F_7, and F_{10}; and F_{10} feeds on F_3, F_5, and F_7. Thus two species, one of which feeds on the other, must be shipped in different aquariums. Model this problem as a graph-coloring problem and determine the smallest number of aquariums needed.

115. Suppose the edges of K_6 are painted red or green. Prove that there is a subgraph of K_6 that is a red triangle or a green triangle.

Chapter 11 Review Exercises

116. Let $V = \{a,b,c,d\}$ and $E = \{\{a,b\},\{a,c\},\{a,d\},\{c,d\}\}$. Draw a picture that represents the simple graph $G = (V,E)$.

117. Let $G = (V,E)$ be the simple graph represented by the following figure.

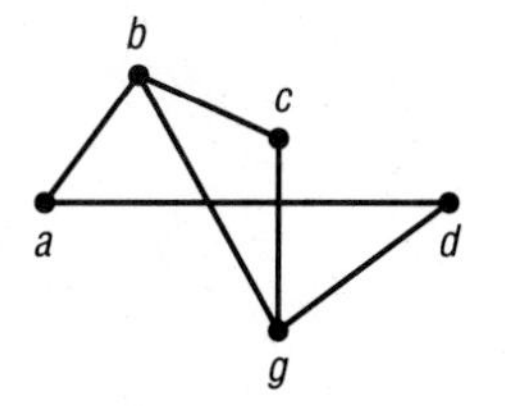

a. Find V.

b. Find E.

c. Find the degree of each vertex.

d. Which vertices are adjacent to b?

e. Which edges are incident with b?

118. Show that every complete simple graph is regular.

119. Find the number of vertices, the number of edges, and the degree of each vertex in each of the following representations of graphs. Identify the simple graphs and the graphs without loops.

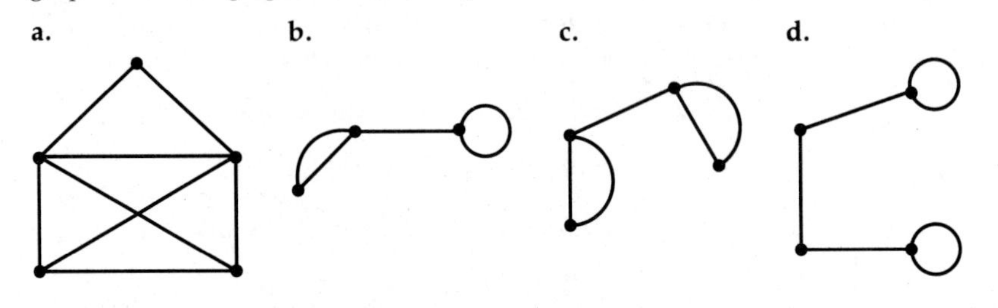

120. Find the sum of the degrees of the vertices of each graph in Exercise 119. Compare this number and the number of edges.

121. For each of the following graphs, determine if it is possible to begin a walk at vertex a, cross each edge exactly once, and end the walk at vertex a.

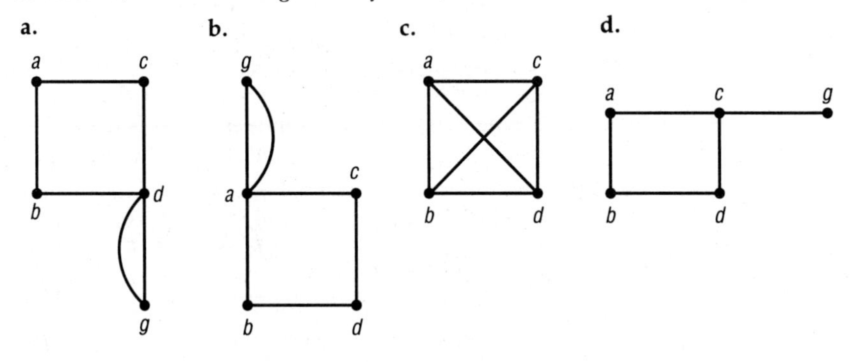

122. Let $G = (V,E,f)$ be a graph with five vertices. Suppose that the degrees of these vertices are 3, 2, 2, 2, and 1. How many edges does G have?

123. Why can there not be a graph on five vertices such that the degrees of these five vertices are 3, 2, 2, 1, and 1?

124. Find the underlying simple graph of the graph represented by the following picture.

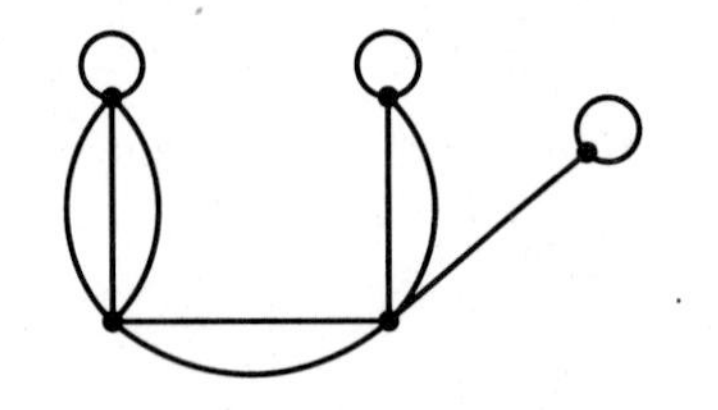

125. For each of the graphs represented by the following pictures, find the adjacency matrix and the incidence matrix using the natural order of the alphabet and natural numbers to determine the order of the rows and columns.

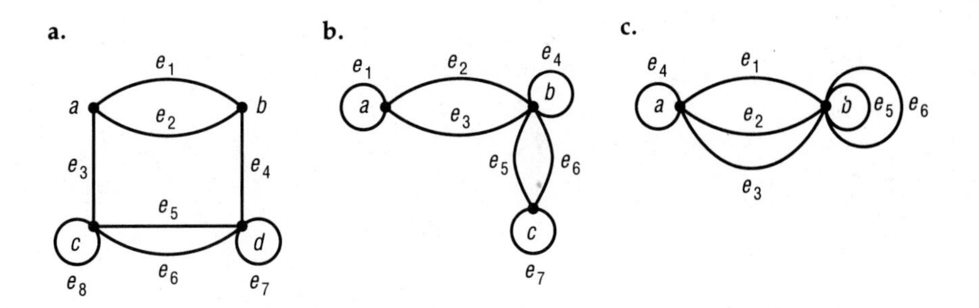

126. Draw a picture of the graph that is represented by each of the following adjacency matrices.

a.
$$\begin{array}{c} \\ a \\ b \\ c \end{array}\begin{array}{c} \begin{array}{ccc} a & b & c \end{array} \\ \begin{bmatrix} 0 & 1 & 1 \\ 1 & 3 & 0 \\ 1 & 0 & 0 \end{bmatrix} \end{array}$$

b.
$$\begin{array}{c} \\ a \\ b \\ c \end{array}\begin{array}{c} \begin{array}{ccc} a & b & c \end{array} \\ \begin{bmatrix} 2 & 1 & 1 \\ 1 & 2 & 1 \\ 1 & 1 & 2 \end{bmatrix} \end{array}$$

c.
$$\begin{array}{c} \\ a \\ b \\ c \end{array}\begin{array}{c} \begin{array}{ccc} a & b & c \end{array} \\ \begin{bmatrix} 2 & 2 & 2 \\ 2 & 0 & 0 \\ 2 & 0 & 0 \end{bmatrix} \end{array}$$

127. Draw a picture of the graph with the following incidence matrix.

$$\begin{array}{c} \\ a \\ b \\ c \\ d \end{array}\begin{array}{c} \begin{array}{ccccccc} e_1 & e_2 & e_3 & e_4 & e_5 & e_6 & e_7 \end{array} \\ \begin{bmatrix} 1 & 0 & 1 & 0 & 1 & 0 & 0 \\ 1 & 0 & 0 & 1 & 0 & 1 & 2 \\ 0 & 1 & 1 & 0 & 0 & 1 & 0 \\ 0 & 1 & 0 & 1 & 1 & 0 & 0 \end{bmatrix} \end{array}$$

128. Which of the following pairs of simple graphs are isomorphic?

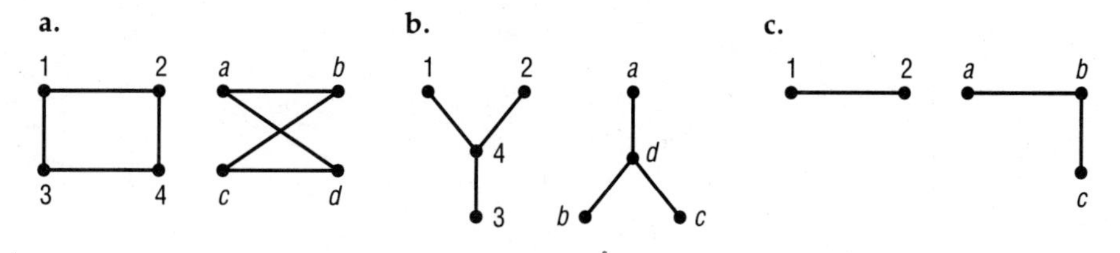

129. Show that the following two graphs are isomorphic.

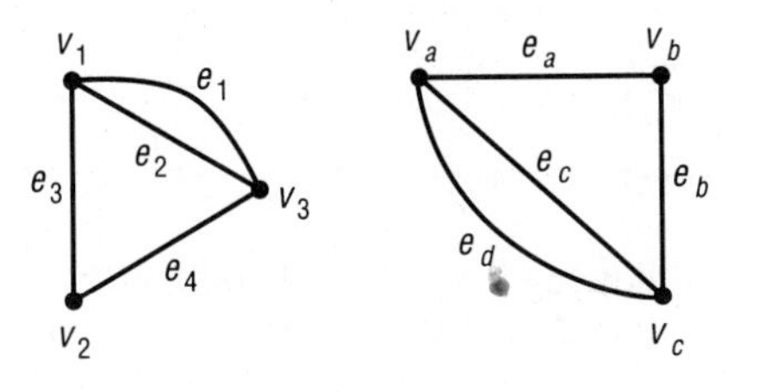

130. Find a property that shows that the following two graphs are not isomorphic.

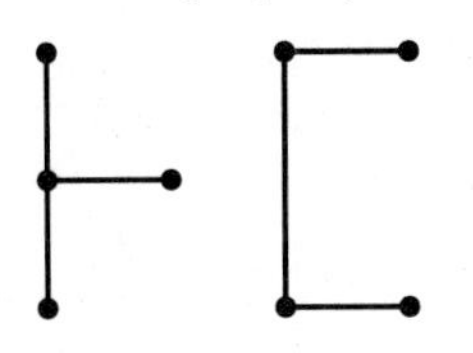

131. How many components does each of the following graphs have? Which graphs are connected?

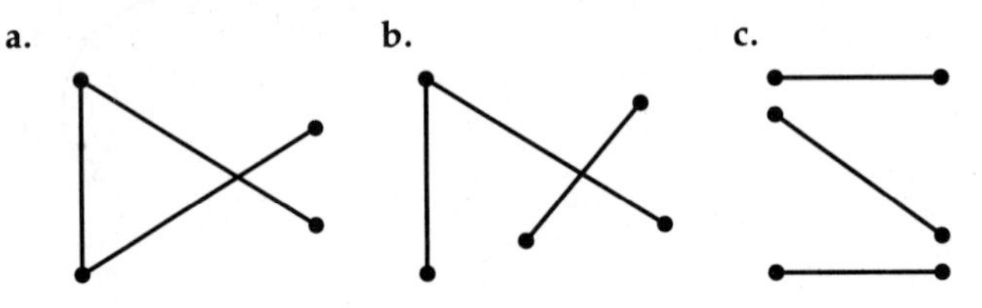

132. In the following graphs, find the number of paths between a and g of length less than or equal to 4. What is the shortest length of a path from a to g?

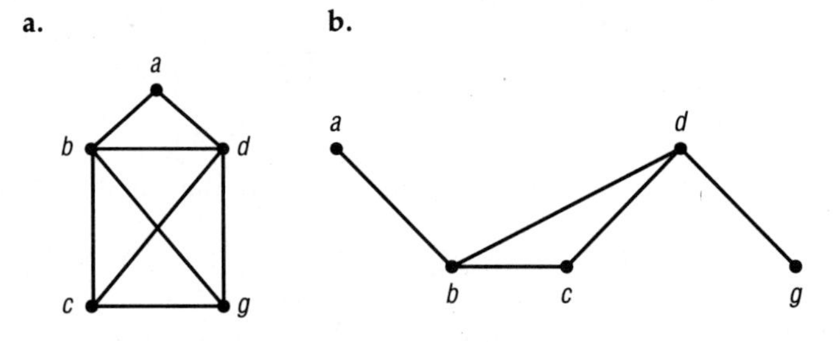

133. Determine whether each of the following graphs is eulerian. Construct an Euler tour when one exists.

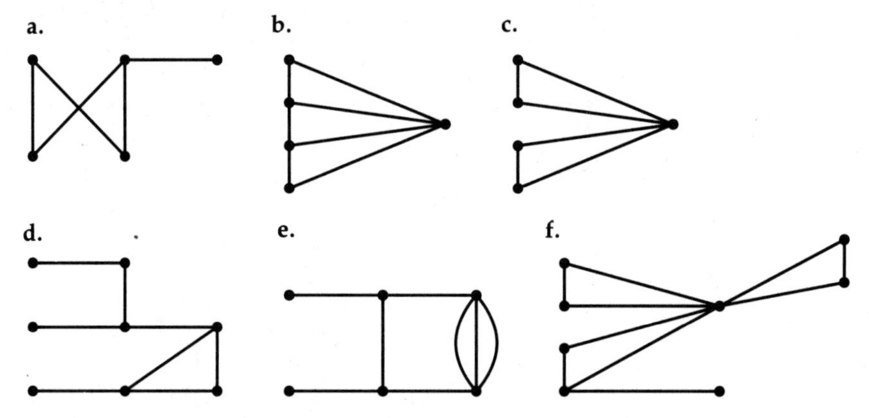

134. Which of the following graphs are hamiltonian. For those that are, find a Hamilton cycle. For those that are not, explain why?

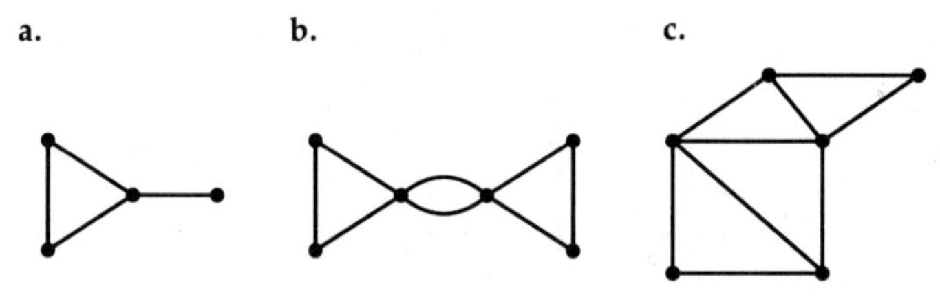

135. Use Dijkstra's algorithm to find the length of a shortest path from u_0 to all other vertices in each of the following weighted graphs.

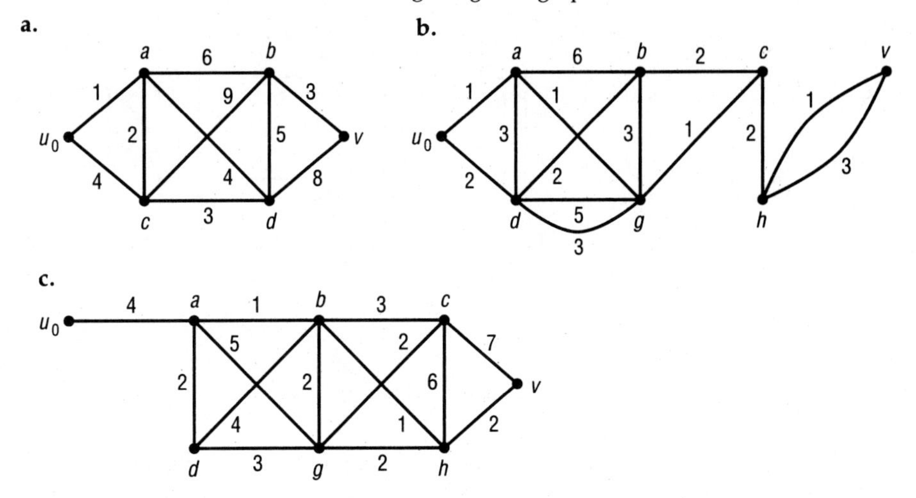

136. Use Warshall's algorithm to find the minimum weight of a path between each pair of vertices of the following weighted graph.

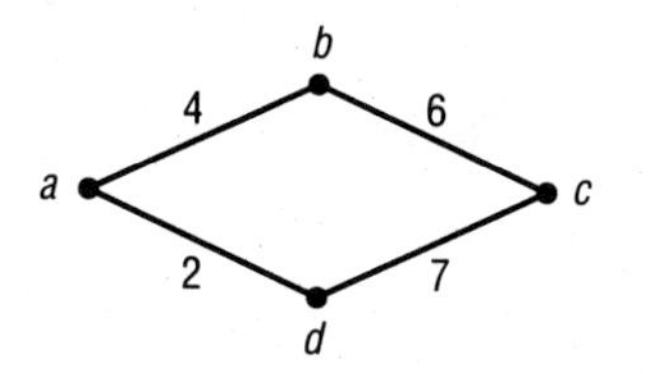

137. Draw planar graphs that represent

a. K_4 **b.** $K_{2,2}$ **c.** K_3 **d.** $K_{2,3}$

138. Draw planar graphs of the following.

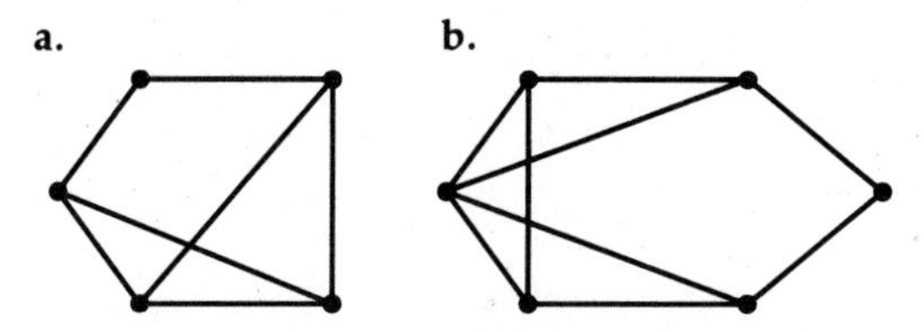

139. Which of the following graphs are planar?

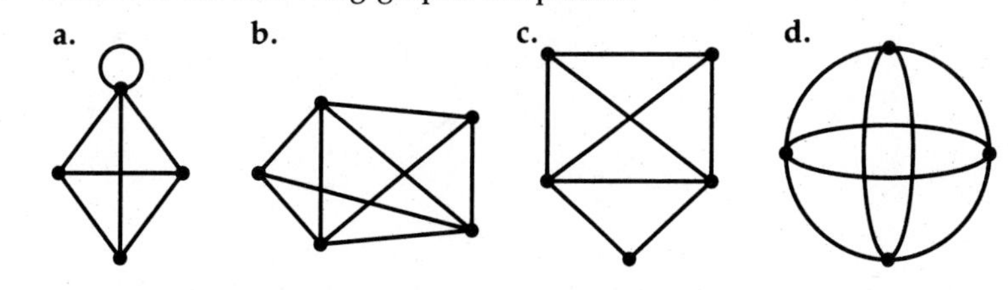

140. What is the minimum number of colors needed to color each of the following maps so that no two adjacent regions have the same color?

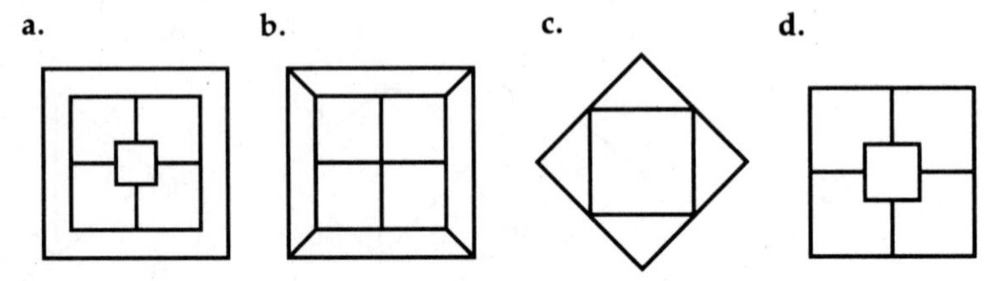

141. Find the chromatic number of each of the following graphs.

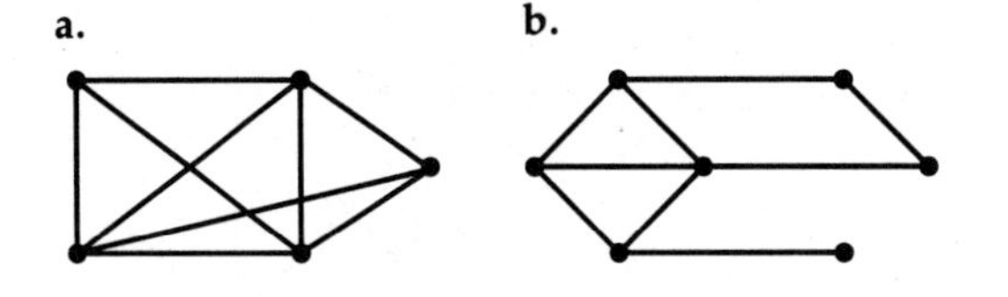

142. A faculty organization has five committees, C_1, C_2, C_3, C_4, and C_5, and there is one time slot from 11 to 12 o'clock on Thursday set aside for meetings of the committees. The membership of each committee is as follows:

C_1 = {Jones, Smith, Patty, Hoyle, Fletcher} C_2 = {Apple, Smith, Hoyle}

C_3 = {Jones, Frank, Patty, Fletcher} C_4 = {Frank, Chen, Adrien}

C_5 = {Harvey, Jones, Adrien}

Use graph coloring to determine the least number of weeks needed for all committees to meet once.

Chapter

12

TREES

As we indicated in the previous chapter, a tree is a connected graph with no cycles. Since a tree does not have cycles, it cannot contain loops or multiple edges. Therefore it is a simple graph. Trees may be the most important type of graph used in computer science. They are used in a variety of areas including sorting, searching, data organization, computer design, artificial intelligence, and estimation of errors in numerical calculations. One of the origins of the mathematical study of trees was the study of hydrocarbons. In 1857, Arthur Cayley studied these chemical compounds formed from hydrogen and carbon atoms.

12.1

BASIC PROPERTIES OF TREES

We begin this section by illustrating an application of trees.

Example 1

Suppose that we are planning to construct a new college, which will initially have six buildings. We want to plan a telephone network for the college. We can build a line between every pair of buildings, but this is not very efficient. It is important that any two buildings be linked, but it is not necessary that there be a direct link between any two particular buildings. If we represent the buildings by vertices of a graph and possible telephone lines by edges, then the complete graph on six vertices represents the possibilities for the telephone lines. We want to select a set of edges that gives us a telephone line between any two buildings, but we want the set as small as possible. One such set is the tree illustrated in Figure 12.1. ■

Figure 12.1

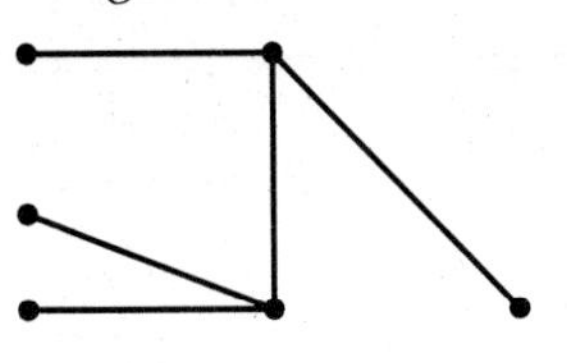

In Chapter 11, we promised to prove that if G is a tree then $\varepsilon = \nu - 1$. Our first goal is to prove this result, and we begin with an illustration.

Example 2

Draw the trees that have six vertices.

Analysis

We begin by constructing a tree none of whose vertices has degree greater than 2. Then we construct the trees with one vertex of degree 3 and continue in this manner until we have constructed all trees that have six vertices, as shown in Figure 12.2.

Figure 12.2

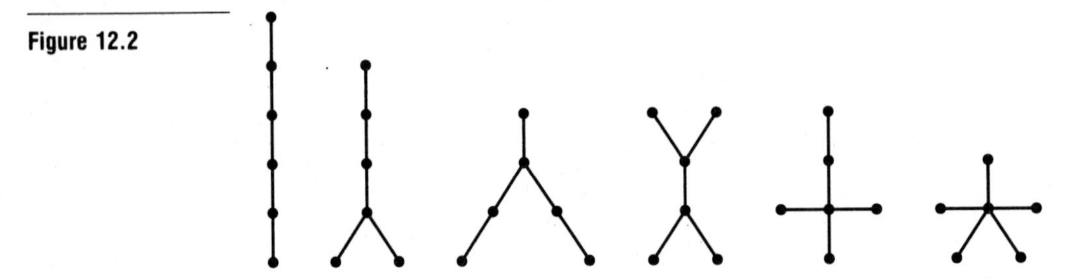

Notice that the number of edges of each tree is five. ■

We obtain the result promised in the previous chapter as a corollary of the following two theorems.

Theorem 12.1

> If $G = (V,E)$ is a tree and u and v are distinct members of V, then there is a unique (u,v)-path.

Proof

Since G is connected, if $u,v \in V$, there is a (u,v)-path. We need to prove that this path is unique. The proof is by contradiction. Suppose that G is a tree and that u and v are vertices of G with two distinct (u,v)-paths P_1 and P_2 in G. Since $P_1 \neq P_2$, there is an edge e of P_1 that is not a member of P_2. By Theorem 11.18, the subgraph $(P_1 \cup P_2) - \{e\}$ is connected. Hence it contains a (u,v)-path P. Consequently, $P \cup \{e\}$ is a cycle in G. This is a contradiction, since G is a tree. □

In Exercise 1 we ask you to label the vertices of each of the six trees in Figure 12.2, to choose two vertices u and v of degree 1 in each tree, and to construct the (u,v)-path.

Theorem 12.2

> Let $G = (V,E)$ be a simple graph. If for each pair u and v of distinct vertices of G there is a unique (u,v)-path, then G is connected and $\varepsilon = \nu - 1$.

Proof

Suppose that for each pair u and v of distinct vertices of G there is a unique (u,v)-path. Then it is clear that G is connected. The proof that $\varepsilon = \nu - 1$ is by induction on the number of vertices of G. Let

$$\begin{aligned} S = \{n \in \mathbb{N}\text{: } & \text{if } G \text{ is a simple graph with } n \text{ vertices such} \\ & \text{that for each pair } u \text{ and } v \text{ of distinct vertices of } G \\ & \text{there is a unique } (u,v)\text{-path, then } \varepsilon = \nu - 1\} \end{aligned}$$

If G is a simple graph with only one vertex, then G does not have any edges. Therefore $1 \in S$.

Suppose $n \in \mathbb{N}$ and $\{1,2, \ldots ,n\} \subseteq S$. Let $G = (V,E)$ be a simple graph with $n + 1$ vertices which also has the property that if u and v are distinct members of V then there is a unique (u,v)-path. Let $e = \{a,b\}$ be a member of E. Since aeb is the unique (a,b)-path, there is no (a,b)-path in $G - \{e\}$. Thus $G - \{e\}$ is not connected. By Theorem 11.14, $\omega(G - \{e\}) = 2$. Let G_1 and G_2 be the components of $G - \{e\}$. Then for each $i = 1,2$ and each pair of vertices c and d of G_i there is a unique (c,d)-path in G_i. (If there are two (c,d)-paths in G_i, then there are two (c,d)-paths in G.) Since for each $i = 1,2$ the number of vertices of G_i is less than or equal to n and $\{1,2, \ldots ,n\} \subseteq S$, $\varepsilon(G_i) = \nu(G_i) - 1$. Therefore

$$\begin{aligned}\varepsilon(G) &= \varepsilon(G_1) + \varepsilon(G_2) + 1\\ &= [\nu(G_1) - 1] + [\nu(G_2) - 1] + 1\\ &= \nu(G) - 1\end{aligned}$$

Hence $n + 1 \in S$, and by the second principle of mathematical induction $S = \mathbb{N}$. □

Corollary 12.3 If G is a tree, then $\varepsilon = \nu - 1$.

Proof

Let G be a tree. Then, by Theorem 12.1, for each pair u and v of distinct vertices of G there is a unique (u,v)-path. Thus, by Theorem 12.2, $\varepsilon = \nu - 1$. □

We pause at this point to give a simple example of a tree. The 1988 Major League playoffs and World Series results are shown in Figure 12.3.

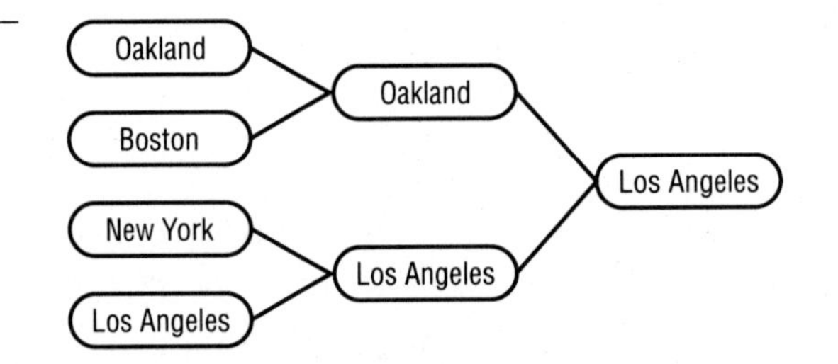

In the American League playoffs Oakland eliminated Boston, and in the National League playoffs Los Angeles eliminated New York. Then the Los Angeles Dodgers won the World Series by beating Oakland in a seven-game series. These playoffs can be represented by the tree shown in Figure 12.4a, which when rotated (Figure 12.4b) bears some resemblance to a tree with root v.

Figure 12.4

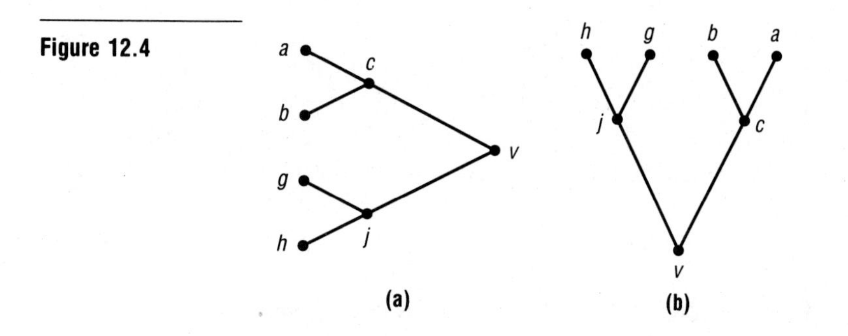

(a) (b)

Once a root r of a tree T has been designated, we can assign a direction to each edge of the tree as follows: for each vertex v ($v \neq r$) of T, there is a unique (r,v)-path, so we choose a direction for each edge by directing it away from the root.

Definition

The directed graph that is obtained from a tree by choosing a root and directing the edges so that all paths lead away from the root is called a **rooted tree.**

It is customary to draw rooted trees so that the top vertex is the root and all the arrows point down. In this case, we leave off the arrows. If we choose v to be the root of the tree in Figure 12.4b, the rooted tree that we obtain is shown in Figure 12.5 (with arrows in (a) and without arrows in (b)).

Figure 12.5

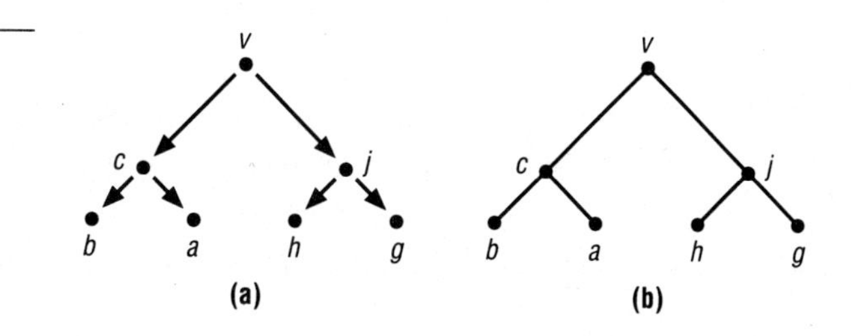

(a) (b)

The proof of the following theorem uses the result of Corollary 12.3 that if G is a tree then $\varepsilon = \nu - 1$.

Theorem 12.4 If $G = (V,E)$ is a nonempty tree, then G has at least two vertices of degree 1.

Proof

Let $G = (V,E)$ be a nonempty tree. Since G is connected and $E \neq \emptyset$, V has at least two members and the degree of each vertex is at least 1. By Theorem 11.1, $\Sigma\{\deg(v): v \in V\} = 2\varepsilon$. By Corollary 12.3, $2\varepsilon = 2\nu - 2$. So $\Sigma\{\deg(v): v \in V\}$ is two less than twice the number of vertices. Therefore there are at least two vertices of degree 1. □

Definition

Vertices of degree 1 of a tree are called **leaves.**

We now give a sufficient condition for a simple graph to be a tree.

Theorem 12.5 If G is a connected simple graph such that $\varepsilon = \nu - 1$, then G is a tree.

Proof

The proof is by contradiction. Let G be a connected simple graph such that $\varepsilon(G) = \nu(G) - 1$, and suppose that G contains a cycle C_1. Let e_1 be an edge in C_1. Then $G - \{e_1\}$ is connected, $\nu(G - \{e_1\}) = \nu(G)$, and $\varepsilon(G - \{e_1\}) = \varepsilon(G) - 1$. If $G - \{e_1\}$ contains a cycle C_2, choose an edge e_2 of C_2 and observe that $G - \{e_1,e_2\}$ is connected, $\nu(G - \{e_1,e_2\}) = \nu(G)$, and $\varepsilon(G - \{e_1,e_2\}) = \varepsilon(G) - 2$. If we continue this process, we eventually obtain a connected simple graph H with ν vertices that does not contain a cycle and has fewer than $\nu - 1$ edges. This is a contradiction because a connected simple graph without cycles is a tree and, by Corollary 12.3, $\varepsilon = \nu - 1$. □

Theorems 12.1, 12.2, and 12.5 yield the equivalence of the following statements in a simple graph $G = (V,E)$:

a. G is a tree.
b. If $u,v \in V$ and $u \neq v$, then there is a unique (u,v)-path.
c. G is connected and $\varepsilon = \nu - 1$.

Definition

Let G be a connected graph. A subgraph T of G is a **spanning tree** if T is a tree and T is a spanning subgraph of G.

Notice that if $T = (V_1,E_1)$ is a spanning tree of $G = (V_2,E_2,f_2)$ then $V_1 = V_2$.

Theorem 12.6 Every connected graph has a spanning tree.

Proof

Let $G = (V,E,f)$ be a connected graph. Suppose that $G' = (V',E')$ is a connected subgraph of G with the property that $V' = V$ and if any edge of G' is removed then G' is not connected. (Of necessity, G' is a simple graph because it is clear that loops and multiple edges can be removed from a connected graph without disconnecting the graph.) If G' contains a cycle C, then by Theorem 11.18 an edge e of C can be removed without disconnecting G'. Therefore G' cannot contain any cycles. Hence it is a tree, so it is clear that it is a spanning tree. □

Example 3

Find a spanning tree for the graph G in Figure 12.6.

Figure 12.6

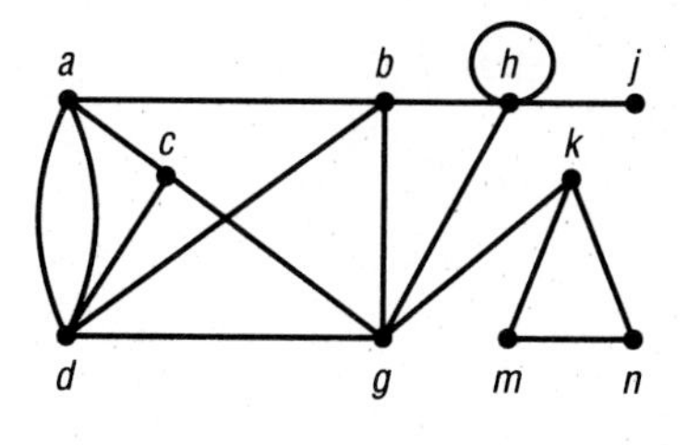

Analysis

We remove edges of G that are members of cycles until no cycles are left. First observe that the loop at h is a cycle, so we remove this loop. The multiple edges whose vertices are a and d constitute a cycle, so we remove one of these edges. Now $\{b,g\}\{g,h\}\{h,b\}$ is a cycle, so we remove $\{g,h\}$. Continuing, we can remove $\{b,d\}$, $\{a,c\}$, $\{d,g\}$, and $\{m,n\}$. Finally, we remove the remaining edge incident with a and d. We are left with the spanning tree shown in Figure 12.7.

Figure 12.7

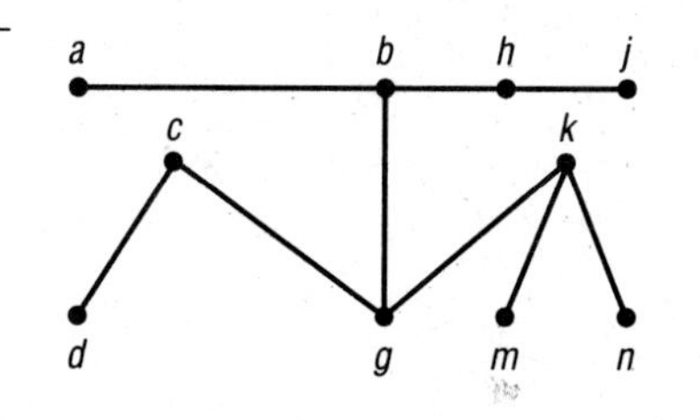

■

A graph may have more than one spanning tree. Indeed, in Exercise 8 we ask you to construct a spanning tree of the graph in Figure 12.6 different from that in Figure 12.7.

Example 4

An airline whose current schedule is represented by the graph in Figure 12.8 must reduce its flight schedule. Which flights can be discontinued and still retain service between each pair of cities?

Figure 12.8

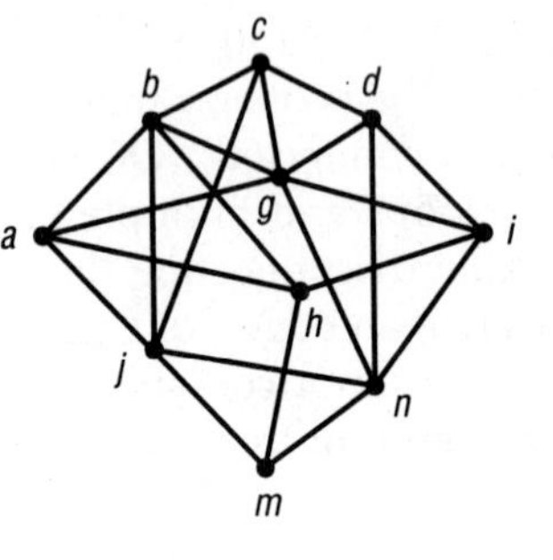

Analysis

We can answer this question by constructing a spanning tree (Figure 12.9) of the graph. All other flights can be discontinued.

Figure 12.9

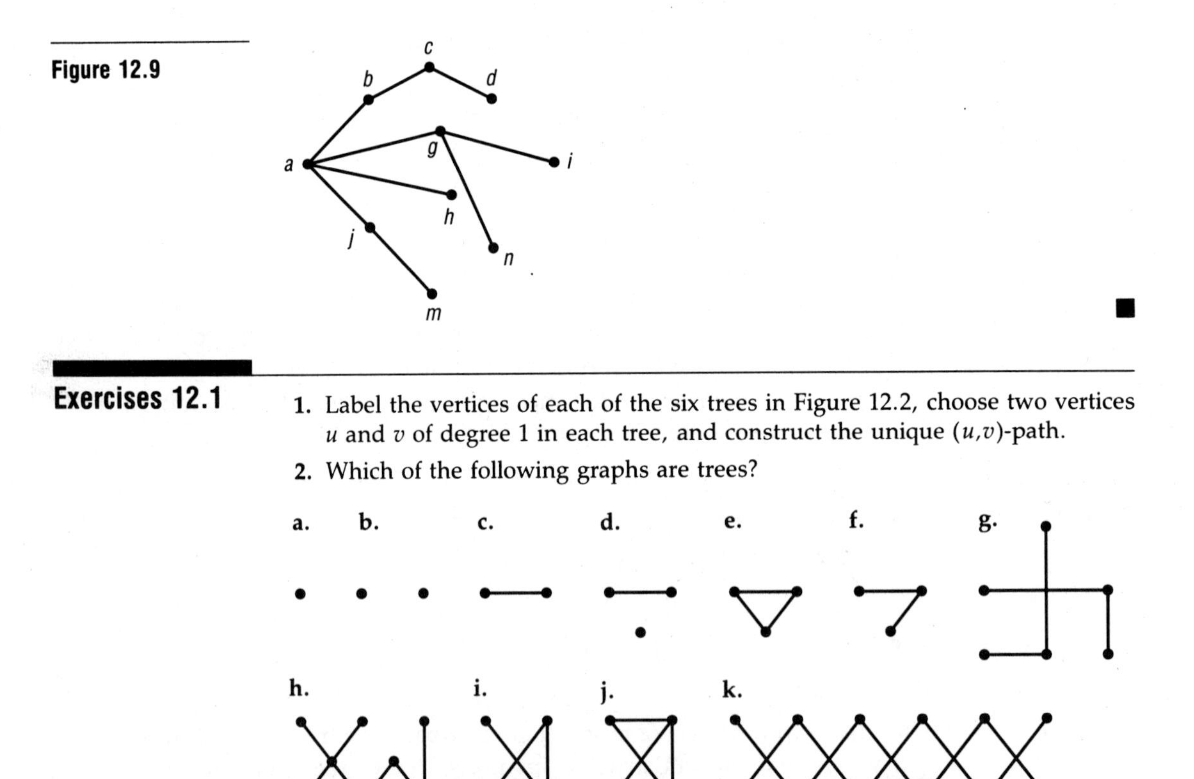

■

Exercises 12.1

1. Label the vertices of each of the six trees in Figure 12.2, choose two vertices u and v of degree 1 in each tree, and construct the unique (u,v)-path.
2. Which of the following graphs are trees?

a. b. c. d. e. f. g.

h. i. j. k.

3. How many edges are there in a tree with eleven vertices?
4. How many vertices are there in a tree with seventeen edges?
5. Draw the trees that have seven vertices.
6. How many vertices of degree 1 are there in a tree with five vertices of degree 3, two vertices of degree 2, and no vertices of degree more than 3?
7. How many paths are there in a tree with six vertices?
8. Construct a spanning tree for the graph in Figure 12.6 that is different from the one in Figure 12.7.
9. Construct a spanning tree for the following graph.

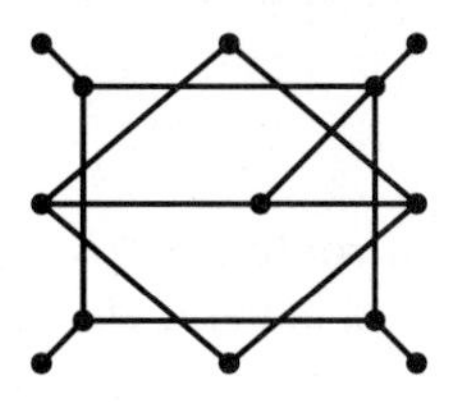

10. If $n \geqslant 2$, what is the smallest number of colors needed to color a tree with n vertices?
11. If $n \geqslant 2$, how many paths are there in a tree with n vertices?
12. Let T be a tree. Prove that if $k = \max\{\deg(v)\colon v \text{ is a vertex of } T\}$ then there are at least k vertices of degree 1.
13. For which m and n is $K_{m,n}$ a tree?

12.2 ROOTED TREES

In Exercise 21 of Chapter 5 we asked you to apply a binary search to the sorted list of natural numbers 1,2,3, . . . ,14,15. This binary search can be pictured as a rooted tree, as shown in Figure 12.10.

Figure 12.10

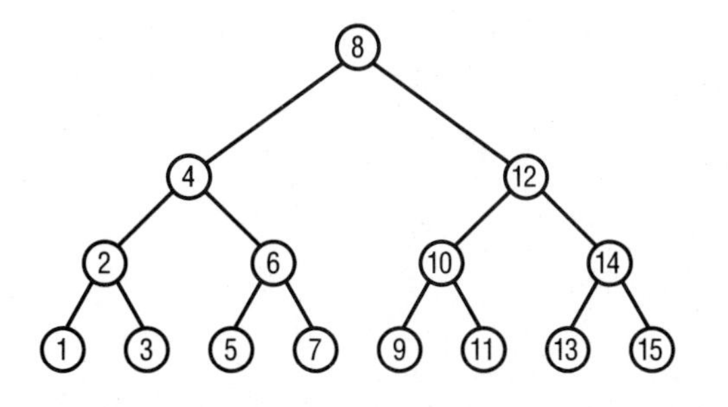

A given number p in the list 1,2,3, . . . ,14,15 is first compared to 8. If $p = 8$, the search is over. If $p < 8$, then it is compared to 4. If $p > 8$, then it is compared to 12. One proceeds in this manner down the tree until p is found.

In the preceding section we defined a leaf of a tree to be a vertex of degree 1. There is one exception to this in rooted trees. We do not call the root a leaf even when it has degree 1. The origins of the terminology for trees are genealogical as well as botanical.

Definition

If uv is a directed edge in a rooted tree, then u is a **parent** of v and v is a **child** of u. An **internal vertex** of a rooted tree is a vertex that has children.

Notice that every vertex of a rooted tree except the root has exactly one parent, but a vertex may have several children.

Definition

If u and v are vertices of a rooted tree, then v is a **descendant** of u provided that $v \neq u$ and u is a vertex of the unique path from the root r to v. Distinct vertices with the same parent are called **siblings.** If u is a vertex of a rooted tree, the **subtree with root *u*** is the tree consisting of u, the descendants of u, all the directed edges from u to the descendants of u, and all the directed edges from one descendant of u to another.

Example 5

Find the parent of c, the children of c, the siblings of c, the descendants of c, and the subtree with root c in the rooted tree illustrated in Figure 12.11.

Figure 12.11

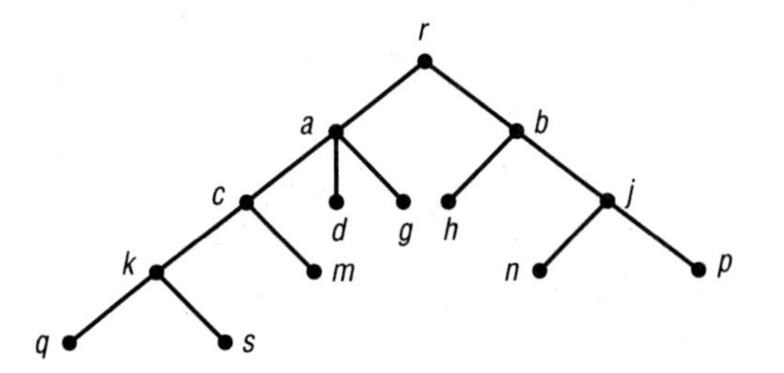

Analysis

The parent of c is a. The children of c are k and m. The siblings of c are d and g, and the subtree with root c is shown in Figure 12.12.

Figure 12.12

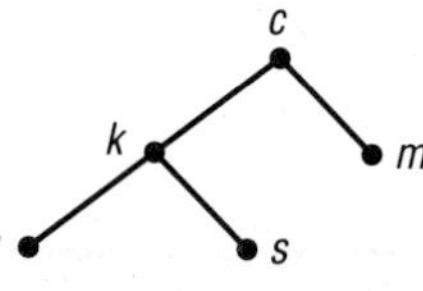

■

Definition

If $\overrightarrow{uv}$ is a directed edge in a directed graph, then u is called the **initial vertex** and v is called the **terminal vertex** of this directed edge. If v is a vertex of a directed graph G, the **indegree** of v is the number of directed edges of G that have v as terminal vertex and the **outdegree** of v is the number of directed edges of G that have v as initial vertex.

Example 6

Find the indegree and the outdegree of each vertex of the directed graph in Figure 12.13.

Figure 12.13

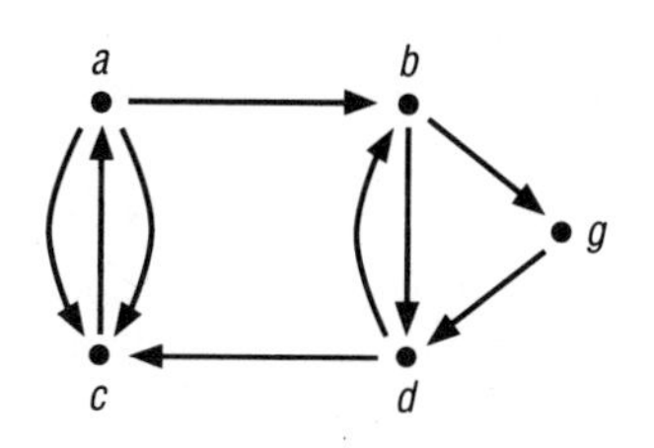

Analysis

The indegree of a is 1, the indegree of b is 2, the indegree of c is 3, the indegree of d is 2, and the indegree of g is 1. The outdegree of a is 3, the outdegree of b is 2, the outdegree of c is 1, the outdegree of d is 2, and the outdegree of g is 1. ■

Definition

A rooted tree in which each vertex with positive outdegree corresponds to a decision is called a **decision tree.** The subtree with root at each such vertex represents a **possible outcome** of the decision.

Example 7

The chain of command in a company can be modeled by a rooted tree. Each position in the company is represented by a vertex, and an edge from one vertex to another indicates that the person represented by the initial vertex is the direct supervisor of the person represented by the terminal vertex (see Figure 12.14).

Figure 12.14

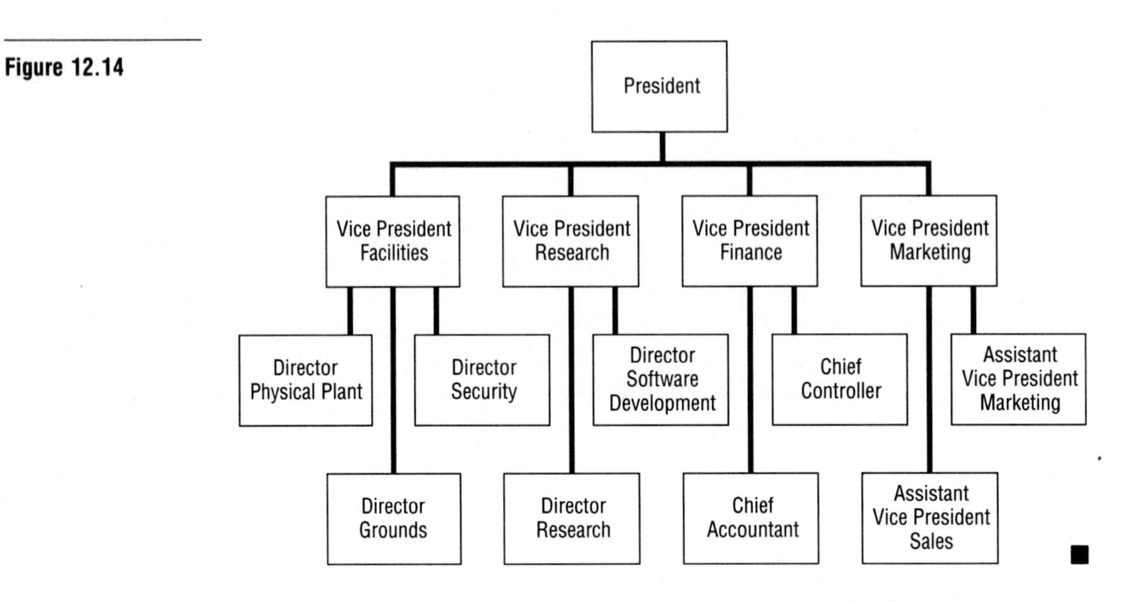

The following theorem gives a characterization of rooted trees.

Theorem 12.7

A directed tree T is a rooted tree if and only if T has one vertex of indegree 0 and all the other vertices of T are of indegree 1.

We do not give a proof of Theorem 12.7, but observe that a tree is rooted if and only if it has a root r whose indegree is necessarily 0. Given any other vertex v, there is a unique path from r to v, and this unique path contains the unique directed edge with terminal vertex v.

Definition

Let $m \in \mathbb{N}$. A rooted tree T is an ***m*-ary tree** if the outdegree of each of its vertices is less than or equal to m. The 2-ary trees are called **binary trees.** A **regular *m*-ary tree** is an m-ary tree such that the outdegree of each of its vertices that is not a leaf is m. Similarly, a **regular binary tree** is a binary tree such that the outdegree of each of its vertices that is not a leaf is 2.

Observe that the rooted tree in Figure 12.10 is a regular binary tree.

Example 8

For which natural numbers m is the rooted tree T in Figure 12.15 an m-ary tree?

Figure 12.15

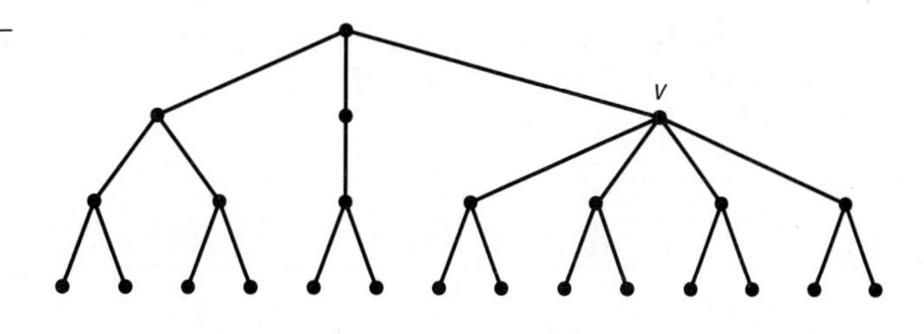

Analysis

The outdegree of v is 4, and the outdegree of any other vertex is less than 4. Therefore T is a 4-ary tree. It is also the case that if $m > 4$ then T is an m-ary tree. ■

Example 9

Construct a regular 3-ary tree.

Analysis

We must construct a rooted tree in which the outdegree of each vertex is 0 or 3. One such tree is constructed in Figure 12.16.

Figure 12.16

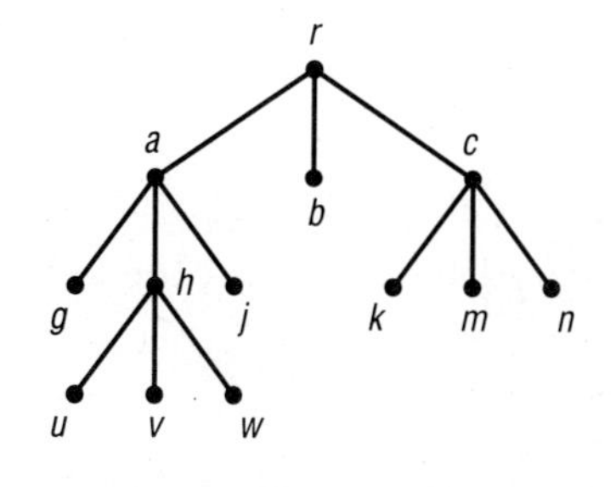

■

Definition

Let T be a rooted tree with root r. The **level number** of r is 0, and the **level number** of a vertex v ($v \neq r$) of T is the length of the unique path from r to v. The **height** of T is the maximum of the level numbers of the vertices of T.

Example 10

Find the level number of each vertex of the rooted tree in Figure 12.16.

Analysis

The level number of r is 0. The level number of a, b, and c is 1. The level number of g, h, j, k, m, and n is 2, and the level number of u, v, and w is 3. Therefore the height of T is 3. ■

Observe that, if v is a vertex of a rooted tree T whose level number is the height of T, then v is a leaf.

Theorem 12.8

Let $m \in \mathbb{N}$. If T is an m-ary tree of height h, then T has at most m^h leaves.

Proof

The proof uses the second principle of mathematical induction. Let $S = \{h \in \mathbb{N}$: if T is an m-ary tree of height h, then T has at most m^h leaves$\}$. Let T be an m-ary tree of height 1 with root r. Then each vertex v $(v \neq r)$ of T is a child of r, and r has at most m children. Therefore T has at most $m = m^1$ leaves. Thus $1 \in S$.

Suppose $h \in \mathbb{N}$ and $\{1,2, \ldots ,h\} \subseteq S$. Let T be an m-ary tree of height $h + 1$ with root r. Let $v_1,v_2, \ldots ,v_n$ denote the children of r. Since T is m-ary, $n \leqslant m$. For each $i = 1,2, \ldots ,n$, let T_i be the subtree of T with root v_i. Then, for each $i = 1,2, \ldots ,n$, T_i is m-ary and the height h_i of T_i is less than or equal to h. Since $h_i \in S$ for each $i = 1,2, \ldots ,n$, T_i has at most m^{h_i} leaves. Since $m^{h_i} \leqslant m^h$, the number of leaves of T is at most $nm^h \leqslant mm^h = m^{h+1}$. Therefore $h + 1 \in S$, and by the second principle of mathematical induction, $S = \mathbb{N}$. □

Definition

Let $m \in \mathbb{N}$. A **full m-ary tree (full binary tree)** is a regular m-ary tree (regular binary tree) such that the level number of each leaf is the height of the tree.

The rooted tree in Figure 12.15 is a full 4-ary tree, whereas the regular 3-ary tree in Figure 12.16 is not full since there are leaves whose level numbers are less than 3.

Theorem 12.9

Let $m \in \mathbb{N}$. If T is a full m-ary tree of height h, then T has m^h leaves, $(m^h - 1)/(m - 1)$ parents, and $(m^{h+1} - 1)/(m - 1)$ vertices.

Proof

Let T be a full m-ary tree of height h with root r. Since T is regular m-ary, there are m vertices, $v_1,v_2, \ldots ,v_m$, with level number 1. Likewise for each $i = 1,2, \ldots ,m$, v_i has m children. Thus there are m^2 vertices with level number 2. Continuing in this manner, we can prove by induction that for each $n \leq h$ the number of vertices with level number n is m^n. Hence the number of leaves of T is m^h and the total number of vertices of T is $1 + m + m^2 + \cdots + m^h$. By Example 6 of Chapter 4,

$$\sum_{i=0}^{h} m^i = \frac{m^{h+1} - 1}{m - 1}$$

Finally, the number of parents is

$$\frac{m^{h+1} - 1}{m - 1} - m^h = \frac{m^{h+1} - 1 - m^h(m - 1)}{m - 1}$$

$$= \frac{m^{h+1} - 1 - m^{h+1} + m^h}{m - 1} = \frac{m^h - 1}{m - 1}$$

□

An *ordered rooted tree* is a rooted tree in which the children of each parent are linearly ordered. Ordered rooted trees are drawn so that the children of each parent are shown in order from left to right. When drawing an ordered binary tree, we draw a left child below and to the left of its parent and a right child below and to the right of its parent.

Definition

If a parent in an ordered binary tree T has two children, the first child is called the **left child** and the second is called the **right child.** The subtree of T whose root is the left child u of a parent is called the **left subtree of** u**,** and the subtree of T whose root is the right child v of a parent is called the **right subtree of** v**.**

Example 11

Find the left and right children of a and the left and right subtrees of b in the rooted tree in Figure 12.17.

Figure 12.17

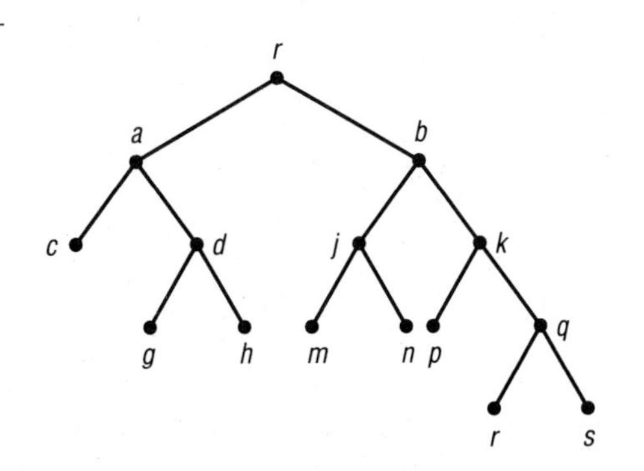

Analysis

The left child of a is c, and the right child of a is d. The left subtree of b is shown in Figure 12.18a and the right subtree of b is shown in Figure 12.18b.

Figure 12.18

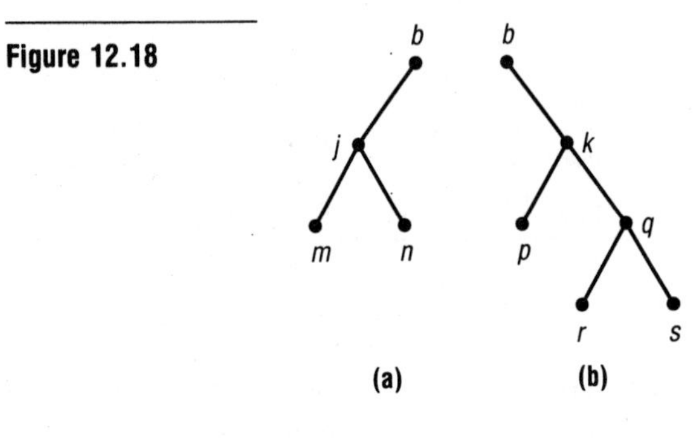

■

For purposes of organizing and storing information, it is convenient to label the vertices of an ordered m-ary tree. One method of doing this is to let $\Sigma = \{0,1,2, \ldots ,m - 1\}$ and use words from Σ^*. The ordered children of the root are labeled $0,1,2, \ldots$, and the ordered children of i (i is the label for one of the children of the root) are labeled $i0,i1,i2, \ldots$. In general, if w is a label for a vertex, then the ordered children of w are labeled $w0,w1,w2, \ldots$.

Example 12

Use the procedure that we have just explained to label the vertices of the ordered rooted tree shown in Figure 12.19.

Figure 12.19

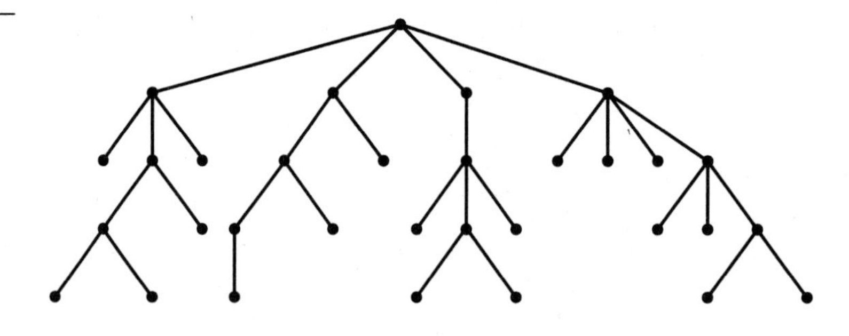

Analysis

All vertices except the root must be labeled. The ordered rooted tree shown in Figure 12.19 is a 4-ary tree. Therefore we use words from Σ^*, where $\Sigma = \{0,1,2,3\}$. There are four children of the root, so we label these four children 0, 1, 2, and 3. Then we label the children of these four children (the grandchildren of the root) 00, 01, 02, 10, 11, 20, 30, 31, 32, and 33. We continue in this manner and label the remaining children. The result is shown in Figure 12.20.

Figure 12.20

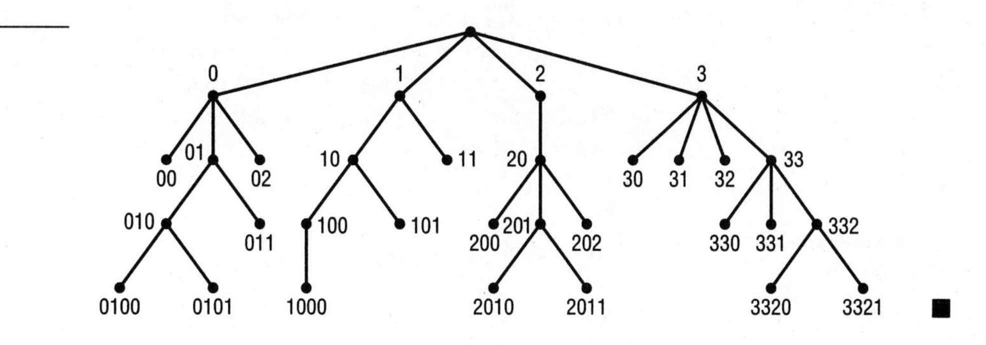

■

The method of labeling in Example 12 permits us to readily locate a particular vertex. For example, by Theorem 12.8, a 3-ary tree of height 8 may have as many as $3^8 = 6{,}561$ leaves. In spite of this, we can retrieve information stored at one of these leaves in eight steps.

Example 13

Let T be an ordered 3-ary tree of height 8. Describe the exact location of the vertex labeled 021102.

Analysis

We can easily locate the vertex labeled 021102 by drawing the subtree of T shown in Figure 12.21. We can describe the location of this vertex. It is the third child of the vertex labeled 02110, which in turn is the first child of the vertex labeled 0211. The latter child is the second child of the vertex labeled 021, which in turn is the second child of the vertex labeled

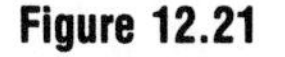

Figure 12.21

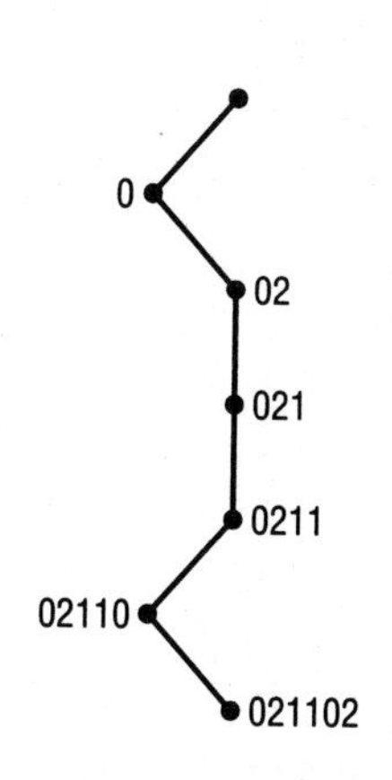

02. The latter child is the third child of the vertex labeled 0, which in turn is the first child of the root. ■

Definition

The **Fibonacci trees** T_n are defined recursively as follows: T_1 and T_2 are rooted trees consisting of a single vertex. For each $n \geqslant 3$, T_n is the rooted tree with root r such that r has left child u and right child v, T_{n-1} is the left subtree of u, and T_{n-2} is the right subtree of v.

Example 14

Draw the Fibonacci trees T_3 and T_4.

Analysis

The Fibonacci tree T_3 has root r with two children, and these children are T_2 and T_1. The Fibonacci tree T_4 has root r with left child u and right child v, u is the root of T_3, and v is the root of T_2.

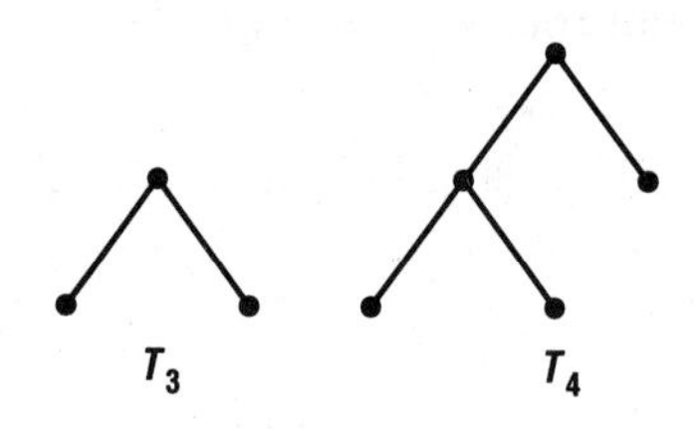

■

Exercises 12.2

14. a. Draw the following tree so that r is the root.

b. What is the level number of vertex c?

c. What is the height of the tree?

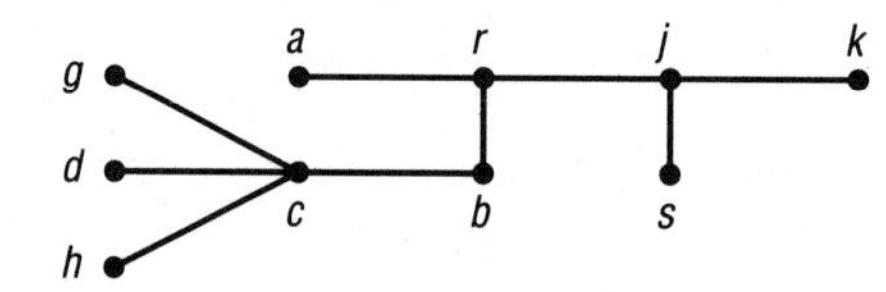

15. For each of the following rooted trees, find the level numbers of the vertices and the height of the tree.

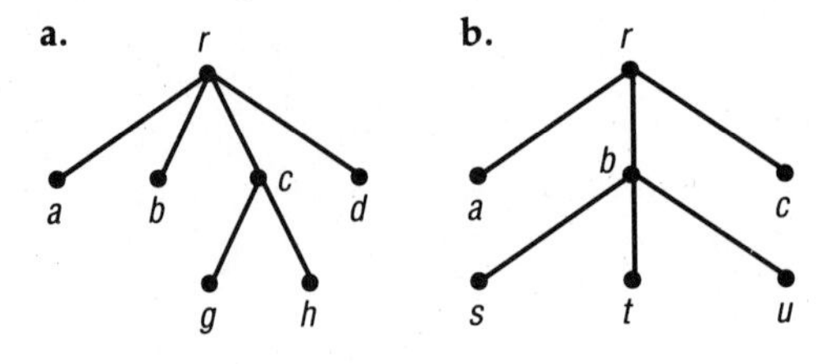

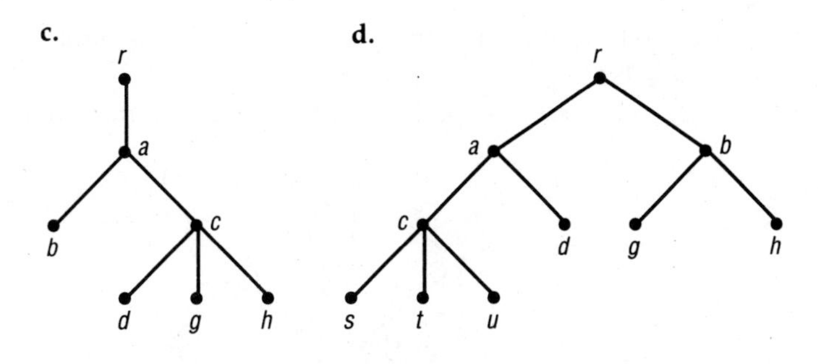

16. **a.** Draw a 3-ary tree of height 4 that is neither a regular 3-ary tree nor a binary tree.
    b. How many leaves does this tree have?
    c. How many parents does it have?
    d. How many vertices does it have?
17. **a.** Draw a regular 3-ary tree of height 2 that is not a full 3-ary tree.
    b. How many leaves does this tree have?
    c. How many parents does it have?
    d. How many vertices does it have?
18. **a.** Draw a full 3-ary tree of height 2.
    b. How many leaves does this tree have?
    c. How many parents does it have?
    d. How many vertices does it have?
19. Is it possible to draw a 3-ary tree of height 2 with eight leaves? Explain your answer.
20. Are the following rooted trees regular m-ary trees for some natural number m? If the answer is yes, what are the possible values for m?

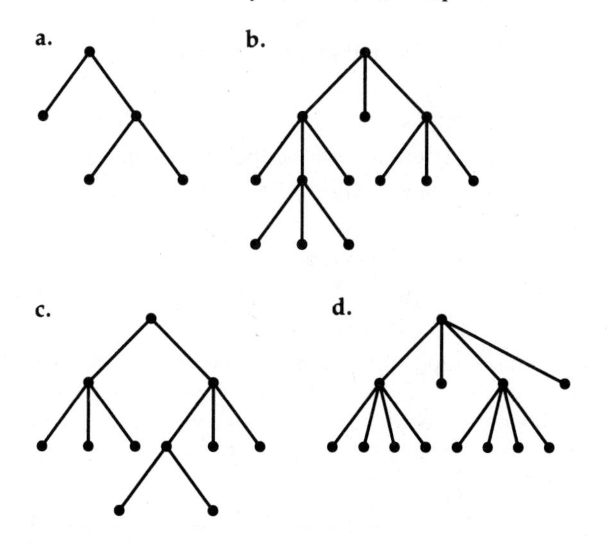

21. Which (if any) of the trees in Exercise 20 are full m-ary trees? Explain your answer.

22. The questions and tasks in this exercise pertain to the following rooted tree.

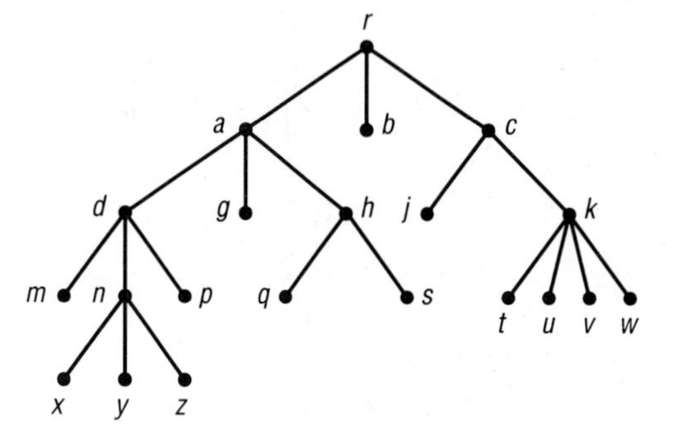

a. Which vertex is the root?

b. Which vertices are leaves?

c. Which vertices are parents?

d. Which vertices are siblings of d?

e. Which vertices are children of d?

f. Which vertices are descendants of d?

g. Draw the subtree with root d.

h. Draw the subtree with root a.

i. What is the level number of each vertex?

j. What is the height of the tree?

23. Draw the Fibonacci trees T_5, T_6, and T_7.

24. Let $n \in \mathbb{N}$ and let T_n be the nth Fibonacci tree.

a. What is the height of T_n? Explain your answer.

b. How many vertices does T_n have? Explain your answer.

c. How many leaves does T_n have? Explain your answer.

25. Let T be a tree. A *center* of T is a vertex v such that the choice of v as a root yields a rooted tree of minimal height. Show that T can have at most two centers. (*Hint:* Prove that if u and v are distinct centers of T then u and v are adjacent.)

26. Let T be a regular m-ary tree. Prove the following propositions.

a. If T has i internal vertices, then T has $mi + 1$ vertices.

b. If T has i internal vertices, then T has $(m - 1)i + 1$ leaves.

c. If T has λ leaves, then T has $(\lambda - 1)/(m - 1)$ internal vertices and $(m\lambda - 1)/(m - 1)$ vertices.

d. If T has n vertices, then T has $(n - 1)/m$ internal vertices and $[n(m - 1) + 1]/m$ leaves.

27. An organization consists of 261 people. A telephone chain is set up so that a leader calls a given set of five people. Each of these five calls a given set of five other people. This process continues until all people are called. How many people have to make calls for all people to be called? (*Hint:* Use Exercise 26.)

28. Let G be a simple graph that is a disjoint union of p trees. If G has n vertices, how many edges does it have?

LABELED TREES AND SPANNING TREES

The two labeled trees illustrated here are distinct because $\{b,c\}$ is an edge of one but not the other:

a b c   c a b

Example 15

Find the number of distinct trees with three labeled vertices.

Analysis

Using the six different arrangements of the letters a, b, and c, we can label trees with three vertices as in Figure 12.22.

Figure 12.22

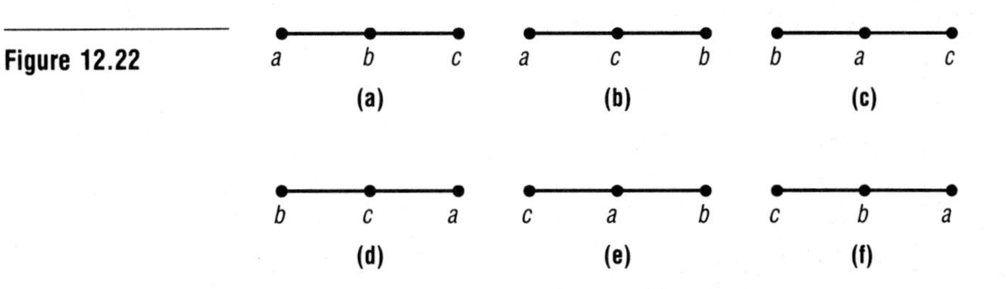

Trees (a) and (f) have the same edges, $\{a,b\}$ and $\{b,c\}$, so they are the same labeled tree. Likewise, trees (b) and (d) are the same labeled tree, and (c) and (e) are the same. Thus there are three distinct trees with three labeled vertices. ■

The problem of finding the distinct trees with four labeled vertices is more complicated. There is, however, an algorithm that allows us to count the number of distinct trees with n ($n \in \mathbb{N}$) labeled vertices. This algorithm, known as Prüfer's algorithm, establishes a one-to-one correspondence between trees with labeled vertices $1,2,\ldots,n$ and a list $a_1,a_2,\ldots,a_{n-2}$, where $1 \leq a_i \leq n$ for each $i = 1,2,\ldots,n-2$. The algorithm itself tells us how to get this list from a tree whose vertices are labeled $1,2,\ldots,n$ ($n > 2$). We summarize Prüfer's algorithm as follows:

Prüfer's Algorithm

1. Let T be a given labeled tree and let $k = 1$.
2. Select the vertex u of degree 1 in T that has the smallest label.

3. Let e denote the edge that is incident with u, let v be the vertex adjacent to u, and let a_k be the label on v.
4. Delete edge e and vertex u from T and let T' denote the resulting tree.
5. If T' has exactly two vertices, stop. The list $a_1, a_2, \ldots, a_{n-2}$ is the desired list.
6. Replace T by T', replace k by $k + 1$, and go to step 2.

Prüfer's algorithm is named in honor of Heinz Prüfer, who in 1918 published the first explicit correspondence demonstrating a connection between labeled trees and a list.

Example 16

Use Prüfer's algorithm to find the list a_1, a_2, a_3, a_4, a_5 associated with the tree in Figure 12.23.

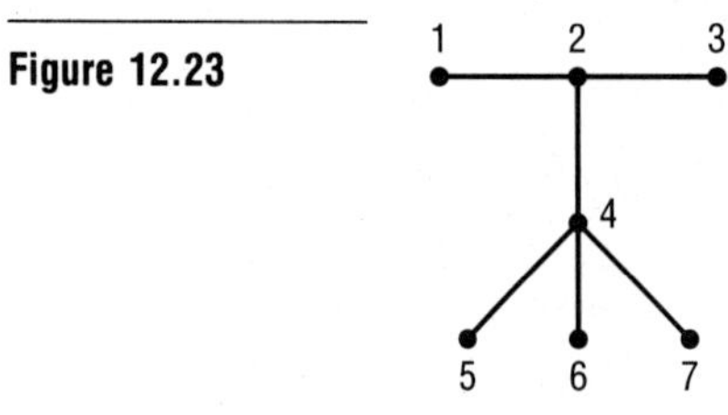

Figure 12.23

Analysis

1. Let T be the tree shown in Figure 12.23 and let $k = 1$.
2. u is the vertex labeled 1.
3. $e = \{1,2\}$, so $a_1 = 2$.
4. T' is the tree shown in Figure 12.24.

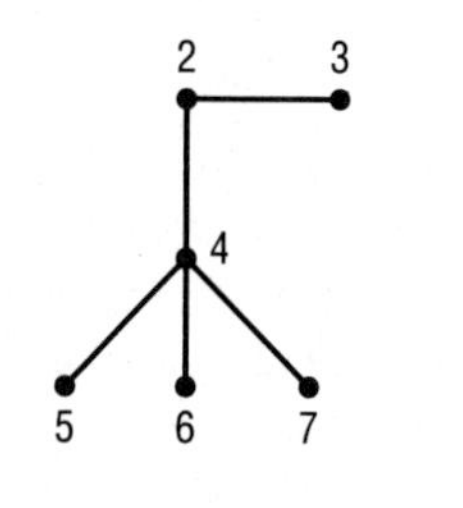

Figure 12.24

5. T' has more than two vertices.
6. Replace T by T', let $k = 2$, and go to step 2.
2. u is the vertex labeled 3.
3. $e = \{3,2\}$, so $a_2 = 2$.
4. T' is the tree shown in Figure 12.25.

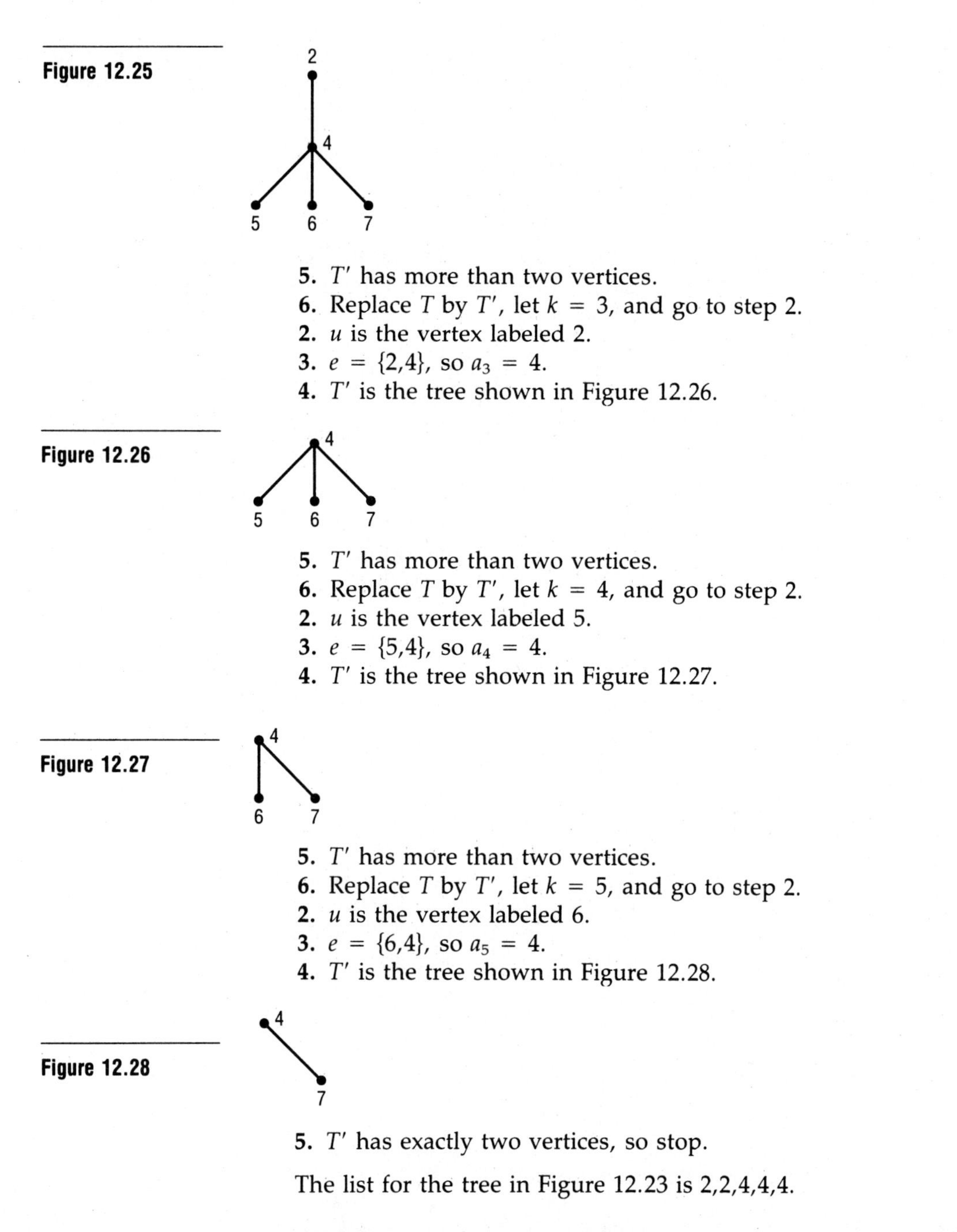

Figure 12.25

5. T' has more than two vertices.
6. Replace T by T', let $k = 3$, and go to step 2.
2. u is the vertex labeled 2.
3. $e = \{2,4\}$, so $a_3 = 4$.
4. T' is the tree shown in Figure 12.26.

Figure 12.26

5. T' has more than two vertices.
6. Replace T by T', let $k = 4$, and go to step 2.
2. u is the vertex labeled 5.
3. $e = \{5,4\}$, so $a_4 = 4$.
4. T' is the tree shown in Figure 12.27.

Figure 12.27

5. T' has more than two vertices.
6. Replace T by T', let $k = 5$, and go to step 2.
2. u is the vertex labeled 6.
3. $e = \{6,4\}$, so $a_5 = 4$.
4. T' is the tree shown in Figure 12.28.

Figure 12.28

5. T' has exactly two vertices, so stop.

The list for the tree in Figure 12.23 is 2,2,4,4,4. ■

We have seen how to start with a tree whose vertices are labeled $1,2, \ldots ,n$ $(n > 2)$ and construct a sequence $a_1,a_2, \ldots ,a_{n-2}$ such that

$1 \leqslant a_i \leqslant n$ for each i. We consider the opposite direction and give an algorithm to construct a tree whose vertices are labeled $1,2,\ldots,n$ from a list L of $n - 2$ natural numbers, where each member of L is less than or equal to n. We refer to this algorithm as the *converse of Prüfer's algorithm*, and we summarize it as follows:

Converse of Prüfer's Algorithm

1. Let $j = 1$, let $A = \{1,2,\ldots,n\}$, and let L be the list $a_1,a_2,\ldots,a_{n-2}$.
2. Select the smallest member i_j of A that is not in list L and let a_j be the first number in L.
3. Construct the edge $\{i_j,a_j\}$.
4. Let L' be the list obtained from L by deleting a_j and let A' be the set obtained from A by deleting i_j.
5. If L' is the empty list, construct the edge that consists of the two members of A' and stop. The resulting tree is the desired tree.
6. Replace L by L', A by A', j by $j + 1$, and go to step 2.

Example 17

Use the converse of Prüfer's algorithm to construct the tree whose vertices are labeled 1,2,3,4,5,6 from the list L: 2,3,1,6.

Analysis

1. Let $j = 1$, let $A = \{1,2,3,4,5,6\}$, and let L be the list 2,3,1,6.
2. 4 is the smallest member of A that is not in list L, and 2 is the first number in L.
3. Construct edge $\{4,2\}$ (see Figure 12.29).

Figure 12.29

4 2

4. Let L' be the list 3,1,6, and let $A' = \{1,2,3,5,6\}$.
5. There are numbers in L'.
6. Replace L by L', A by A', j by $j + 1$, and go to step 2.
2. 2 is the smallest member of A that is not in L, and 3 is the first number in L.
3. Construct edge $\{2,3\}$ (see Figure 12.30).

Figure 12.30

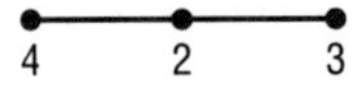

4 2 3

4. Let L' be the list 1,6, and let $A' = \{1,3,5,6\}$.
5. There are members of L'.
6. Replace L by L', A by A', j by $j + 1$, and go to step 2.
2. 3 is the smallest member of A that is not in L and 1 is the first member of L.
3. Construct edge $\{3,1\}$ (see Figure 12.31).

Figure 12.31

4 2 3 1

4. Let L' be the list 6, and let $A' = \{1,5,6\}$.
5. There are members of L'.
6. Replace L by L', A by A', j by $j + 1$, and go to step 2.
2. 1 is the smallest member of A that is not in L, and 6 is the first member of L.
3. Construct edge $\{1,6\}$ (see Figure 12.32).

Figure 12.32

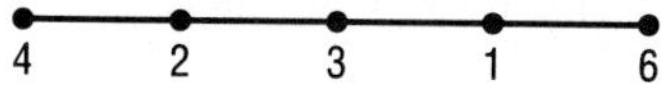

4 2 3 1 6

4. Let L' be the list obtained from L by deleting 6, and let $A' = \{5,6\}$.
5. There are no members of L', so construct edge $\{5,6\}$ and stop (see Figure 12.33).

Figure 12.33

4 2 3 1 6 5

■

In Prüfer's algorithm, all but two of the vertices are labeled once. At most n operations are required to locate the first vertex that is deleted, at most $n - 1$ operations are required to locate the second vertex that is deleted, and so on. Thus the number of operations needed to locate the vertices that are deleted (see Example 5 in Chapter 4) is

$$\begin{aligned} n + (n-1) + \cdots + 3 &= \frac{n(n+1)}{2} - 3 \\ &= \frac{n(n+1) - 6}{2} = \frac{n^2 + n - 6}{2} \end{aligned}$$

Therefore the complexity of Prüfer's algorithm is $O(n^2)$.

In the converse of Prüfer's algorithm, for each member of A we must search the list L, so at the first step at most $n(n - 2)$ operations are required. At the second step at most $(n - 1)(n - 3)$ operations are required. Continuing, at the $n - 2$ step, we find that at most 3(1) operations are required. So the total number of operations (see Exercise 16 of Chapter 4) is at most

$$1(3) + 2(4) + \cdots + (n-3)(n-1) + (n-2)n$$

Since

$$\begin{aligned} 1(3) + 2(4) &+ \cdots + (n-3)(n-1) + (n-2)n \\ &\leq 2(3) + 3(4) + \cdots + (n-2)(n-1) + (n-1)n \\ &\leq \frac{(n-1)n(n+1)}{3} \end{aligned}$$

The complexity of the converse of Prüfer's algorithm is $O(n^3)$.

Recall the breadth-first search algorithm in Section 11.8 for finding a shortest path from one vertex to another. The same procedure can be used to find a spanning tree for a simple graph $G = (V,E)$. The breadth-first search spanning tree algorithm finds a spanning tree, if it exists, in a simple graph with n vertices ($n \in \mathbb{N}$). In the algorithm, L denotes a set of labeled vertices, U denotes a set of unlabeled vertices, and T denotes a set of edges that connect the vertices in L. We summarize the algorithm as follows:

The Breadth-First Search Spanning Tree Algorithm

1. Select a vertex u, assign the label 0 to u, let $L = \{u\}$, let $T = \varnothing$, and let $k = 0$.
2. If $L = V$, stop. The edges in T and the vertices in L form a spanning tree for G.
3. ($L \neq V$) Let k be the largest label that has been used, and find the vertices U in $V - L$ that are adjacent to vertices in L that bear the label k. If there are no such vertices, go to step 4. Assign the label $k + 1$ to the vertices in U and put them in L. For each vertex p with label $k + 1$, choose an edge e that connects p and a vertex with label k and place e in T. Go to step 2.
4. G has no spanning tree. Stop.

Example 18

Use the breadth-first search spanning tree algorithm to find a spanning tree for the simple graph in Figure 12.34.

Figure 12.34

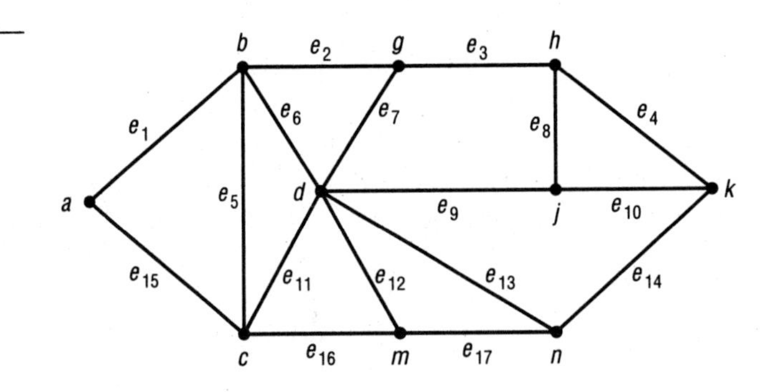

Analysis

1. Select vertex a, assign the label 0 to a, let $L = \{a\}$, let $T = \varnothing$, and let $k = 0$.
2. $L \neq V$.
3. There are two unlabeled vertices b and c that are adjacent to a. Assign the label 1 to b and c, let $L = \{a,b,c\}$, and let $T = \{e_1,e_{15}\}$.

2. $L \neq V$.
3. There are three unlabeled vertices d, g, and m that are adjacent to vertices in L that bear the label 1. Assign the label 2 to d, g, and m, let $L = \{a,b,c,d,g,m\}$, and let $T = \{e_1,e_{15},e_6,e_2,e_{16}\}$.
2. $L \neq V$.
3. There are three unlabeled vertices h, j, and n that are adjacent to vertices in L that bear the label 2. Assign the label 3 to h, j, and n, let $L = \{a,b,c,d,g,m,h,j,n\}$ and let $T = \{e_1,e_{15},e_6,e_2,e_{16},e_3,e_9,e_{13}\}$.
2. $L \neq V$.
3. There is one unlabeled vertex k that is adjacent to vertices in L that bear the label 3. Assign the label 4 to k, let $L = \{a,b,c,d,g,m,h,j,n,k\}$, and let $T = \{e_1,e_{15},e_6,e_2,e_{16},e_3,e_9,e_{13},e_4\}$.
2. $L = V$. Stop. The edges in T and the vertices in L form a spanning tree for G (see Figure 12.35).

Figure 12.35

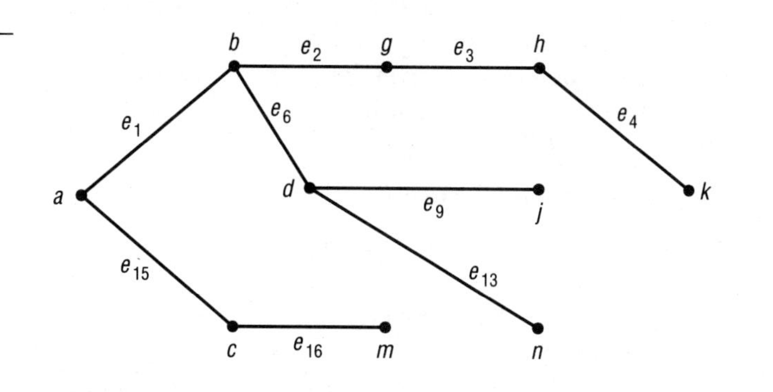

■

With the breadth-first search spanning tree algorithm, after one vertex has been selected we examine the edges to see which edges connect vertices in V with vertices not in V. The number of edges in a simple graph with n vertices is at most

$$(n - 1) + (n - 2) + \cdots + 1 = \frac{n(n-1)}{2} \quad \text{(see Example 5 of Chapter 4)}$$

At the next step there is one less edge to examine because one edge has been placed in T and we do not have to examine the edges in T. So the number of edges that must be examined at the next step is

$$(n - 2) + (n - 3) + \cdots + 1 = \frac{(n-1)(n-2)}{2}$$

Therefore the total number of operations involving edges is

$$\frac{n(n-1)}{2} + \frac{(n-1)(n-2)}{2} + \cdots + \frac{1 \cdot 2}{2}$$

$$= \frac{[n(n-1) + (n-1)(n-2) + \cdots + 2 \cdot 1]}{2}$$

$$= \frac{(n-1)n(n+1)}{6}.$$

The number of operations involving vertices cannot exceed

$$(n-1) + (n-2) + \cdots + 1 = \frac{n(n-1)}{2}$$

Therefore the complexity of the breadth-first search spanning tree algorithm is $O(n^3)$.

Another algorithm for finding a spanning tree of a simple graph $G = (V,E)$ with n vertices ($n \in \mathbb{N}$) is the depth-first search algorithm. Again L denotes a set of labeled vertices and T denotes a set of edges that connect the vertices in L. With this algorithm, only one vertex is added to L at each step until L contains all vertices. For each $k = 1,2, \ldots ,n$, we have exactly one vertex labeled k, and at each step if $k < n$ we add to L one vertex that is adjacent to a vertex in L. We summarize the algorithm as follows:

The Depth-First Search Algorithm

1. Select a vertex u, assign the label 1 to u, let $L = \{u\}$, let $T = \varnothing$, and let $k = 1$.
2. If $L = V$, stop. The edges in T and the vertices in L form a spanning tree for G.
3. ($L \neq V$) If there is a vertex in L that is adjacent to a vertex not in L, go to step 5.
4. ($L \neq V$) G has no spanning tree. Stop.
5. Let v be the vertex in L with largest label that is adjacent to a vertex not in L. Select a vertex w not in L that is adjacent to v. Assign the label $k + 1$ to w, put w in L, and put edge $\{v,w\}$ in T. Replace k by $k + 1$, and go to step 2.

Example 19

Use the depth-first search algorithm to find a spanning tree for the simple graph in Figure 12.36.

Analysis

1. Select vertex a, assign the label 1 to a, let $L = \{a\}$, let $T = \varnothing$, and let $k = 1$.
2. $L \neq V$.
3. There is a vertex in L that is adjacent to a vertex not in L.
5. There are three vertices, b, d, and g, not in L that are adjacent to a. Select b, assign the label 2 to b, let $L = \{a,b\}$, let $T = \{e_1\}$, and let $k = 2$.

Figure 12.36

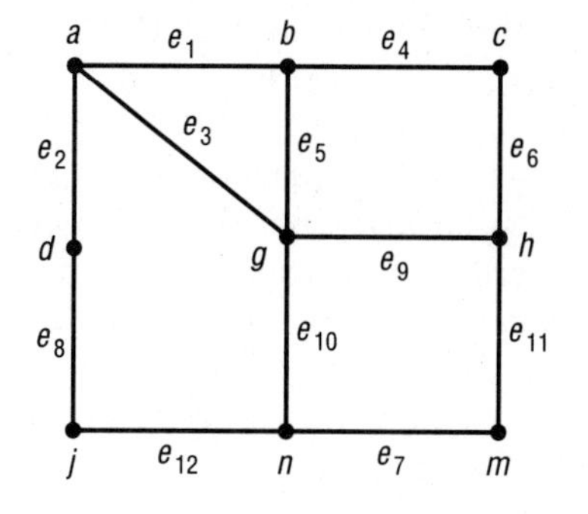

2. $L \neq V$.
3. There is a vertex in L that is adjacent to a vertex not in L.
5. The vertex in L with largest label that is adjacent to a vertex not in L is b. There are two vertices, c and g, not in L that are adjacent to b. Select c, assign the label 3 to c, let $L = \{a,b,c\}$, let $T = \{e_1,e_4\}$, and let $k = 3$.
2. $L \neq V$.
3. There is a vertex in L that is adjacent to a vertex not in L.
5. The vertex in L with largest label that is adjacent to a vertex not in L is c. There is only one vertex, h, not in L that is adjacent to c. Assign the label 4 to h, let $L = \{a,b,c,h\}$, let $T = \{e_1,e_4,e_6\}$, and let $k = 4$.
2. $L \neq V$.
3. There is a vertex in L that is adjacent to a vertex not in L.
5. The vertex in L with largest label that is adjacent to a vertex not in L is h. There are two vertices, g and m, not in L that are adjacent to h. Select g, assign the label 5 to g, let $L = \{a,b,c,h,g\}$, let $T = \{e_1,e_4,e_6,e_9\}$, and let $k = 5$.
2. $L \neq V$.
3. There is a vertex in L that is adjacent to a vertex not in L.
5. The vertex in L with largest label that is adjacent to a vertex not in L is g. There is only one vertex, n, not in L that is adjacent to g. Assign the label 6 to n, let $L = \{a,b,c,h,g,n\}$, let $T = \{e_1,e_4,e_6,e_9,e_{10}\}$, and let $k = 6$.
2. $L \neq V$.
3. There is a vertex in L that is adjacent to a vertex not in L.
5. The vertex in L with largest label that is adjacent to a vertex not in L is n. There are two vertices, j and m, not in L that are adjacent to n. Select j, assign the label 7 to j, let $L = \{a,b,c,h,g,n,j\}$, let $T = \{e_1,e_4,e_6,e_9,e_{10},e_{12}\}$, and let $k = 7$.
2. $L \neq V$.
3. There is a vertex in L that is adjacent to a vertex not in L.
5. The vertex in L with largest label that is adjacent to a vertex not in L is j. There is only one vertex, d, not in L that is adja-

cent to j. Assign the label 8 to d, let $L = \{a,b,c,h,g,n,j,d\}$, let $T = \{e_1,e_4,e_6,e_9,e_{10},e_{12},e_8\}$, and let $k = 8$.

2. $L \neq V$.

3. There is a vertex in L that is adjacent to a vertex not in L.

5. The vertex in L with largest label that is adjacent to a vertex not in L is n. There is only one vertex, m, not in L that is adjacent to n. Assign the label 9 to m, let $L = \{a,b,c,h,g,n,j,d,m\}$, let $T = \{e_1,e_4,e_6,e_9,e_{10},e_{12},e_8,e_7\}$, and let $k = 9$.

2. $L = V$. The edges in T and the vertices in L form a spanning tree for G (see Figure 12.37).

Figure 12.37

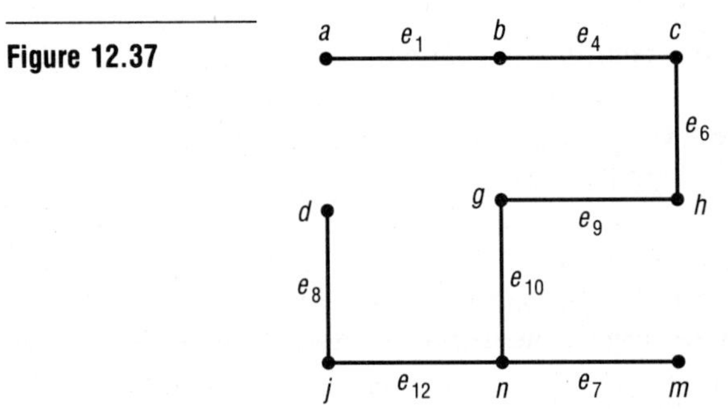

■

In the depth-first search algorithm, each vertex is labeled at most once and each edge is used at most twice, once in step 3 and once in step 5. So if the graph has n vertices and ε edges, the number of operations is at most $n + 2\varepsilon$. Since $\varepsilon \leq n(n-1)/2$, the number of operations is at most $n + n(n-1)/2$. Therefore the complexity of the depth-first search algorithm is $O(n^2)$.

Exercises 12.3

29. Use Prüfer's algorithm to find the list $a_1,a_2, \ldots ,a_{n-2}$ associated with each of the following trees.

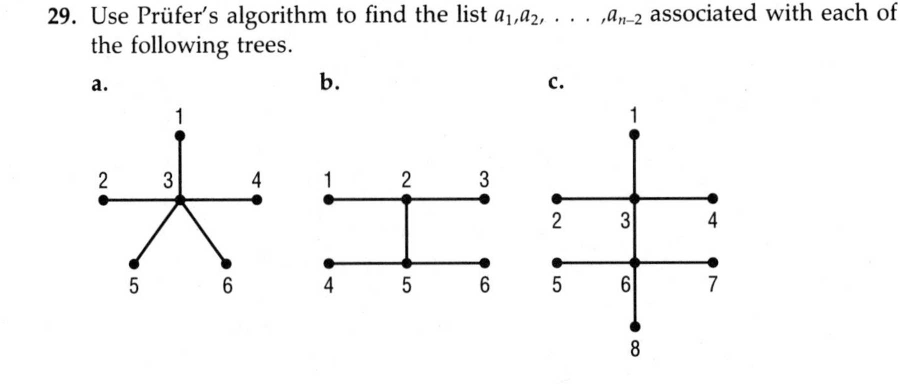

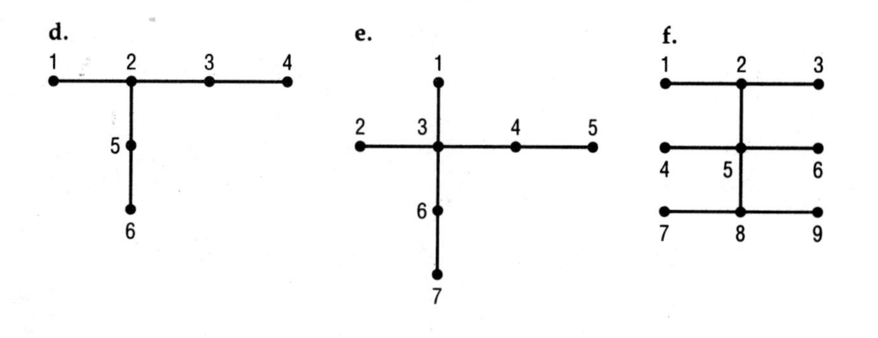

30. Use the converse of Prüfer's algorithm to construct each of the following trees.

a. the tree whose vertices are labeled 1,2,3,4,5 from the list L: 2,1,2

b. the tree whose vertices are labeled 1,2,3,4,5 from the list L: 2,2,2

c. the tree whose vertices are labeled 1,2,3,4,5,6 from the list L: 3,4,2,3

d. the tree whose vertices are labeled 1,2,3,4,5,6 from the list L: 3,5,5,3

e. the tree whose vertices are labeled 1,2,3,4,5,6,7 from the list L: 6,7,6,2,7

f. the tree whose vertices are labeled 1,2,3,4,5,6,7 from the list L: 4,4,4,1,1

31. Draw the sixteen distinct trees whose vertices are labeled 1,2,3,4.

32. Use the breadth-first search spanning tree algorithm to find a spanning tree for each of the following simple graphs.

a.

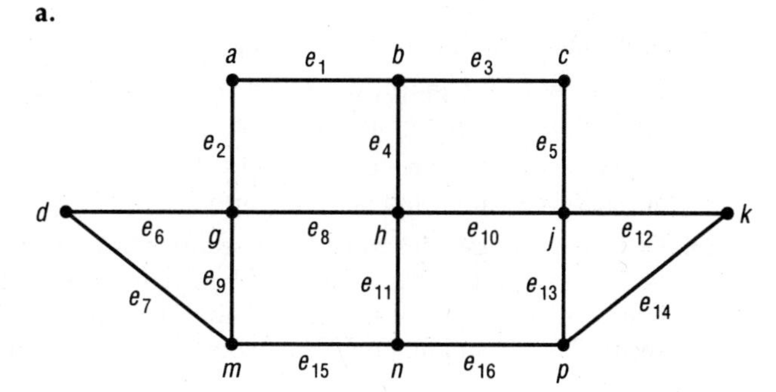

b.

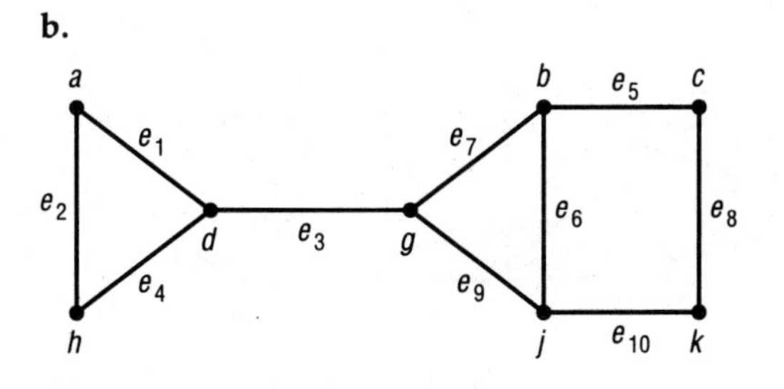

c.

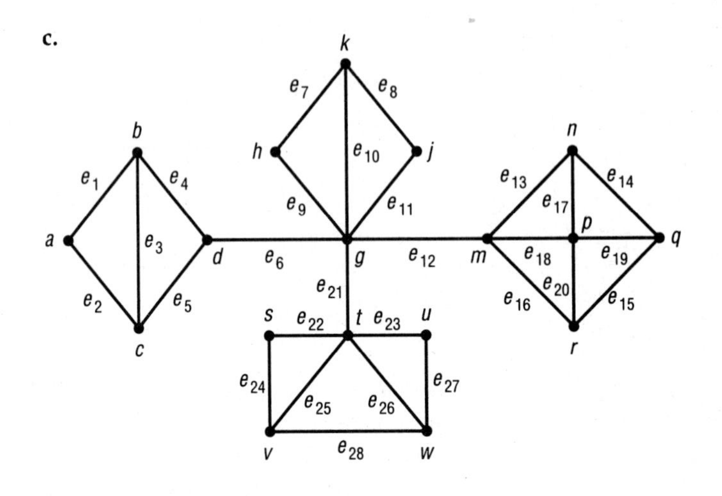

33. Use the depth-first search algorithm to find a spanning tree for each of the simple graphs in Exercise 32.
34. Explain why the graph produced by the breadth-first search spanning tree algorithm when applied to a simple graph is a tree.
35. Explain why the tree produced by the breadth-first search spanning tree algorithm when applied to a connected simple graph is a spanning tree.
36. Explain by use of an example why the tree produced by the breadth-first search spanning tree algorithm when applied to a simple graph with two components is not a spanning tree.
37. Explain why the graph produced by the depth-first search algorithm when applied to a simple graph is a tree.
38. Explain why the tree produced by the depth-first search algorithm when applied to a connected simple graph is a spanning tree.
39. Government agencies must have secure computer communications, but it is not necessary that all their lines be secure. Given the following computer lines between agencies, indicate how a minimum number of lines can be secured while still having secure communications between any two agencies.

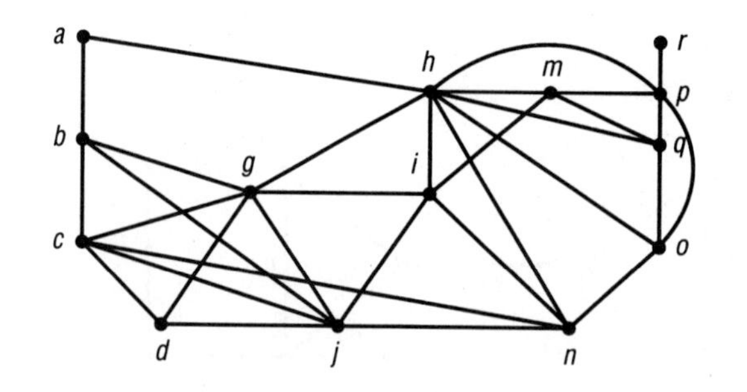

40. Suppose that T_1 and T_2 are spanning trees of a connected simple graph G. Do T_1 and T_2 necessarily have an edge in common? If so, give a proof, and if not, give a counterexample.

***41.** Show that a connected simple graph with only one spanning tree is a tree.

MINIMAL SPANNING TREES

In the previous section we gave two algorithms for finding a spanning tree in a simple graph. The problem is more interesting if the graph is weighted. In this case we want to find a spanning tree whose weight is less than or equal to the weight of any other spanning tree. Such a spanning tree is called a *minimal spanning tree*. In this section we give two algorithms for finding a minimal spanning tree. Both proceed by successively adding edges of smallest weight from those edges that have not been used, and they are examples of what are known as *greedy algorithms*. A greedy algorithm is a procedure that makes an optimal choice at each step.

The first algorithm we discuss was discovered by Joseph Kruskal in 1956. Kruskal's algorithm constructs a minimal spanning tree for a connected weighted simple graph $G = (V,E)$ with n vertices ($n \in \mathbb{N}$ and $n \geqslant 2$) whose edges $e_1, e_2, \ldots, e_\varepsilon$ have been sorted so that $w(e_1) \leqslant w(e_2) \leqslant \cdots \leqslant w(e_\varepsilon)$. In the algorithm, T denotes a set of edges and $U(T)$ denotes the set of vertices of G that are vertices of edges in T. We summarize the algorithm as follows:

Kruskal's Algorithm for Minimal Spanning Trees

1. Let $T = \varnothing$ and $U(T) = \varnothing$.
2. Let k be the smallest natural number such that e_k is not in T and the graph $(U(T \cup \{e_k\}), T \cup \{e_k\})$ does not contain a cycle.
3. Replace T by $T \cup \{e_k\}$ and $U(T)$ by $U(T \cup \{e_k\})$.
4. If T contains $n - 1$ edges, stop. $(U(T), T)$ is a minimal spanning tree.
5. Go to step 2.

Example 20

Use Kruskal's algorithm to find a minimal spanning tree for the weighted graph shown in Figure 12.38.

Analysis

To sort edges $e_1, e_2, \ldots, e_{12}$ so that $w(e_1) \leqslant w(e_2) \leq \cdots \leq w(e_{12})$, let $e_1 = \{b, g\}$, $e_2 = \{n,m\}$, $e_3 = \{a,d\}$, $e_4 = \{j,n\}$, $e_5 = \{d,g\}$, $e_6 = \{c,h\}$, $e_7 = \{g,n\}$, $e_8 = \{a,b\}$, $e_9 = \{g,h\}$, $e_{10} = \{b,c\}$, $e_{11} = \{d,j\}$, and $e_{12} = \{h,m\}$.

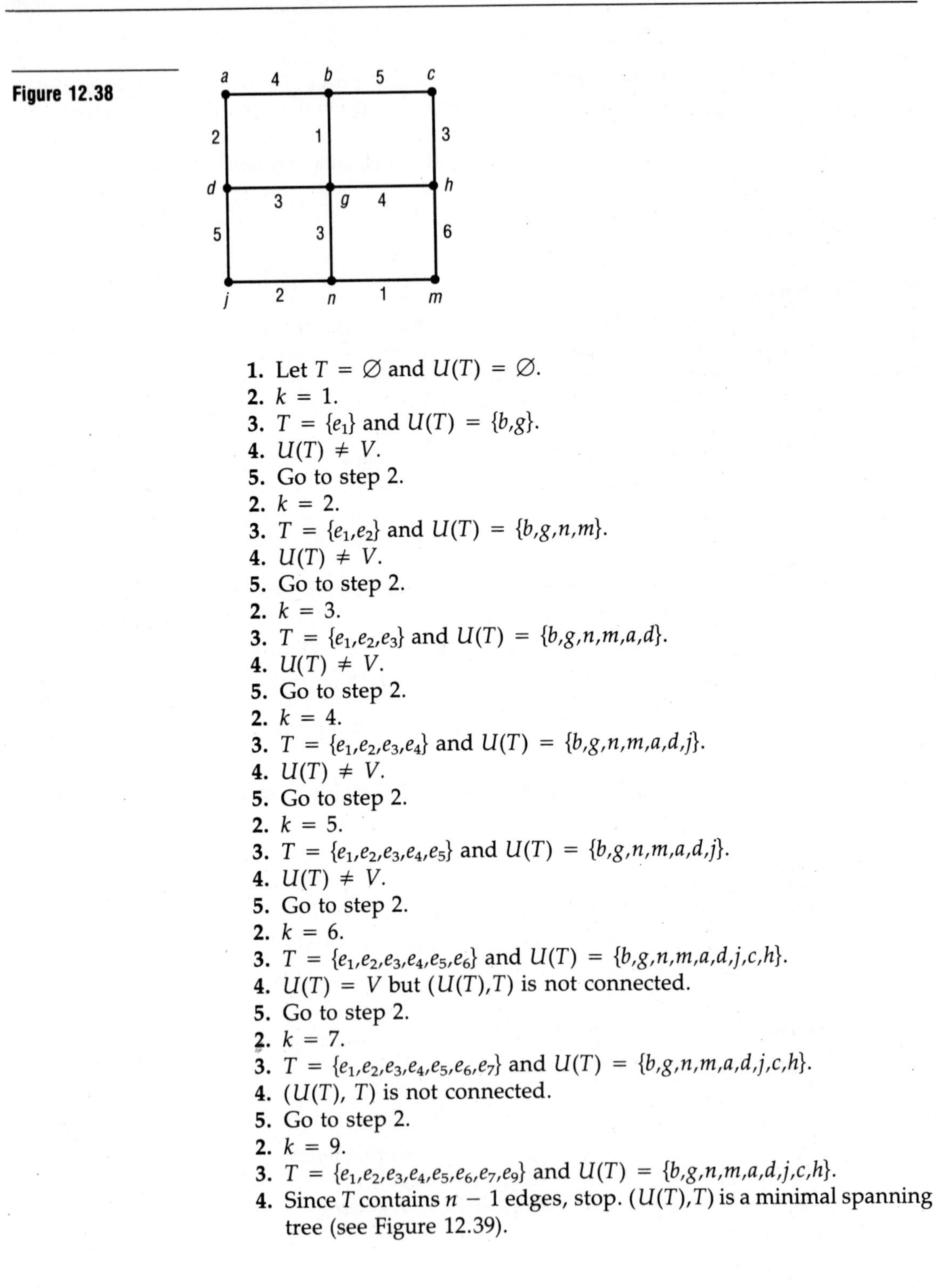

Figure 12.38

1. Let $T = \varnothing$ and $U(T) = \varnothing$.
2. $k = 1$.
3. $T = \{e_1\}$ and $U(T) = \{b,g\}$.
4. $U(T) \neq V$.
5. Go to step 2.
2. $k = 2$.
3. $T = \{e_1,e_2\}$ and $U(T) = \{b,g,n,m\}$.
4. $U(T) \neq V$.
5. Go to step 2.
2. $k = 3$.
3. $T = \{e_1,e_2,e_3\}$ and $U(T) = \{b,g,n,m,a,d\}$.
4. $U(T) \neq V$.
5. Go to step 2.
2. $k = 4$.
3. $T = \{e_1,e_2,e_3,e_4\}$ and $U(T) = \{b,g,n,m,a,d,j\}$.
4. $U(T) \neq V$.
5. Go to step 2.
2. $k = 5$.
3. $T = \{e_1,e_2,e_3,e_4,e_5\}$ and $U(T) = \{b,g,n,m,a,d,j\}$.
4. $U(T) \neq V$.
5. Go to step 2.
2. $k = 6$.
3. $T = \{e_1,e_2,e_3,e_4,e_5,e_6\}$ and $U(T) = \{b,g,n,m,a,d,j,c,h\}$.
4. $U(T) = V$ but $(U(T),T)$ is not connected.
5. Go to step 2.
2. $k = 7$.
3. $T = \{e_1,e_2,e_3,e_4,e_5,e_6,e_7\}$ and $U(T) = \{b,g,n,m,a,d,j,c,h\}$.
4. $(U(T), T)$ is not connected.
5. Go to step 2.
2. $k = 9$.
3. $T = \{e_1,e_2,e_3,e_4,e_5,e_6,e_7,e_9\}$ and $U(T) = \{b,g,n,m,a,d,j,c,h\}$.
4. Since T contains $n - 1$ edges, stop. $(U(T),T)$ is a minimal spanning tree (see Figure 12.39).

Figure 12.39

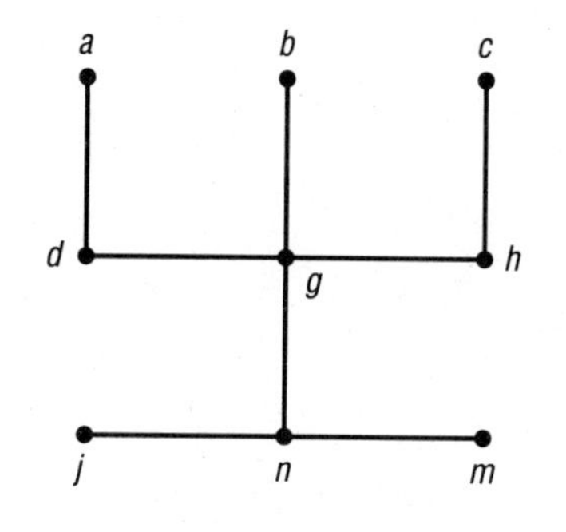

Since the weight of a tree is the sum of the weights of its edges, the weight of the tree in Figure 12.39 is

$$1 + 1 + 2 + 2 + 3 + 3 + 3 + 4 = 19$$ ■

Kruskal's algorithm involves a test of edges to see if they belong to cycles. Thus we do not analyze the complexity. However, one method of making such a test is to use a variation of Dijkstra's algorithm. Since the complexity of Dijkstra's algorithm is $O(n^2)$, if we use the variation of Dijkstra's algorithm to test acyclicity in Kruskal's algorithm, the complexity of Kruskal's algorithm is also $O(n^2)$.

The next algorithm we discuss was discovered by Robert Prim in 1957. Prim's algorithm produces at each stage a subgraph that is a tree. At each stage it adds an edge of smallest weight incident with a vertex that is already in the constructed subgraph. We assume that $G = (V,E)$ is a connected weighted simple graph with n vertices, and in the algorithm T denotes a set of edges and U denotes a set of vertices. We summarize the algorithm as follows:

Prim's Algorithm for Minimal Spanning Trees

1. Select a vertex u, let $U = \{u\}$, and let $T = \varnothing$.
2. If $U = V$, stop. The tree (U,T) is a minimal spanning tree for G.
3. $(U \neq V)$ Select an edge e of smallest weight that is incident with one vertex u in U and one vertex v not in U. Put e in T, put v in U, and go to step 2.

Example 21

Use Prim's algorithm to find a minimal spanning tree for the weighted graph $G = (V,E)$ in Figure 12.40.

Analysis

Let $e_1 = \{a,b\}$, $e_2 = \{a,c\}$, $e_3 = \{a,d\}$, $e_4 = \{a,g\}$, $e_5 = \{b,g\}$, $e_6 = \{c,d\}$, $e_7 = \{c,h\}$, $e_8 = \{c,j\}$, $e_9 = \{h,j\}$, $e_{10} = \{d,g\}$, $e_{11} = \{d,j\}$, and $e_{12} = \{g,j\}$.

Figure 12.40

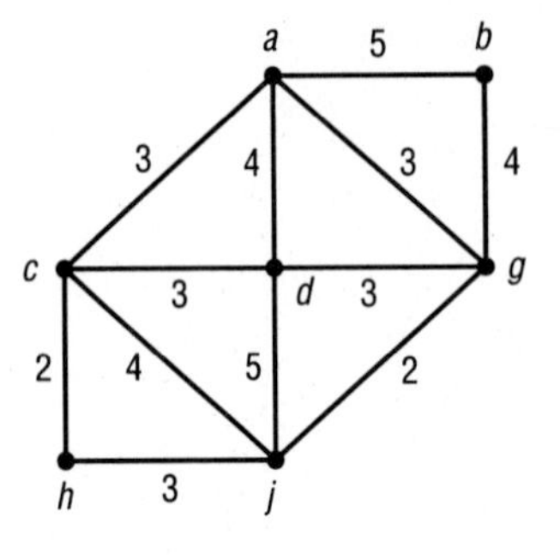

1. Let $U = \{a\}$ and $T = \emptyset$.
2. $U \neq V$.
3. There are two edges, e_2 and e_4, of smallest weight that are incident with a. Select e_2, put it in T, and put c in U. So $U = \{a,c\}$ and $T = \{e_2\}$.
2. $U \neq V$.
3. The edge of smallest weight that is incident with one vertex in U and one vertex not in U is e_7. So $U = \{a,c,h\}$ and $T = \{e_2,e_7\}$.
2. $U \neq V$.
3. There are three edges, e_4, e_6, and e_9, of smallest weight that are incident with one vertex in U and one vertex not in U. Select e_4, so $U = \{a,c,h,g\}$ and $T = \{e_2,e_7,e_4\}$.
2. $U \neq V$.
3. The edge of smallest weight that is incident with one vertex in U and one vertex not in U is e_{12}. So $U = \{a,c,h,g,j\}$ and $T = \{e_2,e_7,e_4,e_{12}\}$.
2. $U \neq V$.
3. There are two edges, e_6 and e_{10}, of smallest weight that are incident with one vertex in U and one vertex not in U. Select e_6, so $U = \{a,c,h,g,j,d\}$ and $T = \{e_2,e_7,e_4,e_{12},e_6\}$.
2. $U \neq V$.
3. The edge of smallest weight that is incident with one vertex in U and one vertex not in U is e_5, so $U = \{a,c,h,g,j,d,b\}$ and $T = \{e_2,e_7,e_4,e_{12},e_6,e_5\}$.
2. $U = V$. The tree (U,T) is a minimal spanning tree for G (see Figure 12.41).

The weight of the tree in Figure 12.41 is $3 + 2 + 3 + 2 + 3 + 4 = 17$. ■

In Prim's algorithm, there are $n - 1$ iterations. Each iteration involves finding an edge (and a vertex of this edge) with a certain property, and this can be accomplished with an algorithm whose complexity is $O(n)$. So the complexity of Prim's algorithm is $O(n^2)$.

Figure 12.41

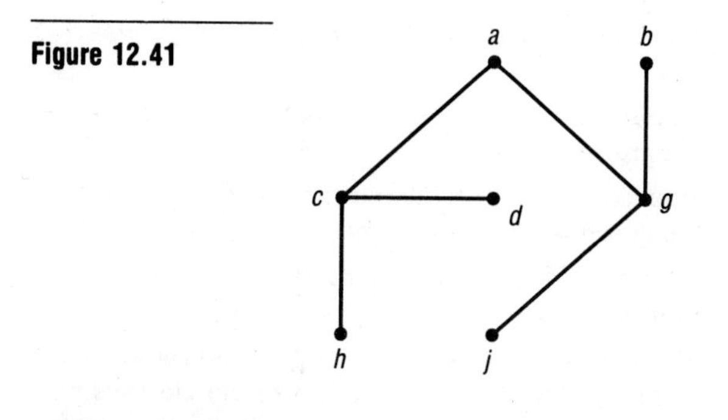

Exercises 12.4

42. Use Kruskal's algorithm to find a minimal spanning tree for each of the following weighted graphs.

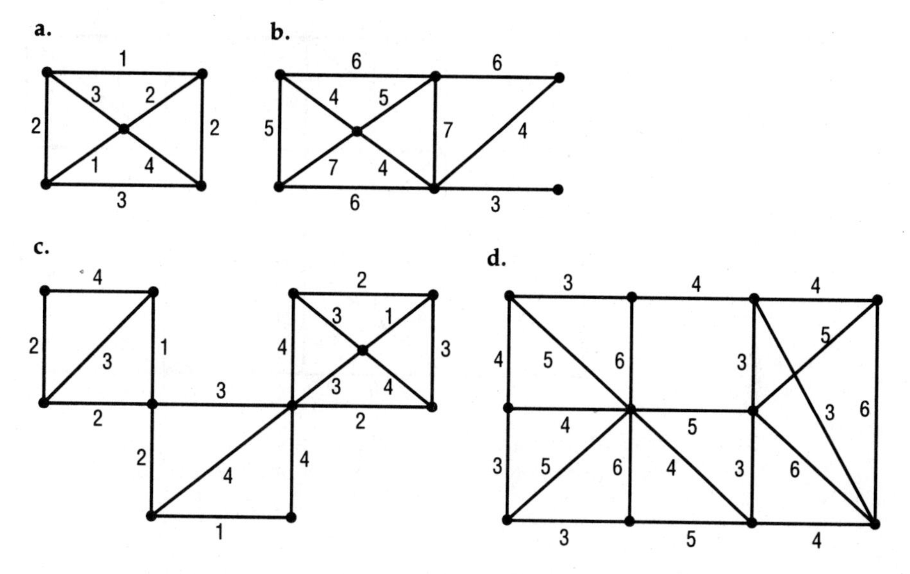

43. Find the weights of the minimal spanning trees that you found in Exercise 42.

44. Use Prim's algorithm to find a minimal spanning tree for each of the weighted graphs in Exercise 42.

45. Find the weights of the minimal spanning trees that you found in Exercise 44.

46. Explain why the graph produced by Prim's algorithm when applied to a connected weighted simple graph is a tree.

47. Explain why the tree produced by Prim's algorithm when applied to a connected weighted simple graph is a spanning tree.

*48. Prove that Prim's algorithm produces a minimal spanning tree of a connected weighted simple graph.

49. Explain why the graph produced by Kruskal's algorithm when applied to a connected weighted simple graph is a tree.

50. Explain why the tree produced by Kruskal's algorithm when applied to a connected weighted simple graph is a spanning tree.

*51. Prove that Kruskal's algorithm produces a minimal spanning tree of a connected weighted simple graph.

52. An electric company built too many lines in a housing subdivision. In the following graph, vertices represent houses and edges represent electric lines. The weights of the edges represent the lengths of the lines. The electric company wants to ensure that each pair of houses is connected by a path of electric lines, but it wants to remove excess lines so that the sum of the lengths of the remaining lines is as small as possible. Show how this can be done.

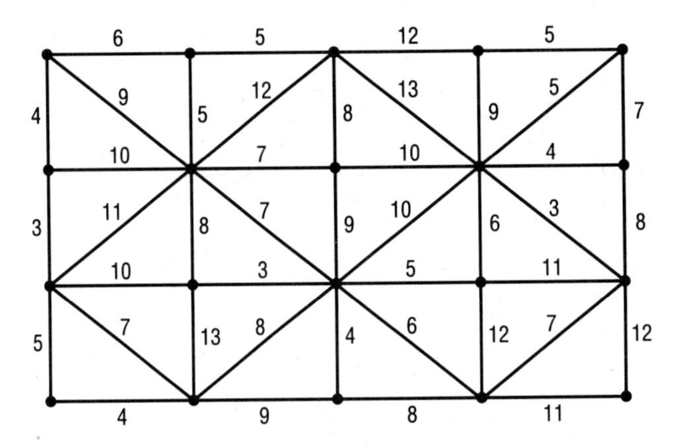

*53. Prove that, if the weights in a connected weighted simple graph G are all different, then G has exactly one minimal spanning tree.

12.5

TREE TRAVERSAL ALGORITHMS

An ordered binary tree can be used to represent such things as arithmetic expressions, compound propositions, and set combinations. The internal vertices are used to represent the operations, and the leaf vertices are used to represent the variables. Each operation operates on its left and right subtrees in that order, and the root denotes the final operation. An ordered binary tree that is used in this manner is called an *expression tree.* Expression trees are built from the bottom up. We illustrate the use of expression trees in an example.

Example 22

Let a, b, c, and d denote real numbers. Draw the expression trees for the following expressions:

a. ab
b. $a + bc$
c. $a + b(c - d)$
d. $(a + b)^2 + (c - d)/3$

Analysis

a. We have two variables, a and b, and only one operation, multiplication, so a and b are represented by leaves and $\cdot$ is represented by the root of the ordered binary tree. The expression ab is represented by the binary tree shown here.

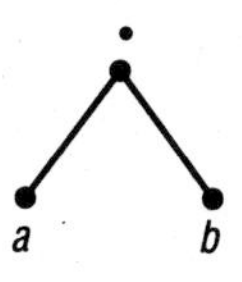

b. We have three variables, a, b, and c, and two operations, addition and multiplication, so the ordered binary tree that represents this expression has three leaves and two internal vertices. The last operation to be performed is addition, so $+$ represents the root of the ordered binary tree. As indicated, we build the expression tree from the bottom up. We construct the subtree for bc, and from this subtree we build the binary tree that represents $a + bc$ as shown here.

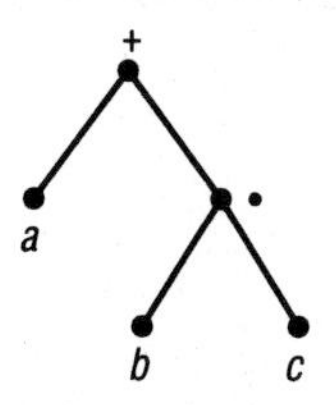

c. We have four variables, a,b,c, and d, and three operations, addition, multiplication, and subtraction, so the ordered binary tree that represents this expression has four leaves and three internal vertices. The last operation to be performed is addition, so $+$ represents the root of the expression tree. Starting from the bottom, we first build the subtree that represents $c - d$, then we build the subtree that represents $b(c - d)$, and finally we build the expression tree for $a + b(c - d)$ as shown here.

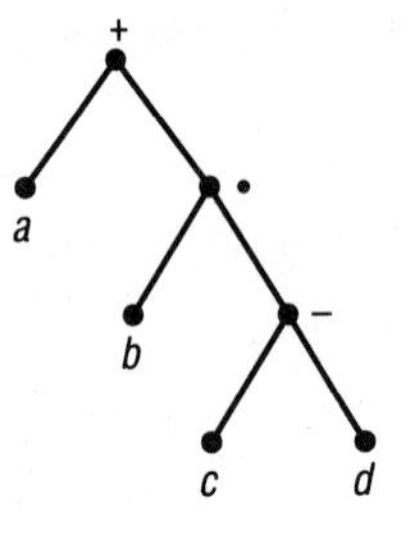

d. We have four variables, a, b, c, and d, and four operations, addition, exponentiation (which we denote by $\wedge$), subtraction, and division. The last operation to be performed is addition, so $+$ represents the root of the expression tree. This time we need to build nontrivial left *and* right subtrees of the root, and the binary tree that represents $(a + b)^2 + (c - d)/3$ is shown here.

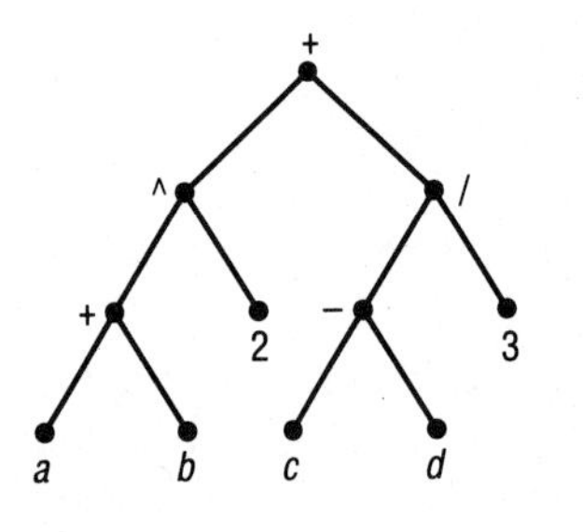

■

Notice that we must be careful to distinguish between the left and right subtrees, because if the operation is not commutative we get different answers. For example, the left expression tree of the following figure represents $a - b$, whereas the right tree represents $b - a$.

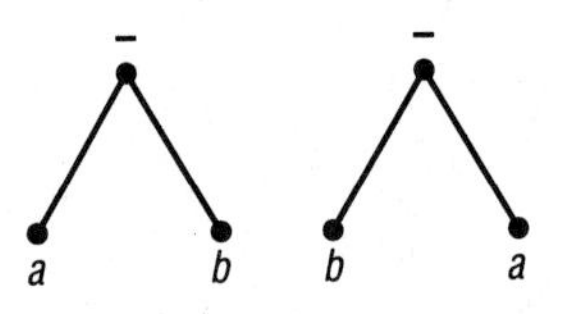

Once we have the expression tree, we must process it to obtain an evaluation of the expression. We want a systematic way to examine each vertex exactly once.

A *tree traversal algorithm* is a procedure for listing (or visiting, or searching) each vertex of an ordered rooted tree exactly once. If the tree is a regular binary tree, then a tree traversal algorithm is used in the computer to evaluate expressions.

The three tree traversal algorithms we consider here are the preorder traversal, the postorder traversal, and the inorder traversal algorithms. It is difficult to present these tree traversal algorithms by pseudocode, and we make no such attempt. Instead, we simply illustrate them. In particular, we indicate which vertex is listed first. Then, after a given vertex has been listed, we indicate which vertex is listed next.

The *preorder traversal algorithm* is a special case of the depth-first search algorithm. With the preorder traversal algorithm we always list the root of an ordered rooted tree first, and in general we list a parent before its children and we list the children according to the linear order (that is, from left to right). The preorder traversal algorithm is different from the algorithms we have previously considered in that it refers to itself, or, in programming terminology, it calls itself. In mathematical terminology, the algorithm is defined recursively.

The easiest way to understand the preorder traversal algorithm is to look at an example. Let T be the ordered rooted tree shown in Figure 12.42.

Figure 12.42

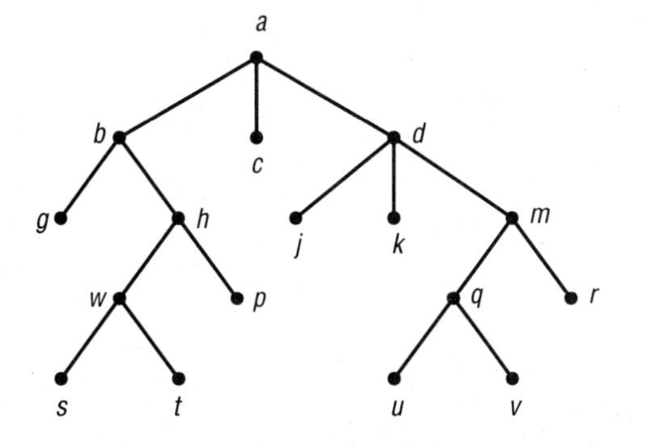

Application of the Preorder Traversal Algorithm

1. List the root a of T.
2. The ordered children of a are b, c, and d.
3. List b.
4. The ordered children of b are g and h.
5. List g.
6. Note that g has no children.
7. List h.
8. The ordered children of h are w and p.
9. List w.
10. The ordered children of w are s and t.
11. List s.
12. Note that s has no children.
13. List t.
14. Note that t has no children.
15. List p.

16. Note that p has no children.
17. List c.
18. Note that c has no children.
19. List d.
20. The ordered children of d are j, k, and m.
21. List j.
22. Note that j has no children.
23. List k.
24. Note that k has no children.
25. List m.
26. The ordered children of m are q and r.
27. List q.
28. The ordered children of q are u and v.
29. List u.
30. Note that u has no children.
31. List v.
32. Note that v has no children.
33. List r.
34. Note that r has no children.
35. Stop.

The order of the vertices is shown in Figure 12.43.

Figure 12.43

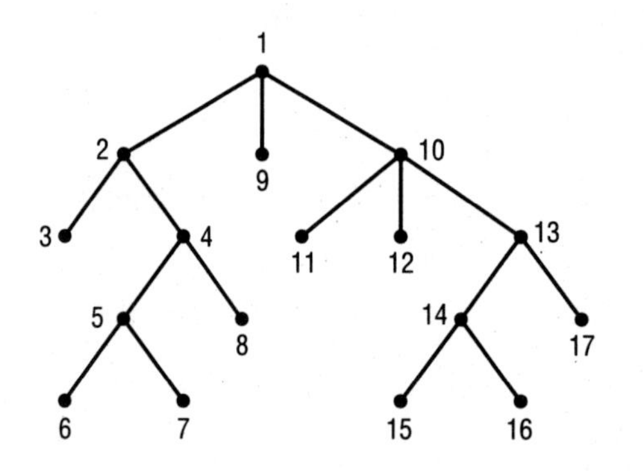

Observe the procedure we follow to determine which vertex is listed after a given vertex x. If x has a child, we next list the first child of x. If x does not have a child but has a sibling that has not been listed, we next list the first such sibling of x. If x does not have a child and does not have a sibling that has not been listed, we go back up the tree until we find a parent who has a child that has not been listed, and we list the first such child.

Definition

When we apply the preorder traversal algorithm to an expression tree, the resulting list of variables and operations is called the **prefix form** of the expression, and expressions written in prefix form are said to be written in **Polish notation.**

Polish notation was named in honor of the logician Jan Lukasiewicz (who was actually Ukrainian rather than Polish).

Example 23

Give the Polish notation for the expression $(a + b)^2 + (c - d)/3$.

Analysis

The expression tree for $(a + b)^2 + (c - d)/3$ and the order that results when we apply the preorder traversal algorithm to this expression tree are shown in Figure 12.44. Thus the Polish notation for the expression $(a + b)^2 + (c - d)/3$ is $+ \wedge + a\, b\, 2 / - c\, d\, 3$. If we think of the binary operations $+, -, *, /, \wedge$ as functions, then expressions written in Polish notation are easy to evaluate. For example, $+(1,3)$ means $1 + 3$, and $-(5,2)$ means $5 - 2$.

Figure 12.44

■

Example 24

Evaluate the following expression (written in Polish notation):

$+ \wedge + 1\ 3\ 2 / -5\ 2\ 3$

Analysis

We first add parentheses and commas:

$$+ \wedge + 1\ 3\ 2 / -5\ 2\ 3 = +(\wedge(+(1,3),2),/(-(5,2),3))$$
$$= +(\wedge(4,2),/(3,3)) = +(16,1) = 17$$ ■

With the *postorder traversal algorithm* we list the root last, and in general we list the children before the parent according to the linear order (from left to right). This algorithm also refers to itself. As with the preorder traversal algorithm, the easiest way to understand the postorder traversal algorithm is to look at an example. Again let T be the ordered rooted tree in Figure 12.42.

Application of the Postorder Traversal Algorithm

1. The ordered children of the root a are b, c, and d.
2. The ordered children of b are g and h.
3. Note that g has no children.
4. List g.
5. The ordered children of h are w and p.
6. The ordered children of w are s and t.
7. Note that s has no children.
8. List s.
9. Note that t has no children.
10. List t.
11. Note that we have listed all the children of w.
12. List w.
13. Note that p has no children.
14. List p.
15. Note that we have listed all the children of h.
16. List h.
17. Note that we have listed all the children of b.
18. List b.
19. Note that c has no children.
20. List c.
21. The ordered children of d are j, k, and m.
22. Note that j has no children.
23. List j.
24. Note that k has no children.
25. List k.
26. The ordered children of m are q and r.
27. The ordered children of q are u and v.
28. Note that u has no children.
29. List u.
30. Note that v has no children.
31. List v.
32. Note that we have listed all the children of q.
33. List q.
34. Note that r has no children.
35. List r.
36. Note that we have listed all the children of m.
37. List m.

38. Note that we have listed all the children of d.
39. List d.
40. Note that we have listed all the children of a.
41. List a.
42. Stop.

The order of the vertices is shown in Figure 12.45.

Figure 12.45

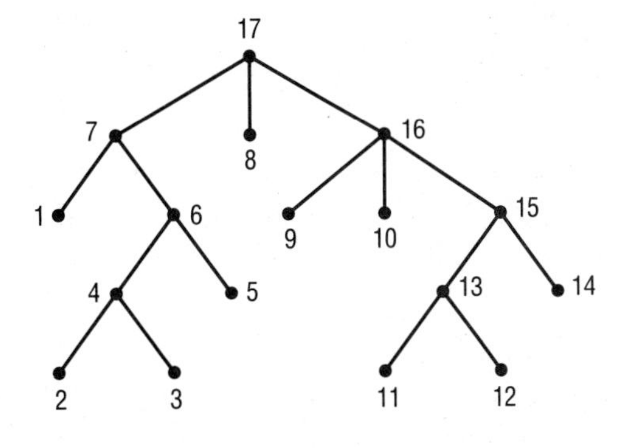

Observe the procedure we follow to determine the first vertex to list. We consider the first child of the root, then we consider the first child of the first child, and we continue this process until we obtain a vertex x with no children. Then we list x. Also observe the procedure we follow to determine which vertex to list after a given vertex y. If all the children of the parent of y have been listed, we next list the parent of y; if y has no parent, then y is the root and we are finished. If the parent of y has a child that has not been listed, we consider the first such child. Then we consider the first child of this first child, and we continue this process until we obtain a vertex z with no children. Then we list z.

Definition

When we apply the postorder traversal algorithm to an expression tree, the resulting list of variables and operations is called the **postfix form** of the expression, and expressions written in postfix form are said to be written in **reverse Polish notation.**

This notation was also introduced by Lukasiewicz.

Example 25

Give the reverse Polish notation for the expression $(a + b)^2 + (c - d)/3$.

Analysis

The expression tree for $(a + b)^2 + (c - d)/3$ and the order that results when we apply the postorder traversal algorithm to this expression tree are shown in Figure 12.46. Thus the reverse Polish notation for the expression $(a + b)^2 + (c - d)/3$ is $a\ b\ +\ 2 \wedge c\ d\ -\ 3\ /\ +$.

Figure 12.46

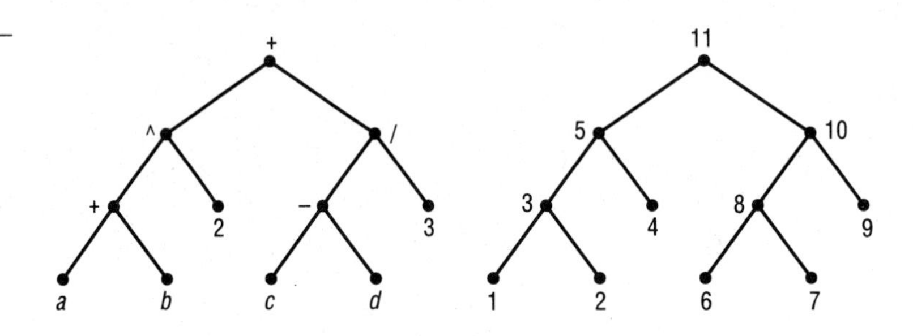

Think of entering in order the symbols $a\ b\ +\ 2 \wedge c\ d\ -\ 3\ /\ +$ into a computer. We first enter the real numbers a and b. Then the $+$ means that we add the two real numbers previously entered. Then we enter a 2, and the $\wedge$ means that we square the number we already have in the computer. Next we enter, in order, c and d, and the $-$ means that we subtract. The 3 followed by $/$ means that we divide the number that we have in the computer by 3. Finally, the $+$ means that we add the two numbers that we have in the computer. Indeed, some hand calculators require that algebraic expressions be entered according to reverse Polish notation. ■

We begin the discussion of the *inorder traversal algorithm* by explaining how it works in an ordered binary tree. In this case, it lists the left child c first, then the parent p of c, and finally the right child of p. As with the previous two cases, the easiest way to understand the inorder traversal algorithm is to look at an example. Again let T be the ordered rooted tree in Figure 12.42.

Application of the Inorder Traversal Algorithm

1. The ordered children of the root a are b, c, and d.
2. The ordered children of b are g and h.
3. Note that g has no children.
4. List g.
5. List b (the parent of g).
6. The ordered children of h are w and p.
7. The ordered children of w are s and t.
8. Note that s has no children.
9. List s.

10. List w (the parent of s).
11. Note that t has no children.
12. List t.
13. Note that the parent w of t has been listed and that there are no children of w that have not been listed.
14. List h (the parent of w).
15. Note that p has no children.
16. List p.
17. Note that the parent h of p has been listed and that there are no children of h that have not been listed.
18. Note that the parent b of h has been listed and that there are no children of b that have not been listed.
19. List a (the parent of b).
20. Note that c has no children.
21. List c.
22. The ordered children of d are j, k, and m.
23. Note that j has no children.
24. List j.
25. List d (the parent of j).
26. Note that k has no children.
27. List k.
28. Note that the parent d of k has been listed.
29. The ordered children of m are q and r.
30. The ordered children of q are u and v.
31. Note that u has no children.
32. List u.
33. List q (the parent of u).
34. Note that v has no children.
35. List v.
36. Note that the parent q of v has been listed and that there are no children of q that have not been listed.
37. List m (the parent of q).
38. Note that r has no children.
39. List r.
40. Note that the parent m of r has been listed and that there are no children of m that have not been listed.
41. Note that the parent d of m has been listed and that there are no children of d that have not been listed.
42. Note that the parent a of m has been listed and that there are no children of a that have not been listed.
43. Since a is the root, stop.

The order of vertices is shown in Figure 12.47.

Observe the procedure we follow to determine the first vertex to list. We consider the first child of the root, then we consider the first child of

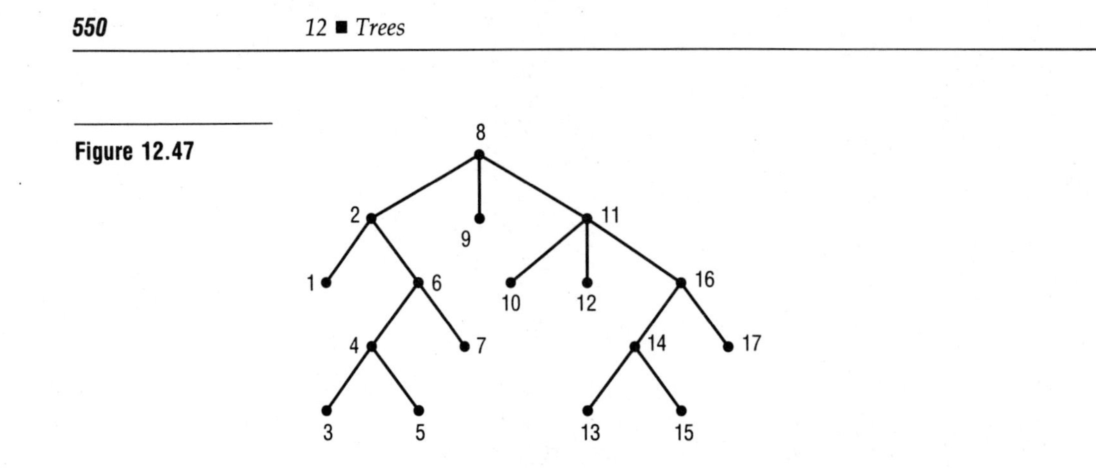

Figure 12.47

this first child, and we continue this process until we obtain a vertex x with no children. Then we list x. Also observe the procedure we follow to determine which vertex to list after a given vertex y. If y has a parent z that has not been listed, we next list z. If the parent of y has been listed, but there are siblings of y that have not been listed, we consider the first such sibling of y. Then we consider the first child of this first sibling, and we continue this process until we obtain a vertex z with no children. Then we list z. If the parent of y and all the siblings of y have been listed, then we consider the parent of the parent of y. We continue this process and we know that we are finished when we reach the root and find that it and all its children have been listed.

Example 26

List the variables and operations that result when the inorder traversal algorithm is applied to the expression tree for $(a + b)^2 + (c - d)/3$.

Analysis

The expression tree for $(a + b)^2 + (c - d)/3$ and the order that results when we apply the inorder traversal algorithm to this expression tree are shown in Figure 12.48. Thus the list of variables and operations that result when the inorder traversal algorithm is applied to the expression tree for $(a + b)^2 + (c - d)/3$ is $a + b \wedge 2 + c - d \,/\, 3$.

Figure 12.48

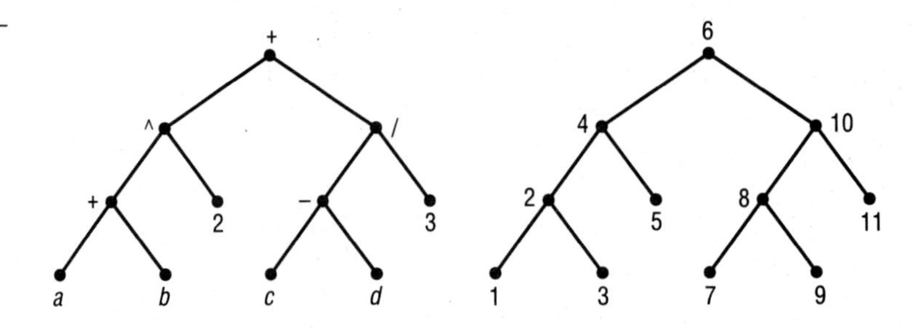

Notice that with Polish notation the operation signs precede the variables, whereas with reverse Polish notation the operation signs follow the variables. In the list of variables and operations that results when the inorder traversal algorithm is applied to an expression tree, the operation signs are between variables; we must, however, insert parentheses to get the correct result. ■

There is a simple trick to labeling the vertices of a tree using the preorder traversal algorithm, which shows that this algorithm really is a search algorithm. Think of the tree as a system of corridors and imagine that you have been blindfolded and placed at its root. Put your right hand on the back wall and travel through the entire system of corridors without ever taking your hand off the wall. List each vertex the first time you come to it.

As the terminology suggests, the trick to labeling the vertices of a tree using the postorder traversal algorithm is the old trick done backward. Start at the root, but this time put your left hand on the back wall. List each vertex the first time you come to it, but this time count down: The order of T, the order of T minus one, the order of T minus two . . . , as if you were planning on launching a rocket.

We can offer no hand-on-the-wall trick for the inorder traversal algorithm. Perhaps this is because both Polish notation and reverse Polish notation are more natural than the notation we are all used to.

Exercises 12.5

54. Draw an expression tree for each of the following expressions.

a. $(a + b)c$ **b.** $(a + b)(c + d)$

c. $(a + b)/3$ **d.** $(a + b)^4$

e. $[(a + b)/2](c + d/3)$ **f.** $a(b(c(d + 2)))$

55. In which order does the preorder traversal algorithm list the vertices of each of the following ordered rooted trees?

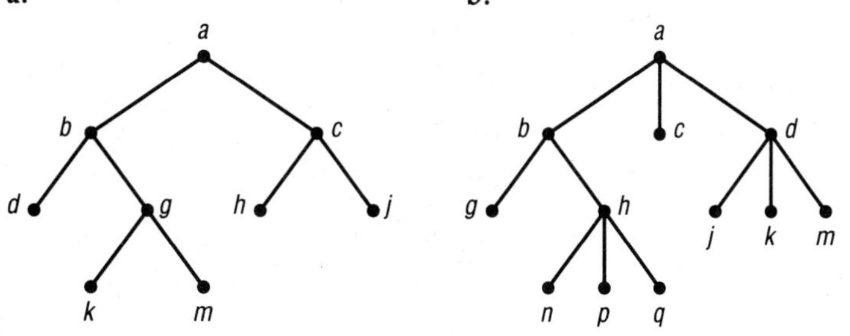

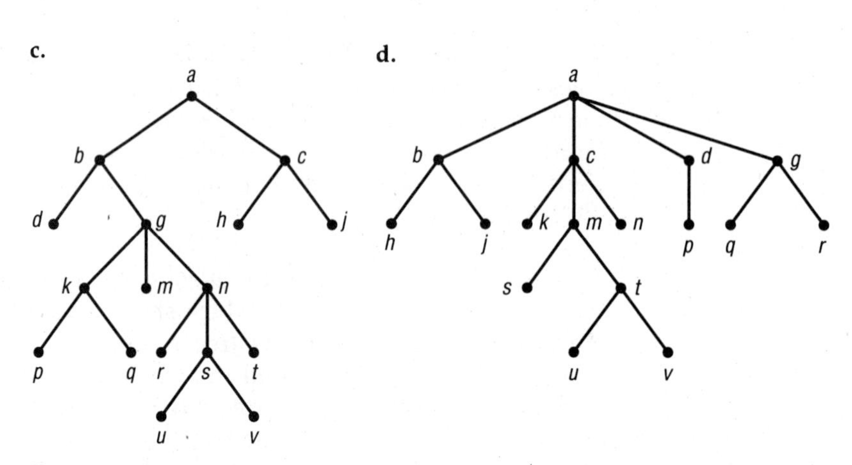

56. Construct the expression tree for $ab + (c - d)/3 + e^5$ and apply the preorder traversal algorithm to obtain the Polish notation for this expression.
57. Construct the expression tree for $(a - b)^4 + (cd)/e$ and apply the preorder traversal algorithm to obtain the Polish notation for this expression.
58. In which order does the postorder traversal algorithm list the vertices of each of the ordered rooted trees in Exercise 55.
59. Apply the postorder traversal algorithm to the expression tree constructed in Exercise 56 to obtain the reverse Polish notation for the expression $ab + (c - d)/3 + e^5$.
60. Apply the postorder traversal algorithm to the expression tree constructed in Exercise 57 to obtain the reverse Polish notation for the expression $(a - b)^4 + (cd)/e$.
61. In which order does the inorder traversal algorithm list the vertices of each of the ordered rooted trees in Exercise 55.
62. List the variables and operations that result when the inorder traversal algorithm is applied to the expression tree for $(a + b)/5 - (c + d)^3e$.

12.6

OPTIMAL BINARY TREES

Computers use strings of 0s and 1s to represent symbols, and each 0 or 1 is called a *bit*. In ASCII (American Standard Code for Information Interchange), each symbol is represented by a string of eight bits. For example, the ASCII code for A, B, C, D, and E is

A: 10100001

B: 10100010

C: 10100011

D: 10100100

E: 10100101

Thus we write *BAD* as 101000101010000110100100. As you can see, we write the ASCII code for *B*, immediately follow it by the ASCII code for *A*, and immediately follow this by the ASCII code for *D*. To translate a long string of bits into its ASCII symbols, we find the ASCII symbol represented by the first eight bits, the ASCII symbol represented by the second eight bits, and so on. Thus the translation of 101001001010010110100011 is *DEC*.

ASCII is a fixed-length representation; that is, each symbol is represented by a code word of the same length. In situations involving large-volume storage, it is more efficient to use code words of variable lengths, where symbols that are used frequently have shorter code words. In the English language, *S, E, T, O, A,* and *I* are used more frequently than *V, K, J, Q, X,* and *Z*. Thus an efficient way to build a code using these twelve letters is the following:

S: 0	*V*: 000
E: 1	*K*: 001
T: 00	*J*: 010
O: 01	*Q*: 100
A: 10	*X*: 101
I: 11	*Z*: 011

Alas, with this assignment of code words we have a problem with decoding. For example, suppose we have 01101011. This could be *ZSXE* or it could be *OAAI*. In general, should we start by looking at the first digit, the first two digits, or the first three digits? The following code, using eleven letters, avoids this problem:

S: 01	*D*: 00100
A: 00000	*H*: 00111
C: 00011	*T*: 11
G: 00110	*B*: 00001
E: 10	*F*: 00101
I: 00010	

Any time we see two 0s, we know that these two 0s together with the next three bits represent a letter. Otherwise, a letter is represented by two bits. So, for example, 0100000011010 represents *SAGE*.

Definition

A set of code words is said to have the **prefix property** if no code word is the first part of any other code word.

The set of code words just illustrated has the prefix property; the set preceding it does not. Observe the similarity between a set of code words having the prefix property and the choice of alphabets in Chapter 4. To decode a string of bits, where the set of code words has the prefix property, we read one digit at a time until the string of digits represents a code word.

As we have seen, an efficient method of coding should use a set of code words such that (a) symbols used frequently have shorter code words than symbols used infrequently, and (b) the set of code words has the prefix property.

We can use any regular binary tree T to construct a set of code words with the prefix property. We assign 0 to each edge that connects a parent to its left child and 1 to each edge that connects a parent to its right child. Then, given any leaf t of T, we use the string of bits along the unique path from the root r to t as a code word. The set of all such strings is a set of code words with the prefix property, because each string ends at a leaf and hence cannot be part of another code word.

Example 27

Use the regular binary tree shown in Figure 12.49 to construct a set of code words with the prefix property.

Figure 12.49

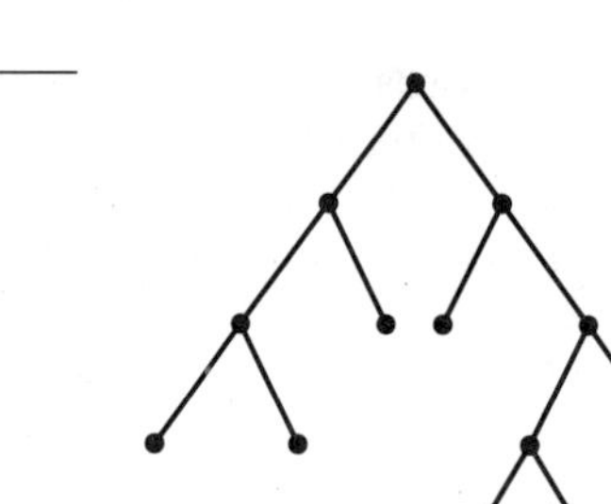

Analysis

We begin by assigning 0s and 1s to the edges as previously explained. This assignment is shown in Figure 12.50. Next we list the code words using the unique path from the root to each leaf. This procedure determines a set of code words: 000, 001, 01, 10, 1100, 1101, and 111. There are seven code words because the tree has seven leaves and this set of

Figure 12.50

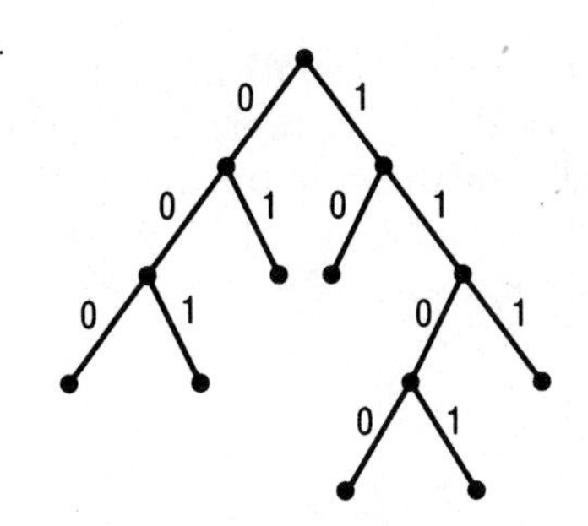

code words has the prefix property. To complete the task, we assign the most frequently used symbols to the code words associated with the leaves closest to the root. ■

Definition

> Let $k \in \mathbb{N}$ and let $w_1, w_2, \ldots, w_k$ be nonnegative real numbers. A regular binary tree with k leaves labeled $w_1, w_2, \ldots, w_k$ is called a **binary tree for the weights** $w_1, w_2, \ldots, w_k$. For each $i = 1, 2, \ldots, k$, let d_i be the length of the unique path from the root to w_i. Then $\sum_{i=1}^{k} d_i w_i$ is the **weight** of the binary tree for the weights $w_1, w_2, \ldots, w_k$.

Example 28

Construct a binary tree for the weights 1, 3, 5, 7, and 8, and find the weight of this binary tree.

Analysis

We have five weights, so the binary tree must have five leaves. We begin by constructing a regular binary tree with five leaves as shown in Figure 12.51a. Then we use the weights to label the leaves, as shown in Figure 12.51b. The weight of this tree is

$$2 \cdot 1 + 2 \cdot 3 + 2 \cdot 5 + 3 \cdot 7 + 3 \cdot 8 = 2 + 6 + 10 + 21 + 24 = 63$$

Figure 12.51

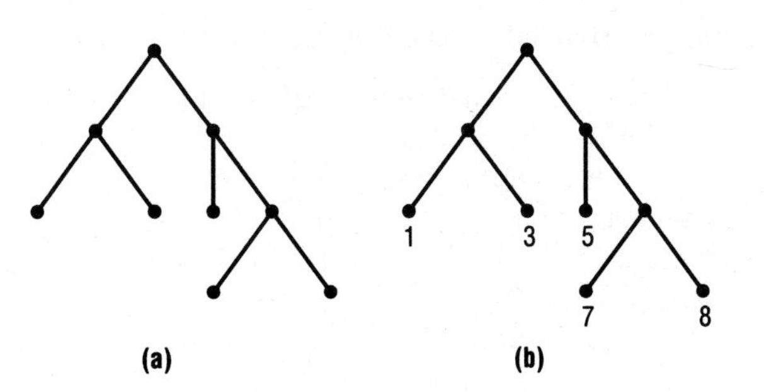

Given a set of weights, there are several binary trees for these weights; for example, in Figure 12.52 we draw two more binary trees for the weights 1, 3, 5, 7, and 8. The weight of the tree on the left is $1 \cdot 5 + 2 \cdot 7 + 3 \cdot 8 + 4 \cdot 1 + 4 \cdot 3 = 5 + 14 + 24 + 4 + 12 = 59$, and the weight of the tree on the right is $2 \cdot 1 + 2 \cdot 3 + 3 \cdot 5 + 3 \cdot 7 + 2 \cdot 8 = 2 + 6 + 15 + 21 + 16 = 60$.

Figure 12.52

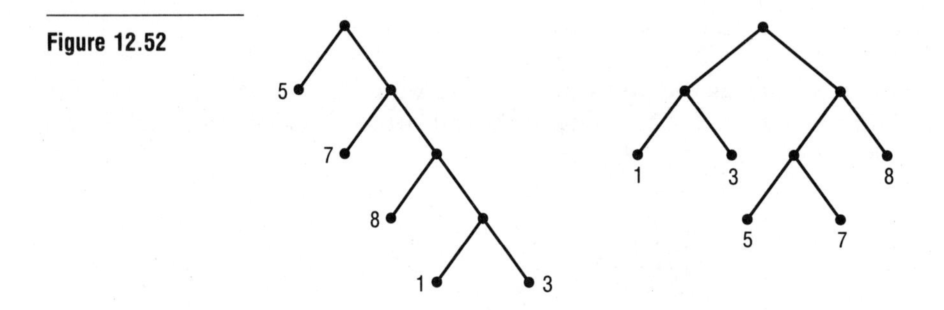

■

To construct codes, we want a regular binary tree with the smallest weight, in which the weights represent the frequencies of the symbols.

Definition

If $k \in \mathbb{N}$ and $w_1, w_2, \ldots, w_k$ are nonnegative real numbers, an **optimal binary tree for the weights** $w_1, w_2, \ldots, w_k$ is a regular binary tree for the weights $w_1, w_2, \ldots, w_k$ whose weight is as small as possible.

Let $k \in \mathbb{N}$ and let $w_1, w_2, \ldots, w_k$ be nonnegative real numbers. One algorithm for finding an optimal binary tree is due to David A. Huffman. At each stage, the two smallest weights, say u and v, are replaced by their sum, $u + v$, and a regular binary tree is constructed. We summarize the algorithm as follows:

Huffman's Optimal Binary Tree Algorithm

1. If there are at least two weights on the list, select two, u and v, of smallest value and go to step 3.
2. Stop. The tree we have constructed is an optimal binary tree for the weights $w_1, w_2, \ldots, w_k$.
3. Construct a regular binary tree whose root is assigned the label $u + v$, whose left child is assigned the label u, and whose right child is assigned the label v. If any regular binary tree whose roots are assigned the labels u and v have been constructed, include

these in the regular binary tree currently being constructed. Replace u and v in the list by $u + v$, and go to step 1.

Example 29

Use Huffman's optimal binary tree algorithm to construct an optimal binary tree for the weights 1, 3, 5, 7, and 8.

Analysis

1. Select 1 and 3.

3. Construct a binary tree whose root is assigned the label $4 = 1 + 3$, whose left child is assigned the label 1, and whose right child is assigned the label 3 (see Figure 12.53). Replace 1 and 3 on the list by 4; the list of weights is now 4, 5, 7, and 8. Go to step 1.

Figure 12.53

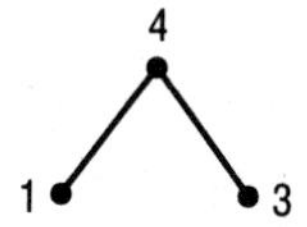

1. Select 4 and 5.

3. Construct a binary tree whose root is assigned the label $9 = 4 + 5$, whose left child is assigned the label 4, and whose right child is assigned the label 5 (see Figure 12.54). Since 4 is the label assigned the root of the tree shown in Figure 12.53, include this

Figure 12.54

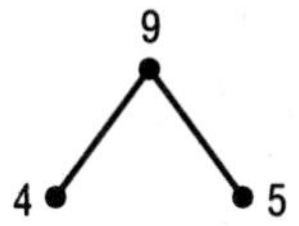

tree in the construction. The result is the tree shown in Figure 12.55. Replace 4 and 5 on the list by 9; the list of weights is now 9, 7, and 8. Go to step 1.

Figure 12.55

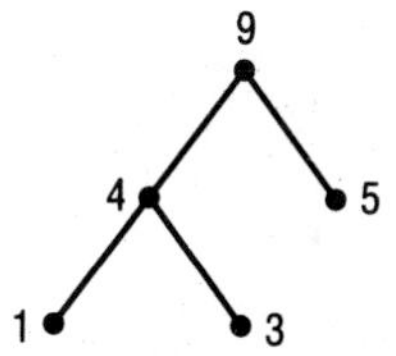

1. Select 7 and 8.

3. Construct a binary tree whose root is assigned the label $15 = 7 + 8$, whose left child is assigned the label 7, and whose right child is assigned the label 8 (see Figure 12.56). Replace 7 and 8 on the list by 15; the list of weights is now 9 and 15. Go to step 1.

Figure 12.56

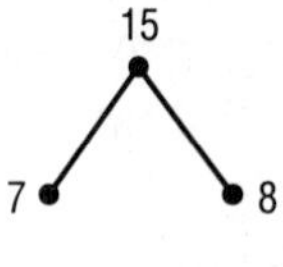

1. Select 9 and 15.
3. Construct a binary tree whose root is assigned the label 24 = 9 + 15, whose left child is assigned the label 9, and whose right child is assigned the label 15 (see Figure 12.57). Since 9 is the label assigned the root of the tree shown in Figure 12.55, include this

Figure 12.57

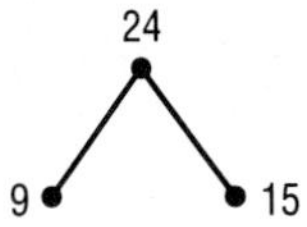

tree in the construction. Since 15 is the label assigned the root of the tree shown in Figure 12.56, include this tree in the construction. The result is the tree shown in Figure 12.58. Replace 9 and 15 on the list by 24, and go to step 1.
2. There is only one weight on the list, so stop. The tree in Figure 12.58 is an optimal binary tree for the weights 1, 3, 5, 7, and 8.

Figure 12.58

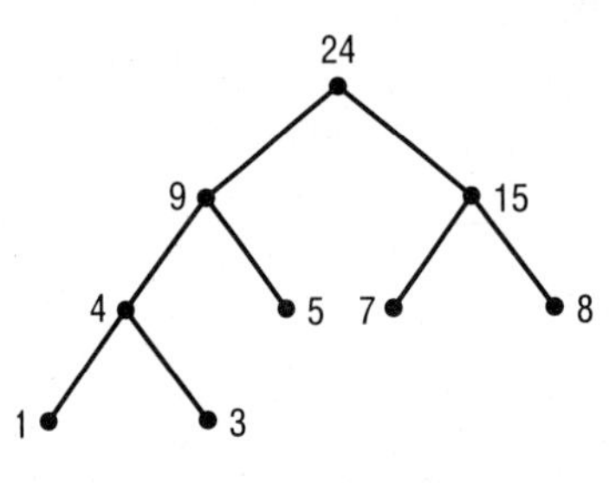

The weight of this tree is $3 \cdot 1 + 3 \cdot 3 + 2 \cdot 5 + 2 \cdot 7 + 2 \cdot 8 = 3 + 9 + 10 + 14 + 16 = 52$. ■

It is a corollary of the following theorem that Huffman's optimal binary tree algorithm provides an optimal binary tree for the weights $w_1, w_2, \ldots, w_k$.

Theorem 12.10

Let $k \in \mathbb{N}$ and let $w_1, w_2, \ldots, w_k$ be nonnegative real numbers such that $w_l \leqslant w_2 \leqslant \cdots \leqslant w_k$.

a. If T is an optimal binary tree for the weights $w_1, w_2, \ldots, w_k$ and i and j are natural numbers such that $w_i < w_j$, then the distance from the root of T to w_i is greater than or equal to the distance from the root of T to w_j.

b. There is an optimal binary tree for the weights $w_1, w_2, \ldots, w_k$ such that w_1 and w_2 are children of the same parent.

c. If T is an optimal binary tree for the weights $w_1 + w_2, w_3, \ldots, w_k$, then the tree obtained from T by replacing the leaf labeled $w_1 + w_2$ by a binary tree whose root is labeled $w_1 + w_2$, whose left child is a leaf labeled w_1, and whose right child is a leaf labeled w_2 is an optimal binary tree for the weights $w_1, w_2, \ldots, w_k$.

To find a set of code words with the prefix property such that the symbols used frequently have shorter code words than the symbols used infrequently, we construct an optimal binary tree with the frequencies of the symbols as its weights. Then we assign 0s and 1s to the edges of this tree as described just prior to Example 27. In the remainder of this section, we use the following list of frequencies (Table 12.1) of the letters of the alphabet:

Table 12.1

A:	73	*G:*	16	*L:*	35	*Q:*	3	*V:*	13
B:	9	*H:*	35	*M:*	25	*R:*	77	*W:*	16
C:	30	*I:*	74	*N:*	78	*S:*	63	*X:*	5
D:	44	*J:*	2	*O:*	74	*T:*	93	*Y:*	19
E:	130	*K:*	3	*P:*	27	*U:*	27	*Z:*	1
F:	28								

Example 30

Use the frequencies of Table 12.1 as weights and construct a set of code words for the first nine letters of the English alphabet from Huffman's optimal binary tree algorithm.

Analysis

1. Select 9 and 16.
3. Construct a binary tree whose root is assigned the label 25 = 9 + 16, whose left child is assigned the label 9, and whose right child is assigned the label 16 (see Figure 12.59). Replace 9 and 16 on the list by 25; the list of weights is now 25, 28, 30, 35, 44, 73, 74, 130. Go to step 1.

Figure 12.59

25

9 16

1. Select 25 and 28.
3. Construct a binary tree whose root is assigned the label 53 = 25 + 28, whose left child is assigned the label 25, and whose right child is assigned the label 28 (see Figure 12.60).

Figure 12.60

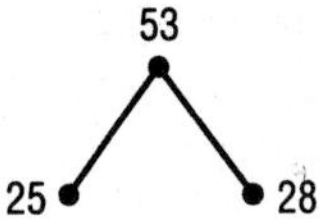

Since 25 is the label assigned the root of the tree shown in Figure 12.59, include this tree in the construction. The result is the tree shown in Figure 12.61. Replace 25 and 28 on the list by 53; the list of weights is now 53, 30, 35, 44, 73, 74, 130. Go to step 1.

Figure 12.61

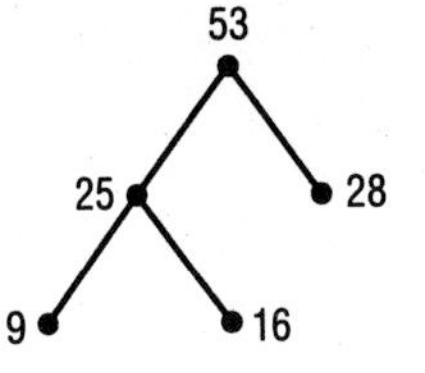

1. Select 30 and 35.
3. Construct a binary tree whose root is assigned the label 65 = 30 + 35, whose left child is assigned the label 30, and whose right child is assigned the label 35 (see Figure 12.62). Replace 30 and 35 on the list by 65; the list of weights is now 53, 65, 44, 73, 74, and 130. Go to step 1.

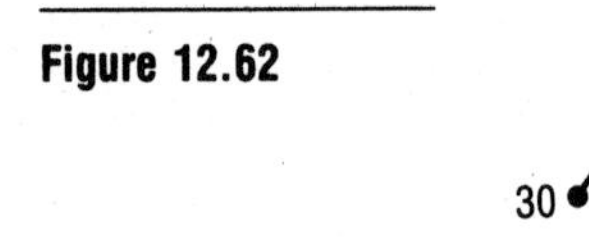

Figure 12.62

65

30 35

1. Select 44 and 53.
3. Construct a binary tree whose root is assigned the label 97 = 44 + 53, whose left child is assigned the label 44, and whose right child is assigned the label 53 (see Figure 12.63).

Figure 12.63

97

44 53

Since 53 is the label assigned the root of the tree shown in Figure 12.61, include this tree in the construction. The result is the tree shown in Figure 12.64. Replace 44 and 53 on the list by 97; the list of weights is now 97, 65, 73, 74, 130. Go to step 1.

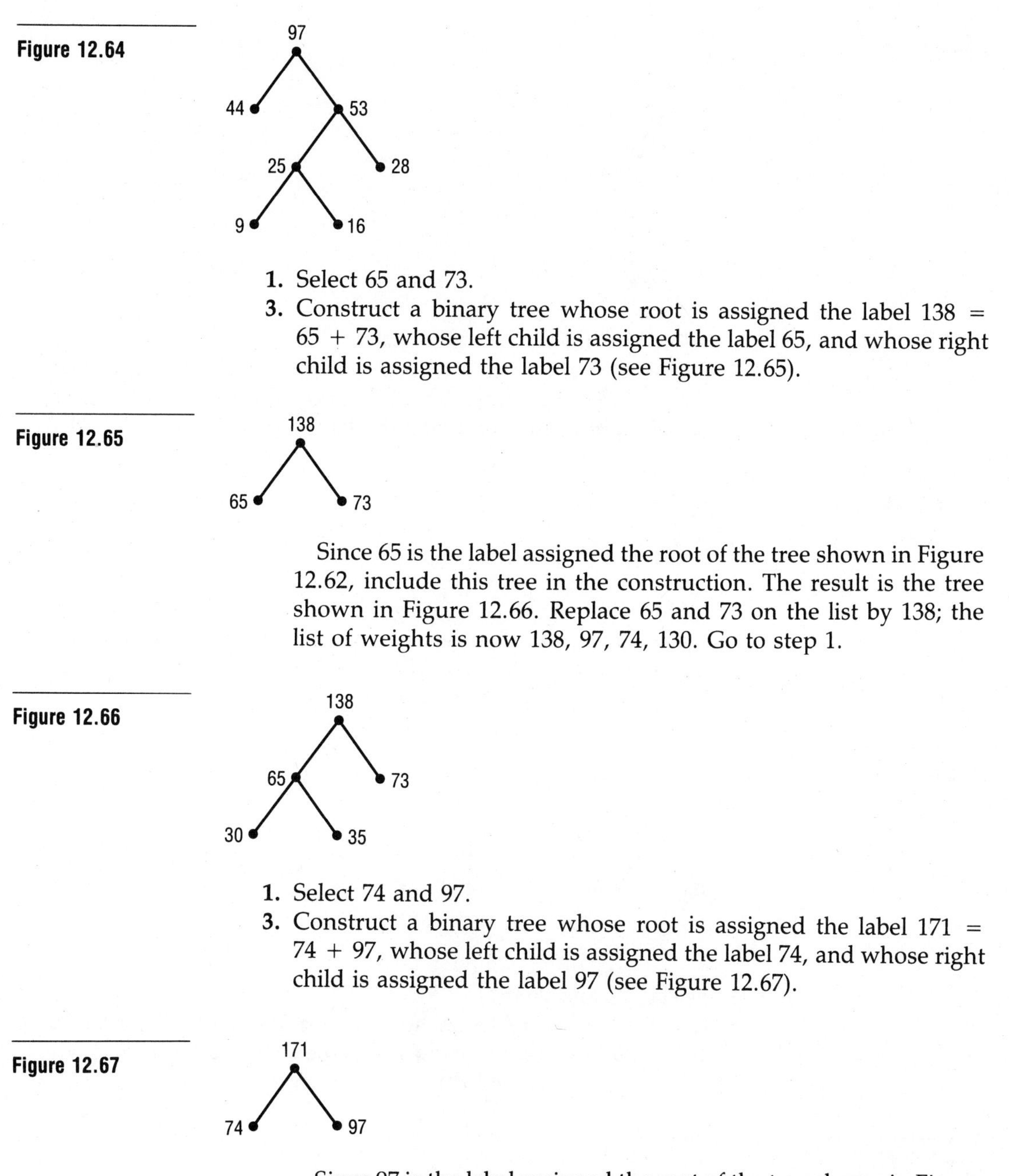

Figure 12.64

1. Select 65 and 73.
3. Construct a binary tree whose root is assigned the label 138 = 65 + 73, whose left child is assigned the label 65, and whose right child is assigned the label 73 (see Figure 12.65).

Figure 12.65

Since 65 is the label assigned the root of the tree shown in Figure 12.62, include this tree in the construction. The result is the tree shown in Figure 12.66. Replace 65 and 73 on the list by 138; the list of weights is now 138, 97, 74, 130. Go to step 1.

Figure 12.66

1. Select 74 and 97.
3. Construct a binary tree whose root is assigned the label 171 = 74 + 97, whose left child is assigned the label 74, and whose right child is assigned the label 97 (see Figure 12.67).

Figure 12.67

Since 97 is the label assigned the root of the tree shown in Figure 12.64, include this tree in the construction. The result is the tree shown in Figure 12.68. Replace 74 and 97 on the list by 171; the list of weights is now 171, 138, 130. Go to step 1.

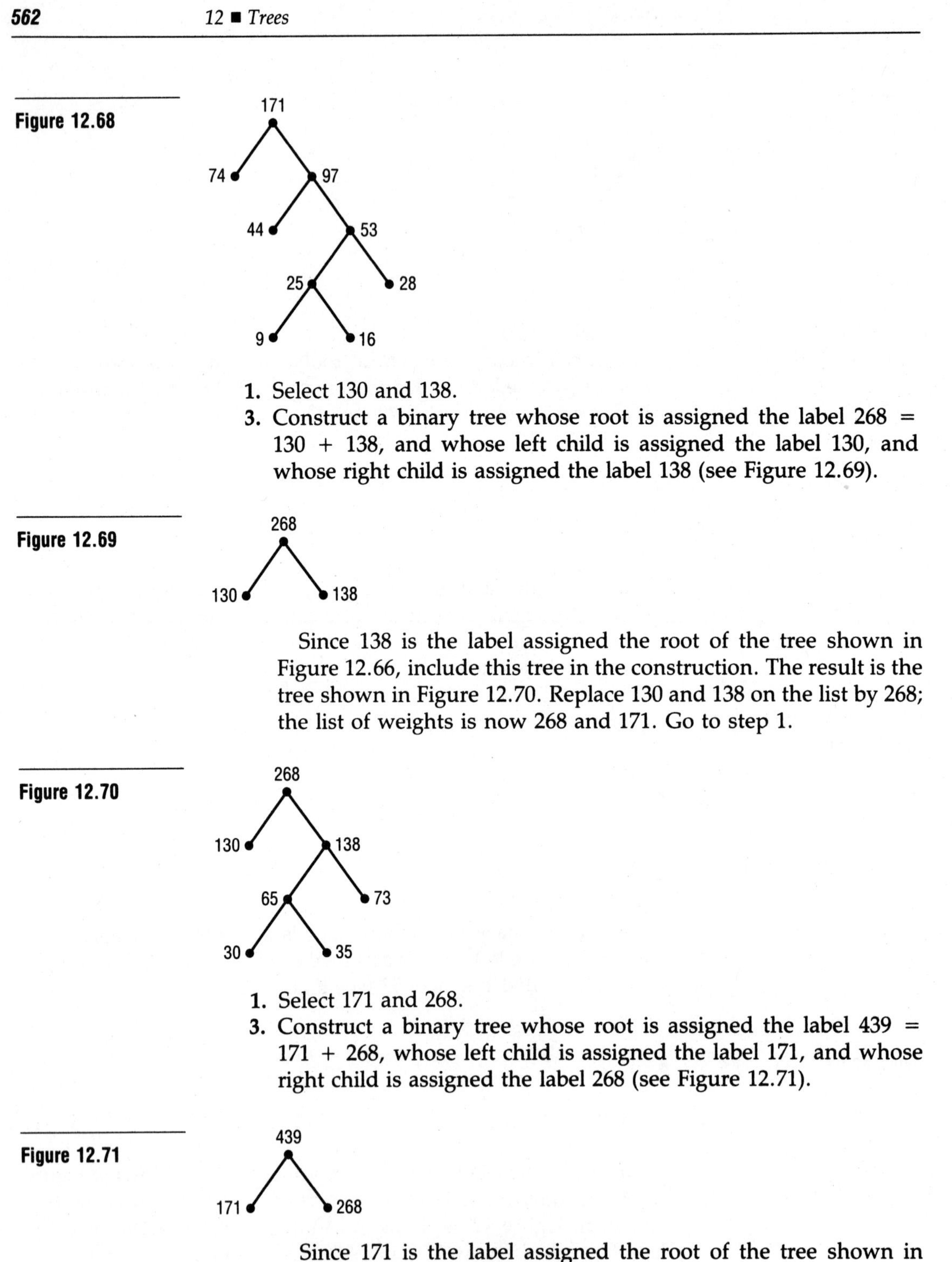

Figure 12.68

1. Select 130 and 138.
3. Construct a binary tree whose root is assigned the label 268 = 130 + 138, and whose left child is assigned the label 130, and whose right child is assigned the label 138 (see Figure 12.69).

Figure 12.69

Since 138 is the label assigned the root of the tree shown in Figure 12.66, include this tree in the construction. The result is the tree shown in Figure 12.70. Replace 130 and 138 on the list by 268; the list of weights is now 268 and 171. Go to step 1.

Figure 12.70

1. Select 171 and 268.
3. Construct a binary tree whose root is assigned the label 439 = 171 + 268, whose left child is assigned the label 171, and whose right child is assigned the label 268 (see Figure 12.71).

Figure 12.71

Since 171 is the label assigned the root of the tree shown in Figure 12.68 and 268 is the label assigned the root of the tree shown in Figure 12.70, include these trees in the construction. The result is the tree shown in Figure 12.72. Replace 171 and 268 on the list by 439. Go to step 2.

Figure 12.72

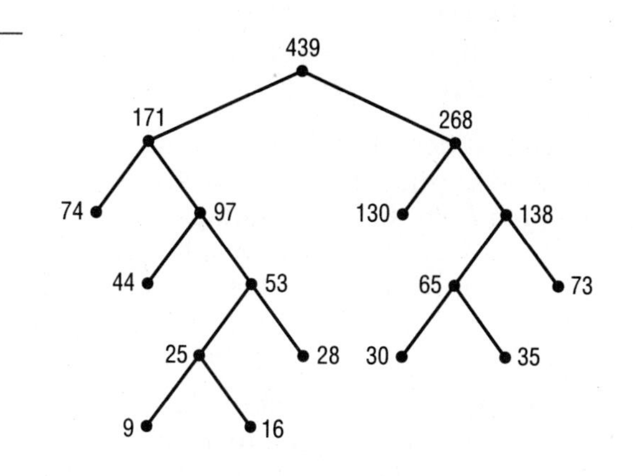

2. There is only one weight on the list, so stop. The tree in Figure 12.72 is an optimal binary tree for the weights 9, 16, 28, 30, 35, 44, 73, 74, 130.

To construct code words, we first assign 0s and 1s to the edges of the tree in Figure 12.72 as shown in Figure 12.73. Next we list the code

Figure 12.73

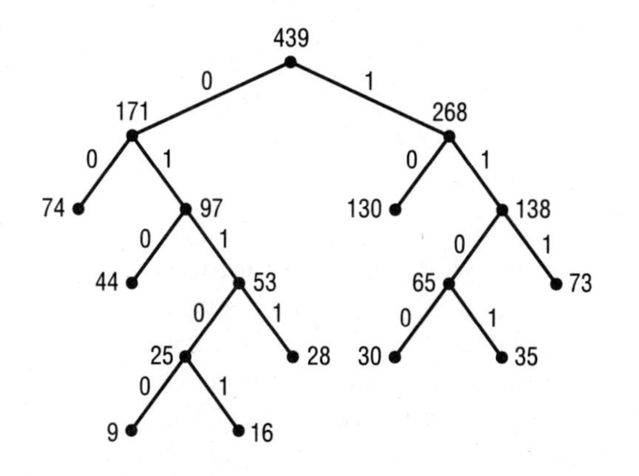

words by using the unique path from the root to each leaf. The set of code words is as follows:

00	010	01100	01101	0111
10	1100	1101	111	

The most frequently used letter is E, so we want to assign E a shortest code word. We have two choices, 00 or 10. Suppose we choose 00. The next most frequently used letter is I, so we assign it the code word 10.

The next most frequently used letter is A, so we have two choices, 010 or 111. Suppose we choose 010. We continue in this manner and obtain the following list of code words for the first nine letters of the alphabet:

A: 010	D: 111	G: 01100
B: 01101	E: 00	H: 0111
C: 1100	F: 1101	I: 10

■

A binary tree that describes the code words can be used to decode a string of bits. Given a string of bits, we start at the root of the binary tree and follow the path from the root indicated by the bits until we reach a leaf. Then the string of bits that we have followed is decoded by the code word at the leaf. Then we go back to the root and start over with the next digit.

Example 31

Suppose that the binary tree in Figure 12.74 describes our code words (the labels on the leaves refer to the frequencies in Table 12.1, p. 559). Decode the string 011100101111100011.

Figure 12.74

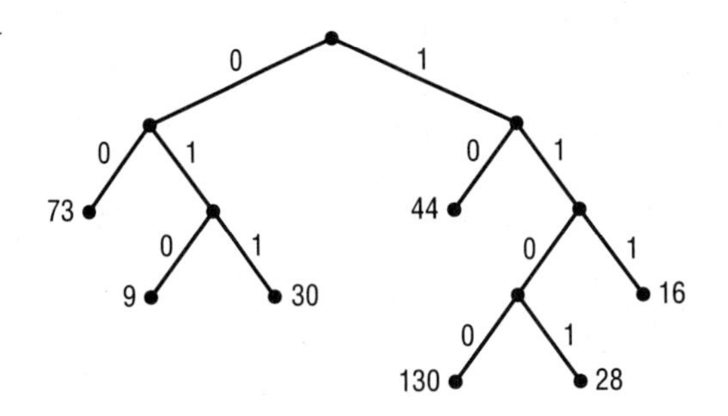

Analysis

We begin at the root, travel along the edges labeled 0, 1, and 1, and reach the leaf labeled 30. Then we begin at the root again, travel along the edges labeled 1 and 0, and reach leaf labeled 44. We begin again at the root, this time travel along the edges labeled 0, 1, and 0, and arrive at the leaf labeled 9. Then we begin at the root, travel along the edges labeled 1, 1, and 1, and arrive at the leaf labeled 16. Next we begin at the root, travel along the edges labeled 1, 1, 0, and 0, and arrive at the leaf labeled 130. Finally, we begin at the root, travel along the edges labeled 0, 1, and 1, and arrive at the leaf labeled 30. Thus our frequency labels are 30, 44, 9, 16, 130, and 30. Using Table 12.1, we see the message is *CDBGEC*. ■

Recall that, if L_1 and L_2 are sorted lists of numbers with n_1 and n_2 members, respectively, then L_1 and L_2 can be merged into one sorted list with at most $2n_1 + 2n_2 - 1$ comparisons. Suppose that we have more than two sorted lists of numbers. By merging two lists at a time, how can we minimize the number of comparisons needed to merge all the sorted lists into one sorted list? Huffman's optimal binary tree algorithm can be used to answer this question. The procedure has already been explained, so we simply give an example.

Example 32

Suppose that we have six sorted lists of 25, 30, 40, 55, 75, and 100 numbers. Use Huffman's optimal binary tree algorithm to find the minimum number of comparisons needed to merge these six sorted lists into one sorted list.

Analysis

We begin by merging the two sorted lists with 25 and 30 members. We obtain one sorted list with 55 members, and $2 \cdot 25 + 2 \cdot 30 - 1 = 109$ comparisons are needed.

Then we merge the two sorted lists with 40 and 55 members. We obtain one sorted list with 95 members, and $2 \cdot 40 + 2 \cdot 55 - 1 = 189$ comparisons are needed. Since we have a binary tree with root 55, we include it in the tree we are constructing.

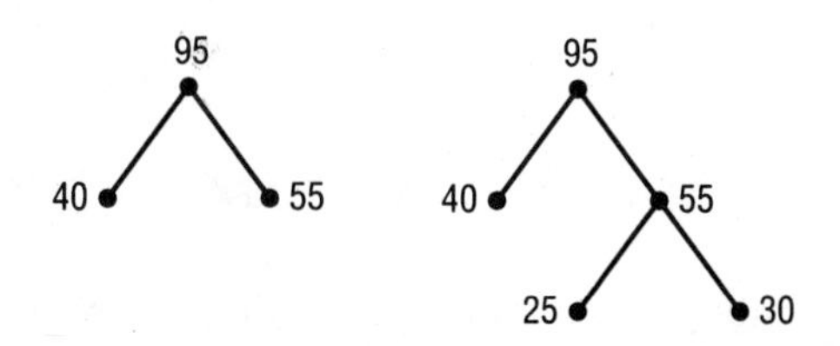

Next we merge the two sorted lists with 55 and 75 members. We obtain one sorted list with 130 members, and $2 \cdot 55 + 2 \cdot 75 - 1 = 259$ comparisons are needed. We have already merged the tree with root 55 in the tree with root 95. Thus we have the trees shown here.

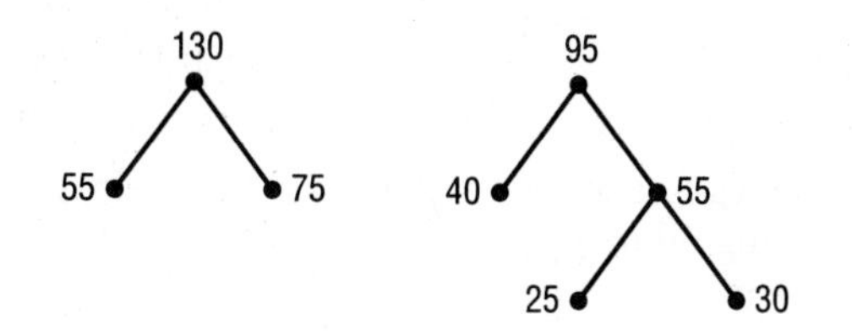

We merge the two sorted lists with 95 and 100 members and obtain one sorted list with 195 members, and $2 \cdot 95 + 2 \cdot 100 - 1 = 389$ comparisons are needed. Since we have a binary tree with root 95, we include it in the tree we are constructing. We have the two trees shown here on the right.

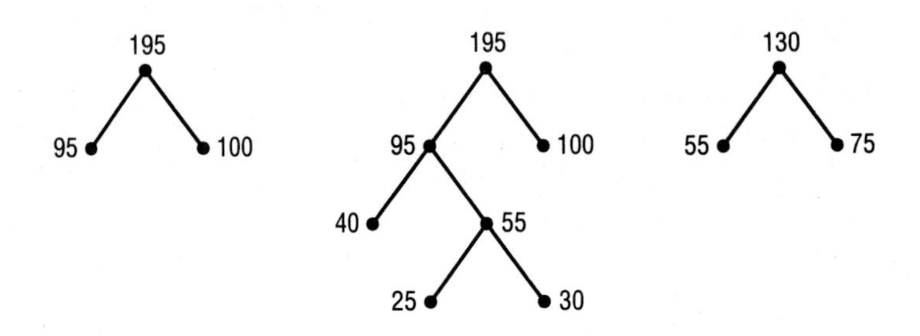

Finally, we merge the two sorted lists with 130 and 195 members and obtain one sorted list with 325 members, and $2 \cdot 130 + 2 \cdot 195 - 1 = 649$ comparisons are needed. Since we have binary trees with roots 130 and 195, we include them in the tree we are constructing. The result is the tree shown here on the right.

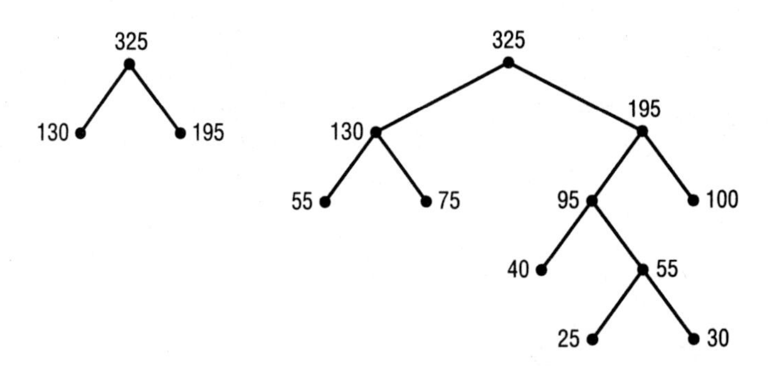

The total number of comparisons is $109 + 189 + 259 + 389 + 649 = 1{,}595$. ■

Exercises 12.6

63. Determine whether each of the following sets of code words has the prefix property.

a. {0,11,10,01}
b. {0,10,11,101}
c. {0, 11,100,101}
d. {00,11,100,011,010}
e. {01,10,000,001,110, 111}
f. {010,011,110,001}

64. What is the largest set of code words with the prefix property that can be constructed using no more than three digits, where each digit is a bit?

65. What is the largest set of code words with the prefix property that can be constructed using no more than four digits, where each digit is a bit?

66. Use each of the following regular binary trees to construct a set of code words with the prefix property.

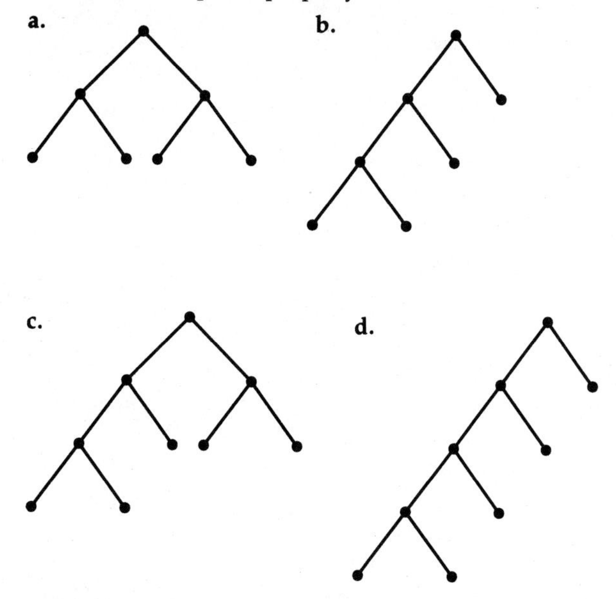

67. Without using Huffman's optimal binary tree algorithm, construct two regular binary trees for the weights 2, 3, 5, and 7 and find the weights of these trees.

68. Without using Huffman's optimal binary tree algorithm, construct two regular binary trees for the weights 4, 5, 7, 9, 11, and 13 and find the weights of these trees.

69. Construct a regular binary tree that generates a set of six code words with the prefix property. What are these six code words?

70. Construct a regular binary tree that generates a set of eleven code words with the prefix property. What are these eleven code words?

71. Given the assignment *E*: 11, *H*: 000, *L*: 10, *O*: 001, and *Y*: 01, decode the message 000001011011.

72. Given the assignment *A*: 000, *L*: 001, *O*: 01, *S*: 10, *U*: 110, and *Y*: 111, decode the message 10000000100111100101110.

73. Using the following regular binary tree and the frequencies in Table 12.1 (p. 559), decode the following messages. (*Note:* 74 could be *I* or *O*. Choose the one you feel is more appropriate.)

a. 000100011100000101

b. 1111010101000110000010010101000111111

c. 00001000111110011100001000000100101 01

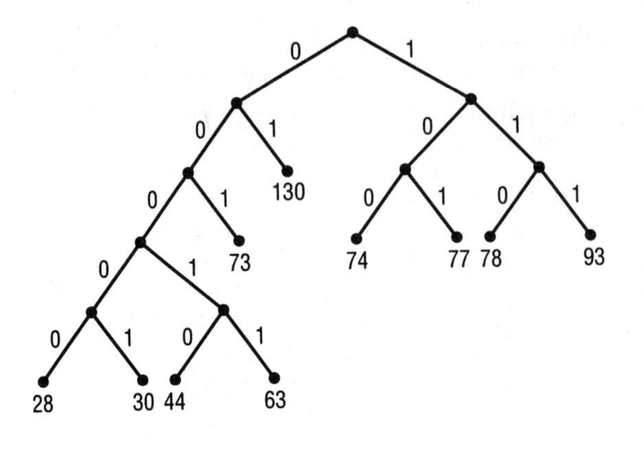

74. Use Huffman's optimal binary tree algorithm to construct an optimal binary tree for each of the following sets of weights.

a. 2,3,5,7 **b.** 8,9,11,13,17 **c.** 4,5,7,9,11,13

75. Use the frequencies in Table 12.1 (p. 559) and Huffman's optimal binary tree algorithm to construct a set of code words for the letters A, B, K, N, R, and T (by using these frequencies as weights).

76. Use Huffman's optimal binary tree algorithm to find the minimum number of comparisons needed to merge sorted lists with the given number of members into one sorted list.

a. 30,73,101 **b.** 8,31,47,61 **c.** 5,17,31,41,91

77. Prove that for each natural number n greater than or equal to 2 there is a regular binary tree with exactly n leaves.

12.7

BINARY SEARCH TREES

State University, with an enrollment of 25,000 students, maintains a list of students registered for Math 101. If, on the fourth day of classes, Tom Jones submits a request to register for Math 101, the registrar needs to know if Tom is already registered for the course, and if not she needs to add Tom's name to the roll. Later, if Kim Aaron drops the course, the registrar needs to remove Kim's name from the roll. How should the university maintain such a list on the computer? If the university simply maintains the list according to the order in which students register for Math 101, it will be necessary to search every name to see if Tom Jones is registered. If the university maintains the list in alphabetical order, it will be necessary to reposition almost all the names after Kim Aaron drops the course. Another method of storing the list is to store names at the vertices of a binary tree, where the binary tree is arranged so that, if a vertex v belongs to the left subtree of a vertex u, then v precedes u in alphabetical order, and if a vertex w belongs to the right subtree of a vertex u, then u precedes w in alphabetical order. Figure 12.75 gives an example of a binary tree that is labeled and arranged in this manner.

Figure 12.75

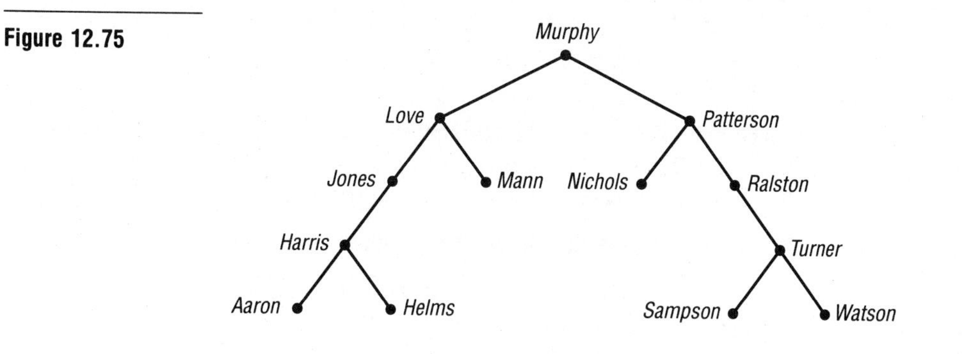

Definition

Suppose that we have a list of L of distinct real numbers or words and that the list is ordered by the usual "less than" or by alphabetical order. In either case, we use the symbol $<$ to denote the order. A **binary search tree** for L is a binary tree in which each vertex is labeled by a member of the list so that (a) each member of the list is used to label exactly one vertex, (b) a vertex v belongs to the left subtree of a vertex u if and only if $v < u$, and (c) a vertex w belongs to the right subtree of a vertex u if and only if $u < w$.

The tree in Figure 12.75 is a binary search tree for the list *Aaron, Harris, Helms, Jones, Love, Mann, Murphy, Nichols, Patterson, Ralston, Sampson, Turner, Watson.*

Example 33

Construct a binary search tree for the list 2, 3, 5, 7, 9, 11, 13.

Analysis

A binary search tree for a list is certainly not unique. In particular, we can choose any member of the list for the root. In this case, suppose we choose 5. Then 2 and 3 must belong to the left subtree of 5, and 7, 9, 11, and 13 must belong to the right subtree of 5. With this choice, the binary search tree for the list 2, 3, 5, 7, 9, 11, 13 is shown in Figure 12.76.

Figure 12.76

■

There are algorithms for constructing binary search trees. In the algorithm we discuss, we assume that $a_1, a_2, \ldots, a_n$ are distinct members of a linearly ordered set. We summarize the algorithm as follows:

Binary Search Tree Construction Algorithm

1. Construct a root and label it a_1. If $n = 1$, stop.
2. Let $v = a_1$ and $k = 2$.
3. If $v < a_k$, go to step 8.
4. ($a_k < v$) If v has a left child, u, go to step 7.
5. ($a_k < v$ and v has no left child) Construct a left child for v and label it a_k. If $k = n$, stop.
6. Replace k by $k + 1$, replace v by a_1, and go to step 3.
7. ($a_k < v$ and v has a left child) Replace v by u and go to step 3.
8. ($v < a_k$) If v has a right child w, go to step 11.
9. ($v < a_k$ and v has no right child) Construct a right child for v and label it a_k. If $k = n$, stop.
10. Replace k by $k + 1$, replace v by a_1, and go to step 3.
11. ($v < a_k$ and v has a right child) Replace v by w and go to step 3.

Example 34 illustrates the binary search tree construction algorithm.

Example 34

Use the binary search tree construction algorithm to construct a binary search tree for the list 7, 5, 3, 9, 11, 6, 8.

Analysis

1. Construct a root and label it 7.
2. Let $v = 7$ and $k = 2$.
3. $7 \nless 5$.
4. 7 has no left child.
5. Construct a left child for 7 and label it 5 (see Figure 12.77).

Figure 12.77

6. $k = 3$ and $v = 7$.
3. $7 \not< 3$.
4. 7 has a left child 5.
7. $v = 5$.
3. $5 \not< 3$.
4. 5 has no left child.
5. Construct a left child for 5 and label it 3 (see Figure 12.78).

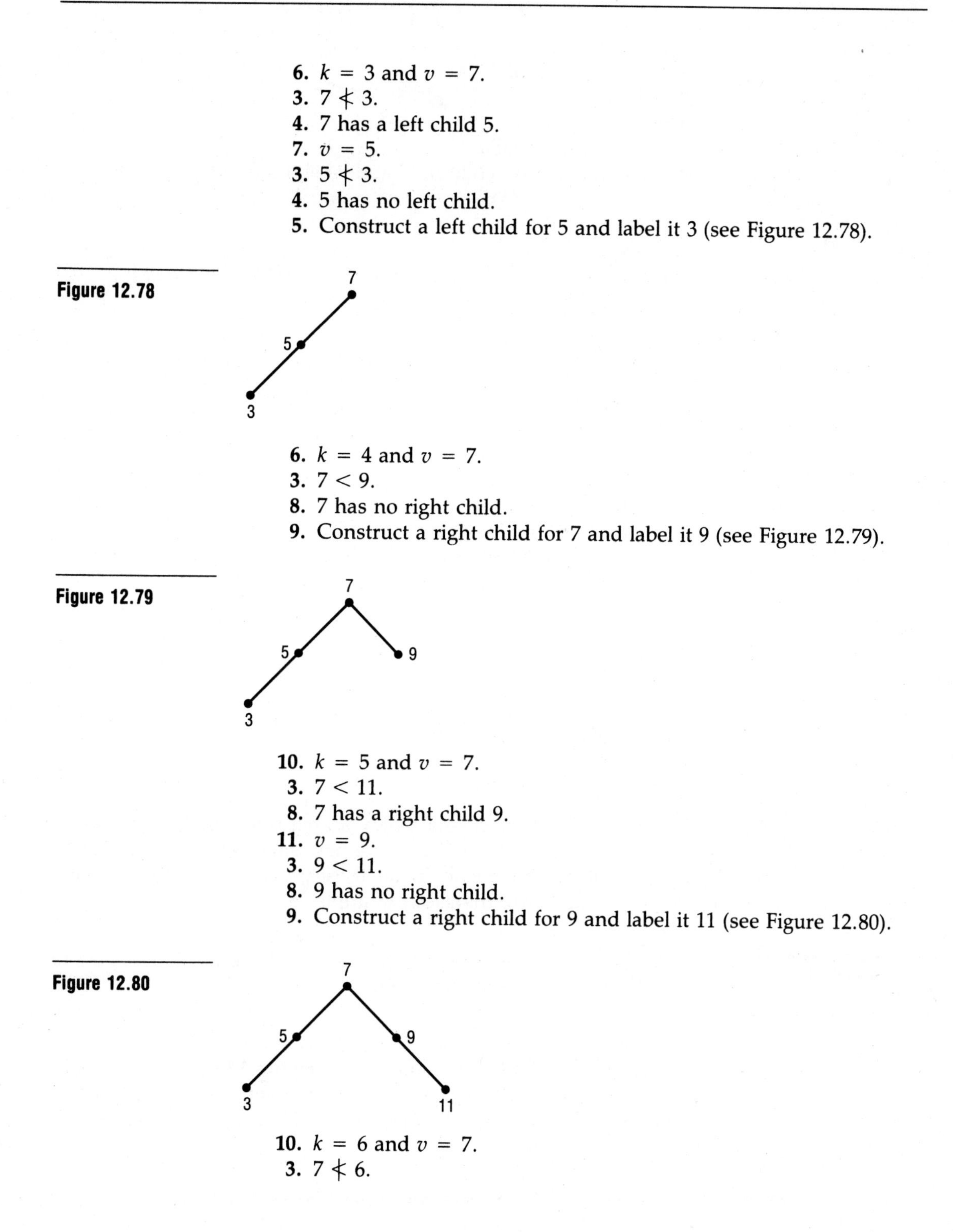

Figure 12.78

6. $k = 4$ and $v = 7$.
3. $7 < 9$.
8. 7 has no right child.
9. Construct a right child for 7 and label it 9 (see Figure 12.79).

Figure 12.79

10. $k = 5$ and $v = 7$.
3. $7 < 11$.
8. 7 has a right child 9.
11. $v = 9$.
3. $9 < 11$.
8. 9 has no right child.
9. Construct a right child for 9 and label it 11 (see Figure 12.80).

Figure 12.80

10. $k = 6$ and $v = 7$.
3. $7 \not< 6$.

4. 7 has a left child 5.
7. $v = 5$.
3. $5 < 6$.
8. 5 has no right child.
9. Construct a right child for 5 and label it 6 (see Figure 12.81).

Figure 12.81

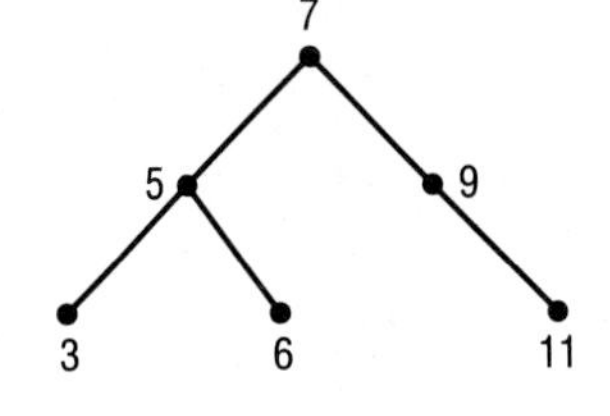

10. $k = 7$ and $v = 7$.
3. $7 < 8$.
8. 7 has a right child 9.
11. $v = 9$.
3. $9 \nless 8$.
4. 9 has no left child.
5. Construct a left child for 9 and label it 8 (see Figure 12.82). $k = n$, so stop.

Figure 12.82

■

Adding a new member to a binary search tree is easy, because we need only add one new vertex and one new edge. We illustrate this by showing how to add a name to the binary search tree in Figure 12.75 and how to add a number to the list in Example 33.

Example 35

Add *Obenshain* to the binary search tree in Figure 12.75 and add 8 to the binary search tree in Figure 12.76.

Analysis

With respect to the list determined by Figure 12.75, *Obenshain* falls between *Nichols* and *Patterson*. So we want the vertex labeled *Obenshain* to belong to the left subtree of *Patterson* and the right subtree of *Nichols*. This is shown in the tree in Figure 12.83.

Figure 12.83

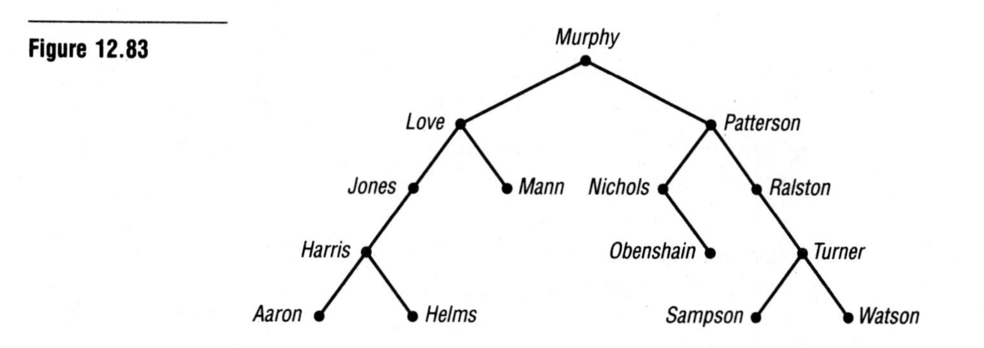

With respect to Example 33, 8 falls between 7 and 9, so we want the vertex labeled 8 to belong to the left subtree of 9 and the right subtree of 7. This is shown in Figure 12.84.

Figure 12.84

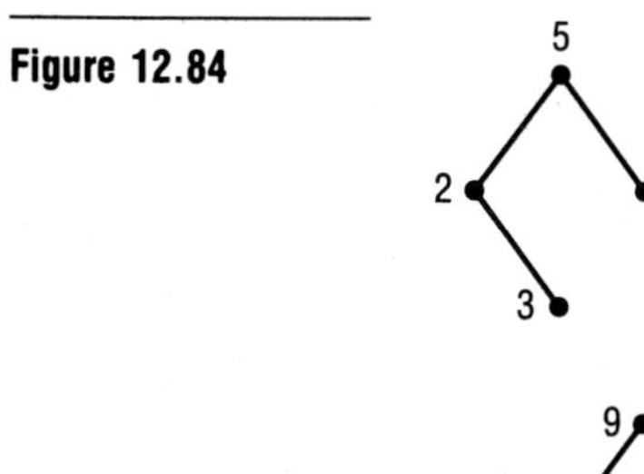

■

The procedure for determining if an item belongs to a binary search tree and for locating it if it does belong is similar to the binary search algorithm mentioned in Chapter 5. We assume that all objects under discussion belong to some linearly ordered set so that comparisons can be made. We begin by comparing the item with the root. We go left if it is less than the root and right if it is greater than the root. We continue until the item is located or we reach a leaf (different from the item), in which case the item does not belong to the binary search tree. (Note that, if we want to add the item to the binary search tree, we add it at this leaf.) We summarize an algorithm for determining if a given item a belongs to a binary search tree below:

Binary Search Tree Search Algorithm

1. Let v be the root of the binary search tree.
2. If $v = a$, stop. Item a has been located in the binary search tree.
3. If $v < a$, go to step 6.
4. If there is a left child u of v, replace v by u and go to step 2.

5. Stop. Item a is not in the binary search tree.
6. If there is a right child w of v, replace v by w and go to step 2.
7. Stop. Item a is not in the binary search tree.

We conclude this section with an illustration of the use of the binary search tree search algorithm.

Example 36

Use the binary search tree search algorithm to search for *fail* in the binary search tree shown in Figure 12.85.

Figure 12.85

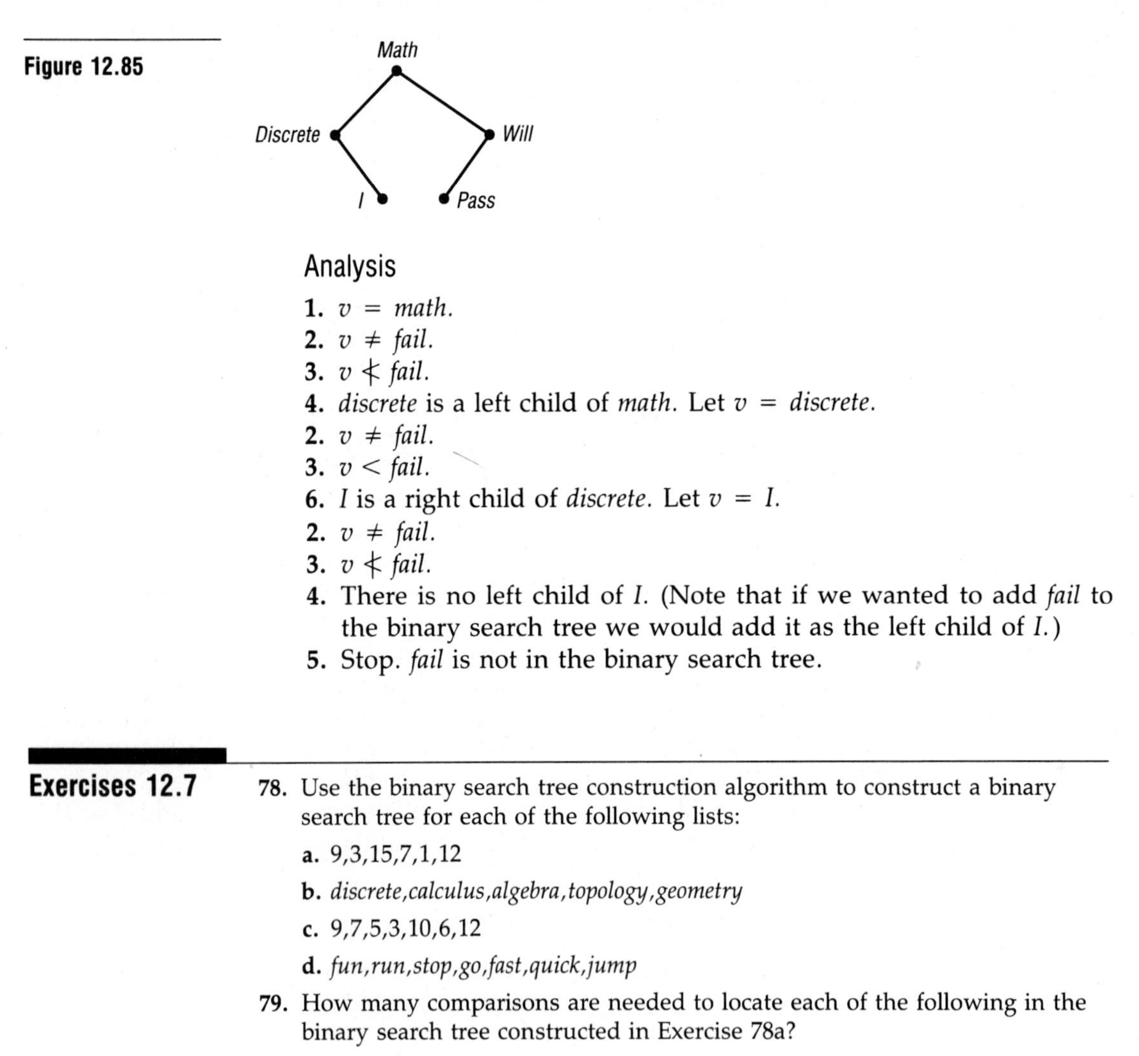

Analysis

1. $v = math$.
2. $v \neq fail$.
3. $v \nless fail$.
4. *discrete* is a left child of *math*. Let $v = discrete$.
2. $v \neq fail$.
3. $v < fail$.
6. I is a right child of *discrete*. Let $v = I$.
2. $v \neq fail$.
3. $v \nless fail$.
4. There is no left child of I. (Note that if we wanted to add *fail* to the binary search tree we would add it as the left child of I.)
5. Stop. *fail* is not in the binary search tree.

Exercises 12.7

78. Use the binary search tree construction algorithm to construct a binary search tree for each of the following lists:

    a. 9,3,15,7,1,12

    b. *discrete,calculus,algebra,topology,geometry*

    c. 9,7,5,3,10,6,12

    d. *fun,run,stop,go,fast,quick,jump*

79. How many comparisons are needed to locate each of the following in the binary search tree constructed in Exercise 78a?

    a. 1 b. 15 c. 12

80. How many comparisons are needed to locate each of the following in the binary search tree constructed in Exercise 78d?

a. *stop* **b.** *fast* **c.** *jump*

81. Explain the use of the binary search tree search algorithm to search for 9 in the binary search tree in Figure 12.84.

82. Explain the use of the binary search tree search algorithm to search for *Fenner* in the binary search tree in Figure 12.83.

83. Add *Rutland* to the binary search tree in Figure 12.83.

84. Add *calculus* and *topology* to the binary search tree in Figure 12.85.

85. Delete *Mann* from the binary search tree in Figure 12.83.

86. Delete *Jones* from the binary search tree in Figure 12.83.

87. Delete *Turner* from the binary search tree in Figure 12.83.

Chapter 12 Review Exercises

88. Which of the following graphs are trees?

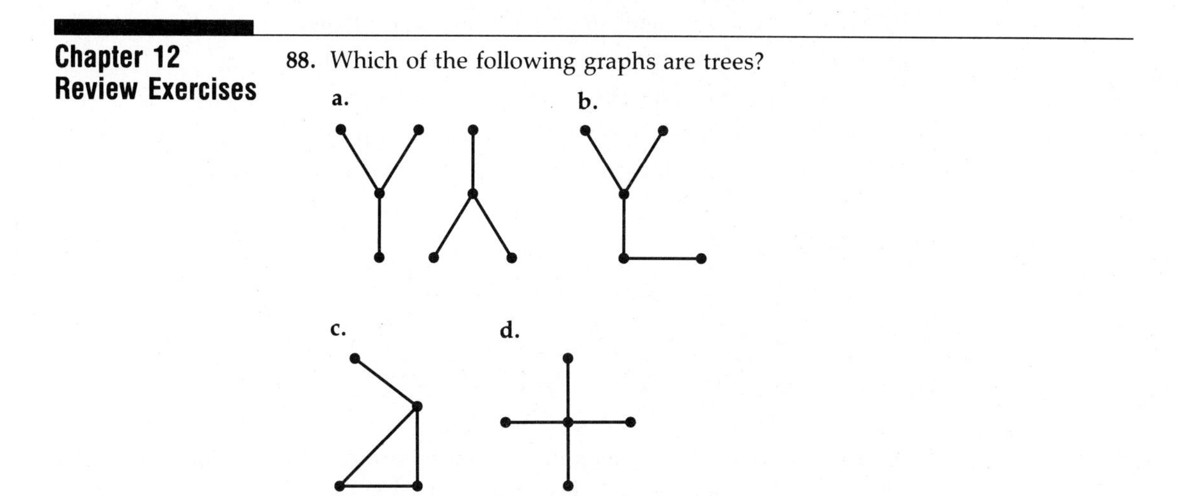

89. Draw all trees that have four edges.

90. How many paths are there in a tree with ten edges?

91. **a.** Construct a spanning tree for the following connected graph.

b. How many different spanning trees are there for this graph?

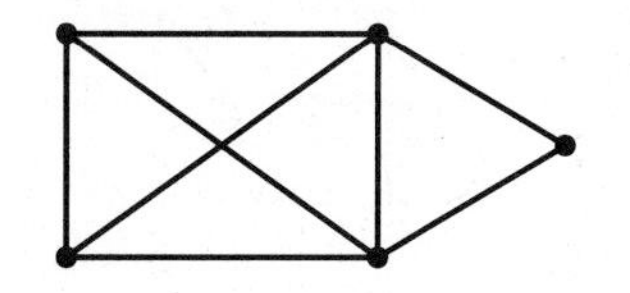

92. Prove that if $n > 1$ then the number of distinct trees with n vertices is n^{n-2}. (*Hint:* Use Prüfer's algorithm.)

93. For each of the following rooted trees, find the level numbers of each vertex and the height of the tree.

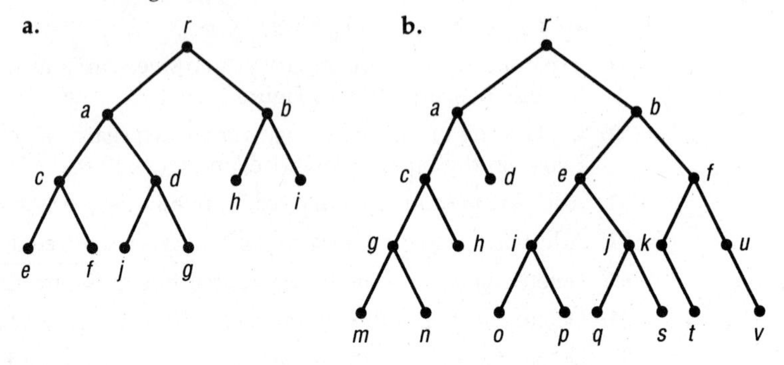

94. How many leaves does each tree in Exercise 93 have?

95. How many descendants does the vertex b have in each tree in Exercise 93?

96. **a.** Draw the Fibonacci tree T_8. **b.** What is the height of T_8?

c. How many leaves does T_8 have?

d. How many descendants does the root of T_8 have?

97. Use Prüfer's algorithm to find the list $a_1, a_2, \ldots, a_{n-2}$ associated with each of the following trees.

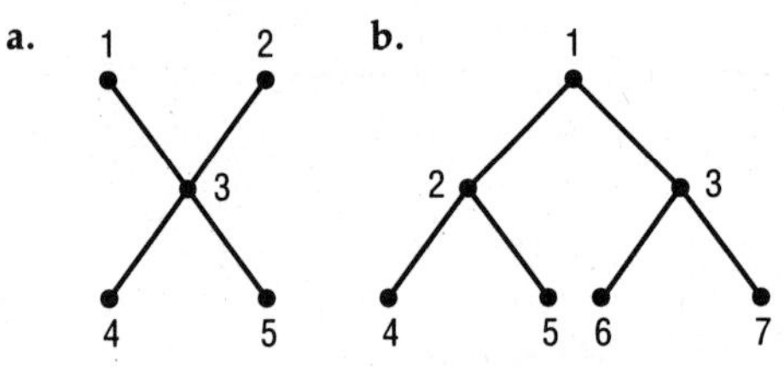

98. Use the converse of Prüfer's algorithm to construct each of the following trees.

a. the tree whose vertices are labeled 1,2,3,4,5 from the list L: 1,2,2

b. the tree whose vertices are labeled 1,2,3,4,5 from the list L: 2,2,1

99. Use the breadth-first search spanning tree algorithm to find a spanning tree for the following simple graph.

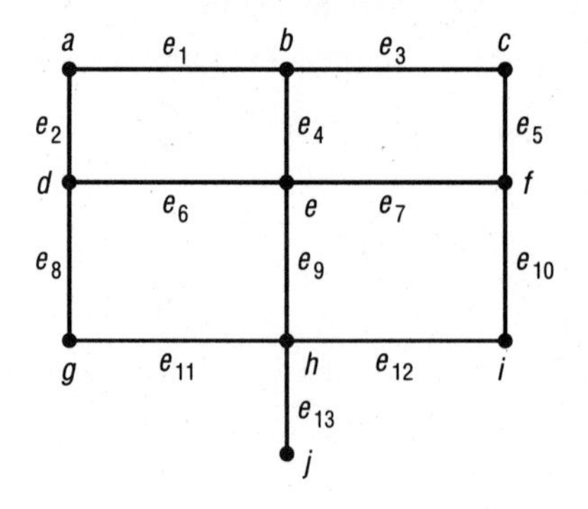

100. Use the depth-first search algorithm to find a spanning tree for the simple graph in Exercise 99.

101. Use Kruskal's algorithm to find a minimal spanning tree for each of the following weighted graphs.

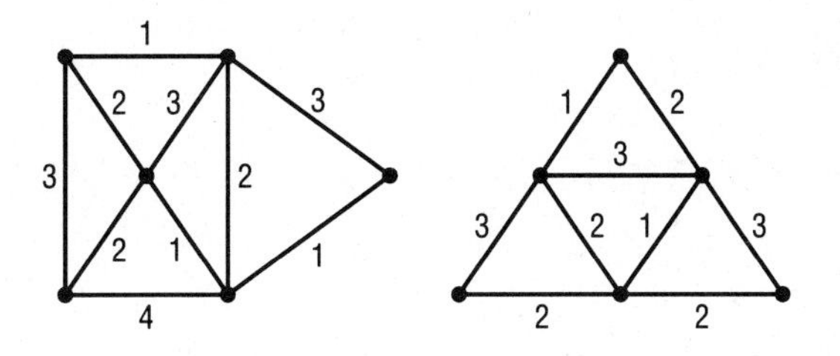

102. Find the weights of the minimal spanning trees that you found in Exercise 101.

103. Use Prim's algorithm to find a minimal spanning tree for each of the weighted graphs in Exercise 101.

104. Find the weights of the minimal spanning trees that you found in Exercise 103.

105. Draw an expression tree for each of the following expressions.

a. $(1 + 2 \cdot 3) \cdot 4$ **b.** $1 \cdot 3 + 2 \cdot 4$

106. Write the preorder list for each tree in Exercise 105.

107. Write the postorder list for each tree in Exercise 105.

108. Write the inorder list for each tree in Exercise 105.

109. Determine whether each of the following sets of code words has the prefix property.

a. {00,01,100,101} **b.** {01,00100,00101,01000} **c.** {000,001,010,011,1}

110. Use the binary search tree construction algorithm to construct a binary search tree for each of the following lists.

a. 7,3,1,2,5

b. *John, Mary, Ralph, Bill, Henry, Jane*

111. Add 6 to the tree you constructed in Exercise 110a.

112. The World Series is a seven-game series between the National League champion and the American League champion, and the first term that wins four games wins the series. Using a tree diagram, find the number of ways that the World Series can occur.

113. An algorithm is composed of a series of six conditionals. The algorithm terminates when a false conditional is reached. The algorithm proceeds to the next conditional when a true conditional is reached. Find the number of paths through the algorithm by using a tree diagram.

Chapter

13

BOOLEAN ALGEBRA REVISITED

This chapter continues the study of Boolean algebra that we began in Chapter 7. You may wish to review the axioms of a Boolean algebra given in Table 7.5. These axioms together with de Morgan's laws, $(a \vee b)' = a' \wedge b'$ and $(a \wedge b)' = a' \vee b'$, are the principle tools of this chapter.

We adopt some simplified notation. From now on, the statement that B is a Boolean algebra is shorthand for saying that $(B,\vee,\wedge,',0,1)$ is a Boolean algebra. For $a, b \in B$ we write a $\leqslant b$ provided that $a \wedge b = a$. Recall (Exercise 22 of Chapter 7) that the resulting relation is a partial order on B such that, for all $b \in B$, $0 \leq b \leq 1$. Also, if B is a Boolean algebra and a and b belong to B, we usually denote $a \wedge b$ by ab. Since meet plays a role in B similar to the role of multiplication in ordinary arithmetic, our second abbreviation is in the spirit of the familiar substitution of ab for $a \cdot b$ in ordinary arithmetic. For this reason, we refer to the "product" rather than the "meet" of elements of a Boolean algebra, and we refer to Boolean expressions such as $wxy \vee xyz \vee wxz \vee wx$ as a "join of products."

13.1

BOOLEAN EXPRESSIONS AND BOOLEAN FUNCTIONS

Let B be a Boolean algebra and let V be an alphabet of variables. It is understood that V is a nonempty set that contains none of the symbols $\vee$, $\wedge$, $'$, (,). For each variable w, both w and w' are called *literals*. The set E_B of all Boolean expressions over B is defined recursively. The members of E_B are *finite* sequences. The basis for the recursive definition of E_B is that all one-element sequences of literals belong to E_B. If it were not for the need for parentheses, we could state the inductive clause of the definition of a Boolean expression as follows: If α and β are finite sequences belonging to E_B, so are $\alpha \vee \beta$, $\alpha\beta$, α', and β'.

Unfortunately, we need to distinguish between expressions such as $(p \vee (q'))$ and $(p \vee q)'$. For this reason, we state the inductive clause using parentheses, even though in practice the required parentheses are often omitted and understood from context. For example, $p \vee q'$ means $(p \vee q')$ and not $((p \vee q)')$, and $p \vee q$ means $(p \vee q)$.

Definition

Let B be a Boolean algebra and let V be an alphabet of variables. A **Boolean expression** over B is a member of the set E_B defined recursively as follows:

a. All one-element sequences of literals belong to E_B. **(basis)**
b. If α and β are elements of E_B, so are $(\alpha \vee \beta)$, $(\alpha\beta)$, and (α'). **(inductive clause)**
c. A Boolean expression must be obtainable from the basis and the inductive clause in finitely many steps. **(extremal condition)**

Because of the extremal condition, only finitely many literals can appear in any Boolean expression, and each literal that does appear can appear only finitely many times. These restrictions hold even in the case that the alphabet of variables is infinite.

Boolean expressions, as finite sequences, are ordered, so $p \vee q$ and $q \vee p$ are two different expressions. We would like to be able to define a notion of equivalent Boolean expressions that makes sense for expressions over an arbitrary Boolean algebra equipped with an arbitrary allowable set of variables. Of course, we want our notion of equivalent expressions to square with the notion we already have in the specific case that we are considering expressions of logic or logic networks. In that specific case, two expressions are equivalent if they have the same output no matter what values from {0,1} are substituted for their variables. This idea motivates the definition of a *Boolean function of degree n,* which is the key to defining equivalent Boolean expressions in a general setting.

Definition

Let B be a Boolean algebra and let n be a natural number. A function from B^n, the set of all ordered n-tuples of members of B, into B is called a **Boolean function of degree n**. The set of all Boolean functions of degree n is denoted by $\mathbb{F}_{B,n}$, or by $\mathbb{F}_n$ if the underlying Boolean algebra is clear from context.

Exercise 30 of Chapter 7 asks if the set B of positive divisors of 15 forms a Boolean algebra when ab is defined to be the greatest common divisor of a and b and $a \vee b$ is defined to be the least common multiple of a and b. Indeed $B = \{1,3,5,15\}$ is a Boolean algebra under these definitions of product and join, and we consider Boolean functions over this Boolean algebra in the examples of this section. We promote the Boolean algebra of divisors of 15 solely because it is a small Boolean algebra different from {0,1}. Since 1 is the zero of B and 15 is its one, for all $b \in B$, $1b = 1$ and $15b = b$.

Example 1

Let B be the Boolean algebra of divisors of 15. Let x_1 and x_2 be variables. Table 13.1 defines a Boolean function of degree 2.

Table 13.1

x_1	15	15	15	15	5	5	5	5	3	3	3	3	1	1	1	1
x_2	15	5	3	1	15	5	3	1	15	5	3	1	15	5	3	1
$f((x_1, x_2))$	15	3	15	15	3	5	1	5	5	3	1	3	15	1	1	1

According to Table 13.1, $f((3,15)) = 5$ whereas $f((15,3)) = 15$. Since we already know that the domain of f is $B \times B$ and that the codomain of f is B, the function f is defined by listing the members of its graph. It is customary to list these members in tabular form, and hereafter we define Boolean functions by listing the elements of their graphs in tables such as the table of this example. ■

Example 2

Let B be the Boolean algebra of positive divisors of 15 and let x_1 and x_2 be variables. Then $x_1 \vee x_2'$ is a Boolean expression, and this expression defines a Boolean function $f: B \times B \to B$, where for all $(x_1,x_2) \in B \times B$, $f((x_1,x_2)) = x_1 \vee x_2'$. Because de Morgan's laws hold in any Boolean algebra, this same Boolean function is defined by the Boolean expression $(x_1'x_2)'$. The values of this Boolean function f of degree 2 are given in Table 13.2.

Table 13.2

x_1	15	15	15	15	5	5	5	5	3	3	3	3	1	1	1	1
x_2	15	5	3	1	15	5	3	1	15	5	3	1	15	5	3	1
$f((x_1, x_2))$	15	15	15	15	5	15	5	15	3	3	15	15	1	3	5	15

■

As Example 2 illustrates, the difference between Boolean functions and Boolean expressions is like the difference between ordinary functions and algorithms. Each Boolean expression in n variables provides an algorithm that defines a Boolean function of degree n, but there are many different Boolean expressions of degree n that define the same Boolean function. As we see in the next section, there are some Boolean functions that cannot be represented by Boolean expressions. This situation does not arise, however, if we look only at Boolean functions over the smallest Boolean algebra $B = \{0,1\}$.

Given the set $\mathbb{F}_n$ of Boolean functions of degree n over a Boolean algebra B, we may define the join, product, and complement for the members f and g of $\mathbb{F}_n$ as follows:

$f \vee g: B^n \to B$ is defined by
$$f \vee g((b_1,b_2,\ldots,b_n)) = f((b_1,b_2,\ldots,b_n)) \vee g((b_1,b_2,\ldots,b_n))$$

$f \wedge g: B^n \to B$ is defined by
$$f \wedge g((b_1,b_2,\ldots,b_n)) = f((b_1,b_2,\ldots,b_n)) \wedge g((b_1,b_2,\ldots,b_n))$$

$f': B^n \to B$ is defined by
$$f'((b_1,b_2,\ldots,b_n)) = (f((b_1,b_2,\ldots,b_n)))'$$

With these definitions, which we illustrate in Example 3, $\mathbb{F}_n$ itself becomes a Boolean algebra whose zero is the constant function that assigns each ordered n-tuple the value 0 of B and whose 1 is the constant function that assigns each ordered n-tuple the value 1 of B.

Example 3

Let B be the Boolean algebra of positive divisors of 15 and let x_1 and x_2 be variables. Then the functions of Examples 1 and 2, which we now call f_1 and f_2, are both members of $\mathbb{F}_2$, as are the functions $f_1 \vee f_2$, $f_1 \wedge f_2$, f_1', and f_2'. Moreover, for each $b \in B$, the constant function $\underline{b}$: $B \times B \rightarrow B$ belongs to $\mathbb{F}_2$ and $\underline{0}$ and $\underline{1}$ are the zero and one of the Boolean algebra $\mathbb{F}_2$. Although we have chosen the Boolean algebra B for its smallness and are considering Boolean functions of only two variables, the number of members of $\mathbb{F}_2$ is gargantuan (see Exercise 5). The Boolean functions f_1, f_2, $f_1 \vee f_2$, $f_1 \wedge f_2$, f_1', $\underline{0}$, and $\underline{1}$ are defined by Table 13.3.

Table 13.3

x_1	x_2	f_1	f_2	$f_1 \vee f_2$	$f_1 \wedge f_2$	f_1'	$\underline{0}$	$\underline{1}$
15	15	15	15	15	15	1	1	15
15	5	3	15	15	3	5	1	15
15	3	15	15	15	15	1	1	15
15	1	15	15	15	15	1	1	15
5	15	3	5	5	3	5	1	15
5	5	5	15	15	5	3	1	15
5	3	1	5	5	1	15	1	15
5	1	5	15	15	5	3	1	15
3	15	5	3	5	3	3	1	15
3	5	3	3	3	3	5	1	15
3	3	1	15	15	1	15	1	15
3	1	3	15	15	3	5	1	15
1	15	15	1	15	1	1	1	15
1	5	1	3	3	1	15	1	15
1	3	1	5	5	1	15	1	15
1	1	1	15	15	1	15	1	15

■

We now use Boolean functions to define an equivalence relation on the set E_B of all Boolean expressions over B in variables from some given alphabet V of variables.

Definition

Let B be a Boolean algebra, let V be a set of variables, and let α and β belong to E_B. We say that α and β are equivalent expressions provided that, if α determines a Boolean function f_α in n variables, β determines a Boolean function f_β in these same n variables, and these n variables include all the variables appearing in either of the expressions α and β, then the functions f_α and f_β are equal. Since f_α and f_β have the same domain and the same codomain, $f_\alpha = f_\beta$ if and only if, whenever $(b_1,b_2, \ldots, b_n)$ is an n-tuple of members of B, $f_\alpha((b_1,b_2,\ldots,b_n)) = f_\beta((b_1,b_2,\ldots,b_n))$.

Exercises 13.1

1. Let $B = \{0,1\}$ and let x_1 and x_2 be variables over B. Write the table that defines each of the following Boolean functions:

   a. f_1: $B \times B \rightarrow B$ defined by $f_1((x_1,x_2)) = x_1 \wedge x_2$

   b. f_2: $B \times B \rightarrow B$ defined by $f_2((x_1,x_2)) = x_1' \vee x_2$

   c. $f_1 \vee f_2$: $B \times B \rightarrow B$

   d. f_1': $B \times B \rightarrow B$

2. Let B be the Boolean algebra of positive divisors of 15 and let f: $B^3 \rightarrow B$ be the Boolean function defined by $f((b_1,b_2,b_3)) = b_1b_2'b_3$. Which of the following are true for all $(b_1,b_2,b_3) \in B^3$? Give a counterexample to each false statement.

   a. $(f(b_1,b_2,b_3))' = f'((b_1,b_2,b_3))$ b. $f \wedge f' = (f \wedge f)'$

   c. $f \wedge f'$ is a constant function

   d. $f((b_1,b_2,b_3)) \vee f((b_1',b_2',b_3')) = f((15,15,15))$

   e. $f'((b_1',b_2',b_3')) = f((b_1,b_2,b_3))$

3. Let $B = \{0,1\}$. Then $\mathbb{F}_2$ has sixteen members. We have listed three of these functions in the table below.

x_1	x_2	f_1	f_2	f_3	f_4	f_5	f_6	f_7	f_8	f_9	f_{10}	f_{11}	f_{12}	f_{13}	f_{14}	f_{15}	f_{16}
1	1	1	0	1													
1	0	1	1	0													
0	1	1	1	1													
0	0	1	1	1													

   a. List the remaining members.

   b. Evaluate $f_1 \vee f_3'((1,0))$, $f_1 \wedge f_2((0,1))$, and $f_1' \wedge f_2 \wedge f_3'((1,1))$.

4. Let B be the Boolean algebra of Example 1. Then $B = \{1,3,5,15\}$.

   a. How many members does $\mathbb{F}_{B,2}$ have?

   b. How many members of $\mathbb{F}_{B,2}$ are constant functions?

5. Let B be a Boolean algebra with 2^m members and let n be a positive integer. Then $\mathbb{F}_{B,n}$ has 2^x members. Give a formula for x in terms of m and n. Evaluate x in the specific case that B is the Boolean algebra of subsets of a ten-element set and $n = 15$.

6. Let $B = \{0,1\}$ and let f_1 and f_2 be Boolean functions belonging to $\mathbb{F}_{B,3}$ defined by the accompanying table.

x_1	1	1	1	1	0	0	0	0
x_2	1	1	0	0	1	1	0	0
x_3	1	0	1	0	1	0	1	0
$f_1((x_1, x_2, x_3))$	1	0	0	1	1	1	0	1
$f_2((x_1, x_2, x_3))$	0	1	1	0	1	0	1	1

Evaluate

a. $f_1 \vee f_2((1,1,1))$ **b.** $f_1 \vee (f_2')((1,1,1))$

c. $(f_1 \wedge f_2)'((1,0,1))$ **d.** $(f_1)' \wedge (f_2)((1,0,1))$

7. Does there exist a nonconstant function $f \in \mathbb{F}_{B,3}$, where $B = \{0,1\}$, such that $f((x_1',x_2,x_3)) = f((x_1,x_2',x_3)) = f((x_1,x_2,x_3'))$ for all $(x_1,x_2,x_3) \in B^3$? If so, exhibit such a function.

8. Let $B = \{0,1\}$. Since $\mathbb{F}_{B,3}$ is also a Boolean algebra, we may define a partial order $\leq$ on $\mathbb{F}_{B,3}$ in the usual way: $f \leq g$ provided that $f \wedge g = f$.

a. For f_1 and f_2 as defined in Exercise 6, is $f_1 \leq f_2$? Is $f_2 \leq f_1$?

b. Give an example of a function $f_3 \in \mathbb{F}_{B,3}$ such that $f_3 \leq f_2$, $f_3 \leq f_1$, and $f_3 \neq \underline{0}$.

c. Give an example of a function $f \in \mathbb{F}_{B,3}$ such that $f \neq \underline{0}$ and, if g is any member of $\mathbb{F}_{B,3}$ for which $g < f$, then $g = \underline{0}$.

9. Let f be the Boolean function defined in Table 13.1. How many functions g belonging to $\mathbb{F}_{B,2}$ have the property that $f \wedge g$ is the zero of the Boolean algebra $\mathbb{F}_{B,2}$?

13.2 JOIN-NORMAL FORM

In this section we are interested primarily in Boolean expressions and Boolean functions over the simplest Boolean algebra $B = \{0,1\}$. Before turning to this specific Boolean algebra, we compare Boolean functions and expressions over an arbitrary Boolean algebra with polynomial functions and polynomial expressions over the set of all real numbers.

Suppose that we are given two quadratic polynomial functions, $f(x) = a_1x^2 + b_1x + c_1$ and $g(x) = a_2x^2 + b_2x + c_2$. If $f(3) = g(3), f(11) = g(11)$, and $f(17) = g(17)$, then 3, 11, and 17 are the three roots of the polynomial function

$$(f - g)(x) = (a_1 - a_2)x^2 + (b_1 - b_2)x + (c_1 - c_2)$$

Since the only quadratic function with three roots is $h(x) = 0x^2 + 0x + 0$, $f = g$. In general, any two polynomial functions of degree n that have $n + 1$ roots in common are equal. There is a remarkable analogue of this result, sometimes referred to as the verification theorem, which holds for two Boolean functions that can be represented by Boolean expressions:

Theorem 13.1

Verification Theorem Let B be a Boolean algebra and let f and g be two Boolean functions of degree n *both of which can be represented by Boolean expressions in n variables.* Suppose that for each of the 2^n n-tuples of 0s and 1s, $(b_1,b_2,\ldots,b_n)$, $f((b_1,b_2,\ldots,b_n)) = g((b_1,b_2,\ldots,b_n))$. Then $f = g$.

Example 4

Let $U = \{a,b,c\}$ be a set with three members and let B be the Boolean algebra of subsets of U (see Theorem 7.5). Let members f and g of $\mathbb{F}_{B,2}$ be defined by $f((x,y)) = x \vee y'$ and $g((x,y)) = (yx')'$. Note that f and g have the same domain $B \times B$, which has 64 members. According to Table 13.4, $f((b_1,b_2)) = g((b_1,b_2))$ as long as $b_1, b_2 \in \{0,1\} = \{\varnothing,U\}$. Thus the verification theorem verifies that $f = g$. For example, given $x = \{a,b\}$ and $y = \{b,c\}$, we already know from Theorem 13.1 and Table 13.4 that $f((x,y)) = g((x,y))$. In fact, $f((x,y)) = \{a,b\} \cup \{b,c\}^{\sim} = \{a,b\} \cup \{a\} = \{a,b\}$ and $g((x,y)) = (\{b,c\} \cap \{a,b\}^{\sim})^{\sim} = (\{b,c\} \cap \{c\})^{\sim} = \{c\}^{\sim} = \{a,b\}$.

Table 13.4

x	y	$f((x,y))$	$g((x,y))$
U	U	U	U
U	$\varnothing$	U	U
$\varnothing$	U	$\varnothing$	$\varnothing$
$\varnothing$	$\varnothing$	U	U

It is noteworthy that to verify that $f = g$ without using Theorem 13.1 we would have to check that $f((x,y)) = g((x,y))$ for all 64 ordered pairs of subsets of U. ■

Let B be the Boolean algebra of Example 4 and let E be the collection of all Boolean functions of degree 2 that can be represented by Boolean expressions over B in two variables x and y. For each Boolean function f belonging to E we may select a Boolean expression σ representing f, and according to Theorem 13.1 f is determined by Table 13.5. Since we are using only U and $\varnothing$ for x and y, it is clear that, no matter what expression $\sigma((x,y))$ may be, the right-hand column of Table 13.5 has only the values U and $\varnothing$. Thus there are two choices for each entry of the table for a total of 2^4 possible tables. We conclude that there can be at most 2^4 Boolean functions belonging to E. (In fact, it follows from Theorem G.2 (Appendix G) that there are exactly 2^4 Boolean functions that can be represented by Boolean expressions in two variables x and y.) But since the Boolean algebra of subsets of a three-element set has 2^3 members, the table defining a typical Boolean function of degree 2 has 64 entries. To count the members of $\mathbb{F}_{B,2}$, we must realize that each of the 64 lines of a table defining a typical Boolean function of degree 2 may be given eight different values

Table 13.5

x	y	$\sigma((x,y))$
U	U	$\sigma((U,U))$
U	$\varnothing$	$\sigma((U,\varnothing))$
$\varnothing$	U	$\sigma((\varnothing,U))$
$\varnothing$	$\varnothing$	$\sigma((\varnothing,\varnothing))$

in its right-hand column. Thus $\mathbb{F}_{B,2}$ has $8^{64} = 2^{192}$ functions, and only 16 of these functions can be represented by Boolean expressions in variables x and y.

As we promised at the beginning of this section, we restrict our attention to the Boolean algebra $B = \{0,1\}$, and from now on B always denotes this Boolean algebra. The study of Boolean expressions and functions over B is vastly simpler than the study of these expressions and functions over an arbitrary Boolean algebra. There are $2^{(2^n)}$ Boolean functions in $\mathbb{F}_n$, and the tables that define Boolean functions f and g from $\mathbb{F}_n$ are the same as the tables used in the verification theorem to decide if $f = g$. Thus things are so much simpler that the verification theorem is useless. That is the bad news. The good news is that every function belonging to $\mathbb{F}_n$ can be represented by a Boolean expression in n variables.

Theorem 13.2

Let $f \in \mathbb{F}_n$. Then there is a Boolean expression σ in n variables such that for each n-tuple $(b_1,b_2,\ldots,b_n)$ in B^n, $f((b_1,b_2,\ldots,b_n)) = \sigma((b_1,b_2,\ldots,b_n))$.

We set about the task presented by Theorem 13.2. Given a particular function $f \in \mathbb{F}_n$, how can we find a Boolean expression such as the one promised by this theorem? The task is surprisingly simple. Assume for the moment that f is not the constant function $\underline{0}$, which is the zero of the Boolean algebra $\mathbb{F}_n$. This assumption is reasonable because there is no difficulty in finding a Boolean expression in n variables that represents $\underline{0}$. For example, $\sigma(x_1,x_2,\ldots,x_n) = x_1x_1'x_2x_2'\ldots x_nx_n'$ will do. Under the assumption that f is not the constant function $\underline{0}$, the trick illustrated in Example 5 always finds a Boolean expression that represents f.

Example 5

Let $f \in \mathbb{F}_3$ be the Boolean function whose table is given in Table 13.6. Find a Boolean expression that represents f.

Table 13.6

x	1	1	1	1	0	0	0	0
y	1	1	0	0	1	1	0	0
z	1	0	1	0	1	0	1	0
$f((x,y,z))$	0	1	1	1	0	0	1	0
Selected product of literals	—	xyz'	$xy'z$	$xy'z'$	—	—	$x'y'z$	—

Then $f((x,y,z)) = xyz' \vee xy'z \vee xy'z' \vee x'y'z$ (see Exercise 14). ■

Why does the trick illustrated in Example 5 work? It hinges on the following definition.

Definition

Given n variables $\{x_1,x_2,\ldots,x_n\}$, a product $v_1v_2v_3\ldots v_n$ of n literals is called a **minterm,** or **complete product,** of the variables $x_1,x_2,\ldots,x_n$ provided that for every literal v_i either $v_i = x_i$ or $v_i = x_i'$.

Like any Boolean expression, each minterm defines a Boolean function, but the Boolean functions defined by minterms are special.

Example 6

Let $\sigma_1(x,y,z) = xyz'$, $\sigma_2(x,y,z) = xy'z$, $\sigma_3(x,y,z) = xy'z'$, and $\sigma_4(x,y,z) = x'y'z$. For $i = 1,2,3,4$, let f_i be the function defined by σ_i. Then each f_i assumes the value 1 for exactly one ordered triple in B^3. Consequently, if f is the function defined by Table 13.7, $f = f_1 \vee f_2 \vee f_3 \vee f_4$. Before recording these results, let us consider the first minterm xyz'. If $x = 0$ or $y = 0$ or $z = 1$, then $f_1((x,y,z)) = xyz' = 0$. Consequently, if $f_1((x,y,z)) \neq 0$, then $x \neq 0$, *and* $y \neq 0$, *and* $z \neq 1$; that is, if $f_1((x,y,z)) = 1$, then $x = 1$, $y = 1$, and $z = 0$. Conversely, $f_1((1,1,0)) = 1$. We have seen that $f_1((1,1,0)) = 1$ and f_1 of anything else is 0. With this hint, you should be able to fill in Table 13.7 with ease (see Exercise 15)

Table 13.7

x	y	z	$f_1((x,y,z))$	$f_2((x,y,z))$	$f_3((x,y,z))$	$f_4((x,y,z))$	$f_1 \vee f_2 \vee f_3 \vee f_4$	f
1	1	1	0					0
1	1	0	1	0	0	0	1	1
1	0	1	0					1
1	0	0	0					1
0	1	1	0					0
0	1	0	0					0
0	0	1	0					1
0	0	0	0					0

■

Definition

Let f be a member of $\mathbb{F}_n$. If f is represented by a Boolean expression that is a join of minterms, we say that f is written in **join-normal form**. (Some texts, especially those concerned with logic, call join-normal form "disjunctive-normal form.")

We can make the following observations about join-normal form:

a. The constant function $\underline{0}$ is the only Boolean function that cannot be written in join-normal form.
b. Any non-zero Boolean function can be written in join-normal form using the procedure illustrated in Examples 5 and 6.

The word "normal" in the phrase "join-normal form" means "canonical" or "standard," and so this terminology would be inappropriate if any Boolean function could be written in join-normal form in two different ways. Because a minterm takes on the value 1 for exactly one n-tuple from B^n, no function has two join-normal forms. For each minterm that appears in the join-normal form of a Boolean function f there is exactly one n-tuple $(b_1,b_2,\ldots,b_n)$ in the domain of f for which $f((b_1,b_2,\ldots,b_n)) = 1$. If we throw away some minterm, there is no chance of recovering the value 1 from some other minterm, because the other minterms have the value 1 for other n-tuples (and no minterm takes on the value 1 twice). Suppose then that A and B are join-normal forms for the same Boolean function f. We have just seen that each minterm in A is needed in B and each minterm in B is needed in A. In other words, A and B are the join of exactly the same minterms, and so A is B.

We state the previous observations as a theorem, which can be thought of as an analogue for Boolean functions of the fundamental theorem of arithmetic for natural numbers.

Theorem 13.3

Every Boolean function in $\mathbb{F}_n$ other than $\underline{0}$ can be written in join-normal form and, except for the order in which the minterms are joined and the order in which the literals appear within each minterm, the expression of a Boolean function in join-normal form is unique.

Example 7

Let $f: B^3 \rightarrow B$ be defined by Table 13.8. Find a Boolean expression in these variables that represents f.

Table 13.8

x	y	z	f
1	1	1	1
1	1	0	0
1	0	1	1
1	0	0	1
0	1	1	1
0	1	0	1
0	0	1	0
0	0	0	1

Analysis

It is of course possible to express f in join-normal form as

$$f((x,y,z)) = xyz \vee xy'z \vee xy'z' \vee x'yz \vee x'yz' \vee x'y'z'$$

But there is a second approach. We may express f uniquely in meet-normal form:

$$f((x,y,z)) = (x' \vee y' \vee z) \wedge (x \vee y \vee z')$$

Here $x' \vee y' \vee z$ and $x \vee y \vee z'$ are *maxterms*, so named because they take on the value 0 exactly once. The details of this example are given in Exercise 22. ■

A non-zero element a of a Boolean algebra is called an *atom* provided that if x belongs to the Boolean algebra and $ax = x$ then $x = 0$ or $x = a$. The term *atom* arose long before the Manhattan Project and is meant to suggest the indivisible members of a Boolean algebra (other than 0). In the Boolean algebra of subsets of a nonempty set, the singleton sets are the atoms. In the Boolean algebra of positive divisors of 15, 3 and 5 are the atoms. We ask you to find the atoms of $\mathbb{F}_3$ in Exercise 29.

Exercises 13.2

10. Let $\mathcal{P}(S)$ be the Boolean algebra of subsets of $S = \{1,2,3,4\}$. Assume that f and g are both represented by Boolean expressions in four variables w, x, y, and z. Using Theorem 13.1, how many 4-tuples of subsets of S would you have to check to verify that f and g are equal? If f and g are of degree 4 but are not known to be represented by Boolean expressions, how many 4-tuples of subsets of S would you have to check?

11. Let B be the Boolean algebra of subsets of $\{1,2,3,4\}$.

a. Use Theorem 13.1 to verify that f: $B^3 \to B$ defined by $f((x,y,z,)) = ((x \vee y) \wedge z) \vee ((x \wedge y) \vee z) \vee (x' \wedge y')$ and g: $B^3 \to B$ defined by $g((x,y,z)) = (x' \vee y \vee z) \wedge (x \vee y' \vee z)$ are equal.

b. Check to see that $f((\{1,3\},\{2,3\},\{1,4\})) = g((\{1,3\},\{2,3\},\{1,4\}))$.

12. Let f: $B^2 \to B$ be defined by $f((x,y)) = x \vee (xy)$ and g: $B^2 \to B$ be defined by $g((x,y)) = x$. Use Theorem 13.1 to show that $x \vee (xy)$ and x are equivalent Boolean expressions.

13. a. Use Theorem 13.1 to test whether the Boolean expressions $(x \wedge y) \vee (x' \wedge y')$ and $(x' \vee y \vee z) \wedge (x \vee y' \vee z)$ represent the same Boolean function f: $B^3 \to B$, where B is the Boolean algebra of subsets of a 2,000-element set S.

14. Let f: $B^3 \rightarrow B$ be the function defined in Table 13.6 of Example 5. Show that for the following ordered triples (a,b,c), $abc' \vee ab'c \vee ab'c' \vee a'b'c = f((a,b,c))$.

a. $(a,b,c) = (1,1,1)$ **b.** $(a,b,c) = (0,1,1)$

c. $(a,b,c) = (0,1,0)$ **d.** $(a,b,c) = (1,0,0)$

15. Fill in Table 13.7 with the values that define f_1, f_2, f_3, f_4, and $f_1 \vee f_2 \vee f_3 \vee f_4$.

16. Let $B = \{0,1\}$ and let f_1, f_2, f_3, and f_4 be the functions mapping B^2 into B defined in the following table.

a. Write f_1 in join-normal form. **b.** Write f_2 in join-normal form.

c. Write f_3 in join-normal form. **d.** Write f_4 in join-normal form.

x	y	$f_1((x,y))$	$f_2((x,y))$	$f_3((x,y))$	$f_4((x,y))$
1	1	1	1	0	1
1	0	0	0	0	1
0	1	0	1	1	1
0	0	1	1	1	1

17. Let $B = \{0,1\}$ and let f_1, f_2, and f_3 be the functions mapping B^3 into B defined in the following table.

a. Write f_1 in join-normal form. **b.** Write f_1' in join-normal form.

c. Write $f_1 \wedge f_2$ in join-normal form.

d. Find an expression that determines f_3.

e. Find an expression that determines f_3'.

x	y	z	$f_1((x,y,z))$	$f_2((x,y,z))$	$f_3((x,y,z))$
1	1	1	1	0	0
1	1	0	0	1	0
1	0	1	0	1	0
1	0	0	1	1	0
0	1	1	1	0	0
0	1	0	0	0	1
0	0	1	1	1	0
0	0	0	0	1	1

18. Find a Boolean expression in join-normal form that is equivalent to $(x \wedge y) \vee z$.

19. Find a Boolean expression in join-normal form that is equivalent to $(x' \wedge y') \vee z'$.

20. Find a Boolean expression in join-normal form that is equivalent to $(x \wedge y \wedge z') \vee (x \wedge y' \wedge z) \vee (x' \wedge y')$.

21. Find a Boolean expression in three variables x, y, and z in join-normal form that is equivalent to $x \vee y$.

22. Let $f: B^3 \to B$ be the function defined in Table 13.8 of Example 7.
    a. Write f' in join-normal form.
    b. Using de Morgan's laws, write $(f')'$ in meet-normal form.
23. a. A certain Boolean function in B^3 is written in join-normal form as $f((x,y,z)) = x'yz \vee xyz' \vee xyz \vee x'y'z' \vee x'y'z$. Write f in meet-normal form.
    b. A certain Boolean function g in B^3 is written in meet-mormal form as $g((x,y,z)) = (x \vee y \vee z)(x \vee y' \vee z)(x' \vee y' \vee z)(x \vee y' \vee z')$. Write g in join-normal form.
24. Let $\mathcal{P}(S)$ be the Boolean algebra of subsets of some nonempty set S.
    a. Argue that if $c \neq d$ then $\{c,d\}$ is not an atom of $\mathcal{P}(S)$.
    b. Argue that $\{c\}$ is an atom of $\mathcal{P}(S)$.
25. If S is a nonempty set with n elements, how many atoms does $\mathcal{P}(S)$ have?
26. Let B^* be the set of subsets of a ten-element set U but let join $\vee^*$ be intersection and meet $\wedge^*$ be union (so that $0^* = U$ and $1^* = \varnothing$).
    a. Describe the atoms of $(B^*,\vee^*,\wedge^*)$.
    b. How many atoms does $(B^*,\vee^*,\wedge^*)$ have?
    c. How many members does B^* have?
27. Prove that a non-zero element $a \in B$ is an atom of B if and only if 0 and a are the only members b of B for which $b \leqslant a$.
28. Let B be the Boolean algebra of positive divisors of 210 (we promise that B is a Boolean algebra). List the atoms of B.
29. Let $B = \{0,1\}$. List the atoms of $\mathbb{F}_{B,3}$.

13.3

SIMPLIFYING BOOLEAN EXPRESSIONS

Although join-normal form provides a canonical way to express each non-zero Boolean function, it does not provide a method of simplification. Figure 13.1a shows the circuit of an expression in join-normal form, but (as we ask you to verity in Exercise 30) this circuit is equivalent to the much simpler circuit of Figure 13.1b.

What we mean by "simplifying" a circuit is not so clear as Figure 13.1 seems to indicate. Because of physical considerations, circuits requiring more than a certain number of inputs may have to be assembled from circuits with fewer inputs. This restriction, known as *fan-in limitation*, raises the problem of minimizing the area of a circuit on a computer chip rather than minimizing the number of inputs. We consider here only the problem of simplifying joins of products. Even in this restricted setting we cannot always provide a best simplification.

Figure 13.1

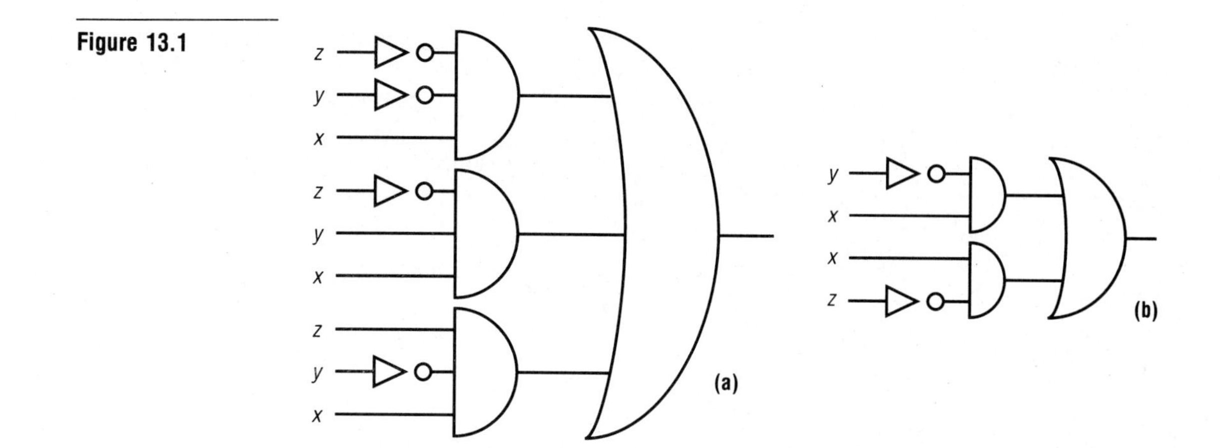

Definition

Let σ and τ be equivalent Boolean expressions both of which are joins of products. We say that σ is **as simple as** τ if σ is not the join of more products than τ and the number of literals in σ is less than or equal to the number of literals in τ. If σ is as simple as τ and σ has fewer products than τ or σ has fewer literals than τ, then we say that σ is **simpler than** τ. If τ is the join of products and there is no simpler join of products, we way that τ is a **minimal join of products**. A minimal join of products equivalent to a Boolean expression σ is said to **simplify** σ.

As Example 8 shows, a Boolean expression can have two essentially different minimal joins of products that simplify it.

Example 8

Let f: $B^3 \to B$ be the Boolean function defined in Table 13.9 and let σ and τ be the Boolean expressions $\sigma((x,y,z)) = xy' \vee x'z \vee yz'$ and $\tau((x,y,z)) = xz' \vee x'y \vee y'z$. Then σ and τ are equivalent expressions and if ρ is the join-normal form of f then σ and τ both simplify ρ.

Table 13.9

x	y	z	$f((x,y,z)) = \sigma((x,y,z)) = \tau((x,y,z))$
1	1	1	0
1	1	0	1
1	0	1	1
1	0	0	1
0	1	1	1
0	1	0	1
0	0	1	1
0	0	0	0

■

Exercises 13.3

30. Show that the circuits of Figure 13.1 define the same Boolean function $f: B^3 \to B$ by writing each circuit's Boolean expression in join-normal form.
31. For each pair of Boolean expressions, state whether the left expression is simpler than the right expression, as simple as the right expression, or neither of these.
    a. $wx'z$, $wx'y'z \vee wx'yz$
    b. $wx'z \vee wx'yz$, $wx'y'z \vee wx'yz$
    c. $x'yz' \vee x'y \vee xz' \vee y'z \vee xy' \vee yz' \vee x'z$,
    $xyz' \vee x'yz \vee x'yz' \vee x'y'z \vee xy'z \vee xy'z'$
    d. $y'z \vee x'y \vee xz'$, $xy' \vee yz' \vee x'z$
    e. $xy' \vee yz' \vee x'z$, $y'z \vee x'y \vee xz'$
32. Write the function f of Example 8 in join-normal form.
33. a. Verify that, if ρ is the expression in join-normal form that represents the function of Example 8, then the expression σ simplifies ρ and the expression τ simplifies ρ.
    b. Does σ simplify τ or does τ simplify σ?

13.4

MARQUAND MATRICES AND THE QUINE–McCLUSKEY METHOD OF SIMPLIFICATION

How can we simplify a given Boolean expression in join-normal form?[1] One standard answer to this question is to use the Quine–McCluskey method and Karnaugh maps. Our approach, which is nonstandard, is to replace the Karnaugh maps with Marquand matrices.

The method of simplification known as the Quine–McCluskey method is due to W. V. O. Quine and E. J. McCluskey, Jr. The twin pillars of this method (as with other methods of simplification) are the distributive law $a(b \vee c) = ab \vee ac$ and the law that $b \vee b' = 1$. These laws guarantee that the join of two minterms that differ in exactly one variable can be combined to form an equivalent product in which the differing variable does not appear. For example, $wx'y'z \vee wx'yz$ is the join of two products

[1]Our answer to this question has been changed by conversations with Morton Nadler, Professor of Electrical Engineering and Computer Science at Virginia Polytechnic Institute and State University, who pointed out to us the usefulness of Marquand matrices.

that differ only with respect to the variable y. Using the laws previously cited (as well as commutativity and associativity, which we take for granted) we have $wx'y'z \vee wx'yz \sim wx'z(y' \vee y) \sim wx'z(1) \sim wx'z$.

The following definition is natural and not new to those who have worked Exercise 37 of Chapter 2.

Definition

Let σ and τ be Boolean expressions that are products of n literals. If $n - 1$ of these literals agree and the remaining variable appears complemented in one of the expressions and uncomplemented in the other, then σ and τ are called **neighbors.**

Example 9

The minterm $pq'r'swxyz'$ in eight variables has eight neighbors. They are $p'q'r'swxyz'$, $pqr'swxyz'$, $pq'rswxyz'$, $pq'r's'wxyz'$, $pq'r'sw'xyz'$, $pq'r'swx'yz'$, $pq'r'swxy'z'$, and $pq'r'swxyz$. A minterm in n variables has exactly n neighbors. ■

The following method of charting minterms and their neighbors was first presented by A. Marquand in 1881.

Example 10

The Marquand matrix for a minterm of two variables is illustrated in Figure 13.2.

Figure 13.2 Two-variable Marquand matrix

		y
	00	01
x	10	11

The boxes are labeled in base 2 with the numbers zero through three. Reading from left to right, we read off the four minterms corresponding to the binary entries in the two-variable Marquand matrix. They are $x'y'$ (00), $x'y$ (01), xy' (10), and xy (11).

The Marquand matrix for a minterm of three variables is illustrated in Figure 13.3. The eight minterms corresponding to the binary entries

Figure 13.3 Three-variable Marquand matrix

			y	y
		z		z
	000	001	010	011
x	100	101	110	111

in the three-variable matrix are (reading from left to right) $x'y'z'$, $x'y'z$, $x'yz'$, $x'yz$, $xy'z'$, $xy'z$, xyz', and xyz.

Figure 13.4 begins the Marquand matrix for four variables. We could continue counting out the binary entries, but we have deliberately left some entries unaccounted. Suppose that we want to know the binary entry in the lower left-hand corner without having to count to get it. This corner is covered by both w and x but not by either y or z. (To see this, simply observe the bars outside the Marquand matrix.) Thus the minterm in the lower left-hand corner is $wxy'z'$, whose base 2 representation is 1100_2.

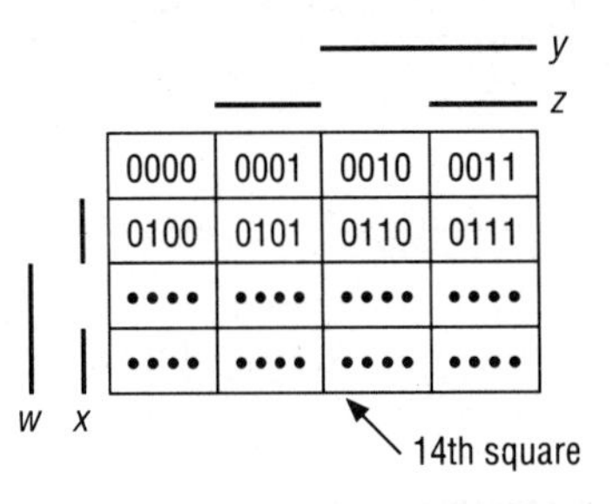

Figure 13.4
Four-variable Marquand matrix

The fourteenth square (1110_2) is $wxyz'$, but remember that we start counting at zero so that the fourteenth square is the next-to-last square.

If we call the variables x_0, x_1, x_2, and so forth, it is natural to label the Marquand matrix so that the kth digit of binary entry indicates the state of the kth variable. We illustrate such a labeling in Figure 13.5, but

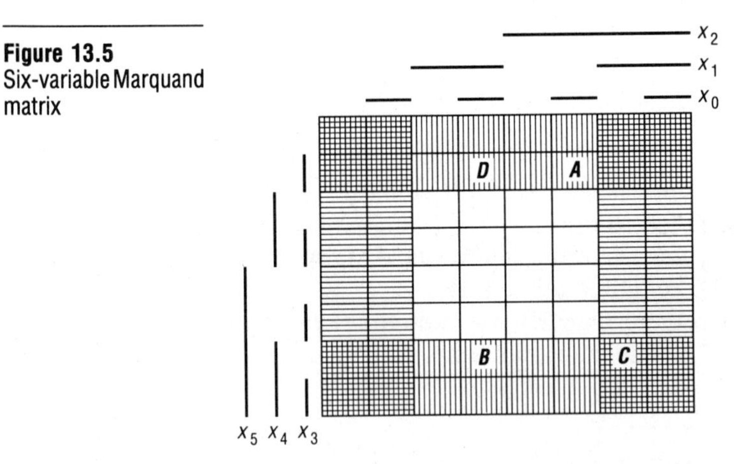

Figure 13.5
Six-variable Marquand matrix

the choice of order of the variables is simply a matter of taste. The binary entries and corresponding minterms for the squares labeled A, B, C, and D are

A $x_5'x_4'x_3x_2x_1'x_0$ (001101_2)
B $x_5x_4x_3'x_2'x_1x_0$ (110011_2)
C $x_5x_4x_3'x_2x_1x_0'$ (110110_2)
D $x_5'x_4'x_3x_2'x_1x_0$ (001011_2)

We have shaded this 8×8 Marquand matrix representing the six-variable minterms to indicate that this matrix consists of sixteen 2×2 submatrices. A 2×2 submatrix is shaded vertically if it is an outside square in the vertical direction and shaded horizontally if it is an outside square in the horizontal direction. The four middle squares are not shaded horizontally or vertically because they are inside squares in both horizontal and vertical directions. ■

A great advantage of the Marquand matrix of n-variable minterms is that we can read off mechanically from it all the neighbors of any given minterm. Suppose we are given the minterm σ appearing in the ith row and jth column of a Marquand matrix. Any minterm appearing off the ith row differs from σ in at least one of the literals indicated to the left of the ith row, and any minterm appearing off the jth column differs from σ in at least one of the literals listed above the jth column. Thus a minterm appearing in neither the ith row nor the jth column differs in at least two literals from σ and cannot be one of its neighbors. Theorem 13.4 states these observations in a positive way.

Theorem 13.4

Let σ be an n-variable minterm appearing in the ith row and jth column of a Marquand matrix M. All n of the neighbors of σ appear in either the ith row or jth column of M.

Theorem 13.4 makes finding the neighbors of a two-variable minterm a mindless task, as we show in the next example.

Example 11

The neighbors of the minterm $A = xy$ are $x'y$ and xy', which we indicate here by small a's.

	y
	a
a	A

(rows: bottom row x) ■

Finding the neighbors of a three-variable minterm is not quite mindless, because we must keep in mind that in the horizontal direction outside squares have inside neighbors (and inside squares have outside neighbors).

Example 12

The minterm labeled A in the three-variable matrix shown here appears in an *outside* square. There are two inside squares in the row in which A appears; both of these represent neighbors, and the remaining neighbor lies in the one remaining square allowed by Theorem 13.4. Similarly, the minterm labeled B appears in an *inside* square. There are two outside squares in the row in which B appears; both represent neighbors of B, and the third neighbor of B lies in the one remaining allowable square.

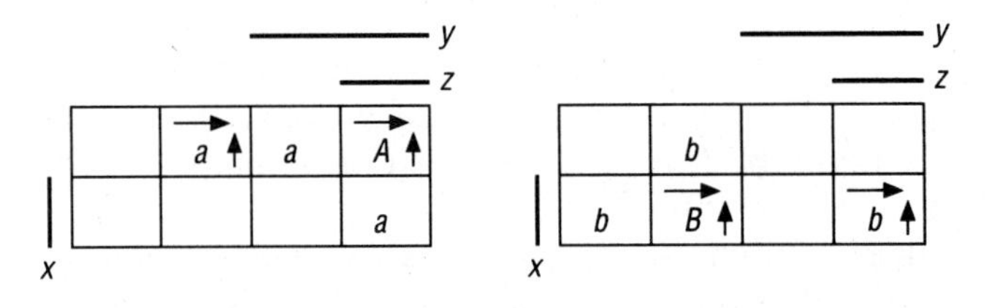

The arrows point out a pattern that makes it easy to find those neighbors that do not belong to the 2×2 submatrix containing the given minterm. If you do not see this pattern, don't worry. Look for it again in Example 13. As for the neighbors that do belong to the 2×2 submatrix containing the given minterm, their position is just as in Example 11. ■

For four-variable minterms, we must keep track of both outside and inside in both the horizontal and vertical directions, but the same rules apply. Outside minterms have inside neighbors and inside minterms have outside neighbors. And, of course, Theorem 13.4 still applies.

Example 13

The neighbors of A and B have been indicated by the corresponding small letters in the four-variable Marquand matrix illustrated here.

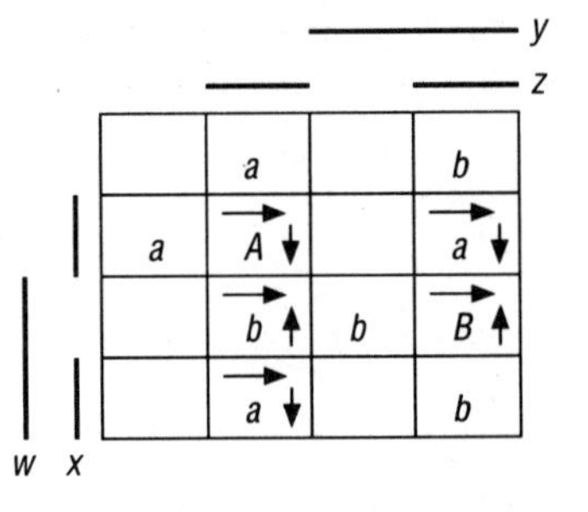

Note that the 4×4 case mimics the trivial 2×2 case in the following way. By considering the four 2×2 submatrices that form M, we see that three of these have neighbors and one does not. Take the minterm A, for example. We check those submatrices that have a neighbor of A:

$$\begin{bmatrix} \checkmark & \checkmark \\ \checkmark & \times \end{bmatrix}$$

The neighbors contained in the first submatrix occupy the same position as in the 2×2 case

	a
a	*A*

The other two submatrices each contain one neighbor, and that neighbor occupies the same relative position in its submatrix that A occupied. The arrows around A indicate that it is in the lower right-hand corner of its submatrix. The two lowercase letter a's with arrows around them are also in the lower right-hand corner of their submatrices. Note that this same pattern holds in Example 12. ■

Although we do not consider simplifying expressions in more than six variables, we assure you that the repetitive pattern indicated in Examples 12 and 13 persists for all Marquand matrices. For us the six-variable Marquand matrix constitutes a "worst case," and finding the neighbors of A, B, C, and D comprises all the difficulties of extending the pattern.

Example 14

We have indicated the neighbors of A and B in the six-variable matrix illustrated here. Note that those neighbors of A contained in the 4×4 submatrix containing A occupy the same positions as the neighbors of A in Example 13. Compare the lower left-hand 4×4 submatrix containing B with the Marquand matrix of Example 13.

					a		
	a			*a*	*A*		*a*
			b				
		D			*a*		
			b				
					a		
	b	*b*	*B*			*C*	*b*
			b				

Exercises 40 and 41 ask for the positions and binary numbers of the neighbors of C and D. Since C belongs to an outside-outside 2×2 submatrix and D belongs to an inside-inside 2×2 submatrix, taken together the minterms represented by A, C, and D are typical of the problems involved in searching out neighbors in the six-variable Marquand matrix. ■

We are now ready for the Quine–McCluskey method of simplifying Boolean expressions. Since each non-zero Boolean function can be represented by an expression in join-normal form, we always begin with an expression in join-normal form.

Example 15

Show how the Quine–McCluskey method is used to simplify the expression

$$\sigma(w,x,y,z) = wxyz \vee w'xyz \vee wx'yz \vee wz'yz' \vee wx'y'z \vee w'x'y'z \vee w'xy'z' \vee wx'y'z'$$

Analysis

We begin by filling in the Marquand matrix with each minterm appearing in σ:

y

z

	0001		
0100			0111
1000	1001	1010	1011
			1111

w x

Next we pair up each minterm with each of its neighbors. Most people prefer to use the base 10 names of the minterms to indicate the pairs:

(1,9), (7,15), (8,10), (9,11), (8,9), (10,11), (11,15)

These are really unordered pairs, for A and B are neighbors if and only if B and A are neighbors. Unordered pairs are two-element sets, and so we should write {1,9} in place of (1,9), and so on. We should, but we won't. Under step 1 of Table 13.10 we list the pairs of neighbors and the resulting simplification in which a dash replaces the variable that has been subsumed. For example, (1,9) = (0001,1001) yields _001. It is instructive to see the mathematics that justifies replacing (1,9) by _001 (see Exercise 42).

Table 13.10

Step 1 Pair	Term	Step 2
(1,9)	_001	
(7,15)	_111	
(8,10)	10_0	(8,9,10,11)10_ _
(9,11)	10_1	
(8,9)	100_	(8,9,10,11)10_ _
(10,11)	101_	
(11,15)	1_11	

In step 2 we combine all terms which have a dash *in the same position* and which differ in exactly one literal. At the same time, we record all minterms that have played a role in producing these offspring. We could continue to step 3 if there were any hope of combining products with two dashes to yield a product with one literal and three dashes, but in this example our hopes are dashed. We continue, therefore, to the last step. We form Table 13.11, whose columns are labeled by the eight minterms of our original expression. The rows of this table are labeled by all *hermits* (those minterms of our original expression that had no neighbors appearing in the expression), by all products that did not have a hand in forming our final offspring, and by all final offspring.

Hermits: 4 = 0100_2 $w'xy'z'$

Final offspring: 10_ _ wx'

Products not involved in forming wx':

1_11 wyz

_111 xyz

_001 $x'y'z$

We place an X in a given row and column if the minterm naming the column played a role in producing the product naming the row. For example, $wxyz$ = 15 = 1111_2 played a role in producing both wyz (1_11) and xyz (_111). If an X appears in the ith row and jth column, the product

Table 13.11

	(15) $wxyz$	(7) $w'xyz$	(11) $wx'yz$	(10) $wx'yz'$	(9) $wx'y'z$	(1) $wx'y'z'$	(4) $w'xy'z'$	(8) $wx'y'z'$
$w'xy'z'$							X	
wx'			X	X	X			X
wyz	X		X					
xyz	X	X						
$x'y'z$					X	X		

t

naming the ith row is said to cover the minterm naming the jth column; thus wx' covers 8, 9, 10, and 11. We select products naming the rows in such a way that the selected products cover *all* original minterms, which label the columns. We must select $w'xy'z'$ (and in general every hermit must be selected). We must select wx', since it is the only product that covers 8. We must select $x'y'z$, since it is the only product that covers 1. Also, we must select xyz, since it is the only product that covers 7. These four products cover all eight minterms, and so we are through: $w'xy'z' \vee wx' \vee x'y'z \vee xyz$ is a simplification of the original expression. ■

A given expression may have more than one simplification, and a valuable aspect of the Quine–McCluskey method is that it allows us to determine all possible simplifications. The final example of this section illustrates the point.

Example 16

Use the Quine-McCluskey method to simplify the Boolean expression

$$\sigma(x,y,z) = (xyz') \vee (x'yz) \vee (x'yz') \vee (x'y'z) \vee (xy'z) \vee (xy'z')$$

Find two different simplifications.

Analysis

First we draw the Marquand matrix of σ:

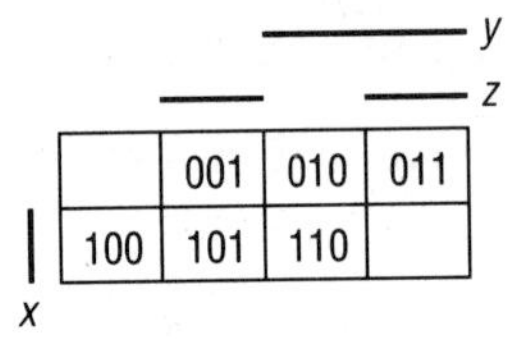

	y		
	z		z
	001	010	011
x: 100	101	110	

The neighbors are (1,5), (1,3) (2,3), (2,6), (4,5), and (4,6). We list these, as in Example 15, and simplify where possible. There are no hermits. Step 2 fails to produce any new offspring.

Step 1		Step 2
Pair	*Term*	
(1,5)	_01	
(1,3)	0_1	
(2,3)	01_	
(2,6)	_10	
(4,5)	10_	
(4,6)	1_0	

Table 13.12 shows that no one product listed on the left has to be used. If you wish to simplify σ and avoid the term xy', it can be done:

$$xz' \vee x'y \vee y'z$$

Table 13.12

	(3) $x'yz$	(6) xyz'	(2) $x'yz'$	(4) $xy'z'$	(5) $xy'z$	(1) $x'y'z$
$y'z$					X	X
$x'z$	X					X
$x'y$	X		X			
yz'		X	X			
xy'				X	X	
xz'		X		X		

If you wish to avoid xz', that is also possible:

$$xy' \vee yz' \vee x'z$$

Of course, $\rho(x,y,z) = xz' \vee x'y \vee y'z$ and $\tau(x,y,z) = xy' \vee yz' \vee x'z$ are equivalent expressions, since they both represent the same Boolean function, namely, the function represented by σ. For the nonce, we call two Boolean expressions *super-equivalent* if one can be obtained from the other using only the commutative and associative laws of Boolean algebra. Then ρ and τ are not super-equivalent, and in this sense ρ and τ are the only two essentially different simplifications of σ. ■

Exercises 13.4

34. Give the minterms indicated by A, B, C, and D in the following Marquand matrices.

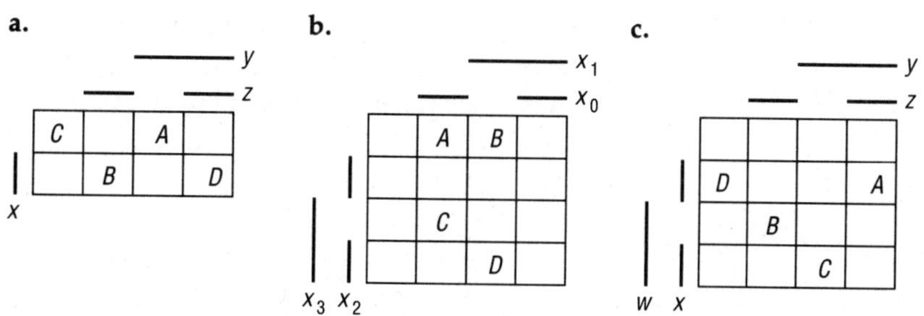

35. **a.** Give the binary numbers associated with A,B,C, and D of Exercise 34a.

b. Give the binary numbers associated with A,B,C, and D of Exercise 34b.

c. Give the binary numbers associated with A,B,C, and D of Exercise 34c.

36. Identify the products of literals of the following "base 2 numbers." Assume that the variables are listed w, x, y, and z.

a. 1 1 0 1 **b.** 1 1 1 1 **c.** 1 _ _1 **d.** 1 _1 0

e. 1 _0 _ **f.** 1 0 _0 **g.** _ _ _ 0 **h.** _ _ _ 1

37. Let σ be the n-variable minterm all of whose variables are complemented and let τ be the n-variable minterm none of whose variables are complemented.

a. Where in the n-variable Marquand matrix does σ appear?

b. Where in the n-variable Marquand matrix does τ appear?

38. Give the dimensions of the n-variable Marquand matrix when

a. $n = 5$ **b.** $n = 7$ **c.** $n = 8$ **d.** $n = 10$

39. List all the neighbors of $x_5x_4'x_3x_2x_1x_0'$.

40. Give the position and binary numbers of the neighbors of the minterm labeled C in the following diagram.

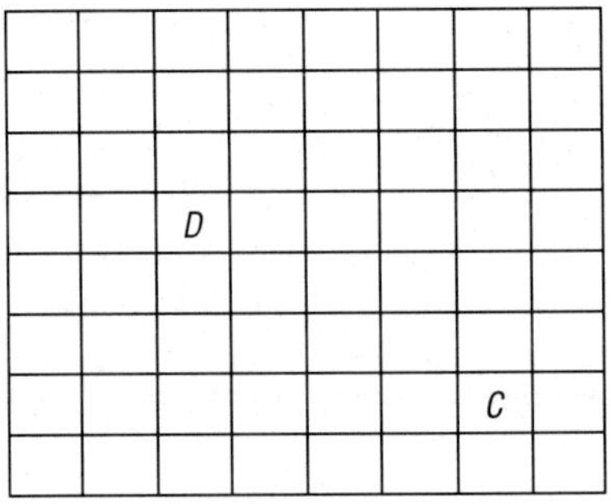

<table>
<tr><td></td><td></td><td></td><td></td><td></td><td></td><td></td><td></td></tr>
<tr><td></td><td></td><td></td><td></td><td></td><td></td><td></td><td></td></tr>
<tr><td></td><td></td><td></td><td></td><td></td><td></td><td></td><td></td></tr>
<tr><td></td><td></td><td>D</td><td></td><td></td><td></td><td></td><td></td></tr>
<tr><td></td><td></td><td></td><td></td><td></td><td></td><td></td><td></td></tr>
<tr><td></td><td></td><td></td><td></td><td></td><td></td><td></td><td></td></tr>
<tr><td></td><td></td><td></td><td></td><td></td><td></td><td>C</td><td></td></tr>
<tr><td></td><td></td><td></td><td></td><td></td><td></td><td></td><td></td></tr>
</table>

41. Give the position and binary numbers of the neighbors of the minterm labeled D in the preceding diagram.

42. In the four-variable Marquand matrix, (1,9) represents $w'x'y'z \vee wx'y'z$, and _001 represents $x'y'z$. Using the laws of Boolean algebra and justifying each step with one of these laws, show that $w'x'y'z \vee wx'y'z \sim x'y'z$.

43. Is super-equivalence, as defined in Example 16, an equivalence relation on the set of all Boolean expressions over $B = \{0,1\}$? Explain.

44. Fill in the four-variable Marquand matrix with the minterms appearing in the expression $\sigma(w,x,y,z) = wxyz \vee wxy'z' \vee wx'yz \vee wx'yz' \vee wx'y'z' \vee w'x'y'z$.

45. List the neighbors of

a. $wxyz$ **b.** $wxy'z'$ **c.** $wx'yz$ **d.** $wx'yz'$ **e.** $wx'y'z'$ **f.** $w'x'y'z$

Which if any of these minterms are hermits?

46. Using Exercises 44 and 45 and the Quine–McCluskey method, find a minimal join of products equivalent to the Boolean expression σ of Exercise 44.

47. Using the Quine–McCluskey method, find a minimal join of products equivalent to

a. $xyz \vee xy'z' \vee xy'z \vee x'y'z$

b. $xyz' \vee xy'z \vee xy'z'$

48. Using the Quine–McCluskey method, find a minimal join of products equivalent to

a. $wxy'z' \vee wxy'z \vee wxyz \vee w'xyz \vee w'x'y'z \vee w'x'yz$

b. $wxyz \vee w'xy'z' \vee wx'y'z' \vee w'x'y'z' \vee w'x'y'z$

c. $wxyz' \vee w'xy'z \vee wx'yz \vee wx'y'z' \vee w'x'yz' \vee w'x'y'z \vee wx'yz' \vee wxyz$

49. Use the Quine–McCluskey method to simplify the circuit of Figure 13.1a.

50. Use the Quine–McCluskey method to simplify the following circuits. (*Hint:* Your first need is an expression in join-normal form.)

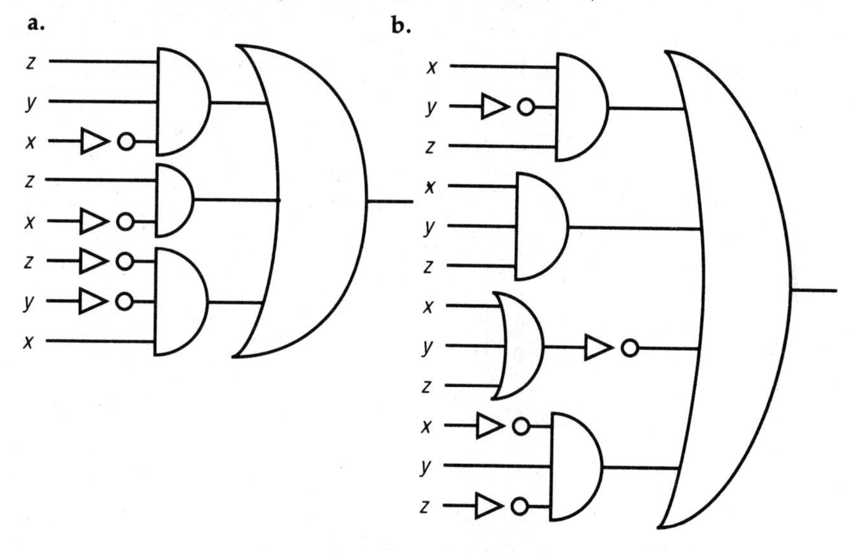

51. a. Design a circuit with five variables such that the output is 1 if and only if an odd number of the variables have the value 1.

b. If possible, simplify your circuit using the Quine–McCluskey method.

52. a. Design a circuit with five variables such that the output is 1 if and only if at least three of the variables have the value 0.

b. If possible, simplify your circuit using the Quine–McCluskey method.

Chapter 13 Review Exercises

53. Let $B = \{0,1\}$. We say that two expressions σ and τ in n variables take on opposite values provided that if $\sigma(x_1,x_2,\ldots,x_n) = 0$ then $\tau(x_1,x_2,\ldots,x_n) = 1$, and if $\sigma(x_1,x_2,\ldots,x_n) = 1$ then $\tau(x_1,x_2,\ldots,x_n) = 0$. Let $\sigma(x,y,z) = x'y \vee xy \vee z$. Find an expression τ such that σ and τ take on opposite values.

54. Let $B = \{0,1\}$. A Marquand matrix M in n variables ($n \geq 2$) is called an n-way switch provided that if σ is any one of the minterms appearing in M then σ and each neighbor of σ take on opposite values.

a. Write a Marquand matrix for a 3-way switch in which the minterm $\sigma(x,y,z) = xyz$ takes on the value 1, and write the Boolean function defined by the matrix in join-normal form.

b. Write a Marquand matrix for a 3-way switch in which the minterm $\sigma(x,y,z) = xyz$ takes on the value 0.

c. Are there any other 3-way switches?

55. Write a Marquand matrix of a 4-way switch, and write the Boolean function defined by the matrix in join-normal form.

56. Let f and g be the Boolean functions of degree 3 over $B = \{0,1\}$ that are defined by $f(x,y,z) = x'yz \vee xyz \vee xy'z \vee x'y'z$ and $g(x,y,z) = xy'z' \vee x'y'z' \vee xyz'$.

a. Fill in the following table, which defines f, g, $f \vee g$, fg, and f'.

			f	g	$f \vee g$	fg	f'
1	1	1					
1	1	0					
1	0	1					
1	0	0					
0	1	1					
0	1	0					
0	0	1					
0	0	0					

b. Use the Quine–McCluskey method to simplify $f \vee g$.

Chapter

14

ADVANCED COUNTING

In this chapter we study some problems that cannot be handled easily by the counting techniques of Chapter 6. One such problem is to find the number of integral solutions of an equation in more than one variable subject to certain constraints, and the principle of inclusion–exclusion can be used to solve such a problem. To solve other problems, permutations or combinations need to be generated, not just counted. Still another problem is to find the solutions of recursion relations, and we develop methods for finding explicit formulas for the terms of sequences that satisfy certain recursion relations. Generating functions can be used to solve recursion relations, and they can also be used to model counting problems.

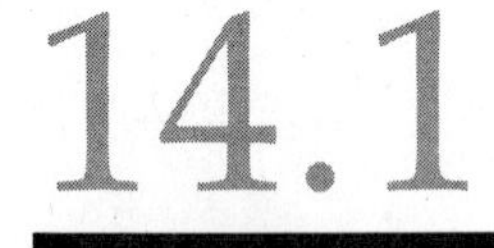

14.1

PRINCIPLE OF INCLUSION–EXCLUSION

By Theorem 2.7, if A and B are finite sets then $n(A \cup B) = n(A) + n(B) - n(A \cap B)$. With this theorem we can find the number of elements in the union of two finite sets. Recall that we touted this theorem not as a tool but as a guide. One reason we were reluctant to peddle it as a useful result is that the theorem is just a special case of the principle of inclusion–exclusion, which provides a more general formula for arbitrary finite unions of finite sets.

Before considering an arbitrary finite union, let us find a formula for the number of members in the union of three sets. Let A, B, and C be finite sets. The problem is easy if $\{A,B,C\}$ is a pairwise disjoint collection of sets. In this case, $n(A \cup B \cup C) = n(A) + n(B) + n(C)$. So let us assume that $\{A,B,C\}$ is not a pairwise disjoint collection. The most complicated situation arises when the intersection of any two of these sets is nonempty, as we illustrate in Figure 14.1. Observe that $n(A) + n(B) + n(C)$ counts elements that are in all three sets three times and counts elements that are in exactly two of the sets two times.

Figure 14.1

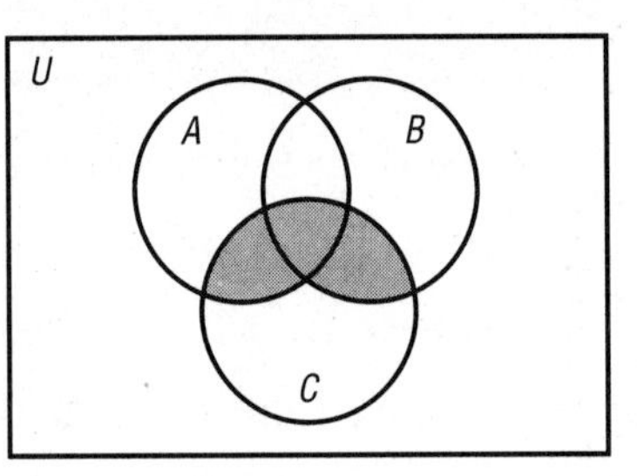

Example 1

Suppose that 6,847 students at State U. are enrolled in a mathematics course, 3,241 are enrolled in a computer science course, and 1,845 are enrolled in an English course. How many students are enrolled in at least one of these three courses?

Analysis

Notice that we cannot answer this question without additional information. If we simply add 6,847 and 3,241 and 1,845, each student who is enrolled in exactly one course is counted exactly once, each student who is enrolled in exactly two courses, say, mathematics and computer science, is counted twice, and each student who is enrolled in all three courses is counted three times. Suppose we have the additional information that there are 1,980 students enrolled in mathematics and computer science, 1,113 students enrolled in mathematics and English, and 589 students enrolled in computer science and English. As previously stated, each student enrolled in exactly two courses is counted twice when we add 6,847 and 3,241 and 1,845. So is 6,847 + 3,241 + 1,845 − 1,980 − 1,113 − 589 = 8,251 the answer to our question? No; each student who is enrolled in exactly one course is still counted exactly once, each student who is enrolled in exactly two courses is now counted exactly once, but a student who is enrolled in all three courses is not counted at all. To answer the question we have posed, we need to know the number of students who are enrolled in all three courses and we need to add this number. If 127 students are enrolled in all three courses, then the number of students enrolled in at least one of the three courses is 6,847 + 3,241 + 1,845 − 1,980 − 1,113 − 589 + 127 = 8,378 ■

Example 1 illustrates the following result.

Theorem 14.1

If A, B, and C are finite sets, then $n(A \cup B \cup C) = n(A) + n(B) + n(C) - n(A \cap B) - n(A \cap C) - n(B \cap C) + n(A \cap B \cap C)$.

Proof

By Theorem 2.1c, if A, B, and C are sets, then $A \cap (B \cup C) = (A \cap B) \cup (A \cap C)$. We use Theorems 2.7 and 2.1c in the following proof of the theorem:

$$\begin{aligned} n(A \cup B \cup C) &= n(A \cup (B \cup C)) \\ &= n(A) + n(B \cup C) - n(A \cap (B \cup C)) \\ &= n(A) + n(B) + n(C) - n(B \cap C) \\ &\quad - n((A \cap B) \cup (A \cap C)) \\ &= n(A) + n(B) + n(C) - n(B \cap C) \\ &\quad - n(A \cap B) - n(A \cap C) \\ &\quad + n((A \cap B) \cap (A \cap C)) \\ &= n(A) + n(B) + n(C) - n(B \cap C) \\ &\quad - n(A \cap B) - n(A \cap C) \\ &\quad + n(A \cap B \cap C) \end{aligned}$$ □

We introduce the following notation to generalize Theorem 14.1 to an arbitrary number of finite sets. Let $A_1, A_2, \ldots, A_n$ be a list of finite sets, and for each $r = 1, 2, \ldots, n$ let m_r be the sum of the cardinalities of all possible intersections of r sets chosen without repetition from $A_1, A_2, \ldots, A_n$.

We illustrate the definition of m_r by considering two special cases, $n = 3$ and $n = 4$. Note that in each case m_r is found by considering all possible intersections involving r sets. Thus, for example, if $n = 3$, we have three sets A_1, A_2, A_3, and

$$m_1 = n(A_1) + n(A_2) + n(A_3)$$

$$m_2 = n(A_1 \cap A_2) + n(A_1 \cap A_3) + n(A_2 \cap A_3)$$

$$m_3 = n(A_1 \cap A_2 \cap A_3)$$

So, by Theorem 14.1, $n(A_1 \cup A_2 \cup A_3) = m_1 - m_2 + m_3$. Likewise, if $n = 4$, we have four sets A_1, A_2, A_3, A_4, and

$$m_1 = n(A_1) + n(A_2) + n(A_3) + n(A_4)$$

$$m_2 = n(A_1 \cap A_2) + n(A_1 \cap A_3) + n(A_1 \cap A_4) + n(A_2 \cap A_3) + n(A_2 \cap A_4) + n(A_3 \cap A_4)$$

$$m_3 = n(A_1 \cap A_2 \cap A_3) + n(A_1 \cap A_2 \cap A_4) + n(A_1 \cap A_3 \cap A_4) + n(A_2 \cap A_3 \cap A_4)$$

$$m_4 = n(A_1 \cap A_2 \cap A_3 \cap A_4)$$

The principle of inclusion–exclusion, which we give as Theorem 14.2, tells us that $n(A_1 \cup A_2 \cup A_3 \cup A_4) = m_1 - m_2 + m_3 - m_4$. We first give an example that illustrates the use of this formula.

Example 2

A bridge hand consists of 13 cards chosen from a standard 52-card deck, and the number of different bridge hands is $C(52,13)$ (see Chapter 6, Example 29). How many of these hands are void in a suit, that is, contain no cards in a suit?

Analysis

Let A denote the set of all bridge hands that are void in spades, let B denote the set of all hands void in hearts, let C denote the set of all hands void in diamonds, and let D denote the set of all hands void in clubs. Then by the preceding formula $n(A \cup B \cup C \cup D) = m_1 - m_2 + m_3 - m_4$, and we must determine m_i for each $i = 1,2,3,4$.

Since a hand void in spades consists of 13 cards chosen from the 39 cards that are not spades, $n(A) = C(39,13)$. It is also true that $n(B) = n(C) = n(D) = C(39,13)$. Therefore

$$m_1 = 4 \times C(39,13) = 4 \times 8{,}122{,}425{,}444 = 32{,}489{,}701{,}776$$

A hand void in spades and hearts consists of 13 cards chosen from the 26 cards that are diamonds and clubs. Therefore $n(A \cap B) = C(26,13)$. Also $n(A \cap C) = n(A \cap D) = n(B \cap C) = n(B \cap D) = n(C \cap D) = C(26,13)$, so

$$m_2 = 6 \times C(26,13) = 6 \times 10{,}400{,}600 = 62{,}403{,}600$$

A hand void in spades, hearts, and diamonds consists of the 13 cards that are clubs, so there is only one such hand. Since there are four suits,

$$m_3 = 4 \times C(13,13) = 4 \times 1 = 4$$

Finally, a hand cannot be void in spades, hearts, diamonds, and clubs, so $m_4 = 0$. Therefore the number of different bridge hands that are void in a suit is

$$32{,}489{,}701{,}776 - 62{,}403{,}600 + 4 - 0 = 32{,}427{,}298{,}180$$ ■

Theorem 14.2

The Principle of Inclusion–Exclusion Let $A_1, A_2, \ldots, A_n$ be a list of finite sets and for each $r \leq n$ let m_r be the sum of the cardinalities of all possible intersections of r sets chosen without repetition from $A_1, A_2, \ldots, A_n$. Then

$$n\left(\bigcup_{r=1}^{n} A_r\right) = \sum_{r=1}^{n} (-1)^{r+1} m_r$$

Analysis

Before proving this theorem, let us make sure that we understand $\sum_{r=1}^{n} (-1)^{r+1} m_r$. As we have seen in the two special cases $n = 3$ and $n = 4$, m_1 is $\sum_{r=1}^{n} n(A_r)$. Likewise, m_2 is the sum of all $n(A_i \cap A_j)$, where $i,j = 1,2, \ldots, m$ and $i \neq j$, and m_3 is the sum of all $n(A_i \cap A_j \cap A_k)$, where $i,j,k = 1,2, \ldots, m$, $i \neq j$, $i \neq k$, and $j \neq k$. In general, for each $r = 1,2, \ldots, m$, m_r is the sum of all $n(A_i \cap A_j \cap \cdots \cap A_p)$, where the intersection $A_i \cap A_j \cap \cdots \cap A_p$ involves r distinct sets and all possible combinations of r sets are used.

Proof

We want to prove the equation

$$n\left(\bigcup_{r=1}^{n} A_r\right) = \sum_{r=1}^{n} (-1)^{r+1} m_r$$

Let x be an element of $\bigcup_{r=1}^{n} A_r$. Then $n(\bigcup_{r=1}^{n} A_r)$ counts this element once. We must make sure that $\sum_{r=1}^{n} (-1)^{r+1} m_r$ also counts x once. Let us suppose that x belongs to k of the sets $A_1, A_2, \ldots, A_n$. Let $B = \{j \in \mathbb{N}: x$ belongs to $A_j\}$; that is, B is the set whose members are those subscripts j such that x belongs to A_j. Since we are assuming that x belongs to k of the sets $A_1, A_2, \ldots, A_m$, B has k members, and the first term m_1 of $\Sigma_{r=1}^{n} (-1)^{r+1} m_r$ counts x k times. How many times does the second term m_2 of $\Sigma_{r=1}^{n} (-1)^{r+1} m_r$ count x? The second term counts x (negatively) just as many times as there are two-element subsets of the set B, namely, $C(k,2)$. Then the third term adds back $C(k,3)$, but the fourth term subtracts $C(k,4)$, and so forth. Summing and remembering that $k = C(k,1)$, we find that the number of times x is counted in $\Sigma_{r=1}^{n} (-1)^{r+1} m_r$ is $C(k,1) - C(k,2) + C(k,3) - \cdots \pm C(k,k)$. It remains to show that this sum equals 1. By the binomial theorem,

$$\begin{aligned} 0 &= (1 + (-1))^k \\ &= C(k,0)(1)^k(-1)^0 + C(k,1)(1)^{k-1}(-1)^1 + C(k,2)(1)^{k-2}(-1)^2 \\ &\quad + \cdots + C(k,k)(1)^0(-1)^k \\ &= 1 - C(k,1) + C(k,2) - C(k,3) + \cdots + C(k,k)(-1)^k \end{aligned}$$

Subtracting 1 from both sides and multiplying through by -1, we have

$$1 = C(k,1) - C(k,2) + C(k,3) - \cdots \pm C(k,k)$$

as was required. □

Example 3

How many natural numbers less than 10,000 are divisible by 7, 13, or 17?

Analysis

Let $A = \{m \in \mathbb{N}: m < 10{,}000$ and m is divisible by $7\}$, let $B = \{m \in \mathbb{N}: m < 10{,}000$ and m is divisible by $13\}$, and let $C = \{m \in \mathbb{N}: m < 10{,}000$ and m is divisible by $17\}$. Then $A \cup B \cup C$ is the set of natural numbers less than 10,000 that are divisible by 7, 13, or 17. By the principle of inclusion–exclusion,

$$n(A \cup B \cup C) = n(A) + n(B) + n(C) - n(A \cap B) - n(A \cap C) - n(B \cap C) + n(A \cap B \cap C)$$

Now $n(A) = \lfloor 10{,}000/7 \rfloor = 1{,}428$, $n(B) = \lfloor 10{,}000/13 \rfloor = 769$, and $n(C) = \lfloor 10{,}000/17 \rfloor = 588$. Since 7 and 13, 7 and 17, and 13 and 17 are relatively prime, $n(A \cap B) = \lfloor 10{,}000/91 \rfloor = 109$, $n(A \cap C) = \lfloor 10{,}000/119 \rfloor = 84$, and $n(B \cap C) = \lfloor 10{,}000/221 \rfloor = 45$. Likewise, 91 and 17 are relatively prime, so $n(A \cap B \cap C) = \lfloor 10{,}000/1{,}547 \rfloor = 6$. Therefore, by the principle of inclusion–exclusion,

$$n(A \cup B \cup C) = 1{,}428 + 769 + 588 - 109 - 84 - 45 + 6 = 2{,}553$$

■

Definition

If S is a nonempty set, a **derangement of S** is a permutation α on S such that, for each $x \in S$, $\alpha(x) \neq x$.

A derangement of {1,2,3,4} is shown in Figure 14.2.

Figure 14.2

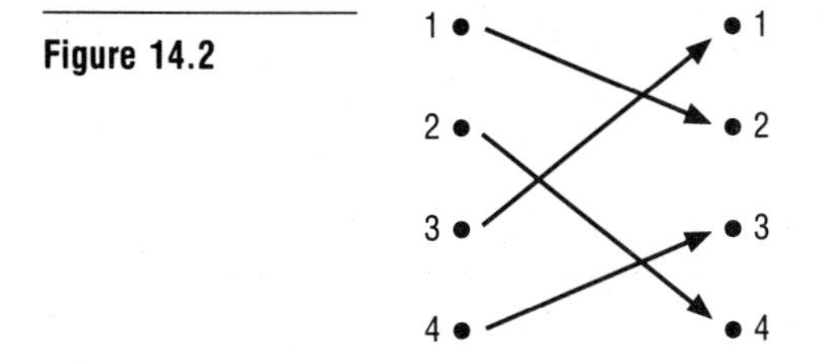

Example 4

The permutation $\alpha = (12)(345)$ is a derangement of {1,2,3,4,5}, whereas (12)(35)(4) is not a derangement of {1,2,3,4,5}. ■

Example 5

How many derangements of {1,2,3,4,5} are there?

Analysis

Let U denote the set of all permutations on {1,2,3,4,5}. Then the number of members of U is $P(5,5) = 5! = 120$. For each $i = 1,2,3,4,5$, let $A_i = \{\alpha \in U: \alpha(i) = i\}$. Then $U - \cup_{i=1}^{5} A_i$ is the set of all derangements of {1,2,3,4,5}, so we can answer the question we posed by calculating $n(\cup_{i=1}^{5} A_i)$ and subtracting this number from 120. For each $i = 1,2,3,4,5$, $n(A_i) = P(4,4) = 4! = 24$. If $i \neq j$, then $A_i \cap A_j = \{\alpha \in U: \alpha(i) = i$ and $\alpha(j) = j\}$. Therefore, for each $i,j = 1,2,3,4,5$ $(i \neq j)$, $n(A_i \cap A_j) = P(3,3) = 3! = 6$. Likewise, for each $i,j,k = 1,2,3,4,5$ $(i \neq j,\ i \neq k$, and $j \neq k)$, $A_i \cap A_j \cap A_k = \{\alpha \in U: \alpha(i) = i,\ \alpha(j) = j$, and $\alpha(k) = k\}$, so $n(A_i \cap A_j \cap A_k) = P(2,2) = 2! = 2$. Also, for each $i,j,k,m = 1,2,3,4,5$ $(i \neq j,\ i \neq k,\ i \neq m,\ j \neq k,\ j \neq m$, and $k \neq m)$, $A_i \cap A_j \cap A_k \cap A_m = \{\alpha \in U: \alpha(i) = i,\ \alpha(j) = j,\ \alpha(k) = k$, and $\alpha(m) = m\}$, so $n(A_i \cap A_j \cap A_k \cap A_m) = P(1,1) = 1! = 1$. Finally, $\cap_{i=1}^{5} A_i = \{\alpha \in U: \alpha(j) = j$ for each $j = 1,2,3,4,5\}$, so $n(\cap_{i=1}^{5} A_i) = 1$. We use the principle of inclusion–exclusion and conclude that

$$n\left(\bigcup_{i=1}^{5} A_i\right) = 5P(4,4) - 10P(3,3) + 10P(2,2) - 5P(1,1) + 1$$

$$= 5(24) - 10(6) + 10(2) - 5(1) + 1 = 76$$

Hence the number of derangements of {1,2,3,4,5} is 120 − 76 = 44. ■

Example 5 illustrates the following theorem.

Theorem 14.3

The number D_n of derangements of a set with exactly n members is given by the formula

$$D_n = n!\left[1 - \frac{1}{1!} + \frac{1}{2!} - \frac{1}{3!} + \cdots + \frac{(-1)^n}{n!}\right]$$

Proof

Let $A = \{1,2, \ldots ,n\}$, let U denote the set of all permutations on A, and for each $i = 1,2, \ldots ,n$ let $A_i = \{\alpha \in U: \alpha(i) = i\}$. Then $U - \bigcup_{i=1}^{n} A_i$ is the set of all derangements of $\{1,2, \ldots ,n\}$. Now $n(U) = n!$, so we can calculate the number of derangements of $\{1,2, \ldots ,n\}$ by calculating $n(\bigcup_{i=1}^{n} A_i)$ and subtracting this number from $n!$. For each $i = 1,2, \ldots ,n$, let m_i be the sum of the cardinalities of all possible intersections of i sets chosen without repetition from $A_1,A_2, \ldots ,A_n$. By the principle of inclusion–exclusion,

$$n\left(\bigcup_{i=1}^{n} A_i\right) = \sum_{i=1}^{n} (-1)^{i+1} m_i$$

Let $\alpha \in A_1$. Then $\alpha(1) = 1$, there are $n - 1$ choices for $\alpha(2)$, $n - 2$ choices for $\alpha(3)$, and so on. Therefore the number of members of A_1 is $(n - 1)!$. Likewise, for each $i = 2,3, \ldots ,n$, the number of members of A_i is $(n - 1)!$. Therefore $m_1 = n(n - 1)! = C(n,1)(n - 1)!$.

Let $\alpha \in A_1 \cap A_2$. Then $\alpha(1) = 1$, $\alpha(2) = 2$, there are $n - 2$ choices for $\alpha(3)$, $n - 3$ choices for $\alpha(4)$, and so on. Therefore the number of members of $A_1 \cap A_2$ is $(n - 2)!$. Similarly, for each $i,j = 1,2, \ldots ,n$, $i \neq j$, the number of members of $A_i \cap A_j$ is $(n - 2)!$. Since the number of ways of choosing two distinct members of $A_1,A_2, \ldots ,A_n$ is $C(n,2)$, $m_2 = C(n,2)(n - 2)!$.

In general, if $\bigcap_{j=1}^{p} A_{ij}$ denotes the intersection of p distinct members of $A_1,A_2, \ldots ,A_n$ and $\alpha \in \bigcap_{j=1}^{p} A_{ij}$, then α keeps p members of A fixed, and so the number of members of $\bigcap_{j=1}^{p} A_{ij}$ is $(n - p)!$. Since the number of ways of choosing p distinct members of $A_1,A_2, \ldots ,A_n$ is $C(n,p)$, $m_p = C(n,p)(n - p)!$. Hence

$$\sum_{i=1}^{n} (-1)^{i+1} m_i = \sum_{i=1}^{n} (-1)^{i+1} C(n,i)(n-i)!$$

and so

$$\begin{aligned}
D_n &= n! - C(n,1)(n-1)! + C(n,2)(n-2)! - C(n,3)(n-3)! \\
&\quad + \cdots + (-1)^n C(n,n)(n-n)! \\
&= n! - \frac{n!}{1!(n-1)!}(n-1)! + \frac{n!}{2!(n-2)!}(n-2)! \\
&\quad - \frac{n!}{3!(n-3)!}(n-3)! + \cdots + (-1)^n \frac{n!}{n!0!} \cdot 0! \\
&= n! - n! + \frac{n!}{2!} - \frac{n!}{3!} + \cdots + \frac{(-1)^n n!}{n!} \\
&= n!\left[1 - 1 + \frac{1}{2!} - \frac{1}{3!} + \cdots + \frac{(-1)^n}{n!}\right]
\end{aligned}$$

□

Example 6

How many solutions does $x_1 + x_2 + x_3 = 17$ have, where x_1, x_2, and x_3 are nonnegative integers with $x_1 \leqslant 4$, $x_2 \leqslant 6$, and $x_3 \leqslant 8$?

Analysis

A solution of $x_1 + x_2 + x_3 = 17$ corresponds to a way of selecting seventeen items from a set with three types of members so that x_1 items of type one are chosen, x_2 items of type two are chosen, and x_3 items of type three are chosen. Hence, by Theorem 6.13, the number of solutions of $x_1 + x_2 + x_3 = 17$ is $C(3 + 17 - 1,17) = C(19,17) = (19 \cdot 18)/(1 \cdot 2) = 171$. Thus, if $U = \{(x_1,x_2,x_3) \in \mathbb{Z} \times \mathbb{Z} \times \mathbb{Z}: x_1 + x_2 + x_3 = 17\}$, then $n(U) = 171$.

A solution of $x_1 + x_2 + x_3 = 17$ subject to the constraint $x_1 \geqslant 5$ corresponds to choosing five items of type x_1 and then selecting twelve additional items. Again by Theorem 6.13, the number of solutions of $x_1 + x_2 + x_3 = 17$ subject to this constraint is $C(3 + 12 - 1,12) = C(14,12) = (14 \cdot 13)/(1 \cdot 2) = 91$. Therefore, if $A = \{(x_1,x_2,x_3) \in U: x_1 \geqslant 5\}$, then $n(A) = 91$. Similarly, if $B = \{(x_1,x_2,x_3) \in U: x_2 \geqslant 7\}$, then $n(B) = C(3 + 10 - 1,10) = C(12,10) = (12 \cdot 11)/(1 \cdot 2) = 66$, and if $C = \{(x_1,x_2,x_3) \in U: x_3 \geqslant 9\}$, then $n(C) = C(3 + 8 - 1,8) = C(10,8) = (9 \cdot 10)/(1 \cdot 2) = 45$.

A solution of $x_1 + x_2 + x_3 = 17$ subject to the constraint $x_1 \geqslant 5$ and $x_2 \geqslant 7$ corresponds to choosing five items of type x_1 and seven items of type x_2 and then selecting five additional items. So $n(A \cap B) =$

$C(3+5-1,5) = C(7,5) = (7 \cdot 6)/(1 \cdot 2) = 21$. In the same way, $n(A \cap C) = C(3+3-1,3) = C(5,3) = (5 \cdot 4)/(1 \cdot 2) = 10$ and $n(B \cap C) = C(3+1-1,1) = C(3,1) = (3 \cdot 2)/(2 \cdot 1) = 3$. Finally, $A \cap B \cap C = \emptyset$, so $n(A \cap B \cap C) = 0$.

By the principle of inclusion–exclusion, the number of solutions of $x_1 + x_2 + x_3 = 17$ subject to the constraints x_1, x_2, and x_3 are nonnegative integers, $x_1 \leqslant 4$, $x_2 \leqslant 6$, and $x_3 \leqslant 8$ is

$$n(U) - n(A) - n(B) - n(C) + n(A \cap B) + n(A \cap C)$$
$$+ n(B \cap C) - n(A \cap B \cap C)$$
$$= 171 - 91 - 66 - 45 + 21 + 10 + 3 - 0 = 3$$

Since there are only three solutions satisfying the given constraints, we should be able to find them. Notice that, if we choose each of x_1, x_2, and x_3 to be the maximum, then their sum, $4 + 6 + 8$, is 18. Consequently we have selected one item too many. We can easily remedy this and obtain the three solutions by reducing, in turn, x_1, x_2, and x_3 by 1. So the three solutions are (3,6,8), (4,5,8), and (4,6,7). ■

Before considering the following example, review Theorem 3.7, which is used repeatedly.

Example 7

How many primes are there that do not exceed 100?

Analysis

Recall that a composite m is divisible by a prime not exceeding the square root of m. Thus a composite that does not exceed 100 has a prime divisor that does not exceed 10. The only primes less than 10 are 2,3,5, and 7. The remaining primes that do not exceed 100 are those natural numbers that are greater than 1, less than 101, and not divisible by 2, 3, 5, or 7. Let

$$A_1 = \{m \in \mathbb{N}: m \leqslant 100 \text{ and } m \text{ is divisible by } 2\}$$

$$A_2 = \{m \in \mathbb{N}: m \leqslant 100 \text{ and } m \text{ is divisible by } 3\}$$

$$A_3 = \{m \in \mathbb{N}: m \leqslant 100 \text{ and } m \text{ is divisible by } 5\}$$

$$A_4 = \{m \in \mathbb{N}: m \leqslant 100 \text{ and } m \text{ is divisible by } 7\}$$

Then the number of primes that do not exceed 100 is $4 + [99 - n(\cup_{i=1}^{4} A_i)]$. By the principle of inclusion–exclusion, if i, j, and k are distinct natural numbers,

$$n\left(\bigcup_{i=1}^{4} A_i\right) = \sum_{i=1}^{4} n(A_i) - \sum_{\substack{i,j=1\\ i\neq j}}^{4} n(A_i \cap A_j) + \sum_{\substack{i,j,k=1\\ i\neq j, j\neq k, i\neq k}}^{4} n(A_i \cap A_j \cap A_k)$$

$$- n\left(\bigcap_{i=1}^{4} A_i\right)$$

$$= \lfloor 100/2 \rfloor + \lfloor 100/3 \rfloor + \lfloor 100/5 \rfloor + \lfloor 100/7 \rfloor - \lfloor 100/(2\cdot 3) \rfloor$$

$$- \lfloor 100/(2\cdot 5) \rfloor - \lfloor 100/(2\cdot 7) \rfloor - \lfloor 100/(3\cdot 5) \rfloor - \lfloor 100/(3\cdot 7) \rfloor$$

$$- \lfloor 100/(5\cdot 7) \rfloor + \lfloor 100/(2\cdot 3\cdot 5) \rfloor + \lfloor 100/(2\cdot 3\cdot 7) \rfloor$$

$$+ \lfloor 100/(2\cdot 5\cdot 7) \rfloor + \lfloor 100/(3\cdot 5\cdot 7) \rfloor - \lfloor 100/(2\cdot 3\cdot 5\cdot 7) \rfloor$$

$$= 50 + 33 + 20 + 14 - 16 - 10 - 7 - 6 - 4 - 2$$

$$+ 3 + 2 + 1 + 0 - 0$$

$$= 117 - 45 + 6 - 0 = 78$$

Therefore the number of primes that do not exceed 100 is $4 + 99 - 78 = 25$. ■

Example 8

If A is a set with seven members and B is a set with three members, how many functions are there that map A onto B?

Analysis

We first calculate the number of functions that map A into B. For each of the seven members of A there are three choices in B. Therefore the number of functions that map A into B is 3^7.

Let $B = \{b_1, b_2, b_3\}$, and for each $i = 1,2,3$ let $F_i = \{f\colon A \to B\colon b_i \notin f(A)\}$. Note that f maps A onto B if and only if $f\colon A \to B$ and $f \notin \bigcup_{i=1}^{3} F_i$. Therefore the number of functions that map A onto B is $3^7 - n(\bigcup_{i=1}^{3} F_i)$. By the principle of inclusion–exclusion, if i, j, and k are distinct natural numbers, then

$$n\left(\bigcup_{i=1}^{3} F_i\right) = \sum_{i=1}^{3} n(F_i) - \sum_{\substack{i,j=1\\ i\neq j}}^{3} n(F_i \cap F_j) + n\left(\bigcap_{i=1}^{3} F_i\right)$$

Let $f\colon A \to B$. For each $i = 1,2,3$, $f \in F_i$ if and only if $b_i \notin f(A)$. Therefore for each $a \in A$ there are two choices for $f(a)$. Hence $n(F_i) = 2^7$. For each $i, j = 1,2,3$, $i \neq j$, $f \in F_i \cap F_j$ if and only if b_i and b_j do not belong to

$f(A)$. Therefore there is only one choice for $f(a)$ for each $a \in A$. Hence $n(F_i \cap F_j) = 1$. Finally $\cap_{i=1}^{3} F_i = \varnothing$. Hence $n(\cup_{i=1}^{3} F_i) = 2^7 + 2^7 + 2^7 - 1 - 1 - 1 = 3(128) - 3 = 381$. Therefore the number of functions that map A onto B is $3^7 - 381 = 1{,}806$. ■

Example 9

As they enter the bookstore at State U., students are required to check the books they are carrying. During the course of an hour on a dull afternoon in the middle of the semester, six students check books. While they are in the bookstore, the electricity goes off, and each of the six students simply selects a stack of books at random from the six stacks at the entrance. What is the probability that no one receives the correct stack of books?

Analysis

By Theorem 14.3, the number of derangements of a set with six members is

$$D_6 = 6!\left(1 - \frac{1}{1!} + \frac{1}{2!} - \frac{1}{3!} + \frac{1}{4!} - \frac{1}{5!} + \frac{1}{6!}\right)$$

$$= 720\left(\frac{360 - 120 + 30 - 6 + 1}{720}\right) = 265$$

The total number of permutations is $6! = 720$, so the probability that no one receives the correct stack of books is $265/720 = 53/144 \approx .36806$. ■

Example 9 illustrates a general problem, usually known as the hatcheck problem.

Example 10

The Hatcheck Problem The person who is checking hats at the 1921 annual ball forgets to put claim numbers on the hats. When the dance ends, the checker gives out hats at random. If m hats are checked, what is the probability that no one receives the correct hat?

Analysis

The number of permutations of the hats is $m!$, and the number of derangements of the hats is D_m. Thus the probability that no one receives the correct hat is

$$\frac{D_m}{m!} = 1 - \frac{1}{1!} + \frac{1}{2!} - \frac{1}{3!} + \cdots + \frac{(-1)^m}{m!}$$

An interesting fact about the hatcheck problem is that $D_5/5!$ and $D_{100}/100!$ are very close together; that is, the probabilities that no one gets the correct hat when five hats are checked and when one hundred hats are checked are virtually the same. In fact, it can be shown that $D_m/m!$ is approximately 0.368 for sufficiently large m. ■

We conclude this section with an introduction to Sterling numbers of the second kind. These numbers are named in honor of the British mathematician James Sterling (1692–1770).

Definition

Suppose $m,p \in \mathbb{N}$ with $p \geqslant m$. Let $\left\{ p \atop m \right\}$ denote the number of ways to partition a set with exactly p members into m nonempty disjoint subsets. The number $\left\{ p \atop m \right\}$ is called a **Sterling number of the second kind.**

Notice that for each $m \in \mathbb{N}$, $\left\{ m \atop 1 \right\} = \left\{ m \atop m \right\} = 1$.

Example 11

Let $A = \{1,2,3\}$. Then the partitions of A that consist of two nonempty disjoint subsets are $\{\{1,2\},\{3\}\}$, $\{\{1,3\},\{2\}\}$, and $\{\{2,3\},\{1\}\}$

Thus $\left\{ 3 \atop 2 \right\} = 3$. ■

Exercises 14.1

1. In a survey, 68 people like golf and 45 like baseball. If 21 of these people like both golf and baseball, how many like golf or baseball?
2. List all the derangements of $\{1,2,3,4\}$.
3. How many derangements are there of a set with eight members?
4. Find the number of solutions of the equation $x_1 + x_2 + x_3 = 13$ when, for each $i = 1,2,3$, $x_i \in \mathbb{Z}$ and $0 \leqslant x_i \leqslant 6$.
5. Find the number of solutions of the equation $x_1 + x_2 + x_3 + x_4 = 19$ when x_1, x_2, x_3, and x_4 are nonnegative integers with $x_1 \leqslant 3$, $x_2 \leqslant 5$, $x_3 \leqslant 7$, and $x_4 \leqslant 8$.
6. In how many ways can Diane distribute a duck, a rabbit, a basket, a bear, a pumpkin, and an Easter basket to Sibyle, Janet, and Linda if each person gets at least one item.

7. A freshman dorm has 400 residents, and 289 of the residents are taking mathematics and 187 are taking computer science. If 100 are taking both mathematics and computer science, how many are taking neither?
8. In a group of investors it is found that 42 own stock in GM, 23 own stock in IBM, 59 own stock in AT&T, 9 own stock in GM and IBM, 19 own stock in GM and AT&T, 12 own stock in IBM and AT&T, 6 own stock in GM, IBM, and AT&T, and 83 do not own any of the three stocks. How many investors are in the group?
9. How many five-card poker hands contain at least one card in each suit?
10. If A is a set with six members and B is a set with four members, how many functions are there that map A onto B?
11. In how many ways can eight different jobs be assigned to five different employees so that each employee is assigned at least one job?
12. How many integers between 1 and 1,003 are divisible by 2,3,5, or 7?
13. Use the principle of inclusion–exclusion to find the number of primes that do not exceed 200.
14. Express the answers to this question in terms of $m!$ and D_m; in other words, do *not* perform the calculations. If the hatcheck person returns hats at random, what is the probability that in a set of 50 hats
    a. no hat is returned to the proper person?
    b. exactly one hat is returned to the proper person?
    c. exactly 49 hats are returned to the proper person?
    d. all hats are returned to the proper person?
15. What is the probability that a thirteen-card bridge hand contains at least one ace, at least one king, at least one queen, and at least one jack?
16. Let $m,p \in \mathbb{N}$ with $m \geqslant p$, let A be a set with exactly m members, and let B be a set with exactly p members. Prove that the number of functions that map A onto B is $C(p,0)(p-0)^m - C(p,1)(p-1)^m + C(p,2)(p-2)^m - \cdots + (-1)^{p-1}C(p,p-1) \cdot 1^m$.
17. Prove that $D_{m+1} = (m+1)D_m + (-1)^{m+1}$ for each $m \geqslant 1$.
18. Use Exercise 17 to prove that $D_m = (m-1)(D_{m-1} + D_{m-2})$ for each $m \geqslant 2$.
19. Let $A = \{1,2,3,4\}$.
    a. List the partitions of A that consist of two nonempty disjoint subsets.
    b. What is $\left\{ {4 \atop 2} \right\}$?
    c. List the partitions of A that consist of three nonempty disjoint subsets.
    d. What is $\left\{ {4 \atop 3} \right\}$?
20. Prove that $\left\{ {m \atop m-1} \right\} = C(m,2)$ for each $m \in \mathbb{N}$ with $m \geqslant 2$.

21. Prove that $\left\{ {m \atop 2} \right\} = 2^{m-1} - 1$ for each $m \in \mathbb{N}$ with $m \geqslant 2$.

*22. Prove that if $m,p \in \mathbb{N}$ with $p \geqslant m$ then $\left\{ {p + 1 \atop m} \right\} = m\left\{ {p \atop m} \right\} + \left\{ {p \atop m - 1} \right\}$.

14.2

GENERATING PERMUTATIONS AND COMBINATIONS

We described several methods for counting permutations and combinations in Chapter 6. Unfortunately, there are many problems, such as the knapsack problem, for which no efficient solution is known and the only method of solution is to check all possibilities. Consequently, we need a systematic method of listing the possibilities; that is, we need to generate permutations or combinations in a systematic manner.

We first consider the problem of generating all the permutations of a set with exactly n members. For convenience, we assume that the set consists of the first n natural numbers. The number of permutations of this set is $n!$, and many algorithms for generating these permutations have been developed. We describe an algorithm that lists the permutations in *lexicographic*, or *dictionary*, *order*. Throughout this section, we use the notation for permutations given in Section 6.3. Thus 231 indicates the permutation of {1,2,3} that sends 1 to 2, 2 to 3, and 3 to 1.

Definition

Let $p = p_1p_2 \ldots p_n$ and $q = q_1q_2 \ldots q_n$ be different permutations of $\{1,2, \ldots ,n\}$. Since $p \neq q$, there is an integer i such that $p_i \neq q_i$. Let k denote the smallest such integer. We say that p **precedes** q if $p_k < q_k$.

If $n = 5$, $p = 23451$ precedes $q = 23514$, since these permutations agree in the first two positions but the number 4 in the third position of p is less than the number 5 in the third position of q.

In Figure 14.3, we use a tree to list all the permutations of {1,2,3} in lexicographic order. Notice that if we think of 1,2,3 as the first three letters of an alphabet then the last column lists permutations in alphabetical order.

Except for the last word, every word in a dictionary has a unique successor. Hence we make the following definition.

Definition

Let p be a permutation of $\{1,2, \ldots ,n\}$ other than the last permutation $n(n-1)(n-2) \ldots (3)(2)(1)$. A permutation q of $\{1,2, \ldots ,n\}$ is the **successor** of p provided that p precedes q, and, if r is a permutation of $\{1,2,3, \ldots ,n\}$ different from q such that p precedes r, then q precedes r.

To construct an algorithm for listing the permutations on $\{1,2, \ldots ,n\}$ in lexicographic order, we must know how to find the successor of a permutation if it exists and we must stop when no successor exists.

Figure 14.3

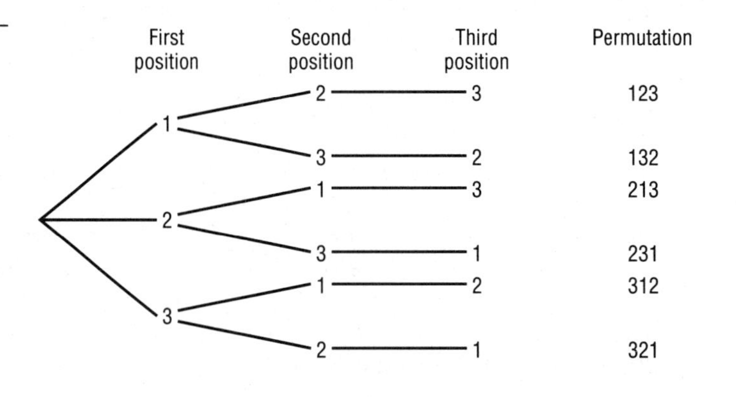

Example 12

Find the successor of the permutation $p = 7632541$ of $\{1,2,3,4,5,6,7\}$.

Analysis

Since $4 \nless 1$, we cannot find a permutation r of $\{1,2,3,4,5,6,7\}$ such that p precedes r by interchanging the numbers in the last two positions. Similarly, since $5 \nless 4$, we cannot find a permutation r of $\{1,2,3,4,5,6,7\}$ such that p precedes r by rearranging the numbers in the last three positions. However, $2 < 5$, so we can find a permutation r of $\{1,2,3,4,5,6,7\}$ such that p precedes r by rearranging the numbers in the last four positions. For example, we can simply interchange 2 and 5 to obtain the permutation $r = 7635241$. However, r is *not* the successor of p. To obtain the successor of p, we put the smallest of the numbers 5, 4, and 1 that is larger than 2 in the fourth position and place the remaining three numbers in the last three positions in increasing order. Since 4 is the smallest such number, we place 4 in the fourth position and arrange the numbers 2, 5, and 1 in increasing order and obtain the permutation $q = 7634125$ as the successor of p.

Since p and q agree in the first three positions and $2 < 4$, p precedes q. If $r = r_1r_2 \ldots r_7$ is a permutation of $\{1,2,3,4,5,6,7\}$ different from q such that p precedes r, how do we know that q precedes r? Since 7 occupies the first position in p, $r_1 = 7$. In the same way, $r_2 = 6$. Then r_3 is 3, 4, or 5. If r_3 is 4 or 5, then q precedes r, so suppose $r_3 = 3$. Then $r_4 = 2, 4$, or 5. If $r_4 = 5$, then q precedes r. If $r_4 = 2$, then we cannot arrange the numbers in the last three positions so that p precedes r. So suppose $r_4 = 4$. Then q precedes r. We have shown that indeed q precedes r. ■

To list the $n!$ permutations of $\{1,2, \ldots ,n\}$ in lexicographic order, we begin with the permutation $123 \ldots n$ and apply the procedure explained in Example 12. But before we present the general algorithm, we consider another example.

Example 13

List the permutations of $\{1,2,3,4\}$ in lexicographic order.

Analysis

We begin with 1234. Since $3 < 4$, the successor to 1234 is 1243. Since $2 < 4$, the successor to 1243 is obtained by putting a 3 (the smaller of 3 and 4) in the second position and then arranging 2 and 4 in increasing order. So the successor to 1243 is 1324. Since $2 < 4$, the successor to 1324 is 1342. Since $4 \nless 2$ and $3 < 4$, the successor to 1342 is obtained by placing 4 in the second position and then arranging 2 and 3 in increasing order. So the successor to 1342 is 1423. Since $2 < 3$, the successor to 1423 is 1432. Therefore the permutations of $\{1,2,3,4\}$ with a 1 in the first position are, in lexicographic order,

1234 1243 1324 1342 1423 1432

We leave as Exercise 23 the problem of placing the remaining permutations of $\{1,2,3,4\}$ in lexicographic order. ■

The algorithm that we have discussed for listing the permutations of $\{1,2, \ldots ,n\}$ in lexicographic order follows:

Algorithm for the Lexicographic Ordering of Permutations

1. For each $i = 1,2, \ldots ,n$, let $p_i = i$.
2. Print $p_1p_2p_3 \ldots p_n$.
3. Find the largest k such that $p_k < p_{k+1}$. If no such k exists, stop.
4. Find the smallest entry p_{k+j} to the right of p_k that is larger than p_k.
5. Interchange p_k and p_{k+j} (change the labels also).
6. Reverse the order of $p_{k+1}p_{k+2} \ldots p_n$ (change the labels also).
7. Go to step 2.

Example 14

List the permutations of $\{1,2,3,4,5\}$ in lexicographic order.

Analysis

1. Let $p_1 = 1$, $p_2 = 2$, $p_3 = 3$, $p_4 = 4$, and $p_5 = 5$.
2. Print 12345.
3. The largest k such that $p_k < p_{k+1}$ is 4.
4. The smallest (only) entry to the right of p_4 that is larger than p_4 is $p_5 = 5$.
5. Interchange 4 and 5.
6. $p_{4+1} = p_5 = 4$, so there is only one entry.
7. Go to step 2.
2. Print 12354.
3. The largest k such that $p_k < p_{k+1}$ is 3.
4. The smallest entry to the right of p_3 is $p_5 = 4$.
5. Interchange 3 and 4 to obtain 12453.
6. Reverse the order of $p_{k+1} = p_4 = 5$ and $p_{k+2} = p_5 = 3$ to obtain 12435.
7. Go to step 2.
2. Print 12435.
3. The largest k such that $p_k < p_{k+1}$ is 4.
4. The only entry to the right of p_4 that is larger than p_4 is 5.
5. Interchange 3 and 5 to obtain 12453.
6. $p_{4+1} = p_5 = 3$, so there is only one entry.
7. Go to step 2.
8. Print 12453.

We continue this process until the $n! = 5! = 120$ permutations of $\{1,2,3,4,5\}$ have been listed in lexicographic order. ■

In Exercise 30 we ask you to explain why the preceding algorithm lists the permutations of $\{1,2, \ldots ,n\}$ in lexicographic order.

The lexicographic order is the most natural ordering of permutations of a finite set, but determining the successor of a given permutation requires several comparisons. Consequently, listing the permutations in a different order may be more efficient.

We present two approaches to listing combinations. In the first approach we write the combination as a "string" of zeros and ones and work with this string.

Recall that a *bit* is a zero or a one. We call a finite sequence of bits a *bit string* and the number of bits in the bit string the *length* of the bit string. Also recall that, if $\{a_1,a_2, \ldots ,a_n\}$ is a set with exactly n members and $0 \leq r \leq n$, then an r-combination of $\{a_1,a_2, \ldots ,a_n\}$ is simply a subset of $\{a_1,a_2, \ldots ,a_n\}$ with r members. We can associate with each such subset of $\{a_1,a_2, \ldots ,a_n\}$ a unique bit string of length n: For each $k = 1,2, \ldots ,n$, the bit string has a 1 in position k if a_k belongs to the subset and a 0 in position k otherwise. So, for example, the subset $\{2,4,7\}$ of $\{1,2,3,4,5,6,7,8\}$

has associated with it the bit string 01010010. Finally, recall that a bit string of length n is the binary representation of an integer between 0 and $2^n - 1$, so the bit strings can be listed in order of their increasing size as integers. Then these bit strings (which yield r-combinations) can be converted to r-combinations as indicated in the preceding paragraph, and so we have the r-combinations ordered according to the bit strings associated with them. We consider an example to illustrate this idea.

To produce all bit strings of length n (that is, all binary representations of integers of length n), we start with the bit string 000 . . . 00 of n zeros and successively find the binary representation of the next integer until we reach the bit string 111 . . . 11 of n ones. At each stage, the binary representation of the next integer is found by locating the first position from the right that is a 0, changing it to 1, and changing all the 1s to the right of this position to 0s.

Example 15

Use the idea of bit strings to order the 3-combinations of {1,2,3,4,5}.

Analysis

The bit strings that represent 3-combinations of {1,2,3,4,5} are those bit strings of length 5 that consist of exactly three 1s and two 0s, and we want to order these according to their size as integers. We simply choose bit strings that have three 1s and two 0s and write them in numerical order. When we interpret the 3-combinations of {1,2,3,4,5} that are represented by these bit strings, we get this list:

00111	{3,4,5}	10101	{1,3,5}
01011	{2,4,5}	10110	{1,3,4}
01101	{2,3,5}	11001	{1,2,5}
01110	{2,3,4}	11010	{1,2,4}
10011	{1,4,5}	11100	{1,2,3}

■

Now we consider a second approach. We list the r-combinations of $\{1,2, \ldots ,n\}$ in lexicographic order and represent an r-combination of $\{1,2, \ldots ,n\}$ by a sequence that contains the members of the r-combinations in increasing order. We begin with $\{1,2, \ldots ,r\}$, and the procedure for finding the next r-combination after $a_1a_2 \ldots a_r$ is given by the following algorithm.

Algorithm for Finding the Next r-Combination in Lexicographic Order

1. Find the last member a_i of the sequence $a_1a_2 \ldots a_r$ such that $n - r + i \neq a_i$. Stop when a_i does not exist.
2. Replace a_i by $a_i + 1$.

3. For each $j = i + 1, i + 2, \ldots, r$, replace a_j by $a_{j-1} + 1$.
4. Go to step 1.

We consider an example using this procedure.

Example 16

List the 3-combinations of {1,2,3,4,5} in lexicographic order.

Analysis

We begin with {1,2,3}. We are listing the 3-combinations, so $a_1 = 1$, $a_2 = 2$, and $a_3 = 3$. Now $n - r = 2$, so $n - r + 3 = 5$, and $5 \neq a_3$. Therefore we replace a_3 with $a_3 + 1 = 4$ and obtain {1,2,4} as the next 3-combination. We have $a_1 = 1$, $a_2 = 2$, and $a_3 = 4$, so $n - r + 3 = 5$, and $5 \neq a_3$. Therefore we replace a_3 with $a_3 + 1 = 5$ and obtain {1,2,5} as the next 3-combination. We have $a_1 = 1$, $a_2 = 2$, and $a_3 = 5$. Since $n - r + 3 = 5$ and $5 = a_3$, but $n - r + 2 = 4$ and $4 \neq a_2$, we replace a_2 by $a_2 + 1$ and a_3 by $a_2 + 1 = 3 + 1 = 4$ and obtain {1,3,4} as the next 3-combination. Continuing in this manner, we obtain the following ordering of the 3-combinations.

{1,2,3} {1,2,4} {1,2,5} {1,3,4} {1,3,5}

{1,4,5} {2,3,4} {2,3,5} {2,4,5} {3,4,5}

Note that this is the reverse order of the order we obtained by using the idea of bit strings and the order (as integers) of bit strings. ■

In Exercise 31 we ask you to explain why the preceding algorithm produces the next r-combination in lexicographic order.

Exercises 14.2

23. Complete Example 13 by listing all the permutations of {1,2,3,4} in lexicographic order.
24. For each of the following pairs p and q of permutations, determine whether p precedes q or q precedes p.

    a. $p = 321$, $q = 213$ b. $p = 45231$, $q = 54132$

    c. $p = 12435$, $q = 12354$ d. $p = 253146$, $q = 253461$
25. Find the successor of each of the following permutations.

    a. 1432 b. 25431 c. 41532 d. 45123 e. 54123 f. 7614235
26. Place the following permutations of {1,2,3,4,5} in lexicographic order: 54321, 43512, 15423, 23145, 15432, 23451, 43251, 31452, 54123, 14523.
27. Place the following 5-combinations of {1,2,3,4,5,6,7} in lexicographic order: {6,4,5,7,1}, {3,4,1,7,2}, {5,4,3,2,1}, {1,5,6,2,7}, {2,6,5,4,7}, {3,1,6,5,4}, {6,2,3,4,5}, {2,6,1,3,5}.

28. Use bit strings to order the 3-combinations of {1,2,3,4,5,6}.

29. List the 4-combinations of {1,2,3,4,5,6} in lexicographic order.

30. Explain why the algorithm given in this section for listing the permutations of $\{1,2, \ldots ,n\}$ in lexicographic order does indeed list the permutations in lexicographic order.

31. Explain why the algorithm given in this section for finding the next r-combination in lexicographic order after $a_1a_2 \ldots a_r$ does indeed find this r-combination.

SOLVING RECURSION RELATIONS

In Chapter 4 we studied several sequences that were defined recursively. In computer science, sorting algorithms involves recursion. For example, the time required to sort a list of n items depends on the time required to sort a list of $n - 1$ items and the time required to put the nth item in its proper place. In this section, we discuss methods of finding explicit formulas for sequences that have been defined recursively. In our first method, *iteration,* we use repeatedly the given recursion relation, each time for a different value of n. We begin with an example.

Example 17

Joe invests $1,000 at 8% compounded annually. If A_n represents the amount at the end of n years, what is the relation between A_n and A_{n-1}? Determine an explicit formula for A_n.

Analysis

Since A_{n-1} represents the amount at the end of $n - 1$ years and A_n represents the amount at the end of n years,

$$A_n = A_{n-1} + 0.08A_{n-1} = 1.08A_{n-1}$$

We have a recursion relation whose initial value is $A_0 = 1000$. We also have

$$\begin{aligned} A_n &= 1.08A_{n-1} \\ &= 1.08(1.08A_{n-2}) \\ &= (1.08)^2A_{n-2} \\ &= (1.08)^2(1.08A_{n-3}) \\ &= (1.08)^3A_{n-3} \end{aligned}$$

$$= (1.08)^3(1.08A_{n-4})$$
$$= (1.08)^4A_{n-4}$$

Continuing in this manner, we obtain the explicit formula

$$A_n = (1.08)^nA_0 = (1.08)^n(1000)$$

This explicit formula can be proved by induction. ■

We have already seen one example of an explicit solution of a recursion relation—in Example 16 of Chapter 4. We had the recursion relation $s_{n+1} = 2s_n + 1$ with initial value $s_1 = 1$, where s_n was the number of moves required to move the n disks from the first peg to the third peg according to the rules of the Tower of Hanoi, and we found that an explicit formula for s_n was $2^n - 1$.

Example 18

We say that a set of straight lines in the plane is in *general position* if no two of the lines are parallel and no three of the lines go through the same point. If we have a set of n lines in the plane in general position, into how many distinct regions does this set of lines divide the plane?

Analysis

One line divides the plane into two distinct regions, and two lines divide the plane into four regions:

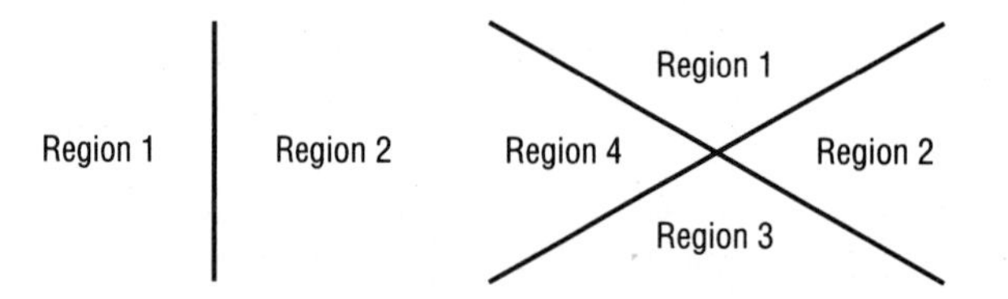

What about three lines in general position?

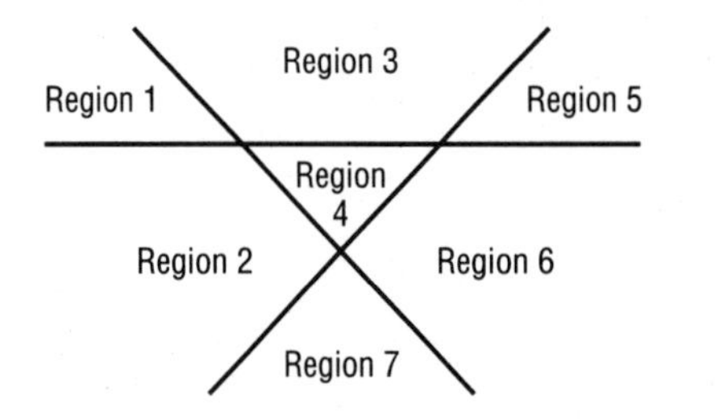

What about four lines in general position?

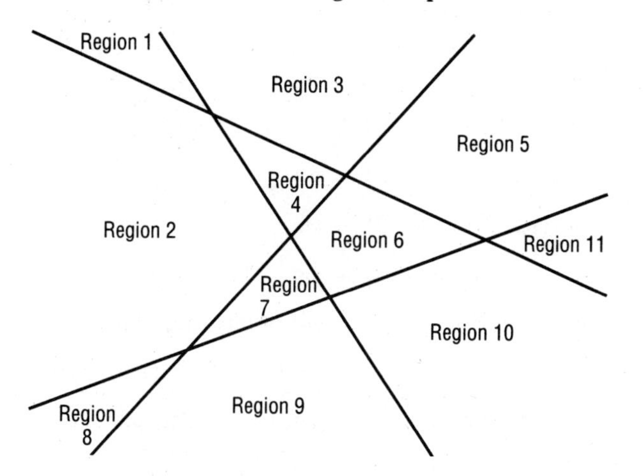

Suppose that n lines in general position divide the plane into p_n distinct regions. We have seen that

$$p_1 = 2 \qquad p_2 = 4 \qquad p_3 = 7 \qquad p_4 = 11$$

Consider a set of $n - 1$ lines in general position. These lines divide the plane into p_{n-1} distinct regions. Now add an nth line to these $n - 1$ lines so that the set of n lines is in general position (see Figure 14.4). The new line intersects each of the old lines at points $p_1, p_2, \ldots, p_{n-1}$. Choose a point on the new line that does not lie between any two of these points. Move along the new line toward the points $p_1, p_2, \ldots, p_{n-1}$. As soon as the new line meets the first of the old lines, it has divided the region where we started into two regions. When it meets the second of the old lines, it has divided this region into two regions. When it meets the last of the old lines, it has divided each of $n - 1$ regions into two regions. After it has crossed all the old lines, it divides one more region into two regions, so $p_n = p_{n-1} + n$.

Figure 14.4

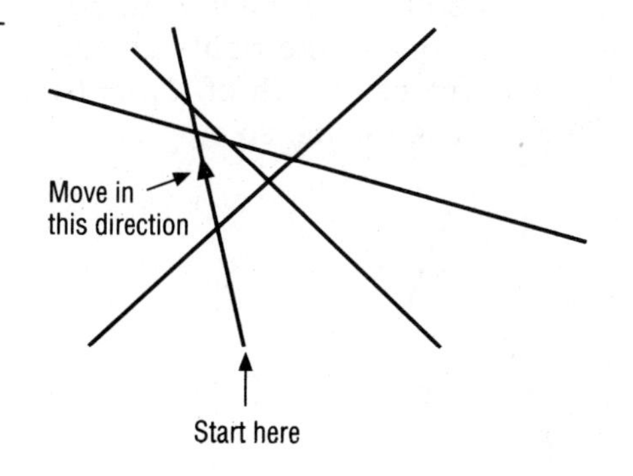

We have a recursion relation $p_n = p_{n-1} + n$ with initial condition $p_1 = 2$. Therefore

$$\begin{aligned} p_n &= p_{n-1} + n \\ &= (p_{n-2} + n - 1) + n \\ &= p_{n-2} + (n - 1) + n \\ &= (p_{n-3} + n - 2) + (n - 1) + n \\ &= p_{n-3} + (n - 2) + (n - 1) + n \end{aligned}$$

Continuing in this manner, we obtain the explicit formula

$$p_n = p_1 + 2 + 3 + \cdots + (n-2) + (n - 1) + n$$

(Once again, this explicit formula can be proved by induction.) Since $p_1 = 2$ and $1 + 2 + 3 + \cdots + (n - 2) + (n - 1) + n = n(n + 1)/2 = C(n + 1,2)$ (see Example 5 in Chapter 4), $p_n = C(n + 1,2) + 1$. ■

In the following example we develop a recursion relation for the number of binary trees with n vertices.

Example 19

For each $n \in \mathbb{N}$, let a_n denote the number of binary trees with n vertices. Since a binary tree with one vertex has no edges, $a_1 = 1$. There are two binary trees with two vertices, so $a_2 = 2$:

Two vertices

We construct the binary trees with three vertices as follows. One vertex must be the root r. There are three possibilities for the remaining two vertices: (1) they can both be in the left subtree of r, (2) one can be in the left subtree of r and one in the right subtree of r, or (3) they can both be in the right subtree of r. Each of (1) and (3) gives rise to two binary trees and (2) gives rise to one, so $a_3 = 5$:

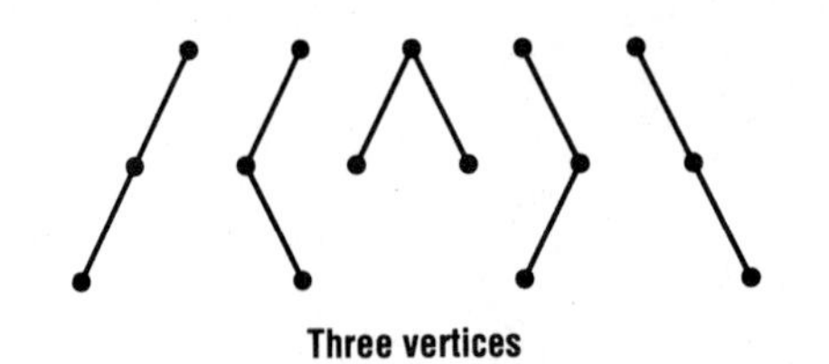

Three vertices

If a binary tree has four vertices, then one vertex must be the root r, and the remaining three vertices can be positioned as follows.

Position	*A*	*B*	*C*	*D*
Vertices in left subtree	0	1	2	3
Vertices in right subtree	3	2	1	0

Since $a_3 = 5$, there are five binary trees in each of positions A and D. Since $a_1 = 1$ and $a_2 = 2$, there are two binary trees in each of positions B and C. Therefore $a_4 = 5 + 2 + 2 + 5 = 14$.

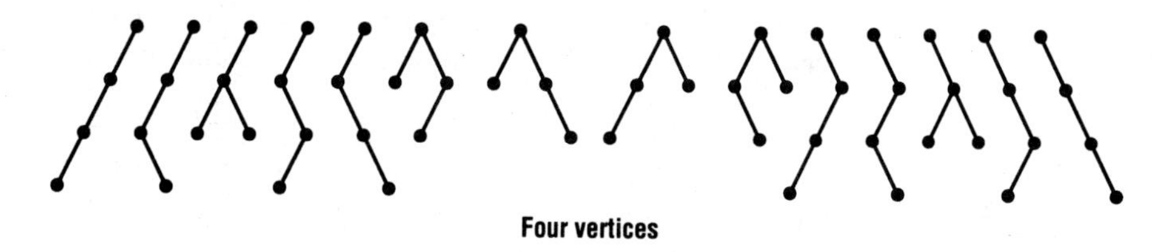

Four vertices

Suppose that $n \in \mathbb{N}$ with $n > 4$ and we know how to construct all binary trees with fewer than n vertices. Then we can construct the binary trees with n vertices by putting k vertices ($0 \leqslant k \leqslant n - 1$) in the left subtree of the root r and $n - k - 1$ vertices in the right subtree of r. For each k the number p_k of such trees with k vertices in the left subtree of r is as follows.

k	0	1	2	3	$\cdots$	$n-2$	$n-1$
p_k	$1 \cdot a_{n-1}$	$1 \cdot a_{n-2}$	$2 \cdot a_{n-3}$	$5 \cdot a_{n-4}$	$\cdots$	$a_{n-2}a_1$	$a_{n-1} \cdot 1$

Thus

$$a_n = \sum_{k=0}^{n-1} p_k$$

By taking $a_0 = 1$, we can write

$$a_n = \sum_{k=0}^{n-1} a_k a_{n-k-1}$$

It is beyond the scope of this text, but it can be shown that $a_n = C(2n, n)/(n + 1)$. ■

The following example is a simplified model to show how estimations of animal population use recursion relations.

Example 20

Suppose that deer reproduce each spring and that each doe of at least one year of age gives birth to exactly two fawns. Assume also that does and bucks occur in equal numbers, that p_1 is the percentage of deer that survive to the first year, that p_2 is the percentage of deer that survive to the second year, and that no deer survive to the third year. Find a recursion relation that gives the deer population in the spring of the nth year.

Analysis

For the nth year, let A_n denote the number of newborn fawns, let B_n denote the number of 1-year-old deer, let C_n denote the number of 2-year-old deer, and let S_n denote the total deer population. Then we have

$$S_n = A_n + B_n + C_n,\ B_n = p_1A_{n-1}, \text{ and } C_n = p_2B_{n-1} = p_2p_1A_{n-2}$$

Also, since half the deer are does and each doe produces two offspring, the number of newborn fawns in the nth year is

$$A_n = 2(1/2)B_n + 2(1/2)C_n = B_n + C_n$$

Thus $S_n = A_n + A_n$, or $A_n = (1/2)S_n$. Also, $A_n = B_n + C_n = p_1A_{n-1} + p_2p_1A_{n-2}$. Therefore

$$(1/2)S_n = p_1(1/2)S_{n-1} + p_2p_1(1/2)S_{n-2}$$

so $S_n = p_1S_{n-1} + p_2p_1S_{n-2}$ ■

We can use the formula obtained in Example 20 to estimate the deer population for any given year in the future.

Example 21

Suppose that the deer population, measured in millions, is 1.6 in 1990 and 1.7 in 1991. Suppose further that in Example 20, $p_1 = 0.75$ and $p_2 = 0.8$. Estimate the deer population in the year 1993.

Analysis

We are assuming that $S_1 = 1.6$ and $S_2 = 1.7$, and we want to find S_5. We use the recursion relation found in Example 20, so

$$\begin{aligned} S_5 &= (0.75)S_4 + (0.8)(0.75)S_3 \\ &= (0.75)[(0.75)S_3 + (0.8)(0.75)S_2] \\ &\quad + (0.8)(0.75)[(0.75)S_2 + (0.8)(0.75)S_1] \\ &= (0.75)^2S_3 + 2(0.8)(0.75)^2S_2 + (0.8)^2(0.75)^2S_1 \end{aligned}$$

$$= (0.75)^2[(0.75)S_2 + (0.8)(0.75)S_1] + 2(0.8)(0.75)^2S_2 + (0.8)^2(0.75)^2S_1$$

$$= [(0.75)^3 + 2(0.8)(0.75)^2]S_2 + [(0.8)(0.75)^3 + (0.8)^2(0.75)^2]S_1$$

$$= (0.75)^2(0.75 + 1.6)S_2 + (0.8)(0.75)^2[0.75 + 0.8]S_1$$

$$= (0.75)^2(2.35)S_2 + (0.8)(0.75)^2(1.55)S_1$$

$$= 3.3631875$$ ■

The need to solve this recursion relation is obvious because the calculation required to estimate the deer population in the year 2010, for example, is enormous. Thus we turn our attention to some general results regarding solutions of recursion relations.

Definition

Let $k \in \mathbb{N}$. A **linear homogeneous recursion relation of order k with constant coefficients** is a recursion relation of the form

$$a_n = c_1a_{n-1} + c_2a_{n-2} + \cdots + c_ka_{n-k}$$

where $c_1, c_2, \ldots, c_k$ are real numbers and $c_k \neq 0$.

The recursion relation in Example 17 is a linear homogeneous recursion relation of order 1. The recursion relation $s_n = 2s_{n-1} + 1$ and the recursion relation in Example 18 are not linear homogeneous recursion relations of order k (for any k).

The recursion relation in the preceding definition is *linear* since the right-hand side is a sum of multiples of the previous terms of the sequence. The recursion relation in this definition is *homogeneous* since every term on the right-hand side is a multiple of previous terms of the sequence. Thus $s_n = 2s_{n-1} + 1$ and $p_n = p_{n-1} + n$ fail to be homogeneous since 1 and n are not multiples of previous terms of the sequence. The recursion relation $a_n = na_{n-1}$ is not homogeneous since it does not have constant coefficients, and the recursion relation $a_n = (a_{n-1})^2$ is not linear since a_{n-1} is squared.

The recursion relation $f_n = f_{n-1} + f_{n-2}$ used to define the Fibonacci sequence is a linear homogeneous recursion relation of order 2 with constant coefficients (the coefficients are both 1).

We turn our attention to the problem of finding a general method of solving linear homogeneous recursion relations of order 2 with constant coefficients. Once we have done this, we will be able to solve the recursion relation $f_n = f_{n-1} + f_{n-2}$.

You are entitled to wonder at this point just what it means to "solve" a recursion relation. Roughly speaking, it means to find a way to express a sequence that has been defined recursively in a form that is easier to think about. This form, which is called *closed form*, may itself be defined recursively. In Example 17 we wrote the recursion relation $\{A_n\} = \{1.08A_{n-1}\}$ in the closed form $\{A_n\} = \{(1.08)^n A_0\}$, but recall that the sequence $\{(1.08)^n\}$ is itself defined by recursion. As we see in Example 26, the closed form of the Fibonacci sequence is more difficult to work with both for man and machine, so in this case the advantage of the closed form over the original recursion relation is theoretical rather than computational. Often, however, the form we discover by solving a recursion relation is easier to work with as well as being easier to think about.

Suppose that we have two sequences of real numbers both of which are solutions of a recursion relation of the form $a_n = c_1a_{n-1} + c_2a_{n-2}$. The following theorem tells us how to obtain an additional solution of the recursion relation using any two real numbers a and b and the two sequences that were already promised to be solutions.

Theorem 14.4

Let $a_1, a_2, c_1, c_2 \in \mathbb{R}$. If $\{p_n\}$ and $\{q_n\}$ are real-number solutions of the recursion relation $a_n = c_1a_{n-1} + c_2a_{n-2}$ and $b_1, b_2 \in \mathbb{R}$, then $\{s_n\}$, where $s_n = b_1p_n + b_2q_n$ for each $n \in \mathbb{N}$, is also a solution of the recursion relation.

Proof

Let $\{p_n\}$ and $\{q_n\}$ be real-number solutions of $a_n = c_1a_{n-1} + c_2a_{n-2}$. Then $p_n = c_1p_{n-1} + c_2p_{n-2}$ and $q_n = c_1q_{n-1} + c_2q_{n-2}$. Multiplying the first equation by b_1 and the second one by b_2, we obtain $b_1p_n = b_1c_1p_{n-1} + b_1c_2p_{n-2}$ and $b_2q_n = b_2c_1q_{n-1} + b_2c_2q_{n-2}$. Therefore

$$\begin{aligned} s_n &= b_1p_n + b_2q_n \\ &= c_1(b_1p_{n-1} + b_2q_{n-1}) + c_2(b_1p_{n-2} + b_2q_{n-2}) \\ &= c_1s_{n-1} + c_2s_{n-2} \end{aligned}$$

Hence $\{s_n\}$ is a solution of the recursion relation. □

Example 22

The functions $\{p_n\}$ and $\{q_n\}$, where for each $n \in \mathbb{N}$, $p_n = 2^n$ and $q_n = n2^n$, are both solutions of the recursion relation $a_n = 4a_{n-1} - 4a_{n-2}$ (see Exercise 36). By Theorem 14. 4, the sum $\{p_n + q_n\}$ is also a solution of the recursion relation. Of course, for each $n \in \mathbb{N}$, $p_n + q_n = 2^n + n2^n$. ■

Suppose that we are given a linear homogeneous recursion relation of order k with constant coefficients: $a_n = c_1a_{n-1} + c_2a_{n-2} + \cdots + c_ka_{n-k}$. Now suppose that by a stroke of good luck there is a non-zero real number r such that $\{a_n\} = \{r^n\}$ (that is, $r^n = a_n$ for each $n \in \mathbb{N}$). Our supposition is not too fanciful, because $r = 2$ satisfies our supposition for the linear homogeneous recursion relation of Example 22. Anyway, if r is such a number, then for each natural number n

$$r^n = a_n = c_1a_{n-1} + c_2a_{n-2} + \cdots + c_ka_{n-k}$$

$$r^n = c_1r^{n-1} + c_2r^{n-2} + \cdots + c_kr^{n-k}$$

and so dividing through by r^{n-k} we have

$$r^k = c_1r^{k-1} + c_2r^{k-2} + \cdots + c_{k-1}r + c_k \tag{14.1}$$

Conversely, if r is a non-zero real number satisfying Equation (14.1), then the sequence $\{r^n\}$ is a solution to the recursion relation $\{a_n\}$.

Definition

For each $i = 1,2, \ldots ,k$, let $a_i,c_i \in \mathbb{R}$. The equation

$$r^k - c_1r^{k-1} - c_2r^{k-2} - \cdots - c_{k-1}r - c_k = 0$$

is called the **characteristic equation** of the linear homogeneous recursion relation $a_n = c_1a_{n-1} + c_2a_{n-2} + \cdots + c_ka_{n-k}$.

The observations preceding the definition of the characteristic equation are so useful that they deserve to be incorporated as a theorem.

Theorem 14.5

For each $i = 1,2, \ldots ,k$, let $a_i,c_i \in \mathbb{R}$. The non-zero number r_1 is a root of the characteristic equation

$$r^k - c_1r^{k-1} - c_2r^{k-2} - \cdots - c_{k-1}r - c_k = 0$$

of $a_n = c_1a_{n-1} + c_2a_{n-2} + \cdots + c_ka_{n-k}$ if and only if the sequence $\{p_n\}$ defined by $p_n = (r_1)^n$ is a solution of the recursion relation.

Proof

Suppose that r_1 is a root of

$$r^k - c_1r^{k-1} - c_2r^{k-2} - \cdots - c_{k-1}r - c_k = 0$$

Then

$$c_1(r_1)^{n-1} + c_2(r_1)^{n-2} + \cdots + c_k\,(r_1)^{n-k}$$

$$= (r_1)^{n-k}(c_1(r_1)^{k-1} + c_2(r_1)^{k-2} + \cdots + c_k)$$
$$= (r_1)^{n-k}(r_1)^k = (r_1)^n$$

Thus the sequence $\{p_n\}$ defined by $p_n = (r_1)^n$ is a solution of the recursion relation.

Suppose that the sequence $\{p_n\}$ defined by $p_n = (r_1)^n$ is a solution of the recursion relation. Then

$$(r_1)^n = c_1(r_1)^{n-1} + c_2(r_1)^{n-2} + \cdots + c_k(r_1)^{n-k}$$

Hence

$$(r_1)^n - c_1(r_1)^{n-1} - c_2(r_1)^{n-2} - \cdots - c_k(r_1)^{n-k} = 0$$

Dividing both sides by r_1^{n-k}, we obtain

$$(r_1)^k - c_1(r_1)^{k-1} - c_2(r_1)^{k-2} - \cdots - c_k = 0$$

Thus r_1 is a root of the characteristic equation. □

Example 23

Find the roots of the characteristic equation of $a_n = -a_{n-1} + 2a_{n-2}$ and use Theorem 14.5 to find the solutions of this recursion relation.

Analysis

The characteristic equation of $a_n = -a_{n-1} + 2a_{n-2}$ is $r^2 + r - 2 = 0$. We solve this quadratic equation by factoring $(r + 2)(r - 1) = 0$. Thus -2 and 1 are the roots of the characteristic equation of the recursion relation. By Theorem 14.5, the sequences $\{p_n\}$ and $\{q_n\}$, where for each $n \in \mathbb{N}$, $p_n = (-2)^n$ and $q_n = 1^n$, are solutions of the recursion relation $a_n = -a_{n-1} + 2a_{n-2}$. ■

In the next theorem and the examples that follow, it is convenient to assume that our sequences $\{a_n\}$ begin with a_0 rather than a_1.

Theorem 14.6

Let $D_0, D_1, c_1, c_2 \in \mathbb{R}$ and suppose that $r^2 - c_1 r - c_2 = 0$ has distinct real roots r_1 and r_2. Then a sequence $\{p_n\}$ is a solution of the recursion relation $a_n = c_1 a_{n-1} + c_2 a_{n-2}$ with initial values $a_0 = D_0$ and $a_1 = D_1$ if and only if there are real numbers b_1 and b_2 such that $p_n = b_1(r_1)^n + b_2(r_2)^n$ for each nonnegative integer n.

Proof

Suppose that there are real numbers b_1 and b_2 such that $p_n = b_1(r_1)^n + b_2(r_2)^n$ for each nonnegative integer n. Since r_1 and r_2 are roots of

$r^2 - c_1r - c_2 = 0$, $(r_1)^2 = c_1r_1 + c_2$ and $(r_2)^2 = c_1r_2 + c_2$. Therefore

$$\begin{aligned} c_1p_{n-1} + c_2p_{n-2} &= c_1(b_1(r_1)^{n-1} + b_2(r_2)^{n-1}) + c_2(b_1(r_1)^{n-2} + b_2(r_2)^{n-2}) \\ &= c_1b_1(r_1)^{n-1} + c_1b_2(r_2)^{n-1} + c_2b_1(r_1)^{n-2} \\ &\quad + c_2b_2(r_2)^{n-2} \\ &= b_1(r_1)^{n-2}(c_1r_1 + c_2) + b_2(r_2)^{n-2}(c_1r_2 + c_2) \\ &= b_1(r_1)^{n-2}(r_1)^2 + b_2(r_2)^{n-2}(r_2)^2 \\ &= b_1(r_1)^n + b_2(r_2)^n \\ &= p_n \end{aligned}$$

Therefore $\{p_n\}$ is a solution of the recursion relation $a_n = c_1a_{n-1} + c_2a_{n-2}$.

Now suppose that $\{p_n\}$ is a solution of the recursion relation $a_n = c_1a_{n-1} + c_2a_{n-2}$. Let $p_0 = D_0$ and $p_1 = D_1$ be the initial values. We need to show that there are real numbers b_1 and b_2 such that $p_n = b_1(r_1)^n + b_2(r_2)^n$ for each nonnegative integer n. This sequence has the same initial values. Therefore $p_0 = D_0 = b_1 + b_2$ and $p_1 = D_1 = b_1r_1 + b_2r_2$. Hence $b_2 = D_0 - b_1$ and

$$D_1 = b_1r_1 + (D_0 - b_1)r_2 = b_1r_1 + D_0r_2 - b_1r_2 = b_1(r_1 - r_2) + D_0r_2$$

Therefore $b_1(r_1 - r_2) = D_1 - D_0r_2$. So

$$b_1 = \frac{D_1 - D_0r_2}{r_1 - r_2} \text{ and}$$

$$\begin{aligned} b_2 = D_0 - b_1 &= D_0 - \frac{D_1 - D_0r_2}{r_1 - r_2} \\ &= \frac{D_0(r_1 - r_2) - D_1 + D_0r_2}{r_1 - r_2} = \frac{D_0r_1 - D_1}{r_1 - r_2} \end{aligned}$$

Since the initial conditions and the recursion relation uniquely determine the sequence, $p_n = b_1(r_1)^n + b_2(r_2)^n$ for each nonnegative integer n. □

Notice that in the proof of Theorem 14.6, we divide by $r_1 - r_2$, so it is necessary for r_1 and r_2 to be distinct roots of $r^2 - c_1r - c_2 = 0$.

We now use Theorem 14.6 to find the solutions of a couple of linear homogeneous recursion relations of order 2 with constant coefficients.

Example 24

Solve the recursion relation $a_n = -a_{n-1} + 2a_{n-2}$ with initial values $a_0 = -1$ and $a_1 = 11$.

Analysis

The characteristic equation of the recursion relation $a_n = -a_{n-1} + 2a_{n-2}$ is $r^2 + r - 2 = 0$, and its roots are $r_1 = -2$ and $r_2 = 1$. Therefore the sequence $\{p_n\}$ is a solution of the recursion relation if and only if there are real numbers b_1 and b_2 such that $p_n = b_1(-2)^n + b_2 \cdot 1^n$ for each nonnegative integer n. To find b_1 and b_2, we use the initial values $p_0 = -1$ and $p_1 = 11$ and obtain

$$-1 = b_1(-2)^0 + b_2(1)^0 = b_1 + b_2$$

$$11 = b_1(-2)^1 + b_2(1)^1 = -2b_1 + b_2$$

Therefore $b_2 = -1 - b_1$ and hence $-2b_1 + (-1 - b_1) = 11$, or $-3b_1 = 12$. Thus $b_1 = -4$ and $b_2 = 3$. Hence a solution is the sequence $\{p_n\}$, where $p_n = (-4)(-2)^n + 3(1)^n = (-4)(-2)^n + 3$ for each nonnegative integer n. ■

Example 25

Solve the recursion relation $a_n = 2a_{n-1} + 3a_{n-2}$ with initial values $a_0 = 1$ and $a_1 = 2$.

Analysis

The characteristic equation of the recursion relation $a_n = 2a_{n-1} + 3a_{n-2}$ is $r^2 - 2r - 3 = 0$, and its roots are $r_1 = 3$ and $r_2 = -1$. Therefore the sequence $\{p_n\}$ is a solution of the recursion relation if and only if there are real numbers b_1 and b_2 such that $p_n = b_1 3^n + b_2(-1)^n$ for each nonnegative integer n. To find b_1 and b_2, we use the initial values $p_0 = 1$ and $p_1 = 2$ and obtain

$$1 = p_0 = b_1 3^0 + b_2(-1)^0 = b_1 + b_2$$

$$2 = p_1 = b_1 3^1 + b_2(-1)^1 = 3b_1 - b_2$$

Therefore $b_2 = 1 - b_1$ and hence $3b_1 - (1 - b_1) = 2$, or $4b_1 = 3$. Thus $b_1 = 3/4$ and $b_2 = 1/4$. Hence a solution is the sequence $\{p_n\}$, where

$$p_n = \frac{3}{4} \cdot 3^n + 14 \cdot (-1)^n$$

for each nonnegative integer n. ■

As promised earlier, we now find an explicit solution for the Fibonacci sequence.

Example 26

Solve the recursion relation $f_n = f_{n-1} + f_{n-2}$ with initial values $f_0 = f_1 = 1$.

Analysis

The characteristic equation of the recursion relation $f_n = f_{n-1} + f_{n-2}$ is $r^2 - r - 1 = 0$, and the roots of this characteristic equation are $r_1 = (1 + \sqrt{1 + 4})/2$ and $r_2 = (1 - \sqrt{1 + 4})/2$. Therefore the sequence $\{g_n\}$ is a solution of the recursion relation if and only if there are real numbers b_1 and b_2 such that

$$g_n = b_1\left(\frac{1 + \sqrt{5}}{2}\right)^n + b_2\left(\frac{1 - \sqrt{5}}{2}\right)^n$$

for each nonnegative integer n. To find b_1 and b_2 we use the initial values $g_0 = g_1 = 1$ and obtain

$$1 = b_1\left(\frac{1 + \sqrt{5}}{2}\right)^0 + b_2\left(\frac{1 - \sqrt{5}}{2}\right)^0 = b_1 + b_2$$

$$1 = b_1\left(\frac{1 + \sqrt{5}}{2}\right)^1 + b_2\left(\frac{1 - \sqrt{5}}{2}\right)^1$$

Therefore $b_2 = 1 - b_1$ and hence

$$1 = b_1\left(\frac{1 + \sqrt{5}}{2}\right) + (1 - b_1)\left(\frac{1 - \sqrt{5}}{2}\right)$$

$$2 = (1 + \sqrt{5})b_1 + 1 - \sqrt{5} + b_1(\sqrt{5} - 1)$$

$$1 + \sqrt{5} = 2\sqrt{5}b_1$$

$$b_1 = \frac{1 + \sqrt{5}}{2\sqrt{5}}$$

Thus

$$b_2 = 1 - \frac{1 + \sqrt{5}}{2\sqrt{5}} = \frac{2\sqrt{5} - 1 - \sqrt{5}}{2\sqrt{5}} = \frac{\sqrt{5} - 1}{2\sqrt{5}}$$

Hence a solution is the sequence $\{g_n\}$ where

$$g_n = \left(\frac{1 + \sqrt{5}}{2\sqrt{5}}\right)\left(\frac{1 + \sqrt{5}}{2}\right)^n + \left(\frac{\sqrt{5} - 1}{2\sqrt{5}}\right)\left(\frac{1 - \sqrt{5}}{2}\right)^n$$

$$= \frac{1}{\sqrt{5}}\left(\frac{1 + \sqrt{5}}{2}\right)^{n+1} - \frac{1}{\sqrt{5}}\left(\frac{1 - \sqrt{5}}{2}\right)^{n+1}$$

■

As we mentioned, Theorem 14.6 cannot be used to solve a linear homogeneous recursion relation of order 2 with constant coefficients unless the characteristic equation of the recursion relation has distinct roots. We now state, without proof, a theorem that can be used to solve a system whose characteristic equation has a multiple root.

Theorem 14.7

Let $c_1, c_2, D_0, D_1 \in \mathbb{R}$ with $c_2 \neq 0$ and suppose that $r^2 - c_1 r - c_2 = 0$ has one root r_0. Then the sequence $\{p_n\}$ is a solution of the recursion relation $a_n = c_1 a_{n-1} + c_2 a_{n-2}$ with initial values $a_0 = D_0$ and $a_1 = D_1$ if and only if there are real numbers b_1 and b_2 such that $p_n = b_1(r_0)^n + b_2 n(r_0)^n$ for each nonnegative integer n.

Example 27

Solve the recursion relation $a_n = 10a_{n-1} - 25a_{n-2}$ with initial values $a_0 = 1$ and $a_1 = 2$.

Analysis

The characteristic equation of the recursion relation $a_n = 10a_{n-1} - 25a_{n-2}$ is $r^2 - 10r + 25 = 0$, and its only root is $r_0 = 5$. Therefore the sequence $\{p_n\}$ is a solution of the recursion relation if and only if there are real numbers b_1 and b_2 such that $p_n = b_1 5^n + b_2 n 5^n$. To find b_1 and b_2 we use the initial values $p_0 = 1$ and $p_1 = 2$ and obtain

$$1 = p_0 = b_1 5^0 + b_2 \cdot 0 \cdot 5^n = b_1$$

$$2 = p_1 = b_1 5^1 + b_2 \cdot 1 \cdot 5^1 = 5b_1 + 5b_2$$

Thus $5b_2 = 2 - 5b_1 = 2 - 5 = -3$ and hence $b_2 = -3/5$. Thus a solution is the sequence $\{p_n\}$, where

$$p_n = 5^n - \frac{3}{5} n 5^n$$

for each nonnegative integer n. ■

We state, without proof, a generalization of Theorem 14.6 and give an example to illustrate its use.

Theorem 14.8

Let $k \in \mathbb{N}$ with $k > 2$, and for each $i = 1, 2, \ldots, k$ let $c_i \in \mathbb{R}$. Suppose that the equation $r^k - c_1 r^{k-1} - \cdots - c_k = 0$ has distinct real roots $r_1, r_2, \ldots, r_k$. Then a sequence $\{p_n\}$ is a solution of the recursion relation $a_n = c_1 a_{n-1} + c_2 a_{n-2} + \cdots + c_k a_{n-k}$ if and only if there are real numbers $b_1, b_2, \ldots, b_k$ such that $p_n = b_1(r_1)^n + b_2(r_2)^n + \cdots + b_k(r_k)^n$ for each nonnegative integer n.

Example 28

Solve the recursion relation $a_n = 2a_{n-1} + a_{n-2} - 2a_{n-3}$ with initial values $a_0 = 3$, $a_1 = 2$, and $a_2 = 6$.

Analysis

The characteristic equation of the recursion relation $a_n = 2a_{n-1} + a_{n-2} - 2a_{n-3}$ is $r^3 - 2r^2 - r + 2 = 0$, and, since $r^3 - 2r^2 - r + 2 = (r - 1)(r + 1)(r - 2)$, its roots are $r_1 = 1$, $r_2 = -1$, and $r_3 = 2$. Therefore the sequence $\{p_n\}$ is a solution of the recursion relation if and only if there are real numbers b_1, b_2, and b_3 such that $p_n = b_1 \cdot 1^n + b_2(-1)^n + b_3 2^n$ for each nonnegative integer n. To find b_1, b_2, and b_3, we use the initial values $p_0 = 3$, $p_1 = 2$, and $p_2 = 6$ and obtain

$$3 = p_0 = b_1 \cdot 1^0 + b_2(-1)^0 + b_3 2^0 = b_1 + b_2 + b_3$$

$$2 = p_1 = b_1 \cdot 1^1 + b_2(-1)^1 + b_3 2^1 = b_1 - b_2 + 2b_3$$

$$6 = p_2 = b_1 \cdot 1^2 + b_2(-1)^2 + b_3 2^2 = b_1 + b_2 + 4b_3$$

Therefore $b_3 = 3 - b_1 - b_2$, so $2 = b_1 - b_2 + 2(3 - b_1 - b_2)$, $b_1 + 3b_2 = 4$, or $b_1 = 4 - 3b_2$. Hence

$$\begin{aligned} 6 &= (4 - 3b_2) + b_2 + 4(3 - b_1 - b_2) \\ &= 4 - 2b_2 + 12 - 4b_1 - 4b_2 \\ &= 16 - 6b_2 - 4(4 - 3b_2) = 6b_2 \end{aligned}$$

Thus $b_2 = 1$, $b_1 = 1$, and $b_3 = 1$. Therefore a solution is the sequence $\{p_n\}$, where $p_n = 1 + (-1)^n + 2^n$ for each nonnegative integer n. ■

We conclude this section with a brief discussion of nonhomogeneous recursion relations.

Example 29

Solve the recursion relation $a_n = a_{n-1} + 5$ with initial value $a_0 = 0$.

Analysis

Note that $a_0 = 0$, $a_1 = 5$, $a_2 = 10$, $a_3 = 15$, and $a_4 = 20$. Thus a reasonable guess is that $a_n = 5n$. We prove this by induction. Let $S = \{n \in \mathbb{N} \cup \{0\}: a_n = 5n\}$. Since $a_0 = 0 = 5 \cdot 0$, $0 \in S$. Suppose $n \in S$. Then $a_n = 5n$. Therefore $a_{n+1} = a_n + 5 = 5n + 5 = 5(n + 1)$. Hence $n + 1 \in S$. By the principle of mathematical induction, $S = \mathbb{N} \cup \{0\}$. ■

The recursion relation in Example 29 is a nonhomogeneous, linear recursion relation of order 1. We state the formal definition of such equations and give theorems that provide the unique solutions of some of these equations.

Definition

> **A linear nonhomogeneous recursion relation of order 1 with constant coefficients** is a recursion relation of the form $a_{n+1} = ca_n + b_n$.

If the sequence b in the recursion relation $a_{n+1} = ca_n + b_n$ is a constant sequence, then the recursion relation becomes $a_{n+1} = ca_n + d$ for some constant d. Furthermore, suppose $a_0 = e$. Then

$$a_1 = ca_0 + d = ce + d$$

$$a_2 = ca_1 + d = c(ce + d) + d = c^2e + cd + d$$

$$a_3 = ca_2 + d = c(c^2e + cd + d) + d = c^3e + c^2d + cd + d$$

$$a_4 = ca_3 + d = c(c^3e + c^2d + cd + d) + d = c^4e + c^3d + c^2d + cd + d$$

Thus it seems reasonable to guess that

$$\begin{aligned} a_n &= c^ne + c^{n-1}d + c^{n-2}d + \cdots + cd + d \\ &= c^ne + d(c^{n-1} + c^{n-2} + \cdots + c + 1) \\ &= \begin{cases} c^ne + d\left(\dfrac{1 - c^n}{1 - c}\right), & \text{if } c \neq 1 \\ e + dn, & \text{if } c = 1 \end{cases} \end{aligned}$$

We use the principle of mathematical induction to prove this fact.

Theorem 14.9

> The recursion relation $a_{n+1} = ca_n + d$ with initial value $a_0 = e$ has the unique solution
>
> $$a_n = \begin{cases} c^ne + d\left(\dfrac{1 - c^n}{1 - c}\right), & \text{if } c \neq 1 \\ e + dn, & \text{if } c = 1 \end{cases}$$

Proof

First suppose $c \neq 1$, and let

$$S = \left\{n \in \mathbb{N} \cup \{0\}: a_n = c^ne + d\left(\frac{1 - c^n}{1 - c}\right)\right\}$$

Since $c^0e + d\left(\dfrac{1 - c^0}{1 - c}\right) = e$, $0 \in S$.

Suppose $n \in S$. Then $a_n = c^ne + d\left(\dfrac{1 - c^n}{1 - c}\right)$ and hence

$$\begin{aligned}
a_{n+1} &= ca_n + d = c\left[c^ne + d\left(\frac{1 - c^n}{1 - c}\right)\right] + d \\
&= c^{n+1}e + \frac{cd - c^{n+1}d}{1 - c} + d \\
&= c^{n+1}e + \frac{cd - c^{n+1}d + d - cd}{1 - c} \\
&= c^{n+1}e + \frac{d - c^{n+1}d}{1 - c} \\
&= c^{n+1}e + d\left(\frac{1 - c^{n+1}}{1 - c}\right)
\end{aligned}$$

Therefore $n + 1 \in S$. By the principle of mathematical induction, $S = \mathbb{N} \cup \{0\}$. The proof that $a_n = e + dn$ where $c = 1$ is left as Exercise 39.

□

Example 30

Solve the recursion relation $a_n = 2a_{n-1} + 3$ with initial value $a_0 = 0$.

Analysis

By Theorem 14.9, with $c = 2$, $d = 3$, and $e = 0$, $a_n = 0 \cdot 2^n + 3(1 - 2^n)/(1 - 2) = 3(2^n - 1)$. ■

Now suppose that we consider a linear nonhomogeneous recursion relation $a_{n+1} = ca_n + b_n$ and the sequence b is not a constant sequence. If the initial value a_0 of the recursion relation $a_{n+1} = ca_n + b_n$ is e, we observe that

$$\begin{aligned}
a_1 &= ca_0 + b_0 = ce + b_0 \\
a_2 &= ca_1 + b_1 = c(ce + b_0) + b_1 = c^2e + cb_0 + b_1 \\
a_3 &= ca_2 + b_2 = c(c^2e + cb_0 + b_1) + b_2 = c^3e + c^2b_0 + cb_1 + b_2 \\
a_4 &= ca_3 + b_3 = c(c^3e + c^2b_0 + cb_1 + b_2) + b_3 \\
&= c^4e + c^3b_0 + c^2b_1 + cb_2 + b_3
\end{aligned}$$

This time a reasonable guess is that, for $n \geq 1$,

$$a_n = c^n e + c^{n-1}b_0 + c^{n-2}b_1 + \cdots + cb_{n-2} + b_{n-1}$$

$$= c^n e + \sum_{i=1}^{n} c^{n-i} b_{i-1}$$

Theorem 14.10 The recursion relation $a_{n+1} = ca_n + b_n$ with initial value $a_0 = e$ has the unique solution $a_n = c^n e + \sum_{i=1}^{n} c^{n-i} b_i$.

The principle of mathematical induction is also used to prove Theorem 14.10, and the proof is left as Exercise 40.

Example 31

Solve the recursion relation $a_n = 2a_{n-1} + n$ with initial value $a_0 = 3$.

Analysis

By Theorem 14.10, with $c = 2$, $e = 3$, and $b_i = i$ for each i, $a_n = 2^n \cdot 3 + \sum_{i=1}^{n} (2^{n-i} i)$. ■

Exercises 14.3

32. Solve each of the following recursion relations with the given initial value

a. $a_n = na_{n-1}$, $a_0 = 1$ **b.** $a_n = 2na_{n-1}$, $a_0 = 1$

c. $a_n = a_{n-1} + 2^n$, $a_0 = 1$ **d.** $a_n = a_{n-1} + 3^{n-1}$, $a_0 = 1$

e. $a_n = a_{n-1} + n^2$, $a_0 = 1$ **f.** $a_n = 3a_{n-1} + 4$, $a_0 = 2$

33. Tell whether each of the following recursion relations is a linear homogeneous recursion relation with constant coefficients. What is the order of each linear homogeneous recursion relation with constant coefficients?

a. $a_n = 4a_{n-1}$ **b.** $a_n = 2a_{n-1} + 3a_{n-2}$

c. $a_n = 3(n - 1)a_{n-1}$ **d.** $a_n = a_{n-1} + 2$

e. $a_n = 2a_{n-1}a_{n-2}$ **f.** $a_n = (a_{n-1})^2 + 2a_{n-2}$

g. $a_n = 2a_{n-2}$ **h.** $a_n = 2a_{n-1} - 3a_{n-2} + 4a_{n-3}$

34. Solve each of the following recursion relations with the given initial values

a. $a_n = 3a_{n-1} - 2a_{n-2}$, $a_0 = 1$, $a_1 = 2$

b. $a_n = a_{n-1} + 5a_{n-2}$, $a_0 = 1$, $a_1 = 2$

c. $a_n = -4a_{n-1} + 4a_{n-2}$, $a_0 = -2$, $a_1 = 2$

d. $a_n = -6a_{n-1} - 9a_{n-2}$, $a_0 = 3$, $a_1 = -3$

e. $a_n = 2\sqrt{2}a_{n-1} - 2a_{n-2}$, $a_0 = \sqrt{2}$, $a_1 = \sqrt{2}$

f. $a_n = -4a_{n-1} + 5a_{n-2}$, $a_0 = 2$, $a_1 = 8$.

35. Prove that $a_n = -2^{n+1}$ is a solution of the recursion relation $a_n = 3a_{n-1} + 2^n$.

36. Show that the functions $\{p_n\}$ and $\{q_n\}$, where for each $n \in \mathbb{N}$, $p_n = 2^n$ and $q_n = n2^n$, are both solutions of the recursion relation $a_n = 4a_{n-1} - 4a_{n-2}$.

37. Solve each of the following recursion relations with the given initial value
    a. $a_n = 3a_{n-1} + 5$, $a_0 = 1$
    b. $a_n = 3a_{n-1} + 5$, $a_0 = 0$
    c. $a_n = a_{n-1} + 4$, $a_0 = 0$
    d. $a_n = a_{n-1} + 4$, $a_0 = 3$
    e. $a_n = -2a_{n+1} - 3$, $a_0 = -1$
    f. $a_n = -3a_{n+1} + 2$, $a_0 = -1$

38. Solve each of the following recursion relations with the given initial value.
    a. $a_n = 3a_{n-1} + n$, $a_0 = 1$
    b. $a_n = 3a_{n-1} + (n - 1)$, $a_0 = 1$
    c. $a_n = a_{n-1} + n^2$, $a_0 = 0$
    d. $a_n = a_{n-1} + 2n$, $a_0 = 0$
    e. $a_n = 2a_{n-1} + (n + 1)$, $a_0 = 3$
    f. $a_n = 5a_{n-1} + (n + 1)$, $a_0 = 2$

39. Prove that the recursion relation $a_{n+1} = a_n + d$ with initial value $a_0 = e$ has the unique solution $e + dn$.

40. Prove that the recursion relation $a_{n+1} = ca_n + b_n$ with initial value $a_0 = e$ has the unique solution $a_n = c^n e + \Sigma_{i=1}^{n} c^{n-i} b_i$.

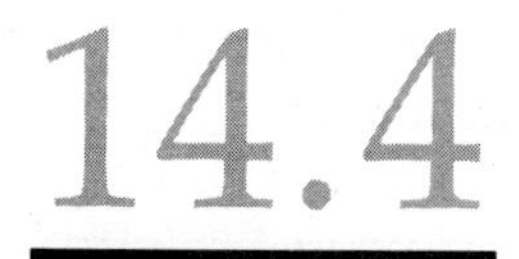

14.4 GENERATING FUNCTIONS

Generating functions provide yet another important tool for solving counting problems. Suppose we have a sequence $a_0, a_1, a_2, \ldots$ of real numbers. Then we can set up an infinite sum in terms of a "parameter" x:

$$G(x) = a_0 + a_1x + a_2x^2 + \cdots + a_nx^n + \cdots$$

We have no reason to expect that this infinite sum exists for any number other than $x = 0$, but we can see if it gives us useful information about the sequence $\{a_n\}$ if it does exist.

Definition

The **generating function** for a sequence $a_0, a_1, a_2, \ldots$ of real numbers is the function defined by the infinite sum

$$G(x) = a_0 + a_1x + a_2x^2 + \cdots + a_nx^n + \cdots$$

The generating function of a finite sequence $a_0, a_1, a_2, \ldots, a_n$ is defined by taking $a_m = 0$ for all $m > n$. Hence the generating function for $a_0, a_1, a_2, \ldots, a_n$ is $G(x) = a_0 + a_1x + a_2x^2 + \cdots + a_nx^n$.

Example 32

What is the generating function for the finite sequence 1,1,1,1,1?

Analysis

Since $(x^5 - 1)/(x - 1) = 1 + x + x^2 + x^3 + x^4$, $G(x) = (x^5 - 1)/(x - 1)$ is the generating function for the sequence 1,1,1,1,1. ■

Example 33

What is the generating function for the sequence 1,7,21,35,35,21,7,1?

Analysis

Since $C(7,0) = 1$, $C(7,1) = 7$, $C(7,2) = 21$, $C(7,3) = 35$, $C(7,4) = 35$, $C(7,5) = 21$, $C(7,6) = 7$, and $C(7,7) = 1$, the generating function for the sequence 1,7,21,35,35,21,7,1 is the function defined by

$$G(x) = C(7,0) + C(7,1)x + C(7,2)x^2 + C(7,3)x^3 + C(7,4)x^4 + C(7,5)x^5 + C(7,6)x^6 + C(7,7)x^7$$

Thus by the binomial theorem $G(x) = (1 + x)^7$. ■

Theorem 14.11

Let $c_1, c_2 \in \mathbb{R}$. If G_1 is the generating function for the sequence $a_0, a_1, a_2, \ldots$ and G_2 is the generating function for the sequence $b_0, b_1, b_2, \ldots$, then $c_1G_1 + c_2G_2$ is the generating function for $c_1a_0 + c_2b_0, c_1a_1 + c_2b_1, c_1a_2 + c_2b_2, \ldots$.

Proof

Let G_1 be the generating function for the sequence $a_0, a_1 a_2, \ldots$, and let G_2 be the generating function for the sequence $b_0, b_1, b_2, \ldots$. Then $G_1(x) = a_0 + a_1x + a_2x^2 + \cdots + a_nx^n + \cdots$ and $G_2(x) = b_0 + b_1x + b_2x^2 + \cdots + b_nx^n + \ldots$. Therefore

$$(c_1G_1 + c_2G_2)(x) = (c_1a_0 + c_2b_0) + (c_1a_1 + c_2b_1)x + (c_1a_2 + c_2b_2)x^2 + \cdots + (c_1a_n + c_2b_n)x^n + \ldots \quad \square$$

Example 34

If G is the generating function for the sequence $a_0, a_1, a_2, \ldots$, what is the generating function for the sequence $0, a_0, a_1, a_2, \ldots$?

Analysis

$$G(x) = a_0 + a_1x + a_2x^2 + \cdots + a_nx^n + \ldots, \quad \text{so}$$

$$xG(x) = a_0x + a_1x^2 + a_2x^3 + \cdots + a_nx^{n+1} + \ldots$$

Therefore the function F defined by $F(x) = xG(x)$ is the generating function for $0, a_0, a_1, a_2, \ldots$. ■

We state the following theorem, which is a generalization of Example 34, without proof.

Theorem 14.12

If G is the generating function for the sequence $a_0,a_1,a_2, \ldots$ and $n \in \mathbb{N}$, then the function F defined by $F(x) = x^nG(x)$ is the generating function for the sequence whose first n terms are zero and whose mth term $(m > n)$ is a_{m-n-1}.

According to Theorem 14.12, for example, if G is the generating function for the sequence $a_0,a_1,a_2, \ldots$, then the function F defined by $F(x) = x^4G(x)$ is the generating function for the sequence $0,0,0,0,a_0,a_1,a_2, \ldots$.

We have observed, in the proof of Theorem 14.11, that if

$$f(x) = a_0 + a_1x + a_2x^2 + \cdots + a_nx^n + \ldots \quad \text{and}$$

$$g(x) = b_0 + b_1x + b_2x^2 + \cdots + b_nx^n + \ldots, \quad \text{then}$$

$$f(x) + g(x) = (a_0 + b_0) + (a_1 + b_1)x + (a_2 + b_2)x^2 + \cdots + (a_n + b_n)x^n + \cdots$$

We state, without proof, two more facts about infinite sums in the following two theorems.

Theorem 14.13

If $f(x) = a_0 + a_1x + a_2x^2 + \cdots + a_nx^n + \cdots$ and $g(x) = b_0 + b_1x + b_2x^2 + \cdots + b_nx^n + \cdots$, then $f(x)g(x) = a_0b_0 + (a_0b_1 + a_1b_0)x + (a_0b_2 + a_1b_1 + a_2b_0)x^2 + \cdots + (a_0b_n + a_1b_{n-1} + a_2b_{n-2} + \cdots + a_{n-2}b_2 + a_{n-1}b_1 + a_nb_0)x^n + \cdots$

Theorem 14.14

If $|x| < 1$, then $\dfrac{1}{1-x} = 1 + x + x^2 + \cdots$

Example 32 can be generalized to show that if $n \in \mathbb{N}$ then the function G defined by $G(x) = (x^n - 1)/(x - 1)$ is the generating function for the finite sequence $1,1, \ldots ,1$ consisting of n 1s. Notice that according to

Theorem 14.14 the function G defined by $G(x) = 1/(1 - x)$ is the generating function for the constant sequence $1,1,1, \ldots$.

Corollary 14.15

> The function G defined by $G(x) = 1/(1 - x)^2$, for each $x \in \mathbb{R}$ such that $|x| < 1$, is the generating function for the sequence $1,2,3,4, \ldots$.

Proof

Since $|x| < 1$, by Theorem 14.14

$$\frac{1}{1-x} = 1 + x + x^2 + \cdots + x^n + \cdots$$

By Theorem 14.13

$$\frac{1}{(1-x)^2} = \frac{1}{1-x} \cdot \frac{1}{1-x} = 1 + (1 \cdot 1 + 1 \cdot 1)x + (1 \cdot 1 + 1 \cdot 1 + 1 \cdot 1)x^2 + \cdots + \underbrace{(1 \cdot 1 + 1 \cdot 1 + \cdots + 1 \cdot 1)}_{n\,+\,1 \text{ times}} x^n \ldots$$

□

Theorem 14.16

> If a is a non-zero real number and $|x| < 1/|a|$, then
>
> $$\frac{1}{1-ax} = 1 + ax + a^2x^2 + \cdots + a^nx^n + \cdots$$

Notice that according to Theorem 14.16 the function G defined by $G(x) = 1/(1 - ax)$, for each $x \in \mathbb{R}$ such that $|x| < 1/|a|$, is the generating function for the sequence $1,a,a^2,a^3, \ldots ,a^n, \ldots$.

Generating functions can be used to solve problems such as the one in Example 6. We first consider a simpler problem.

Example 35

How many solutions does $e_1 + e_2 = 13$ have, where e_1 and e_2 are integers with $3 \leqslant e_1 \leqslant 8$ and $6 \leqslant e_2 \leqslant 9$?

Analysis

We choose this problem first because we can see that the solutions are

$e_1 = 4$ and $e_2 = 9$ $\qquad$ $e_1 = 5$ and $e_2 = 8$

$e_1 = 6$ and $e_2 = 7$ $\quad$ $e_1 = 7$ and $e_2 = 6$

We want to see how generating functions can be used to find the number of solutions. It is easy to see that the number of solutions with the given constraints is the coefficient of x^{13} in the expansion of

$$(x^3 + x^4 + x^5 + x^6 + x^7 + x^8)(x^6 + x^7 + x^8 + x^9)$$

This is the case because we obtain a term involving x^{13} in the product of these two generating functions by selecting a term x^{e_1} in the first function and a term x^{e_2} in the second function such that $e_1 + e_2 = 13$. Therefore the term involving x^{13} is $x^4x^9 + x^5x^8 + x^6x^7 + x^7x^6$ and the coefficient is 4, which indicates that there are four solutions to the original equation subject to the given constraints. ■

Now we consider a nontrivial example.

Example 36

How many solutions does $e_1 + e_2 + e_3 + e_4 = 23$ have, where e_1, e_2, e_3, and e_4 are integers with $3 \leqslant e_1 \leqslant 7$, $4 \leqslant e_2 \leqslant 8$, $5 \leqslant e_3 \leqslant 9$, and $6 \leqslant e_4 \leqslant 10$?

Analysis

The number of solutions with the given constraints is the coefficient of x^{23} in the expansion of

$$(x^3 + x^4 + x^5 + x^6 + x^7)(x^4 + x^5 + x^6 + x^7 + x^8)(x^5 + x^6 + x^7 + x^8 + x^9)(x^6 + x^7 + x^8 + x^9 + x^{10})$$

For each term x^{e_1} in the first function, each term x^{e_2} in the second function, each term x^{e_3} in the third function, and each term x^{e_4} in the fourth function such that $e_1 + e_2 + e_3 + e_4 = 23$, we obtain one solution. Hence the number of solutions is the coefficient of x^{23} in the product. By listing the possible ways of obtaining a 23 as the exponent, we see that the coefficient of x^{23} in the product is 52. Thus the number of solutions of the given equation subject to the given constraints is 52. ■

Note that in Example 36 we found that the coefficient of x^{23} is 52 by listing the possible ways of obtaining a 23 as exponent. Hence we did not save any work. We may as well have enumerated all the solutions of the given equation subject to the given constraints. The method can, however, be used to solve other counting problems.

Example 37

In how many different ways can ten identical golf balls be distributed among three golfers if each golfer receives at least two but not more than four golf balls?

Analysis

For each $n \in \mathbb{N}$, let a_n be the number of ways to distribute n golf balls among the three golfers. Since each golfer receives at least two golf balls but no more than four, the generating function for a_n is $x^2 + x^3 + x^4$. Since we have three golfers, the number of solutions is the coefficient of x^{10} in the expansion of $(x^2 + x^3 + x^4)^3$. The term involving x^{10} is

$$x^2x^4x^4 + x^3x^3x^4 + x^3x^4x^3 + x^4x^2x^4 + x^4x^3x^3 + x^4x^4x^2$$

so the number of ways of distributing the golf balls is 6. ■

We next illustrate the use of generating functions in solving recursion relations. We solve the recursion relation by finding an explicit formula for its generating function.

Example 38

Solve the recursion relation $a_n = 4a_{n-1}$ with initial value $a_0 = 1$.

Analysis

Let G be the generating function for the sequence $\{a_n\}$. Then

$$G(x) = a_0 + a_1x + a_2x^2 + \cdots + a_nx^n + \cdots, \quad \text{so}$$

$$xG(x) = a_0x + a_1x^2 + \cdots + a_{n-1}x^n + \cdots, \quad \text{and}$$

$$4xG(x) = 4a_0x + 4a_1x^2 + \cdots + 4a_{n-1}x^n + \cdots$$

Therefore

$$G(x) - 4xG(x) = a_0 + (a_1 - 4a_0)x + (a_2 - 4a_1)x^2 + \cdots + (a_n - 4a_{n-1})x^n + \cdots$$

Since $a_k = 4a_{k-1}$ for each $k \in \mathbb{N}$, the coefficient of x^k is 0 for each $k \in \mathbb{N}$. Thus $G(x) - 4xG(x) = a_0$. So $G(x)(1 - 4x) = a_0$ and hence $G(x) = a_0/(1 - 4x) = 1/(1 - 4x)$. By Theorem 14.16,

$$G(x) = 1 + 4x + 4^2x^2 + \cdots + 4^nx^n + \cdots$$

Therefore a solution of the recursion relation is the sequence $\{a_n\}$, where $a_n = 4^n$ for each nonnegative integer n. ■

Example 39

Solve the recursion relation $a_n = a_{n-1} + 2$ with initial value $a_0 = 1$.

Analysis

Let G be the generating function for the sequence $\{a_n\}$. Then

$$G(x) = a_0 + a_1x + a_2x^2 + \cdots + a_nx^n + \cdots$$

Multiplying both sides of the recursion relation by x^n, we obtain $a_nx^n = a_{n-1}x^n + 2x^n$. Therefore, since $a_0 = 1$,

$$\begin{aligned} G(x) - 1 &= (a_0x + 2x) + (a_1x^2 + 2x^2) \\ &\quad + \cdots + (a_{n-1}x^n + 2x^n) + \cdots \\ &= a_0x + a_1x^2 + \cdots + a_{n-1}x^n + \cdots + 2x + 2x^2 \\ &\quad + \cdots + 2x^n + \cdots \\ &= x(a_0 + a_1x + \cdots + a_{n-1}x^{n-1} + \cdots) \\ &\quad + 2(x + x^2 + \cdots + x^n + \cdots) \end{aligned}$$

We may assume that $|x| < 1$. Then, by Theorem 14.14,

$$G(x) - 1 = xG(x) + 2\left(\frac{1}{1 - x} - 1\right)$$

Therefore $G(x) - xG(x) = 2/(1 - x) - 1$. So

$$(1 - x)G(x) = \frac{2 - (1 - x)}{1 - x} = \frac{1 + x}{1 - x}$$

and, by Corollary 14.15,

$$\begin{aligned} G(x) &= \frac{1 + x}{(1 - x)^2} \\ &= \frac{1}{(1 - x)^2} + \frac{x}{(1 - x)^2} \\ &= (1 + 2x + 3x^2 + \cdots + (n + 1)x^n + \cdots) \\ &\quad + (x + 2x^2 + 3x^3 + \cdots + nx^n + \cdots) \\ &= 1 + 3x + 5x^2 + \cdots + (2n + 1)x^n + \cdots \end{aligned}$$

Therefore a solution is the sequence $\{a_n\}$, where $a_n = 2n + 1$ for each nonnegative integer n. ■

Example 40

Solve the recursion relation $a_n = 2a_{n-1} - a_{n-2}$ with initial values $a_0 = 1$ and $a_1 = 1$.

Analysis

Our goal is to obtain a function defined by an infinite sum in which the coefficient of x^n is 0 for all $n \geq 2$. Since $a_n - 2a_{n-1} + a_{n-2} = 0$, we can accomplish this by obtaining an infinite sum in which the coefficient of x^n is $a_n - 2a_{n-1} + a_{n-2}$. Let G be the generating function for the sequence $\{a_n\}$. Then

$$G(x) = a_0 + a_1x + a_2x^2 + \cdots + a_nx^n + \cdots, \qquad \text{so}$$

$$-2xG(x) = -2a_0x - 2a_1x^2 - 2a_2x^3 - \cdots - 2a_{n-1}x^n \cdots \qquad \text{and}$$

$$x^2G(x) = a_0x^2 + a_1x^3 + \cdots + a_{n-2}x^n + \cdots$$

Therefore

$$G(x) - 2xG(x) + x^2G(x) = a_0 + (a_1 - 2a_0)x + (a_2 - 2a_1 + a_0)x^2 + \cdots + (a_n - 2a_{n-1} + a_{n-2})x^n + \cdots$$

and indeed the coefficient of x^n is 0 for all $n \geq 2$. Therefore

$$G(x)(1 - 2x + x^2) = a_0 + (a_1 - 2a_0)x$$

We may assume that $|x| < 1$. Then, by Theorem 14.14,

$$G(x) = \frac{1 - x}{(1 - x)^2} = \frac{1}{1 - x} = 1 + x + x^2 + \cdots + x^n + \cdots$$

Therefore a solution is the sequence $\{a_n\}$, where $a_n = 1$ for each nonnegative integer n. ■

Exercises 14.4

41. Find the generating function of each of the following finite sequences.

a. 3,3,3,3,3,3 **b.** 1,3,9,27,81,243

c. 1,4,16,64 **d.** 0,1,0,1,0,1,0,1

42. Find the generating function of each of the following infinite sequences $\{a_n\}$, where for each $n = 0,1,2,\ldots$

a. $a_n = 2$ **b.** $a_n = 5^n$ **c.** $a_n = n + 1$ **d.** $a_n = 2^n + 1$

43. For each of the following find a sequence with the given function G as its generating function.

a. $G(x) = (1 + x)^3$ **b.** $G(x) = (2 + x)^4$

c. $G(x) = (x^2 + 1)^4$ **d.** $G(x) = 5/(1 - 3x)$

44. Find a generating function for each of the following recursion relations and use this generating function to solve the recursion relation with the given value(s).

a. $a_n = 5a_{n-1}$, $a_0 = 2$ **b.** $a_n = -4a_{n-1}$, $a_0 = -1$

c. $a_n = a_{n-1} + 3$, $a_0 = 1$ **d.** $a_n = 3a_{n-1} + 2$, $a_0 = 1$

e. $a_n = 2a_{n-1} - a_{n-2}$, $a_0 = 0$, $a_1 = 1$

f. $a_n = -2a_{n-1} - a_{n-2}$, $a_0 = 0$, $a_1 = 1$

g. $a_n = 3a_{n-1} + 4^{n-1}$, $a_0 = 1$

h. $a_n = 5a_{n-1} - 6a_{n-2}$, $a_0 = 2$, $a_1 = 5$

i. $a_n = a_{n-1} + 2(n - 1)$, $a_0 = 2$

j. $a_n = 3a_{n-1} + 3a_{n-2} + a_{n-3}$, $a_0 = 0$, $a_1 = 0$, $a_2 = 1$

Chapter 14 Review Exercises

45. Find the number of natural numbers less than or equal to 50 that are divisible by none of 3, 5, and 7.

46. Find the number of natural numbers less than 90 that are relatively prime to 90.

47. Nine different letters are placed in nine mailboxes. Find the number of ways in which

a. no letter is placed in the proper mailbox.

b. exactly one letter is placed in the proper mailbox.

c. at least two letters are placed in the proper mailboxes.

(*Hint:* Express your answers in terms of the number of derangements of a set.)

48. Find the number of natural numbers between 1 and 900 that are neither perfect squares, perfect cubes, nor perfect fourth powers.

49. A leap year is a year that is divisible by 4 and has the property that if it is divisible by 100 then it is divisible by 400. Find the number of leap years between year 1 and year 2500.

50. For each of the following pairs p and q of permutations determine, with respect to lexicographic order, whether p precedes q or q precedes p.

a. $p = 4132$, $q = 3124$ **b.** $p = 25143$, $q = 25314$

51. Find the successor, with respect to lexicographic order, of each of the following permutations.

a. 3124 **b.** 25143 **c.** 12354 **d.** 651432

52. List the 4-combinations of {1,2,3,4,5} in lexicographic order.

53. Use bit strings to order the 4-combinations of {1,2,3,4,5,6}.

54. Solve each of the following recursion relations with the given initial value.

a. $a_n = a_{n-1} + 3$, $a_0 = 1$ **b.** $a_n = 3a_{n-1}$, $a_0 = 1$

c. $a_n = 4a_{n-1}$, $a_0 = 4$ **d.** $a_n = 2a_{n-1} + 1$, $a_0 = 1$

55. Solve each of the following recursion relations with the given initial values.

a. $a_n = 3a_{n-1} - 2a_{n-2}$, $a_0 = 1$, $a_1 = 1$

b. $a_n = a_{n-1} + 2a_{n-2}$, $a_0 = 1$, $a_1 = 0$

c. $a_n = 6a_{n-1} + a_{n-2}$, $a_0 = 0$, $a_1 = 1$

d. $a_n = 7a_{n-1} - 6a_{n-2}$, $a_0 = 1$, $a_1 = 2$

e. $a_n = 6a_{n-1} - 11a_{n-2} + 6a_{n-3}$, $a_0 = 1$, $a_1 = -1$, $a_2 = 1$

f. $a_n = 4a_{n-2} - 3a_{n-3}$, $a_0 = 1$, $a_1 = 1$, $a_2 = 2$

56. Determine the number of integer solutions of $e_1 + e_2 + e_3 + e_4 = 18$ subject to the conditions $e_1 \geqslant 0$, $e_2 \geqslant 3$, $1 \leqslant e_3 \leqslant 5$, and $e_4 \geqslant 0$.

57. Find the generating function of each of the following infinite sequences $\{a_n\}$, where for each $n \in \mathbb{N} \cup \{0\}$

a. $a_n = (-1)^n(n + 2)$

b. $a_n = n(n + 1)(n + 2)$

58. Let $n \in \mathbb{N}$ and define an order relation $\leqslant$ on the set of all permutations on $\{1,2, \ldots ,n\}$ by agreeing that $f \leqslant g$ means $f < g$ with respect to the lexicographic order or f is g.

a. Is $\leqslant$ a partial order?

b. Is $\leqslant$ a linear order?

c. Is $\leqslant$ a well order?

Chapter

15

ALGEBRAIC STRUCTURES AND FORMAL LANGUAGES

Formal languages were developed as a result of research in the automatic translation of one language to another. They are specified by a well-defined set of rules of syntax. We study the syntax of a class of languages called *phrase structure grammars*. Finite-state machines can be used to describe the effect of computer programs. They are useful in the study of formal languages and are often found in compilers and interpreters for various computer programming languages. Regular expressions play an important role in the study of formal languages and finite-state machines, and semigroups are used in the study of finite-state machines.

Recall from Chapter 4 the definition of an alphabet Σ and the set Σ^* of all words over Σ. Also recall that Σ^* contains the empty word ε. The members of Σ^* are also called strings. If $w_1 = a_1a_2 \ldots a_m$ and $w_2 = b_1b_2 \ldots b_n$ are members of Σ^*, we define the concatenation of w_1 with w_2, w_1w_2, to be the sequence $a_1a_2 \ldots a_mb_1b_2 \ldots b_n$. Notice that if $w = c_1c_2 \ldots c_p \in \Sigma^*$ then

$$\varepsilon w = \varepsilon c_1c_2 \ldots c_p = c_1c_2 \ldots c_p = c_1c_2 \ldots c_p\varepsilon = w\varepsilon = w$$

15.1 REGULAR EXPRESSIONS

The concept of *regular expression* plays an important role in the study of phrase structure grammar and finite-state automata. We begin with a definition of concatenation of sets of words.

Definition

Let L_1 and L_2 be sets whose members are words. Then the **concatenation of L_1 with L_2**, denoted L_1L_2, is defined by $L_1L_2 = \{w_1w_2: w_1 \in L_1 \text{ and } w_2 \in L_2\}$.

Example 1

Let $L_1 = \{a,aca,bac,\varepsilon\}$ and $L_2 = \{ba,bc\}$. Then $L_1L_2 = \{aba,abc,acaba,acabc,bacba,bacbc,ba,bc\}$. ■

If L is a set whose members are words, powers of L designate the concatenation of L with itself the appropriate number of times.

Definition

Let L be a set whose members are words. We define $L^0 = \{\varepsilon\}$, $L^1 = L$, and for $n > 1$, L^n is defined by the recursion relation $L^n = L^{n-1}L$.

Example 2

Let $\Sigma = \{a,b\}$ and $L = \{a,ab\}$. Then $L^0 = \{\varepsilon\}$, $L^1 = \{a,ab\}$, $L^2 = \{aa,aab,aba,abab\}$, and $L^3 = \{aaa,aaab,aaba,aabab,abaa,abaab,ababa,ababab\}$. ■

Definition

Let Σ be an alphabet and let $L \subseteq \Sigma^*$. The **Kleene closure** L^* of L is defined to be $\bigcup_{i=0}^{\infty} L^i$.

Notice that L^* is the set consisting of ε and all words that can be obtained by the concatenation of L with itself any number of times.

Example 3

If $L = \Sigma = \{0,1\}$, then L^* consists of ε and all words that can be formed using 0s and 1s. For example, 0010100111 is a member of L^*. ■

The regular expressions over an alphabet are defined recursively in much the same way that Boolean expressions are defined in Chapter 13.

Definition

Let Σ be an alphabet. A **regular expression** over Σ is a string that can be constructed from the members of Σ and the symbols (,), $\vee$, $*$, and ε according to the following rules:

a. ε is a regular expression.
b. If $a \in \Sigma$, then a is a regular expression.
c. If α and β are regular expressions, then $\alpha \vee \beta$ is a regular expression.
d. If α and β are regular expressions, then $\alpha\beta$ is a regular expression.
e. If α is a regular expression, then α^* and (α) are regular expressions.

Notice that (a) and (b) provide initial regular expressions, while (c), (d), and (e) define regular expressions in terms of those already defined. Parentheses may be essential in some cases and may be used in other cases for clarity. In the absence of parentheses, the hierarchical order is as follows: $*$ is applied first, concatenation second, and $\vee$ third; thus if $\Sigma = \{a,b\}$, then $a \vee ab$ means $a \vee (ab)$, ab^* means $a(b^*)$, and $a \vee b^*$ means $a \vee (b^*)$.

Example 4

Show that the expression $0^*(1 \vee 0)^*$ is a regular expression over $\Sigma = \{0,1\}$.

Analysis

By part (b) of the definition, 1 and 0 are regular expressions. Thus, by part (c), $1 \vee 0$ is a regular expression. Then, by part (e), 0^* and $(1 \vee 0)^*$ are regular expressions. Finally, by part (d), $0^*(1 \vee 0)^*$ is a regular expression. ■

With each regular expression over an alphabet Σ, we associate a subset of Σ^* according to the following rules:

a. The subset of Σ^* associated with ε is $\{\varepsilon\}$.
b. If $a \in \Sigma$, the subset of Σ^* associated with a is $\{a\}$.

c. If α and β are regular expressions and A and B are the subsets of Σ^* associated with α and β, the subset of Σ^* associated with $\alpha \vee \beta$ is $A \cup B$.

d. If α and β are regular expressions and A and B are the subsets of Σ^* associated with α and β, then the subset of Σ^* associated with $\alpha\beta$ is AB (the concatenation of A with B).

e. If α is a regular expression and A is the subset of Σ^* associated with α, then the subset of Σ^* associated with α^* is A^* (the Kleene closure of A).

Definition

Let Σ be an alphabet. A **regular set** over Σ is the empty set or a set that is associated with a regular expression over Σ.

Example 5

Let $\Sigma = \{a,b\}$. In the left column of Table 15.1, we list some regular expressions over Σ, and in the right column we list the set associated with each regular expression.

Table 15.1

Regular expression	Associated set
a	$\{a\}$
b	$\{b\}$
$a \vee b$	$\{a,b\}$
ab	$\{ab\}$
a^*	$\{a\}^* = (\varepsilon,a,aa,aaa, \ldots\}$
$(ab)^*$	$\{ab\}^* = \{\varepsilon,ab,abab,ababab, \ldots\}$
a^*b	$\{a\}^*\{b\} = \{b,ab,aab,aaab, \ldots\}$
ab^*	$\{a\}\{b\}^* = \{a,ab,abb,abbb, \ldots\}$
$a^* \vee b$	$\{a\}^* \cup \{b\} = \{b, \varepsilon, a,aa,aaa, \ldots\}$
$a^* \vee b^*$	$\{a\}^* \cup \{b\}^* = \{\varepsilon,a,aa,aaa, \ldots ,b,bb,bbb, \ldots\}$

■

Example 6

Let $\Sigma = \{a,b,c\}$. Then some regular sets over Σ are $\varnothing$, $\{\varepsilon\}$, $\{a\}$, $\{b\}$, $\{c\}$, $\{\varepsilon,a,aa,aaa, \ldots\}$, $\{ac\}$, $\{b,c\}$, $\{c,ac,aac,aaac, \ldots\}$, and $\{a,ac,acc,accc, \ldots\}$. ■

Exercises 15.1

1. Let $\Sigma = \{0,1\}$. Describe the set associated with each of the following regular expressions. **a.** 0^*1^* **b.** $0^*(1^* \vee 0^*)$

2. Explain why each of the following expressions is a regular expression over $\Sigma = \{0,1\}$.

   a. $0 \vee 1$ **b.** 01 **c.** $(0 \vee 1)^*$ **d.** $(01)^*$

   e. $(00)^*$ **f.** $0^*(0 \vee 1)^*$ **g.** $(0^*1)0^*$

3. Describe the set that is associated with each of the regular expressions in Exercise 2.
4. Let $\Sigma = \{1,2,3\}$. Each of the following sets is associated with a regular expression. In each case, find that regular expression.
   a. $\{1,2,3\}$
   b. $\{1,12,122,1222, \ldots\}$
   c. $\{\varepsilon,12,1212,121212, \ldots\}$
   d. $\{\varepsilon,2,3,33,333, \ldots\}$
   e. $\{\varepsilon,1,2,3,12,1212,121212, \ldots\}$
   f. the set consisting of all words that contain an even number of 1s
5. Let $\Sigma = \{0,1\}$. Find the set that is associated with each of the following regular expressions.
   a. $(00^*)[(0 \vee 1)^*1]$
   b. $(01)^*[(01) \vee 1^*]$

15.2

FORMAL LANGUAGES

Let Σ be an alphabet. Our first thought is that Σ probably consists of digits or letters. But Σ can be a collection of words. Thus Σ^* may be regarded as the collection of all possible sentences that can be formed by using the words in Σ. Of course, the so-called sentences do not have to make sense. For example, suppose Σ = {*hat, ran, sit, tree, up, the, green*}. Then *ran tree green hat sit* is a member if Σ^* (provided that it is agreed that we may use a blank space to indicate the empty word), but it is not a properly constructed English sentence. On the other hand, *the green tree ran up the hat* is a properly constructed English sentence that is meaningless. We may think of a language as requiring three specifications. First, there must be a set of words. Second, a subset of Σ^* that consists of the properly constructed sentences must be specified, and finally, a method of determining which of the properly constructed sentences have meaning must be specified.

The body of rules governing the construction of a language is called the *syntax* of the language. When we learn to program in a programming language, we learn the syntax of the language, and if we make a mistake in syntax in a language such as FORTRAN the compiler detects the mistake and generates an appropriate error message. In this section we study the syntax of a class of languages called *phrase structure grammar*. Phrase structure grammar was introduced by Noam Chomsky in 1954 in his effort to describe the natural languages.

Definition

A **phrase structure grammar** G is an ordered 4-tuple (N,T,S,P), where

a. $N \cap T = \varnothing$;
b. T is a finite set, called the **terminal alphabet,** whose members are symbols that make up the resulting words;
c. N is a finite set, called the **nonterminal alphabet,** whose members are used to generate the word patterns;
d. S is a specific member of N, called the **start symbol;**
e. P is a finite set of rules, called **productions,** that govern the generation of the word patterns. Each production is of the form $\alpha \to \beta$, where α is a string that contains at least one member of N, and $\alpha \to \beta$ indicates that α can be replaced by β.

The **language** $L(G)$ generated by G is the set of all words in T^* that can be derived from S by a finite sequence of productions. A sequence of productions that terminate with a word in T^* is called a **derivation.**

This definition can best be explained by considering some examples, but first we explain productions. A production is a replacement relation in the sense that $\alpha \to \beta$ indicates that we may replace α by β whenever the string α occurs, either alone or as a substring. If $G = (N,T,S,P)$ is a phrase structure grammar, then P usually consists of more than one production. Thus we use the notation $\alpha \to_a \beta$ to provide a name for the production. This allows the reader to determine readily which production is being used.

Example 7

Let $N = \{S,A,B,C\}$ and $T = \{0,1\}$, and suppose that P consists of the productions $S \to_a AS$, $S \to_b ACB$, $A \to_c 0$, $B \to_d 0$, and $C \to_e 1$. Let $G = \{N,T,S,P)$. List three members of $L(G)$.

Analysis

The idea is to begin with S and successively apply productions to determine a word that contains only members of T. Any word obtained in this manner is in $L(G)$. So we start with S. Notice that there are two productions we can use. We choose one of the two and continue using productions until we obtain a word in T^*. We list three derivations and obtain three members of $L(G)$.

$$S \rightarrow_a AS \rightarrow_b AACB \rightarrow_c A0CB \rightarrow_d A0C0 \rightarrow_e A010 \rightarrow_c 0010$$

$$S \rightarrow_b ACB \rightarrow_c 0CB \rightarrow_d 0C0 \rightarrow_e 010$$

$$S \rightarrow_a AS \rightarrow_a AAS \rightarrow_b AAACB \rightarrow_c 0AACB \rightarrow_c 00ACB$$

$$\rightarrow_c 000CB \rightarrow_d 000C0 \rightarrow_e 00010$$ ■

In the next example, we illustrate the use of rooted trees to show the derivation of a specific sentence in the language generated by a phrase structure grammar. The start symbol is taken as the root of the tree. The vertices of level number 1 are labeled by the members of $N \cup T$ that are on the right side of productions whose left sides consist of the start symbols. The vertices of level number 2 are labeled by the members of $N \cup T$ that are on the right side of productions whose left sides consist of vertices of level number 1. The process continues. The resulting tree is called a *derivation tree*. Derivation trees can be used whenever the left sides of the productions consist of a single member of N.

Example 8

Let

N = {S, noun, adjective, verb, adverb, phrase},

T = {*rapidly, boy, runs, the, car, slowly, walked*}, and

P = {$S \rightarrow_a$ (phrase adverb), phrase $\rightarrow_b$ (adjective phrase),

adjective $\rightarrow_c$ *the*, adverb $\rightarrow_d$ *rapidly*, adverb $\rightarrow_e$ *slowly*,

phrase $\rightarrow_f$ (noun verb), noun $\rightarrow_g$ *boy*, noun $\rightarrow_h$ *car*,

verb $\rightarrow_i$ *runs*, verb $\rightarrow_j$ *walked*}

Show that the sentence *the boy walked slowly* is a member of the language generated by $G = (N,T,S,P)$.

Analysis

We must start with S and find a derivation that yields the desired sentence. We construct a tree with root S whose leaves are labeled *the, boy, walked,* and *slowly*. To draw this tree, we use the following productions, in the order listed: $S \rightarrow_a$ (phrase adverb), phrase $\rightarrow_b$ (adjective phrase), adjective $\rightarrow_c$ *the*, phrase $\rightarrow_f$ (noun verb), noun $\rightarrow_g$ *boy*, verb $\rightarrow_j$ *walked*, adverb $\rightarrow_e$ *slowly*. The construction of the tree is shown in Figure 15.1

Figure 15.1

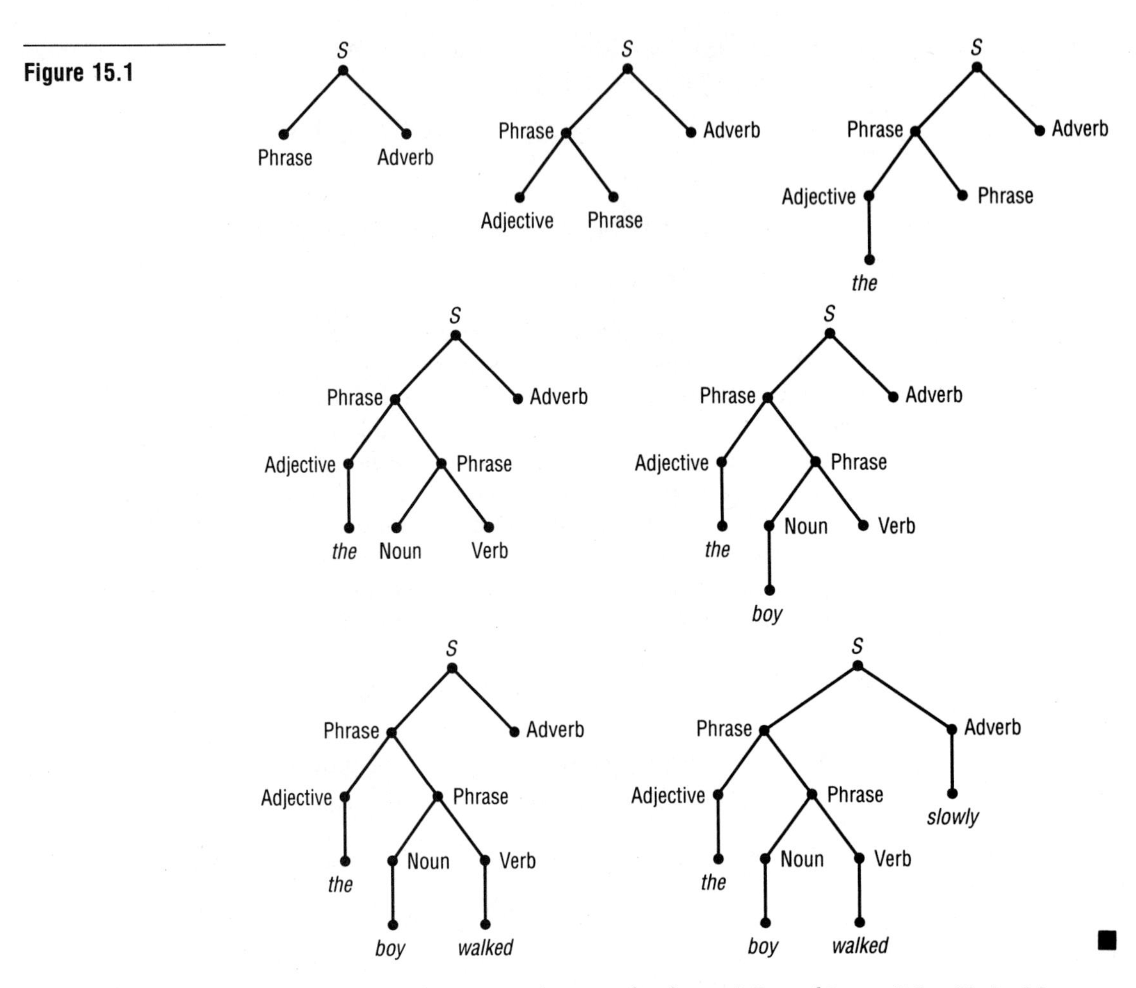

■

Example 9

Let $N = \{S, May, Jim\}$, $T = \{0,1\}$, and $P = \{S \rightarrow_a (May\ Jim),\ May \rightarrow_b 1,\ Jim \rightarrow_c (0\ Jim),\ Jim \rightarrow_d 0\}$. Describe the language $L(G)$ generated by $G = (N,T,S,P)$.

Analysis

We must begin every derivation with $S \rightarrow_a (May\ Jim)$. We must also use the production $May \rightarrow_b 1$ at some stage of every derivation. Thus every derivation tree must contain the following tree:

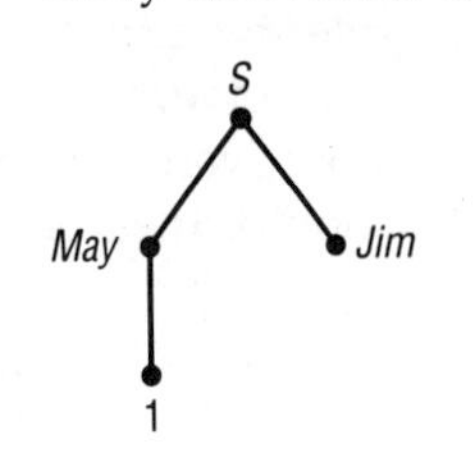

If we next use the production $Jim \rightarrow_d 0$, the derivation ends and we have 10 in $L(G)$. On the other hand, if we next use the production $Jim \rightarrow_c (0\ Jim)$ followed by the production $Jim \rightarrow_d 0$, we obtain 100 as a member of $L(G)$. By using the production $Jim \rightarrow_c (0\ Jim)$ as many times as we wish, we can obtain as many 0s as we wish in a member of $L(G)$. Each such string of 0s will, however, be preceded by a 1. Thus each member of $L(G)$ consists of a 1 followed by a string of 0s. Therefore if A denotes the set associated with the regular expression 10^*, then $L(G) = A - \{1\}$. ■

The set of productions for a phrase structure grammar $G = (N,T,S,P)$ may contain a production of the form $\alpha \rightarrow \varepsilon$. Such a production erases the string α, and consequently it is called an *erasing production*.

Example 10

Let $N = \{S,A,B,C\}$, $T = \{0,1\}$, and $P = \{S \rightarrow_a A0,\ A \rightarrow_b 1B,\ A \rightarrow_c 1C,\ B \rightarrow_d 1C0,\ C \rightarrow_e \varepsilon\}$. Describe the language $L(G)$ generated by $G = (N,T,S,P)$.

Analysis

We must begin every derivation with $S \rightarrow_a A0$. Then we have two choices: we can use the production $A \rightarrow_b 1B$ or the production $A \rightarrow_c 1C$. If we use the production $A \rightarrow_b 1B$, then the derivation is $S \rightarrow_a A0 \rightarrow_b 1B0 \rightarrow_d 11C00 \rightarrow_e 1100$. If we use the production $A \rightarrow_c 1C$, then the derivation is $S \rightarrow_a A0 \rightarrow_c 1C0 \rightarrow_e 10$. Therefore $L(G) = \{1100,10\}$. ■

We have considered several examples in which we start with a phrase structure grammar G and describe the language generated by G. In the following example, we start with a particular language and find a phrase structure grammar that generates the language. As the example indicates, there may be more than one phrase structure grammar that generates the language.

Example 11

Let $\Sigma = \{1\}$ be an alphabet and let L be the subset of Σ^* consisting of all words with an even number of 1s. Find two phrase structure grammars that generate L.

Analysis

Since each word in L uses only the symbol 1, let $T = \{1\}$. We know that N must contain the start symbol S, so let $N = \{S\}$. We want an even number of 1s, so we can let one production be $S \rightarrow_a 11S$. By repeated use of this production we can obtain as many 1s as we want, but we always obtain an even number of 1s. Since an ending S is always required, we add an erasing production $S \rightarrow_b \varepsilon$. Therefore, if $P = \{S \rightarrow_a 11S,\ S \rightarrow_b \varepsilon\}$, then $G = (N,T,S,P)$ is a phrase structure grammar that generates L.

To find another phrase structure grammar that generates L, we again let $T = \{1\}$, but this time we let $N = \{S,A,B\}$ and $P = \{S \rightarrow_a 1A,\ A \rightarrow_b 1,\ A \rightarrow_c 1B,\ B \rightarrow_d 1A\}$. Then $G = (N,T,S,P)$ is the desired phrase structure grammar. ■

We give two more examples in which we start with a phrase structure grammar and describe the language generated by the grammar.

Example 12

Let $N = \{S,A\}$, $T = \{0,1,2\}$, and $P = \{S \rightarrow_a 22A,\ A \rightarrow_b 0A,\ A \rightarrow_c 1\}$. Describe the language $L(G)$ generated by $G = (N,T,S,P)$.

Analysis

We must start with the production $S \rightarrow_a 22A$. Then we have two choices. If we use the production $A \rightarrow_b 0A$, we have $S \rightarrow_a 22A \rightarrow_b 220A$. Once again we have two choices. We add a 0 each time we use $A \rightarrow_b 0A$, and the derivation is complete as soon as we use $A \rightarrow_c 1$. Thus some members of $L(G)$ are 221, 2201, 22001, and 220001. Therefore $L(G)$ is the set associated with the regular expression 220^*1. The derivation tree for 22001 is shown in Figure 15.2.

Figure 15.2

■

Example 13

Let $N = \{S,A\}$, $T = \{0,1,2\}$, and $P = \{S \rightarrow_a 1S2,\ S2 \rightarrow_b 2A,\ 12A \rightarrow_c 0\}$. Describe the language $L(G)$ generated by $G = (N,T,S,P)$.

Analysis

We must begin with the production $S \rightarrow_a 1S2$, and we can use this production as many times as we want. To eliminate S, we must eventually use $S2 \rightarrow_b 2A$. At this point we must use $12A \rightarrow_c 0$ to eliminate A. Repeated use of $S \rightarrow_a 1S2$ results in a string of 1s, followed by S, followed by a string of 2s, with the string of 2s the same length as the string of 1s. Then, when we use $S2 \rightarrow_b 2A$, we have a string of 1s, followed by 2, followed by A, followed by a string (perhaps the empty string) of 2s, with the length of the string of 2s one less than the length of the string of 1s. Finally, when we use $12A \rightarrow_c 0$, we have a string of 1s (perhaps the empty string), followed by 0, followed by a string of 2s, with the string of 2s the same length as the string of 1s. Thus a typical member of $L(G)$ is 111102222. ■

In Example 13, derivation trees cannot be used to describe the derivations because the left side of some of the productions consists of more than a single symbol. Also, $L(G)$ cannot be associated with a regular expression. Because of examples of this type (as well as other considerations), Chomsky devised a classification of phrase structure grammars.

Definition

Let $G = (N,T,S,P)$ be a phrase structure grammar. Then G is a **type 0 grammar** if no restrictions are placed on the members of P. G is a **type 1 grammar** if the length of the left side of each production is less than or equal to the length of the right side. G is a **type 2 grammar** if the left side of each production is a single member of N. G is a **type 3 grammar** if the left side of each production is a single member of N, the right side of each production contains no more than one member of N, and any member of N occurring on the right side of a production occurs on the extreme right of the string.

In each of the four types, we allow the inclusion of the production $S \rightarrow \varepsilon$ even though it violates the definition of types 1, 2, and 3. This is done to avoid special cases. Note that Chomsky's classification system is a hierarchy because every type 3 grammar is a type 2 grammar, every type 2 grammar is a type 1 grammar, and every type 1 grammar is a type 0 grammar.

Definition

A language L is a **type** $\boldsymbol{i}$ ($i = 0,1,2,3$) language if there is a type i phrase structure grammar that generates L.

The phrase structure grammar in Example 12 is a type 3 grammar, the phrase structure grammar in Example 13 is a type 0 grammar that is not type 1, and the phrase structure grammar in Example 9 is a type 2 grammar that is not type 3. In Exercise 12, we ask you to give an example of a type 1 phrase structure grammar that is not type 2.

Exercises 15.2

6. Find five members of the language generated by each of the following phrase structure grammars.
   a. $G = (N,T,S,P)$, where $N = \{S,A,B,C\}$, $T = \{0,1\}$, and
   $P = \{S \rightarrow_a 0A,\ A \rightarrow_b 1A,\ A \rightarrow_c 0B,\ A \rightarrow_d C,\ B \rightarrow_e 0B,\ B \rightarrow_f C,\ C \rightarrow_g 1\}$
   b. $G = (N,T,S,P)$, where $N = \{S,A,B\}$, $T = \{0,1\}$, and
   $P = \{S \rightarrow_a 1A,\ A \rightarrow_b 1AB,\ A \rightarrow_c 1,\ B \rightarrow_d 0\}$
   c. $G = (N,T,S,P)$, where $N = \{S,A,B\}$, $T = \{0,1\}$, and
   $P = \{S \rightarrow_a A0B,\ A \rightarrow_b SA,\ SA \rightarrow_c 11B,\ B \rightarrow_d 01,\ A \rightarrow_e \varepsilon,\ S \rightarrow_f \varepsilon,\ B \rightarrow_g \varepsilon\}$
   d. $G = (N,T,S,P)$, where $N = \{S,A,B\}$, $T = \{0,1\}$, and
   $P = \{S \rightarrow_a 1BA,\ BA \rightarrow_b 0A,\ A \rightarrow_c 0,\ B \rightarrow_d 1A,\ B \rightarrow_e 1AB\}$
   e. $G = (N,T,S,P)$, where $N = \{S,A,B\}$, $T = \{0,1,2\}$, and
   $P = \{S \rightarrow_a 2BA,\ B \rightarrow_b 0,\ A \rightarrow_c 0,\ A \rightarrow_d BA,\ BA \rightarrow_e 1,\ BA \rightarrow_f 0A\}$
7. Describe the language generated by each of the following phrase structure grammars.
   a. $G = (N,T,S,P)$, where $N = \{S\}$, $T = \{1\}$, and $P = \{S \rightarrow_a 11,\ S \rightarrow_b 11S\}$
   b. $G = (N,T,S,P)$, where $N = \{S,A\}$, $T = \{1\}$, and
   $P = \{S \rightarrow_a A,\ A \rightarrow_b 111,\ A \rightarrow_c 111A\}$
   c. $G = (N,T,S,P)$, where $N = \{S,A\}$, $T = \{0,1,2\}$, and
   $P = \{S \rightarrow_a 0S,\ S \rightarrow_b 1A,\ A \rightarrow_c 1A,\ A \rightarrow_d 2\}$
   d. $G = (N,T,S,P)$, where $N = \{S,A,B,C\}$, $T = \{1\}$, and
   $P = \{S \rightarrow_a 1A,\ A \rightarrow_b 1B,\ B \rightarrow_c 1C,\ C \rightarrow_d \varepsilon,\ C \rightarrow_e 1B\}$
   e. $G = (N,T,S,P)$, where $N = \{S,A,B\}$, $T = \{0,1,2\}$, and
   $P = \{S \rightarrow_a BA,\ A \rightarrow_b A10,\ A \rightarrow_c \varepsilon,\ B \rightarrow_d 20B,\ B \rightarrow_e 20\}$
8. Let G be the phrase structure grammar defined in Exercise 6a. Which of the following strings belong to $L(G)$? Explain your answers.
   a. 01 **b.** 10 **c.** 111 **d.** 000 **e.** 1000 **f.** 0111
9. Let G be the phrase structure grammar defined in Exercise 6b. Which of the following strings belong to $L(G)$? Explain your answers.
   a. 11 **b.** 00 **c.** 110 **d.** 001 **e.** 0001 **f.** 1110
10. Let G be the phrase structure grammar defined in Exercise 6c. Which of the following strings belong to $L(G)$? Explain your answers.
   a. ε **b.** 1 **c.** 11 **d.** 01 **e.** 001 **f.** 001001
11. Find a phrase structure grammar that generates each of the following languages
   a. the language of all words over $\{0,1\}$ that consist of n 1s followed by n 0s, where $n \in \mathbb{N}$
   b. the language of all words over $\{0,1,2\}$ that consist of n 1s followed by n 0s followed by n 2s, where $n \in \mathbb{N}$.
   c. The language of all words over $\{0,1,2\}$ that consist of an even number of 1s followed by two 0s followed by an even number of 2s.

12. Give an example of a type 1 phrase structure grammar that is not type 2.
13. Draw a derivation tree for 202010 in the language generated by the phrase structure grammar of Exercise 7e.
14. Draw a derivation tree for 11111 in the language generated by the phrase structure grammar of Exercise 7d.
15. Recall that a *palindrome* (see Exercise 15 of Chapter 6) is a string that reads the same from right to left as it does from left to right. Find a phrase structure grammar that generates the set of all palindromes on {0,1}.

SEMIGROUPS

A Boolean algebra is an example of an algebraic system with important applications in information processing and switching theory. In this section we study an algebraic system with important applications in formal languages and automata theory.

Definition

A **semigroup** is an ordered pair $(S,*)$, where S is a nonempty set and $*$ is an associative binary operation on S. A semigroup $(S, *)$ is **commutative** if $*$ is commutative.

Our first three examples involve concepts with which you are familiar, so we leave the analysis of these examples as Exercises 16, 20, and 21.

Example 14

Let S be a set. Then $(\mathcal{P}(S),\cup)$ is a commutative semigroup. ■

Example 15

Let S be a nonempty set and let S^S denote the set of all functions $f: S \to S$. Then $(S^S, \circ)$ is a semigroup that is not commutative. ■

Example 16

Let $n \in \mathbb{N}$, let $\Sigma = \{1,2, \ldots ,n\}$ be an alphabet, and let $\cdot$ denote concatenation on Σ^*. Then $(\Sigma^*,\cdot)$ is a semigroup, and it is called the *free semigroup generated by* Σ. ■

Definition

An element e of a semigroup $(S,*)$ is called an **identity element** of S if $e*a = a*e = a$ for all $a \in S$.

The empty set is the identity element of the semigroup in Example 14, the identity function is the identity element of the semigroup in Example 15, and the empty word is the identity element of the semigroup in Example 16.

Example 17

The commutative semigroup $(\mathbb{N}, +)$ has no identity element. ■

Theorem 15.1

If a semigroup $(S,*)$ has an identity, that identity is unique.

Proof

Suppose that e and e' are identity elements of $(S,*)$. Since e is an identity, $e*e' = e'$. Since e' is an identity, $e*e' = e$. Therefore $e = e'$. □

Definition

A semigroup that has an identity is called a **monoid.**

Example 18

$(\mathbb{Z}, +)$ and $(\mathbb{Z}, \cdot)$ are commutative monoids. ■

Definition

Let $(S,*)$ be a semigroup. If T is a nonempty subset of S that is closed under the operation $*$, then $(T,*)$ is a semigroup, called a **subsemigroup** of $(S,*)$. The operation $*$ in $(T,*)$ is really the restriction of the operation $*$ in $(S,*)$ to T.

Definition

Let $(S,*)$ be a monoid with identity e. If T is a subset of S that contains e and is closed under the operation $*$, then $(T,*)$ is a monoid, called a **submonoid** of $(S,*)$.

Note that a semigroup is a subsemigroup of itself. Note also that if $(S,*)$ is a monoid with identity e then $(S,*)$ and $(\{e\},*)$ are submonoids of $(S,*)$.

Definition

Let $(S,*)$ be a semigroup and let $a \in S$. We define the powers of a by $a^1 = a$ and $a^n = a^{n-1} * a$ for each natural number n greater than 1. If $(S,*)$ is a monoid with identity e, we also define $a^0 = e$.

Example 19

If $(S,*)$ is a semigroup, $a \in S$, and $T = \{a^n: n \in \mathbb{N}\}$, then $(T,*)$ is a subsemigroup of $(S,*)$. If $(S,*)$ is a monoid, $a \in S$, and $T = \{a^n: n = 0$ or $n \in \mathbb{N}\}$, then $(T,*)$ is a submonoid of $(S,*)$.

Analysis

It can be shown that, if a is a member of a semigroup $(S,*)$ and $m,n \in \mathbb{N}$, then $a^m * a^n = a^{m+n}$. The results in this example are an immediate consequence of this fact. ■

Example 20

If T is the set of even integers, then $(T,\cdot)$ is a subsemigroup of the monoid $(\mathbb{Z},\cdot)$, but it is not a submonoid since $1 \notin T$. ■

Definition

Let $(S,*)$ and $(T,\#)$ be semigroups. A function $f: S \to T$ such that $f(a * b) = f(a) \# f(b)$ for all $a,b \in S$ is called a **homomorphism.** If there is a homomorphism that maps S onto T, then $(T,\#)$ is said to be the **homomorphic image** of $(S,*)$.

Example 21

Let $\Sigma = \{0,1\}$. Then (as seen in Example 16) $(\Sigma^*,\cdot)$ is a semigroup. Also, if $\oplus$ is defined by the following table, then $(\mathbb{Z}_2,\oplus)$ is a semigroup. Show that $(\mathbb{Z}_2,\oplus)$ is the homomorphic image of $(\Sigma^*,\cdot)$.

$\oplus$	[0]	[1]
[0]	[0]	[1]
[1]	[1]	[0]

Analysis

Define $f: \Sigma^* \to \mathbb{Z}_2$ by $f(\alpha) = [0]$ if α has an even number of 1s and $f(\alpha) = [1]$ if α has an odd number of 1s. Then it is easy to see that $f(\alpha\beta) = f(\alpha) \oplus f(\beta)$, and hence f is a homomorphism (see Exercise 23). ■

Theorem 15.2

Let $(S,*)$ and $(T,\#)$ be monoids with identities e and e', respectively, and let $f\colon S \to T$ be a homomorphism of S onto T. Then $f(e) = e'$.

Proof

Let $b \in T$. Since f maps S onto T, there exists $a \in S$ such that $f(a) = b$. Since e is the identity of S, $a * e = e * a = a$. Since f is a homomorphism, $b = f(a) = f(a * e) = f(a) \# f(e) = b \# f(e)$, and $b = f(a) = f(e * a) = f(e) \# f(a) = f(e) \# b$. Therefore $f(e)$ is an identity element of T. Hence, by Theorem 15.1, $f(e) = e'$. □

Theorem 15.3

Let $(S,*)$ and $(T,\#)$ be semigroups, let $(A,*)$ be a subsemigroup of $(S,*)$, and let $f\colon S \to T$ be a homomorphism. Then $(f(A),\#)$ is a subsemigroup of $(T,\#)$.

Proof

It is sufficient to show that, if $c,d \in f(A)$, then $c \# d$ is a member of $f(A)$, so let $c,d \in f(A)$. By definition of $f(A)$, there exist $a,b \in A$ such that $f(a) = c$ and $f(b) = d$. Since $(A,*)$ is a subsemigroup of $(S,*)$, $a*b \in A$. By definition of $f(A)$, $f(a*b) \in f(A)$. Since f is a homomorphism, $f(a * b) = f(a) \# f(b) = c\#d$. Therefore $c\#d \in f(A)$ and hence $(f(A),\#)$ is a subsemigroup of $(T,\#)$. □

A special case of Theorem 15.3 is that, if $(S,*)$ and $(T,\#)$ are semigroups and $f\colon S \to T$ is a homomorphism, then $(f(S),\#)$ is a subsemigroup of $(T,\#)$.

Corollary 15.4

Let $(S,*)$ and $(T,\#)$ be monoids, let $(A,*)$ be a submonoid of $(S,*)$, and let $f\colon S \to T$ be a homomorphism of S onto T. Then $(f(A),\#)$ is a submonoid of $(T,\#)$.

The proof of Corollary 15.4 is left as Exercise 24.

Theorem 15.5

Let $(S,*)$ and $(T,\#)$ be semigroups and let $f\colon S \to T$ be a homomorphism. If $(S,*)$ is commutative, then $(f(S),\#)$ is also commutative.

The proof of Theorem 15.5 is left as Exercise 25. A special case of Theorem 15.5 is that if, f maps S onto T and $(S,*)$ is commutative, then $(T,\#)$ is also commutative.

Definition

Let $(S,*)$ and $(T,\#)$ be semigroups. A one-to-one homomorphism $f: S \to T$ that maps S onto T is called an **isomorphism.** If an isomorphism $f: S \to T$ exists, then $(S,*)$ and $(T,\#)$ are said to be **isomorphic**.

Example 22

Let $S = \{a,b,c\}$ and $T = \{x,y,z\}$. Define binary operations $*$ and $\#$ on S and T, respectively, by the following tables:

*	a	b	c
a	a	c	c
b	c	a	a
c	c	a	a

#	x	y	z
x	y	x	y
y	x	y	x
z	y	x	y

It is a straightforward exercise (see Exercise 26) to verify that $(S,*)$ and $(T,\#)$ are semigroups. Show that $(S,*)$ and $(T,\#)$ are isomorphic.

Analysis

Define $f: S \to T$ by $f(a) = y$, $f(b) = z$, and $f(c) = x$. It is clear that f is one-to-one and that f maps S onto T. We leave as Exercise 27 the proof that f is a homomorphism. ■

Example 23

Define $\oplus$ on $\mathbb{Z}_2$ as in Example 21 and define $\oplus$ on $\mathbb{Z}_4$ by the following table. Show that there is a homomorphism from $\mathbb{Z}_4$ onto $\mathbb{Z}_2$ but that $(\mathbb{Z}_4,\oplus)$ and $(\mathbb{Z}_2,\oplus)$ are not isomorphic.

⊕	[0]	[1]	[2]	[3]
[0]	[0]	[1]	[2]	[3]
[1]	[1]	[2]	[3]	[0]
[2]	[2]	[3]	[0]	[1]
[3]	[3]	[0]	[1]	[2]

Analysis

Since $\mathbb{Z}_4$ has four members and $\mathbb{Z}_2$ has only two members, it is clear that there is no one-to-one function f mapping $\mathbb{Z}_4$ onto $\mathbb{Z}_2$. However, the function f: $\mathbb{Z}_4 \to \mathbb{Z}_2$ defined by $f([0]) = f([2]) = [0]$ and $f([1]) = f([3]) = [1]$ is a homomorphism (see Exercise 28). ■

Exercises 15.3

16. Let S be a set. Show that $(\mathcal{P}(S), \cup)$ is a commutative semigroup.

17. Let $A = \{0,1\}$. Which of the following tables define an operation $*$ on A such that $(A,*)$ is a semigroup? Which of the following tables define an operation $*$ on A such that $(A,*)$ is a monoid?

a.

*	0	1
0	0	0
1	0	0

b.

*	0	1
0	0	0
1	0	1

c.

*	0	1
0	0	0
1	1	0

d.

*	0	1
0	0	1
1	1	0

e.

*	0	1
0	1	1
1	0	0

f.

*	0	1
0	1	0
1	0	1

18. Determine whether $(A,*)$ is a semigroup under the following conditions.
    a. $A = \mathbb{N}$ and $a * b = \max\{a,b\}$
    b. $A = \mathbb{N}$ and $a * b = \text{lcm}\{a,b\}$
    c. $A = \mathbb{N}$ and $a * b = \gcd\{a,b\}$
    d. $A = \mathbb{N}$ and $a * b = b$
    e. $A = \mathcal{P}(\mathbb{N})$ and $a * b = a \cap b$

19. Let $S = \{0,1\}$. Find the members of S^S, and exhibit the table for the binary operation $\circ$ on S^S.

20. Let S be a nonempty set. Show that $(S^S, \circ)$ is a semigroup that is not commutative. (Hint: Recall some theorems about the composition of functions.)

21. Let $n \in \mathbb{N}$ and let $\Sigma = \{1,2, \ldots ,n\}$. Show that $(\Sigma^*,\cdot)$, where $\cdot$ is concatenation, is a semigroup.

22. Let $S = \{0,1\}$. Find the members of $\mathcal{P}(S)$ and exhibit the table for the binary operation $\cup$ on $\mathcal{P}(S)$.

23. Let $\Sigma = \{0,1\}$, and let $\cdot$ be concatenation on Σ^*. Define $\oplus$ on $\mathbb{Z}_2$ as in Example 21. Define f: $\Sigma^* \to \mathbb{Z}_2$ by $f(\alpha) = [0]$ if α has an even number of 1s and $f(\alpha) = [1]$ if α has an odd number of 1s. Show that f is a homomorphism.

24. Let $(S,*)$ and $(T,\#)$ be monoids, let $(A,*)$ be a submonoid of $(S,*)$, and let f: $S \to T$ be a homomorphism of S onto T. Prove that $(f(A),\#)$ is a submonoid of $(T,\#)$.

25. Let $(S,*)$ and $(T,\#)$ be semigroups and let f: $S \to T$ be a homomorphism. Show that if $(S,*)$ is commutative then $(f(S),\#)$ is also commutative.

26. Verify that $(S,*)$ and $(T,\#)$ as defined in Example 22 are semigroups.

27. Let f: $S \to T$, where f is defined as in the analysis of Example 22. Show that f is a homomorphism.

28. Verify that the function f: $(\mathbb{Z}_4,\oplus) \to (\mathbb{Z}_2,\oplus)$ defined by $f([0]) = f([2]) = [0]$ and $f([1]) = f([3]) = [1]$ is a homomorphism.

29. Let $(R,@)$, $(S,*)$ and $(T,\#)$ be semigroups and let $f: R \to S$ and $g: S \to T$ be homomorphisms. Prove that $g \circ f: R \to T$ is a homomorphism.

30. Let $\Sigma = \{0,1\}$ and let $S = \{\alpha \in \Sigma^*: \alpha \text{ has an odd number of 1s}\}$. Let $\cdot$ be the binary operation of concatenation. Is $(S,\cdot)$ a subsemigroup of $(\Sigma^*,\cdot)$? Verify your answer.

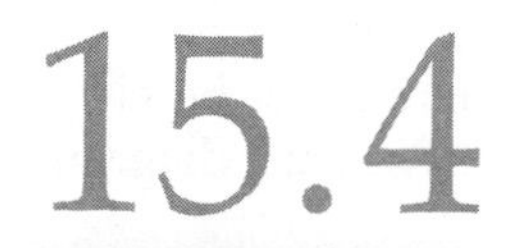

FINITE-STATE MACHINES

A finite-state machine (or finite-state automaton) is a mathematical model of how a computer functions while running a program. The internal memory of a computer is a system that can exist in only a finite number of different states. There are several types of finite-state machines, but we discuss only the most elementary versions. The study of finite-state machines is an important application of graph theory to computer science.

Definition

A **finite-state machine** M is a 5-tuple (S,Σ,d,s_0,F), where

a. S is a finite set whose members are called the **states** of the machine;

b. Σ is a finite set whose members are the **inputs** of the machine (the set Σ is called the **alphabet** of the machine);

c. $d: S \times \Sigma \to S$ is the **transition function** of the machine;

d. s_0 is an element of S called the **initial state** of the machine;

e. F is a subset of S whose members are called the **final states** of the machine.

The transition function d describes how the machine changes from one state to another. If the machine is in state s and receives an input σ, then the state of the machine becomes $d(s,\sigma)$.

A directed graph, called the *state diagram* of the machine, can be used to represent a finite-state machine. The states of the machine are represented by vertices, the inputs of the machine are represented by edges, and there is a directed edge σ from the vertex s_i to the vertex s_j if and only if $d(s_i,\sigma) = s_j$. It is customary to use small circles to denote the vertices in a state diagram of a machine, with the initial state designated by →○ and the final states by ◎.

Example 24

Let $S = \{s_0, s_1, s_2, s_3\}$, let $\Sigma = \{\sigma_1, \sigma_2\}$, let $F = \{s_2, s_3\}$, and let $d: S \times \Sigma \to S$ be defined by $d(s_0,\sigma_1) = s_2$, $d(s_0,\sigma_2) = s_1$, $d(s_1,\sigma_1) = s_2$, $d(s_1,\sigma_2) = s_2$, $d(s_2,\sigma_1) = s_3$, $d(s_2,\sigma_2) = s_1$, $d(s_3,\sigma_1) = s_0$, and $d(s_3,\sigma_2) = s_3$. (We always let s_0 denote the initial state of the machine.) Draw the state diagram of the finite-state machine (S,Σ,d,s_0,F).

Analysis

The state diagram has vertices s_0, s_1, s_2, and s_3 and edges σ_1 and σ_2. Since there is an edge for each member of $S \times \Sigma$, there must be eight edges. Consequently, several edges bear the same label. The state diagram of (S,Σ,d,s_0,F) is shown in Figure 15.3.

Figure 15.3

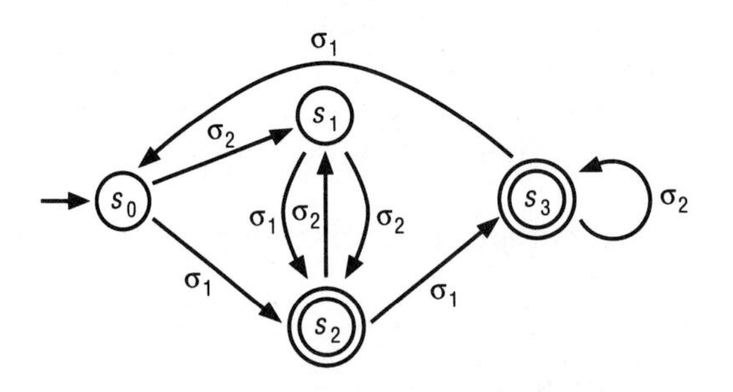

■

A table, called the *state table* of the machine, gives another way of describing a finite-state machine. The state table is really a matrix whose rows are labeled by the states (the initial state is the label of the first row) and whose columns are labeled by the inputs. The entry in row s_i and column σ_j is $d(s_i,\sigma_j)$. Since a state table specifies the states, the inputs, the transition function, and the initial state of a finite-state machine, a complete description of a finite-state machine is given by its state table and the set of final states.

Example 25

Construct the state table of the finite-state machine whose state diagram is shown in Figure 15.3.

Analysis

We construct a table whose rows are labeled s_0, s_1, s_2, and s_3, whose columns are labeled σ_1 and σ_2, and whose entries are $d(s_i,\sigma_j)$ for all $i = 0,1,2,3$ and $j = 1,2$.

	σ_1	σ_2
s_0	s_2	s_1
s_1	s_2	s_2
s_2	s_3	s_1
s_3	s_0	s_3

■

Example 26

Specify the finite-state machine $M = (S,\Sigma,d,s_0,F)$ associated with the state diagram in Figure 15.4.

Figure 15.4

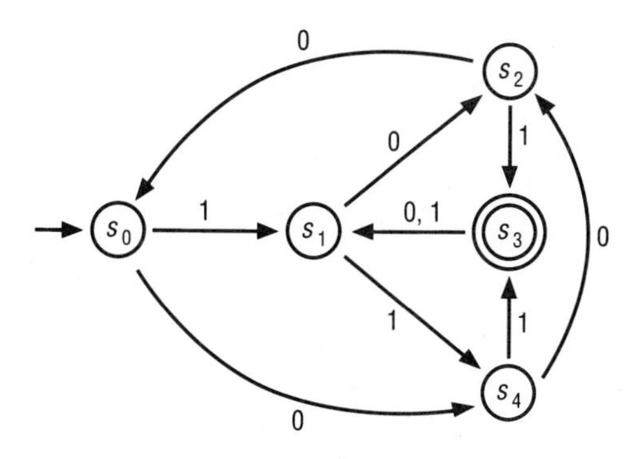

Analysis

We can easily see that $S = \{s_0,s_1,s_2,s_3, s_4\}$, $\Sigma = \{0,1\}$, $F = \{s_3\}$, and $d\colon S \times \Sigma \to S$ is the function defined by $d(s_0,0) = s_4$, $d(s_0,1) = s_1$, $d(s_1,0) = s_2$, $d(s_1,1) = s_4$, $d(s_2,0) = s_0$, $d(s_2,1) = s_3$, $d(s_3,0) = s_1$, $d(s_3,1) = s_1$, $d(s_4,0) = s_2$, and $d(s_4,1) = s_3$. ■

Definition

Let $M = (S,\Sigma,d,s_0,F)$ be a finite-state machine, and let R be an equivalence relation on S. Then R is a **machine congruence** on M if, for each $s, t \in S$, $(s,t) \in R$ implies $(d(s,\sigma),d(t,\sigma)) \in R$ for all $\sigma \in \Sigma$.

If R is a machine congruence on M, we let $\overline{S} = S/R$ (the set of equivalence classes of S with respect to R), and we let $\overline{d}\colon \overline{S} \times \Sigma \to \overline{S}$ be the function defined by $\overline{d}([s] \times \sigma) = [d(s,\sigma)]$. To see that $\overline{d}$ is a function, we must show that if $[s] = [t]$ then $[d(s,\sigma)] = [d(s,\sigma)]$. Suppose $[s] = [t]$. Then $(s,t) \in R$, and hence $(d(s,\sigma),d(t,\sigma)) \in R$. But this means that $[d(s,\sigma)] = [d(s,\sigma)]$. We also let $\overline{F} = \{[s]\colon s \in F\}$. Then $\overline{M} = (\overline{S},\Sigma,\overline{d},s_0, \overline{F})$ is a finite-state machine, and it is called the *quotient of M with respect to R.*

Example 27

Let $F = \{s_1,s_2\}$, let $M = (S,\Sigma,d,s_0,F)$ be the finite-state machine whose state table is given as follows, and let R be the relation whose associated matrix is M_R.

	0	1
s_0	s_4	s_3
s_1	s_4	s_2
s_2	s_2	s_0
s_3	s_3	s_1
s_4	s_0	s_3

$$M_R = \begin{array}{c} \\ s_0 \\ s_1 \\ s_2 \\ s_3 \\ s_4 \end{array} \begin{array}{c} \begin{array}{ccccc} s_0 & s_1 & s_2 & s_3 & s_4 \end{array} \\ \begin{bmatrix} 1 & 1 & 0 & 0 & 0 \\ 1 & 1 & 0 & 0 & 0 \\ 0 & 0 & 1 & 1 & 0 \\ 0 & 0 & 1 & 1 & 0 \\ 0 & 0 & 0 & 0 & 1 \end{bmatrix} \end{array}$$

Show that R is a machine congruence on M and construct the state table of the quotient of M with respect to R.

Analysis

It is easily seen from the matrix M_R that R is an equivalence relation on S and that $\overline{S} = S/R = \{[s_0],[s_2],[s_4]\}$, where $[s_0] = \{s_0,s_1\}$, $[s_2] = \{s_2,s_3\}$, and $[s_4] = \{s_4\}$. For R to be a machine congruence on M, it must be the case that each of the following sets belongs to a member of $\overline{S}$: $\{d(s_0,0),d(s_1,0)\}$, $\{d(s_0,1),d(s_1,1)\}$, $\{d(s_2,0),d(s_3,0)\}$, and $\{d(s_2,1),d(s_3,1)\}$. It is easy to see that this is true. The state table of $\overline{M}$ is shown here.

	0	1
$[s_0]$	$[s_4]$	$[s_2]$
$[s_2]$	$[s_2]$	$[s_0]$
$[s_4]$	$[s_0]$	$[s_2]$

■

In general, the quotient of a finite-state machine M with respect to a machine congruence is simpler than M. In the next section we discuss the language of a finite-state machine. We conclude this section by stating, without proof, a theorem that permits us to simplify a finite-state machine and obtain a finite-state machine with the same language as the original machine. But first we need some terminology.

Definition

Let $M = (S,\Sigma,d,s_0,F)$ be a finite-state machine. For each $\sigma \in \Sigma$, let $d_\sigma = d|(S \times \{\sigma\})$, and for each $\alpha = \sigma_0\sigma_1 \ldots \sigma_n \in \Sigma^*$, define $d_\alpha = d_{\sigma_n} \circ d_{\sigma_{n-1}} \circ \cdots \circ d_{\sigma_1} \circ d_{\sigma_0}$. The function d_α: $S \rightarrow S$ is called the **transition function corresponding to α.**

Definition

Let $M = (S,\Sigma,d,s_0,F)$ be a finite-state machine, and let $\sigma \in \Sigma^*$. We say that members s and t of S are **σ-compatible** if $d_\sigma(s)$ and $d_\sigma(t)$ both belong to F or neither belongs to F.

Theorem 15.6

> Let $M = (S,\Sigma,d,s_0,F)$ be a finite-state machine and define a relation R on S by $(s,t) \in R$ if and only if s and t are σ-compatible for all $\sigma \in \Sigma^*$. Then R is a machine congruence on M.

It can be shown that the quotient of a finite-state machine with respect to the machine congruence given by Theorem 15.6 has the same language as the original machine.

Exercises 15.4

31. Specify the finite-state machine $M = (S,\Sigma,d,s_0,F)$ associated with each of the following state diagrams.

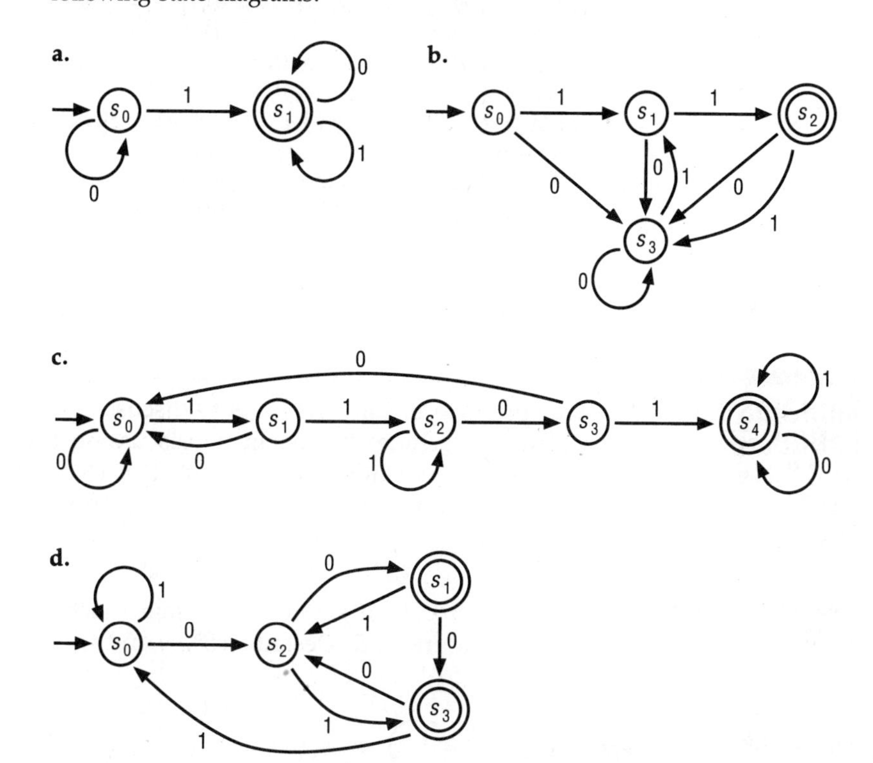

32. Draw the state diagram of each of the following finite-state machines.

a. Let $S = \{s_0,s_1,s_2\}$, $\Sigma = \{0,1\}$, $F = \{s_0,s_2\}$, and define $d: S \times \Sigma \to S$ by $d(s_0,0) = s_0$, $d(s_0,1) = s_1$, $d(s_1,0) = s_1$, $d(s_1,1) = s_2$, $d(s_2,0) = s_0$, and $d(s_2,1) = s_2$.

b. Let $S = \{s_0,s_1,s_2,s_3,s_4\}$, $\Sigma = \{0,1\}$, $F = \{s_3,s_4\}$, and define $d: S \times \Sigma \to S$ by $d(s_0,0) = s_2$, $d(s_0,1) = s_4$, $d(s_1,0) = s_3$, $d(s_1,1) = s_4$, $d(s_2,0) = s_1$, $d(s_2,1) = s_0$, $d(s_3,0) = s_4$, $d(s_3,1) = s_3$, $d(s_4,0) = s_3$, and $d(s_4,1) = s_4$.

c. Let $S = \{s_0,s_1,s_2,s_3\}$, $\Sigma = \{0,1,2\}$, $F = \{s_3\}$, and define $d\colon S \times \Sigma \to S$ by $d(s_0,0) = s_0$, $d(s_0,1) = s_0$, $d(s_0,2) = s_2$, $d(s_1,0) = s_2$, $d(s_1,1) = s_3$, $d(s_1,2) = s_2$, $d(s_2,0) = s_1$, $d(s_2,1) = s_0$, $d(s_2,2) = s_3$, $d(s_3,0) = s_3$, $d(s_3,1) = s_2$, and $d(s_3,2) = s_0$.

33. Construct the state table of each of the finite-state machines in Exercise 32.

34. Let $S = \{s_0,s_1,s_2,s_3,s_4,s_5,s_6\}$, $\Sigma = \{0,1\}$, $F = \{s_4,s_5\}$, and define $d\colon S \times \Sigma \to S$ by $d(s_0,0) = s_3$, $d(s_0,1) = s_0$, $d(s_1,0) = s_0$, $d(s_1,1) = s_6$, $d(s_2,0) = s_0$, $d(s_2,1) = s_3$, $d(s_3,0) = s_4$, $d(s_3,1) = s_2$, $d(s_4,0) = s_1$, $d(s_4,1) = s_0$ $d(s_5,0) = s_3$, $d(s_5,1) = s_1$, $d(s_6,0) = s_4$, and $d(s_6,1) = s_1$. Let $M = (S,\Sigma, d,s_0,F)$ and define a relation R on S by $(s,t) \in R$ if and only if s and t are σ-compatible for all $\sigma \in \Sigma^*$. By Theorem 15.6, R is a machine congruence on M. Find the quotient of M with respect to R.

35. Let $S = \{s_0,s_1,s_2,s_3,s_4,s_5,s_6,s_7\}$, $\Sigma = \{0,1\}$, $F = \{s_5,s_6,s_7\}$, and define $d\colon S \times \Sigma \to S$ by $d(s_0,0) = s_0$, $d(s_0,1) = s_7$, $d(s_1,0) = s_1$, $d(s_1,1) = s_6$, $d(s_2,0) = s_1$, $d(s_2,1) = s_2$, $d(s_3,0) = s_7$, $d(s_3,1) = s_2$, $d(s_4,0) = s_0$, $d(s_4,1) = s_3$, $d(s_5,0) = s_1$, $d(s_5,1) = s_6$, $d(s_6,0) = s_3$, $d(s_6,1) = s_2$, $d(s_7,0) = s_7$, and $d(s_7,1) = s_2$. Let $M = \{S,\Sigma,d,s_0,F\}$, and let R be the equivalence relation associated with the partition $\mathcal{P} = \{\{s_0,s_4\},\{s_1,s_5\},\{s_2\},\{s_3,s_6,s_7\}\}$.
a. Verify that R is a machine congruence on M.
b. Find the quotient of M with respect to R.

15.5 MONOIDS AND LANGUAGES OF FINITE-STATE MACHINES

Our purpose in this section is to discuss the languages accepted by a finite-state machine. We begin by discussing the monoids associated with a given finite-state machine. Let $M = (S,\Sigma,d,s_0,F)$ be a finite-state machine. According to the discussion in Section 15.3, $(S^S, \circ)$ and $(\Sigma^*,\cdot)$ are monoids.

Example 28

Let $M = (S,\Sigma,d,s_0,F)$ be the finite-state machine whose state diagram is shown in Figure 15.5. Describe $(S^S, \circ)$ and $(\Sigma^*,\cdot)$.

Figure 15.5

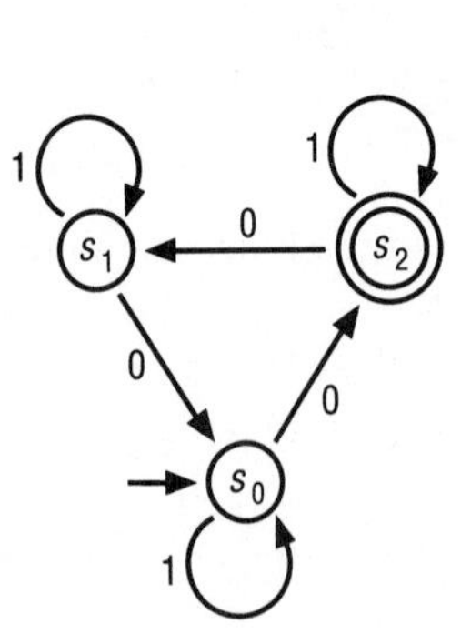

Analysis

Notice that $S = \{s_0,s_1,s_2\}$. In defining the members of S^S, we have three choices for the image of s_0, three choices for the image of s_1, and three choices for the image of s_2. Consequently, there are $3 \times 3 \times 3 = 27$ members of S^S. The binary operation $\circ$ is composition, and we can see that $\circ$ is noncommutative by considering the functions $f_1,f_2: S \to S$ defined by $f_1(s_0) = f_1(s_1) = f_1(s_2) = s_0$ and $f_2(s_0) = f_2(s_1) = f_2(s_2) = s_1$. The function $f_2 \circ f_1$ is equal to f_2, and the function $f_1 \circ f_2$ is f_1. Thus $f_2 \circ f_1 \neq f_1 \circ f_2$. Since $\Sigma = \{0,1\}$, Σ^* consists of all strings of 0s and 1s. Thus Σ^* is an infinite set, and hence there is no one-to-one correspondence between Σ^* and S^S. Therefore $(\Sigma^*,\cdot)$ and $(S^S, \circ)$ cannot be isomorphic. The operation of concatenation simply "joins" two words, so $(\Sigma^*,\cdot)$ is a noncommutative monoid. ■

Definition

Let $M = (S,\Sigma,d,s_0,F)$ be a finite-state machine. Let $\alpha = \sigma_0\sigma_1 \ldots \sigma_n \in \Sigma^*$, and for each $i = 0,1, \ldots ,n$ let $s_{i+1} = d(s_i,\sigma_i)$. If $s_{n+1} \in F$, we say that α is **accepted by** M, and if $s_{n+1} \notin F$ we say that α is **rejected by** M.

Example 29

Let $M = (S,\Sigma,d,s_0,F)$ be the finite-state machine whose state diagram is shown in Figure 15.5. Let $\alpha = 101$ and $\beta = 010$. Then $\alpha,\beta \in \Sigma^*$. Are α and β accepted by M?

Analysis

To decide whether α is accepted by M we calculate as follows: $d(s_0,1) = s_0$, $d(s_0,0) = s_2$, and $d(s_2,1) = s_2$. Since $s_2 \in F$, α is accepted by M. Likewise, we calculate for β: $d(s_0,0) = s_2$, $d(s_2,1) = s_2$, and $d(s_2,0) = s_1$. Since $s_1 \notin F$, β is not accepted by M. ■

Notice that, if d_α is the transition function corresponding to α and $d_\alpha(s_0) \in F$, then α is accepted by M and otherwise α is rejected by M. The *language* of M is defined to be the set of all $\alpha \in \Sigma^*$ such that $d_\alpha(s_0) \in F$, and it is denoted by $L(M)$.

Example 30

Let $M = (S,\Sigma,d,s_0,F)$ to be finite-state machine whose state diagram is shown in Figure 15.6. Describe $L(M)$.

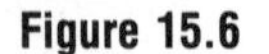

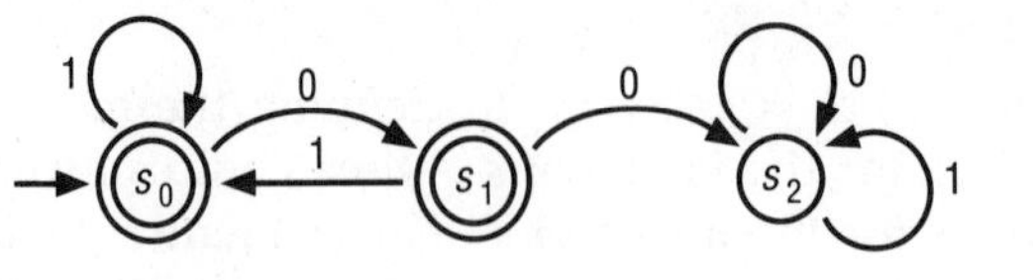

Figure 15.6

Analysis

Since $S - F = \{s_2\}$ and once the state s_2 is entered there is no way for a computation to leave s_2, we can describe the subset A of Σ^* that consists of those words that do not belong to $L(M)$. Thus $L(M)$ is $\Sigma^* - A$. We therefore consider this question: For which $\alpha \in \Sigma^*$ is $d_\alpha(s_0) = s_2$? The state s_2 can be reached only when we are in state s_1 and the input is 0. The state s_1 can be reached only when we are in state s_0 and the input is 0. Thus the word $00 \in A$. Also, if $\beta \in \Sigma^*$ leads to the initial state s_0, then $\beta 00 \in A$. If we are not already in state s_2, then a 1 leads to state s_0. Since we start at s_0, we conclude that any member of Σ^* of the form $\beta 00 \gamma$, where β is either ε or ends in a 1 and γ is any member of Σ^*, belongs to A. Therefore any member of Σ^* with two consecutive 0s belongs to A. Suppose $\alpha \in \Sigma^*$ and α does not contain two consecutive 0s. Then we can never reach state s_2. Consequently, $\alpha \notin A$. Thus $L(M)$ consists of all members of Σ^* that do not contain two consecutive 0s. ■

Let $M = (S,\Sigma,d,s_0,F)$. Define $H: \Sigma^* \to S^S$ as follows: $H(\varepsilon)$ is the identity function. Let α be any member of Σ^* different from ε. Then there exists $\sigma_0,\sigma_1, \ldots ,\sigma_n \in \Sigma$ such that $\alpha = \sigma_0\sigma_1 \ldots \sigma_n$. Define $H(\alpha) = d_{\sigma_0} \circ d_{\sigma_1} \circ \cdots \circ d_{\sigma_n}$.

Theorem 15.7 The function $H: \Sigma^* \to S^S$ is a homomorphism from $(\Sigma^*,\cdot)$ into $(S^S, \circ)$.

Proof

Let $\alpha,\beta \in \Sigma^*$. Then there exists $\sigma_0,\sigma_1, \ldots ,\sigma_m,\tau_0,\tau_1, \ldots ,\tau_n \in \Sigma$ such that $\alpha = \sigma_0\sigma_1 \ldots \sigma_m$ and $\beta = \tau_0\tau_1 \ldots \tau_n$. Therefore

$$\begin{aligned} H(\alpha\beta) &= H(\sigma_0\sigma_1 \ldots \sigma_m\tau_0\tau_1 \ldots \tau_n) \\ &= d_{\sigma_0} \circ d_{\sigma_1} \circ \cdots \circ d_{\sigma_m} \circ d_{\tau_0} \circ d_{\tau_1} \circ \cdots \circ d_{\tau_n} \\ &= (d_{\sigma_0} \circ d_{\sigma_1} \circ \cdots \circ d_{\sigma_m}) \circ (d_{\tau_0} \circ d_{\tau_1} \circ \cdots \circ d_{\tau_n}) \\ &= H(\alpha) \circ H(\beta) \end{aligned}$$

□

Definition

Since the identity function on S is a member of $H(\Sigma^*)$, $(H(\Sigma^*), \circ)$ is a submonoid of $(S^S, \circ)$. The monoid $(H(\Sigma^*), \circ)$ is called the **monoid** of M.

Example 31

Let $M = (S,\Sigma,d,s_0,F)$ be the finite-state machine whose state diagram is shown in Figure 15.7. Compute the functions d_{012}, d_{111}, and d_{012111}, and verify that $d_{012111} = d_{111} \circ d_{012}$.

Figure 15.7

Analysis

Since $d_{012} = d_2 \circ d_1 \circ d_0$,

$$d_{012}(s_0) = (d_2 \circ d_1 \circ d_0)(s_0) = (d_2 \circ d_1)(s_1) = d_2(s_1) = s_2$$

$$d_{012}(s_1) = (d_2 \circ d_1 \circ d_0)(s_1) = (d_2 \circ d_1)(s_0) = d_2(s_0) = s_2 \text{ and}$$

$$d_{012}(s_2) = (d_2 \circ d_1 \circ d_0)(s_2) = (d_2 \circ d_1)(s_0) = d_2(s_0) = s_2$$

Since $d_{111} = d_1 \circ d_1 \circ d_1$,

$$d_{111}(s_0) = (d_1 \circ d_1 \circ d_1)(s_0) = (d_1 \circ d_1)(s_0) = d_1(s_0) = s_0$$

$$d_{111}(s_1) = (d_1 \circ d_1 \circ d_1)(s_1) = (d_1 \circ d_1)(s_1) = d_1(s_1) = s_1 \text{ and}$$

$$d_{111}(s_2) = (d_1 \circ d_1 \circ d_1)(s_2) = (d_1 \circ d_1)(s_1) = d_1(s_1) = s_1$$

Thus $(d_{111} \circ d_{012})(s_0) = d_{111}(s_2) = s_1$, $(d_{111} \circ d_{012})(s_1) = d_{111}(s_2) = s_1$, and $(d_{111} \circ d_{012})(s_2) = d_{111}(s_2) = s_1$. Since $d_{012111} = d_1 \circ d_1 \circ d_1 \circ d_2 \circ d_1 \circ d_0$,

$$d_{012111}(s_0) = (d_1 \circ d_1 \circ d_1 \circ d_2 \circ d_1 \circ d_0)(s_0) = (d_1 \circ d_1 \circ d_1 \circ d_2 \circ d_1)(s_1)$$
$$= (d_1 \circ d_1 \circ d_1 \circ d_2)(s_1) = (d_1 \circ d_1 \circ d_1)(s_2) = (d_1 \circ d_1)(s_1) = d_1(s_1) = s_1$$

$$d_{012111}(s_1) = (d_1 \circ d_1 \circ d_1 \circ d_2 \circ d_1 \circ d_0)(s_1) = (d_1 \circ d_1 \circ d_1 \circ d_2 \circ d_1)(s_0)$$
$$= (d_1 \circ d_1 \circ d_1 \circ d_2)(s_0) = (d_1 \circ d_1 \circ d_1)(s_2) = (d_1 \circ d_1)(s_1) = d_1(s_1) = s_1$$

and

$$d_{012111}(s_2) = (d_1 \circ d_1 \circ d_1 \circ d_2 \circ d_1 \circ d_0)(s_2) = (d_1 \circ d_1 \circ d_1 \circ d_2 \circ d_1)(s_0)$$
$$= (d_1 \circ d_1 \circ d_1 \circ d_2)(s_0) = (d_1 \circ d_1 \circ d_1)(s_2) = (d_1 \circ d_1)(s_1) = d_1(s_1) = s_1$$

Therefore $d_{012111} = d_{111} \circ d_{102}$. ■

We conclude this section by stating, without proof, Kleene's theorem.

Theorem 15.8 Let Σ be an alphabet and let $L \subseteq \Sigma^*$. Then L is a type 3 language if and only if there is a finite-state machine M such that $L = L(M)$.

Exercises 15.5

36. Let $M = (S,\Sigma,d,s_0,F)$ be the finite-state machine whose state diagram is shown here. Which of the following strings are accepted by M? Justify your answers.

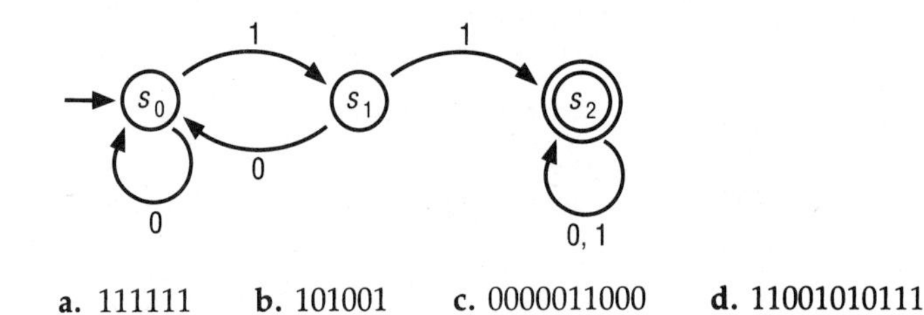

a. 111111 **b.** 101001 **c.** 0000011000 **d.** 11001010111

37. What property is required of a string for it to be accepted by the finite-state machine in Exercise 36?

38. Let $F = \{s_2\}$ and let $M = (S,\Sigma,d,s_0,F)$ be the finite-state machine whose state transition table is shown here. Which of the following strings are accepted by M? Justify your answers.

a. 012013 **b.** 01102201

c. 01010101 **d.** 3010301

	0	1	2	3
s_0	s_1	s_0	s_0	s_2
s_1	s_1	s_1	s_1	s_2
s_2	s_2	s_2	s_2	s_2

39. What property is required of a string for it to be accepted by the finite-state machine in Exercise 38?

40. Let $M = (S,\Sigma,d,s_0,F)$ be the finite-state machine whose state diagram is shown here. List the values of d_{01001} and d_{11100}.

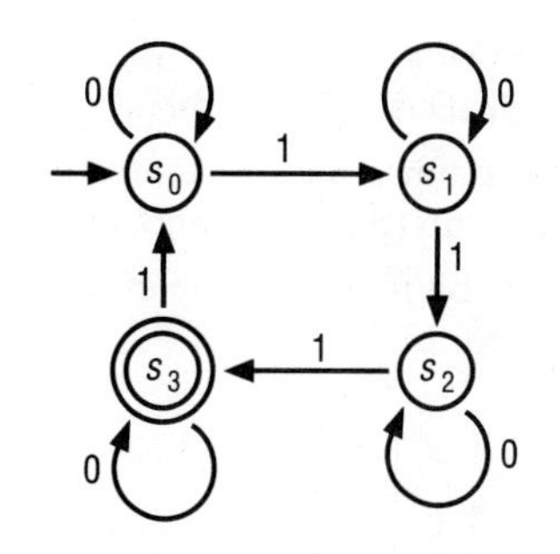

41. With respect to the finite-state machine in Exercise 40, describe the strings α of 0s and 1s which have the property that $d_\alpha(s_0) = s_0$.

42. With respect to the finite-state machine in Exercise 40, describe the strings α of 0s and 1s which have the property that $d_\alpha(s_0) = s_2$.

43. Let $F = \{s_4\}$ and let $M = (S,\Sigma,d,s_0,F)$ be the finite-state machine whose state transition table is shown here. Describe $L(M)$.

	0	**1**
s_0	s_0	s_1
s_1	s_1	s_2
s_2	s_2	s_3
s_3	s_3	s_4
s_4	s_4	s_0

44. Let $M = (S,\Sigma,d,s_0,F)$ be the finite-state machine whose diagram is shown here. Describe $L(M)$.

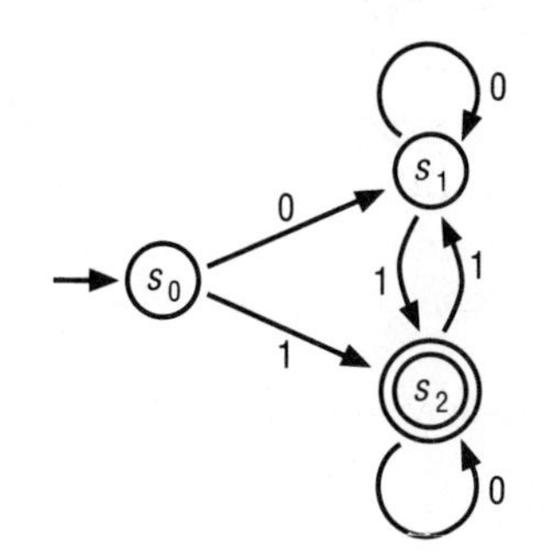

Chapter 15 Review Exercises

45. Let $\Sigma = \{a,b\}$. Find the set associated with each of the following regular expressions.

a. $(a^* \vee b^*)b$ **b.** $a^*(a \vee b)^*b$ **c.** $(ab)^*(ab \vee b)$

46. Each of the following sets is associated with a regular expression. In each case, find the regular expression.

a. $\{\varepsilon, ba, baba, bababa, \ldots\}$

b. $\{\varepsilon, b, bb, bbb, \ldots, a, ab, abb, abbb, \ldots, aa, aab, aabb, aabbb, \ldots, aaa, aaab, aaabb, aaabbb, \ldots\}$

47. Let $G = (N,T,S,P)$, where $N = \{S,A\}$, $T = \{0,1\}$, and $P = \{S \rightarrow_a AS,\ S \rightarrow_b A,\ A \rightarrow_c 0\}$. Describe the members of $L(G)$.

48. Find a phrase structure grammar that generates the language of all words over $\{0,1\}$ which consist of alternating 0s and 1s.

49. Let $A = \{0,1,2\}$. Which of the following tables define an operation $*$ on A such that $(A,*)$ is a semigroup?

a.

*	0	1	2
0	0	0	1
1	0	1	2
2	1	2	0

b.

*	0	1	2
0	0	0	1
1	0	1	0
2	0	0	0

c.

*	0	1	2
0	1	2	0
1	2	0	1
2	0	1	2

d.

*	0	1	2
0	2	1	0
1	1	2	1
2	0	1	2

e.

*	0	1	2
0	0	2	1
1	2	1	0
2	1	0	2

f.

*	0	1	2
0	2	0	1
1	0	1	2
2	2	1	0

50. Which of the tables in Exercise 49 define an operation $*$ on A such that $(A,*)$ is a monoid?

51. Recall that $(\mathbb{Z}_4,\oplus)$ is defined in Example 23. Define $\oplus$ on $\mathbb{Z}_3$ by the following table.

$\oplus$	[0]	[1]	[2]
[0]	[0]	[1]	[2]
[1]	[1]	[2]	[0]
[2]	[2]	[0]	[1]

Define f: $(\mathbb{Z}_4,\oplus) \rightarrow (\mathbb{Z}_3,\oplus)$ by $f([0]) = f([3]) = [0]$, $f([1]) = [1]$, and $f([2]) = [2]$. Is f a homomorphism? Prove your answer.

52. Recall that $(\mathbb{Z}_2,\oplus)$ is defined in Example 21. Let X denote the set of all permutations on $\{1,2,3\}$. Define a homomorphism that maps $(X, \circ)$ onto $(\mathbb{Z}_3,\oplus)$.

53. Find the finite-state machine $M = (S,\Sigma,d,s_0,F)$ associated with each of the following state diagrams.

a. b.

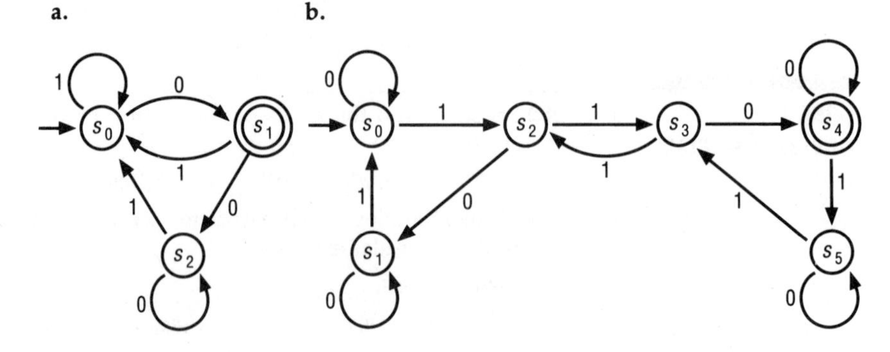

54. Draw the state diagram of each of the following finite-state machines.

a. Let $S = \{s_0,s_1,s_2\}$, $\Sigma = \{0,1\}$, $F = \{s_2\}$, and define d: $S \times \Sigma \to S$ by $d(s_0,0) = s_1$, $d(s_0,1) = s_0$, $d(s_1,0) = s_2$, $d(s_1,1) = s_0$, $d(s_2,0) = s_2$, and $d(s_2,1) = s_0$.

b. Let $S = \{s_0,s_1,s_2,s_3\}$, $\Sigma = \{0,1,2\}$, $F = \{s_1,s_3\}$, and define d: $S \times \Sigma \to S$ by $d(s_0,0) = s_3$, $d(s_0,1) = s_0$, $d(s_0,2) = s_2$, $d(s_1,0) = s_3$, $d(s_1,1) = s_2$, $d(s_1,2) = s_3$, $d(s_2,0) = s_1$, $d(s_2,1) = s_0$, $d(s_2,2) = s_2$, $d(s_3,0) = s_1$, $d(s_3,1) = s_3$, and $d(s_3,2) = s_2$.

55. In each of the following, construct finite-state machines that accept the strings described and no others.

a. $\Sigma = \{0,1\}$, strings where the number of 0s is a non-zero multiple of 3

b. $\Sigma = \{0,1\}$, strings that end with 1100

c. $\Sigma = \{0,1\}$, strings that contain 10 and end with 000

d. $\Sigma = \{0,1\}$, strings that end with 1010

e. $\Sigma = \{0,1\}$, strings that do not contain successive 0s

56. Let $M = (S,\Sigma,d,s_0,F)$ be the finite-state machine illustrated here. Let R be the equivalence relation on S with the given matrix. Find the quotient of M with respect to R.

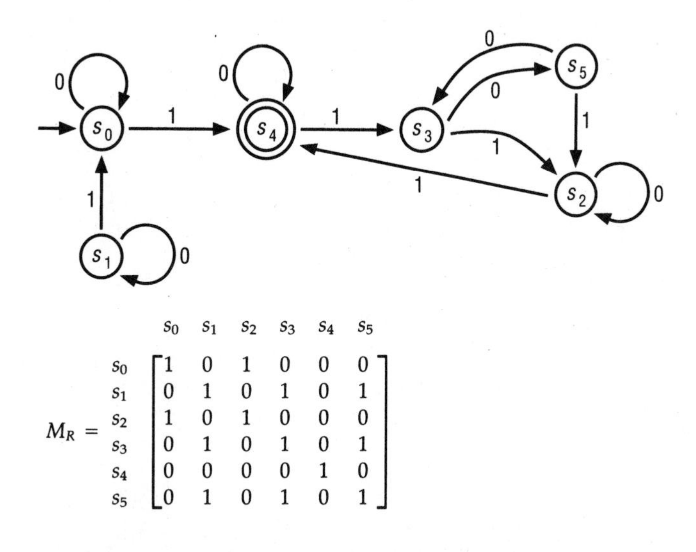

$$M_R = \begin{matrix} & s_0 & s_1 & s_2 & s_3 & s_4 & s_5 \\ s_0 & 1 & 0 & 1 & 0 & 0 & 0 \\ s_1 & 0 & 1 & 0 & 1 & 0 & 1 \\ s_2 & 1 & 0 & 1 & 0 & 0 & 0 \\ s_3 & 0 & 1 & 0 & 1 & 0 & 1 \\ s_4 & 0 & 0 & 0 & 0 & 1 & 0 \\ s_5 & 0 & 1 & 0 & 1 & 0 & 1 \end{matrix}$$

57. Let S be a set, and for $A,B, \in \mathcal{P}(S)$ define $A \oplus B$ to be $(A \cup B) - (A \cap B)$. Is $(\mathcal{P}(S),\oplus)$ a commutative monoid? Explain.

Appendix A

THE COMPUTER

The modern electronic digital computer is such a complicated tangle of wires and flashing lights that its machinations are comprehensible only to those who have spent years studying mathematics, electrical engineering, and computer science. It performs millions of calculations a second and in doing so it rarely makes more than one or two small miscalculations. Unfortunately, these few mistakes each second pile up after a while, and so occasionally a computer may mistakenly stop a subscription to a favorite magazine, empty a bank account, or cause a collapse of a major international currency. In short, the computer is the cause of both minor annoyance and major woe; it is an object to be feared.

Opinions such as this, though popular, are of course erroneous. It is possible to operate a computer without knowing any discrete mathematics (or electrical engineering either, for that matter), and in the years to come millions of people will operate computers while knowing little or nothing of the mathematics in this book. Futhermore, there is no need to seek protection from the whims of computer error; almost all computer errors are in reality man-made. Finally, the notion that a digital computer is a complicated and mysterious black box containing an incredible tangle of electric circuits is, at best, a half-truth. We begin by considering this half-truth, because much of the mathematics in this book is motivated by this mysterious black box we call a digital computer.

What is a digital computer? We are as uneasy with this question as we were with the question considered in the introduction—what is discrete mathematics? We start with a stopgap answer: A digital computer is a collection of on–off switches connected in interesting ways. It is easy to see that there is something wrong with this definition. It is too vague. What one person finds interesting, another may find dull.

Under our admittedly vague interpretation of a computer, the first computers were structures used by the ancients to calculate events of the calendar. The most famous of these structures is at Stonehenge, but the Mayans had such a computer at Chichén Itzá (Mexico), and the Indians of North America calculated astronomical events using computers that blended so well into the environment that their purpose was not understood by modern man for centuries. In these ancient structures, the on–

off switches were simply windows so cleverly arranged that they allowed light to pass only when the sun or moon occupied a significant position in the heavens.

The history of computers reflects man's long struggle with calculating sums and products. One example of a digital computer, which often goes unnoticed, is the ordinary checkerboard. The use of the checkerboard to calculate products dates to the early seventeenth century, and the only vestige of this use of the checkerboard lies in such financial terms as "checkbook" and "Chancellor of the Exchequer." In Figures A.1 and A.2 we illustrate the calculation of 13 × 6 on a checkerboard.

Figure A.1

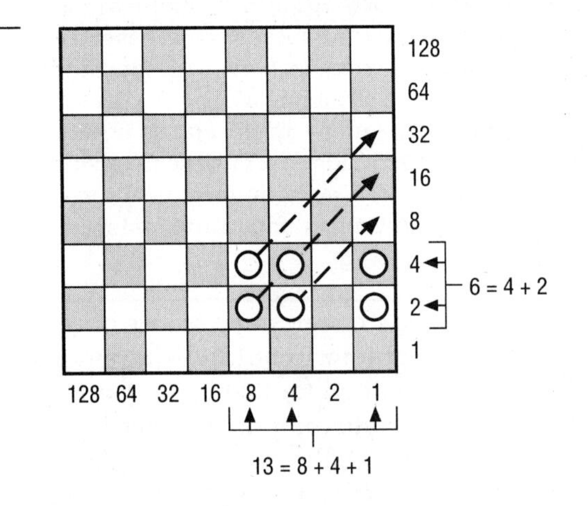

The first step is to enter 13 by choosing columns that add up to 13. Notice that a given column is chosen or is not chosen. Thus the data is being entered in discrete units: yes the column is chosen, or no the column is not chosen. Similarly, 6 is entered by choosing rows that add up to 6. Checkers are placed wherever the chosen rows and columns have a common square. There are six such squares, and, although the placing of the checkers is not done by machine, the operator is not required to have any mathematical skills or even any understanding of multiplication. Next all checkers are slid diagonally to the far right column. At that stage, if two checkers occupy the same square they are removed and a single checker is added to the square directly above the square from which the checkers have been taken. This process is continued until no two checkers occupy the same square (see Figure A.2). The answer can now be read off by finding the sum of the numbers with a checker beside them. Thus the answer is 64 + 8 + 4 + 2 = 78.

The checkerboard has the two essential features of a computer: it accepts data in discrete units and it processes data without requiring any further mental effort on the part of its operator. Note also that the answer

Figure A.2

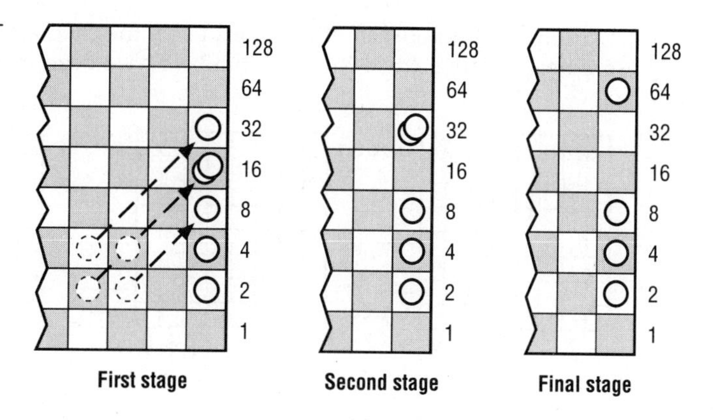

is given in discrete units. This example, however, is somewhat misleading, because it gives the impression that a computer must be able to make a mathematical calculation. In fact, a computer processes information of all kinds, and the use of a computer as a word processor is just as typical as its use as a calculator.

We have been considering a digital computer because it motivates or makes practical much of discrete mathematics. We mention in passing that there is a second type of computer known as an *analog* computer. An analog computer works by making use of a physical situation that mimics the desired mathematics in its behavior. Technically, the slide rule is such a computer, because the sliding of a slide rule's two ruled sticks is the physical analogue of adding numbers. If you want to see a computer scientist wince, you have only to cite this example, because the slide rule is to analog computers what Stonehenge is to digital computers, an ancient relic. What is usually meant by an analog computer is an electronic computer in which flow of current is the physical situation that provides the analogue of mathematics. Because the electronic analog computer relies on *measuring* physical quantities such as voltage or amperage, its use is more restricted than that of the digital computer, but in some settings there are compensating advantages. From our point of view, however, the analog computer is not of much interest, because both its input and output may be continuous rather than discrete and so the mathematics involved with such a computer is not likely to be discrete mathematics.

Exercises Appendix A

1. Choose a specific automobile. Does it have any computers on board? If so, which are digital and which are analog? Among other items, consider the car's clock, speedometer, and odometer.
2. A compact disk player is playing a digital recording of "Swan Lake." Is it a digital computer? Briefly justify your answer.

3. Perform the indicated multiplications using the checkerboard. Indicate in ordinary notation (base 10) what numbers have been multiplied.

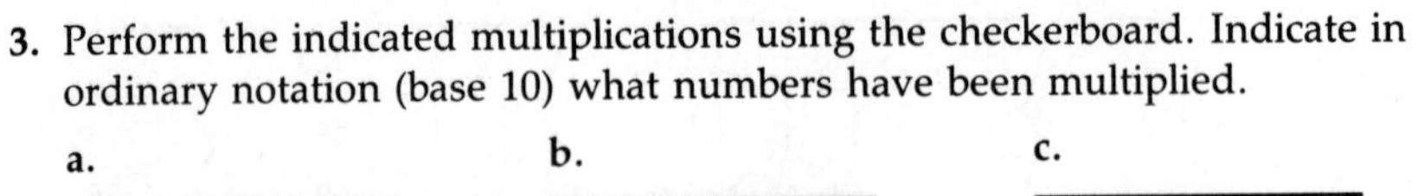

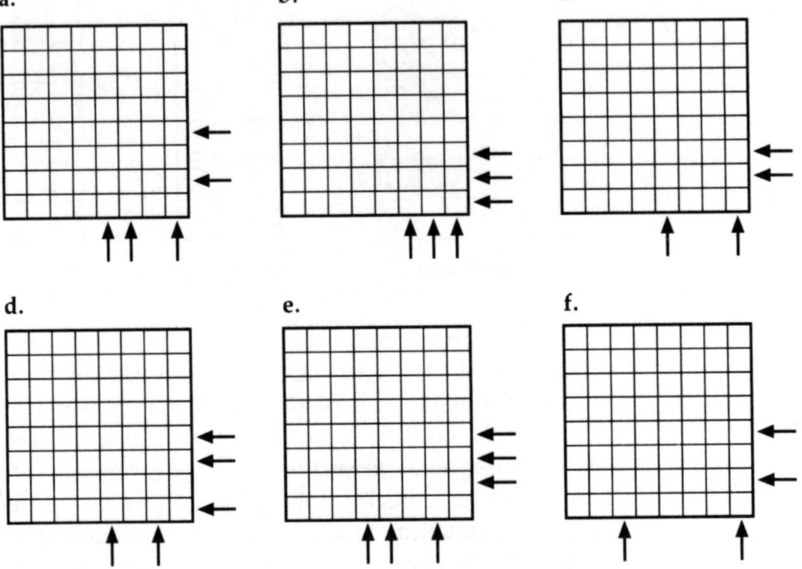

4. Are there two numbers both greater than 1 such that their product on the checkerboard is as follows? If so, indicate two such numbers in checkerboard style. (Remark: It is an unsolved problem to determine all stacks of checkers for which the answer to this question is yes.)

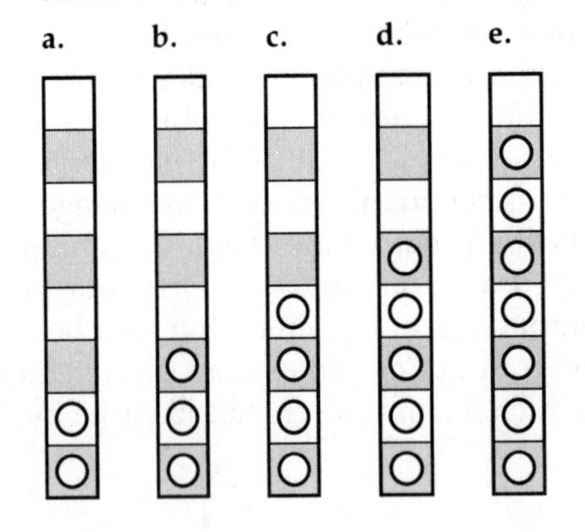

*5. a. What is the largest product that the right-hand column of an 8×8 checkerboard can accommodate?

b. Find two numbers which when multiplied yield this product.

Appendix

B

FLOWCHARTS

Flowcharts have been used for many purposes. Our purpose in using them is to provide a graphic representation of steps to be performed inside a computer. The standard flowchart symbols are shown here:

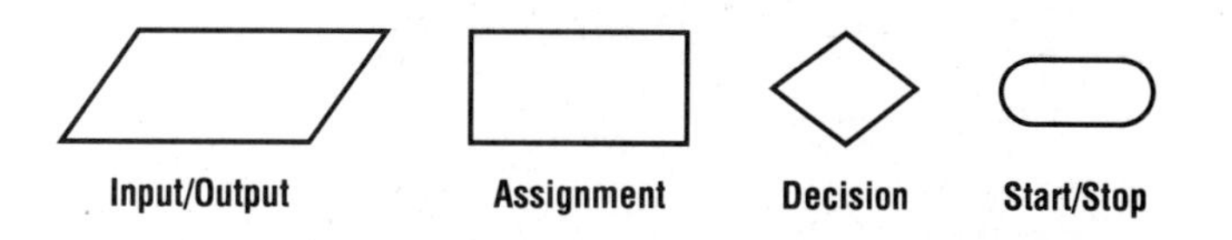

The input/output symbol is used, as the name suggests, to indicate what is fed into the computer (input) and what the computer supplies to the operator (output). Examples of input, output, and assignment are shown here:

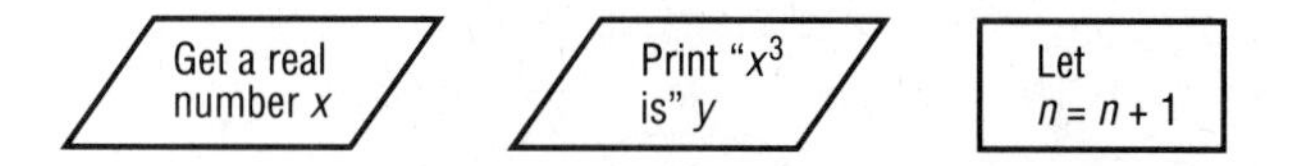

A word of explanation regarding the arithmetic instruction in the assignment box is in order. To the mathematician, $n = n + 1$ is nonsense, and $y = ny$ is not true unless $n = 1$ or $y = 0$. But to the computer scientist this notation means something. In the phrase $n = n + 1$, for example, n is the name of a storage location and $n = n + 1$ means go to storage location labeled n, take the number stored in this location, add 1 to it, and put the result back in the storage location labeled n.

The decision symbol indicates operations of logic or comparison, and its English equivalent structure is IF . . . THEN . . . or IF . . . THEN . . . ELSE . . . . Each decision operation has one entrance path and at least two exit paths, as shown here:

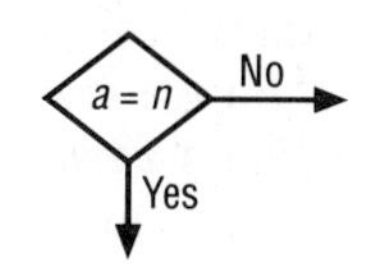

As indicated by this decision symbol, the exit paths are usually determined by a yes or no answer to some question. In a programming language or pseudocode, the statement IF P THEN Q means if P is logically true then perform the action Q, otherwise go to the next statement; whereas IF P THEN Q ELSE R means if P is logically true then perform the action Q, otherwise perform the action R, and in either case go to the next statement. The start/stop symbol is simply used to indicate the beginning and the end of a procedure.

We illustrate the use of these symbols with the simple flowchart in Figure B.1. The arrows in the flowchart tell us the direction. We analyze two cases, $n = 1$ and $n = 4$.

Figure B.1

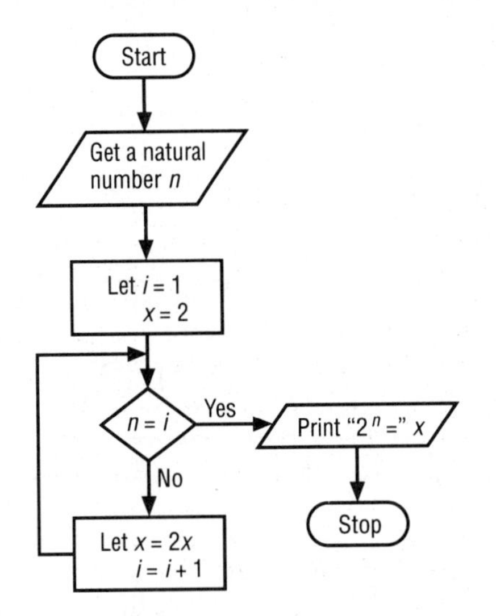

Case 1: $n = 1$. Suppose that as our first step we get the natural number 1. We make the assignments $i = 1$ and $x = 2$. When we reach the decision symbol, we note that indeed n is equal to i (n and i are each 1), so we exit along the Yes arrow and print $2^n = 2$. Then the procedure stops. Note that $2^n = 2$ is true, since $n = 1$.

Case 2: $n = 4$. This time, as our first step we get the natural number 4. Again we make the assignments $i = 1$ and $x = 2$. When we reach the decision symbol, we note that n is not equal to i ($n = 4$ and $i = 1$), so we exit along the No arrow, replace x by $2x$ (so x is now 4), and replace i by $i + 1$ (so i is now 2). We make a loop back to the decision symbol and again note that n is not equal to i. So we exit along the No arrow, replace x by $2x$ (so x is now 8), and replace i by $i + 1$ (so i is now 3). Again we loop back to the decision symbol and note that n is not equal to i. So we continue and exit along the No arrow, replace x by $2x$ (so x

is now 16), and replace i by $i + 1$ (so i is now 4). We loop back to the decision symbol, note with relief that n is finally equal to i, and so exit along the Yes arrow and print $2^n = 16$. Then the procedure stops. Again note that $2^n = 16$ is true since $n = 4$.

Figure B.2

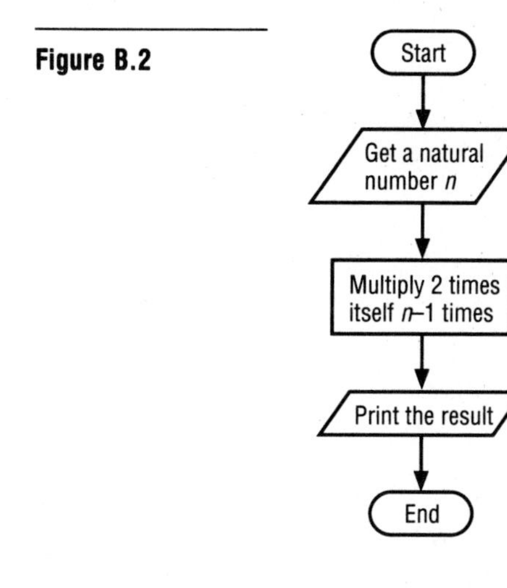

Why did we complicate the problem by drawing a flowchart with a loop rather than drawing the simple flowchart in Figure B.2? The truth is that this simpler flowchart illustrates how the problem would be solved if it were given to a computer programmer. But we wished to illustrate the use of a loop in a simple setting. We can summarize the verbal description for computing 2^4 by using the flowchart in Figure B.1 as follows:

n	i	x
4	1	2
	2	4
	3	8
	4	16
		Print x

Notice that n is 4, the initial value for i is 1, the initial value for x is 2, i increases by 1 at each stage until it reaches n, and x doubles at each stage.

Recall from Section 4.2 that a sequence of numbers in which each term is obtained from the preceding one by adding a constant d is called an *arithmetic progression;* for example, if we begin a sequence with 1 and obtain each term from the preceding one by adding 2, the sequence we obtain is the sequence 1,3,5,. . ., consisting of the odd natural numbers in their usual order. Now let a and d be real numbers. Figure B.3 gives a flowchart that has as input a natural number n and prints the nth term

Figure B.3

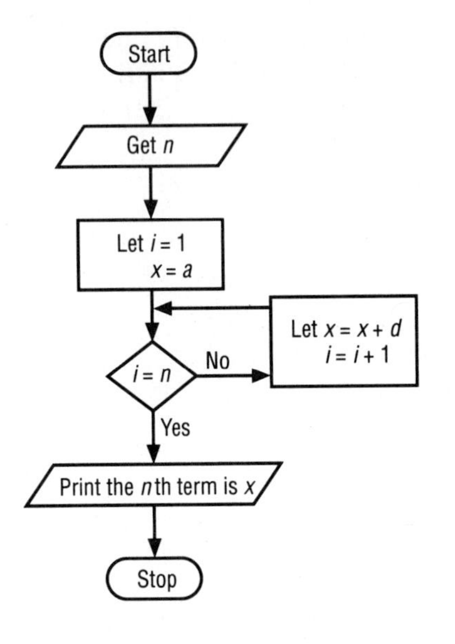

of the arithmetic progression $a, a + d, a + 2d, \ldots$. We analyze this flowchart in the case that $a = 2$, $d = 3$, and $n = 5$. As our first step, we make the assignments $i = 1$ and $x = 2$. When we reach the decision symbol, we note that n is not equal to i ($n = 5$ and $i = 1$). We exit along the No arrow, replace i by $i + 1$, and replace x by $x + d$ (i is now 2 and x is 5). We make the loop back to the decision symbol and again note that n is not equal to i. We exit along the No arrow and again replace i by $i + 1$ and x by $x + d$ (i is now 3 and x is 8). We loop back to the decision symbol and note that n is not equal to i. We continue and exit along the No arrow, replace i by $i + 1$ and replace x by $x + d$ (i is now 4 and x is 11). Yet again we loop back to the decision symbol, note that i is not equal to n, and replace i by $i + 1$ and x by $x + d$ (i is now 5 and x is 14). This time we note that i is equal to n, so we print 14 as the fifth term of the arithmetic progression.

The following chart summarizes the preceding paragraph. Notice that the initial value of i is 1, the initial value of x is $a = 2$, i increases by 1 at each stage until it reaches $n = 5$, and x increases by $d = 3$ at each stage.

$n = 5$ $a = 2$ $d = 3$	i	x
	1	2
	2	5
	3	8
	4	11
	5	14

Print 14

Example 1

Using the intuitive concept of *operation* (namely, that it is either a comparison or a computation), find the number of operations involved in printing the fifth term of the arithmetic progression whose first term is 2 and whose nth term is obtained from the preceding term by adding 3.

Analysis

Our computation is based on the flowchart in Figure B.3. We go around the No loop in the following figure until i is 5; that is, we go around this loop four times.

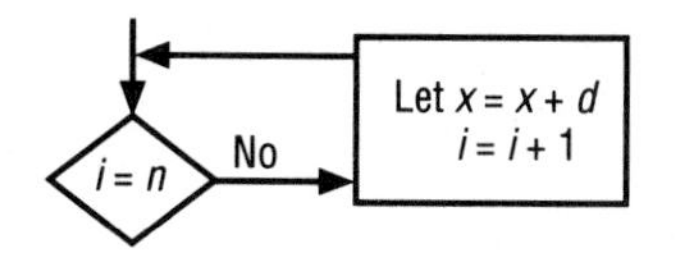

Note that each time we go around the loop there are three operations involved. Then one additional comparison is made. Therefore there are a total of thirteen operations. ■

Exercises Appendix B

1. Find the seventh term of the arithmetic progression 4,2,0, ...
2. Recall from Section 4.2 that a sequence of real numbers in which each term is obtained from the preceding one by multiplying by a constant r is called a *geometric progression*. Let a and r be real numbers and consider the geometric progression $a, ar, ar^2, \ldots$. Draw a flowchart that has as input the natural number n and prints the nth term of this geometric progression.
3. The flowchart in Figure B.1 describes a process to find 2^n when n is a natural number. Using the intuitive concept of *operation* (namely, that it is either a comparison or a computation), find the number of operations involved when $n = 4$ and when $n = 10$.
4. With respect to the flowchart in Figure B.3, find the number of operations involved when $n = 6$ and when $n = 9$.
5. Given the flowchart you drew in Exercise 2, find the number of operations involved when $n = 5$ and when $n = 12$.

Appendix

C

DISCRETE PROBABILITY ON AN INFINITE SET

In this appendix we consider problems of probability in which the sample space, though discrete, is infinite. As you might expect, the switch from finite to infinite can play havoc.

Suppose that the cards of an ordinary bridge deck are turned over one at a time and you are asked to predict the color of each card before it is turned over. The cards are shuffled, and the first 25 cards dealt are either clubs or spades (i.e., all black). Assuming a fair deal, what color do you expect the next card to be? Twenty-seven cards remain, and all but one of them is red. Naturally, you bet on red and win a free trip to Monte Carlo. Congratulations. Once there, you head to the roulette table, where, to your surprise, for 25 straight times the little ball falls into a black slot. If you assume the croupier is honest, what color do you expect on the next play? This is a much harder question than the previous one, so we give you a suggestion. We all know that after say 2,000 spins the ball will very probably have dropped into the red holes nearly 1,000 times, so it has some catching up to do. Just as in the card game, red seems an especially good bet.

But it isn't. The reasoning we have given is fallacious; in fact, this reasoning is called, for the obvious reasons, the *Monte Carlo fallacy*. Our mistake smacks of believing that a sequence that begins 1,2,4,8,16 must continue with 32. But the fallacy is more than this. If, as we should expect, during the next 1,975 spins the ball falls into a red slot roughly half the time, the small disturbance at the beginning will have negligible effect. The catching up we spoke of happens because, when we divide by larger and larger numbers n, $(25 + n/2)/n$ and $(n/2)/n$ $(= 1/2)$ become ever closer values, and there is no need to tamper with the law of averages to make this happen.

In distinguishing between finite and infinite probability, there is also a distinction to be made about the meaning of "probability 0." In finite probability, an event A has probability 0 if and only if $A = \varnothing$; in other words, the only events with probability 0 are impossible. But in infinite probability, a possible (= nonempty) event may be so unlikely that it has probability 0. For example, imagine flipping a coin infinitely many times. The probability that someone could guess each flip correctly is 0, but it could happen just the same. Our example is not very convincing because, after all, you cannot flip a coin infinitely many times. Nicholas Bernoulli (1695–1726) and his brother Daniel Bernoulli (1700–1782), who is Bernoulli of the principle that keeps airplanes up, have a much better example, called the *Petersburg paradox*. Since the Petersburg paradox makes use of the idea of mathematical expectation, we first illustrate this concept.

The *mathematical expectation* of an event is the product of its probability and its payoff. If you are to roll a die and are to be paid 24¢ if you roll a 5, then the mathematical expectation of the event {5} is (1/6)(24¢) = 4¢. You should be willing to pay 4¢ to roll the die, because on average you will pay out 24¢ for six rolls and will win 24¢ once. We illustrate the usefulness of mathematical expectation by analyzing a sucker bet:

Example 1

Three dice are rolled. If one 5 appears, the sucker gets $1.00; if two 5s appear, the sucker gets $1.50; and if all three 5s appear, the sucker gets $6.00. What is the mathematical expectation of this wager?

Analysis

The sample space $I_6 \times I_6 \times I_6$ has $216 = 6^3$ ordered triples. Of these, 75 have exactly one 5, 15 have exactly two 5s, and 1 has all three 5s. The mathematical expectation is

$$\frac{75}{216}(\$1.00) + \frac{15}{216}(\$1.50) + \frac{1}{216}(\$6.00) = \$\left(\frac{207}{432}\right)$$

or about 48¢. But the sucker reasons as follows: The probability of rolling a 5 on the first die is 1/6, on the second die it is 1/6, and on the third die it is 1/6. So I should win $1.00 at least half the time. Therefore I am willing to pay 50¢. ■

We are now ready to consider Nicholas Bernoulli's Petersburg paradox, which illustrates the difference between finite and infinite probability. Here is Bournoulli's wager. We flip a coin. If it comes up heads, you win $2.00 and are allowed to play again. If the second flip is heads, you win $4.00 *more* and are allowed to play again. In general, on the nth flip you win 2^n *more* dollars if the flip is heads. As soon as the coin comes up tails, you may keep what you have won, but the game is over. Clearly, you should be willing to pay something to play this game. How much? In mathematical parlance, what is the mathematical expectation?

This is a hard question, so we give a suggestion. Let us consider a *finite* version of Bernoulli's wager in which the game is over after ten flips no matter what. Then the mathematical expectation is

$$\frac{1}{2}(\$2) + \frac{1}{4}(\$4) + \frac{1}{8}(\$8) + \cdots + \frac{1}{2^{10}}(\$2^{10}) = \$10$$

We hope our suggestion has been helpful.

Appendix D

CHANGING BASES

Let b be a natural number greater than 1. Then each natural number n can be written uniquely in the form

$$n = a_k b^k + a_{k-1} b^{k-1} + \cdots + a_1 b + a_0$$

where $a_k \neq 0$ and $a_0, a_1, \ldots, a_k$ are all nonnegative integers less than b. The numeral $(a_k a_{k-1} \ldots a_1 a_0)_b$ denotes this representation of n and is called the *standard base b expansion of n*. The standard base 2 expansion of n is called the *binary expansion* of n, and the standard base 10 expansion of n is called the *decimal expansion* of n. When n is denoted by $(a_k a_{k-1} \ldots a_1 a_0)_b$, n is said to be *written in base b*. We often write $a_k a_{k-1} \ldots a_1 a_0$ in place of $(a_k a_{k-1} \ldots a_1 a_0)_{10}$, and if no base is indicated the numeral is presumed to be a decimal expansion, unless another base is clear from context. In particular, the number b indicating the base of a standard base b expansion is always written in its decimal expansion. We do not prove that each natural number n has a unique standard base b expansion for each $b > 1$; instead, we illustrate a procedure for writing a natural number n in base b which makes repeated use of the division algorithm.

Example 1

Write 1042 in base 8.

Analysis

From the division algorithm, we have the following equations:

$$1042 = 8(130) + 2$$

$$130 = 8(16) + 2$$

$$16 = 8(2) + 0$$

$$2 = 8(0) + 2$$

Therefore

$$\begin{aligned} 1042 &= 8(130) + 2 \\ &= 8[8(16) + 2] + 2 \end{aligned}$$

$$= 8\{8[8(2) + 0] + 2\} + 2$$
$$= 2 \cdot 8^3 + 0 \cdot 8^2 + 2 \cdot 8^1 + 2 \cdot 8^0$$

By definition, $1042 = (2022)_8$. ■

Notice that the base 8 expansion of 1042 in Example 1 is formed by writing the remainders from the division algorithm in the reverse order in which they are obtained. We can make use of this fact to simplify the procedure for finding a base b representation of a number that is written in its decimal representation.

Example 2

Write 1042 in base 8.

Analysis

8 ⌊ 1042		
8 ⌊ 130	Remainder 2	↑
8 ⌊ 16	Remainder 2	
8 ⌊ 2	Remainder 0	
0	Remainder 2	Read up

$1042 = (2022)_8$ ■

Given a number in its base b expansion, we have a simple procedure for determining its decimal expansion.

Example 3

Write $(2022)_8$ in base 10.

Analysis

By definition, $(2022)_8$ denotes

$$2(8^3) + 0(8^2) + 2(8^1) + 2(8^0) = 1024 + 0 + 16 + 2 = 1042$$ ■

As we mentioned previously, for those working in discrete mathematics the most important base is base 2, and it is worthwhile becoming adept at calculating in this base. We begin modestly by counting to fifteen.

1 one	110 six	1011 eleven
10 two	111 seven	1100 twelve
11 three	1000 eight	1101 thirteen
100 four	1001 nine	1110 fourteen
101 five	1010 ten	1111 fifteen

Addition and multiplication tables for the two one-digit binary numbers 0 and 1 are given by

+	0	1
0	0	1
1	1	10

x	0	1
0	0	0
1	0	1

Using these tables, we can add and multiply any two binary numbers as in Example 4.

Example 4

$$\begin{array}{r} 10110 \\ +11011 \\ \hline 110001 \end{array} \qquad \begin{array}{r} 10110 \\ \times\ 101 \\ \hline 10110 \\ 00000 \\ 10110 \\ \hline 1101110 \end{array}$$

■

With a little practice, it is easy to read off the decimal expansion of a number that is written in base 2. We give all the particulars in the examples that follow, but of course the idea is to do the calculations in your head.

Example 5

Write $(1010001010)_2$ in its decimal expansion.

Analysis

$$\begin{aligned}(1010001010)_2 &= (\underline{1010}000000)_2 + (1010)_2 \\ &= (10 \times 64) + 10 \\ &= 650\end{aligned}$$

■

Example 6

Write $(110010011010)_2$ in its decimal expansion.

Analysis

$$\begin{aligned}(110010011010)_2 &= (\underline{11}0000000000)_2 + (\underline{1001}0000)_2 + (1010)_2 \\ &= (3)(2^{10}) + 9(2^4) + 10 \\ &= 3(1024) + 9(16) + 10 \\ &= 3072 + 144 + 10 \\ &= 3226\end{aligned}$$

■

There is nothing wrong with determining the decimal expansion of numbers such as those of Examples 4 and 5, but most people would need paper and pencil.

For any base b, we need b one-digit numerals to write all the number representations in that base. As long as $1 < b \leq 10$, we can use the familiar base 10 numerals. If $b > 10$, we must invent numerals. Luckily there is only one base greater than 10 that is of interest, and that is base 16, which requires 16 one-digit numerals. Usually these numerals are taken to be 0,1,2,3,4,5,6,7,8,9,A,B,C,D,E, and F, where the numerals A through F represent the numbers 10 through 15. Base 16 numerals are of interest because binary numerals, especially 8-digit and 16-digit binary numerals which occur naturally in computer science, can easily be written in base 16. For the same reason, base 8 numerals are also useful. In the terminology of computer science, a *bit* is a single on–off switch, a *byte* is eight bits, and a *word* is 16 bits. It is common to let "on" be represented by the number 1 and "off" by the number 0. The term *bit* is used to denote both a single on–off switch and the one-digit binary numeral 0 or 1 that represents the state of that switch. Thus a bit can be 0 or 1, a byte can be any binary numeral from 00000000 (base 10 number 0) to 11111111 (base 10 number 255), and a word can be any binary numeral from 0000000000000000 (base 10 number 0) to 1111111111111111 (base 10 number 65535). Most computers have byte- or word-size memory locations. As you can see, 16-digit binary numerals are hard to write and hard to distinguish. The difference between 1011111100000100 and 1011111100001000 is not apparent unless these numerals are lined up directly underneath each other. When the base is small, the addition and multiplication tables are simple, but the names of the numbers are unwieldy. The computer is right at home with the binary system; it never forgets a 16-digit binary numeral and it has no difficulty in distinguishing between two such numerals. But humans have a hard time considering even relatively small binary numerals.

Since it is hard for us to deal with binary numerals, we use larger bases to talk about the contents of bytes and words in the computer. This is where octal and hexadecimal come into play. *Octal* means base 8, and *hexadecimal* (frequently just called *hex*) means base 16. Table D.1 lists the first 17 numbers in the decimal, octal, hex, and binary systems.

Notice that the rules to switch from base 16 to base 10 are somewhat different from the rules we have had for switching bases so far. Take the numeral AB9C in hex. This stands for

$$(A \times 10^3) + (B \times 10^2) + (9 \times 10) + C$$

in hex. To find the number in base 10 (our usual decimal system), not only do we need to replace 10 by 16, we also need to replace each of the letters by their appropriate decimal equivalent. Hence we need to calculate

$$[10 \times (16)^3] + [11 \times (16)^2] + [9 \times 16] + 12$$

which is 43,932 in base 10.

Table D.1

Decimal	Octal	Hex	Binary
0	0	0	0
1	1	1	1
2	2	2	10
3	3	3	11
4	4	4	100
5	5	5	101
6	6	6	110
7	7	7	111
8	10	8	1000
9	11	9	1001
10	12	A	1010
11	13	B	1011
12	14	C	1100
13	15	D	1101
14	16	E	1110
15	17	F	1111
16	20	10	10000
17	21	11	10001

Switching from base 10 to hex also requires a slight modification. When using the "divide-by-16" process, any remainders that are ten or larger must be changed to their hex single-digit names.

Changing from Decimal to Hex

Start with 3128 in base 10.

Division	Remainder	Hex digits
16 ⌊ 3128		
16 ⌊ 195	8	8
16 ⌊ 12	3	3
0	12 ⟶	C

$3128_{10} = C38_{16}$.

It is remarkably easy to switch back and forth between base 8 and base 2. To switch from octal to binary, simply replace each octal digit with its binary equivalent from Table D.1. For example, 725 in octal is

111 010 101 in binary. To switch from binary to octal, group the digits in sets of three, starting at the right-most digit of the numeral. If you need an extra digit or two at the left end of the numeral, add one or two zeros. Now replace each group of three binary digits with its equivalent octal digit from Table D.1.

Changing from Binary to Octal

Start with the binary numeral 10001110.

Group in threes.	10	001	110
Add 0s to the left.	010	001	110
Change to octal digits.	2	1	6

$10001110_2 = 216_8$.

The ease in switching back and forth between binary and octal is worth the trouble of using an unfamiliar base.

The relationship between base 2 and base 16 is similar to the relationship between base 2 and base 8. To switch from hex to binary, simply replace each hex digit with its binary equivalent from Table D.1. For example, AB9 in hex is 1010 1011 1001 in binary. To switch from binary to hex, group the digits in sets of four, starting at the right-most digit of the numeral. If you need an extra digit or so at the left end of the numeral, add one, two, or three zeros. Now replace each group of four binary digits with its equivalent hex digit from Table D.1.

Changing from Binary to Hex

Start with the binary numeral 1101010111011.

Group in fours.	1	1010	1011	1011
Add 0s to the left.	0001	1010	1011	1011
Change to hex digits.	1	A	B	B

$1101010111011_2 = 1ABB_{16}$.

As you can see, it is just as easy to switch back and forth between binary and hex as it is to switch between binary and octal.

Both the octal and hex systems are used extensively to describe memory contents and addresses within a computer. The octal system was the main system used when the electronic digital computer was first introduced, because this system was easier to explain than hex. In the past ten years, the hex system has prevailed.

It turns out that there is a small problem with using the octal system to describe memory contents and addresses. The easiest way to see this problem is to look at a memory address that is word size and so consists of two 8-bit bytes. The problem arises in changing the address to octal. There are two ways to proceed. One way to change each 8-bit byte to octal is to place the two octal numbers side by side and separate them with a ".". This is called *split octal* notation. The other method is to change the entire 16 bits to octal in one step and give the usual octal notation. These two methods of changing an address to octal notation give different results. If we use hex notation no such problem arises, because we change four bits at a time instead of three, and 4 divides 8 and 16 but 3 does not. Example 7 provides a comparison of hex with octal and split octal.

Example 7

Write the number expressed as the binary numeral 1011110111111010 in hex, octal, and split octal.

Analysis

a. Change to hex.

1011 1101 1111 1010

B D F A

The answer in hex is BDFA.

b. Change to octal.

1 011 110 111 111 010

001 011 110 111 111 010

1 3 6 7 7 2

The answer in octal is 136772.

c. Change to split octal.

10 111 101 . 11 111 010

010 111 101 . 011 111 010

2 7 5 . 3 7 2

The answer in split octal is 275.372. ■

Because of the discrepancy between octal and split octal, hex has become more popular in recent years, even though there is some work involved in defining the digits A, B, C, D, E, and F.

In this appendix we have concentrated on bases 2, 8, and 16. We have a primary interest in base 2, but our interest in bases 8 and 16 is derived solely from the ease with which we can switch back and forth between these bases and base 2 and the fact that long base 2 numbers are hard for computer programmers to handle. It may seem reasonable to ask if we can change arbitrary bases, say, from base 9 to base 7. Of course, the answer is yes, but such an exercise is of no interest to those who work with computers.

Exercises Appendix D

1. Find the binary, decimal, octal, split octal, and hex representation of each of the following numbers.

   a. 1011011 in binary **b.** 3727 in octal

   c. 459A in hex **d.** 9812 in decimal

   e. 177777 in octal **f.** 111.111 in split octal

2. Each of the numbers in this exercise is in binary. Perform each of the indicated operations in binary.

   a. $\begin{array}{r} 11011 \\ +11111 \\ \hline \end{array}$ **b.** $\begin{array}{r} 1101 \\ \times 1011 \\ \hline \end{array}$ **c.** $\begin{array}{r} 11000000 \\ -1101111 \\ \hline \end{array}$ **d.** $1101 \overline{)1000001}$ **e.** $\begin{array}{r} 1101 \\ 1110 \\ 1011 \\ 1111 \\ +1001 \\ \hline \end{array}$

3. Each of the numbers in this problem is in octal. Perform each of the indicated operations in octal.

   a. $\begin{array}{r} 777 \\ +65 \\ \hline \end{array}$ **b.** $\begin{array}{r} 672 \\ \times 600 \\ \hline \end{array}$ **c.** $\begin{array}{r} 3307 \\ -670 \\ \hline \end{array}$ **d.** $76 \overline{)37174}$ **e.** $(55 \times 43) + (26 \times 43)$

4. Each of the numbers in this problem is in hex. Perform each of the indicated operations in hex.

   a. $\begin{array}{r} AB36 \\ +C49D \\ \hline \end{array}$ **b.** $\begin{array}{r} D4F \\ \times AF \\ \hline \end{array}$ **c.** $\begin{array}{r} 3759 \\ -979 \\ \hline \end{array}$

5. Write the addition and multiplication tables for octal notation.

6. **a.** Using the method of Example 2, write 43 in base 2.

   b. Multiply 101101_2 by 101101_2 using binary arithmetic.

Appendix E

NUMERATION OF RATIONAL AND IRRATIONAL NUMBERS

In Appendix D we consider the use of binary notation to denote integers, but like the decimal system the binary system can also be used to denote real numbers. Although there is a general theory that pertains to numerals of an arbitrary base b, in this section we consider only decimal and binary expansions of real numbers. We begin by considering our familiar decimal system.

The easiest decimal numbers to understand, and the only decimal numbers that a computer can fathom, are those with finitely many digits. These numbers are called *terminating decimals.* Any terminating decimal can be written in *scientific notation.* A terminating decimal is said to be written in scientific notation if it is written in the form $a \times 10^p$, where a is a terminating decimal, $1 \leq |a| < 10$, and p is an integer. We can write a terminating decimal in this form as follows:

1. Beginning on the left, locate the first non-zero digit; put the decimal point immediately to the right of this digit.
2. Let p denote the number of places the decimal point must be moved to put it back in its original position.
3. Multiply the number in (1) by 10^p if the decimal point must be moved to the right to put it back in its original position and by 10^{-p} if it must be moved to the left to put it back in its original position.

Example 1

We write 36,700,000 in scientific notation by first placing the decimal point immediately to the right of 3: 3.6700000. We must move the decimal point seven places to the right to put it back in its original position, so the scientific notation for 36,700,000 is 3.67×10^7. ■

Example 2

We write 0.000285 in scientific notation by first placing the decimal point immediately to the right of 2: 2.85. We must move the decimal point four places to the left to put it back in its original position, so the scientific notation for 0.000285 is 2.85×10^{-4}. ■

Recall that a positive fraction a/b is an indicated division of an integer a by an integer b, where either both a and b are positive or both a and b are negative. Each positive terminating decimal represents infinitely many different positive fractions, and fractions convey more information than their corresponding decimal representations. For example, it is one thing to sport a .750 batting average after the first game of the season, but that same average is a lot more impressive after 80 times at bat. Recall (see Example 14 of Chapter 10) that two fractions a/b and c/d are equivalent (written $a/b \sim c/d$) provided that $ad = bc$. Two equivalent fractions have the same decimal representation, and $\sim$ is an equivalence relation on the set of all fractions (see Exercise 36 of Chapter 10). As we soon prove, some fractions, such as 1/3, have no terminating decimal representations, but the fault with these fractions arises only because there is no power of 10 that their denominator divides. In base 3, the terminating decimal 0.1_3 when written in scientific notation is $1. \times 3^{-1}$, which is 1/3.

If n is a natural number and c is a natural number that is not divisible by 10 and is greater than 10^n, then there exists a nonnegative integer k such that

$$c/10^n = a_k 10^k + a_{k-1}10^{k-1} + \cdots + a_0 10^0 + b_1 10^{-1} + b_2 10^{-2} + \cdots + b_n 10^{-n}$$

where each a_i and b_j is one of the digits from 0 to 9. Hence $c/10^n$ can be expressed as the terminating decimal $a_k a_{k-1} \ldots a_0.b_1 b_2 \ldots b_n$.

Example 3

Using the preceding terminology, let $c = 8347$ and $n = 3$. Then $c > 10^n$ and the fraction

$$\begin{aligned} 8347/10^3 &= 8000/1000 + 300/1000 + 40/1000 + 7/1000 \\ &= 8 + 3/10 + 4/100 + 7/1000 \\ &= 8 + 3(1/10) + 4(1/100) + 7(1/1000) \\ &= (8 \times 10^0) + (3 \times 10^{-1}) + (4 \times 10^{-2}) + (7 \times 10^{-3}) \\ &= 8.347 \end{aligned}$$

■

If n is a natural number and c is a natural number that is not divisible by 10 and is less than 10^n, then

$$c/10^n = b_1 10^{-1} + b_2 10^{-2} + \cdots + b_n 10^{-n}$$

where each b_i is one of the digits from 0 to 9. Hence $c/10^n$ can be expressed as the terminating decimal $0.b_1b_2 \ldots b_n$.

Example 4

Using the preceding terminology, let $c = 346$ and $n = 5$. Then $c < 10^n$ and the fraction

$$\begin{aligned}346/10^5 &= 300/100000 + 40/100000 + 6/100000\\ &= 3/1000 + 4/10000 + 6/100000\\ &= 0(1/10) + 0(1/100) + 3(1/1000) + 4(1/10000) + 6(1/100000)\\ &= (0 \times 10^{-1}) + (0 \times 10^{-2}) + (3 \times 10^{-3}) + (4 \times 10^{-4})\\ &\quad + (6 \times 10^{-5})\\ &= 0.00346\end{aligned}$$

■

If a/b is any fraction that is equivalent to $c/10^n$ for some natural numbers c and n, then the procedures illustrated in Examples 3 and 4 can be used to write a/b in decimal notation.

Example 5

Write 7/40 in decimal notation.

Analysis

We first find an equivalent fraction whose denominator is a power of 10. Since $7/40 = (7/40)(m/m)$, we need a natural number m such that $40m$ is a power of 10. Since

$$40m = 2 \cdot 2 \cdot 2 \cdot 5m = 2 \cdot 2 \cdot 10m$$

we can obtain a power of 10 in the denominator by replacing m by $5 \cdot 5 = 25$. So we have

$$\begin{aligned}7/40 &\sim (7 \cdot 25)/(40 \cdot 25) = 175/1000\\ &= 100/1000 + 70/1000 + 5/1000\\ &\sim 1/10 + 7/100 + 5/1000\\ &= 0.175\end{aligned}$$

■

Example 6

Write 3/250 in decimal notation.

Analysis

We use the same procedure as in Example 5. We first find a natural number m such that $250m$ is a power of 10. Since

$$250m = 2 \cdot 5 \cdot 5 \cdot 5m = 10 \cdot 5 \cdot 5m$$

we can obtain a power of 10 by replacing m by $2 \cdot 2 = 4$. So we have

$$\begin{aligned} 3/250 &\sim (3 \cdot 4)/(4 \cdot 250) = 12/1000 \\ &= 10/1000 + 2/1000 \\ &\sim 1/100 + 2/1000 \\ &= 0.012 \end{aligned}$$

■

Examples 5 and 6 suggest the following theorem.

Theorem E.1

> If a/b is a nonnegative fraction in lowest terms and b does not have any prime factors except 2 or 5, then the decimal expansion of a/b terminates.

Proof

It is sufficient to prove that a/b is equivalent to a fraction whose denominator is a power of 10. We are given that there exist nonnegative integers m and n such that $a/b = a/(2^m \cdot 5^n)$. Therefore

$$\begin{aligned} a/b &\sim (a \cdot 2^n \cdot 5^m)/(2^m \cdot 5^n \cdot 2^n \cdot 5^m) \\ &= (a \cdot 2^n \cdot 5^m)/(2^{m+n} \cdot 5^{m+n}) \\ &= (a \cdot 2^n \cdot 5^m)/10^{m+n} \end{aligned}$$

□

Example 7

We prove that $1/3$ cannot be expressed as a terminating decimal by proving that there do not exist natural numbers c and n such that $1/3 \sim c/10^n$. The proof is by contradiction. Suppose that there are natural numbers c and n such that $1/3 \sim c/10^n$. Then $1 \cdot 10^n = 3c$. Now 3 divides $3c$ and hence 3 divides 10^n. This contradicts the fundamental theorem of arithmetic, since $10^n = 2^n \cdot 5^n$. ■

The proof of the following theorem, which we leave as an exercise, is similar to the preceding proof.

Theorem E.2

> If a/b is a nonnegative fraction in lowest terms and b has a prime factor different from 2 and 5, then a/b cannot be expressed as a terminating decimal.

It is a relatively simple matter to find a fraction represented by a given terminating decimal. Finding such a fraction is called *converting* a decimal to a fraction.

Example 8

Convert 0.3714 to a fraction.

Analysis

By definition

$$\begin{aligned} 0.3714 &= 3(1/10) + 7(1/100) + 1(1/1000) + 4(1/10000) \\ &= 3/10 + 7/100 + 1/1000 + 4/10000 \\ &= (3000 + 700 + 10 + 4)/10000 \\ &= 3714/10000 \end{aligned}$$

As we have already noted, each decimal represents many different fractions, and the choice of 3714/10000 rather than 1857/5000 or any other equivalent fraction is simply a matter of taste. ■

We now illustrate how to find a decimal representation of those fractions that cannot be expressed as a terminating decimal. Consider, for example, the fraction 11/14, which according to Theorem E.2 cannot be expressed as a terminating decimal:

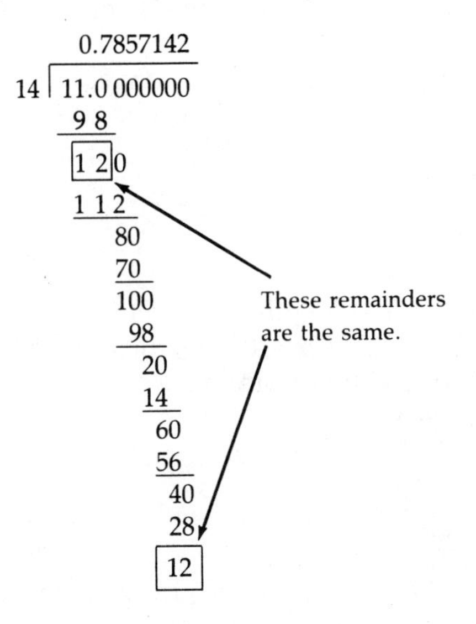

When a remainder is repeated, the decimal repeats because the same succession of digits occurs over and over. To indicate that the digits continue to repeat, the decimal representation is written with a bar over the repeating part:

$$11/14 = 0.7\overline{857142}$$

In this illustration the divisor is 14, so at each stage in the division the remainder is a member of $\{n \in \mathbb{Z}: 0 \leq n \leq 13\}$. If the remainder is 0, the process ends and the number is a terminating decimal. If the remainder is never 0, then there are only thirteen possibilities, and hence the repetition must start after twelve divisions. The number is twelve (rather than thirteen) because as far as the repetition is concerned we have a remainder of 11 before we start. A similar argument can be used to prove the following theorem.

Theorem E.3

> If the decimal that represents a nonnegative fraction does not terminate, it must repeat. If a/b is a nonnegative fraction in lowest terms whose decimal representation does not terminate, then the maximum length of the repeating cycle is $b - 1$.

The maximum length determined in Theorem E.3 is actually attained by some fractions, for example, $1/7 = 0.\overline{142857}$, which is of length 6, and $1/17 = 0.\overline{0588235294117647}$, which is of length 16. It is an unsolved problem, first posed by Gauss, whether there are infinitely many such fractions. The corresponding repeating decimals of such fractions have remarkable properties (see Exercise 4).

We have already seen how to convert a terminating decimal to a fraction. A repeating decimal can also be converted to a fraction.

Example 9

Convert $0.\overline{729}$ to a fraction.

Analysis

Let $x = 0.\overline{729}$. The length of the repeating cycle is 3, so we multiply the equation $x = 0.\overline{729}$ by 10^3 and obtain $1000x = 729.\overline{729}$. Now we subtract the first equation from the second

$$\begin{aligned} 1000x &= 729.\overline{729} \\ x &= 0.\overline{729} \\ 999x &= 729 \end{aligned}$$

Therefore $x = 729/999$. Other, equivalent fractions, such as $81/111$ or $27/37$, are also represented by the repeating decimal $0.\overline{729}$. ■

Example 10

Convert $6.325\overline{4817}$ to a fraction.

Analysis

Let $x = 6.325\overline{4817}$. Since the length of the repeating cycle is 4, we multiply the equation by 10^4 and obtain $10000x = 63254.817\overline{4817}$. Subtracting the first equation from the second, we obtain

$$10000x = 63254.8174\overline{817}$$
$$x = 6.3254\overline{817}$$
$$9999x = 63248.492$$

Since there are three digits to the right of the decimal, we multiply the equation by 10^3 and obtain $9999000x = 63248492$. Therefore $x = 63248492/9999000$. ■

We have seen that any repeating or terminating decimal represents a fraction and that any fraction can be represented by such a decimal. Thus the rational numbers are the decimals that repeat or terminate and the irrational numbers are the decimals that neither terminate nor repeat. Here is the beginning of an irrational number: 0.102003000400005 . . . . If you have guessed correctly the decimal that we have in mind, it should be clear that this decimal is irrational; the decimal we have in mind does not terminate and it cannot repeat, since each segment has one more zero than the previous segment.

So much for base 10; what about base 2? The theorems for this base are much the same as the theorems we have already established for base 10, and, while we admit that working in a less familiar base might prove more difficult, the required mathematical arguments remain the same. We therefore just state the results corresponding to Theorems E.1, E.2, and E.3.

Theorem E.4

If a/b is a nonnegative fraction in lowest terms and b has no prime factor other than 2 (that is, b is a power of 2), then the binary expansion of a/b terminates.

Theorem E.5

If a/b is a nonnegative fraction in lowest terms and b has a prime factor other than 2 (that is, b is not a power of 2), then the binary expansion of a/b does not terminate.

Theorem E.6

If the binary expansion of a nonnegative fraction does not terminate, it must repeat. If a/b is a nonnegative fraction in lowest terms whose binary expansion does not terminate, then the maximum length of the repeating cycle is $b - 1$.

The following examples illustrate the previous three theorems.

Example 11

$101_2/100000_2$ (alias 5/32) is a fraction whose denominator is a power of 2. According to Theorem E.4, the binary expansion of this fraction terminates.

The binary expansion of $101_2/100000_2$ is obtained by moving the "decimal" point of $101._2$ five places to the left. Thus $101_2/100000_2$ is 0.00101_2, which obviously terminates. ■

In the remaining examples of this section we work entirely in base 2, and throughout these examples we omit the subscript 2 that indicates the use of binary numerals.

Example 12

Convert 0.1101 to a binary fraction.

Analysis

$$\begin{aligned} 0.1101 &= 1(1/10) + 1(1/100) + 0(1/1000) + 1(1/10000) \\ &= 1/10 + 1/100 + 1/10000 \\ &= 1000/10000 + 100/10000 + 1/10000 \\ &= 1101/10000 \end{aligned}$$

■

Example 13

Find the binary expansion of 1011/1110 (alias 11/14).

Analysis

According to Theorem E.5 the binary expansion of this fraction does not terminate, and so according to Theorem E.6 the repeating cycle of the binary expansion has thirteen or fewer digits. In fact 1011/1110 $= 0.11\overline{001}$. ■

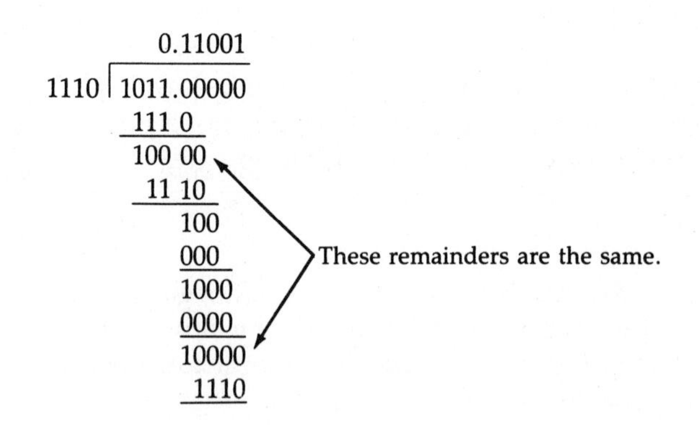

■

Example 14

Convert $0.11\overline{001}$ to a fraction.

Analysis

Let $x = 0.11\overline{001}$. Then

$$
\begin{aligned}
1000x &= 110.01\overline{001} \\
x &= 0.11\overline{001} \\
111x &= 101.10
\end{aligned}
$$

Therefore $11100x = 10110$ and $x = 10110/11100$. Of course, $1011/1110$ is a fraction that is equivalent to $10110/11100$. ■

In Exercise 10, we ask you to convert binary numerals having a binary point to base 10 numbers having a decimal point. The easiest way to do this is first to write the binary numerals as binary fractions and then to convert the numerators and denominators of these fractions to base 10 numerals. For example, $0.11\overline{001}_2 = 1011_2/1110_2 = 11/14 = 0.7\overline{857142}$.

Exercises Appendix E

1. Write each of the following numbers in scientific notation.
   a. 35092 **b.** 0.0006431 **c.** 0.304002
   d. 35.092 **e.** 3040.02 **f.** 304.00200
2. Which of the following fractions have terminating decimal expansions?
   a. $7/16$ **b.** $8/15$ **c.** $3/15$ **d.** $1/10000550$
3. **a.** Is $0.\overline{9}$ less than 1? **b.** Does $3 \times 0.\overline{3} = 0.\overline{9}$? **c.** Does $3 \times 0.\overline{3} = 1$?
4. As we have already observed, $1/17 = 0.\overline{0588235294117647}$ and $1/7 = 0.\overline{142857}$.
   a. Write $16/17$ as a repeating decimal. (*Hint:* Use Exercise 3 and the equation $16 + 1 = 17$.)
   b. Write $6/7$ as a repeating decimal.
   c. Multiply the positive integer 142857 by 2, by 4, and by 24.
   d. Explain what is unusual about the way the integer 588235294117647 behaves under multiplication.
5. **a.** If a fraction has a terminating binary expansion, must it also have a terminating octal expansion?
   b. If a fraction has a terminating octal expansion, must it also have a terminating binary expansion?
6. Convert each of the following decimals to fractions.
   a. $2.3\overline{4}$ **b.** $0.\overline{24}$ **c.** $3.13\overline{12}$
   d. $5.48\overline{163}$ **e.** $0.3456\overline{651}$ **f.** $8.3\overline{45167}$
7. Prove that, if a/b is a nonnegative fraction in lowest terms and b has a prime factor different from 2 and 5, then a/b cannot be expressed as a terminating decimal.

8. Find the binary expansions of each of the following binary fractions.
   a. 101/1001 b. 10010/101 c. 1101/110
9. Convert each of the following to a binary fraction.
   a. $0.0110\overline{1}_2$ b. $101.01\overline{101}_2$ c. $0.\overline{01100}_2$ d. $0.\overline{1101}_2$
10. Convert each of the following binary numerals to base 10 numerals.
    a. 0.1111 b. 0.1101 c. $0.\overline{1111}$
    d. $0.\overline{1101}$ e. 101.101 f. 10.0001

*11. We have already observed that the collection of all rational numbers is closed under both addition and multiplication.
    a. Is the collection of all irrational numbers closed under addition?
    b. Is the collection of all irrational numbers closed under multiplication?
    Demonstrate that your answers are correct.

Appendix

F

COMPUTER ARITHMETIC

In this appendix we discuss some problems one encounters in performing arithmetic operations on a computer. Of course, before arithmetic operations can be performed, numbers must be entered into a computer. In some computer languages, Pascal for example, variables are given data types, and if a variable X is declared to be of integer data type, then X can be assigned the value 7 but cannot be assigned the value 7. or 7.0 since decimal points indicate that the data type is real numbers.

We have seen how to write numbers in scientific notation in Appendix E. Many computer languages, including FORTRAN, allow real numbers to be printed in scientific notation; for example, 3.478 E 05 means 3.478×10^5. A *normalized number* in exponential notation is the same as the number in scientific notation except that the decimal point immediately precedes the first non-zero digit of the number. Thus 0.2368×10^3 is the normalized form of 236.8, whereas 2.368×10^2 is the scientific notation for this number. The numeral 3 is the *exponent*, and the numeral 2368 is the *mantissa*. (The word *mantissa*, which is one of the few words in the English language of Celtic origin, first meant an addition of trifling value.)

The use of *floating point* storage of numbers in the computer is similar to the use of scientific notation. The major differences involve using base 2 rather than base 10 and storing only what is needed. Suppose, for example, that we want to use floating point storage to store the number $26\frac{2}{3}$. The number is first converted to base 2. Using the techniques already described in this book, we see that $26\frac{2}{3}$ is approximately 11010.101010101010. In base 2, a normalized form of this number is $0.1101010101010101 \times 10^{101}$. It is customary to allow a byte for the exponent and its sign and two bytes for the mantissa and its sign. Since all binary numerals in normalized form begin with 0.1, the first 1 need not be stored, and some computers thereby gain an extra bit of storage space. In any case, a 0 in the first bit means that the number is nonnegative and a 1 in the first bit means that the number is negative. This choice of meanings may appear arbitrary or even backward, but as we see in this section the choice is natural and is motivated by addition in $\mathbb{Z}_n$. Since the exponent indicates the size of a number, it is customary to store the exponent first. Thus $26\frac{2}{3}$ is stored as follows:

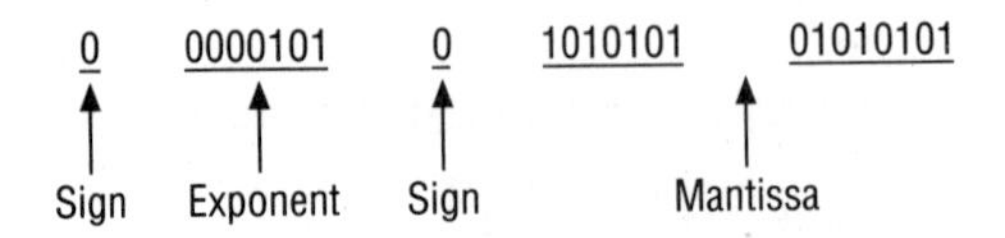

Note that the decimal numbers stored in these three bytes are 5, 85, and 85. Base 10 numbers with as many as 38 digits can be stored in this manner, and these numbers are accurate up to about five decimal places. Doubling the number of digits stored from 16 to 32, usually referred to as *double precision arithmetic,* increases the accuracy to ten decimal places.

When a number is written in normalized form, the digits of the mantissa are called *significant digits*. The significant digits of 00347800 are 3,4,7, and 8, since the normalized form of 00347800 is 0.3478×10^6. The significant digits of 07.040200 are 7,0,4,0, and 2, since the normalized form of 07.040200 is 0.70402×10^1. When a number is written in normalized form, the number of significant digits is called its *precision,* and the power of 10 that multiplies the mantissa is called its *magnitude*. The precision of 0.002900 is 2 and its magnitude is 10^{-2}.

Suppose that the answer to an arithmetic problem is 0.341278×10^5, but also that the computer we are using has a precision of 4. In this case, the computer usually *truncates,* or chops off, the extra digits and gives 0.3412×10^5. Thus an error of 0.000078×10^5 has been introduced. This is the *absolute error,* or difference between the result and the correct answer. The *relative error* is the ratio of the absolute error and the correct answer; in this case, the relative error is $(0.000078 \times 10^5)/34127.8 \approx 0.2285 \times 10^{-3}$.

In the following discussion, we use the symbols $+$, $-$, $*$, $/$, and $\wedge$ to denote the arithmetic operations of addition, subtraction, multiplication, division, and exponentiation. When several operations are present in an expression, such as $2 + 3 * 4 + 2 \wedge 5$, some priority must be established for the order in which these operations are performed. One simple rule is to execute the operations from left to right, and in the machine-dependent programming language, *assembly language,* this is what is usually done. Most computer languages adopt the priority listed in Figure F.1, and we use it in this text as well, but it is always a good idea to check the manual of the particular programming language you are using.

Figure F.1

Hierarchy of Operations

1. Parentheses override the following precedences.
2. Exponentiation.
3. Unary minus; that is, negation of a number.
4. Multiplication and division.
5. Addition and subtraction.

The hierarchy of operations for evaluating $2 + 3 * 4 + 2 \wedge 5$ yields

$$2 + 3 * 4 + 2 \wedge 5 = 2 + 3 * 4 + 32$$
$$= 2 + 12 + 32$$
$$= 46$$

Observe that according to the hierarchy of operations $-3 \wedge 2 = -9$, whereas $(-3) \wedge 2 = 9$.

A byte can contain one of 256 different binary numbers. Computers, keyboards, CRT terminals, printers, and disk drives must all communicate with each other and with humans. Since the invention of the electronic digital computer, various codes have been invented for this communication. Today, the most popular code used is the American Standard Code for Information Interchange (ASCII). When you enter a number on a calculator keypad or on a computer keyboard, the actual 8-bit code generated by each of the digits in the number is the ASCII code for this single digit. Converting the sequence of 8-bit codes to a single decimal number is an interesting exercise. The following table gives the ASCII codes for the digits 0 through 9:

Digit	0	1	2	3	4	5	6	7	8	9
ASCII code in decimal	48	49	50	51	52	53	54	55	56	57

As you can see, the ASCII codes for the single-digit numbers have been chosen craftily. To convert the ASCII code for a digit to the actual digit, all you need do is subtract the decimal 48 from the ASCII code.

The algorithm or repetitive process for entering a decimal number into a computer is shown in the flowchart in Figure F.2.

Figure F.2

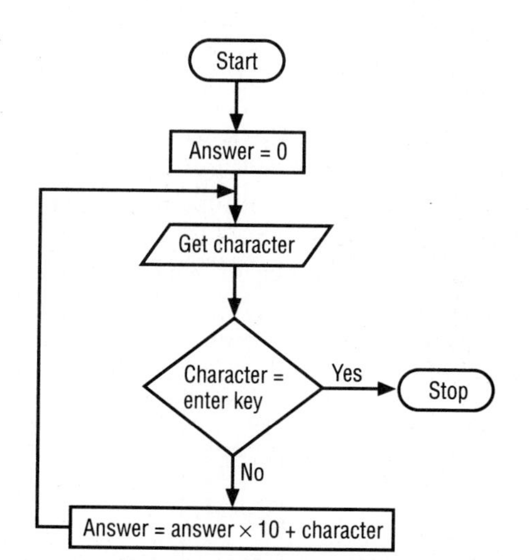

Suppose we type the number 6354 on the keyboard of a computer and push the Enter key. Answer starts as 0. When the 6 is typed, we multiply 0 by 10 and add 6; thus Answer becomes 6. When the 3 is typed, we multiply 6 by 10 and add 3; thus Answer becomes 63. When the 5 is typed, we multiply 63 by 10 and add 5; thus Answer becomes 635. When the 4 is typed, we multiply 635 by 10 and add 4; thus Answer becomes 6354. When the Enter key is pushed, the process stops.

Having considered some of the rudimentary problems involved in entering numbers into a computer and in establishing a pecking order for the usual arithmetic operations, we turn to the problem of calculating sums, differences, and products. There are really two problems involved with each of these arithmetic operations: What is an efficient algorithm to perform the arithmetic operation, and given a suitable algorithm how can we design the circuitry needed to implement that algorithm? We sample both types of problems, but our samples are chosen for their simplicity rather than their efficiency and do not represent the solutions used today.

Suppose that we wish to add two numbers that are written in base 2. We all know the addition algorithm to be used, which is even simpler than the addition algorithm we learned in elementary school for addition in base 10. The two bits at the extreme right of our binary numerals are particularly easy to add, since there is no possibility that a 1 has been carried. The circuit to add these bits is called a *half adder*, so named because it does not cope with a carry of 0 or 1 from a previous addition (see Figure F.3).

Figure F.3
The Half Adder

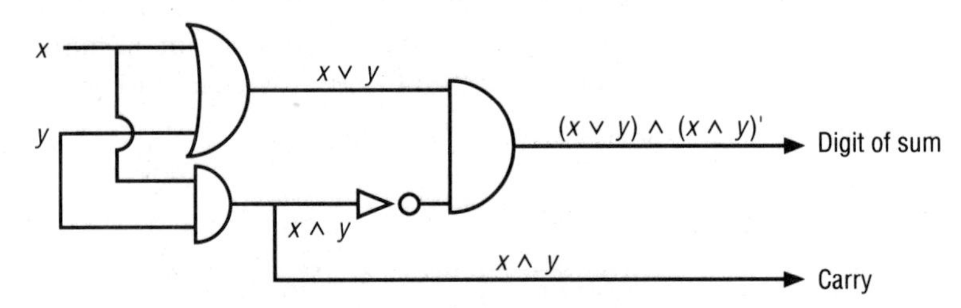

The half adder is not a gate, because it has two outputs (a gate can have only one). The carry output $x \wedge y$ is 1 exactly when both x and y are 1, and this is the one instance in which we would like to carry 1. What about the output $(x \vee y) \wedge (x \wedge y)'$? As the following truth table shows, the resulting output is a 1 only when $x = 0$ and $y = 1$, or vice versa:

x	y	$x \vee y$	$(x \wedge y)'$	$(x \vee y) \wedge (x \wedge y)'$
1	1	1	0	0
1	0	1	1	1
0	1	1	1	1
0	0	0	1	0

All is well. Using a half adder, we can add the last two bits of our binary numerals. Perhaps it would be comforting to write the truth tables of $x \wedge y$ and $(x \vee y) \wedge (x \wedge y)'$ sideways

x	1	1	0	0
	+	+	+	+
y	1	0	1	0
$(x \vee y) \wedge (x \wedge y)'$	0	1	1	0
$x \wedge y$ **(carry)**	1	0	0	0

The two outputs of the half adder are easier to work with if we adopt the notation of juxtaposition to denote logical product (writing xy for $x \wedge y$) and recall that $(x \vee y) \wedge (x \wedge y)'$ is equivalent to the symmetric difference of x and y (which we have agreed to denote by $x \oplus y$; see Section 2.2). Then the carry output is xy and the digit output is $x \oplus y$.

To add the next pair of digits, x and y, or for that matter any pair of digits other than the pair we have just added, we need a *full adder* (see Figure F.4). In our illustration, the input labeled c indicates the value (0 or 1) of the carry from the (half) adder that added the previous pair of digits, and x and y indicate the values of the digits we are adding.

Figure F.4
The Full Adder

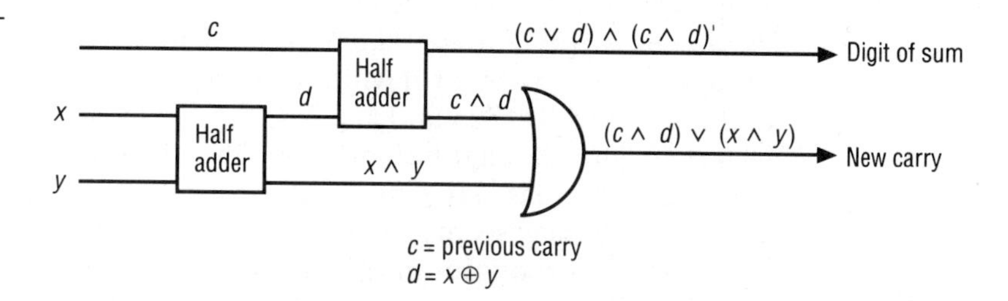

In Exercise 17, we ask you to show that the digit output of the full adder is equivalent to $x \oplus y \oplus c$. (In light of Theorem 2.4, the notation $x \oplus y \oplus c$ is unambiguous.) No matter what notation we use to denote the outputs of the full adder, it is complicated enough to merit the truth table supplied as Table F.1. It will be worth your effort to check the calculations of this table and to rewrite it sideways (see Exercise 18).

In 1642, Blaise Pascal invented an adding machine to assist his father in checking accounts at Rueun. The machine, which was the first desk calculator, carried mechanically and dealt only with numbers that had six digits. Those numbers with more than six digits were beyond the capabilities of the calculator, and if a number had fewer than six digits Pascal simply prefixed enough zeros to make it six digits long. Although the modern calculator is designed on the basis of binary numerals, these numerals had not yet been invented—indeed Leibniz, their inventor, had not yet been born—and so Pascal's machine worked in base 10. The problem Pascal faced was that, while his machine could carry it could not borrow, and his father needed to be able to subtract as well as to add.

Table F.1

x	y	c	$x \wedge y$	$x \vee y$	$(x \vee y) \wedge (x \wedge y)'$ (alias d)	$(c \vee d) \wedge (c \wedge d)'$ (new digit)	$(c \wedge d) \vee (x \wedge y)$ (new carry)
1	1	1	1	1	0	$1 \wedge 1 = 1$	$0 \vee 1 = 1$
1	1	0	1	1	0	$0 \wedge 1 = 0$	$0 \vee 1 = 1$
1	0	1	0	1	1	$1 \wedge 0 = 0$	$1 \vee 0 = 1$
1	0	0	0	1	1	$1 \wedge 1 = 1$	$0 \vee 0 = 0$
0	1	1	0	1	1	$1 \wedge 0 = 0$	$1 \vee 0 = 1$
0	1	0	0	1	1	$1 \wedge 1 = 1$	$0 \vee 0 = 0$
0	0	1	0	0	0	$1 \wedge 1 = 1$	$0 \vee 0 = 0$
0	0	0	0	0	0	$0 \wedge 1 = 0$	$0 \vee 0 = 0$

Definition

Let x be a base 10 numeral with six or fewer digits. Then the **complement** of x is $1{,}000{,}000 - x$.

Let us suppose that Pascal wanted to subtract 9,375 from 86,214. The complement of 9,375 is $1{,}000{,}000 - 9{,}375$, which Pascal would treat as $(999{,}999 - 9{,}375) + 1 = 990{,}625$. Note that Pascal's machine could find the complement without having to borrow. Next Pascal would use his machine to add 990,625 and 086,214, but, as we have said, the machine dealt only with six-digit numbers, so when it added 086,214 and 990,625 it kept only the last six digits of the sum, 076,839. This is the correct answer, which may seem somewhat surprising since in the last calculation the machine introduced an absolute error of 1,000,000. To see what is going on, we must consider arithmetic in $\mathbb{Z}_{10^6}$. In $\mathbb{Z}_{10^6}$, where

$$\begin{aligned}[86{,}214] - [9{,}375] &= [86{,}214] + [0] - [9{,}375] \\ &= [86{,}214] + [1{,}000{,}000] - [9{,}375] \\ &= [86{,}214] + [990{,}625] \\ &\quad (= [86{,}214] + [\text{the complement of } 9{,}375]) \\ &= [1{,}076{,}839] \\ &= [76{,}839]\end{aligned}$$

Once we have the equation $[86{,}214] - [9{,}375] = [76{,}839]$, it follows that $86{,}214 - 9{,}375 = 76{,}839$.

Pascal's trick can easily be adapted for the modern computer. As an illustration, we use byte-size binary numerals in place of Pascal's six-digit decimal numbers. We first keep track of things in base 10, working in $\mathbb{Z}_{256}$. As a notational convenience, for $0 \leq n \leq 255$, we write n to denote

Table F.2 Modulo 256

Positive number	Additive inverse
1	255
2	254
3	253
4	252
.	.
.	.
.	.
125	131
126	130
127	129
128	128
.	.
.	.
.	.
252	4
253	3
254	2
255	1

both the integer n and the equivalence class in $\mathbb{Z}_{256}$ to which n belongs. Notice that, in $\mathbb{Z}_{256}$, $56 + 200 = 0$ and so 200 is the additive inverse of 56. In Table F.2 we list the numbers between 0 and 256 and their additive inverses. As you can see, it is reasonable to consider the numbers 1 through 127 to be positive and the numbers 129 through 255 to be negative. There is a problem with 128 in that it could be considered either positive or negative. But if you look at the binary representation of the numbers from 0 to 255, you see that all the positive numbers through 127 have a 0 as their first binary digit and all the negative numbers from 129 through 255 have a 1 as their first binary digit. Since 128 has a 1 in its first binary digit, it is reasonable to choose 128 to be a negative number. Then an 8-bit binary number is negative if and only if its first binary digit is a 1. This method of determining which numbers are negative is reminiscent of Pascal's easy way to do subtraction. For example, to subtract 55 from 100, find the additive inverse in $\mathbb{Z}_{256}$ of 55, which is 201, and then add 201 to 100 giving 301, which is 45 modulo 256. When we work in base 10, all this manipulation is not fruitful. Technically, we have traded a problem in subtraction for a problem in addition, but finding out that $[-55] = [201]$ and that $[301] = [45]$ involves two subtractions, each of which is more difficult than the subtraction we were given to start with. Using 8-bit binary numerals, however, we gain the same advantage Pascal had when he used complements with respect to 1,000,000 to deal with 6-digit decimal numerals.

The *one's complement* of an 8-bit binary number is obtained by replacing all of its 0s by 1s and all of its 1s by 0s. The *two's complement* of an 8-bit binary number is obtained by adding 1 to the one's complement of the number (see Example 1). Finding the one's and two's complement of an 8-bit binary number is easy, and, as you are asked to show in Exercise 12, the two's complement of an 8-bit binary number in $\mathbb{Z}_{256}$ is its additive inverse.

Example 1

Find the one's complement and two's complement of 101101_2 and verify that in $\mathbb{Z}_{256}$ the two's complement of 101101_2 is its additive inverse.

Analysis

Number	00101101_2
One's complement	11010010_2
	$+1$
Two's complement	11010011_2

To see that $[11010011_2]$ is the additive inverse of $[00101101_2]$, it suffices to observe that

$$[00101101_2] + [11010011_2] = [100000000_2] = [256_{10}] = [0]$$ ■

Example 2

Evaluate $11101110_2 - 101101_2$ using two's-complement arithmetic.

Analysis

$$\begin{aligned}11101110_2 - 101101_2 &= [11101110_2] + [-101101_2] \\ &= [11101110_2] + [\,11010011_2] && \text{(by Example 1)} \\ &= [111000001_2] && \text{(by addition in base 2)} \\ &= [11000001_2] && \text{(by congruence modulo 256)}\end{aligned}$$

Thus $11101110_2 - 101101_2 = 11000001_2$. ■

There are two features of binary arithmetic that allow for the ease of calculating subtractions of numbers. First, the two's complement allows us to find the additive inverse of a number. Second, as in the last step of Example 2, we may replace any 9-digit binary number by the number formed from its last eight digits. This replacement is possible because we are working in $\mathbb{Z}_{256}$. Note that to subtract 55 from 100 we use subtraction to find the additive inverse of 55 and also to find that $[301] = [45]$. But in binary arithmetic, two's complementation allows us to find the additive inverse of a number without subtracting, and subtracting 256 from a number such as 301 is now a simple matter of dropping the leading bit. Although most computers actually perform low-level arithmetic in byte size (that is, in $\mathbb{Z}_{256}$), there is no change in the basic principles of two's complementation provided that we work with a modulus that is a power of 2. Exercise 13 illustrates this point by considering two's-complement subtraction in $\mathbb{Z}_{65536}$.

Multiplication is the last arithmetic operation we consider in this appendix. The classic 8080 microcomputer had no direct multiplication operation, so multiplication with the 8080, even at the one-byte level, is an interesting process. This process is based on the Egyptian method of multiplication. Example 3 indicates in modern notation how the Egyptians multiplied two numbers.

Example 3

Egyptian multiplication of 27×43:

$1 + 2 + 8 + 16 = 27$

√ 1	43	43
√ 2	86	86
4	172	
√ 8	344	344
√16	688	688
		$1161 = 27 \times 43$

Like the checkerboard method of multiplication discussed in Appendix A, we begin Egyptian multiplication by writing one of the numbers as sums of powers of 2. The required powers of 2 are checked in the left-hand column. In the right-hand column the other number is doubled and redoubled as far as is needed. The numbers in the right-hand column corresponding to numbers marked in the left-hand column are added, and the resulting sum is the product of the two numbers. ■

The Egyptian method of multiplication developed into a somewhat improved method known as the *Russian peasants' method*, or the method of *duplation and mediation.* In both the Egyptian method and the Russian peasants' method the multiplicand is listed on the right and a column of its successive doubles is written down. In the Russian peasants' method the multiplier is listed on the left and successively halved. In the halving process we truncate, so that for an odd positive integer $2n + 1$ the subsequent half is n and not $n + 1/2$. Those numbers in the right-hand column that correspond to odd numbers in the left-hand column are then added. In Example 4, we compare the two methods of multiplication.

Example 4

Multiply 35 × 28 using the Egyptian method and the Russian peasants' method:

Analysis

	Egyptian method			Russian peasants' method		
	√ 1	28	28	*35	28	28
	√ 2	56	56	*17	56	56
	4	112		8	112	
	8	224		4	224	
	16	448		2	448	
	√32	896	896	*1	896	896
32 + 2 + 1 = 35			980	*odd numbers		980

□

The advantage of the Russian peasants' method is that the numbers to be summed are chosen automatically, whereas in the Egyptian system we have to figure out which powers of 2 add up to the multiplier. Although it is unlikely that either the Egyptians or the Russian peasants understood why their methods worked, both methods are really multiplications in base 2. To see this, let us perform the multiplication 35 × 28 in base 2:

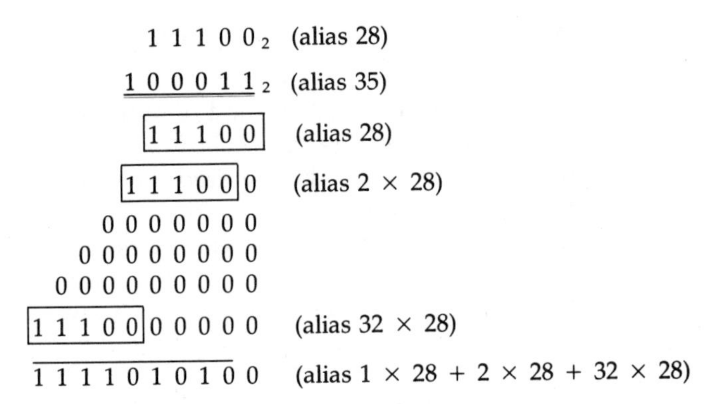

To see why the Russian peasants' method selects the powers of 2 that add to the multiplier, we have only to compare their method of repeatedly halving the multiplier with the method of writing the multiplier in base 2 that we have considered in Example 2 of Appendix D. We again select the number 35 to make this comparison.

Example 5

Two ways to find the standard binary expansion of 35:

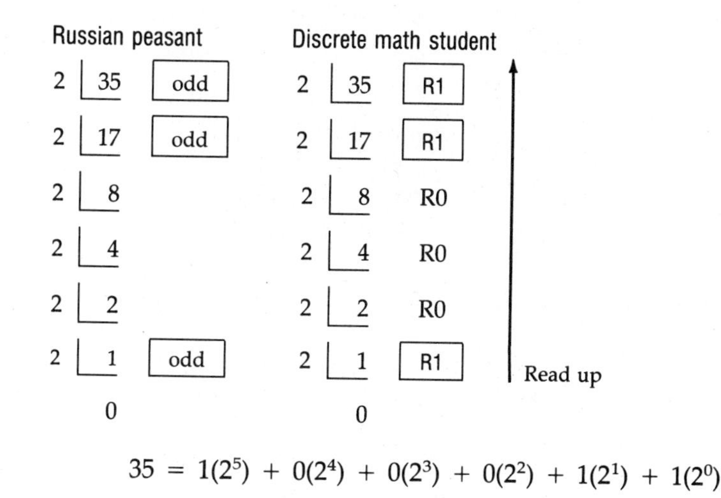

$$35 = 1(2^5) + 0(2^4) + 0(2^3) + 0(2^2) + 1(2^1) + 1(2^0)$$

The crux of the matter is the not-very-profound observation that a number is odd if and only if it leaves a remainder of 1 when divided by 2. ■

The usual multiplication algorithm is a farrago of addition and multiplication unsuitable to feed to a computer, but in both the Egyptian and Russian peasants' methods the multiplications and additions are separated, with the multiplications done first and the additions done last. Moreover, the Russian peasants' method, being completely mechanical, is well suited to the computer.

Although the 8080 microprocessor has no direct multiplication operation, it has the ability to rotate the bits in a byte left or right one position. For example, RLC transforms the byte 00110101 into 01101010:

RLC (rotate left)

00110101   0 0 1 1 0 1 0 1   01101010

The effect of the RLC command is to multiply the binary number in a byte by 2 unless there is a 1 in the most significant digit. Thus RLC doubles the binary number in a byte modulo 256, whereas, RRC halves the binary number in a byte and truncates. Then, to multiply two 8-bit binary numbers in a computer with an 8080 microprocessor, we use the Russian peasants' method. All numbers in the examples that follow are written in base 2.

Example 6

Use the Russian peasants' method to find the product 00000101 * 00001011.

Analysis

	RRC	RLC
odd	00000101	00001011
	00000010	
odd	00000001	00101100
		00110111 ← answer

■

Example 7

Use the Russian peasants' method to find the product 00001111 * 00001111.

Analysis

	RRC	RLC
odd	00001111	00001111
odd	00000111	00011110
odd	00000011	00111100
odd	00000001	01111000
		11100001 ← answer

■

Exercises Appendix F

1. Write each of the following numbers in normalized form.

   a. 13.25 **b.** 0.003 **c.** 14259.379 **d.** 0.000000397

2. Find the mantissa, precision, and magnitude of the numbers given in Exercise 1.

3. Write each of the following in normalized binary form.

   a. 2.4 **b.** 10.3 **c.** $13\frac{1}{3}$

4. Assuming that the hierarchy rule is in effect, evaluate each of the following expressions.

   a. 2 * 3 + 4 * 5 **b.** 2 * 3/2 * 3

   c. −2 ∧ 2 + 2 * (1 + 3) **d.** 1 + (2 − 3) * 5

5. When the computer stops, what is in Answer when 579 Enter is typed on a keyboard of a computer and the computer uses the process indicated in Figure F.2?

6. What are the values in Answer in increasing order when 236 Enter is typed on the keyboard of a computer and the computer uses the process indicated in Figure F.2?

7. Find each of the indicated additions modulo 256.

   a. $\begin{array}{r} 10011001_2 \\ +\ 00110111_2 \\ \hline \end{array}$ **b.** $\begin{array}{r} 160 \\ +\ 100 \\ \hline \end{array}$ **c.** $\begin{array}{r} 255_8 \\ +\ 30_8 \\ \hline \end{array}$ **d.** $\begin{array}{r} 11011110_2 \\ +\ 11101001_2 \\ \hline \end{array}$

8. Find each of the indicated multiplications modulo 256.

a. $\begin{array}{r} 15 \\ \times\ 10 \\ \hline \end{array}$ b. $\begin{array}{r} 30 \\ \times\ 20 \\ \hline \end{array}$ c. $\begin{array}{r} 01010101_2 \\ \times\ 00001011_2 \\ \hline \end{array}$ d. $\begin{array}{r} 11111111_2 \\ \times\ 00001010_2 \\ \hline \end{array}$

9. Find the one's complement of each of the following.

a. 10101010_2 b. 11001101_2 c. 237 d. 796

10. Find the two's complement of each of the following.

a. 00110101_2 b. 11111111_2 c. 240 d. 12

11. Subtract each of the following using two's-complement subtraction.

a. $\begin{array}{r} 10110110_2 \\ -\ 00101011_2 \\ \hline \end{array}$ b. $\begin{array}{r} 11011100_2 \\ -\ 01101111_2 \\ \hline \end{array}$

c. $\begin{array}{r} 11000000_2 \\ -\ 11111111_2 \\ \hline \end{array}$ d. $\begin{array}{r} 100 \\ -\ 50 \\ \hline \end{array}$

12. Argue that the two's complement of an 8-bit binary number is its additive inverse.

13. a. Find the one's complement and two's complement of the word-size binary numerals x and y, where $x = 1110111001001001_2$ and $y = 0000111100001010_2$.

b. Using two's-complement arithmetic, evaluate $x - y$.

c. Using two's-complement arithmetic, evaluate $y - x$.

14. In each of the following, multiply using the Russian peasants' method.

a. 25×13 b. 63×12

c. $00110101_2 \times 00000011_2$ d. $00001011_2 \times 00001101_2$

e. $11111111_2 \times 10101010_2$

*15. One way to store fractions in a computer is to store the numerator in one byte and the denominator in a second byte. For example, $3/5$ can be stored as 00000011 in byte one and 00000101 in byte two. Some fractions cannot be stored in this manner—$400/3$ for example, because $400 > 256$.

a. Find the closest fraction to $400/3$ that can be stored in the described manner.

b. Find the closest fraction to $400/501$ that can be stored in the described manner.

c. Given that x is a number between $1/255$ and 255, is there always a closest fraction to x that can be stored in the described manner?

16. Let $x = 1$, $y = 1$, and $c = 1$. Evaluate the following.

a. $d = x \oplus y$ b. $(c \wedge d) \vee (x \wedge y)$ c. $(c \vee d) \wedge (c \wedge d)'$

17. Let c, x, and y be variables and let d be $(x \vee y) \wedge (x \wedge y)'$. Show that $(c \vee d) \wedge (c \wedge d)'$ and $x \oplus y \oplus c$ are equivalent Boolean expressions.

18. Fill in the table appropriately and verify that the resulting values jibe with the truth values in Table F.1.

c	1	0	1	0	1	0	1	0
x	1	1	0	0	1	1	0	0
+								
y	1	1	1	1	0	0	0	0
New digit								
New carry								

Appendix G

FINITE BOOLEAN ALGEBRAS

In this appendix we establish that, if B is a finite Boolean algebra and A is the set of its atoms, then B is isomorphic to the Boolean algebra of subsets of A. One important consequence of this result is that if B is a finite Boolean algebra then the number of elements of B is a power of 2. Throughout this discussion B denotes a finite Boolean algebra.

According to Exercise 27 of Chapter 13, a non-zero element a of B is an atom if and only if 0 and a are the only members b of B for which $b \leq a$. We list some other facts about atoms and the partial order $\leq$, and the proofs are left as exercises. Let x, y, p, and q be elements of B and let a and a_1 be atoms.

Fact 1. $xy \leq x$ and $xy \leq y$.

Fact 2. If $x \leq p$ and $x \leq q$, then $x \leq pq$.

Fact 3. If $x \leq p$ or $x \leq q$, then $x \leq p \vee q$.

Fact 4. If $a = x \vee y$, then $a = x$ or $a = y$.

Fact 5. If $a \neq a_1$, then $aa_1 = 0$.

We first show that every nonzero element of a finite Boolean algebra is either an atom or the join of atoms. As usual, $x < y$ means $x \leq y$ and $x \neq y$.

Definition

Let B be a Boolean algebra with n elements and for each natural number k with $1 \leq k \leq n$, define an element b of B to be **on level k** provided that k is the largest positive integer for which there exist k distinct non-zero elements $b_1, b_2, \ldots, b_k$ such that $b_1 < b_2 < \cdots < b_k = b$.

The motivation for the terminology *on level* k is clear from the Hasse diagram of any finite Boolean algebra.

Example 1

In the Boolean algebra of subsets of {1,2,3,4,5}, {2,3,5} is on level 3 because $\varnothing < \{2\} < \{2,5\} < \{2,3,5\}$. In the Boolean algebra of divisors of 210, 70 is on level 3 because $1 < 7 < 14 < 70$ (see Exercise 1). ■

Theorem G.1

> Let B be a finite Boolean algebra. Then every nonzero element of B is on some level, and an element b of B is on level 1 if and only if it is an atom.

Proof

Let a be an atom of B. By Exercise 27 of Chapter 13, $0 \leqslant a$, and if $b \leqslant a$ then $b = 0$ or $b =$ a. Therefore a is on level 1. It also follows from Exercise 27 of Chapter 13 that any element of level 1 is an atom (see Exercise 4).

Let b be a non-zero element of B. We wish to show that b is on some level. If b is an atom, we already know that b is on level 1. If b is not an atom, then by Exercise 27 of Chapter 13 there is a nonzero element b_1 such that $0 < b_1 < b_2 = b$. It follows that $k = 2$ is one positive integer such that there exist distinct non-zero elements $b_1, b_2, \ldots, b_k$ of B such that $0 < b_1 < \cdots < b_k = b$. Since B is finite, there must be a largest such integer k (which is greater than or equal to 2), and b is on level k. □

We have already seen that an element b of B is on level 1 if and only if there is exactly one atom x (namely, b itself) for which $x \leqslant b$. It is reasonable to guess that for each natural number k an element b of B is on level k if and only if there are exactly k atoms x of B for which $x \leq b$. Although this guess is correct, for our needs the following simpler result suffices.

Theorem G.2

> Let B be a finite Boolean algebra and let A be its set of atoms. Define $f: B \to \mathcal{P}(A)$ by $f(b) = A_b$, where $A_b = \{a \in A: a \leq b\}$. Then f is a one-to-one correspondence.

Proof

The proof that f is one-to-one is by contradiction. Suppose that $f(x) = f(y)$ and yet $x \neq y$. then $x \nleqslant y$ or $y \nleqslant x$. We assume without loss of generality that $x \nleqslant y$, for if $y \nleqslant x$ we can just switch the names of the elements x and y.

Since $x \nleqslant y$, $xy \neq x$. It follows that $xy' \neq 0$, for otherwise $x = x(1)$

$= x(y \vee y') = xy \vee xy' = xy \vee 0 = xy$. By Theorem G.1, there is a positive integer k such that xy' is on level k, and if we write $0 < b_1 < b_2 < \ldots < b_k = xy'$, then b_1 is an atom (see Exercise 4). Since $b_1 \leq xy' \leq x$ (Fact 1), $b_1 \in A_x = f(x) = f(y) = A_y$. Thus $b_1 \leq y$. But $b_1 \leq xy' \leq y'$, and so $b_1 \leq y$ and $b_1 \leq y'$. By Fact 2, $0 \leq b_1 \leq yy' = 0$. It follows that $b_1 = 0$, and since b_1 is an atom we have reached a contradiction. Therefore f is one-to-one.

It remains to show that f maps B onto $\mathcal{P}(A)$. Evidently $f(0) = \varnothing$. Let $F = \{a_1, a_2, \ldots, a_k\}$ be a nonempty subset of A and let $b = a_1 \vee a_2 \vee \ldots \vee a_k$. To show that $f(b) = F$, we must show that $F \subseteq f(b)$ (that is, if $a \in F$ then $a \leq b$) and that $f(b) \subseteq F$ (that is, if $a \notin F$ then $a \not\leq b$).

Let $a_i \in F$. Then

$$a_ib = a_ia_1 \vee a_ia_2 \vee \ldots \vee a_ia_i \vee a_ia_{i+1} \vee \ldots \vee a_ia_k = a_ia_i \vee 0 \text{ (Fact 5)} = a_i$$

By definition, $a_i \leq b$.

Now let a be an atom not belonging to F. Then $ab = aa_1 \vee aa_2 \vee \ldots \vee aa_k = 0$ (Fact 5). Clearly $ab \neq a$, and so $a \not\leq b$. We have shown that $f(b) = F$, and so f maps B onto $\mathcal{P}(A)$. □

Corollary G.3

Let B be a finite Boolean algebra with n atoms. Then B has 2^n elements.

Proof

By Theorem 4.3, $\mathcal{P}(A)$ has 2^n elements. Since there is a one-to-one correspondence between B and $\mathcal{P}(A)$, B also has 2^n elements. □

Let A be the set of atoms of a finite Boolean algebra B. To assure ourselves that the Boolean algebras B and $\mathcal{P}(A)$ behave alike, we must find a one-to-one correspondence $\Phi: B \to \mathcal{P}(A)$ that preserves the fundamental Boolean operations of meet, join, and complement in the sense that, for all $p,q \in B$,

$$\Phi(p \vee q) = \Phi(p) \vee \Phi(q)$$

$$\Phi(pq) = \Phi(p)\Phi(q) \qquad \text{and}$$

$$\Phi(p') = \Phi(p)'$$

Such a correspondence is called a *Boolean algebra isomorphism*. In the special case we are considering, the join, product, and complement of $\mathcal{P}(A)$ are union, intersection, and set complement, and we use the appropriate set-theoretic notation.

Theorem G.4

Let A be the set of atoms of a finite Boolean algebra B and let $f: B \to \mathcal{P}(A)$ be the one-to-one correspondence of Theorem G.2. Then f is a Boolean algebra isomorphism.

Proof

Let $p,q \in B$.

To show that $f(p \vee q) = f(p) \cup f(q)$, we show that $f(p) \cup f(q) \subseteq f(p \vee q)$ and vice versa. Let a be an atom belonging to $f(p) \cup f(q)$. Then $a \leqslant p$ or $a \leqslant q$ and in either case, by Fact 3, $a \leqslant p \vee q$. Therefore $f(p) \cup f(q) \subseteq f(p \vee q)$.

Let $a \in f(p \vee q)$. Then $a \leqslant p \vee q$. We wish to show that $a \leqslant p$ or $a \leqslant q$. We may assume that neither p nor q is 0 since $f(0) = \varnothing$. This time, we need to use the fact that a is an atom (see Exercise 3b). Since $a \leqslant p \vee q$, $a = a(p \vee q) = ap \vee aq$. By Fact 4, $a = a \wedge p$ or $a = a \wedge q$; that is, $a \leqslant p$ or $a \leqslant q$. Thus $f(p \vee q) = f(p) \cup f(q)$.

The proof that $f(p \wedge q) = f(p) \cap f(q)$ is similar to the argument just given. Let $a \in f(pq)$. Then $a \leqslant pq$. By Fact 1, $a \leqslant p$ and $a \leqslant q$. Thus $f(pq) \subseteq f(p) \cap f(q)$. Now let $a \in f(p) \cap f(q)$. Then $a \leqslant p$ and $a \leqslant q$ and so, by Fact 2, $a \leqslant pq$. Thus $f(p) \cap f(q) \subseteq f(pq)$ and so $f(p) \cap f(q) = f(pq)$.

There is a straightforward proof that $f(p') = f(p)^{\sim}$. We already know that (1) $A = f(1) = f(p \vee p') = f(p) \cup f(p')$ and that (2) $\varnothing = f(0) = f(pp') = f(p) \cap f(p')$. By (1), $f(p)^{\sim} \subseteq f(p')$. By (2), $f(p') \subseteq f(p)^{\sim}$. Therefore $f(p)^{\sim} = f(p')$. □

Theorem G.4 is a fundamental result. Without slogging through details, we mention two of its applications.

1. If two finite Boolean algebras have the same number of elements, they are isomorphic. (Each of them is isomorphic to the Boolean algebra $\mathcal{P}(I_n)$ for some natural number n.)

2. It is clear that, in the Boolean algebra of subsets of a nonempty set A, each non-zero element can be written as the join of atoms (that is, each nonempty subset can be written as the union of singleton sets). It is equally clear that, except for the order in which these atoms are joined, the representation of a non-zero element as the join of atoms is unique. For the finite Boolean algebra $B = \{0,1\}$ and for each natural number n, $\mathbb{F}_{B,n}$ is a finite Boolean algebra and so, by Theorem G.4, must behave like the Boolean algebra of subsets of a nonempty set A. The atoms of $\mathbb{F}_{B,n}$ are the functions in $\mathbb{F}_{B,n}$ that can be represented by minterms. Thus Theorem 13.3, which promises that each non-zero function in $\mathbb{F}_{B,n}$ can be written uniquely in join-normal form, is just a corollary of Theorem G.4.

Exercises Appendix G

1. Draw enough of the Hasse diagrams of the following finite Boolean algebras to show all the elements on levels 1 and 2.
   a. the Boolean algebra of subsets of I_5
   b. the Boolean algebra of divisors of 210
2. How many elements of the Boolean algebra of subsets of a set with ten elements are on level 7? (*Hint*: Consider the binomial theorem.)
3. Let B be the Boolean algebra of divisors of 210.
   a. Evaluate $3 \wedge (5 \vee 7 \vee 2 \vee 3)$ and $3 \wedge (5 \vee 7 \vee 2)$.
   b. Give an example of three members $p,q,r \in B$ such that $r \leq p \vee q$ yet $r \not\leq p$ and $r \not\leq q$.
   c. Give an example of $p,q \in B$ such that $p \not\leq q$ and $q \not\leq p$.
   d. Do there exist distinct elements $p,q \in B$ such that $p \leq q$ and $q \leq p$? Explain how you know.
4. Let B be a finite Boolean algebra and let $b \in B$ on level k. Suppose that $b_1, b_2, \ldots, b_k$ are distinct members of B such that $0 < b_1 < b_2 < \ldots < b_k = b$. Explain why b_1 must be an atom.
5. If possible, give an example of a Boolean algebra with the stated property. If no such Boolean algebra exists, write "no way."
   a. a Boolean algebra with 120 atoms
   b. a Boolean algebra with 120 elements
   c. a finite Boolean algebra with no atoms
   d. a finite Boolean algebra with an odd number of elements
   e. a finite Boolean algebra for which 1 is an atom
6. Show that the divisors of 48 do not form a Boolean algebra.

   For Exercises 7–11, let x, y, p, and q be elements of a Boolean algebra B. These exercises establish the facts of the partial order $\leq$ given on p. 733.

7. Explain why $xy \leq x$ and $xy \leq y$.
8. Show that $xpq = (xp)(xq)$, and explain why it follows from this equation that if $x \leq p$ and $x \leq q$ then $x \leq pq$.
9. Suppose $xp = x$. Prove that $x(p \vee q) = x$. Explain why Fact 3 follows. (*Hint*: $x = x(1)$.)
10. Prove that if a and b are distinct atoms of a Boolean algebra then $ab = 0$. (*Hint*: Use Fact 1, which has been established in Exercise 7.)
11. Suppose that an atom a of a Boolean algebra is the join of two elements x and y. Prove that $a = x$ or $a = y$. (*Hint*: Using $x = x(1)$ and $y = y(1)$, first show that $x \leq x \vee y$ and $y \leq x \vee y$.)

ANSWERS TO SELECTED EXERCISES

Chapter 1

Exercises 1.1, pages 7–8

1. Neither true nor false

3. a. P' **b.** $P' \wedge Q'$ **c.** $P' \vee Q$ **d.** $P' \to Q$
e. $Q \to P'$

5. a, b, c, and d **7.** b, d, e, and f

Exercises 1.2, Pages 12–13

9. a.

X	Y	$X \wedge Y$	$(X \wedge Y)'$
T	T	T	F
T	F	F	T
F	T	F	T
F	F	F	T

b.

X	Y	X'	Y'	$X' \wedge Y'$
T	T	F	F	F
T	F	F	T	F
F	T	T	F	F
F	F	T	T	T

13. a.

P	Q	$P \to Q$
T	T	T
T	F	F
F	T	T
F	F	T

b. line 1 **c.** Yes

15. a and c

Exercises 1.3, pages 16–17

17. a. FORTRAN is not a language and SUPERCALC is not a branch of mathematics.
c. You can write Pascal programs and you cannot easily learn FORTRAN.

19. a. (contrapositive) If either $2^2 \neq 4$ or Raleigh is not the capital of North Carolina, then IBM is not a museum and $3 \geq 4$. (negation) IBM is a museum or $3 < 4$ and either $2^2 \neq 4$ or Raleigh is not the capital of North Carolina.
c. (contrapositive) If Buffalo is not the capital of New York and Mickey Mantle was not a movie star, then $|-3| \neq |3|$ and the Mississippi is not a river. (negation) $|-3| \neq |3|$ or the Mississippi is a river and Buffalo is not the capital of New York and Mickey Mantle was not a movie star.

21. a, e, and h **23.** c and d

25. "Is the right fork the way to your village?" If the man is a Truth Teller, then a "yes" answer means that the right fork is the correct way and a "no" answer means that the left fork is the correct way. If the man is a Liar, then a "yes" answer once again means that the right fork goes to the Truth Teller's village and a "no" answer means that the left fork goes to the Truth Teller's village.

Exercises 1.4, page 23

27. Since a divides b and c, there are integers p and q such that $b = pa$ and $c = qa$. Then $nb + mc = n(pa) + m(qa) = (np)a + (mq)a = (np + mq)a$. Since $np + mq$ is an integer, a divides $nb + mc$.

29. Assume $x = 2$. Then $x^3 - 2x^2 + 4x - 8 = 2^3 - 2 \cdot 2^2 + 4 \cdot 2 - 8 = 8 - 8 + 8 - 8 = 0$. Assume $x^3 - 2x^2 + 4x - 8 = 0$. Then $x^2(x - 2) + 4(x - 2) = 0$ or $(x^2 + 4)(x - 2) = 0$. Since $x^2 + 4 \neq 0$, $x - 2 = 0$. Thus $x = 2$.

31. Since $x \neq y$, $x - y \neq 0$. Thus $(x - y)^2 > 0$. Then $x^2 - 2xy + y^2 > 0$ or, adding $4xy$ to both sides, $x^2 - 2xy + y^2 + 4xy > 4xy$, or $x^2 + 2xy + y^2 > 4xy$, or $(x + y)^2 > 4xy$. Since x and y are both positive, $x + y > 0$. Thus, dividing by $x + y$, we get $x + y > 4xy/(x + y)$.

33. $n^3 - n = n(n^2 - 1) = (n - 1) \cdot n \cdot (n + 1)$. If $n - 1$ is even, then $n^3 - n$ is even. If $n - 1$ is odd, then n is even and once again $n^3 - n$ is even. Hence $n^3 - n$ is even.

Exercises 1.5, pages 27–29

35. a. There is an integer that is not a real number.
c. No integer is prime.

e. There is a house not made of brick.
g. There is a car that is neither silver nor blue.
i. No integer is either even or a perfect square.
37. a. There is a real number x such that, if y is a real number, then $2^x \neq y$.
c. There is a pair of rational numbers a and b with $a < b$ such that, if c is a number such that $a < c < b$, then c is rational.
41. a, c, and d

Review Exercises for Chapter 1, pages 29–30

43. a. Either you cannot draw a flowchart or you cannot write a COBOL program.
c. The machine is not an IBM and I can operate it.
45. a. If a rainbow does not have five colors, then 1992 does not have 366 days.
c. If either x is not positive or y is not positive, then $|x + y|$ is not equal to $|x| + |y|$.
47. $n^2 + n = n(n + 1)$. If n is even, then $n^2 + n$ is even. If n is odd, then $n + 1$ is even, and once again $n^2 + n$ is even. Hence $n^2 + n$ is even.
49. Assume $x^2 - x - 2 < 0$. Then $(x - 2)(x + 1) < 0$. Thus, either $x - 2 > 0$ and $x + 1 < 0$ or $x - 2 < 0$ and $x + 1 > 0$. Thus, either $x > 2$ and $x < -1$ or $x < 2$ and $x > -1$. Since x cannot be larger than 2 and less than -1 at the same time, there are no numbers that satisfy the first half, and only the second half applies. Thus $x < 2$ and $x > -1$, or $-1 < x < 2$.
51. Black. The third man knows that there are only two white hats, so if he sees two white hats he knows that his hat is black. But he does not know. Hence it is not the case that both of the first two hats are white. Now if the second man sees a white hat, he knows that his and the first man's hat cannot both be white, so he knows that his hat is black. But the second man does not know the color of his hat either. So the first man's hat cannot be white. Hence the first man's hat must be black.

Chapter 2

Exercises 2.1, pages 34–36

1. a. $-4, -3, -2, -1$ **c.** 6, 60, 210, 504
e. $-2, 2$ **g.** $-2, 2$
3. a. 1 **c.** The empty set
5. a, b, d, e, g
7. $\mathcal{P}(\{1\}) = \{\emptyset, \{\emptyset\}\}$,
$\mathcal{P}(\mathcal{P}(\{1\})) = \{\emptyset, \{\emptyset\}, \{\{\emptyset\}\}, \{\emptyset, \{\emptyset\}\}\}$
9. a. $A = \{1, 2, 3\}$, $B = \{4\}$ **c.** $A = \emptyset$, $B = \{1\}$

Exercises 2.2, pages 43–45

11. a. $\{1, 2, 4, 5, 7\}$ **c.** $\{4, 7\}$ **e.** $\{3, 6, 8, 9\}$
g. $\{1\}$ **i.** $\{1\}$ **k.** $\{1, 4, 7\}$
13. a. $(A \cap B \cap C) \cup (B \cap (A \cup C)^{\sim})$
c. $B \cap C^{\sim}$ **e.** $(A \cap B \cap C^{\sim}) \cup (C \cap (A \cup B)^{\sim})$
15. $A = \emptyset$, $B = X$; $A = \{1\}$, $B = \{2, 3, 4\}$; $A = \{2\}$, $B = \{1, 3, 4\}$; $A = \{3\}$, $B = \{1, 2, 4\}$; $A = \{4\}$, $B = \{1, 2, 3\}$; $A = \{1, 2\}$, $B = \{3, 4\}$; $A = \{1, 3\}$, $B = \{2, 4\}$; $A = \{1, 4\}$, $B = \{2, 3\}$
17.

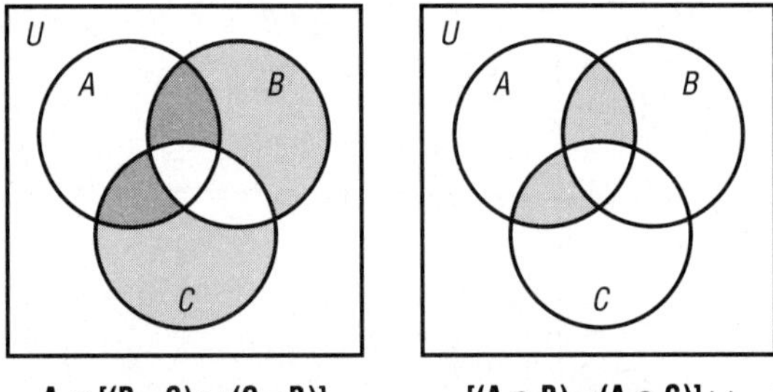

$A \cap [(B - C) \cup (C - B)]$ $[(A \cap B) - (A \cap C)] \cup [(A \cap C) - (A \cap B)]$

19. Let A and B be sets. Let $x \in A$. Either $x \in B$ or $x \notin B$. If $x \in B$, then $x \in A \cap B$ and thus $x \in (A - B) \cup (A \cap B)$. If $x \notin B$, then $x \in A - B$ and thus $x \in (A - B) \cup (A \cap B)$. Hence $(A \cap B) \cup (A \cap B) \supseteq A$. Let $x \in (A - B) \cup (A \cap B)$. Then either $x \in A - B$ or $x \in A \cap B$. In either case, $x \in A$. Hence $(A - B) \cup (A \cap B) \subseteq A$. Therefore $A = (A - B) \cup (A \cap B)$.

Suppose that there is an element $x \in (A \cap B) \cap (A - B)$. Then $x \in A \cap B$ and $x \in A - B$. Thus $x \in B$ and $x \notin B$, which is impossible. Hence $(A \cap B) \cap (A - B) = \emptyset$.
21. The universal set is not chosen generally enough.
23. $A \oplus B = (A - B) \cup (B - A)$, which is the set of elements in A but not in B together with the set of elements in B but not in A. So $A \oplus B$ is the set of elements either in A or in B but not in both.

25. Let A and B be sets. Assume $A \cap B = \emptyset$. Let $x \in A \cup B$. Then $x \in A$ or $x \in B$. If $x \in A$, then $x \notin B$ and $x \in A - B$ and thus $x \in (A - B) \cup (B - A) = A \oplus B$. If $x \in B$, then $x \notin A$ and $x \in B - A$ and thus $x \in (A - B) \cup (B - A) = A \oplus B$. Hence $A \oplus B \supseteq A \cup B$. Let $x \in A \oplus B$. Then $x \in (A - B) \cup (B - A)$. Then either $x \in A - B$ or $x \in B - A$. Thus either $x \in A$ or $x \in B$. Thus $x \in A \cup B$. Hence $A \cup B \supseteq A \oplus B$. Therefore, if $A \cap B = \emptyset$, then $A \cup B = A \oplus B$. Now assume $A \cup B = A \oplus B$. Suppose $A \cap B \neq \emptyset$. Then there is an element $p \in A \cap B$. Then $p \in A$ and $p \in B$. Thus $p \in A \cup B$. But $p \notin A - B$ and $p \notin B - A$. Thus $p \notin (A - B) \cup (B - A) = A \oplus B$. Hence $A \cup B \neq A \oplus B$, which is impossible. Hence A and B are disjoint sets if and only if $A \cup B = A \oplus B$.

Exercises 2.3, pages 48–49

27. $\cup_{n=1}^{5} A_n = \mathbb{N}$; $\cap_{n=1}^{5} A_n = \{k \in \mathbb{N} : k \geq 5\}$; $\cup \{A_n : n \in \mathbb{N}\} = \mathbb{N}$; $\cap \{A_n : n \in \mathbb{N}\} = \emptyset$

29. $\cup_{n=1}^{20} A_n = (-1, 40)$; $\cap_{n=1}^{20} A_n = (-1/20, 2)$; $\cup \{A_n : n \in \mathbb{N}\} = (-1, \infty)$; $\cap \{A_n : n \in \mathbb{N}\} = [0, 2)$

31. Let $m \in \mathbb{N}$. Let $x \in A_m$. By definition, $x \in \cup \{A_n : n \in \mathbb{N}\}$. Thus $\cup \{A_n : n \in \mathbb{N}\} \supseteq A_m$. Let $x \in \cap \{A_n : n \in \mathbb{N}\}$. Since $m \in \mathbb{N}$, $x \in A_m$. Therefore $A_m \supseteq \cap \{A_n : n \in \mathbb{N}\}$.

33. Let $x \in (\cap \{A_n : n \in \mathbb{N}\})^{\sim}$. Then $x \notin \cap \{A_n : n \in \mathbb{N}\}$. So there is an element $k \in \mathbb{N}$ such that $x \notin A_k$. Thus $x \in A_k^{\sim}$. Hence $x \in \cup \{A_n^{\sim} : n \in \mathbb{N}\}$. Therefore $\cup \{A_n^{\sim} : n \in \mathbb{N}\} \supseteq (\cap \{A_n : n \in \mathbb{N}\})^{\sim}$. Let $x \in \cup \{A_n^{\sim} : n \in \mathbb{N}\}$. Then there is a $k \in \mathbb{N}$ such that $x \in A_k^{\sim}$. Then $x \notin A_k$. Thus $x \notin \cap \{A_n : n \in \mathbb{N}\}$. Hence $x \in \{A_n : n \in \mathbb{N}\}^{\sim}$. Therefore $(\cap \{A_n : n \in \mathbb{N}\})^{\sim} \supseteq \cup \{A_n^{\sim} : n \in \mathbb{N}\}$. Hence $(\cap \{A_n : n \in \mathbb{N}\})^{\sim} = \cup \{A_n^{\sim} : n \in \mathbb{N}\}$.

35. $\cup \{A_n : n \in \mathbb{N}\} = (-\infty, 1]$
$\cap \{A_n : n \in \mathbb{N}\} = (-1, 0]$
$\cup \{A_n^{\sim} : n \in \mathbb{N}\} = (-\infty, -1] \cup (0, \infty)$
$\cap \{A_n^{\sim} : n \in \mathbb{N}\} = (1, \infty)$

Exercises 2.4, pages 51–52

37. 8 **39. a.** 40 **b.** 40 **c.** 10

41. b, c, d, e, and f

Exercises 2.5, pages 55–57

43. a. $n(A \cup B) = n(A) + n(B) - n(A \cap B) = 28 + 79 - 3 = 104$

b. $n(A \cap B) = n(A) + n(B) - n(A \cup B) = 35 + 96 - 122 = 9$

45.

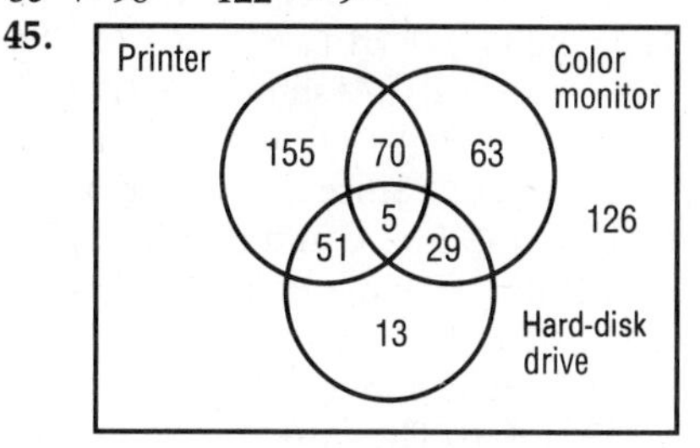

512 computer science majors

There are 126 computer science majors who have only a monochrome monitor and no printer or hard-disk drive.

47. Mrs. Peacock

49.

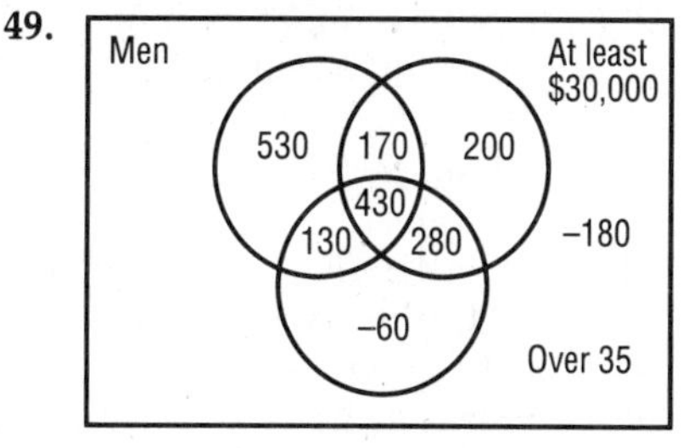

1500 Faculty

No! The region over 35, not male, and under \$30,000 would have to have −60 people in it and the region belonging to none of the three groups would have to have −180 people in it. No region can have a negative number of elements.

51. 257

53. Yes. $n(A \cup B) = n(A) + n(B) - n(A \cap B)$. No.

Review Exercises for Chapter 2, pages 57–58

55. Let $x \in A \cap (B \oplus C)$. Then $x \in A$ and $x \in B \oplus C = (B - C) \cup (C - B)$. Thus either $x \in B$ and $x \notin C$ or $x \in C$ and $x \notin B$. Assume $x \in B$ and $x \notin C$. Then $x \in A \cap B$ and $x \notin A \cap C$. Thus $x \in (A \cap B) - (A \cap C)$ and thus $x \in ((A \cap B) - (A \cap C)) \cup ((A \cap C) - (A \cap B)) = (A \cap B) \oplus (A \cap C)$. Assume $x \in C$ and $x \notin B$. Then

$x \in A \cap C$ and $x \notin A \cap B$. Thus $x \in (A \cap C) - (A \cap B)$ and thus $x \in ((A \cap B) - (A \cap C)) \cup ((A \cap C) - (A \cap B)) = (A \cap B) \oplus (A \cap C)$. Hence $x \in (A \cap B) \oplus (A \cap C)$. Hence $(A \cap B) \oplus (A \cap C) \supseteq A \cap (B \oplus C)$. Let $x \in (A \cap B) \oplus (A \cap C)$. Then either $x \in (A \cap B) - (A \cap C)$ or $x \in (A \cap C) - (A \cap B)$. Assume $x \in (A \cap B) - (A \cap C)$. Then $x \in A \cap B$ and $x \notin A \cap C$. Then $x \in A$ and $x \in B$. Since $x \in A$ and $x \notin A \cap C$, $x \notin C$. Thus $x \in B - C$ and $B \oplus C = (B - C) \cup (C - B) \supseteq B - C$ and $x \in B \oplus C$. Assume $x \in (A \cap C) - (A \cap B)$. Then $x \in A \cap C$ and $x \notin A \cap B$. Then $x \in A$ and $x \in C$. Since $x \in A$ and $x \notin A \cap B$, $x \notin B$. Thus $x \in C - B$ and $B \oplus C = (B - C) \cup (C - B) \supseteq C - B$ and $x \in B \oplus C$. In either case, $x \in B \oplus C$. Hence $x \in A \cap (B \oplus C)$. Thus $A \cap (B \oplus C) \supseteq (A \cap B) \oplus (A \cap C)$. Therefore $A \cap (B \oplus C) = (A \cap B) \oplus (A \cap C)$.

57. $A \times (B \times C) = \{(1, (4, 6)), (1, (5, 6)), (2, (4, 6)), (2, (5, 6)), (3, (4, 6)), (3, (5, 6))\}$
$(A \times B) \times C = \{((1, 4), 6), ((1, 5), 6), ((2, 4), 6), ((2, 5), 6), ((3, 4), 6), ((3, 5), 6)\}$
$A \times B \times C = \{(1, 4, 6), (1, 5, 6), (2, 4, 6), (2, 5, 6), (3, 4, 6), (3, 5, 6)\}$

59. $n(A) = 25$; $n(B) = 41$; $n(C) = 32$; $n(A \cap B) = 16$; $n(A \cap C) = 11$; $n(B \cap C) = 17$; $n(A \cap B \cap C) = 6$. Thus $n(A \cup B \cup C) = n(A) + n(B) + n(C) - n(A \cap B) - n(A \cap C) - n(B \cap C) + n(A \cap B \cap C) = 60$.

61. a. $\{4a, 5b, 5c, 1d\}$ **b.** $\{2b, 3c\}$ **c.** $\{4a, 2c\}$ **d.** $\{3b, 1d\}$ **e.** $\{4a, 7b, 8c, 1d\}$

Chapter 3

Exercises 3.1, pages 63–65

1. a, c, and d **3.** a and d

5. a. $\{(1, 1), (2, 1/2), (3, 1/3), (4, 1/4), (5, 1/5), (6, 1/6)\}$
b. $\{(1, 1), (1/2, 2), (1/3, 3), 3), (1/4, 4), (1/5, 5), (1/6, 6), (1/7, 7)\}$

7. $f(\mathbb{Z})$, odd integers; $f(\mathbb{N})$ odd integers greater than 1; $f(\mathbb{E})$ all integers that are one more than a multiple of 4; $f(\mathbb{R}) = \mathbb{R}$; rng$(f)$, $\mathbb{R}$

9. a. $f(n) = 2$ **b.** $f(n) = 2n$
c. $f(n) = \begin{cases} n/2, & \text{if } n \text{ is even} \\ 3, & \text{if } n \text{ is odd} \end{cases}$

11. a. Yes **b.** No

13. $f(1) = 3(1^2) - 5 = -2 = 3((-1)^2) - 5 = f(-1)$. Thus $f: \mathbb{R} \to \mathbb{R}$ is not a one-to-one function. Suppose that there is an element $x \in \mathbb{R}$ such that $f(x) = -8$. Then $3x^2 - 5 = -8$. Thus $3x^2 = -3$, or $x^2 = -1$, and there is no element x in $\mathbb{R}$ such that $x^2 = -1$. Therefore -8 is not in the range of f.

15. Suppose that $a, b \in A$ such that $f(a) = f(b)$. Then $a^2 + 2a + 5 = b^2 + 2b + 5$, or $a^2 + 2a = b^2 + 2b$, or $a^2 - b^2 + 2(a - b) = 0$, or $(a - b)(a + b + 2) = 0$. Therefore either $a - b = 0$ or $a + b + 2 = 0$. Hence either $a = b$ or $a + b + 2 = 0$. Now $a \geq -1$ and $b \geq -1$. So in order for $a + b + 2$ to be equal to 0 , it must be the case that $a = -1$ and $b = -1$, and hence $a = b$. In either case, $a = b$, so f is a one-to-one function from A into B.

17. Define $f: A \to \mathbb{N}$ by, if $n \in \mathbb{N}$, then $f(n) = n$ and, if $X \in \mathcal{P}(\mathbb{N})$, then $f(X) = 1$. Now $\{1, 2, 3\}$ is both an element of A and a subset of A. Considered as an element of A, $f(\{1, 2, 3\}) = 1$, and considered as a subset of A, $f(\{1, 2, 3\}) = \{1, 2, 3\}$.

Exercises 3.2, pages 69–70

19. 6. $\{(1, -1), (2, 0), (3, 1)\}$; $\{(1, -1), (2, 1), (3, 0)\}$; $\{(1, 0), (-1, 2), (0, 3)\}$; $\{(1, 0), (-1, 3), (0, 2)\}$; $\{(1, 1), (2, -1), (3, 0)\}$; $\{(1, 1), (2, 0), (3, -1)\}$

21. a. $\{(0, 0), (3, -1), (2, 0), (-1, 3)\}$
b. $\{(1, 1), (2, 4), (4, 2), (3, 1)\}$

23. Since $0 < a < 1$, $1 - a > 0$ and $3a/(1 - a) > 0$ and hence $3a/(1 - a) \in (0, \infty)$ and $f(3a/(1 - a)) = a$. Thus rng$(f) = (0, 1)$.

25. a. $(f \circ g)(x) = f(g(x)) = f(x - 3) = 2(x - 3) = 2x - 6$
c. $(f \circ h)(x) = f(h(x)) =$
$$f\left(\begin{cases} 0, & \text{if } x \text{ is odd} \\ 1, & \text{if } x \text{ is even} \end{cases}\right) = \begin{cases} 0, & \text{if } x \text{ is odd} \\ 2, & \text{if } x \text{ is even} \end{cases}$$
e. $(g \circ h)(x) = g(h(x)) =$
$$g\left(\begin{cases} 0, & \text{if } x \text{ is odd} \\ 1, & \text{if } x \text{ is even} \end{cases}\right) = \begin{cases} -3, & \text{if } x \text{ is odd} \\ -2, & \text{if } x \text{ is even} \end{cases}$$
g. $(f \circ (g \circ h))(x) = f((g \circ h)(x)) =$
$$f\left(\begin{cases} -3, & \text{if } x \text{ is odd} \\ -2, & \text{if } x \text{ is even} \end{cases}\right) = \begin{cases} -6, & \text{if } x \text{ is odd} \\ -4, & \text{if } x \text{ is even} \end{cases}$$

27. a. Let $x \in \mathbb{R}$. Then $(f \circ g)(x) = f(g(x))$

$$= f\left(\frac{x}{1+|x|}\right) = \frac{\dfrac{x}{1+|x|}}{1 - \left|\dfrac{x}{1+|x|}\right|} = \frac{\dfrac{x}{1+|x|}}{1 - \dfrac{|x|}{1+|x|}}$$

$$= \frac{\dfrac{x}{1+|x|}}{\dfrac{1+|x|-|x|}{1+|x|}} = \frac{x}{1} = x$$

c. Yes. Yes. Corollary 3.2

29. $f^{-1}(x) = (2x+1)/(x-2)$

Exercises 3.3, pages 74–75

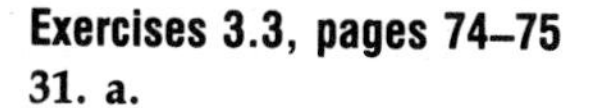

31. a. **e.**

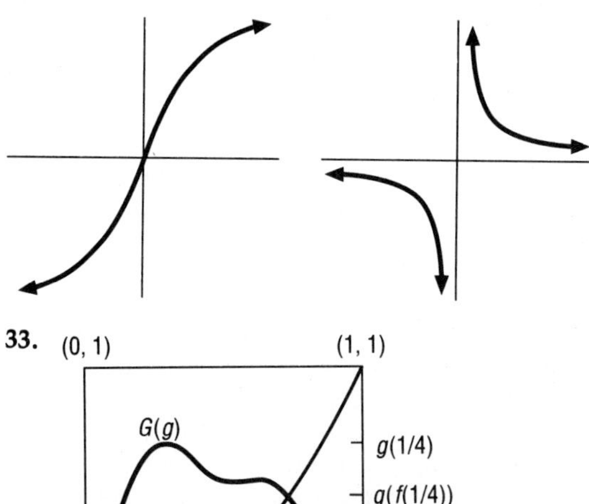

33.

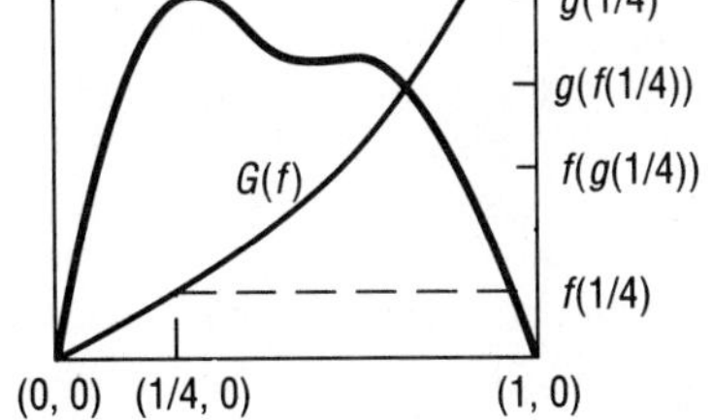

35. a. $q = 32,\ r = 11$ **b.** $q = -33,\ r = 2$ **c.** $q = -11,\ r = 20$

Exercises 3.4, pages 86–88

37. a. $f + g = \{(1, 3), (2, 4), (3, 0), (4, -1)\}$ **c.** $f \cdot g = \{(1, 2), (2, 3), (3, -1), (4, -2)\}$

39. a. 8.5 **c.** 6.5

41. a.

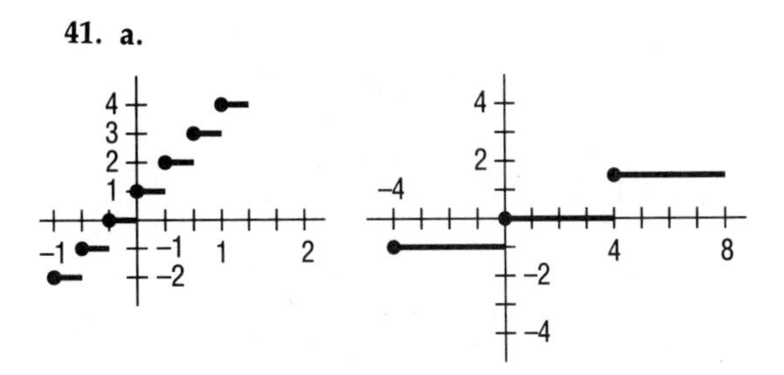

43. No, $f(0)$ and $f(1)$ are both 0. Yes. If $n \in \mathbb{Z}$, then $f(2n) = n$.

45. $(f+g)(x) = \begin{cases} -1 + 2x, & \text{if } x < 0 \\ 1 + 3x, & \text{if } x \geq 0 \end{cases}$

$(f-g)(x) = \begin{cases} -1 - 2x, & \text{if } x < 0 \\ 1 - 3x, & \text{if } x \geq 0 \end{cases}$

$(fg)(x) = \begin{cases} -2x, & \text{if } x < 0 \\ 3x, & \text{if } x \geq 0 \end{cases}$

$(f/g)(x) = \begin{cases} -1/(2x), & \text{if } x < 0 \\ 1/(3x), & \text{if } x \geq 0 \end{cases}$

49. a. $\varnothing$, X **b.** All subsets except $\varnothing$ and X

51. Let $x \in X$. Then $(\chi_A + \chi_{A^\sim})(x) = \chi_A(x) + \chi_{A^\sim}(x) = \begin{cases} 1, & \text{if } x \in A \\ 0, & \text{if } x \notin A \end{cases} + \begin{cases} 0, & \text{if } x \in A \\ 1, & \text{if } x \notin A \end{cases} = \begin{cases} 1, & \text{if } x \in A \\ 1, & \text{if } x \notin A \end{cases} = 1 = \chi_X(x)$.

53. a. **b.**

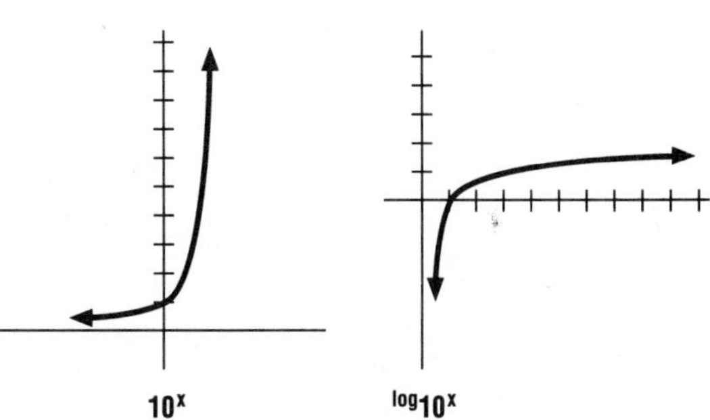

57. $(3 * 2) * 1 = (3+2)^{3+2} * 1 = 5^5 * 1 = 3125 * 1 = (3125+1)^{3125+1} = 3126^{3126}$, and $3 * (2 * 1) = 3 * (2+1)^{2+1} = 3 * 27 = (3+27)^{3+27} = 30^{30}$. Clearly the first number is much larger.

59. Let $f_1 = \{(1, 1), (2, 2)\}$ and $f_2 = \{(1, 2), (2, 1)\}$. Since f_1 is the identity function, $f_1 \circ f_2 = f_2 \circ f_1$ and this is the only nontrivial composition. Thus the binary operation is commutative.
61. a. 24 **b.** 120
c. Let S be a set with n elements. Then $\mathcal{P}(S)$ has 2^n elements and Sym(S) has $n!$ elements. Thus 3 is the largest n can be and have $2^n > n!$.
63. Define $T : A \times B \rightarrow A$ by, if $a \in A$ and $b \in B$, then $T((a,b)) = a$. For instance, T((bat, up)) = bat.

Review Exercises for Chapter 3, pages 88–89

65. a and d **67.** a
69. No. B has six elements and A has only five elements, so at least two elements of B map onto the same element of A.
71. $f^{-1}(x) = (x - 7)/2$
73. $(f \circ g)(x) = f(g(x)) = f(x^2 + 7) = 2(x^2 + 7) + 5 = 2x^2 + 19$; $(g \circ f)(x) = g(f(x)) = g(2x + 5) = (2x + 5)^2 + 7 = 4x^2 + 20x + 32$
75. a. No. $f(-2) = f(0) = f(2)$ **b.** No. Nothing maps into 1.

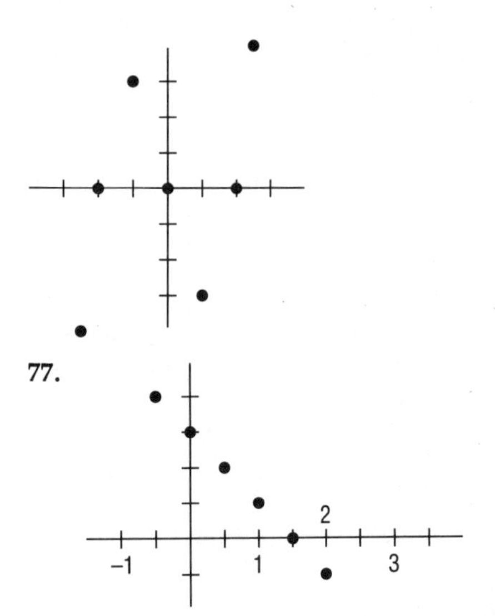

79. graph of $f + g = \{(1, -2), (2, 1), (3, 0), (4, 7), (5, -3), (6, 3), (7, -5/2)\}$
graph of $f - g = \{(1, -2), (2, -1), (3, 2), (4, 3), (5, 2), (6, -3), (7, 7/2)\}$
graph of $fg = \{(1, 0), (2, 0), (3, -1), (4, 10), (5, 2), (6, 0), (7, -3/2)\}$
graph of f/g is not defined since 0 is in the range of g

Chapter 4

Exercises 4.1, pages 93–94

1. $-1; 1; -1; 1; 1; -1$
3. a. $f(1) = 1$; for $n > 1$, $f(n) = nf(n - 1)$
c. 6; 24; 720
5. a. -1; -1 **b.** 4; 32 **c.** 4; 65536 **d.** 50; 1250
e. 1; 675 **f.** 4; 32

Exercises 4.2, pages 101–103

7. a. $a = 2$; $r = 3$; $n = 10$; 177,146
9. a. $1^2 + 2^2 + 3^2 + 4^2 + 5^2 = 55$
c. $1^1 + 2^2 + 3^3 + 4^4 + 5^5 = 3413$
11. a. $\sum_{i=1}^{6} 2i$ **b.** $\sum_{i=1}^{6} (2i - 1)$ **c.** $\sum_{i=1}^{6} i^2$

13. Let A be the set of all natural numbers n such that $3 + 6 + 9 + \cdots + 3n = 3n(n + 1)/2$. Since $3 = 3(1)(1 + 1)/2$, $1 \in A$. Suppose that $k \in A$. Then $3 + 6 + 9 + \cdots + 3k = 3k(k + 1)/2$. Thus $3 + 6 + 9 + \cdots + 3k + 3(k + 1) = 3k(k + 1)/2 + 3(k + 1) = 3(k + 1)((k + 2) + 1 = 3(k + 1)(k/2)/2 = 3(k + 1)((k + 1) + 1)/2$. Thus $k + 1 \in A$. Hence, if $k \in A$, then $k + 1 \in A$. By the principle of mathematical induction, A is the set of all natural numbers and the theorem is proved.
23. Let A be the set of all natural numbers n such that $3^n - 1$ is divisible by 2. Since $3^1 - 1 = 2$, $1 \in A$. Suppose $k \in A$. Then $3^k - 1$ is divisible by 2. Now $3^{k+1} - 1 = 3(3^k - 1) + 3 - 1 = 3(3^k - 1) + 2$, which is divisible by 2 (see Chapter 1, Exercise 27). Thus $k + 1 \in A$. Hence, if $k \in A$, then $k + 1 \in A$. By the principle of mathematical induction, A is the set of all natural numbers and the theorem is proved.
25. a. 1;4;9;16;25 **b.** n^2
c. Let A be the set of all natural numbers n such that $1 + 3 + 5 + \cdots + (2n - 1) = n^2$. Since $1 = 1^2$, $1 \in A$. Let $k \in A$. Then $1 + 3 + 5 + \cdots + (2k - 1) = k^2$. Now $1 + 3 + 5 + \cdots + (2k - 1) + (2(k + 1) - 1) = k^2 + 2(k + 1) - 1) = k^2 + 2k + 1 = (k + 1)^2$. Thus $k + 1 \in A$. Hence,

if $k \in A$, then $k + 1 \in A$. By the principle of mathematical induction, A is the set of all natural numbers and the theorem is proved.

27. Let A be the set of all natural numbers n such that $n \geq 4$ and $n! > 2^n$. Since $4! = 24 > 16 = 2^4$, $4 \in A$. Assume $k \in A$. Then $k \geq 4$ and $k! > 2^k$. Thus $(k + 1)! = (k + 1)k! > 5 \cdot 2^k > 2^{k+1}$. Hence $k + 1 \in A$. Therefore, if $k \in A$, then $k + 1 \in A$. By the principle of mathematical induction, A is the set of all natural numbers greater than or equal to 4. Hence the theorem is proved.

31. Let W be the set of all natural numbers n such that $A \cap (\cup_{i=1}^{n} B_i) = \cup_{i=1}^{n}(A \cap B_i)$. Since $A \cap (\cup_{i=1}^{1} B_i) = A \cap B_1 = \cup_{i=1}^{1} (A \cap B_i)$, $1 \in W$. Assume $k \in W$. Then $A \cap (\cup_{i=1}^{k} B_i) = \cup_{i=1}^{k} (A \cap B_i)$. Recall from Theorem 2.1c that $X \cap (Y \cup Z) = (X \cap Y) \cup (X \cap Z)$. Now $A \cap (\cup_{i=1}^{k+1} B_i) = A \cap (\cup_{i=1}^{k} B_i \cup B_{k+1}) = (A \cap \cup_{i=1}^{k} B_i) \cup (A \cap B_{k+1}) = \cup_{i=1}^{k} (A \cap B_i) \cup (A \cap B_{k+1}) = \cup_{i=1}^{k+1} (A \cap B_i)$. Thus $k + 1 \in W$. Hence, if $k \in W$, then $k + 1 \in W$. Therefore, by the principle of mathematical induction, W is the set of all natural numbers, and the theorem is proved.

Exercises 4.3, page 107

33. Let A be the set of all natural numbers n such that $n \geq 14$ and n is the sum of numbers each of which is either a 3 or an 8. Since $14 = 3 + 3 + 8$, $14 \in A$. Since $15 = 3 + 3 + 3 + 3 + 3$, $15 \in A$. Since $16 = 8 + 8$, $16 \in A$. Let k be a natural number such that $k \geq 16$ and $14, 15, 16, \ldots, k \in A$. Then $k + 1 \geq 17$, so $k - 2 = (k + 1) - 3 \geq 17 - 3 = 14$. Thus $14 \leq k - 2 \leq k$. Therefore $k - 2$ is the sum of numbers each of which is either a 3 or an 8. Hence $k + 1 = (k - 2) + 3$ is the sum of numbers each of which is either a 3 or an 8. Therefore, if $k \in A$, then $k + 1 \in A$. By the second principle of mathematical induction, A is the set of all natural numbers greater than or equal to 14, and the theorem is proved.

37. Let A be the set of all natural numbers n such that $1 + 3 + 5 + \cdots + (2n - 1) = n^2$. Since $1 = 1^2$, $1 \in A$. Let $k \in A$. Then $1 + 3 + 5 + \cdots + (2k - 1) = k^2$. Now $1 + 3 + 5 + \cdots + (2k - 1) + (2(k + 1) - 1) = k^2 + (2(k + 1) - 1) = k^2 + 2k + 1 = (k + 1)^2$. Thus $k + 1 \in A$. Hence, if $k \in A$, then $k + 1 \in A$. By the principle of mathematical induction, A is the set of all natural numbers, and the theorem is proved.

Exercises 4.4, pages 116–118

39. Let A be the set of natural numbers n such that $(1 + 2 + \cdots + n)^2 = 1^3 + 2^3 + \cdots + n^3$. Since $1^2 = 1 = 1^3$, $1 \in A$. Assume $k \in A$. Then $(1 + 2 + \cdots + k)^2 = 1^3 + 2^3 + \cdots + k^3$. Recall (from Example 5) that, if n is a natural number, then $1 + 2 + 3 + \cdots + n = n(n+ 1)/2$. Now $(1 + 2 + \cdots + k + (k + 1))^2 = (1 + 2 + \cdots + k)^2 + 2(1 + 2 + \cdots + k)(k + 1) + (k + 1)^2 = 1^3 + 2^3 + \cdots + k^3 + 2\dfrac{k(k + 1)}{2}(k + 1) + (k + 1)^2 = 1^3 + 2^3 + \cdots + k^3 + k(k + 1)^2 + (k + 1)^2 = 1^3 + 2^3 + \cdots + k^3 + (k + 1)(k + 1)^2 = 1^3 + 2^3 + \cdots + k^3 + (k + 1)^3$. Thus $k + 1 \in A$. Hence, if $k \in A$, then $k + 1 \in A$. Therefore by the principle of mathematical induction, A is the set of all natural numbers and the theorem is proved.

41. $a_2 = 2^2 = 4$; $a_3 = 2^4 = 16$; $a_4 = 2^{16} = 65{,}536$; $a_5 = 2^{65{,}536}$

43. $b_1 = 1$, $b_2 = 2$, $b_3 = 1$, $b_4 = 2$, $b_5 = 1$, $b_6 = 2$, $b_7 = 1$, $b_8 = 2$

45. $f(1) = 1$, $f(n + 1) = f(n) + (n + 1)$

47. Let A be the set of all natural numbers n such that $2n \in S$. Since $2(1) = 2 \in S$, $1 \in A$. Assume $k \in A$. Then $2k \in S$. Now $2 \in S$ and $2k \in S$. Thus $2(k + 1) = 2k + 2 \in S$. Hence $k + 1 \in A$. Therefore, if $k \in A$, then $k + 1 \in A$. Hence by the principle of mathematical induction, A is the set of all natural numbers and the theorem is proved.

55. a. $a_3 = 3$; $a_4 = 4$; $a_5 = 5$ **b.** $a_n = n$
c. Let A be the set of all natural numbers n such that $a_n = n$. Since $a_1 = 1$, $1 \in A$. Assume $1, 2, \ldots, k \in A$. Then $a_{k+1} = 2a_k - a_{k-1} = 2k - (k - 1) = k + 1$. Therefore $k + 1 \in A$. Hence, if $1, 2, \ldots, k \in A$, then $k + 1 \in A$. By the second principle of mathematical induction, A is the set of all natural numbers and the theorem is proved.

61. a. *aaa* **b.** *aaababa* **c.** 5

63. The language consisting of all nonempty words $f{:}I_n \to \Sigma$ ($n \in \mathbb{N}$) for which $f(1) = 2$.

Review Exercises for Chapter 4, pages 118–119

67. $f(n) = (1 + r)^{n-1}$

Let A be the set of all natural numbers n such that $f(n) = (1 + r)^{n-1}$. Since $f(1) = 1 = (1 + r)^0 = (1 + r)^{1-1}$, $1 \in A$. Assume $k \in A$. Then $f(k) = (1 + r)^{k-1}$. Thus $f(k + 1) = f(k)(1 + r) = (1 + r)^{k-1}(1 + r) = (1 + r)^k = (1 + r)^{(k+1)-1}$. Therefore $k + 1 \in A$. Hence, if $k \in A$, then $k + 1 \in A$. Thus, by the first principle of mathematical induction, A is the set of all natural numbers and the theorem is proved.

73. 11 **75.** 98

Chapter 5

Exercises 5.1, pages 124–125

1. 1. Add the two units digits. 2. Write down the units digit of the result. 3. If the tens digit of the result is 1, then set the carry to 1; otherwise set the carry to 0. 4. Add the two tens digits and the carry. 5. Write down the result to the left of the first units digit of the result.

3. The description of the black box might be $(x - 8)(x - 7) \cdots (x - 1)/8! + x$, in which case the output would match for 1–8 but would be 10 when x is 9.

5. It can have two different outputs for the same input.

7. a. 1. Get as input a day of the week. 2. If it is Friday, Saturday, or Sunday, let x = Monday. 3. If it is Monday or Tuesday, let x = Wednesday. 4. If it is Wednesday or Thursday, let x = Friday. 5. Output x.

Exercises 5.2, page 129

9. 1. Get as input fractions a/b and c/d. 2. Let $x = ad - bc$. 3. Let $y = bd$. 4. Output x/y.

11. 2.23

Exercises 5.3, page 133

13. Let A be the set of all natural numbers n such that $n \geq 7$ and $n^4 < n!$. Now $7^4 = 2401$ and $7! = 5040$. Thus $7^4 < 7!$ and $7 \in A$. Now suppose $k \in A$. Then $k^4 < k!$. Thus $(k + 1)^4 = k^4 + 4k^3 + 6k^2 + 4k + 1 < 5k^4 < (k + 1)k^4 < (k + 1)k! = (k + 1)!$. Therefore $k + 1 \in A$. Thus, if $k \in A$, then $k + 1 \in A$. Hence A is the set of all natural numbers greater than or equal to 7 and the theorem is proved.

15. Let $c = 1$, $k = 0$, and $x \geq k$. Then $|f(x)| = |x^2| = x^2 \leq x^2 + x = g(x) = 1 \cdot |g(x)|$. Therefore $f = O(g)$. Let $c = 2$, $k = 1$, and $x \geq k$. Then $|g(x)| = |x^2 + x| = x^2 + x \leq x^2 + x^2 = 2x^2 = 2f(x) = 2|f(x)|$. Therefore $g = O(f)$.

17. a. Let c and k be positive numbers. Any nonzero x has $|f(x)| \geq c|g(x)|$. Thus $f \neq O(g)$.

b. No. Let c and k be positive numbers. Let $x = 9c + k$. Then $x \geq k$ and $|f(x)| = 2(9c + k) > 18c > c|h(x)|$.

Exercises 5.4, page 137

19. a. 10, $O(1)$ **b.** 25, $O(1)$
c. $5n - 1$, $O(1)$ **d.** 7, $O(1)$
e. $2x + 4$, $O(x)$ **f.** $2x$, $O(x)$

Exercises 5.5, pages 141–142

21. a. 8 **b.** 4, 12, **c.** 2, 6, 10, 14
d. 1, 3, 5, 7, 9, 11, 13, 15
e. (a), (b), and (c) only have 7 numbers together, while (d) has 8 numbers.

23.
$$\frac{1}{n}\sum_{m=1}^{n}(3m - 2) = \frac{1}{n}\left(3\sum_{m=1}^{n} m - \sum_{m=1}^{n} 2\right) = \frac{1}{n}\left(3\frac{n(n+1)}{2} - 2n\right) = \frac{3}{2}(n + 1) - 2 = \frac{3n - 1}{2}$$

25. a. best case = 1, worst case = n^3, average = $(1 + n^3)/2$

b.
$$\frac{1}{n}\sum_{i=1}^{n} i^3 = \frac{1}{n}\left[\sum_{i=1}^{n} i\right]^2 = \frac{1}{n}\left[\frac{n(n+1)}{2}\right]^2 = \frac{n(n+1)^2}{4}$$

27. Remember that the number of operations is $(3/2)n - 1/2$. Now $n = 100{,}000{,}000$ and the computer can do 10,000 operations per second. To convert to minutes divide by 60, and to convert to hours divide by 60 again. Thus we get

$$\frac{\frac{3}{2}\cdot(100000000) - \frac{1}{2}}{10000\cdot 60\cdot 60} = 4.16666 \text{ hours}$$

$$= 4 \text{ hours and } 9.9999 \text{ minutes,}$$

which is about 4 hours and 10 minutes.

Exercises 5.6, page 150

29. Remember that the number of operations is $(3/2)n^2 + (1/2)n - 2$. Now $n = 100{,}000{,}000$ and the computer can do 10,000 operations per second. To convert to minutes divide by 60, and to convert to hours divide by 60 again, to convert to days divide by 24, to convert to years divide by $365\frac{1}{4}$, and then to convert to centuries divide by 100. Thus we get

$$\frac{\frac{3}{2}\cdot(100000000)^2 + \frac{1}{2}\cdot(100000000) - 2}{10000\cdot 60\cdot 60\cdot 24\cdot 365\frac{1}{4}\cdot 100}$$

$$= 475.321318 \text{ centuries, or about 475 centuries}$$

31. $$\sum_{i=0}^{n-2}(4i+6) = 4\sum_{i=0}^{n-2} i + \sum_{i=0}^{n-2} 6$$

$$= 4\frac{(n-1)(n-2)}{2} + 6(n-1)$$

$$= 2(n^2 - 3n + 2) + 6(n-1)$$

$$= 2n^2 - 6n + 4 + 6n - 6$$

$$= 2n^2 - 2$$

Exercises 5.7, pages 157–158

35. a. $63 = 1\cdot 35 + 28$
$35 = 1\cdot 28 + 7$
$28 = 4\cdot 7 + 0$ Answer: gcd = 7

c. $396 = 1\cdot 312 + 84$
$312 = 3\cdot 84 + 60$
$84 = 1\cdot 60 + 24$
$60 = 2\cdot 24 + 12$
$24 = 2\cdot 12 + 0$ Answer: gcd = 12

37. $660 = 2\cdot 2\cdot 3\cdot 5\cdot 11$, $1386 = 2\cdot 3\cdot 3\cdot 7\cdot 11$, $14421 = 3\cdot 11\cdot 19\cdot 23$; gcd $= 3\cdot 11 = 33$

39. a. No **b.** Yes **c.** No **d.** Yes **e.** 17

Bill's score can be $18 = 4 + 7 + 7$, $19 = 4 + 4 + 4 + 7$, $20 = 4 + 4 + 4 + 4 + 4$, and $21 = 7 + 7 + 7$. Hence Bill's score can be any score greater than or equal to 18. 17 cannot be written as the sum of 4s and 5s.

43. 1

Review Exercises for Chapter 5, page 158

47.

Pass	S	N	$T = N/S$
1	1	7	7
2	4	7	1.75
3	2.875	7	2.4348
4	2.6549	7	2.6366
5	2.6458	7	2.6458

Pass	$T - S$	Yes/No	$(S + T)/2$
1	6	No	4
2	−2.25	No	2.875
3	−0.4402	No	2.6549
4	−0.0183	No	2.6458
5	0	Yes	

Thus $\sqrt{7} = 2.646$ to the nearest three places.

49. a. Let $c = 1$ and $k = 4$. Yes. **b.** No.

51. 15 seconds

53. a. 4 **b.** 6 **c.** 6

Chapter 6

Exercises 6.1, pages 165–166

1. a. 8 **c.** 12

3. a. $65 + 48 = 113$

5. a. $2^{10} = 1024$

7. a. 6 **c.** 1

e. You are more likely to get a 7.

9. $7\cdot 12 = 84$ **11.** $2^8 = 256$ **13. a.** 20

15. If n is even, then there are $2^{n/2}$ n-letter palindromes. If n is odd, then there are $2^{(n+1)/2}$ n-letter palindrones.

Exercises 6.2, pages 169–171

17. 6 **19.** 13

21. Yes. Connect two of the PCs to each of the remaining PCs and do not make any other connections.

23. $17\cdot 4 + 1 = 69$. You could choose 17 ducks, 17 pitchers, 17 baskets, and 17 bears and still not have 18 objects that are the same. But if you choose one more object, then you must choose one of the four and then you will have 18 of that object.

25. Yes. In the set $\{p_1, p_2, p_3\}$ there must be either two even numbers or two odd numbers. Since the sum of two even numbers is even and the sum of two odd numbers is even, there must be two numbers in the set whose sum is even.

27. Call the six people A, B, C, D, E, and F. Assume that there is not a threesome in which all three pairs are friends or all three pairs are enemies. Then there are 20 groups of three people (ABC, ABD, ABE, ABF, ACD, ACE, ACF, ADE, ADF, AEF, BCD, BCE, BCF, BDE, BDF, BEF, CDE, CDF, CEF, and DEF) that can be made with these six people. In any threesome, there are three pairs of people and, since there are only two states (friends and enemies), two of the pairs must have the same state while the third pair has the other state. If abc is a threesome, then c is called the odd man out if ac and bc have one state while ab has the other state. Since there are only six people and $3 \cdot (6) + 1 = 19 < 20$, there are at least $3 + 1 = 4$ triples with the same odd man out. Call this odd man out d and the four threesomes x_1x_2d, x_3x_4d, x_5x_6d, and x_7x_8d. Since there are only six people, at least two of these triples have a common person other than d. Let xad and yad be these two triples. Now xd and ad have the same state, while xa has the opposite state. Now yd and ad also have the same state and hence xd, yd, and ad all have the same state. If xy has the same state as ad, then xyd is a triple in which all three pairs have the same state. If xy has the opposite state from ad, then xya is a triple in which all three pairs have the same state. Therefore it must be the case that there is a threesome in which all three pairs are all friends or all three pairs are all enemies.

29. Since the midpoint of the line segment joining (x_1, y_1, z_1) and (x_2, y_2, z_2) is $((x_1 + x_2)/2, (y_1 + y_2)/2, (z_1 + z_2)/2)$, the coordinates of the midpoints are all integers if x_1 and x_2 are both even or both odd, y_1 and y_2 are both even or both odd, and z_1 and z_2 are both even or both odd. Assign each of the nine points one of the eight nests (E, E, E), (E, E, O), (E, O, E), (O, E, E), (E, O, O), (O, E, O), (O, O, E), (O, O, O), where, for example, a point P_i is assigned (E, O, E) if its first and last coordinates are even and its second coordinate is odd. Then two points are assigned the same nest and these two points determine a line segment whose midpoint has integral coordinates.

31. There are $n(A)$ pigeons and $n(B)$ nests. Since there are more pigeons than nests, some nest has at least two pigeons.

Exercises 6.3, pages 174–176

37. a. $4 \cdot 4 = 16$
c. $4 \cdot 4 \cdot 4 = 64$
e. $4 \cdot 3 = 12$

39.
a. 123 124 125 132 134 135 142 143 145 152 153 154
213 214 215 231 234 235 241 243 245 251 253 254
312 314 315 321 324 325 341 342 345 351 352 354
412 413 415 421 423 425 431 432 435 451 452 453
512 513 514 521 523 524 531 532 534 541 542 543

41. a. $P(7,4) = 7!/(7 - 4)! = 7!/3! = 840$
c. $P(12,9) = 12!/(12 - 9)! = 12!/3! = 79{,}833{,}600$

43. $5 \cdot 4 = 20$

45. $9 \cdot 8 \cdot 7 \cdot 6 \cdot 5 \cdot 4 \cdot 3 \cdot 2 \cdot 1 = 362{,}880$

47. $26 \cdot 26 \cdot 26 \cdot 9 \cdot 10 \cdot 10 = 15{,}818{,}400$

49. $10 \cdot 10 \cdot 10 \cdot 10 \cdot 10 = 10{,}000$

51. $3 \cdot 4 = 12$

53. a. 16 **c.** 4 **e.** 11

Exercises 6.4, pages 178–179

55. Start down the list:

Exp. Name	Space	Rating
A	1384	6
B	985	5
C	249	4
D	1775	8
E	39	2
	Total rating =	25

List according to ratings:

Exp. Name	Space	Rating
D	1775	8
I	1113	8
G	1200	7
F	860	5
E	39	2
	Total rating =	30

Least space required:

Exp. Name	Space	Rating
E	39	2
C	249	4
H	632	3
F	860	5
J	930	6
B	985	5
I	1113	8
	Total rating =	33

Highest ratio:

Exp. Name	Space	Rating	Ratio
E	39	2	0.0513
C	249	4	0.0161
I	1113	8	0.00719
J	930	6	0.00645
G	1200	7	0.00583
F	860	5	0.00581
	Total rating =	32	

57. 2^{25}; 3.335 seconds

59.

	Weight	Rating	Ratio
Food	15	8	0.5333
Water	15	8	0.5333
Watch	0.1	0.1	1.0

Total weight that can be carried is 30 pounds. Using the ratio method, the watch and the food would be taken for a total rating of 8.1. It is clear the food and water should be taken, for a rating of 16. Not only is the rating higher (16), you won't get thirsty.

Exercises 6.5, pages 182–184

61. a. $52 \cdot 48 = 2{,}496$ **b.** $52 \cdot 3 = 156$
c. $52 \cdot 12 = 624$ **d.** $52 \cdot 25 = 1{,}300$
63. $4 \cdot 4 = 16$
65. $6 \cdot 5 \cdot 4 \cdot 3 = 360$
67. $1{,}024 - 1 - 8 = 1{,}015$. There are $2^{10} = 1{,}024$ ten-digit numbers using only the digits 1 and 2. There is one number with no 2s. There are eight numbers with one 2. Thus there are 1,015 numbers with at least two 2s.
69. $729 - 4 = 725$. There are $9^3 = 729$ three-digit numbers with no 0s. The only numbers whose digits do not add up to at least 5 are 111, 112, 121, and 211.
71. a. $4 \cdot 3 \cdot 2 \cdot 1 = 24$
73. a. $5 \cdot 5 \cdot 4 \cdot 3 \cdot 2 \cdot 1 = 600$
75. $1 \cdot 4! \cdot 3! = 216$. The men and women can be arranged only one way (WMWMWMW), and the men in their places can be arranged 3! ways and the women 4! ways.

Exercises 6.6, pages 189–191

77. 792 **79.** $C(7,5) = 21$
81. a. $C(9,4)C(11,2) = 90 \cdot 55 = 4{,}950$
c. $C(20,6) = 38{,}760$ is the total number of committees. 1,074 is the number of committees with five or six Republicans. Thus the number of committees with at most four Republicans is $38{,}760 - 1{,}074 = 37{,}686$.
83. 55,440
85. a. $C(12,3) + C(12,5) = 220 + 792 = 1{,}012$
b. $C(12,5) + 2C(12,4) = 792 + 990 = 1{,}782$
87. a. $C(13,10) = 286$
c. $C(8,5)C(5,5) + C(8,6)C(5,4) + C(8,7)C(5,3) = 56 \cdot 1 + 28 \cdot 5 + 8 \cdot 10 = 56 + 140 + 80 = 276$

91.
$$C(n,r) = \frac{n!}{r!(n-r)!} = \frac{n(n-1)!}{r(r-1)!(n-r)!} = \frac{n}{r}\,\frac{(n-1)!}{(r-1)!((n-1)-(r-1))!} = \frac{n}{r}C(n-1, r-1)$$

95. a. $\dfrac{8!}{4! \cdot 4!} = 70$

b. $\dfrac{8!}{4! \cdot 4!} - \dfrac{4!}{2! \cdot 2!} \cdot \dfrac{3!}{2! \cdot 1!} = 62$

Exercises 6.7, pages 196–197

97. $21x^5$
99. $C(6,3)(2x)^3(y/2)^3 = 20x^3y^3$
101. $x^3 + 3x^2y + 3xy^2 + y^3$
103. a. $1088640a^8b^6$
105. 0.878
107. $2^n = (1+1)^n = \sum_{r=0}^{n} C(n,r) \cdot 1^n - r \cdot 1^r = C(n,0) + C(n,1) + C(n,2) + \cdots + C(n,n)$

109. Adding the results in parts (a) and (b) and then dividing by 2 gives the desired result.
115. a. 2^{17}
117. a. $C(25,11)$
119. You would be finding the number of two-element subsets of a seven-element set instead of finding the number of five-element subsets of a seven-element set. Notice that these are the same, because for each two-element subset you choose, the set that is left over is the five-element set.

Exercises 6.8, pages 204–206

123. 1,260
125. $C(11,3)C(8,3) = 165 \cdot 56 = 9{,}240$
127. $C(15,5) = 3{,}003$
131. $C(18,5)C(13,6)$ **135.** 78

Exercises 6.9, pages 217–219

137. The area is the sum of the areas of two identical triangles and the area of a rectangle. The triangles have base and height $1/100$ and the rectangle has sides $9\sqrt{2}/100$ and $\sqrt{2}/100$. Thus the area $= 2\,(1/2) \cdot (1/100) \cdot (1/100) + (9\sqrt{2}/100) \cdot (\sqrt{2}/100) = 19/10000$.
139. $1/16$ **141.** $1/36$ **143.** $5/6$
145. a. $P(E_1) = 1/2$, $P(E_2) = 1/2$, $P(E_3) = 1/4$
c. No **147.** $(1/6)^8$ **149.** $13/15$, $2/5$, $2/15$
153. Yes. There is no reason at all that they should be related. Look at the way you calculated the answer to Exercise 152.
155. $253/17325$
157. $9/47$ **159.** $217/595$
163. $\frac{48}{52} \cdot \frac{47}{51} \cdot \frac{46}{50} \cdot \frac{45}{49} \cdot \frac{44}{48} \cdot \frac{43}{47} \cdot \frac{42}{46} \cdot \frac{41}{45} \cdot \frac{40}{44} \cdot \frac{4}{43}$

Review Exercises for Chapter 6, pages 219–221

167. $2n!n!$
169. a. {1,2,3}, {1,2,4}, {1,2,5}, {1,3,4}, {1,3,5}, {1,4,5}, {2,3,4}, {2,3,5}, {2,4,5}, {3,4,5}
171. 60
173. $P(7,7) = 5{,}040$; $P(7,5) = 2{,}520$; $C(7,2) = 21$; $C(7,5) = 21$
175. b **179.** $3/4$ **181. a.** 512 **183. a.** 14!
c. $4! \cdot 3! \cdot 7!$

Chapter 7

Exercises 7.1, pages 228–232

1. a.

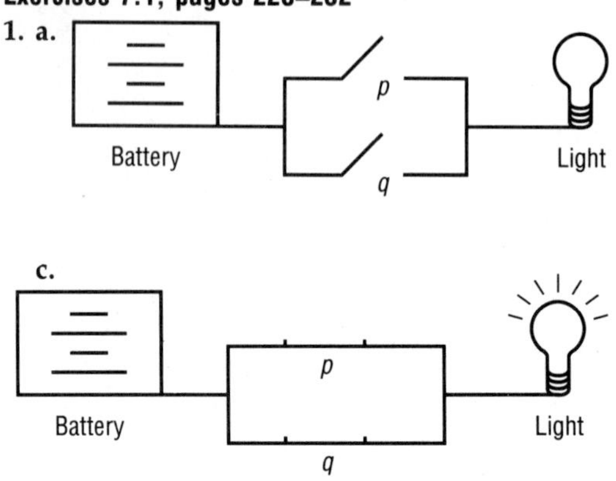

e. If either the switch p or the switch q is closed, then the light bulb lights.

3. a.

p	q	light
closed	closed	lit
closed	open	lit
open	closed	lit
open	open	unlit

b. (p and q) or (p or q)
5. a. not(p and (p or not q))
b. p and not (p or not q) **c.** No

7. a.

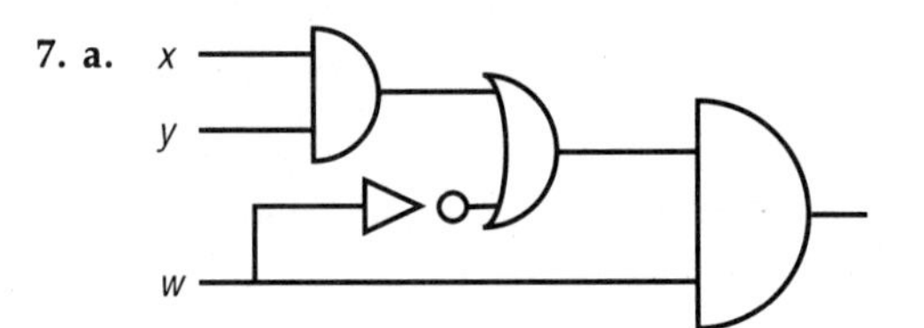

9. a. not (x or y or z)

x	y	z	output
1	1	1	0
1	1	0	0
1	0	1	0
1	0	0	0
0	1	1	0
0	1	0	0
0	0	1	0
0	0	0	1

b. not x and not y and not z

x	y	z	output
1	1	1	0
1	1	0	0
1	0	1	0
1	0	0	0
0	1	1	0
0	1	0	0
0	0	1	0
0	0	0	1

c. Look at the inputs common to both. As long as this set of inputs agrees for both circuits, the output must be the same.

11. Yes or no. Either is acceptable.

Exercises 7.2, pages 235–237

13. a. $3(4 + 5) = 3 \cdot 9 = 27 = 12 + 15 = (3 \cdot 4) + (3 \cdot 5)$

c. $15 \cdot 112 = (10 + 5) \cdot 112 = 10 \cdot 112 + 5 \cdot 112 = 1120 + \frac{1}{2} \cdot (1120) = 1{,}680$ **e.** 1,700

15. a. $x = 1, y = 1, z = 1$

$x + (yz) = 1 + (1 \cdot 1) = 1 + 1 =$

$2(x + y)(x + z) = (1 + 1)(1 + 1) = 2 \cdot 2 = 4$

c. Yes, $x = 1/3, y = 1/3, z = 1/3$

17. Axiom d and Axiom e

19. a. We must show that $1 \vee 0 = 1$ and $1 \wedge 0 = 0$, but these are simply the identity laws. Thus $1' = 0$.

21.

$$
\begin{aligned}
0 \wedge b &= (b' \wedge b) \wedge b && \text{(Axiom e)} \\
&= b' \wedge (b \wedge b) && \text{(Axiom b)} \\
&= b' \wedge b && \text{(Theorem 7.3)} \\
&= 0 && \text{(Axiom e)}
\end{aligned}
$$

23. Suppose $b = b'$. Then

$$
\begin{aligned}
1 &= b \vee b' && \text{(Axiom e)} \\
&= b \vee b && \text{(supposition)} \\
&= b && \text{(Theorem 7.3)} \\
&= b \wedge b && \text{(Theorem 7.3)} \\
&= b \wedge b' && \text{(supposition)} \\
&= 0 && \text{(Axiom e)}
\end{aligned}
$$

But according to Axiom f, $0 \neq 1$ and thus this is impossible. Hence no member of a Boolean algebra is its own complement.

27. a. $2 \vee (3 \vee 5) = 2 \vee 15 = 30$, $(2 \vee 3) \vee 5 = 6 \vee 5 = 30$

29. a. $2 \wedge (3 \wedge 15) = 2 \wedge 15 = 1$; $(2 \wedge 3) \vee (2 \wedge 15) = 1 \vee 1 = 1$

31. Axiom e. 3 has no complement.

33. The cells whose numbers are perfect squares are the only ones whose keys are turned an odd number of times and hence remain unlocked.

Exercises 7.3, pages 240–241

35. a and b

37. a.

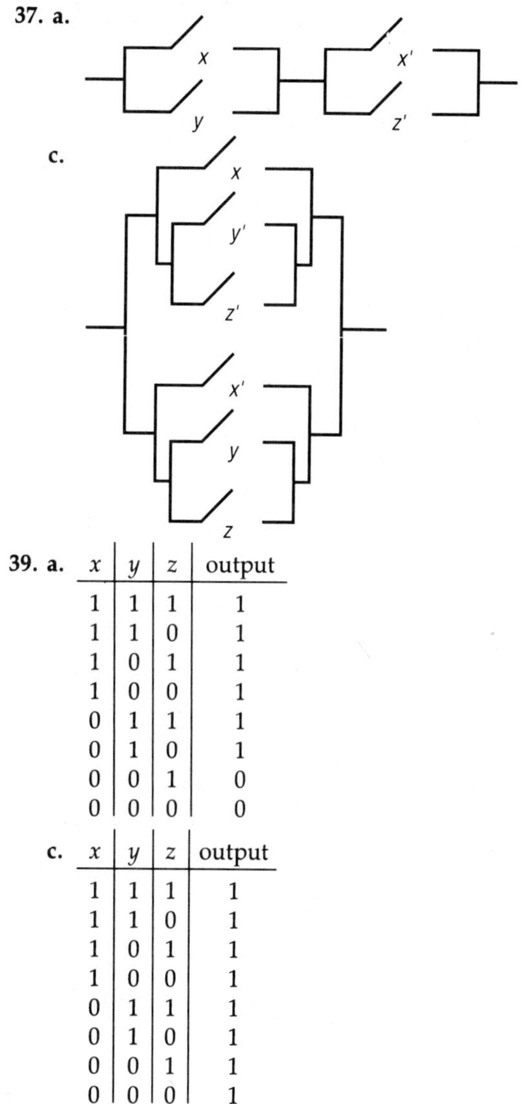

39. a.

x	y	z	output
1	1	1	1
1	1	0	1
1	0	1	1
1	0	0	1
0	1	1	1
0	1	0	1
0	0	1	0
0	0	0	0

c.

x	y	z	output
1	1	1	1
1	1	0	1
1	0	1	1
1	0	0	1
0	1	1	1
0	1	0	1
0	0	1	1
0	0	0	1

41. a. $(x' \wedge (y \vee z))'$, $x \vee (y' \wedge z')$

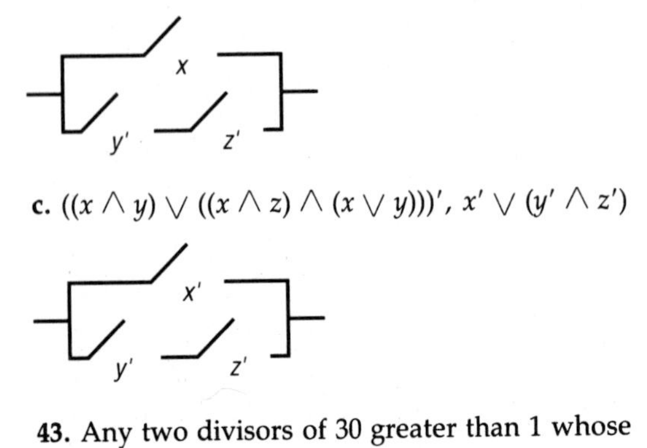

c. $((x \wedge y) \vee ((x \wedge z) \wedge (x \vee y)))'$, $x' \vee (y' \wedge z')$

43. Any two divisors of 30 greater than 1 whose only common divisor is 1 (e.g., 3, 5).
45. Let $x \vee y = x \wedge y$. Then $x = x \wedge (y \vee y') = (x \wedge y) \vee (x \wedge y') = (x \vee y) \vee (x \wedge y') = ((x \vee y) \vee x) \wedge ((x \vee y) \vee y') = (x \vee y) \wedge (x \vee (y \vee y')) = (x \vee y) \wedge (x \vee 1) = (x \vee y) \vee x = x \vee y$. Thus $x = x \vee y = (x \vee y) \vee y = (x \wedge y) \vee y = (x \wedge y) \vee (1 \wedge y) = (x \vee 1) \wedge y = 1 \wedge y = y$.

Exercises 7.4, pages 245–246

47. a. Yes **b.** 1 or and 3 not
c.

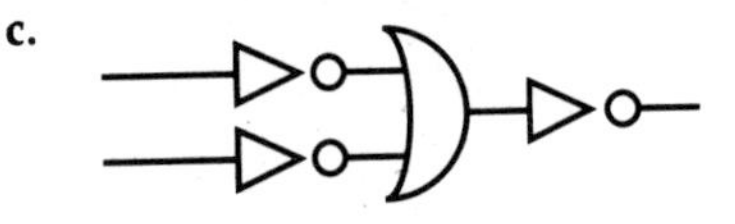

49. a.

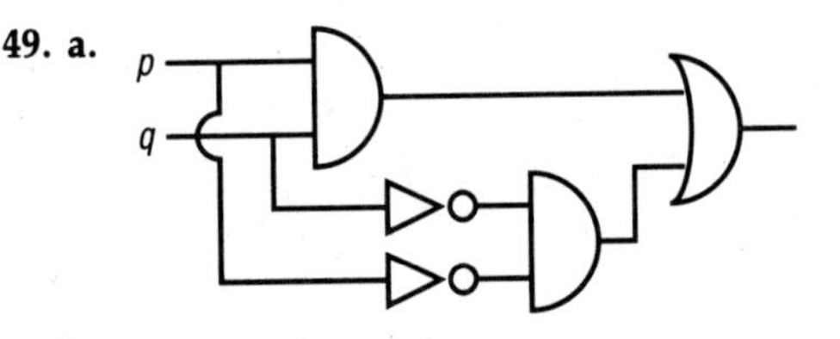

c. **d.**

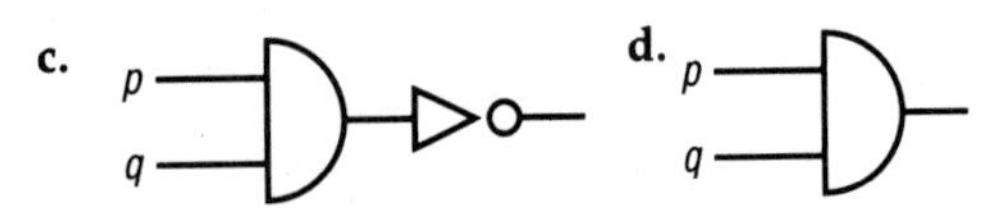

51. (1, 1, 1, 1, 0, 1, 1, 1)
(1, 0, 1, 1, 1, 1, 1, 1)
(0, 0, 1, 1, 0, 1, 1, 1)
(1, 0, 0, 1, 0, 1, 1, 1)
(1, 0, 1, 0, 0, 1, 1, 1)
(1, 0, 1, 1, 0, 0, 1, 1)
(1, 0, 1, 1, 0, 1, 0, 1)
(1, 0, 1, 1, 0, 1, 1, 0)

Review Exercises for Chapter 7, pages 246–247

53. a.

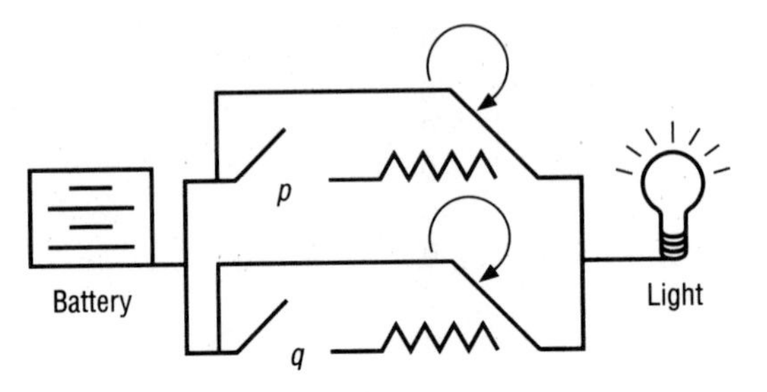

c.

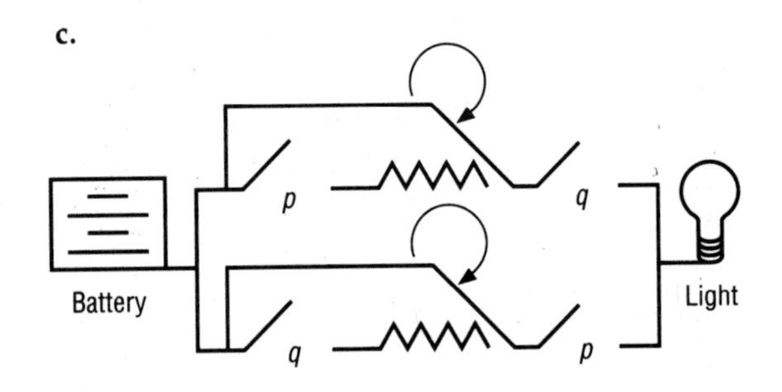

55. Yes. They have the same outputs for equivalent inputs.
57. a. $(5 \wedge 6)' = 1' = 30$, $5' \vee 6' = 6 \vee 5 = 30$
c. $5 \wedge (5 \vee 15) = 5 \wedge 15 = 5$

Chapter 8

Exercises 8.1, pages 260–261

1. a. 3×4 **c.** 1
3. $x = 2$, $y = 2$, $z = 2$
5. $\begin{cases} (x - 2) + 2(y + 1) = 1 \\ 2(x - 2) + 5(y + 1) = 0 \end{cases}$

$$(-2)\begin{cases} x + 2y = 1 \\ 2x + 5y = -1 \end{cases}$$

$x = 7, y = -3$
7. a. 3×4 **b.** 3×5 **e.** 4×5
9. Let $A = [a_{ij}]$, $B = [b_{ij}]$, and $C = [c_{ij}]$ be $m \times n$ matrices. Then $(A + B) + C = A + (B + C)$. Let i and j be natural numbers such that $1 \le i \le m$ and $1 \le j \le n$. The associative law for real numbers tells us that $(a_{ij} + b_{ij}) + c_{ij} = a_{ij} + (b_{ij} + c_{ij})$. Thus $(A + B) + C = A + (B + C)$.

11.

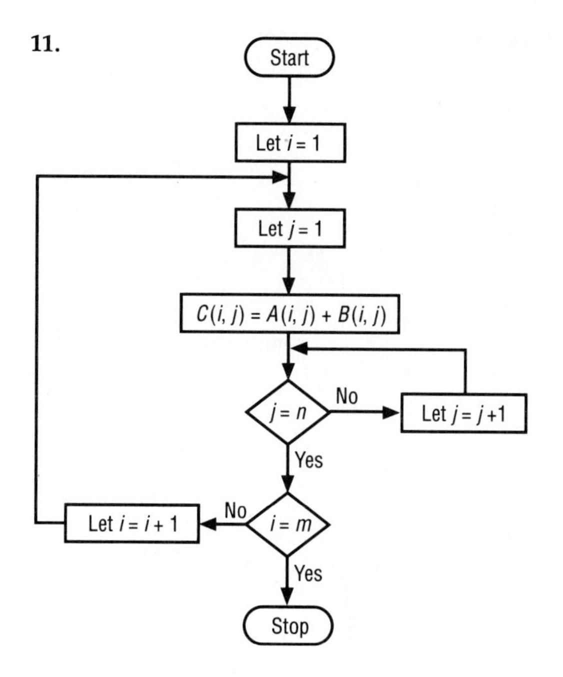

13. $(A \times B) \times C$; $10(80 \cdot 50) + 10(50 \cdot 30) = 55{,}000$; $A \times (B \times C)$ $80(50 \cdot 30) + 10(80 \cdot 30) = 144{,}000$; finding $(A \times B) \times C$ is more efficient.

Exercises 8.2, pages 268–269

15. $\begin{bmatrix} 1 & 0 & 0 & 0 \\ 0 & 2 & 0 & 0 \\ 0 & 0 & 3 & 0 \\ 0 & 0 & 0 & 4 \end{bmatrix}$ **17.** $\begin{bmatrix} 0 & 1 & 1 & 1 \\ 1 & 0 & 1 & 1 \\ 1 & 1 & 0 & 1 \\ 1 & 1 & 1 & 0 \end{bmatrix}$ **19.** $\begin{bmatrix} 1 & 0 & -1 \\ 1 & 0 & -1 \\ 1 & 0 & -1 \end{bmatrix}$

21. For each $n \in \mathbb{N}$, let

$$A_{ij} = \begin{cases} 1, \text{ if } i + j = n + 1 \\ 0, \text{ otherwise} \end{cases}.$$

23. Let $A = [a_{ij}]$ and $B = [b_{ij}]$ be diagonal $n \times n$ matrices and let $C = [c_{ij}]$ be the $n \times n$ matrix such that $C = AB$.
Let i and j be natural numbers such that $1 \le i \le n$ and $1 \le j \le n$. If $i \ne j$, then $a_{ij} = 0$ and $b_{ij} = 0$, and thus $A + B$ is a diagonal matrix. Now $c_{ij} = \sum_{k=1}^{n} a_{ik}b_{kj}$. If $k \ne i$ and $k \ne j$, then $a_{ik} = 0$ and $b_{kj} = 0$. Thus $c_{ij} = a_{ii}b_{ij} + a_{ij}b_{jj} = a_{ii} \cdot 0 + 0 \cdot b_{jj} = 0$. Thus, if i is a natural number, then $c_{ii} = \sum_{k=1}^{n} a_{ik}b_{ki} = a_{ii}b_{ii}$.

25. Let $A = [a_{ij}]$ and $B = [b_{ij}]$ be $m \times n$ zero–one matrices. Let i and j be natural numbers such that $1 \le i \le m$ and $1 \le j \le n$. Then $a_{ij} \vee b_{ij} = b_{ij} \vee a_{ij}$ and $a_{ij} \wedge b_{ij} = b_{ij} \wedge a_{ij}$. Thus $A \vee B = B \vee A$ and $A \wedge B = B \wedge A$.

29. $A = \begin{bmatrix} 1 & 0 & 1 \\ 1 & 0 & 1 \\ 1 & 1 & 1 \end{bmatrix}$ $A^2 = \begin{bmatrix} 1 & 1 & 1 \\ 1 & 1 & 1 \\ 1 & 1 & 1 \end{bmatrix}$

$A^3 = \begin{bmatrix} 1 & 1 & 1 \\ 1 & 1 & 1 \\ 1 & 1 & 1 \end{bmatrix}$ $A^{[n]} = \begin{bmatrix} 1 & 1 & 1 \\ 1 & 1 & 1 \\ 1 & 1 & 1 \end{bmatrix}$

31. $A^{[3n]} = \begin{bmatrix} 1 & 0 & 0 \\ 0 & 1 & 0 \\ 0 & 0 & 1 \end{bmatrix}$ $A^{[3n+1]} = \begin{bmatrix} 0 & 1 & 0 \\ 0 & 0 & 1 \\ 1 & 0 & 0 \end{bmatrix}$

$A^{[3n+2]} = \begin{bmatrix} 0 & 0 & 1 \\ 1 & 0 & 0 \\ 0 & 1 & 0 \end{bmatrix}$

35. $A = \begin{bmatrix} 0 & 0 & 1 \\ 0 & 0 & 0 \\ 1 & 0 & 0 \end{bmatrix}$ $B = \begin{bmatrix} 0 & 0 & 0 \\ 0 & 0 & 0 \\ 1 & 1 & 0 \end{bmatrix}$

$A \odot B = \begin{bmatrix} 1 & 1 & 0 \\ 0 & 0 & 0 \\ 0 & 0 & 0 \end{bmatrix}$

$A \odot B = \begin{bmatrix} 0 & 0 & 0 \\ 0 & 0 & 0 \\ 0 & 0 & 1 \end{bmatrix}$

Exercises 8.3, page 276

39. a. $\begin{cases} x + 2z = 3 \\ 3y = 5 \\ 2x - 3y + 4z = -7 \end{cases}$

c. $\begin{cases} x + 2z - 3w = 5 \\ 2x + 3y + 5z - 4w = 1 \end{cases}$

41. a. $x = 1, y = 2, z = 3$
c. For each real t, $x = 0$, $y = -3t$, $z = t$.

Exercises 8.4, pages 284–285

43. a. $\begin{bmatrix} 4/5 & -1/5 \\ -3/5 & 2/5 \end{bmatrix}$ **c.** $\begin{bmatrix} 0 & 1 \\ 1 & 0 \end{bmatrix}$

45. $\begin{bmatrix} -2 & 1 \\ 3/2 & -1/2 \end{bmatrix}$

47. Let A and B be invertible $n \times n$ matrices. Then $(B^{-1}A^{-1})(AB) = B^{-1}(A^{-1}(AB)) = B^{-1}((A^{-1}A)B) = B^{-1}(I_nB) = B^{-1}B = I_n$.

49. Let A be an invertible $n \times n$ matrix and let k be a real number,. Then $[(1/k)A^{-1}](kA) = (1/k)(A^{-1}(kA)) = 1/k\,(kA^{-1}A) = (1/k)(kI_n) = ((1/k)k)I_n = 1 \cdot I_n = I_n$.

51. a. $\begin{bmatrix} 1&0&0&0 \\ 0&0&0&1 \\ 0&0&1&0 \\ 0&1&0&0 \end{bmatrix}$ **c.** $\begin{bmatrix} 1&0&0&0 \\ 0&1&0&7 \\ 0&0&1&0 \\ 0&0&0&1 \end{bmatrix}$

53. $\begin{bmatrix} 5/28 & -1/28 & 1/4 \\ 3/28 & 5/28 & -1/4 \\ -1/4 & 1/4 & 1/4 \end{bmatrix}$

Exercises 8.5, page 292

55. $7 \cdot \begin{vmatrix} 1 & 3 \\ 0 & 5 \end{vmatrix} - 0 \cdot \begin{vmatrix} 2 & 3 \\ 4 & 5 \end{vmatrix} + (-3)\begin{vmatrix} 2 & 1 \\ 4 & 0 \end{vmatrix} = 7 \cdot 5 - 0 + (-3)(-4) = 47$

57. -170

Exercises 8.6, pages 306–307

59. Let $u = (1, 3, 6)$ and $v = (1, 3, 6)$. Since $3 = 3 \cdot 1$ and $6 = 1 + 5$, u and v are in V. Since $u + v = (2, 6, 12)$ and $12 \neq 2 + 5$, $u + v \notin V$.

61. Suppose that there are numbers a and b such that $(2, -1, 0) = a(1, 1, 1) + b(1, 2, 0)$. Then $a + b = 2$, $a + 2b = -1$, and $a = 0$. But that means that $b = 2$ and $b = -1/2$ which is impossible. Hence $(2, -1, 0)$ is not a linear combination of (1,1,1) and (1, 2, 0).

63. **only a.** $(3, 3, -2) = 3(1, 2, 0) + (-1)(0, 3, 2)$

65. a. $v_1 = 2$, $v_2 = -1$, $v_3 = -2$
c. $v_1 = 9/8$, $v_2 = 17/8$, $v_3 = -7/8$

67. a. det = 19, nonsingular
c. det = 129, nonsingular

69. a. Yes **c.** Yes

71. a. nullity = 1, rank = 2, dimension = 2

Exercises 8.7, pages 314–315

73. a. For eigenvalue 4, the eigenvectors are of the form $\begin{bmatrix} v_1 \\ -v_1 \end{bmatrix}$.
For eigenvalue 9, the eigenvectors are of the form $\begin{bmatrix} v_1 \\ 4v_1 \end{bmatrix}$.

c. For eigenvalue 1, the eigenvectors are of the form $\begin{bmatrix} v_1 \\ -v_1 \end{bmatrix}$.
For eigenvalue 5, the eigenvectors are of the form $\begin{bmatrix} 3v_2 \\ v_2 \end{bmatrix}$.

75. a. 1, 2, -2 **c.** 2

78. a. 1 **c.** 0, 2, 3

Review Exercises for Chapter 8, pages 315–317

87. a. $\begin{bmatrix} 5 & 0 \\ 5 & 10 \end{bmatrix}$ **c.** $\begin{bmatrix} -2 & 1 \\ 0 & 3 \end{bmatrix}$ **e.** $\begin{bmatrix} -2 & 1 \\ -2 & 7 \end{bmatrix}$

89. a. $\begin{bmatrix} 0 & -3 & 1 \\ 1 & 4 & 2 \end{bmatrix}$ **c.** $\begin{bmatrix} -6 & 5 & -3 \end{bmatrix}$

91. Let W be the set of all natural numbers n such that $A^n = \begin{bmatrix} a^n & 0 \\ 0 & 1 \end{bmatrix}$. Since $A^1 = A$ and $a^1 = a$, $1 \in W$. Assume $k \in W$. Then $A^k = \begin{bmatrix} a^k & 0 \\ 0 & 1 \end{bmatrix}$. Thus

$$A^{k+1} = A^k \cdot A = \begin{bmatrix} a^k & 0 \\ 0 & 1 \end{bmatrix}\begin{bmatrix} a & 0 \\ 0 & 1 \end{bmatrix} = \begin{bmatrix} a^{k+1} & 0 \\ 0 & 1 \end{bmatrix}.$$

Therefore $k + 1 \in W$. Hence, if $k \in W$, then $k + 1 \in W$. Thus, by the principle of mathematical induction, W is the set of all natural numbers. Hence the statement is proved.

93. a. $x = 1$, $y = -1$, $z = -1$
c. $x = -1$, $y = -1$, $z = -4$

95. a. $A \wedge B = \begin{bmatrix} 1 & 0 \\ 0 & 0 \end{bmatrix}$, $A \vee B = \begin{bmatrix} 1 & 1 \\ 1 & 1 \end{bmatrix}$

c. $A \wedge B = \begin{bmatrix} 1&0&0 \\ 0&1&0 \\ 1&1&0 \end{bmatrix}$, $A \vee B = \begin{bmatrix} 1&1&1 \\ 0&1&1 \\ 1&1&1 \end{bmatrix}$

97. Let W be the set of all natural numbers n such that $A^{[n]} = A$. Since $A^{[1]} = A$, $1 \in W$. Assume $k \in W$. Then $A^{[k]} = A$. Thus $A^{[k+1]} = A^{[k]} \cdot A = A \cdot A = A$. Therefore $k + 1 \in W$. Hence, if $k \in W$, then $k + 1 \in W$. Thus, by the principle of mathematical induction, W is the set of all natural numbers. Hence the statement is proved.

99. a. $x = 5, y = 2$ **c.** $x = 1, y = -1, z = 2$

101. $v_3 = \begin{bmatrix} 1 \\ 0 \\ 0 \end{bmatrix}$

103. a. No

105. a. 1, 4 **c.** $-2, 2$

107. a. For eigenvalue 1, the eigenspace is $\left\{\begin{bmatrix} 2v \\ v \end{bmatrix} : v \in \mathbb{R}\right\}$ and the geometric multiplicity is 1.

For eigenvalue 4, the eigenspace is $\left\{\begin{bmatrix} v \\ -v \end{bmatrix} : v \in \mathbb{R}\right\}$ and the geometric multiplicity is 1.

c. For eigenvalue -2, the eigenspace is $\left\{\begin{bmatrix} 0 \\ 0 \\ 0 \\ 0 \end{bmatrix}\right\}$ and the geometric multiplicity is 0.

For eigenvalue 4, the eigenspace is $\left\{\begin{bmatrix} u \\ v \\ w \\ -u-v-w \end{bmatrix} : u, v, w \in \mathbb{R}\right\}$ and the geometric multiplicity is 1.

Chapter 9

Exercises 9.1, pages 325–327

1. a. $\{(1, 1), (2, 1), (3, 1), (4, 1), (5, 1), (6, 2), (7, 6)\}$

c. $\{(1, 5), (2, 5), (3, 6), (3, 7), (4,5)\}$

e. $\mathbb{N} \times \mathbb{N}$ **g.** $\{(n\ 17) : n \in \mathbb{N}\}$ **i.** $\mathbb{Z} \times \mathbb{N}$

3. a. $\varnothing$, $\{(1, 2)\}$, $\{(1, 3)\}$, $\{(1, 2), (1, 3)\}$

5. a. Yes

c. $(1, 6), (-1, 6), (3, -\pi), (-2, \pi), (0, 5)$

7. a. $\{3, 5, 6\}$ **c.** $\{3\}$

Exercises 9.2, page 332

11. $\{(1, 1), (1, 2), (1, 3), (7, 1), (2, 5), (4, 4), (4, 5), (5, 3)\}$

13.

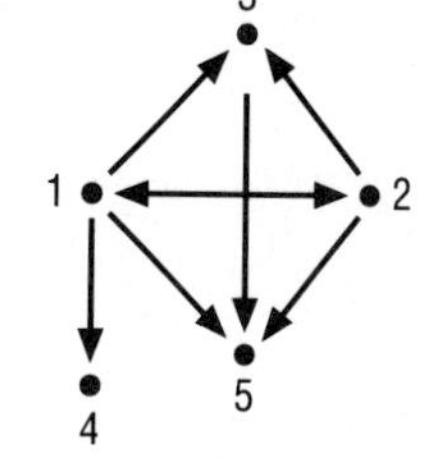

17. a. (1234) **c.** (26435)

Exercises 9.3, page 339

19. a. $\{(x, y) \in \mathbb{R} \times \mathbb{R} : x = 2y^2 + 1\}$

c. $\{(x, y) \in \mathbb{R} \times \mathbb{R} : y = 2x + 7\}$

21. $R^2 = R^3 = R^5 = R^8 = \{1, 2, 3\} \times \{1, 2, 3\}$

23. Let $(x, z) \in (R \circ S) \circ T$. There is a $y \in B$ such that $(x, y) \in T$ and $(y, z) \in R \circ S$. Since $(y, z) \in R \circ S$, there is a $p \in B$ such that $(y, p) \in S$ and $(p, z) \in R$. Since $(x, y) \in T$ and $(y, p) \in S$, $(x,p) \in S \circ T$. Since $(x, p) \in S \circ T$ and $(p, z) \in R$, $(x, z) \in R \circ (S \circ T)$. Thus $R \circ (S \circ T) \supseteq (R \circ S) \circ T$.

Exercises 9.4, pages 344–345

31. a. $\{(1, 1), (1, 2), (2, 2), (2, 3), (3, 1)\}$

c. $\{(1, 3), (2, 2), (3, 1)\}$

33. $M_{R^{-1}} = M^t$

35. a. $\begin{bmatrix} 0 & 1 & 0 \\ 0 & 1 & 1 \\ 1 & 0 & 0 \end{bmatrix}$ **c.** $\begin{bmatrix} 1 & 1 & 1 \\ 1 & 1 & 1 \\ 1 & 1 & 1 \end{bmatrix}$

37. Let $S = \{n \in \mathbb{N} : M_{R^n} = (M_R)^{[n]}\}$. $1 \in S$ because $M_{R^1} = M_R = (M_R)^1$. Suppose that $k \in S$. Then $M_{R^k} = (M_R)^{[k]}$. Thus $M_{R^{k+1}} = M_{R^k} \odot M_R = (M_R)^{[k+1]}$. Thus $k + 1 \in S$ and $S = \mathbb{N}$, and the theorem is proved.

Review Exercises for Chapter 9, pages 345–346

39. a. (3, 2), (7, 1), (17, 5), (5, 4), (5, 1)

41. a. (234) **c.** (14)(23)

43. a. **c.**

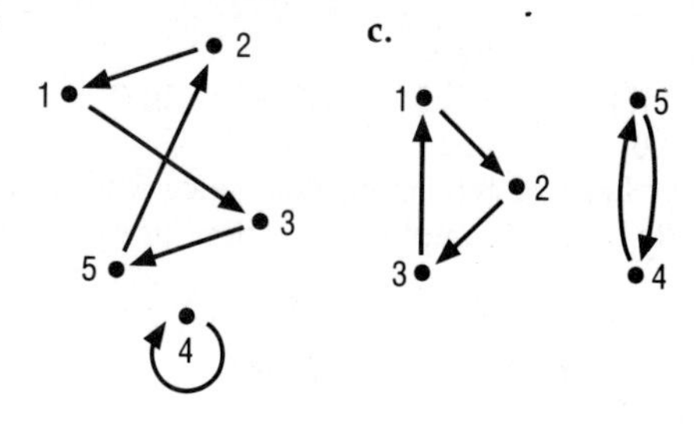

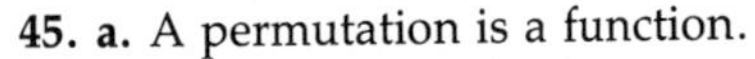

45. a. A permutation is a function.
c.

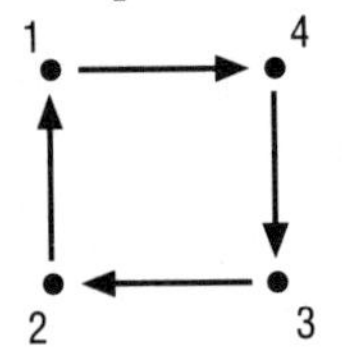

47. a. $\{(2, 1), (4, 3), (3, 2), (1, 3)\}$
c. $\{(1, 2), (2, 4), (2, 1), (4, 3)\}$
e. $\{(1, 3), (2, 1), (3, 3)\}$
49. a. $\{(3, 3), (2, 2), (1, 1), (1, 2)\}$
c. $\{(3, 2), (3, 1), (2, 3), (2, 1), (1, 3), (1, 2)\}$

Chapter 10

Exercises 10.1, pages 354–355

1. a. R_1 is a symmetric but not transitive and not reflexive on A. **c.** R_3 is reflexive on A, symmetric, and transitive. **f.** R_6 is not reflexive on A, not symmetric, and not transitive.
3. a. R_1 is symmetric but not reflexive on $\mathbb{Z}$ and not transitive. **c.** R_3 is reflexive on $\mathbb{Z}$, symmetric, and transitive. **e.** R_5 is symmetric and transitive but not reflexive on $\mathbb{Z}$.
9. Let $(x, y) \in R \cap S$. Then $(x, y) \in R$ and $(x, y) \in S$. Since R and S are symmetric, $(y, x) \in R$ and $(y, x) \in S$. Therefore $(y, x) \in R \cap S$ and hence $R \cap S$ is symmetric.
11. $R = \{(1, 2)\}$ and $S = \{(2, 3)\}$.

Exercises 10.2, pages 359–360

15. a. Let $x \in \mathbb{R}$. Since $x^2 - x^2 = 0$, $(x, x) \in R$ and hence R is reflexive on $\mathbb{R}$. Let $(x, y) \in R$. Then $x^2 - y^2 = 0$, so $y^2 - x^2 = 0$. Hence $(y, x) \in R$ and R is symmetric. Let $(x, y) \in R$ and $(y, z) \in R$. Then $x^2 - y^2 = 0$ and $y^2 - z^2 = 0$. Hence $x^2 - z^2 = (x^2 - y^2) + (y^2 - z^2) = 0 + 0 = 0$ and thus R is transitive.
b. $[5] = \{y \in \mathbb{R} : (5, y) \in R\} = \{y \in \mathbb{R} : 5^2 - y^2 = 0\} = \{-5, 5\}$
19. Let $p \in [y]$. Then yRp. Since R is transitive, xRy and yRp imply xRp. Therefore $p \in [x]$ and $[x] \supseteq [y]$.

Exercises 10.3, pages 366–367

23. 4
27. a. all integers of the form $3n + 1$, where n is an integer
b. $3(1) + 1 = 4$, but 4 is not odd
31. 7
35. Since $a \equiv b \pmod{n}$ and $c \equiv d \pmod{n}$, there exist integers q_1 and q_2 such that $a - b = nq_1$ and $c - d = nq_2$. Therefore $(a + c) - (b + d) = (a - b) + (c - d) = n(q_1 + q_2)$, and hence $a + c \equiv (b + d) \pmod{n}$.
39. Let $c = b - a + n$. Then $(a + c) - b = a + (b - a + n) - b = n$, and hence $a + c \equiv b \pmod{n}$.

Exercise 10.4, pages 373–374

44. a, b, and f **47.** Yes **51.** d
55. a. No. Find partitions $\mathcal{A}$ and $\mathcal{B}$ with the property that there is an $A \in \mathcal{A}$ and a $B \in \mathcal{B}$ such that $A \cap B = \emptyset$. **b.** No. Find partitions $\mathcal{A}$ and $\mathcal{B}$ and a point $p \in X$ such that there are members A_1 and A_2 of $\mathcal{A}$ and B_1 and B_2 of $\mathcal{B}$ with $p \in A_1$, $p \notin A_2$, $p \in B_1$, and $p \notin B_2$. Then $p \in A_1 \cup B_2$ and $p \in A_2 \cup B_1$.

Exercises 10.5, pages 383–385

59. a. $R \cup \{(1, 1), (4, 4), (5, 5)\}$
b. $R \cup \{(2, 1), (3, 2), (1, 3), (1, 4)\}$
63. a. $\mathbb{R} \times \mathbb{R}$ **b.** $\mathbb{R} \times \mathbb{R}$
67. $R \cup \{(1, 3), (2, 2), (3, 3)\}$
71. Let $a \in A$. Since R is reflexive on A, $(a, a) \in R = R^1$. Therefore, by Theorem 10.17, $(a, a) \in R^*$ and hence R^* is reflexive on A.

Exercises 10.6, pages 395–398

75. a and b **79. a.** $\{1, 2\}$ and $\{2, 3\}$
b. 2 and 3 **c.** 3 and 4
83. Let $(a, b) \in A \times B$. Then aRa and bSb. Therefore $(a, b) \leq (a, b)$ and $\leq$ is reflexive on $A \times B$. Suppose $(a, b) \leq (c, d)$ and $(c, d) \leq (a, b)$. Then $(a, c) \in R$ and $(c, a) \in R$. Hence $a = c$. Therefore $(b, d) \in S$ and $(d, b) \in S$, and hence $b = d$. Thus $(a, b) = (c, d)$ and $\leq$ is antisymmetric. Suppose $(a, b) \leq (c, d)$ and $(c, d) \leq (e, f)$. Then $(a, c) \in R$ and $(c, e) \in R$. Hence $(a, e) \in R$. If $a \neq e$, then $(a, b) \leq (e, f)$. Suppose $a = e$. Then $(a, c) \in R$ and $(c, a) \in R$ and hence $a = c$. Therefore $(b, d) \in S$ and $(d, f) \in S$ and hence $(b, f) \in S$. Therefore $(a, b) \leq (e, f)$. Hence $\leq$ is transitive.
87. a. Since a is a greatest element of $(A, \leq)$,

$b \leq a$. Since b is a greatest element of $(A, \leq)$, $a \leq b$. Therefore $a = b$.

93. Let $a, b \in A_1 \cup A_2$. Define $a \leq b$ provided that (1) $a, b \in A_1$ and $a \leq_1 b$, (2) $a, b \in A_2$ and $a \leq_2 b$, or (3) $a \in A_1$ and $b \in A_2$.

Exercises 10.7, page 403–404

95. a. Yes **b.** Yes

97. a. $\{n \in \mathbb{N} : n \leq 25\}$

b. $\mathbb{R}$ **c.** Yes **d.** No

e. 1. Let $n = 1$. 2. Let $x = a[n]$. 3. Let $a[n] = b[n]$. 4. Let $b[n] = x$. 5. If $n = 25$, then stop. 6. Let $n = n + 1$. 7. Go to step 2.

99. 14

101. The mail is on a businessman's desk when he is on vacation. Each day the mail is added to the top of the stack, and when the businessman returns he will look at the top first. Last in, first out.

Review Exercises for Chapter 10, pages 404–407

103. a. R_1 is symmetric and transitive but not reflexive on A.

107. There are two equivalence classes: one is the set of all even integers and the other is the set of all odd integers.

111. a. Some examples of two-element partitions of X are {{1}, {2, 3, 4, 5}} and {{1, 2}, {3, 4, 5}}. There are 5 partitions where one member of the partition is a one-element set, and there are 10 partitions where one member of the partition is a two-element set. Thus there are 15 two-element partitions of X.

115. $R^* = R$

119. a. (1, 1), (1, 2), (1, 3), (1, 4), (2, 1), (2, 2), (2, 3), (2, 4), (3, 1), (3, 2)

123. Friday

Chapter 11

Exercises 11.1, pages 416–417

1.

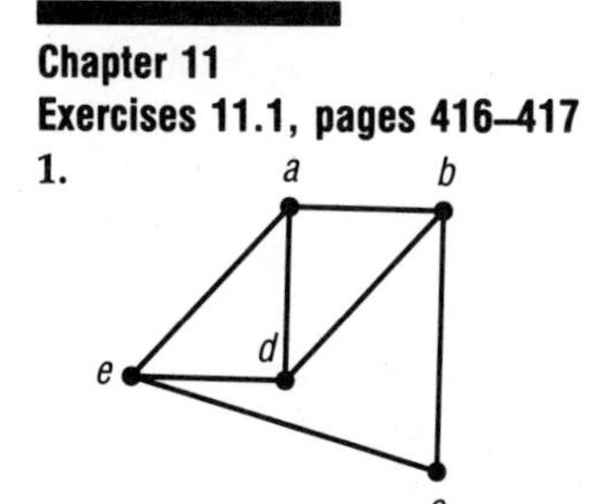

3. a. b and c **b.** c, d, and e **c.** b and c

5.

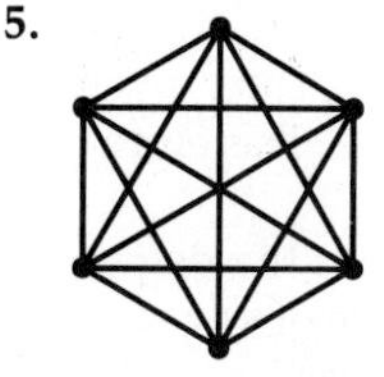

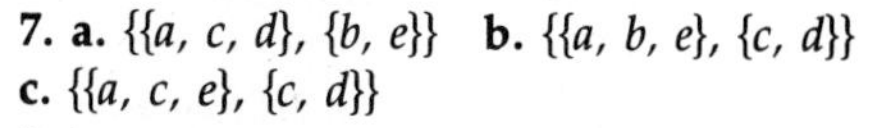

7. a. $\{\{a, c, d\}, \{b, e\}\}$ **b.** $\{\{a, b, e\}, \{c, d\}\}$ **c.** $\{\{a, c, e\}, \{c, d\}\}$

9.

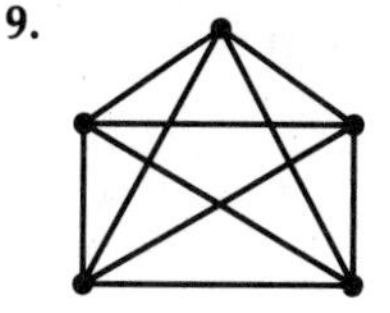

11.

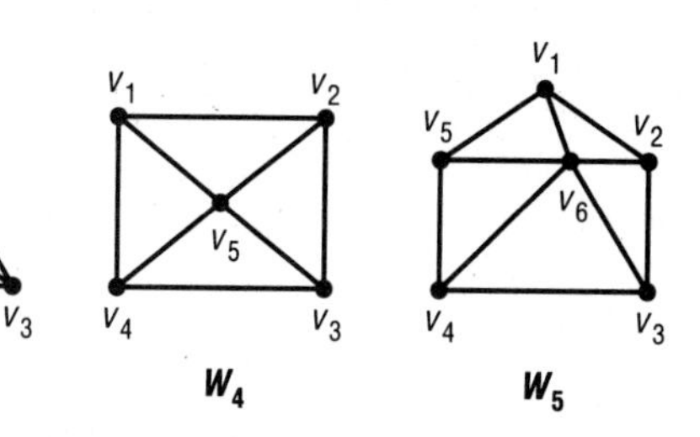

Exercises 11.2, pages 425–427

13. a. six vertices

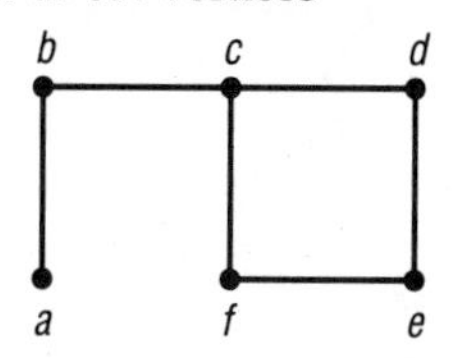

Then deg(a) = 1, deg(b) = 2, deg(c) = 3, deg(d) = 3, deg(e) = 2, and deg(f) = 3 It is a simple graph.

15. a. Since there are five edges, if such a walk were possible it could be represented by a sequence of six letters, where each letter is one of a, b, c, or d, and consecutive letters mean that the edge joining these two vertices has been traversed. Since each edge is to be crossed just once, the letters a and b would appear as consecutive letters once in the sequence. The same is true of the letters b and d, c and d, a and c, and b and c. Since three edges lead to vertex b, the letter b has to appear in the sequence twice, once to denote an entrance to and exit from b and once to indicate an entrance to or exit from b. The same is true of c. The letter d must appear once and the letter a twice, at the beginning and end. Thus seven letters must appear in the sequence. Therefore such a walk is not possible.

16. By Theorem 11.1, the sum of the degrees of the vertices is twice the number of edges. Since the sum of the degrees of the vertices is 14, the number of edges is 7.

17. a. The sum of the degrees of the vertices is 12. Thus the number of edges must be 6. Such a simple graph is

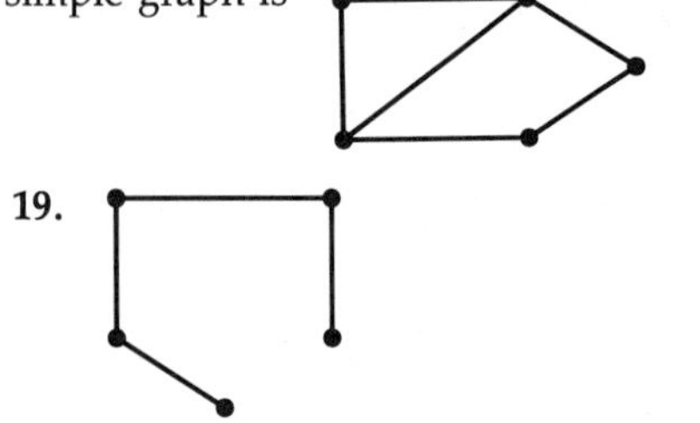

19.

Exercises 11.3, pages 432–433

23. $G_1 \cup G_2 = (V_1 \cup V_2, E_1 \cup E_2, f_1 \cup f_2)$ and $G_1 \cap G_2 = (V_1 \cap V_2, E_1 \cap E_2, f_1 \cap f_2)$

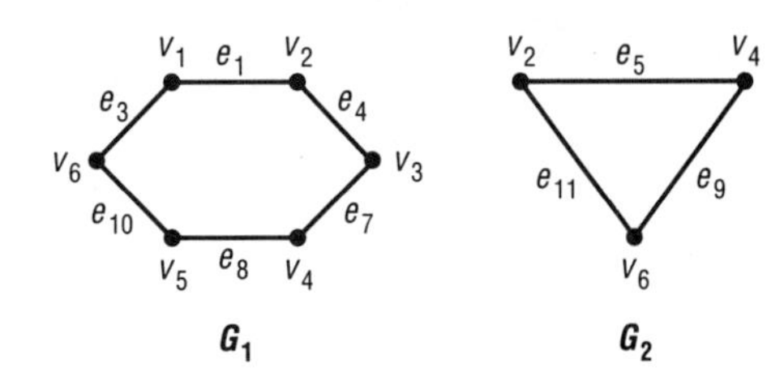

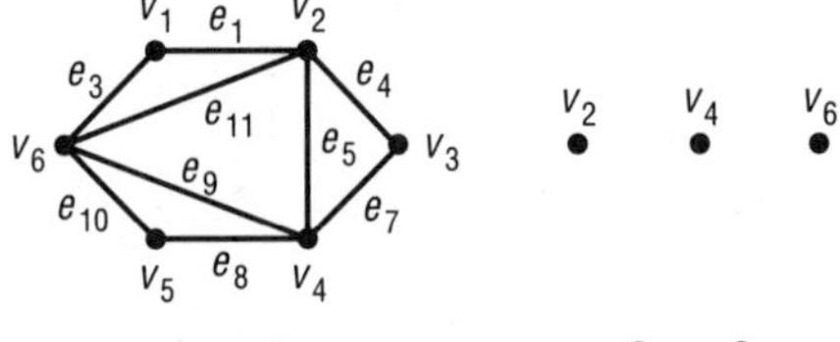

25. a.

$$\begin{array}{c} \\ a \\ b \\ c \\ d \end{array}\begin{array}{c} \begin{array}{cccc} a & b & c & d \end{array} \\ \begin{bmatrix} 0 & 2 & 0 & 1 \\ 2 & 1 & 1 & 0 \\ 0 & 1 & 2 & 2 \\ 1 & 0 & 2 & 0 \end{bmatrix} \end{array}$$

27.

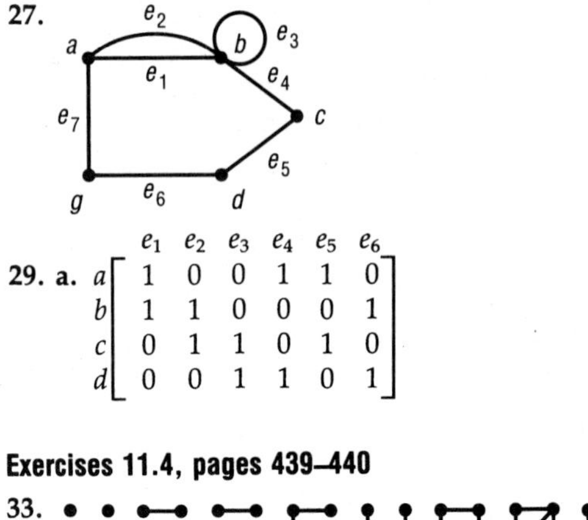

29. a.

$$\begin{array}{c} \\ a \\ b \\ c \\ d \end{array}\begin{array}{c} \begin{array}{cccccc} e_1 & e_2 & e_3 & e_4 & e_5 & e_6 \end{array} \\ \begin{bmatrix} 1 & 0 & 0 & 1 & 1 & 0 \\ 1 & 1 & 0 & 0 & 0 & 1 \\ 0 & 1 & 1 & 0 & 1 & 0 \\ 0 & 0 & 1 & 1 & 0 & 1 \end{bmatrix} \end{array}$$

Exercises 11.4, pages 439–440

33.

35.

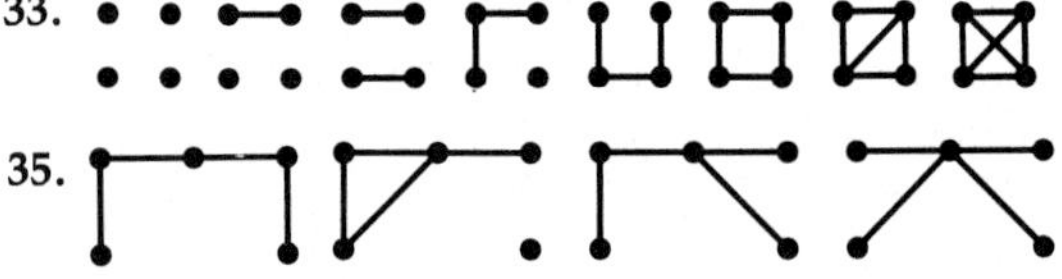

37. Suppose $G = (V_1, E_1)$ and $H = (V_2, E_2)$ are simple graphs and $\theta : V_1 \rightarrow V_2$ is an isomorphism. If $V_1 = \{v_1, v_2, \ldots, v_n\}$, then $V_2 = \{\theta(v_1), \theta(v_2), \ldots, \theta(v_n)\}$ is a permutation of the vertices of H such that G and H have the same adjacency matrix.

Suppose $V_1 = \{u_1, u_2, \ldots, u_n\}$ and $V_2 = \{v_1, v_2, \ldots, v_n\}$ and the vertices of G and H are so that G and H have the same adjacency matrix. Then $\theta : V_1 \rightarrow V_2$ defined by $\theta(u_i) = v_i$ for each $i = 1, 2, \ldots, m$ is an isomorphism.

39. The following two graphs are not isomorphic:

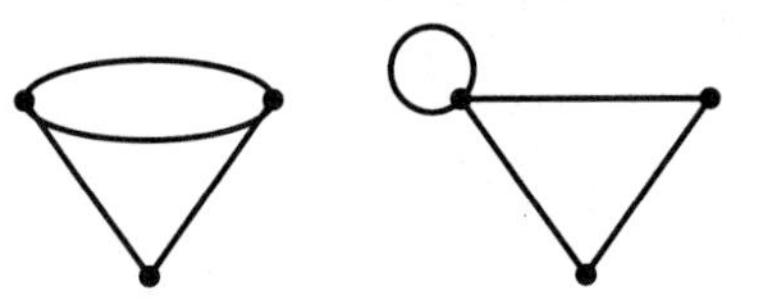

Exercises 11.5, pages 446–448

41. a. (i) $ae_1be_3ce_3be_5g$ (ii) $ae_3ce_6de_7ce_6de_8g$ (iii) $ae_2ke_4ce_5ke_4ce_8de_9he_{11}g$ (iv) $ae_1be_5de_4be_5de_9g$

c. (i) ae_1be_5g (ii) $ae_3ce_6de_8g$ (iii) $ae_1be_7de_9he_{11}g$ (iv) $ae_3de_{10}h$

43. a. is connected

45. a. 2 **b.** 0 **c.** 2 **d.** 0 **47.** 3

49. $K_{4,3}$ $K_{3,4}$ H_1 H_2 H_3 H_4

U_1 U_2 U_3

Exercises 11.6, pages 453–454

56. a. It is not eulerian because it has a vertex of odd degree.

c. It is not eulerian because it has a vertex of odd degree.

e. $v_1e_1v_2e_2v_1e_3v_4e_8v_5e_9v_4e_7v_3e_5v_2e_6v_5e_4v_1$ is an Euler tour.

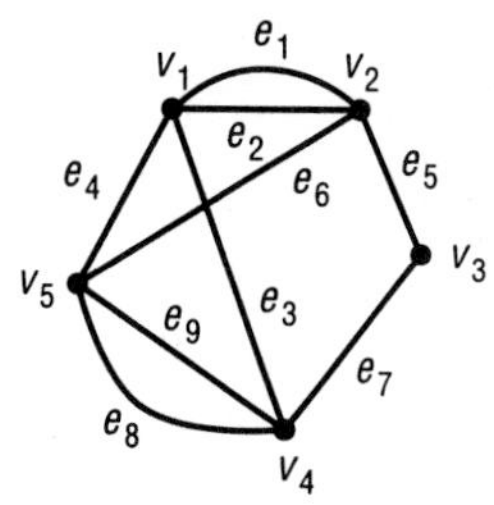

g. It is not eulerian because it is not connected.

i. It is not eulerian because it has a vertex of odd degree.

57. a. $v_1e_1v_2e_2v_3e_5v_4e_6v_5e_4v_2e_3v_4e_7v_6$ is an Euler trail.

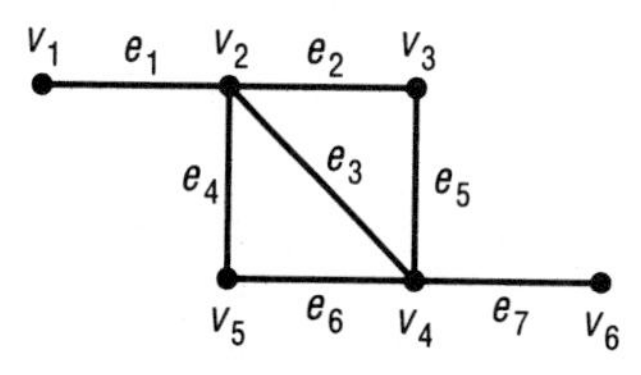

c. Since the graph has three vertices of order 3, it does not contain an Euler trail.

e. The Euler tour in Exercise 56e is an Euler trail.

g. Since the graph is not connected, it does not contain an Euler trail.

i. It does not contain an Euler trail because it has four vertices of odd degree.

61. If $n > 1$ and n is odd, then the degree of each vertex of K_n is even. Therefore K_n is eulerian. If n is even, the degree of each vertex of K_n is odd, so K_n is not eulerian.

Exercises 11.7, pages 461–462

67. b. *abcdgh* is a Hamilton path.

c. The graph is not connected, so no path can contain every vertex.

d. *adcgjhb* is a Hamilton path.

69.

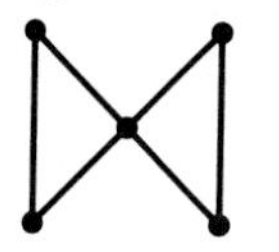

71. Let $\{V_1, V_2\}$ be a bipartition of $K_{n,n}$. Then $V_1 = \{u_1, u_2, \ldots, u_n\}$ and $V_2 = \{v_1, v_2, \ldots, v_n\}$. Then $u_1v_1u_2v_2 \ldots u_nv_nu_1$ is a Hamilton cycle.

73. Suppose C is a Hamilton cycle in the graph G. For each vertex v of G there are exactly two edges incident with v that are members of C. Thus there are three edges incident with d, three edges incident with h, and three edges incident with p that are not members of C. Also there is one edge incident with b, one edge incident with m, one edge incident with r, and one edge incident with t that are not members of C. This leaves only fourteen edges that can be members of C. However, there are sixteen vertices, so every Hamilton cycle must contain sixteen edges. This is a contradiction.

77. Let H be the subgraph of G that results when v and all the edges that are incident with v are removed. Since G is connected but H is not, there are two vertices a and b that are incident with v such that there is no (a, b)-path in H. Since any Hamilton cycle in G must contain v and an (a, b)-path in H, G is not hamiltonian.

Exercises 11.8, pages 474–476

79. a. u_0v **c.** u_0bv **e.** u_0hbckv

85. a. 1. Assign the label 0 to the vertex u and let $L = \{u\}$. 2. $v \notin L$.

3. There are three unlabeled vertices a, b, and c that are adjacent to u. Assign the label 1 to these vertices and let $L = \{u, a, b, c\}$. Vertex u is the predecessor of a, b, and c.

2. $v \notin L$.

3. There are five unlabeled vertices d, g, h, j, and k that are adjacent to vertices in L that bear the label 1. Assign the label 2 to these five vertices and let $L = \{u, a, b, c, d, g, h, j, k\}$. Vertex a is the predecessor of g, vertex b is the predecessor of h, and vertex c is the predecessor of k. Choose a as the predecessor of d and b as the predecessor of j.

2. $v \notin L$.

3. There are three unlabeled vertices m, n, and v that are adjacent to vertices in L that bear the label 2. Assign the label 3 to these three vertices and let $L = \{u, a, b, c, d, g, h, j, k, m, n, v\}$. Vertex k is the predecessor of v. Choose g as the predecessor of m and h as the predecessor of n.

2. $v \in L$. Stop. The distance from u to v is 3, and

a shortest path from u to v is obtained by reversing the order of $vkcu$.

Exercises 11.9, pages 487–488

87. A planar representation of $K_{2,5}$ is

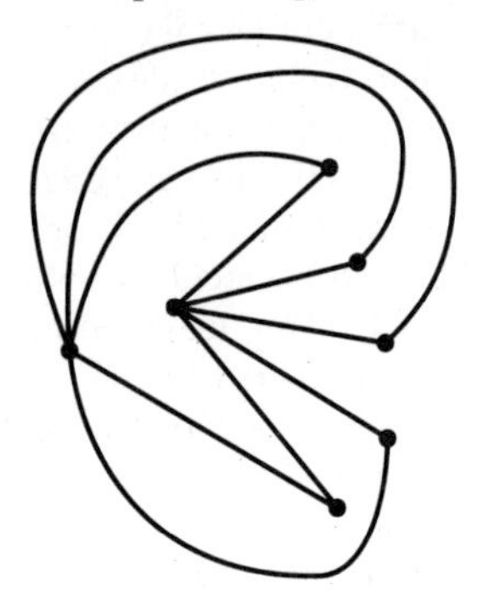

89. a. $\nu = 4$, $\varepsilon = 6$, and $\varnothing = 4$, so $\nu - \varepsilon + \varnothing = 4 - 6 + 4 = 2$

c. $\nu = 5$, $\varepsilon = 9$, and $\varnothing = 6$, so $\nu - \varepsilon + \varnothing = 5 - 9 + 6 = 2$

93. $K_{3,3}$ has six vertices, so $2\nu - 4 = 8$. Since it is a connected simple graph with no cycles of length 3 and $\varepsilon = 9$, by Theorem 11.26 it cannot be planar.

95. Let $v_1, v_2, \ldots, v_7$ denote the vertices of G. Then $\sum_{i=1}^{7} \deg(v_i) = 18$. So by Theorem 11.1 the number of edges is nine. Therefore, by Theorem 11.21, $6 - 9 + \varnothing = 2$. Hence the number of faces of G is five.

99.

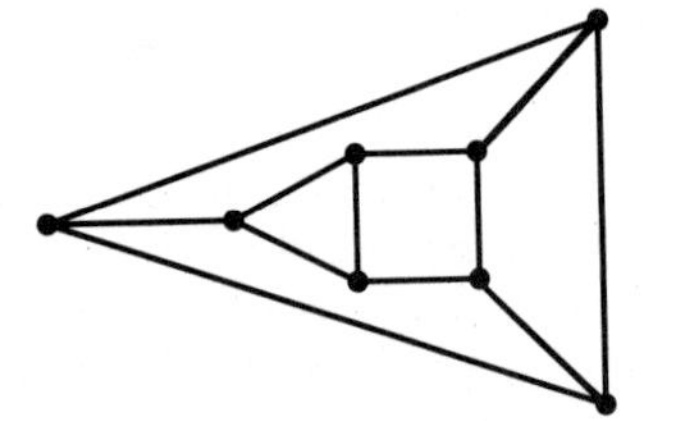

Exercises 11.10, pages 495–497

105. a. **c.**

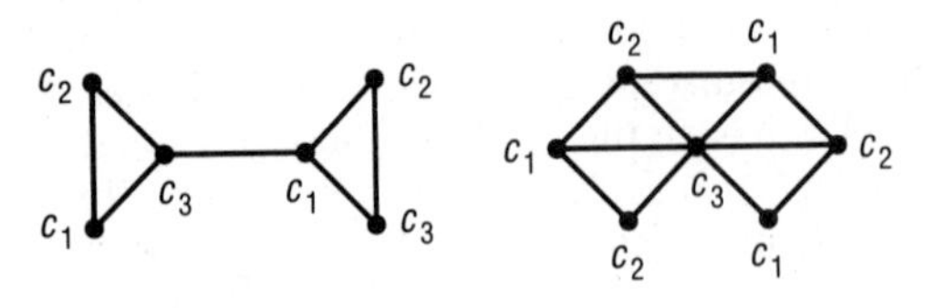

107. 3 if n is even and 4 if n is odd.

109.

111. There are seven committee meetings to arrange. Let each committee be represented by a vertex, and join two vertices with an edge if there is a person on both committees.

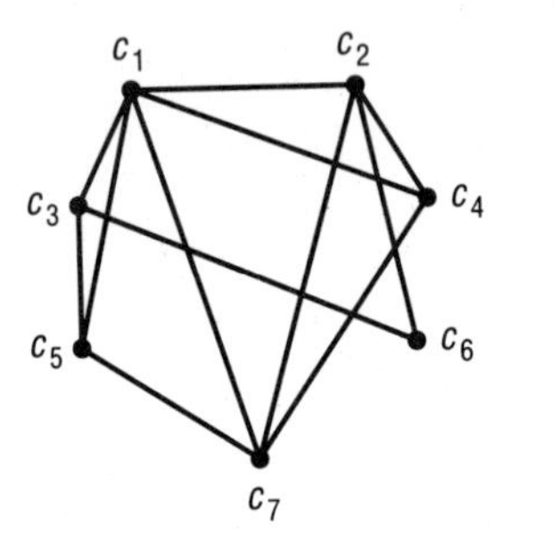

Meetings c_1 and c_6 can be arranged at the same time, c_2 and c_3 can be arranged at the same time, and c_4 and c_5 can be arranged at the same time. Meeting c_7 must meet at a fourth time.

Review Exercises for Chapter 11, pages 497–502

117. a. $V = \{a, b, c, d, g\}$

c. $\deg(a) = \deg(c) = \deg(d) = 2$, $\deg(b) = \deg(g) = 3$

119. a. This graph is simple, and it has five vertices and eight edges. The degree of each vertex is as shown:

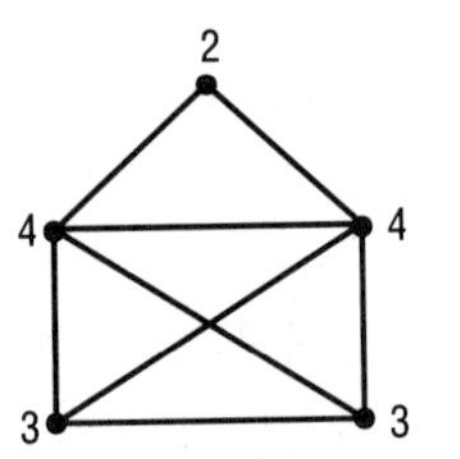

c. This is a graph without loops, but it has multiple edges. It has four vertices and five edges. The degree of each vertex is as shown:

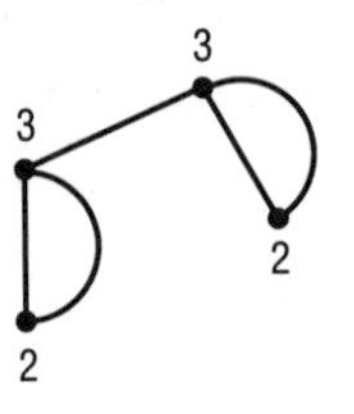

121. a. Label the edges as shown:

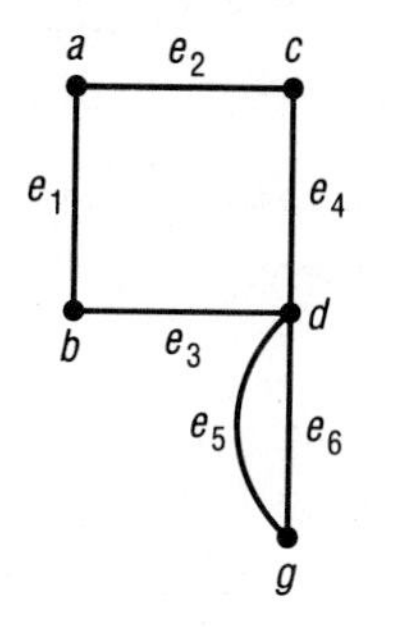

Then $ae_2ce_4de_5ge_6de_3be_1a$ is such a walk.
c. No. There are four vertices of odd degree.
123. $3 + 2 + 2 + 1 + 1 = 9$, but by Theorem 11.1 the sum of the degrees of the vertices must be twice the number of edges.

125. a.

$$A(G) = \begin{array}{c} \\ a \\ b \\ c \\ d \end{array} \begin{array}{c} \begin{array}{cccc} a & b & c & d \end{array} \\ \begin{bmatrix} 0 & 2 & 1 & 0 \\ 2 & 0 & 0 & 1 \\ 1 & 0 & 1 & 2 \\ 0 & 1 & 2 & 1 \end{bmatrix} \end{array}$$

$$M(G) = \begin{array}{c} \\ a \\ b \\ c \\ d \end{array} \begin{array}{c} \begin{array}{cccccccc} e_1 & e_2 & e_3 & e_4 & e_5 & e_6 & e_7 & e_8 \end{array} \\ \begin{bmatrix} 1 & 1 & 1 & 0 & 0 & 0 & 0 & 0 \\ 1 & 1 & 0 & 1 & 0 & 0 & 0 & 0 \\ 0 & 0 & 1 & 0 & 1 & 1 & 0 & 2 \\ 0 & 0 & 0 & 1 & 1 & 1 & 2 & 0 \end{bmatrix} \end{array}$$

c.

$$A(G) = \begin{array}{c} \\ a \\ b \end{array} \begin{array}{c} \begin{array}{cc} a & b \end{array} \\ \begin{bmatrix} 1 & 3 \\ 3 & 2 \end{bmatrix} \end{array}$$

$$M(G) = \begin{array}{c} \\ a \\ b \end{array} \begin{array}{c} \begin{array}{cccccc} e_1 & e_2 & e_3 & e_4 & e_5 & e_6 \end{array} \\ \begin{bmatrix} 1 & 1 & 1 & 2 & 0 & 0 \\ 1 & 1 & 1 & 0 & 2 & 2 \end{bmatrix} \end{array}$$

127.

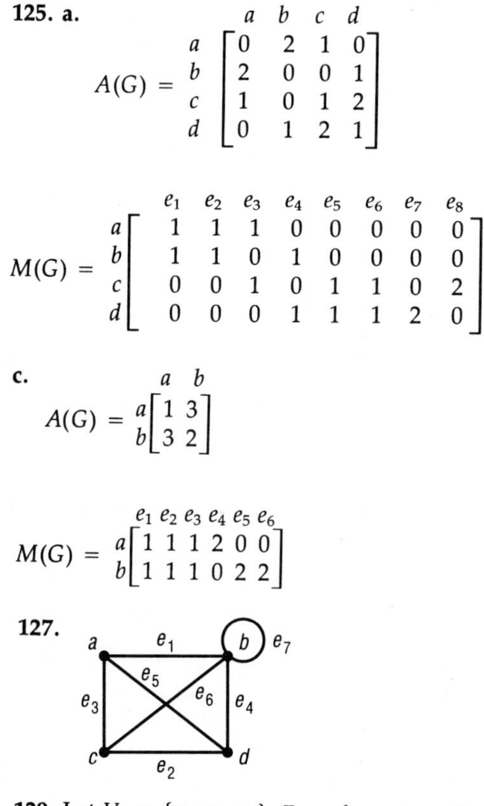

129. Let $V_1 = \{v_1, v_2, v_3\}$, $E_1 = \{e_1, e_2, e_3, e_4\}$, and define $f : E_1 \to \{\{u, v\} : u, v \in V_1\}$ by $f(e_1) = \{v_1, v_3\}$, $f(e_2) = \{v_1, v_3\}$, $f(e_3) = \{v_1, v_2\}$, and $f(e_4) = \{v_2, v_3\}$. Let $V_2 = \{v_a, v_b, v_c\}$, $E_2 = \{e_a, e_b, e_c, e_d\}$, and define $f : E_2 \to \{\{u, v\} : u, v \in V_2\}$ by $f(e_a) = \{v_a, v_b\}$, $f(e_b) = \{v_b, v_c\}$, $f(e_c\} = \{v_a, v_c\}$, and $f(e_d) = \{v_a, v_c\}$. Then one of the graphs is (V_1, E_1, f_1) and the other is (V_2, E_2, f_2). Define a bijection $\theta : V_1 \to V_2$ by $\theta(v_1) = v_a$, $\theta(v_2) = v_b$, $\theta(v_3) = v_c$ and a bijection $\phi : E_1 \to E_2$ by $\phi(e_1) = e_c$, $\phi(e_2) = e_d$, $\phi(e_3) = e_a$, $\phi(e_4) = e_b$. Then (θ, ϕ) is an isomorphism.
131. a. It has one component, and it is connected.
c. It has three components, and it is not connected
133. a. It is not eulerian since it has vertices of odd degree (Corollary 11.17).
c. Label the graph as shown:

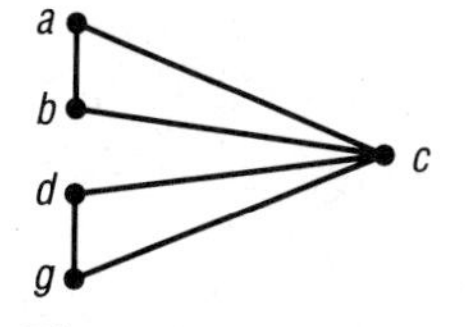

Then *abcdga* is an Euler tour.
e. It is not eulerian since it has vertices of odd degree.
137. a. **c.**

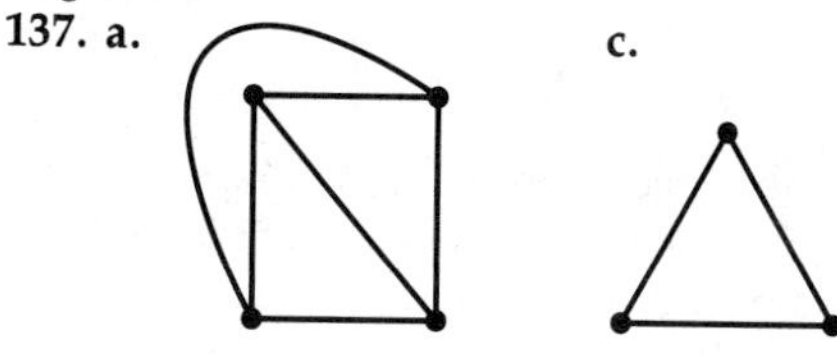

139. They are all planar.
141. a. 4

Chapter 12

Exercises 12.1, pages 510–511

3. 10
7. There is a unique path between each pair of vertices, so the number of paths in a tree with six vertices is $C(6,2) = 15$.
9.

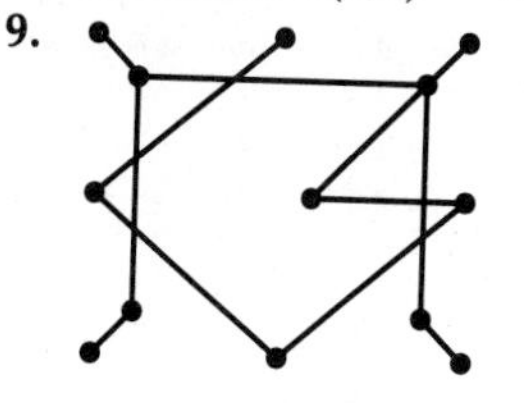

11. $C(n, 2) = n(n - 1)/2$

13. For each $n \in \mathbb{N}$, $K_{1,n}$ and $K_{n,1}$ are trees. Suppose $m, n \in \mathbb{N}$, $m \geq 2$, and $n \geq 2$. Let $\{V_1, V_2\}$ be a bipartition of $K_{m,n}$. If $a, b \in V_1$, and $c, d \in V_2$, then $acbda$ is a cycle in $K_{m,n}$ and hence $K_{m,n}$ is not a tree.

Exercises 12.2, pages 520–523

15. a. The level number of r is 0, the level number of a, b, c, and d is 1, and the level number of g and h is 2. Therefore the height of the tree is 2.
c. The level number of r is 0 and the level number of a is 1. The level number of b and c is 2, and the level number of d, g, and h is 3. Therefore the height of the tree is 3.

17. a.

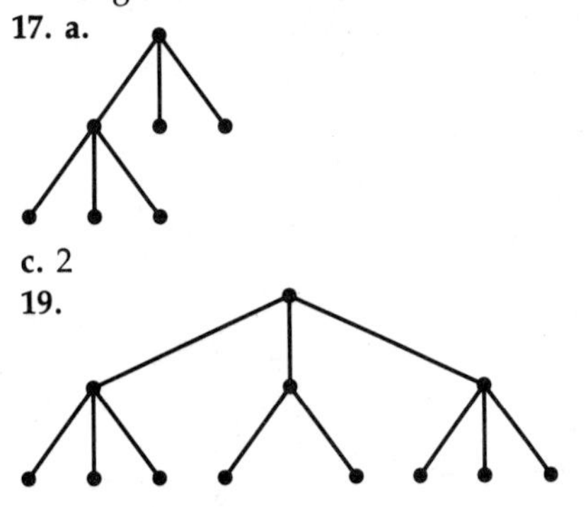

c. 2

19.

Note that this tree is not a regular 3-ary tree.

21. None. The height of (a) is 2 but there is a leaf of level number 1. The height of (b) is 3 but there are leaves of level numbers 1 and 2. Tree (c) is not regular m-ary. The height of (d) is 2 but there are leaves of level number 1.

Exercises 12.3, pages 532–535

31.

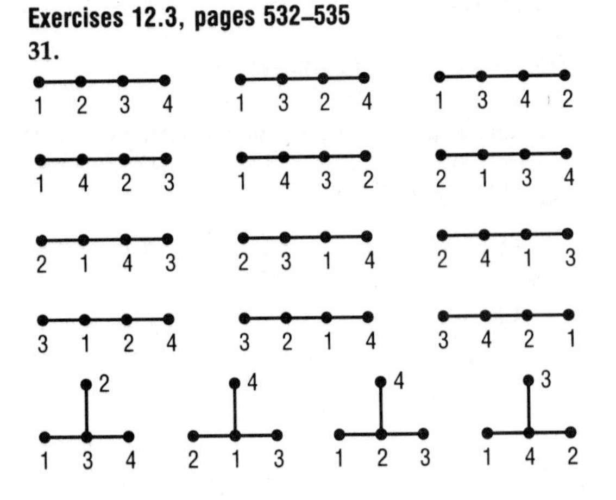

35. We must show that, if $H = (L, T)$ is the tree produced by the breadth-first search spanning tree algorithm, then $L = V$. Let $v \in V$. Since G is connected, there is a (u, v)-path $ue_1a_1e_2 \ldots a_{n-1}e_nv$ in G. By the construction process, v bears a label that is less than or equal to n. Therefore $v \in L$.

37. If $H = (L, T)$ is the graph that results after an iteration of step 5 and v is the vertex that is added during this iteration of step 5, then the construction process guarantees that there is a (u, v)-path in H. Therefore H is connected. Each time an edge e is added to T in step 5, one vertex of e is already in L and the other is not. Thus, adding e to T cannot create any cycles. Therefore H is a tree.

Exercises 12.4, pages 539–540

43. a. $1 + 1 + 2 + 2 = 6$
c. $1 + 1 + 1 + 2 + 2 + 2 + 2 + 2 + 3 + 3 = 19$

45. a. $1 + 2 + 1 + 2 = 6$
c. $2 + 2 + 1 + 2 + 1 + 3 + 2 + 3 + 1 + 1 = 19$

47. Let $v \in V$. Since G is connected, there is a (u,v)-path $ue_1a_1e_2a_2 \ldots a_{n-1}e_nv$ in G. Suppose $v \notin U$. Then no vertex adjacent to v can be in U. In particular, a_{n-1} is not in U. Likewise, for each $i = n - 1, n - 2, \ldots, 2$, if $a_1 \notin U$, then $a_{i-1} \notin U$. Finally, if $a_1 \notin U$, then $u \notin U$. Thus we have a contradiction, so $v \in U$ and hence the tree produced by Prim's algorithm when applied to a connected weighted simple graph is a spanning tree.

48. Let $G = (V, E)$ be a connected weighted simple graph with n vertices. Suppose that the successive edges chosen by Prim's algorithm are $e_1, e_2, \ldots, e_{n-1}$. Let S denote the spanning tree chosen by Prim's algorithm (the edges of S are $e_1, e_2, \ldots, e_{n-1}$). Since a spanning tree of G exists, there is a minimal spanning tree of G. For each $j = 1, 2, \ldots, n - 1$, let S_j be the tree whose edges are $e_1, e_2, \ldots, e_j$, and let S_j' denote the tree whose edges are those edges of S that are not edges of S_j. Let k be the largest integer such that there is a minimal spanning tree of G which contains the first k edges chosen by Prim's algorithm, and let T denote such a tree. It is sufficient to show that $S = T$. Suppose $S \neq T$. Then $k \leq n - 1$, so T contains $e_1, e_2, \ldots, e_k$, but T does not contain e_{k+1}. Let T' be the graph obtained from T by adding the edge e_{k+1}. Then T' has n edges, and hence it cannot be a tree. Since T' is connected, it must contain a cycle C. Since there is no cycle in T, C must contain e_{k+1}. Since S'_k is a tree, there is an edge in C that does not belong to S'_k. Start at the vertex of e_{k+1} that is also a vertex of one of $e_1, e_2, \ldots, e_k$ and follow C until we reach an edge e not in S'_k that has a vertex which is also a vertex of one of the edges $e_1, e_2, \ldots, e_k$. Delete e from T' and obtain a graph T'' with $n - 1$ edges. Since T'' does not contain any cycles, it is a tree. Also, it contains $e_1, e_2, \ldots, e_k, e_{k+1}$. Since e was available when e_{k+1} was chosen by Prim's algorithm, the weight of e_{k+1} is less than or equal to the weight of e. Thus, T'' is a minimal spanning tree of G, and this contradicts the choice of k. Therefore $S = T$.

Exercises 12.5, pages 551–552

55. a. *a b d g k m c h j*

c. *a b d g k p q m n r s u v t c h j*

57.

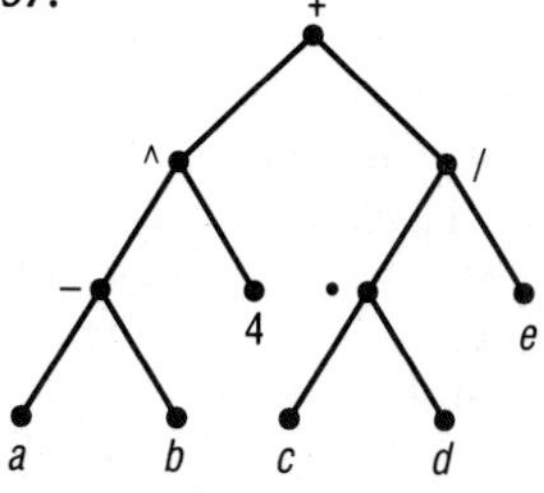

58. a. *d k m g b h j c a*

c. *d p q k m r u v s t n g b h j c a*

59. $a\ b \cdot c\ d\ -\ 3\ /\ +\ e\ 5 \wedge +$

61. a. *d b k g m a h c j*

c. *d b p k q g m r n u s v t a h c j*

Exercises 12.6, pages 566–568

63. a. It does not have the prefix property since 0 is the first part of 01.

c. It has the prefix property.

e. It has the prefix property.

65. The following set of sixteen code words is the largest set: 0000, 0001, 0010, 0100, 1000, 0011, 0101, 0110, 1001, 1010, 1100, 0111, 1011, 1101, 1110, 1111

67.

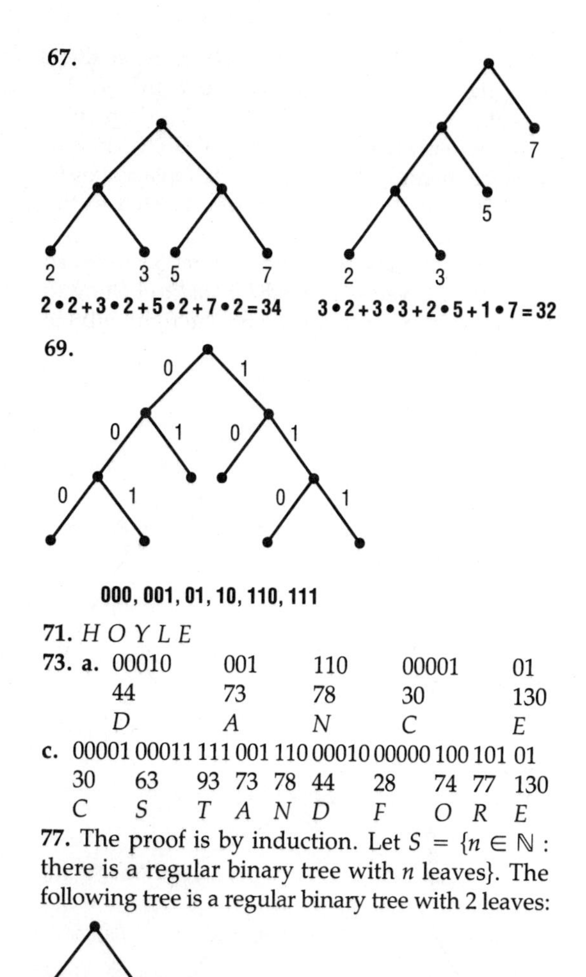

71. *H O Y L E*

73. a.

00010	001	110	00001	01
44	73	78	30	130
D	*A*	*N*	*C*	*E*

c.

00001	00011	111	001	110	00010	00000	100	101	01
30	63	93	73	78	44	28	74	77	130
C	*S*	*T*	*A*	*N*	*D*	*F*	*O*	*R*	*E*

77. The proof is by induction. Let $S = \{n \in \mathbb{N}$: there is a regular binary tree with n leaves$\}$. The following tree is a regular binary tree with 2 leaves:

Therefore $2 \in S$. Suppose $n \in S$, and let T be a regular binary tree with n leaves. Choose one of the leaves of T, call it p. Let T' be the tree obtained from T by adding vertices r and s and edges $\{p, r\}$ and $\{p, s\}$. Then T' is a regular binary tree with $n + 1$ leaves. Therefore $n + 1 \in S$. By the principle of mathematical induction, S is the set of all natural numbers greater than 1.

Exercises 12.7, pages 574–575

79. a. We begin by comparing 1 to 9, then we compare 1 to 3, and finally we compare 1 to 1. So three comparisons are needed.

c. We begin by comparing 12 to 9, then we compare 12 to 15, and finally we compare 12 to 12. So three comparisons are needed.

81. 1. $v = 5$. 2. $v \neq 9$. 3. $v < 9$. 6. Since there is a right child 7 of v, replace v by 7. 2. $v \neq 9$. 3. $v < 9$. 6. Since there is a right child 11 of v, replace v by 11. 2. $v \neq 9$. 3. $v < 9$. 4. Since there is a left child 9 of v, replace v by 9. 2. Since $v = 9$, stop. 9 has been located in the binary search tree.

83. *Rutland* falls between *Ralston* and *Sampson*, so we want the vertex labeled *Rutland* to belong to the left subtree of *Sampson* and the right subtree of *Ralston*:

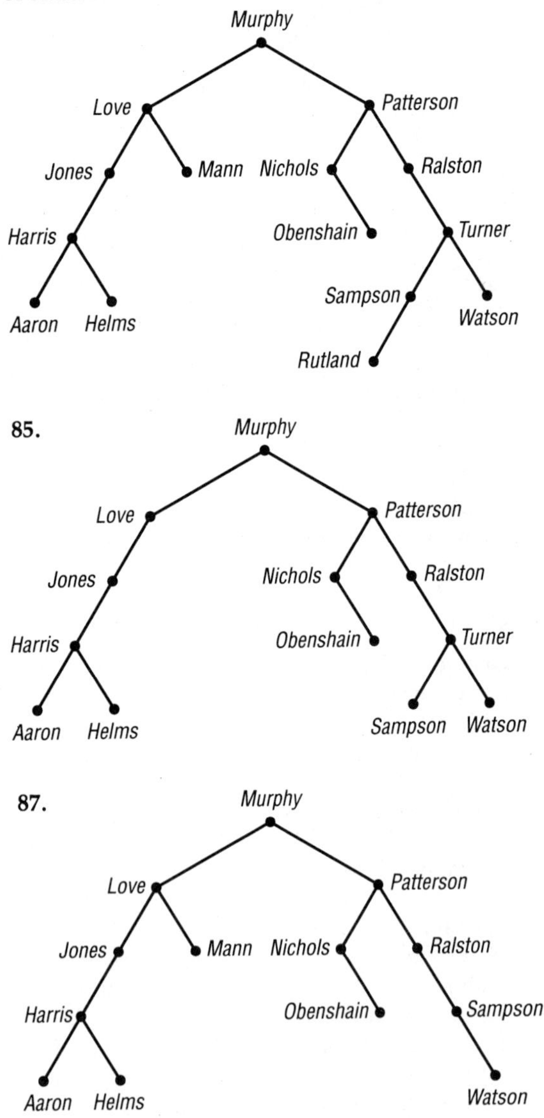

Review Exercises for Chapter 12, pages 575–577

89.

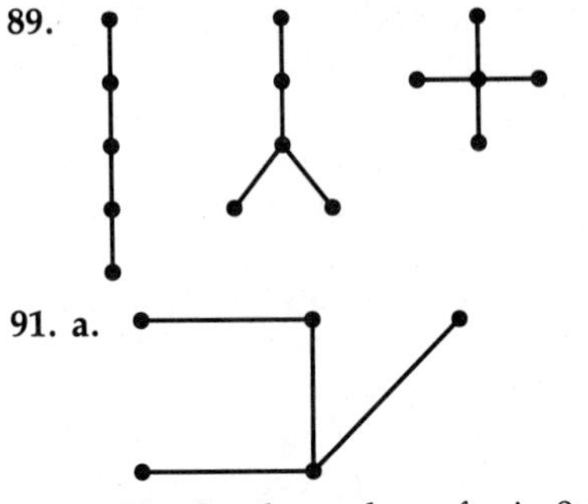

91. a.

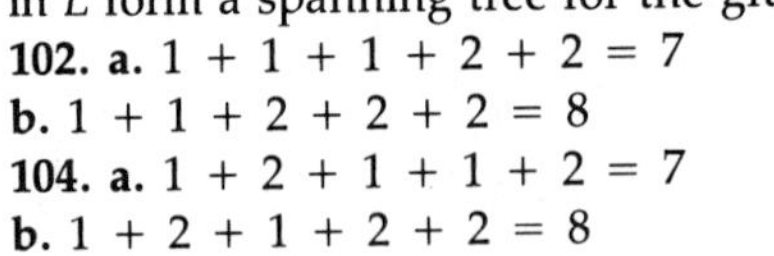

93. a. The level number of r is 0 and the level number of a and b is 1. The level number of c, d, h, and i is 2 and the level number of e, f, j, and g is 3. The height of the tree is 3.

95. a. 2 **b.** 12

99. 1. Select the vertex a, assign the label 0 to a, let $L = \{a\}$, let $T = \varnothing$, and let $k = 0$. 3. There are two unlabeled vertices, b and d, that are adjacent to a. Assign the label 1 to b and d, let $L = \{a, b, c\}$, and let $T = \{e_1, e_2\}$. 3. There are three unlabeled vertices, c, e, and g, that are adjacent to vertices in L that bear the label 1. Assign the label 2 to c, e, and g, let $L = \{a, b, d, c, e, g\}$, and let $T = \{e_1, e_2, e_3, e_4, e_8\}$. 3. There are two unlabeled vertices, f and h, that are adjacent to vertices in L that have been labeled 2. Assign the label 3 to f and h, let $L = \{a, b, d, c, e, g, f, h\}$, and let $T = \{e_1, e_2, e_3, e_4, e_8, e_5, e_9\}$. 3. There are two unlabeled vertices, i and j, that are adjacent to vertices in L that bear the label 3. Assign the label 4 to i and j, let $L = \{a, b, d, c, e, g, f, h, i, j\}$, and let $T = \{e_1, e_2, e_3, e_4, e_8, e_5, e_9, e_{10}, e_{13}\}$. 2. Since $L = V$, stop. The edges in T and the vertices in L form a spanning tree for the graph.

102. a. $1 + 1 + 1 + 2 + 2 = 7$
b. $1 + 1 + 2 + 2 + 2 = 8$

104. a. $1 + 2 + 1 + 1 + 2 = 7$
b. $1 + 2 + 1 + 2 + 2 = 8$

105. a. **b.**

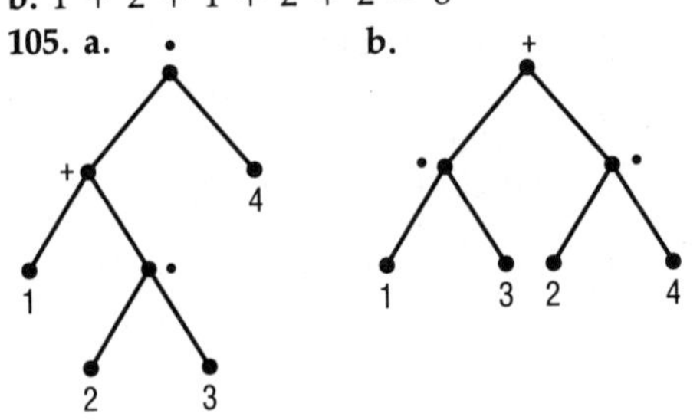

107. a. 1 2 3 · + 4 · **b.** 1 3 · 2 4 · +
109. a. Yes, this set has the prefix property.
c. Yes, this set has the prefix property.
111.

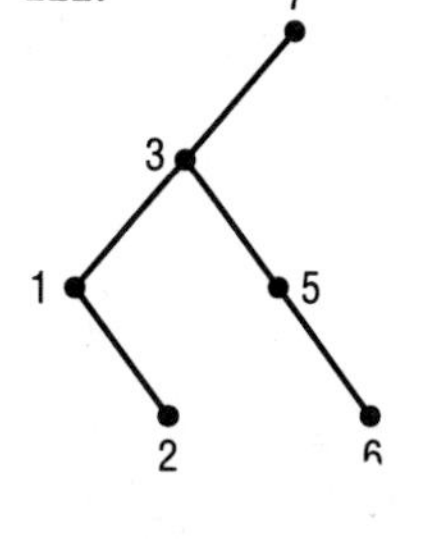

Chapter 13

Exercises 13.1, pages 584–585

1. a.

x_1	1	1	0	0
x_2	1	0	1	0
$f_1(x_1, x_2)$	1	0	0	0

c.

x_1	1	1	0	0
x_2	1	0	1	0
$f_1 \vee f_2$	1	0	1	1

5. $x = m \cdot 2^{mn}$. When $m = 10$ and $n = 10$, then $x = 10 \cdot 2^{150}$. **7.** No

Exercises 13.2, pages 590–592

10. 16, $2^{2^{18}}$
11. a. Let $S = \{1, 2, 3, 4\}$. By the verification theorem, it suffices to verify that $f(x, y, z) = g(x, y, z)$ whenever $x, y, z \in \{\emptyset, S\}$. Thus the table below shows that $f = g$.

x	y	z	f	g
S	S	S	S	S
S	S	$\emptyset$	S	S
S	$\emptyset$	S	S	S
S	$\emptyset$	$\emptyset$	$\emptyset$	$\emptyset$
$\emptyset$	S	S	S	S
$\emptyset$	S	$\emptyset$	$\emptyset$	$\emptyset$
$\emptyset$	$\emptyset$	S	S	S
$\emptyset$	$\emptyset$	$\emptyset$	S	S

13. The expressions do not represent the same function since the expressions do not agree for $(x, y, z) = (1, 0, 1)$.

17. a. $f_1((x, y, z)) = xyz \vee xy'z' \vee x'yz \vee x'y'z$
c. $f_1 \wedge f_2((x, y, z)) = xy'z' \vee x'y'z$
e. $(x'yz' \vee x'y'z')'$
19. $xyz' \vee xy'z' \vee x'yz' \vee x'y'z \vee x'y'z'$
21. $xyz \vee xyz' \vee xy'z \vee xy'z' \vee x'yz \vee x'yz'$
23. a. $(x' \vee y' \vee z) \wedge (x' \vee y \vee z) \wedge (x \vee y' \vee z) \wedge (x \vee y \vee z)$
b. $xyz \vee xy'z \vee xy'z' \vee x'y'z$
25. n atoms
27. Suppose that a is an atom. $0 \leq a$ and $a \leq a$ (since $0(a) = 0$ and $aa = a$). Suppose that $x \leq a$. We must show that $x = 0$ or $x = a$. Since a is an atom, it suffices to show that $ax = x$. But since $x \leq a$, $ax = xa = x$. Now suppose that a is a nonzero element of B for which it is true that 0 and a are the only members of B for which $b \leq a$. Let $x \in B$ such that $ax = x$. Then $x \leq a$ and so $x = 0$ or $x = a$. By definition, a is an atom.
29. $x'y'z'$, $x'y'z$, $x'yz'$, $x'yz$, $xy'z'$, $xy'z$, xyz', and xyz

Exercises 13.3, page 594

31. a. simpler than and as simple as **b.** simpler than and as simple as **c.** neither simpler than nor as simple as **d.** as simple as **e.** as simple as
33. a. The join-normal form has six products and eighteen literals, whereas σ and τ have three products and six literals. **b.** No.

Exercises 13.4, pages 603–605

35. a. A, 110; B, 101; C, 000; D, 111
c. A, 0111; B, 1001; C, 1110; D, 0100
37. a. upper left-hand corner **b.** lower right-hand corner
39. $x_5'x_4'x_3x_2x_1x_0'$, $x_5x_4x_3x_2x_1x_0'$, $x_5x_4'x_3'x_2x_1x_0'$, $x_5x_4'x_3x_2'x_1x_0'$, $x_5x_4'x_3x_2x_1'x_0'$, $x_5x_4'x_3x_2x_1x_0$
41. D is "inside-inside."

			d				
			d				
	d	d	D				d
			d				

001011,010011,011001,011010,011111,111011

43. Yes. The relation is reflexive on the set. (Do nothing.) It is symmetric. Do the string of commutative and associative laws required to show that $\sigma \sim \tau$ in the opposite order starting with τ. The result is σ. Thus, if $\sigma \sim \tau$, $\tau \sim \sigma$. It is transitive. The instructions to get from $\in$ to σ followed by the instructions to get from σ to τ yield a recipe for getting from $\in$ to τ.

45. a. 1011,1110,0111,1101. Since a and c are neighbors, a is not a hermit.
c. 1010,1111,1001,0011. Since c and d are neighbors and c and a are neighbors, c is not a hermit.
e. 1010,1100,1001,0000. Since d and e are neighbors and e and b are neighbors, e is not a hermit.
f. 1001, 0000, 0101, 0011 (f is a hermit)

47. a. $yz' \vee xy' \vee xz$

49. $x \vee x'z$

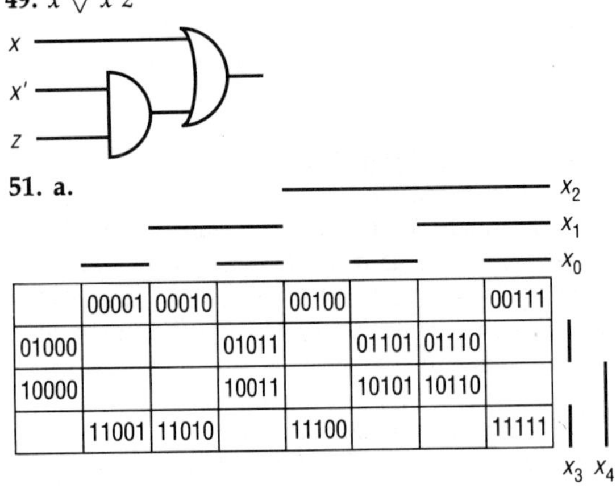

51. a.

	00001	00010		00100			00111
01000			01011		01101	01110	
10000			10011		10101	10110	
	11001	11010		11100			11111

b. Every minterm is a hermit, and so there is no way to simplify the indicated circuit.

Review Exercises for Chapter 13, pages 605–606

53. $\tau(x, y, z) = \sigma'(x, y, z) = (x'y \vee xy \vee z)'$

55.

y
z

1	0	0	1
0	1	1	0
0	1	1	0
1	0	0	1

w x

$\sigma(w, x, y, z) = w'x'y'z' \vee w'x'yz \vee w'xy'z \vee w'xyz' \vee wx'y'z \vee wx'yz' \vee wxy'z' \vee wxyz$

Chapter 14

Exercises 14.1, pages 619–621

1. 92 **3.** 14,833
5. There are four solutions: (2,4,6,7), (3,3,6,7) (3,4,6,6), and (3,4,5,7).
7. 24 **9.** 4,112,784
11. 378,000 **13.** 51
15. $\dfrac{4^4 \cdot 48}{C(52,5)} = .00472804506418$

17. Let S denote the set of nonnegative integers for which the result holds. Clearly $1 \in S$. Suppose $k \geq 1$ and $k \in S$. Then $D_{k+1} =$
$$(k+1)!\left[1 - \frac{1}{1!} + \frac{1}{2!} - \cdots + \frac{(-1)^{k+1}}{(k+1)!}\right]$$
$$= (k+1)\left[k!\left\{1 - \frac{1}{1!} + \frac{1}{2!} - \cdots + \frac{(-1)^k}{k!}\right\}\right]$$
$$+ (k+1)!\frac{(-1)^{k+1}}{(k+1)!} = (k+1)D_k + (-1)^{k+1}.$$ Thus $k + 1 \in S$. Hence $S = \mathbb{N}$ by the principle of mathematical induction.

19. a. {{1}, {2, 3, 4}}, {{2}, {1, 3, 4}}, {{3}, {1, 2, 4}}, {{4}, {1, 2, 3}}, {{1, 2}, {3, 4}}, {{1, 3}, {2, 4}}, {{1, 4}, {2, 3}}
c. {{1}, {2}, {3, 4}}, {{1}, {3}, {2, 4}}, {{1} {4}, {2, 3}}, {{2}, {3}, {1, 4}}, {{2}, {4}, {1, 3}}, {{3}, {4}}, {1, 2}}

21. Let S denote the set of positive integers m such that $\left\{ {m \atop 2} \right\} = 2^{m-1} - 1$. If $m = 2$, then $\left\{ {m \atop 2} \right\} = 1$, since there is only one partition of {1, 2} with two nonempty members. Since $2^{2-1} - 1 = 2 - 1 = 1$, we see $1 \in S$. Suppose now that $k \in S$ for some $k \geq 2$. Let $X = \{1, 2, \ldots, k, k+1\}$. First let $A \cup B = X - \{k+1\}$ and suppose $A \neq \varnothing$, $B \neq \varnothing$, and $A \cap B = \varnothing$. Then $\{A, B\}$ is a two-element partition of $X - \{k+1\}$. This partition "generates" two two-element partitions of X, namely, $\{A \cup \{k+1\}, B\}$ and $\{A, B \cup \{k+1\}\}$. Every two-element partition of X except $\{\{1, 2, \ldots, k\}, \{k+1\}\}$ is generated in this way from partitions of $X - \{k+1\}$. This means that $\left\{ {k+1 \atop 2} \right\} = 2\left\{ {k \atop 2} \right\} + 1 = 2(2^{k-1} - 1) + 1 = 2^k - 2 + 1 = 2^k - 1$. Thus $k + 1 \in S$. Hence $S = \mathbb{N}$ by the principle of mathematical induction.

Exercises 14.2, pages 626–627

23. 1234, 1243, 1324, 1342, 1423, 1432, 2134, 2143, 2314, 2341, 2413, 2431, 3124, 3142, 3214, 3241, 3412, 3421, 4123, 4132, 4213, 4231, 4312, 4321
25. a. 2134 **c.** 42135 **e.** 54132
27. {1,5,6,2,7}, {2,6,1,3,5}, {2,6,5,4,7}, {3,1,6,5,4}, {3,4,1,7,2}, {5,4,3,2,1}, {6,2,3,4,5}, {6,4,5,7,1}
29. {1,2,3,4}, {1,2,3,5}, {1,2,3,6}, {1,2,4,5}, {1,2,4,6}, {1,2,5,6}, {1,3,4,5}, {1,3,4,6}, {1,3,5,6}, {1,4,5,6}, {2,3,4,5}, {2,3,4,6}, {2,3,5,6}, {2,4,5,6}, {3,4,5,6}
31. Regarding each r-combination as an r-digit number (or as r letters in the alphabet), the algorithm adds 1 to the right-most possible digit. This automatically lists the r-combinations in "numerical" or lexicographic order.

Exercises 14.3, pages 644–645

33. a. Yes, order 1 **c.** No
e. No **g.** Yes, order 2
35. By Theorem 14.5, we must show that 2 is a root of the characteristic equation $3(-2^{n-1}) + 2^n$. Let $a_0 = -2$. If $n = 1$, then $a_1 = 3a_0 + 2^1 = 3(-2) + 2 = -4 = -2^{1+1}$. Suppose $a_{k-1} = -2^{k-1+1} = -2^k$. Then $a_k = 3a_{k-1} + 2^k = 3(-2^k) + 2^k = (-3)2^k + 2^k = (-3 + 1)2^k = -2 \cdot 2^k = -2^{k+1}$. So the result follows by induction.
37. a. $a_n = 3n + 5\dfrac{1 - 3^n}{1 - 3}$
c. $a_n = 4n$ **e.** -1
39. When $n = 1$, $a_1 = a_0 + d = e + d \cdot 1$. So the result is true for $n = 1$. Let $S = \{n \in \mathbb{N} : a_n = e + dn\}$. We have shown that $1 \in S$. Suppose $k - 1 \geq 1$ and $k - 1 \in S$. Then $a_k = a_{k-1} + d = e + d(k - 1) + d = e + dk$. Therefore $k \in S$. By the principle of mathematical induction, the theorem follows.

Exercises 14.4, pages 652–653

41. a. $3(x^6 - 1)/(x - 1)$ **c.** $((4x)^4 - 1)/(4x - 1)$
43. a. 1, 3, 3, 1 **c.** 1, 0, 4, 0, 6, 0, 4, 0, 1

Review Exercises for Chapter 14, pages 653–654

45. 23 **47. a.** D_8 **c.** $9! - D_9 - 9D_8$
49. 606 **51. a.** 3142 **c.** 12435
53. {3,4,5,6}, {2,4,5,6}, {2,3,5,6}, {2,3,4,6}, {2,3,4,5}, {1,4,5,6}, {1,3,5,6}, {1,3,4,6}, {1,3,4,5}, {1,2,5,6}, {1,2,4,6}, {1,2,4,5}, {1,2,3,6}, {1,2,3,5}, {1,2,3,4}
55. a. $a_n = 1$
57. a. $G(x) = (2 + x)/(1 + x)^2$

Chapter 15

Exercises 15.1, pages 658–659

1. a. The set associated with 0^* is {ε, 0, 00, 000, . . . }. The set associated with 1^* is {ε, 1, 11, 111, . . . }. Thus the set associated with 0^*1^* is {ε, 1, 11, 111, . . . , 0, 01, 011, . . ., 00, 001, 0011, 00111, . . . , 000, 0001, 00011, 000111, 0001111, . . . , . . . }.
2. a. By part (b) of the definition, 0 and 1 are regular expressions. Thus, by part (c) of the definition, $0 \vee 1$ is a regular expression.
c. By part (a) of this exercise, $0 \vee 1$ is a regular expression. Thus, by part (e) of the definition, $(0 \vee 1)^*$ is a regular expression.
4. a. $1 \vee 2 \vee 3$ **c.** $(12)^*$
e. $1 \vee 2 \vee 3 \vee (12)^*$

Exercises 15.2, pages 666–667

7. a. Each member of $L(G)$ is a string consisting of an even number of 1s.
c. A member of $L(G)$ consists of a string of m 0s followed by a string of n 1s followed by one 2 ($m,n \in \mathbb{N}$).
e. A member of $L(G)$ consists of a string of m 20s followed by a string of n 10s, where $m \in \mathbb{N}$ and $n \in \mathbb{N} \cup \{0\}$.
9. a. 11 is listed in Exercise 6b as being a member of $L(G)$.
c. We must begin with the production $S \rightarrow_a 1A$. If we use $A \rightarrow_c 1$ next, we can never introduce a 0. If we use $A \rightarrow_b 1AB$ next, we must eventually use $A \rightarrow_c 1$ and $B \rightarrow_d 0$. Therefore we will have at least three 1s. Therefore $110 \notin L(G)$.
e. In Exercise 6b we must begin with the production $S \rightarrow_a 1A$. Thus each member of $L(G)$ must begin with 1. Hence $0001 \notin L(G)$.
11. a. $G = (N, T, S, P)$, where $N = \{S\}$, $T = \{0, 1\}$, $P = \{S \rightarrow_a 10, S \rightarrow_b 1S0\}$
c. $G = \{N, T, S, P\}$, where $N = \{S, A, B\}$, $T = \{0, 1, 2\}$, and $P = \{S \rightarrow_a SA, SA \rightarrow_b BS, BS \rightarrow_c 00, B \rightarrow_d 11, A \rightarrow_e 22\}$

13.

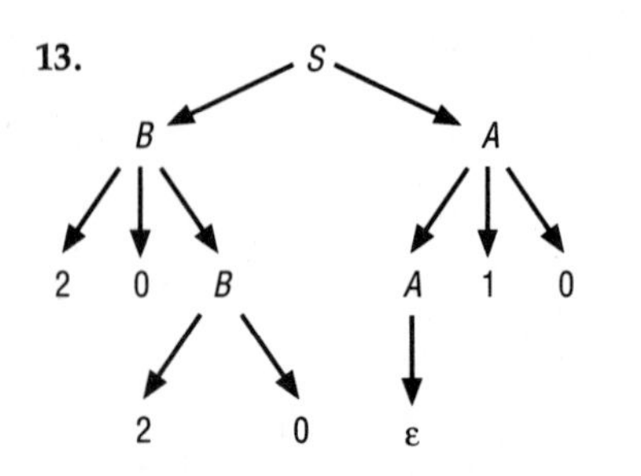

15. $G = \{N, T, S, P\}$, where $N = \{S\}$, $T = \{0, 1\}$, and $P = \{S \rightarrow_a 0S0,\ S \rightarrow_b 1S1,\ S \rightarrow_c \varepsilon,\ S \rightarrow_d 0,\ S \rightarrow_e 1\}$

Exercises 15.3, pages 672–673

17. a. $(A, *)$ is a semigroup; it is not normal.
c. It is not a semigroup because $*$ is not associative.
e. It is not a semigroup because $*$ is not associative.
19. The four members of S^S are the functions e, f, g, and h defined by $e(0) = 0$, $e(1) = 1$, $f(0) = f(1) = 0$, $g(0) = g(1) = 1$, and $h(0) = 1$ and $h(1) = 0$.

*	e	f	g	h
e	e	f	g	h
f	f	f	f	f
g	g	g	g	g
h	h	g	f	e

21. If $w_1, w_2 \in \Sigma^*$, then so is w_1w_2. Hence concatenation is a binary operation on Σ^*. It is also clear that concatenation is associative. Therefore $(\Sigma^*, \cdot)$ is a semigroup.
25. Let $c, d \in f(S)$. Then there exist $a, b \in S$ such that $f(a) = c$ and $f(b) = d$. Since $(S, *)$ is commutative, $a * b = b * a$. Since f is a homomorphism, $c \# d = f(a) \# f(b) = f(a * b) = f(b * a) = f(b) \# f(a) = d \# c$. Therefore $(f(S), \#)$ is commutative.
29. Let $a, b \in R$. Since f is a homomorphism, $f(a @ b) = f(a) * f(b)$. Since g is a homomorphism, $g(f(a @ b)) = g(f(a) * f(b)) = g(f(a)) \# g(f(b))$. Therefore $g \circ f$ is a homomorphism.

Exercises 15.4, pages 677–678

31. a. $S = \{s_0, s_1\}$, $\Sigma = \{0, 1\}$, $F = \{s_1\}$, and $d : S \times \Sigma \rightarrow S$ is the function defined by $d(s_0, 0) = s_0$, $d(s_0, 1) = s_1$, $d(s_1, 0) = s_1$, $d(s_1, 1) = s_1$.

33. a.

	0	1
s_0	s_0	s_1
s_1	s_1	s_2
s_2	s_0	s_2

c.

	0	1	2
s_0	s_0	s_0	s_2
s_1	s_2	s_3	s_2
s_2	s_1	s_0	s_3
s_3	s_3	s_2	s_0

Exercises 15.5, pages 682–683

37. It must contain two consecutive 1s.
39. It must contain a 3.
41. The number of 1s in α must be a multiple of 4.
43. $L(M)$ consists of all $\alpha \in \Sigma^*$ such that the number of 1s in α is of the form $5n + 4$, where n is a natural number.

Review Exercises for Chapter 15, pages 683–685

45. a. $\{b, ab, aab, aaab, \ldots, bb, bbb, bbbb, \ldots\}$
c. $\{ab, abab, ababab, \ldots, b, abb, ababb, abababb, \ldots\}$
47. $L(G)$ consists of all the members of the set associated with o^* except ε.
49. Each table defines a binary operation $*$ on A. Thus, in each case, we must decide whether $*$ is associative. **a.** $*$ is not associative. **c.** $*$ is associative. **e.** $*$ is not associative.
51. f is not a homomorphism.
53. a. $S = \{s_0, s_1, s_2\}$, $\Sigma = \{0, 1\}$, $F = \{s_1\}$, and $d : S \times \Sigma \rightarrow S$ is the function defined by $d(s_0, 0) = s_1$, $d(s_0, 1) = s_0$, $d(s_1, 0) = s_2$, $d(s_1, 1) = s_0$, $d(s_2, 0) = s_2$, and $d(s_2, 1) = s_0$.
57. Yes. If $A, B \in \mathcal{P}(S)$, then $A \oplus B \in \mathcal{P}(S)$. Therefore $\oplus$ is a binary operation on $\mathcal{P}(S)$. By Theorem 2.4, $\oplus$ is associative and commutative. Since $A \oplus \varnothing = (A \cup \varnothing) - (A \cap \varnothing) = A - \varnothing = A$, $\varnothing$ is the identity element.

Exercises Appendix A, pages 689–690

1. Some examples: ignition, either; clock, either; speedometer, analog; odometer, digital; gas gauge, analog; oil light, digital

3. a.

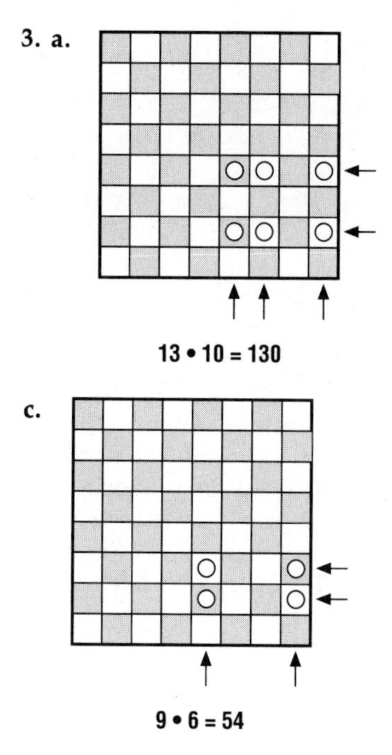

13 • 10 = 130

c.

9 • 6 = 54

e.

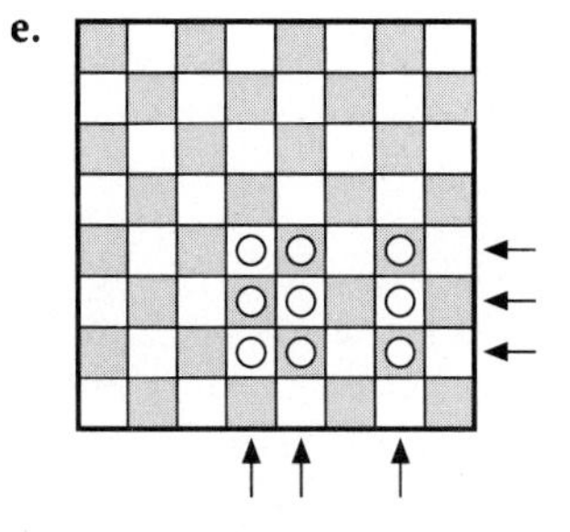

24 • 14 = 108

5. a. 255 **b.** 15 · 17

Exercises for Appendix B, page 696

1. −8

3. n = 4, number of operations = 10; n = 10, number of operations = 28.

5. n = 5, number of operations = 13; n = 12, number of operations = 34.

Exercises for Appendix D, page 708

1. a. binary = 1011011, decimal = 91, octal = 133, split octal = 133, hex = 5B

c. binary = 100010110011010, decimal = 17818, octal = 42632, split octal = 105.232, hex = 459A

e. binary = 1111111111111111, decimal = 65535, octal = 177777, split octal = 377.377, hex = FFFF

3. a. 1064 **c.** 2417 **e.** 4343

5.

+	0	1	2	3	4	5	6	7
0	0	1	2	3	4	5	6	7
1	1	2	3	4	5	6	7	10
2	2	3	4	5	6	7	10	11
3	3	4	5	6	7	10	11	12
4	4	5	6	7	10	11	12	13
5	5	6	7	10	11	12	13	14
6	6	7	10	11	12	13	14	15
7	7	10	11	12	13	14	15	20

*	0	1	2	3	4	5	6	7
0	0	0	0	0	0	0	0	0
1	0	1	2	3	4	5	6	7
2	0	2	4	6	10	12	14	16
3	0	3	6	11	14	17	22	25
4	0	4	10	14	20	24	30	34
5	0	5	12	17	24	31	36	43
6	0	6	14	22	30	36	44	52
7	0	7	16	25	34	43	52	61

Exercises for Appendix E, pages 717–718

1. a. 3.5092×10^4 **c.** 3.04002×10^{-1}

e. 3.04002×10^3

3. a. No **c.** Yes

5. a. Yes

7. Suppose that there are natural numbers c and n such that $a/b = c/10^n$. Then $a10^n = bc$. Let k be a prime factor of b different from 2 and 5. Now k divides bc, but k does not divide a since a and b are relatively prime and does not divide 10 since it is neither 2 nor 5. Thus k does not divide $a10^n$. But this is impossible. Hence a/b cannot be expressed as a terminating decimal.

9. a. $\frac{1101}{100000}$ **c.** $\frac{1100}{11111}$

11. a. No. $\sqrt{2}$ and $-\sqrt{2}$ are irrational and yet their sum is rational.

Exercises for Appendix F, pages 730–732

1. a. 0.1325×10^2 **c.** 0.14259379×10^5
3. a. $0.1\overline{00110} \times 10^{10}$ **c.** $0.1101\overline{01} \times 10^{100}$
5. 579 **7. a.** 11010000_2 **c.** 305_8
9. a. 01010101_2 **c.** 18_{10}
11. a. 10001011_2
c. 11000001_2
13. a. One's complement of x $= 0001000110110110_2$, two's complement of x $= 0001000110110111_2$, one's complement of y $= 1111000011110101_2$, and two's complement of $y = 1111000011110110_2$.
c. $y - x = 0010000011000001_2$
15. a. $133/1$ **c.** Yes. It need not be unique.

Exercises for Appendix G, page 737

1. a.

{1,2} {1,3} {1,4} {1,5} {2,3} {2,4} {2,5} {3,4} {3,5} {4,5}

{1} {2} {3} {4} {5}

∅

3. a. 3, 1 **c.** $p = 6$, $q = 15$
5. a. $\mathcal{P}(I_{120})$ **c.** no way **e.** $B = \{0, 1\}$
7. $xy \le x$ because $(xy)x = x(yx) = x(xy) = (xx)y = xy$. Similarly, $xy \le y$ because $(xy)y = x(yy) = xy$.
9. $x(p \vee q) = xp \vee xq = x \vee xq = x(1) \vee xq = x(1 \vee q) = x(1) = x$. If $x \le p$ then $xp = x$, and as we have just shown it follows that $x(p \vee q) = x$. Thus $x \le p \vee q$. If $x \le q$ then $xq = x$, so $x(p \vee q) = xp \vee xq = xp \vee x = xp \vee x(1) = x(p \vee 1) = x(1) = x$ and $x \le p \vee q$.
11. It is given that $a = x \vee y$. Since $x(x \vee y) = xx \vee xy = x \vee xy = x(1) \vee xy = x(1 \vee y) = x(1) = x$, $x \le x \vee y = a$. Similarly, $y \le x \vee y = a$. Since $x \le a$, $x = 0$ or $x = a$, and since $y \le a$, $y = 0$ or $y = a$. If both $x = 0$ and $y = 0$, then $a = 0 \vee 0 = 0$, which contradicts that a is an atom. Therefore $x = a$ or $y = a$.

INDEX